PLASTICITY TODAY

Modelling, Methods and Applications

The proceedings of an International Symposium on current trends and results in Plasticity, held at the International Centre for Mechanical Sciences, Udine, Italy, 27–30 June, 1983

Symposium dedicated to the memory of

Professor Wacław Olszak

SYMPOSIUM CHAIRMEN

G. Bianchi and A. Sawczuk

HONORARY COMMITTEE

A. V. Turello	*President of CISM, Udine, Italy*
P. Brousse	*Rector of CISM, Paris, France*
W. T. Koiter	*Rector of CISM, Delft, The Netherlands*
W. Nowacki	*Former President of the Polish Academy of Sciences, Warsaw, Poland*
J. H. Argyris	*Copernicus Award Holder of the Polish Academy of Sciences, Stuttgart, FRG*
D. C. Drucker	*Foreign Member of the Polish Academy of Sciences, President of IUTAM, Urbana, Ill., USA*
P. Germain	*Foreign Member of the Polish Academy of Sciences, Paris, France*
P. G. Hodge, Jr	*Minneapolis, Minn., USA*
J. Hult	*Secretary General of IUTAM, Gothenburg, Sweden*
A. J. Ishlinskii	*Foreign Member of the Polish Academy of Sciences, Moscow, USSR*
J. Kravtchenko	*Grenoble, France*
Ch. Massonnet	*Foreign Member of the Polish Academy of Sciences, Liège, Belgium*
F. K. G. Odquist	*Foreign Member of the Polish Academy of Sciences, Djursholm, Sweden*
Y. Ohashi	*Nagoya, Japan*
I. N. Sneddon	*Foreign Member of the Polish Academy of Sciences, Glasgow, UK*

SCIENTIFIC COMMITTEE

D. C. Drucker	*Urbana, USA*	J. Mandel	*Paris, France*
W. Johnson	*Cambridge, UK*	Yu. Rabotnov	*Moscow, USSR*
Th. Lehmann	*Bochum, FRG*	J. R. Rice	*Cambridge, USA*
G. Maier	*Milan, Italy*	O. C. Zienkiewicz	*Swansea, UK*

PLASTICITY TODAY

Modelling, Methods and Applications

Edited by

A. SAWCZUK

International Centre for Mechanical Sciences, Udine, Italy

and

G. BIANCHI

Polytechnic of Milan, Italy

ELSEVIER APPLIED SCIENCE PUBLISHERS
LONDON and NEW YORK

ELSEVIER APPLIED SCIENCE PUBLISHERS LTD
Crown House, Linton Road, Barking, Essex IG 11 8JU, England

Sole Distributor in the USA and Canada
ELSEVIER SCIENCE PUBLISHING CO., INC.
52 Vanderbilt Avenue, New York, NY 10017, USA

British Library Cataloguing in Publication Data

Plasticity today.
1. Plasticity
I. Sawczuk, A. II. Bianchi, G.
531′.3825 QA931

ISBN 0-85334-302-0

WITH 278 ILLUSTRATIONS AND 19 TABLES

The selection and presentation of material and the opinions expressed in this publication are the sole responsibility of the authors concerned

Filmset and printed by The Universities Press (Belfast) Ltd.

Preface

Classical plasticity, as regards its experimental motivation, constitutive modelling, mathematical methods for dealing with its equations, and engineering applications of the theory, is a fairly well founded and established domain of mechanical sciences. The deformation theory of plasticity is properly understood as far as its drawbacks, applicability and usefulness are concerned. The flow theory is endowed with appropriate theorems and mathematical methods for dealing effectively with boundary value problems concerning various technological needs. Its particular field, the limit analysis theory, appears to be applicable and particularly useful in several domains of engineering. To be specific, we may add that classical plasticity is a mature science as to its:

(a) mathematical description of the rate independent material behaviour;
(b) methods of solutions regarding equations involving perfectly plastic material response;
(c) applications of the perfectly plastic model in metal forming and structural engineering, to mention two extremes.

However, the actual requirements of applied research need to be reflected upon and an attempt made to bring together the many facets which go to make up the present state of the domain roughly specified as plasticity. Problems of high pressure and high speed metal forming; of inelastic wave propagation; of the dynamics of vehicles and structures, naval, terrestrial and spatial; of structural behaviour under random excitation and random material response outside the elastic range; of plasticity and failure of structured media; of earthquake and nuclear engineering; of material and structural response when subjected to large configuration changes—if we restrict the list to a few typical examples—force us to look into the field of plasticity and account for the complexities involved. Such an attitude was not needed, in principle, in classical plasticity. However, bordering domains like creep, fissuration, internal damage theories, questions of anisotropy of hardening and softening, of rate dependence, of thermomechanical coupling, of instability and bifurcation of the equilibrium states, of strain localization, of unilateral constraints, of passage from microscale behaviour to macroscopic models and appropriate homogenization techniques, represent examples which cannot be left out of the field of vision of those who study and apply plasticity theories.

All this indicates that an appropriate basis for an objective constitutive modelling of complex material response to external or internal agencies is furnished by non-linear continuum mechanics. This also imposes a need for constructing workable methods of solutions involving new mathematical tools, possibly with an appropriate computer-oriented foundation when the model, well formulated and experimentally confirmed, deserves an engineering implementation.

From such a motivation the symposium Plasticity Today was conceived, although it was clear that the field was as yet only vaguely defined. In order to obtain material for reflection and to bring together the various facets of plasticity an attempt was made to arrive at an objective picture of: (a) the actual state of the plasticity theories accounting for elasticity, damage, hardening, large strains, passage from micro to macro scales, strain localization and instabilities; (b) existing mathematical methods and mathematical possibilities of developing appropriate methods for solving specific and meaningful problems when the plastic response, and thus rate independence, is coupled with other phenomena; (c) current research on plasticity as combined with other domains so as to obtain information with regard to engineering needs and as part of our preparation for serving society in a responsible manner.

To a great extent the objectives of this symposium are the natural consequence of the Warsaw Symposium of 1972 which resulted in the volumes *Foundations of Plasticity* and *Problems in Plasticity*, published shortly after by Noordhoff. Classical plasticity was there assessed, not verbally, but by the scientific contributions contained in these volumes.

The emerging new subdomains and directions were also reported on at that time.

In our attempt to attain the objectives of the present symposium several General Lectures were requested and stimulated, as well as a certain number of Topical Lectures. The aim of the General Lectures was to give the personal views of reliable specialists on the important points of specific domains. The bounds of those domains were, of course, to be established by the Lecturers themselves. The Topical Lectures were intended either to present a résumé of the results obtained by the Lecturer during his years of reflection and creativity or to report on results in newly developing domains or techniques that they had shaped or contributed to. The last group of studies presented at the symposium were the Contributed Papers. These contributions reported on novel results in the sense that solutions to specific questions of developing domains were given.

Although interest in the symposium was particularly extensive only a limited number of proposals could be included in the program. The symposium assembled about 160 participants from 27 countries, stressing its international character, and confirming that no borders exist in the search for truth.

The papers presented at the sessions or accepted for the symposium, even if not actually presented, are collected in two volumes. The present one, entitled *Plasticity Today: Modelling, Methods and Applications,* contains the General and Topical Lectures. The second volume, named *Plasticity Today: New Solutions and Trends in Plasticity,* includes Contributed Papers and is published as the Olszak Memorial Volume in a special issue of *Engineering Fracture Mechanics Journal.*

This is a suitable point to emphasise that the symposium Plasticity Today was dedicated to the memory of Professor Wacław Olszak. A few words about Professor Olszak need to be said here, even though the Memorial Lecture given at the Symposium describes the life path, the research results obtained, and the philosophy of this intellectually mature applied mechanician and structural engineer. This Memorial Lecture opens the Olszak Memorial Volume, which also contains the last research paper by Professor Olszak.

Professor Olszak was born in 1902 and belonged to the Polish nation, to its culture, history, glory and pain. Educated in Austria, Poland and France, he was originally a professional structural engineer. From 1937 he was a University Professor, most of the time with the Technical University of Warsaw and the Institute of Fundamental Technological Research IPPT in Warsaw, the Institute he helped to create within the Polish Academy of Sciences.

His most important contributions concern the plasticity theory, mainly that of non-homogeneous and anisotropic solids and structures, the domain he effectively shaped. A member of the Polish Academy of Sciences and the Institut de France among others, he contributed in an essential manner to the creation of Polish Mechanics after the Second World War. His place in the history and in the culture of my nation is, no doubt, of lasting quality.

His knowledge, research results, culture, and international recognition, being of essential significance for Polish science, were also given to the service of the international scientific community. From 1969 until his passing away in 1980 he was the Resident Rector of the International Centre for Mechanical Sciences, CISM, at Udine, Italy. The symposium Plasticity Today was organized not solely to stress his association with the Centre but also, by the subject and objectives of the meeting, to emphasize his open-mindedness and continued quest for relevance and reliability in research.

A. Sawczuk
Member of the Polish Academy of Sciences
Rector of CISM

Acknowledgement

Many people contributed to the success of the symposium Plasticity Today: the General Lecturers who took time from their own research to prepare their papers; the Topical Lecturers who took pains to render clearly and concisely their studies of specific questions, or who presented and documented new findings; and those whose totally original Contribution Papers added so greatly to the interest of the meeting.

Without the competence and gracious collaboration of the Secretarial Staff of CISM many of the problems of organizing the meeting could not have been solved. Dr Arnilla Bertozzi's initiative and ability were of invaluable assistance, especially during the preparatory phase; and no doubt many of the participants had occasion to appreciate the smiling efficiency of Miss Elsa Venir and Mrs Carla Toros. Miss Bernadette Derderian of the Centre National de la Recherche Scientifique de Marseille also shared in the work while Professor Sawczuk was there. The untimely death of Professor Sawczuk during the course of production of this volume is deeply regretted.

Finally, I am indebted to the University of Udine for their hospitality and technical assistance, and to the publishers, Elsevier Applied Science Publishers, for their efficient help in preparing this edition of the papers.

To all these people I wish to express here my sincere gratitude and thanks.

G.B.

Contents

Section 2: Experimental Plasticity

Section 3: Mathematical Methods of Elastoplasticity

Section 4: Bifurcation and Instability

Section 5: Hardening and Memory Effects

Section 6: Damage and Failure

Section 7: Applications of Plasticity

List of Contributors

M. Achenbach

Hermann-Föttinger-Institut, Technische Universität Berlin, FB-9, Strasse des 17 Juni 1953, 1000 Berlin 12, Federal Republic of Germany

J. M. Alexander

Department of Mechanical Engineering, University College of Swansea, Singleton Park, Swansea SA2 8PP, UK. Present address: University of Surrey, Guildford, Surrey GU2 5XH, UK

J. Argyris

Institut für Statik und Dynamik der Luft- und Raumfahrtkonstruktionen, University of Stuttgart, Pfaffenwaldring 27, 7000 Stuttgart 80, Federal Republic of Germany

A. Baltov

Institute of Mechanics and Biomechanics, Bulgarian Academy of Sciences, Acad. G. Bonchev Street, 1113 Sofia, Bulgaria

Z. P. Bažant

Center for Concrete and Geomaterials, The Technological Institute, Northwestern University, Evanston, Illinois 60201, USA

J. F. Bell

Johns Hopkins University, Baltimore, Maryland 21218, USA

J. F. BESSELING
Laboratory for Engineering Mechanics, Delft University of Technology, Mekelweg 2, 2628 CD Delft, The Netherlands

S. BITTANTI
Department of Electronics, Politecnico di Milano, Piazza L. da Vinci 32, 20133 Milan, Italy

S. R. BODNER
Department of Mechanical Engineering, Technion—Israel Institute of Technology, Haifa 32000, Israel

J. P. BOEHLER
Institut de Mécanique de Grenoble, BP 68, 38402 St Martin d'Hères Cédex, France

Y.-W. CHANG
Department of Theoretical and Applied Mechanics, Cornell University, Ithaca, New York 14853, USA

W. F. CHEN
School of Civil Engineering, Purdue University, West Lafayette, Indiana 47907, USA

R. J. CLIFTON
Division of Engineering, Brown University, Providence, Rhode Island 02912, USA

N. CRISTESCU
Department of Mechanics, University of Bucharest, Str. Academiei 14, Bucharest 70109, Romania

Y. F. DAFALIAS
Department of Civil Engineering, University of California, Davis, California 95616, USA

K. DOLINSKI
Institute of Fundamental Technological Research, Swietokrzyska 21, 00-049 Warsaw, Poland

J. DOLTSINIS
Institut für Statik und Dynamik der Luft- und Raumfahrtkonstruktionen, University of Stuttgart, Pfaffenwaldring 27, 7000 Stuttgart 80, Federal Republic of Germany

D. R. DRUCKER
College of Engineering, University of Illinois, 1308 West Green Street, Urbana, Illinois 61801, USA. Present address: 231 Aerospace Engineering Building, University of Florida, Gainesville, Florida 32611, USA

M. K. DUSZEK
Institute of Fundamental Technological Research, Swietokrzyska 21, 00-049 Warsaw, Poland

G. J. DVORAK
Department of Civil Engineering, University of Utah, Salt Lake City, Utah 84112, USA. Present address: Rensselaer Polytechnic Institute, Troy, New York 12181, USA

J. FAN
Department of Electrical and Computer Engineering, University of Cincinnati, 898 Rhodes Hall, Cincinnati, Ohio 45221, USA

V. FRANCIOSI
Istituto di Scienza delle Costruzioni-Ingegneria, University of Naples, Piazzale Vincenzo Tecchio, 80125 Naples, Italy

H. FROMMER
Institut für Mechanik, ETH, Rämistrasse 101, Zürich 8092, Switzerland

D. J. HAN
School of Civil Engineering, Purdue University, West Lafayette, Indiana 47907, USA

E. W. HART
Department of Theoretical and Applied Mechanics, Cornell University, Ithaca, New York 14853, USA

J. M. Jalinier

Institut National Polytechnique de Grenoble, Génie Physique et Mécanique des Matériaux, Domaine Universitaire, BP 46, 38402 Saint Martin d'Hères, France

W. Johnson

Formerly Department of Engineering, University of Cambridge, Cambridge CB2 1PZ, UK

S. Kaliszky

Department of Mechanics, Technical University of Budapest, Müegyetem rpt. 3, 1521 Budapest, Hungary

M. Kleiber

Institute of Fundamental Technological Research, Swietokrzyska 21, 00-049 Warsaw, Poland. Present address: Department of Civil Engineering, University of California, Berkeley, California 94720, USA

W. Klie

Ingenieurbüro für Maschinendynamik, Hannover, Federal Republic of Germany

J. A. König

Institute of Fundamental Technological Research, Swietokrzyska 21, 00-049 Warsaw, Poland

E. Krempl

Department of Mechanical Engineering, Aeronautical Engineering and Mechanics, Rensselaer Polytechnic Institute, Troy, New York 12181, USA

E. H. Lee

Department of Mechanical Engineering, Aeronautical Engineering and Mechanics, Rensselaer Polytechnic Institute, Troy, New York 12181, USA

K. Leers

Institut für Mechanik, Universität Hannover, Appelstrasse 11, Hannover, Federal Republic of Germany

TH. LEHMANN
Lehrstuhl für Mechanik I, Ruhr-Universität Bochum, Postfach 102148, Universitätstrasse 150, 4630 Bochum, Federal Republic of Germany

T. ŁODYGOWSKI
Institute of Building Structures and Technology, Technical University, Piotrowo 5, 60965 Poznan, Poland

O. MAHRENHOLTZ
Arbeitsgebiet Meerestechnik II, Technische Universität Hamburg-Harburg, Eissendorferstrasse 38, Postfach 901403, 2100 Hamburg 99, Federal Republic of Germany

G. MAIER
Department of Structural Engineering, Politecnico di Milano, Piazza L. da Vinci 32, 20133 Milan, Italy

A. G. MAMALIS
Department of Mechanical Engineering, National Technical University of Athens, Athens, Greece

I. MÜLLER
Hermann-Föttinger-Institut, Technische Universität Berlin, FB-9, Strasse des 17 Juni 1953, 1000 Berlin 12, Federal Republic of Germany

S. MURAKAMI
Department of Energy Engineering, Toyohashi University of Technology, Tempaku-cho, Toyohashi 440, Japan. Present address: Department of Mechanical Engineering, Nagoya University, Chikusa-ku, Nagoya 464, Japan

P. M. NAGHDI
Department of Mechanical Engineering, University of California, Berkeley, California 94720, USA

A. NAPPI
Department of Structural Engineering, Politecnico di Milano, Piazza L. da Vinci 32, 20133 Milan, Italy

S. Nemat-Nasser
Department of Civil Engineering, The Technological Institute, Northwestern University, Evanston, Illinois 60201, USA

Q. S. Nguyen
Laboratoire de Mécanique des Solides, Ecole Polytechnique, 91128 Palaiseau, France

P. Perzyna
Institute of Fundamental Technological Research, Swietokrzyska 21, 00-049 Warsaw, Poland

H. Petryk
Institute of Fundamental Technological Research, Swietokrzyska 21, 00-049 Warsaw, Poland

A. Phillips
Department of Applied Mechanics, Yale University, Becton Center, PO Box 2157, New Haven, Connecticut 06520, USA

C. Polizzotto
Istituto di Scienza delle Costruzioni, Università di Palermo, Viale delle Scienze, I-90128 Palermo, Italy

B. Raniecki
Institute of Fundamental Technological Research, Swietokrzyska 21, 00-049 Warsaw, Poland

J. Raphanel
Institut National Polytechnique de Grenoble, Génie Physique et Mécanique des Matériaux, Domaine Universitaire, BP 46, 38402 Saint Martin d'Hères, France

Y. Sanomura
Department of Mechanical and Electrical Engineering, Tokuyama Technical College, Kume, Tokuyama 745, Japan

M. A. Save
Faculté Polytechnique de Mons, rue de Houdain, Mons 7000, Belgium

M. SAYIR

Institut für Mechanik, ETH, Rämistrasse 101, Zürich 8092, Switzerland

M. SUÉRY

Institut National Polytechnique de Grenoble, Génie Physique et Mécanique des Matériaux, Domaine Universitaire, BP 46, 38402 Saint Martin d'Hères, France

P. M. SUQUET

Université de Montpellier II, Place E Bataillon, 34060 Montpellier Cedex, France

W. SZCZEPIŃSKI

Institute of Fundamental Technological Research, Swietokrzyska 21, 00-049 Warsaw, Poland

J. L. TEPLY

Department of Civil Engineering, University of Utah, Salt Lake City, Utah 84112, USA

V. TVERGAARD

Department of Solid Mechanics, Technical University of Denmark, Building 404, DK-2800 Lyngby, Denmark

K. C. VALANIS

Department of Electrical and Computer Engineering, University of Cincinnati, 898 Rhodes Hall, Cincinnati, Ohio 45221, USA

A. ZAOUI

Laboratoire PMTM Gréco GDE, CNRS, Université Paris-Nord, Avenue J. B. Clément, 93430 Villetaneuse, France

Section 1

MATHEMATICAL MODELLING IN FINITE ELASTOPLASTICITY

1

Elastic–Plastic Materials at Finite Strains

M. KLEIBER* and B. RANIECKI
Institute of Fundamental Technological Research, Warsaw, Poland

ABSTRACT

This paper has basically a review character with the authors' attention focused on the evaluation of different constitutive concepts of plasticity at unrestricted strains. The macroscopic aspects of the theory are discussed. Starting from the Hill–Rice concept and discussing thermostatic properties of elastic–plastic solids we investigate the transformation rules and invariance properties for different quantities in the six-dimensional configurational space. The Mandel–Lee description then follows naturally as a transition from a fixed co-ordinate system to a moving co-ordinate system, the changes of which correspond to the changes in the internal state of the material element. The rate equations are thoroughly presented in the second part of the paper. Some results emerging from the discussion seem to be new, or at least are believed not to have been explicitly stated in the literature.

INTRODUCTION

The aim of this paper is to discuss macroscopic constitutive equations of rate-independent plasticity at unrestricted strains. In particular, a comparison of different existing constitutive descriptions is attempted. Two basic mathematical formalisms are given special attention on

* Present address: Department of Civil Engineering, University of California, Berkeley, USA.

account of their generality and anticipated usefulness. The first one is that proposed in a series of papers by Hill [1–7], Hill and Rice [8] and Rice [9–11] (the approach will further be referred to as the *H–R* formalism) while the second one is due to Mandel [12–17] (*M* formalism). The latter could also be called the *M–L* formalism if one recognized its most significant feature to be a stress-free reference state extensively discussed by Lee [18–21]. Some other significant contributions to finite strain elastoplasticity are also referred to. The paper is divided into two basic parts dealing with thermostatic properties of an elastic–plastic material element and rate equations of the theory, respectively. The major part of the paper is devoted to a unifying presentation of the existing theories and their extensions accounting consistently for thermal effects. At places, some apparently new results are discussed.

1. THERMOSTATIC PROPERTIES OF AN ELASTIC–PLASTIC MATERIAL ELEMENT

1.1. Hill–Rice Formalism

1.1.1. *State Equations of Constrained Equilibrium*

Consider a representative macroscopic sample of a material—a material element $\mathcal{M}_e$ of the unit mass. There exist many possibilities of describing the changes of the shape of such an element under applied load and/or temperature θ. Usually, we choose a reference configuration κ_o and introduce a strain measure $\mathbf{E}$. The reference configuration has not necessarily to be the natural configuration occupied by $\mathcal{M}_e$ at the initial instant of a process of straining and may be taken, in principle, to coincide with any real or fictitious configuration of $\mathcal{M}_e$. Hill [1–3] has distinguished the following family of objective strain tensors

$$\mathbf{E} = \sum_{k=1}^{3} [f(z)]_{z=U_k} \qquad \mathbf{N}_k \otimes \mathbf{N}_k = \bar{\mathbf{f}}(\mathbf{U}) \tag{1.1}$$

where $f(1)=0$, $f'(1)=1$ and U_k, N_k are the principal values (principal stretches) and the unit vectors along the principal directions of a stretch tensor $\mathbf{U}=(\mathbf{F}^T\mathbf{F})^{1/2}$, respectively, $\mathbf{F}$ being the deformation gradient that maps the fixed reference configuration κ_o into a configuration κ (actual configuration) occupied by $\mathcal{M}_e$ at any instant t in the course of straining. The smooth and monotone function f is called the scale function [1–3]. The most commonly used strain tensors belong to the family defined by

$$f(z)=(z^{2n}-1)\,|_{2n}; \qquad \mathbf{E}(n)=\sum_{k=1}^{3}[(U_k^{2n}-1)\,|_{2n}]\mathbf{N}_k\otimes\mathbf{N}_k \tag{1.2}$$

where n may have any value. Distinguishing the class of strain measures (1.1) gave rise to the notion of measure invariance [3, 4]. Some aspects of such an invariance will briefly be discussed in Section 1.2.

Let e_k be an orthonormal triad fixed in physical space (laboratory). The matrix of the components of any tensor with respect to e_k is henceforth denoted by a Roman-type letter as, for instance $\mathbf{A}=\mathrm{A}_{KL}\mathbf{e}_k\otimes\mathbf{e}_L$, $\mathrm{A}=\|\mathrm{A}_{KL}\|$.†

According to Vakulenko [66], Kestin and Rice [38], and Kestin [39], an element of elastic–plastic rate-independent material may be regarded as an elementary thermodynamic system being in constrained equilibrium at any instant of a deformation process. It is assumed for simplicity in illustrating the structure of the *H–R* formalism, that the internal state (or pattern of internal rearrangement—*PIR*, Rice [9]) of $\mathcal{M}_e$ is characterized by a discrete set of internal variables. Their components on $\mathbf{e}_k$ are denoted collectively by H^r, $r=1,2,\ldots,n$. The state of constrained equilibrium of $\mathcal{M}_e$ is characterized by E (matrix of components of $\mathbf{E}$ on $\mathbf{e}_k$), entropy s and H^r. When two neighbouring constrained equilibrium states are considered, the infinitesimal change in the *internal energy* u of $\mathcal{M}_e$ is

$$\mathrm{d}u=\mathrm{tr}\,(T\,\mathrm{d}E)-\pi_r\,\mathrm{d}H^r+\theta\,\mathrm{d}s \tag{1.3}$$

where π_r denotes collectively the components on $\mathbf{e}_k$ of internal thermodynamic forces and T is the symmetric matrix of background components of the work-conjugate stress $\mathbf{T}$. We employ here for convenience the work-conjugate stress 'per unit of mass'. The usual work-conjugate stress per unit of volume can be obtained by multiplying T by ρ_o, $1/\rho_o$ being the volume of $\mathcal{M}_e$ in a fixed reference configuration. Using the polar decomposition $F=RU$ one can derive, cf. [10], the simple relation between E, matrix of background components of the usual Cauchy stress tensor σ and T as

$$T_{IJ}=\frac{1}{\rho}\,\overset{-1}{U}_{MN}(R^T\sigma R)_{NK}\frac{\partial U^{KM}}{\partial E_{IJ}} \tag{1.4}$$

† Throughout the whole paper, the indices run consequently over the following sequences: $\alpha,\beta,\gamma,\delta,\eta=1,2,\ldots,6$; $i,j,k,l,m,n=1,2,\ldots,6$; $\mu,\nu,\sigma=1,2,\ldots,n$; $p,r,s,t=1,2,\ldots,n$; $I,J,K,L,M,N,\ldots=1,2,3$.

where E is related to U by (1.1) and $1/\rho$ is the volume of $\mathcal{M}_e$ at the actual configuration κ. Alternatively, we shall utilize the simplified notation commonly used in the theory of mechanical systems. To this end, we denote by q^i the generalized co-ordinates derived from E, i.e. $q^1 = E_{11}$, $q^2 = E_{22}$, $q^3 = E_{33}$, $q^4 = 2E_{23}$, $q^5 = 2E_{13}$, $q^6 = 2E_{12}$ and by p_i the generalized conjugate variables $p_1 = T_{11}, \ldots, p_4 = T_{23}, \ldots$. The fundamental state equation becomes

$$\mathrm{d}u(\Upsilon^{sq}) = p_i \, \mathrm{d}q^i - \pi_r \, \mathrm{d}H^r + \theta \, \mathrm{d}s \tag{1.5}$$

which allows us to write the increments of the free energy function Φ and the free enthalpy Ψ as

$$\begin{aligned} \Phi &= u - \theta s; \qquad & \mathrm{d}\Phi(\Upsilon^{\theta q}) &= p_i \, \mathrm{d}q^i - s \, \mathrm{d}\theta - \pi_r \, \mathrm{d}H^r \\ \Psi &= \Phi - p_i q^i; \qquad & -\mathrm{d}\Psi(\Upsilon^{\theta p}) &= q^i \, \mathrm{d}p_i + s \, \mathrm{d}\theta + \pi_r \, \mathrm{d}H^r \end{aligned} \tag{1.6}$$

where the sets of variables

$$\Upsilon^{sq} = \{s, q^i, H^r\}, \qquad \Upsilon^{\theta q} = \{\theta, q^i, H^r\}, \qquad \Upsilon^{\theta p} = \{\theta, p_i, H^r\} \tag{1.7}$$

represent the co-ordinates of a point in the three mutually dual thermodynamic representative spaces [25]. Using the H–R formalism one may single out three types of independent differentials corresponding to three different infinitesimal paths (either real or virtual) in a representative space

$$\mathrm{d}_1\Upsilon^{\theta q} = \{\delta\theta, \delta q^i, 0\}; \qquad \mathrm{d}_2\Upsilon^{\theta q} = \{\mathrm{d}\theta, \mathrm{d}q^i, \mathrm{d}H^r\}; \qquad \mathrm{d}_3\Upsilon^{\theta q} = \{0, 0, \mathrm{d}H^r\}.$$

The variation of any state function, say ν, along $\mathrm{d}_1\Upsilon^{\theta q}$, $\mathrm{d}_2\Upsilon^{\theta q}$ and $\mathrm{d}_3\Upsilon^{\theta q}$ will be denoted by $\delta\nu$, $\mathrm{d}\nu$ and $\mathrm{d}^{\mathrm{P}}\nu$, respectively:

$$\delta\nu = \frac{\partial\nu}{\partial\theta}\delta\theta + \frac{\partial\nu}{\partial q^i}\delta q^i; \qquad \mathrm{d}\nu = \frac{\partial\nu}{\partial\theta}\mathrm{d}\theta + \frac{\partial\nu}{\partial q^i}\mathrm{d}q^i + \frac{\partial\nu}{\partial H^r}\mathrm{d}H^r;$$

$$\mathrm{d}^{\mathrm{P}}\nu = \frac{\partial\nu}{\partial H^r}\mathrm{d}H^r \tag{1.8}$$

Here, $\mathrm{d}^{\mathrm{P}}\nu$ denotes the isothermal 'plastic' change in ν, whereas the 'elastic' change in ν is defined as $\mathrm{d}^{\mathrm{E}}\nu = \mathrm{d}\nu - \mathrm{d}^{\mathrm{P}}\nu$. The similar notation is used when Υ^{sq} or $\Upsilon^{\theta p}$ are taken as independent co-ordinates. The d-differentials occurring in eqns. (1.5) and (1.6) are understood in the aforementioned sense.

For rate-independent plastic materials H^r may change only along those path segments of the process that lie on a yield surface. Therefore, the d^P-variation of any quantity can be effected physically only by a finite cycle process that is associated with an infinitesimal change in H^r. The δ-differentials characterize an infinitesimal purely thermoelastic process. Thus, eqns. (1.5), (1.6) written for an elastic process become

$$\delta u = p_i\,\delta q^i + \theta\,\delta s; \qquad \delta\Phi = p_i\,\delta q^i - s\,\delta\theta; \qquad \delta\Psi = -q^i\,\delta p_i - s\,\delta\theta \tag{1.9}$$

From eqns. (1.5), (1.6), there result three sets of equivalent 'equations of state of constrained equilibrium', e.g.

$$p_i(Y^{\theta q}) = \frac{\partial\Phi}{\partial q^i}; \qquad -s(Y^{\theta q}) = \frac{\partial\Phi}{\partial\theta}; \qquad -\pi_r(Y^{\theta q}) = \frac{\partial\Phi}{\partial H^r} \tag{1.10}$$

where $Y^{\theta q}$ is regarded as the set of natural (independent) variables, and

$$-q_i(Y^{\theta p}) = \frac{\partial\Psi}{\partial p_i}; \qquad -s(Y^{\theta p}) = \frac{\partial\Psi}{\partial\theta}; \qquad -\pi_r(Y^{\theta p}) = \frac{\partial\Psi}{\partial H^r} \tag{1.11}$$

where $Y^{\theta p}$ are the natural variables. Equations (1.10), (1.11) are valid for points both at the boundary and inside the elastic region. Applying to eqns. (1.10), (1.11) the d- and δ-differentials, the rate form of state equations is obtained as

$$\left.\begin{aligned}
\mathrm{d}p_i(Y^{\theta q}) &= \underset{\boxed{1.1}}{\frac{1}{\rho_o}\,l_{ij}\,\mathrm{d}q^j} - \underset{\boxed{1.2}}{\frac{1}{\rho_o}\,\beta_i\,\mathrm{d}\theta} + \underset{\boxed{1.3}}{\mathrm{d}^P p_i}, \\
\theta\,\mathrm{d}s(Y^{\theta q}) &= \underset{\boxed{2.1}}{\frac{\theta}{\rho_o}\,\beta_i\,\mathrm{d}q^i} + \underset{\boxed{2.2}}{C_q\,\mathrm{d}\theta} + \underset{\boxed{2.3}}{\theta\,\mathrm{d}^P s(Y^{\theta q})}, \\
\mathrm{d}\pi_r(Y^{\theta q}) &= -\underset{\boxed{3.1}}{\frac{\partial p_j}{\partial H^r}\,\mathrm{d}q^j} + \underset{\boxed{3.2}}{\frac{\partial s(Y^{\theta q})}{\partial H^r}\,\mathrm{d}\theta} + \underset{\boxed{3.3}}{\mathrm{d}^P\pi_r(Y^{\theta q})}.
\end{aligned}\right\} \tag{1.12}$$

$$\delta p_i(Y^{\theta q}) = \frac{1}{\rho_o}\,l_{ij}\,\delta q^j - \frac{1}{\rho_o}\,\beta_i\,\delta\theta; \quad \theta\,\delta s(Y^{\theta q}) = \frac{\theta}{\rho_o}\,\beta_i\,\delta q^i + C_q\,\delta\theta \tag{1.13}$$

$$\left.\begin{aligned}
&\mathrm{d}q^i(\Upsilon^{\theta p}) = \rho_0 m^{ij}\,\mathrm{d}p_j + a^i\,\mathrm{d}\theta + \mathrm{d}^{\mathrm{P}}q^i,\\
&\qquad\qquad \boxed{\bar{1}.\bar{1}} \qquad \boxed{\bar{1}.\bar{2}} \qquad \boxed{\bar{1}.\bar{3}}\\
&\theta\,\mathrm{d}s(\Upsilon^{\theta p}) = \theta a^i\,\mathrm{d}p_j + C_{\mathrm{p}}\,\mathrm{d}\theta + \theta\,\mathrm{d}^{\mathrm{P}}s(\Upsilon^{\theta p}),\\
&\qquad\qquad \boxed{\bar{2}.\bar{1}} \qquad \boxed{\bar{2}.\bar{2}} \qquad \boxed{\bar{2}.\bar{3}}\\
&\mathrm{d}\pi_r(\Upsilon^{\theta p}) = \frac{\partial q^i}{\partial H^r}\,\mathrm{d}p_j + \frac{\partial s(\Upsilon^{\theta p})}{\partial H^r}\,\mathrm{d}\theta + \mathrm{d}^{\mathrm{P}}\pi_r(\Upsilon^{\theta p}),\\
&\qquad\qquad \boxed{\bar{3}.\bar{1}} \qquad \boxed{\bar{3}.\bar{2}} \qquad \boxed{\bar{3}.\bar{3}}
\end{aligned}\right\} \tag{1.14}$$

$$\delta q^i(\Upsilon^{\theta p}) = \rho_o m^{ij}\,\delta p_j + a^i\,\delta\theta; \qquad \theta\,\delta s(\Upsilon^{\theta p}) = \theta a^j\,\delta p_j + C_{\mathrm{p}}\,\delta\theta. \tag{1.15}$$

Here

$$\frac{1}{\rho_o} l_{ij} = \frac{\partial^2\Phi}{\partial q^i\,\partial q^j}, \qquad m^{ij} = -\frac{1}{\rho_o}\frac{\partial^2\Psi}{\partial p_i\,\partial p_j} \tag{1.16}$$

are generalized coefficients of isothermal elastic stiffness and compliance, respectively,

$$\frac{1}{\rho_o}\beta_i = -\frac{\partial p_i(\Upsilon^{\theta q})}{\partial\theta}, \qquad a^j(\Upsilon^{\theta p}) = \frac{\partial q^j}{\partial\theta} \tag{1.17}$$

denote generalized coefficients of elastic thermal stress and elastic thermal strain (thermal expansion) and

$$C_q(\Upsilon^{\theta q}) = \theta\frac{\partial s(\Upsilon^{\theta q})}{\partial\theta}, \qquad C_p(\Upsilon^{\theta p}) = \theta\frac{\partial s(\Upsilon^{\theta p})}{\partial\theta} \tag{1.18}$$

are specific heats at constant q^i and p_i, respectively. Note that $\mathrm{d}^{\mathrm{P}}s(\Upsilon^{\theta q})$ is not equal to $\mathrm{d}^{\mathrm{P}}s(\Upsilon^{\theta p})$; the former is the 'plastic' variation of the entropy at constant temperature and fixed shape of the element (isothermal, isometric plastic change in entropy) whereas the latter is the 'plastic' change in s at constant conjugate stress and θ (isothermal, isostatic plastic change in s). The sets of state equations (1.12) and (1.14) are entirely equivalent, which follows from the identities given in the Appendix. Multiplying (A.4)–(A.6) by $\mathrm{d}H^r$ we obtain the following useful relations

$$\begin{aligned}
&\mathrm{d}^{\mathrm{P}}s(\Upsilon^{\theta q})\,|_{q=q(\Upsilon^{\theta p})} = \mathrm{d}^{\mathrm{P}}s(\Upsilon^{\theta p}) + a^j\,\mathrm{d}^{\mathrm{P}}p_j,\\
&\mathrm{d}H^r\,.\,\mathrm{d}^{\mathrm{P}}\pi_r(\Upsilon^{\theta q})\,|_{q=q(\Upsilon^{\theta p})} = \mathrm{d}^{\mathrm{P}}p_i\,\mathrm{d}^{\mathrm{P}}q^i + \mathrm{d}^{\mathrm{P}}\pi_r(\Upsilon^{\theta p})\,.\,\mathrm{d}H^r,\\
&\mathrm{d}^{\mathrm{P}}p_i\,|_{q=q(\Upsilon^{\theta p})} = -\frac{1}{\rho_o} l_{ij}\,\mathrm{d}^{\mathrm{P}}q^j(\Upsilon^{\theta p}).
\end{aligned} \tag{1.19}$$

It is perhaps worthwhile to stress that eqn. $(1.19)_3$, well-known in isothermal plasticity, remains valid in the general coupled theory of plasticity.

1.1.2. *Coupled Effects. Dissipation*

The terms occurring in eqns. (1.12)–(1.14) represent the so-called 'thermostatic coupling effects' in thermoplasticity. Their interpretation follows directly from the physical meaning of the corresponding partial derivatives. Referring to eqns. (1.12)–(1.14), the following tentative names are assigned to each of the coupling effects: (1.1), $(\bar{1}.\bar{1})$, elasticity; (2.2), $(\bar{2}.\bar{2})$, heat capacity; (3.3), $(\bar{3}.\bar{3})$, mechanics of internal rearrangement; (1.2), elastic thermal stress; $(\bar{1}.\bar{2})$, elastic thermal strain; (2.1), heat of elastic deformation; $(\bar{2}.\bar{1})$, piezocaloric effect; (1.3), isothermal stress relaxation due to changes in *PIR*; $(\bar{1}.\bar{3})$, isothermal variation in strain due to changes in *PIR*; (3.1), strain induced internal forces; $(\bar{3}.\bar{1})$, stress induced internal forces; (2.3), isometric heat of internal rearrangement; $(\bar{2}.\bar{3})$, isostatic heat of internal rearrangement; (3.2), isometric thermally induced internal forces; $(\bar{3}.\bar{2})$, isostatic thermally induced internal forces.

The first law of thermodynamics written for an infinitesimal purely thermoelastic process and an active plastic process reads

$$\delta u = p_i\, \delta q^i - \bar{\bar{\mathrm{d}}}Q, \qquad \mathrm{d}u = p_i\, \mathrm{d}q^i - \bar{\bar{\mathrm{d}}}Q \tag{1.20}$$

Here, $\bar{\bar{\mathrm{d}}}Q$ is an increment of heat transmitted to the surrounding in the course of an infinitesimal process. According to concepts of the classical thermodynamics [25, 26, 38, 39], the dissipation of mechanical work is the difference between the work actually done and the work that would have been done on the system had the process been reversible. Thus, the non-zero increment of energy dissipation is possible only when the process is emanating from a state that lies on the yield surface and its value is

$$-\mathrm{d}^{\mathrm{P}}u = -\mathrm{d}^{\mathrm{P}}\Phi = -\mathrm{d}^{\mathrm{P}}\Psi = \pi_r\, \mathrm{d}H^r \geqslant 0 \tag{1.21}$$

on account of eqns. (1.5), (1.6), (1.20) and the second law of thermodynamics. There is no ambiguity in such a notation since (cf. A.3)

$$\pi_r(\Upsilon^{sq}) = \pi_r(\Upsilon^{\theta q}) = \pi_r(\Upsilon^{\theta p}) \tag{1.22}$$

for all points of mutually dual thermodynamic representative space whose co-ordinates are interrelated by the appropriate equations of state. Eliminating du and ds in eqns. (1.5), (1.20) and $(1.12)_2$ or $(1.14)_2$ one obtains two alternative equations for the increment of temperature

$$\begin{aligned} C_q\,\mathrm{d}\theta &= \pi_r\,\mathrm{d}H^r - \frac{\theta}{\rho_o}\,\beta_i\,\mathrm{d}q^i - \theta\,\mathrm{d}^{\mathrm{P}}s(\Upsilon^{\theta q}) - \bar{\bar{\mathrm{d}}}Q \\ C_p\,\mathrm{d}\theta &= \pi_r\,\mathrm{d}H^r - \theta a^j\,\mathrm{d}p_j - \theta\,\mathrm{d}^{\mathrm{P}}s(\Upsilon^{\theta p}) - \bar{\bar{\mathrm{d}}}Q \end{aligned} \tag{1.23}$$

for a process of active plastic straining, and

$$\begin{aligned} C_q\,\delta\theta &= -\frac{\theta}{\rho_o}\,\beta_i\,\delta q^i - \bar{\bar{\mathrm{d}}}Q \\ C_p\,\delta\theta &= -\theta a^j\,\delta p_j - \bar{\bar{\mathrm{d}}}Q \end{aligned} \tag{1.24}$$

for infinitesimal thermoelastic process.

1.1.3. *Thermodynamic Potentials. Stored Energy*

If at some chosen temperature $\theta = \theta_o$ the surface tractions vanish, $p_i = 0$ then the material element $\mathcal{M}_e$ is said to be in *a natural state.* The natural state at which $\mathcal{M}_e$ is in unconstrained equilibrium is called *the thermodynamical reference state* (trs). At trs we have $p_i = 0$, $\theta = \theta_o$, $H^r = 0$, $\pi_r = 0$ and $u = 0$, $s = 0$ by convention. Consider a material element $\mathcal{M}_e$ in a natural state and suppose that for some range of deformations and histories of deformation for every H^r there exists the unique solution to the following equation

$$\left.\frac{\partial \Phi}{\partial q^i}\right|_{\theta=\theta_o} = 0. \tag{1.25}$$

This requirement corresponds to the situation for which the Gibbs free enthalpy Ψ may be defined. Denote by $\overset{*}{q}{}^i(H^r)$ the solution of eqn. (1.25) and by $\overset{*}{s}(H^r)$ the value of entropy corresponding to $\overset{*}{q}{}^i$

$$\left.\frac{\partial \Phi}{\partial q^i}\right|_{\substack{q^i=\overset{*}{q}{}^i(H^r)\\ \theta=\theta_o}} = 0, \qquad s = -\left.\frac{\partial \Phi}{\partial \theta}\right|_{\substack{q^i=\overset{*}{q}{}^i(H^r)\\ \theta=\theta_o}} = \overset{*}{s}(H^r) \tag{1.26}$$

When the reference configuration coincides with the initial configuration then $\overset{*}{q}{}^i$ is commonly termed 'plastic' (or inelastic) strain. The entropy $\overset{*}{s}(H^r)$ will henceforth be referred to as 'entropy of natural internal rearrangement' or just 'stored entropy' at $\theta = \theta_o$. Without any loss of generality the free energy function, specific heat at $q^i = \text{const.}$, and elastic thermal stress coefficients can be reorganized into the form

$$\Phi(\Upsilon^{\theta q}) = \Phi_1(\Upsilon^{\theta q}) - \theta \overset{*}{s}(H^r) + \overset{*}{u}(H^r),$$
$$\beta_i = \beta_i^o(q^i, H^r) + \Delta\beta_i(\Upsilon^{\theta q}); \qquad \Delta\beta_i\,|_{\theta=\theta_o} = 0, \tag{1.27}$$
$$C_q = C_q^o(\theta, H^r) + \Delta C_q(\Upsilon^{\theta q}); \qquad \Delta C_q\,|_{q^i=\overset{*}{q}^i} = 0,$$

so that

$$\Phi_1 = 0; \qquad \frac{\partial \Phi_1}{\partial q^i} = 0; \qquad \frac{\partial \Phi_1}{\partial \theta} = \theta; \qquad d^P\Phi_1 = 0 \tag{1.28}$$

for $q^i = \overset{*}{q}^i(H^r)$, $\theta = \theta_o$ and every H^r, in view of (1.26). Here, $\overset{*}{u}(H^r)$ denotes the internal energy in the natural state and is usually called 'stored energy' at $\theta = \theta_o$

$$\overset{*}{u}(H^r) = u(\Upsilon^{sq})\Big|_{q^i=\overset{*}{q}^i}^{s=\overset{*}{s}} \tag{1.29}$$

C_q^o is the specific heat measured at zero 'elastic strain' and β_i^o are the elastic thermal stress coefficients measured at the reference temperature θ_o. The quantities $\Delta\beta_i$ and ΔC_q are related by a reciprocal relation of the form

$$\frac{1}{\rho_o}\Delta\beta_i = \int_{\theta_o}^{0} \frac{1}{\theta_1}\frac{\partial\, \Delta C_q}{\partial q^i}\, d\theta_1 \tag{1.30}$$

Using eqns. (1.27), (1.28) and (1.30) the state equations (1.10) can be integrated to give

$$\Phi = \Phi_N(q^i, H^r) - (\theta - \theta_o)s_1(q^i, H^r)$$
$$+ \int_{\theta_o}^{\theta} \left(1 - \frac{\theta}{\theta_1}\right) C_q(\Upsilon^{\theta_1 q})\, d\theta_1 - \theta \overset{*}{s}(H^r) + \overset{*}{u} \tag{1.31}$$

where Φ_N is the usual isothermal elastic potential energy at $\theta = \theta_o$ that vanishes at the stress-free state

$$\Phi_N = 0; \qquad \frac{\partial \Phi_N}{\partial q^i} = 0 \quad \text{for} \quad q^i = \overset{*}{q}^i \tag{1.32}$$

and partial entropy $s_1(q^i, H^r)$ is uniquely defined by

$$\frac{\partial s_1(q^i, H^r)}{\partial q^i} = \frac{1}{\rho_o}\beta_i^o(q^i, H^r) \quad \text{and} \quad s_1 = 0 \quad \text{for} \quad q^i = \overset{*}{q}^i(H^r) \tag{1.33}$$

Similar operations can be carried out for the free enthalpy function Ψ

to give in turn

$$\left.\begin{aligned} a^i &= a^i_o(p_i, H^r) + \Delta a^i(Y^{\theta p}), \qquad \Delta a^i = 0 \quad \text{for} \quad \theta = \theta_o \\ C_p &= C^o_p(\theta, H^r) + \Delta C_p(Y^{\theta p}), \qquad \Delta C_p = 0 \quad \text{for} \quad p_i = 0 \end{aligned}\right\} \tag{1.34}$$

$$\Delta a^i = \int_{\theta_o}^{\theta} \frac{1}{\theta_1} \frac{\partial\, \Delta C_p}{\partial p_j} \,\mathrm{d}\theta_1 \tag{1.35}$$

$$\Psi = \Psi_C(p_j, H^r) - p_i \overset{*}{q}{}^i(H^r) - (\theta - \theta_o)\bar{s}_1(p_i, H^r) + \int_{\theta_o}^{\theta} \left(1 - \frac{\theta}{\theta_1}\right) \Delta C_p(Y^{\theta_1 p})\,\mathrm{d}\theta_1$$
$$+ u^{(s)}(\theta, H^r) - \theta s^{(s)}(\theta, H^r) \tag{1.36}$$

$$\left.\begin{aligned} u^{(s)} &= \overset{*}{u}(H^r) + \int_{\theta_o}^{\theta} C^o_p(\theta_1, H^r)\,\mathrm{d}\theta_1 \\ s^{(s)} &= \overset{*}{s}(H^r) + \int_{\theta_o}^{\theta} \frac{1}{\theta_1} C^o_p(\theta_1, H^r)\,\mathrm{d}\theta_1. \end{aligned}\right\} \tag{1.37}$$

Here, Ψ_C is the usual elastic isothermal $(\theta = \theta_o)$ complementary energy vanishing at $p_i = 0$,

$$\Psi_C = 0; \qquad \frac{\partial \Psi_C}{\partial p_i} = 0 \quad \text{for} \quad p_i = 0 \tag{1.38}$$

$u^{(s)}$ and $s^{(s)}$ are stored internal energy and stored entropy at arbitrary temperature (and stress-free state), respectively, and entropy $\bar{s}_1$ is uniquely defined by

$$\frac{\partial \bar{s}_1(p_i, H^r)}{\partial p_i} = a^i_o(p_i, H^r) \quad \text{and} \quad \bar{s}_1 = 0 \quad \text{for} \quad p_i = 0$$

for arbitrary H^r. Partially integrated forms of the free energies (1.31) and (1.36) enable one to interpret the internal forces in terms of the macroscopic quantities. For isothermal processes such an interpretation was given by Rice [10, 11] who discussed relations of his macroscopic formulation to some deformation mechanisms observed on the microscale. To interpret $u^{(s)}$ consider an isothermal cyclic process in the p_i-space between two stress-free states. In view of (1.20)

$$\oint_{p_i = 0} p_i \,\mathrm{d}q^i - \oint_{p_i = 0} \bar{\bar{\mathrm{d}}}Q = u^{(s)}(\theta, H^r + \Delta H^r) - u^{(s)}(\theta, H^r) \tag{1.39}$$

thus, the increment in stored energy is equal to the difference between the work done on an element and the heat absorbed by a surrounding. The heat which takes part in such experiments is not equal to the energy dissipation though. If (1.39) is combined with (1.5) and (1.21), then

$$\oint_{p_i=0} \pi_r \, \mathrm{d}H^r = \oint_{p_i=0} \bar{\bar{\mathrm{d}}}Q + \theta[s^{(s)}(\theta, H^r + \Delta H^r) - s^{(s)}(\theta, H^r)]. \tag{1.40}$$

While stored energy $\overset{*}{u}(H^r)$ is measurable in macroscopic experiments (cf. [24]) no information is at present available on the role played by the stored entropy $\overset{*}{s}(H^r)$.† This problem arises as a direct consequence of our inability to change reversibly the internal rearrangements at macro-level. Since, during the course of plastic straining, the heat of energy dissipation cannot be separated from the heat of internal rearrangement (2nd term in the rhs of (1.40)), the usefulness of the second law of thermodynamics in the form (1.21) can be questioned. While cold work certainly produces changes in the entropy of metals it is believed that such changes (when multiplied by θ) are small in comparison to changes in $\overset{*}{u}$ [61]. Therefore, in practical applications of the theory the entropy $\overset{*}{s}(H^r)$ is frequently neglected [18]. Another point of view is to consider $\overset{*}{s}(H^r)$ as an additional internal variable that is formally defined by an evolution equation.

1.2. Aspects of Invariance. Transition to M formalism

1.2.1. *Transformation of the Generalized Co-ordinates*

Description of the change of shape of the element $\mathcal{M}_e$ during the deformation process is not unique. Each change of the reference configuration and/or of the strain measure can be interpreted in terms of the transformation of the generalized co-ordinates [3, 6, 7]. Also, the decomposition of the total strain into its elastic and plastic parts may be given a similar interpretation. However, in the latter case, the transformation rule is explicitly dependent upon the internal state and is treated as a transformation from the fixed generalized co-ordinates to a moving set of generalized co-ordinates. Thus‡

$$\begin{aligned} q^\alpha &= q^\alpha(q^i, H^r), \\ q^i &= q^i(q^\alpha, H^r) \end{aligned} \tag{2.1}$$

† Note that the stored entropy $\overset{*}{s}(H^r)$ does not enter the equation for temperature. To prove this, it is sufficient to calculate π_r and $\mathrm{d}^\mathrm{p}s$ by (1.31) and to substitute the result into (1.23).

‡ The convention is consequently used that quantities furnished with Greek letter indices are new co-ordinates whereas those marked with roman letter indices are old co-ordinates.

where

$$\det Q_i^\alpha \neq 0; \qquad Q_i^\alpha = \frac{\partial q^\alpha}{\partial q^i} \tag{2.2}$$

for each H^r. The more general transformation of the form

$$q^\alpha = q^\alpha(q^i, H^r); \qquad H^\sigma = H^\sigma(q^i, H^r) \tag{2.3}$$

could be considered but it is not needed in the present development. (The transformation $(2.3)_2$ is necessary to consider in passing from the micro-level considerations to phenomenological theories based upon certain averaged measures of the structural rearrangement.)

Let κ_o' be a new reference configuration (Fig. 1). The shape of the deformed element can be described by means of the matrix F', $F = F'A$ with A being the matrix that converts the old reference configuration κ_o into the new one κ_o'. Using the polar decomposition $F = RU$, $F' = R'U'$ we have

$$E = \bar{f}(U), \qquad E' = \bar{f}'(U'); \qquad U' = (\overset{-T}{A} U^2 \overset{-1}{A})^{1/2} \tag{2.4}$$

Here, $\bar{f}$ and $\bar{f}'$ are matrix functions defining the deformation measure (cf. eqn. (1.1)). If we wish to consider changes of the deformation measures at fixed reference configuration, then we merely need to put

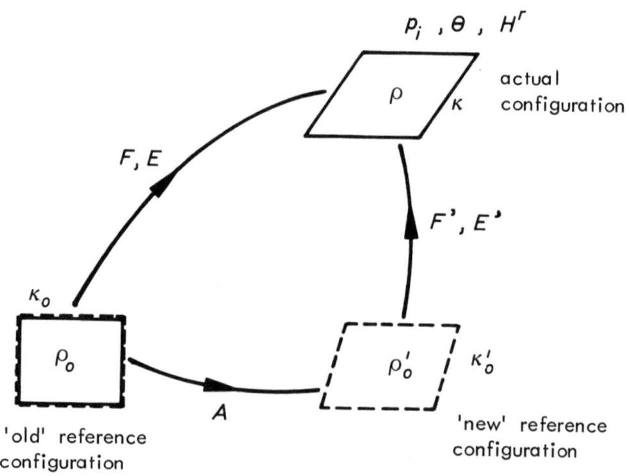

FIG. 1. Change of reference configuration.

$A = I$; changes of the reference configuration for a given deformation measure can be accounted for by putting $\bar{f}' = \bar{f}$. Elimination of U from the relations

$$E = \bar{f}(U), \qquad E' = \bar{f}'[(\overset{-T}{A} U^2 \overset{-1}{A})^{1/2}] \tag{2.5}$$

proves that E' is expressible, at least in principle, as a function of E and vice versa

$$E' = \psi(E); \qquad E = \overset{-1}{\psi}(E'). \tag{2.6}$$

The matrix A is a constant parameter of these transformations. Identifying E' with q^α and E with q^i, eqn. (2.6) can be interpreted as a change of the generalized co-ordinates of the form [3]:

$$q^\alpha = q^\alpha(q^i) \tag{2.7}$$

i.e. the change which does not depend upon H'. The basic connection between E' and E can be derived by expanding the relations (2.5) in a power series with respect to U around the point $U = I$ (I being the unit matrix) and eliminating U between the expansions obtained. The general result is not very stimulative. However, two particular cases are of practical interest:

(i) When the reference configuration is fixed and the strain measure is changed then the expansion is [3]

$$E' = E + (m' - m)E^2 + \dots \tag{2.8}$$

with $2m' = 1 + \left[\dfrac{\mathrm{d}^2 f'}{\mathrm{d}z^2}\right]_{z=1}$; $2m = 1 + \left[\dfrac{\mathrm{d}^2 f}{\mathrm{d}z^2}\right]_{z=1}$, f and f' being two different scale functions from the class (1.1). It follows in terms of generalized co-ordinates

$$Q_i^\alpha \big|_{q^i=0} = \frac{\partial q^\alpha}{\partial q^i}\bigg|_{q^i=0} = \delta_i^\alpha \tag{2.9}$$

for all the strain measures considered, and

$$Q_{ij}^\alpha \big|_{q^i=0} = \frac{\partial^2 q^\alpha}{\partial q^i\, \partial q^j}\bigg|_{q^i=0} = 0 \tag{2.10}$$

for all the strain measures with a common value of m.

(ii) The exact relations (2.6) are particularly simple (linear) when the Green scale function ($n=1$; $m=1$) is assumed for the distinct reference configurations. The same concerns the Almansi scale function ($n=1$; $m=-1$) so that

$$E'(1)=\overset{-T}{A}E(1)\overset{-1}{A}+\tfrac{1}{2}(\overset{-T}{A}\overset{-1}{A}-I); \qquad E(1)=\overset{T}{A}E'(1)A+\tfrac{1}{2}(\overset{T}{A}A-I)$$

$$E'(-1)=AE(-1)\overset{T}{A}+\tfrac{1}{2}(I-A\overset{T}{A}); \quad E(-1)=\overset{-1}{A}E'(-1)\overset{-T}{A}+\tfrac{1}{2}(I-\overset{-1}{A}\overset{-T}{A}) \tag{2.11}$$

where the pairs $E'(1)$, $E(1)$ and $E'(-1)$, $E(-1)$ stand for the Green and Almansi strain measures respectively.

Equations (2.8) and (2.11) are useful in passing from the Lagrangian to Eulerian description of the material properties. In the latter case, the configuration at $t=t_o$ is taken as the new reference configuration. Thus, $F(t)=F'(t)F(t_o)$ for $t\geqslant t_o$ so that $A=F(t_o)$, $F'(t_o)=I$. The transformation of tensorial quantities is particularly simple if the Green (or Almansi) strain measure is employed in the Lagrangian description. The first step of the procedure is then to change the reference configuration according to eqn. (2.11), which is followed by the change of the strain measure by eqn. (2.8).

Note that the origin of the generalized co-ordinates is not altered by (2.7) provided A is an orthogonal matrix. As we mentioned, the decomposition of a total strain into its elastic and plastic parts can be interpreted as a transformation of the generalized co-ordinates. To define elastic strain, one employs the notion of the *instantaneous natural state* (ins) of the constrained equilibrium:

> The instantaneous natural state of $\mathcal{M}_e$ is that state which the material element would have attained, at a given instant, had the surface tractions and temperature been instantaneously reduced to zero and θ_o, respectively, keeping the pattern of internal rearrangement fixed (i.e. $H^r=\text{const.}$).

The configuration $\overset{*}{\kappa}$ of an element being in ins is called the intermediate, relaxed, stress-free or unloaded configuration; it is defined up to an arbitrary rigid-body rotation. The above notions have, of course, a conceptual character, the shape of an element being in ins is accessible to our perception only if the actual configuration κ coincides with $\overset{*}{\kappa}$. In accordance with the notation employed in Section 1.1.3 we shall consistently use the symbol $\overset{*}{q}{}^i(H^r)$ to denote the values of the generalized co-ordinates that describe the shape of $\mathcal{M}_e$ at ins, and the symbol E^{p} for the matrix of the background components of the corresponding plastic strain tensor. The difference in the shape of $\mathcal{M}_e$ at the actual state and at ins can be described by the tensor of the elastic strain defined as

$$E^{\mathrm{e}} = \psi_{\mathrm{o}}(E, E^{\mathrm{p}}) \tag{2.12}$$

where ψ_{o} is an arbitrary (preferably differentiable) function such that $\psi_{\mathrm{o}}(E^{\mathrm{p}}, E^{\mathrm{p}}) = 0$. Identifying E with q^i and E^{e} with q^α the transformation (2.12) can be regarded as a particular type of the transformation (2.1), namely

$$q^\alpha = q^\alpha[q^i, \overset{*}{q}{}^i(H^r)], \qquad q^\alpha\,|_{q^i = \overset{*}{q}{}^i} = 0. \tag{2.13}$$

The geometrical interpretation of E^{e} and E^{p} depends on the form of the function χ_{o}. The intermediate configuration and the multiplicative and polar deformation gradient decompositions offer an attractive means for such an interpretation (Fig. 2)

$$F = F^{\mathrm{e}}F^{\mathrm{p}}; \qquad F^{\mathrm{p}} = R^{\mathrm{p}}U^{\mathrm{p}} = V^{\mathrm{p}}_{(1)}R^{\mathrm{p}}; \qquad F^{\mathrm{e}} = R^{\mathrm{e}}U^{\mathrm{e}} = V^{\mathrm{e}}_{(1)}R^{\mathrm{e}} \tag{2.14}$$

with F, F^{e} and F^{p} being the total, elastic and plastic deformation gradients, respectively. The symbol κ_{o} in Fig. 2 stands for the initial

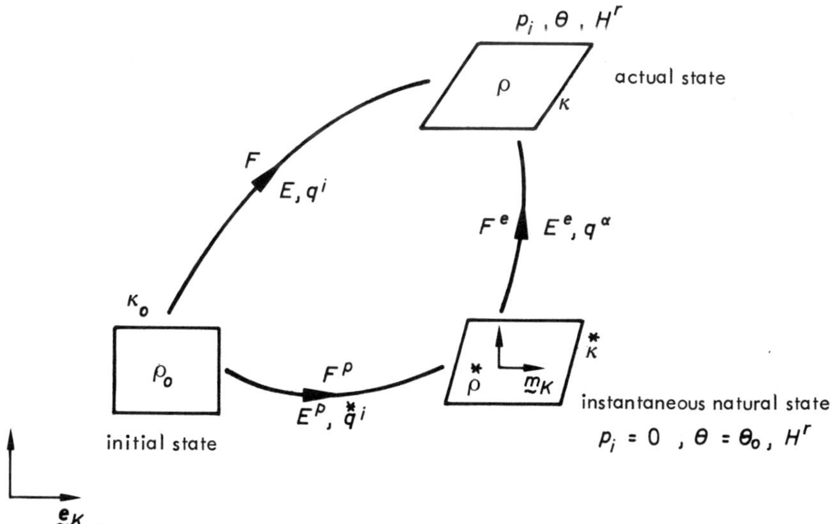

FIG. 2. Decomposition of total strain into elastic and plastic parts.

configuration. One usually adopts the same scale function f (the same strain measure) for the definition of E and E^{p} although different strain measures are also admissible. In refs. 30–32 the simplest function χ_o was proposed leading to the additive decomposition of the total strain

$$E^{\mathrm{e}} = E - E^{\mathrm{p}} \tag{2.15}$$

If the Green scale function is used for the definition of E^{p} then $E^{\mathrm{p}}(1) = \frac{1}{2}(\overset{T}{F^{\mathrm{p}}} F^{\mathrm{p}} - I)$. The decomposition (2.15) has been employed by many researchers.

Another basic relation that leads to eqns. (2.12) and (2.13) with a geometrically interpretable definition of E^{e} arises from the identity

$$2E^{\mathrm{e}} = I - \overset{-1}{U}\overset{2}{U^{\mathrm{p}}}\overset{-1}{U} = R^{\mathrm{T}}(I - \overset{-T}{F^{\mathrm{e}}}\overset{-1}{F^{\mathrm{e}}})R \tag{2.16}$$

Bearing in mind that U and U^{p} are expressible, at least in principle, in terms of E and E^{p}, respectively (cf. eqn. (2.4)), one obtains from (2.16) a group of relations between E^{e}, E^{p} and E [35, 36]. To define the elastic strain Lehmann [45] has used embedded basis and decomposed the metric tensor in the following multiplicative way

$$g^i_k = \mathring{g}^{ij} g_{kj} = \mathring{g}^{im} \overset{*}{g}_{mn} \overset{*}{g}^{nl} g_{lk} \tag{2.17}$$

where $(\mathring{g}^{ij}, \mathring{g}_{kl})$, $(\overset{*}{g}^{ij}, \overset{*}{g}_{kl})$, (g^{ij}, g_{kl}) are pairs of contra- and covariant metric tensors in the configurations κ_o, $\overset{*}{\kappa}$ and κ, respectively (Fig. 2). The elastic strain tensor is defined as

$$_{(e)}\varepsilon^{i}_{.k} = f^i_k({}_{(e)}g^{m}_{.n}), \qquad {}_{(e)}g^{m}_{.n} = \overset{*}{g}^{ml} g_{ln} \tag{2.18}$$

where f^i_k is a representation of an isotropic tensor function. Since ${}_{(e)}g^{l}_{\cdot k}\mathbf{g}_i \otimes \mathbf{g}^k = \overset{T}{\mathbf{F}^{\mathrm{e}}}\mathbf{F}^{\mathrm{e}}$ cf. [58], where $\mathbf{g}^k, \mathbf{g}_l$ are contra- and covariant vectors of the embedded basis in κ we may equivalently rewrite eqn. (2.18) as

$$_{(e)}\boldsymbol{\varepsilon} = \mathbf{f}(\overset{T}{\mathbf{F}^{\mathrm{e}}}\mathbf{F}^{\mathrm{e}}); \qquad {}_{(e)}\boldsymbol{\varepsilon} = {}_{(e)}\varepsilon^{i}_{\cdot k}\mathbf{g}_i \otimes \mathbf{g}^k \tag{2.19}$$

Denoting by ε^{e} the array of background components of the tensor ${}_{(e)}\boldsymbol{\varepsilon}$ and combining eqns. (2.16) and (2.19) we arrive at

$$R^{\mathrm{T}}\varepsilon^{\mathrm{e}}R = f[(1 - 2E^{\mathrm{e}})^{-1}] \tag{2.20}$$

where E^{e} is expressed in terms of E^{p} and E via (2.16). Hence, Lehmann's tensor may essentially be used for isotropic solids only. However, the rotated tensor $'E^{\mathrm{e}} = R^{\mathrm{T}}\varepsilon^{\mathrm{e}}R$ is an example of another elastic strain tensor that belongs to the class (2.12).

The definition (2.12) clearly implies that the tensor E^e is independent of the rotations in the conceptual stress-free configuration. The tensors E, E^e, E^p are frequently employed in the Lagrangian description of the elastic–plastic properties of solids. Such a description can be used to advantage in formulating and solving equations describing many structural mechanics problems (plates, shells). However, it is not advantageous to employ the Lagrangian formalism while developing general macroscopic constitutive equations for elastic–plastic solids. The use of the Lagrangian tensors E, E^e, E^p would complicate any experimental verification of the equations proposed because their components are all dependent upon a certain past configuration κ_o of the element, which is generally unknown. We note that knowledge of κ and $\overset{*}{\kappa}$ is not sufficient to determine E^e even if the actual state is purely elastic! Therefore, it is preferable from the experimental point of view to take as the reference configuration either $\overset{*}{\kappa}$ or κ. According to the basic concept of the macroscopic internal variables the influence of the history of plastic strains and temperature is then accounted for by establishing the actual values of these variables and investigating experimentally the appropriate evolution laws. The advantage here is taken of the experimental evidence that thermoelastic properties (of metallurgically stable ductile metals and alloys) so measured, are basically insensitive to plastic deformations until the grain rotations affect the overall anisotropy of the element.

The stress-free configuration $\overset{*}{\kappa}$ has been accepted as the reference configuration by many authors [12–23, 33, 40–43, 63, 64, 68] starting with the paper by Eckart [29]. The constitutive equations formulated in $\overset{*}{\kappa}$ can later be transformed to κ_o or κ, if required by, say, a solution algorithm.

The mapping $\overset{*}{\kappa}$ changes during plastic straining while the orientation (but not the shape!) in the physical space of the element in ins is arbitrary. This clearly implies that the appropriately invariant $\overset{*}{\kappa}$-description of the material properties may only be accomplished with respect to a frame which is somehow related to the microstructure of $\mathcal{M}_e$ in $\overset{*}{\kappa}$. Three orthogonal unit vectors $\mathbf{m}_k$ representing such a frame were called by Mandel [12–17] the 'director triad'. The orientation of a material element in $\overset{*}{\kappa}$ may thus be characterized by the orthogonal tensor.

$$\overset{*}{\mathbf{R}} = \mathbf{m}_k \otimes \mathbf{e}_k \tag{2.21}$$

The simplest choice of orientation (made here for all further considerations) is such that $\overset{*}{\mathbf{R}} = \text{const.}$ (i.e. $\mathbf{m}_k = \text{const.}$) at each instant of the process. The material element in $\overset{*}{\kappa}$ is rotated at every moment during straining in such a way that the orientation of the director triad with respect to $\mathbf{e}_k$ is preserved. Such a family of the stress-free configurations was called by Mandel [12–17] the family of *isoclinic configurations.* For physical theories accounting for the microstructure of the element (cf. refs. 53 and 65) Mandel's approach seems natural by allowing the description of mutual rotations of the microelements forming the solid macroelement. However, all nine components of the overall plastic deformation gradient F^{p} are then dependent on H^r

$$F^{\mathrm{p}} = F^{\mathrm{p}}(H^r) \tag{2.22}$$

The effective use of Mandel's approach requires three new constitutive equations to be deduced in order to complete the macroscopic description. This somewhat complicates working out appropriate experimental techniques, but, on the other hand, it indicates the physically motivated way of generalizing classical plasticity models.

Putting $A = F^{\mathrm{p}}(H^r)$, $E' = E^{\mathrm{e}}$ into eqn. (2.5) and eliminating U we arrive at a general formula relating the elastic strain E^{e} defined with respect to $\overset{*}{\kappa}$ to the total strain defined with respect to the fixed reference configuration κ_o

$$E^{\mathrm{e}} = \psi_1[E, F^{\mathrm{p}}(H^r)]; \qquad E = \overset{-1}{\psi}_1[E^{\mathrm{e}}, F^{\mathrm{p}}(H^r)]. \tag{2.23}$$

We have again $E^{\mathrm{e}} = 0$ for $E = E^{\mathrm{p}}$. The final form of eqn. (2.23) depends on $F^{\mathrm{p}}(H^r)$. Identifying E^{e} with q^α and E with q^i we obtain the most general transformation of the form (2.1) which satisfies the conditions

$$q^\alpha(q^i, H^r)\,|_{q^i = \overset{*}{q}{}^i} = 0; \qquad q^i(q^\alpha, H^r)\,|_{q^\alpha = 0} = \overset{*}{q}{}^i(H^r) \tag{2.24}$$

Lee [18], Mandel [12–17] and other authors take Green's scale function while defining E^{e}. Then, eqn. (2.23) results directly from (2.11) after putting $A = F^{\mathrm{p}}$, $E^{\mathrm{e}} = E'$. Similarly, for the Almansi strain measure eqn. (2.23) may be deduced directly from (2.11).

1.2.2. *Some Aspects of Invariance*

According to the basic concept of classical thermodynamics [25], the internal energy of $\mathcal{M}_e$ is defined as the work done on the element in an arbitrary adiabatic process beginning at trs. Heat absorbed by $\mathcal{M}_e$ in any particular process between the given states is the difference between the work which would have been done on the system had the process been adiabatic, and the work actually done on $\mathcal{M}_e$ in this process. The process need not involve the infinitesimal changes in state nor be reversible. Therefore, we shall take for granted that internal energy is the basic invariant under a general change (2.1) of the generalized co-ordinates. It also seems expedient to assume that both entropy s and temperature θ are co-ordinate invariant. Hence, free energy Φ is another basic invariant under the change of reference configuration, strain measure and decomposition of total strain into its elastic and plastic contributions,

$$\Phi(\theta, q^i, H^r)\Big|_{\substack{q^i = q^i(q^\alpha, H^\sigma) \\ H^r = H^\sigma \delta^r_\sigma}} = \Phi'(\theta, q^\alpha, H^\sigma) \tag{2.25}$$

Here, the primed symbol indicates the value of a non-indexed function in the new co-ordinate system.

We shall first confine ourselves to considering some invariance properties of functions and relations under the transformation (2.7). This notion has been introduced by Hill [4] (cf. also ref. 3) who in the context of isothermic processes found a fundamental invariant—a differential form that is bilinear in the generalized co-ordinates and their conjugates, and which is interpretable in terms of the work only. Various aspects of invariance in isothermal elastoplasticity are discussed by Hill in refs. 3, 4 and 5. The analysis of the invariance makes it possible, among others, to express some intrinsic co-ordinate invariant properties of a material in the form of inequalities. In the recent paper [7] Hill has extended his investigation to indicate the co-ordinate invariant relationships among state variables in thermoelasticity. Here, we proceed in the similar manner generalizing some of Hill's conclusions to thermoplasticity.

Under the assumption (2.25) not only u and Φ but also their δ-, d^P- and d-differentials are invariant provided u and Φ are expressed in terms of their natural variables Y^{sq} and $Y^{\theta q}$, respectively. A similar remark concerns $s(Y^{\theta q})$. The internal force π_r cannot depend on the way the element shape is described and, therefore, is itself invariant, $\pi_\infty = \pi_r \delta^r_\infty$, which follows directly from eqns. (2.25) and $(1.10)_3$. Thus, all terms occurring in $(1.12)_2$ and $(1.12)_3$ are the examples of co-ordinate invariant quantities.

The expression $q^i p_i$ (or $\operatorname{tr}(TE)$) depends on the strain measure;†

† It is worthwhile to mention that $TE - ET$ is measure invariant as proved in ref. 3.

therefore, the free enthalpy function Ψ is not co-ordinate invariant. This implies that each particular term appearing in the rhs of eqns. $(1.14)_{2,3}$ and $(1.15)_2$ is dependent on the strain measure. In particular, the specific heat at constant conjugate stress C_p is not measure invariant since the same change in temperature at different constant conjugate stresses will, in general, produce different changes in physical dimensions of the element [7]. Let us premultiply both sides of eqn. $(1.14)_1$ by p_i,

$$p_i \, dq^i(\Upsilon^{\theta p}) = \rho_o p_i m^{ij} \, dp_j + p_i a^i \, d\theta + p_i \, d^P q^i \tag{2.26}$$

The lhs of eqn. (2.26) represents a change in the total work and is co-ordinate invariant. However, using the transformation rules to be discussed in Section 1.2.3, it can be shown that none of the individual terms in the rhs of eqn. (2.26) is co-ordinate invariant. In particular, the change in 'plastic work' understood as $p_i \, d^P q^i$ depends on the choice of the strain measure. This is understandable as changes of the element physical dimensions that accompany cyclic changes of p_i are dependent upon the conjugate stress (cf. ref. 9).

Even though the free enthalpy Ψ is not invariant, the expression

$$d^P\Psi(\Upsilon^{\theta p}) = d^P\Phi \tag{2.27}$$

is invariant, on account of (1.22). Further the expression

$$I_1 = \delta \, d^P\Psi = -d^P q^i \, \delta p_i - d^P s(\Upsilon^{\theta p}) \, \delta\theta = \delta \, d^P\Phi = d^P p_i \, \delta q^i - d^P s(\Upsilon^{\theta q}) \, \delta\theta \tag{2.28}$$

does not depend on the choice of the reference configuration and strain measure. From the transformation rules presented in Section 1.2.3 it immediately follows that

$$I_2 = d^P p_i \, \delta q^i, \qquad I_3 = d^P p_i \, dq^i \tag{2.29}$$

are another example of invariants.

Suppose that (in the isothermic case, see ref. 2)

$$I_2 \geqslant 0 \tag{2.30}$$

at any point of a yield surface. Using conventional arguments, one concludes that $d^P p_i$ is co-directional with the inward normal ν_i to the yield surface in the q-space. Clearly, this normality rule is invariant, i.e. when it holds for one conjugate pair it also holds for any other pair. From the identities $(1.19)_3$ it follows that $d^P q^j$ is co-directional with $-m^{ij}\nu_i$ and hence it is also co-directional with the outward normal in the p-space. Thus, the normality rule and its invariance has been shown to remain valid in general coupled thermoplasticity, provided (2.30) is adopted as the basic constitutive inequality.† For isothermic processes Hill [1–3] discussed some further aspects of invariance in elasto–plasticity and stated that:

(a) If the relation between $\mathrm{d}q^i$ and $\mathrm{d}p_i$ allows a potential with one choice of co-ordinates then the rate relation in any other co-ordinates likewise allows a potential;
(b) the usual convexity and hence the usual intrinsic stability conditions are not co-ordinate invariant;
(c) the relative convexity defined in ref. 3 is co-ordinate invariant.

The properties (a) and (c) play the key role in developing variational principles and bifurcation conditions. In coupled thermoplasticity the rate relation in general does not allow a potential [48]. However, if some coupling effects of lesser importance can be neglected [46, 47] the potentiality is preserved and variational principles and relative convexity may be expressed in an invariant manner. (This requires slight generalization of the results presented in refs. 46 and 47 to account for finite strain.) The property (b) results directly from the elastic stiffness transformation rule and remains valid in coupled thermoplasticity.

1.2.3. *Transformation Rules* [3, 60]
The transformation rule (2.7) is considered first. Changing the reference configuration and/or strain measure we have for some of the quantities appearing in eqn. (1.10) and (1.12):

$$\mathrm{d}q^\alpha = Q^\alpha_i\,\mathrm{d}q^i, \tag{2.31}$$

$$\delta q^\alpha = Q^\alpha_i\,\delta q^i, \qquad p_\alpha = Q^i_\alpha p_i, \qquad C'_q = C_q, \tag{2.32}$$

$$\frac{1}{\rho_o}\,l_{\alpha\beta} = \frac{1}{\rho_o}\,Q^i_\alpha l_{ij} Q^j_\beta + Q^i_{\alpha\beta} p_i, \qquad \frac{1}{\rho'_o}\,\beta_\alpha = \frac{1}{\rho_o}\,Q^i_\alpha \beta_i, \tag{2.33}$$

$$\mathrm{d}^{\mathrm{P}} p_\alpha = Q^i_\alpha\,\mathrm{d}^{\mathrm{P}} p_i, \qquad \mathrm{d}^{\mathrm{P}} s'(\Upsilon^{\theta q}) = \mathrm{d}^{\mathrm{P}} s(\Upsilon^{\theta q}) \tag{2.34}$$

where (cf. (2.9) and (2.10))

$$Q^i_\alpha = \frac{\partial q^i}{\partial q^\alpha}; \qquad Q^\alpha_i = \frac{\partial q^\alpha}{\partial q^i}; \qquad Q^\alpha_i Q^i_\beta = \delta^\alpha_\beta; \qquad Q^i_{\alpha\beta} = \frac{\partial^2 q^i}{\partial q^\alpha\,\partial q^\beta} \tag{2.35}$$

† Note that when θ varies eqn. (2.30) is not equivalent to $\mathrm{d}^{\mathrm{P}} q^i\,\delta p_i \leqslant 0$ which is usually adopted in isothermal elasto–plasticity [8]. The latter inequality is not co-ordinate invariant unless $\delta\theta = 0$.

It is noted that the transformation (2.7), which is independent of H^r in the q-space implies in the dual p-space the transformation

$$p_\alpha = p_\alpha(p_i, \theta, H^r) \tag{2.36}$$

which depends on the thermodynamic state. Equation (2.36) results from substituting into the rhs of eqn. $(2.32)_2$ the state equation $(1.11)_1$. The quantities appearing in the rhs of (1.14) have more complex transformation rules. For instance,

$$\mathrm{d}^{\mathrm{P}}q^\alpha = \Gamma_i^\alpha\, \mathrm{d}^{\mathrm{P}}q^i; \qquad a^\alpha = \Gamma_i^\alpha a^i; \qquad \Gamma_i^\alpha = Q_i^\alpha + Q_{ij}^\gamma Q_\beta^j p_\gamma \rho_o' m^{\alpha\beta} \tag{2.37}$$

Because the transformation rules for $\mathrm{d}^{\mathrm{P}}q^i$ and $\mathrm{d}q^i$ are different, $\mathrm{d}^{\mathrm{P}}q^\alpha$ describes different $\mathrm{d}^{\mathrm{P}}q^i$ changes in the physical dimensions of the element. From eqns. $(2.32)_2$ and (2.37) it is seen that plastic work is not invariant ($p_\alpha\, \mathrm{d}^{\mathrm{P}}q^\alpha \neq p_i\, \mathrm{d}^{\mathrm{P}}q^i$). Similar proof of this fact can be found in the recent paper of Hill [6]. This observation contradicts the basic Nemat-Nasser assumption in his controversial paper [50].

The more complex case of the transformation (2.1) is considered now. Even if (2.7) is clearly a particular case of (2.1), it is convenient in the present context to treat the latter transformation as a relation between elastic and total strain only (and not as a transformation of generalized co-ordinates under change of reference state and/or strain measure). The transformation rule for the infinitesimal displacement defined by the d-differential is now not homogeneous

$$\mathrm{d}q^\alpha - \mathrm{d}^{(h)}q^\alpha = Q_i^\alpha\, \mathrm{d}q^i; \qquad \mathrm{d}^{(h)}q^\alpha = \frac{\partial q^\alpha}{\partial H^r}\, \mathrm{d}H^r \tag{2.38}$$

and merely the rules (2.32), (2.33) remain valid. It must be pointed out that the array Q_i^α for the transformation (2.1) has a different meaning than in the case of (2.7). For there is certain arbitrariness in decomposing the total strain into its 'elastic' and 'plastic' parts so that the invariance analysis serves now the purpose of comparing different assumptions and indicating their common points. The most essential feature of the transformation (2.1) is the necessity of the following internal forces transformation

$$\pi_r\, \mathrm{d}H^r = \pi_\sigma\, \mathrm{d}H^\sigma - p_\alpha\, \mathrm{d}^{(h)}q^\alpha \tag{2.39}$$

The forces $-\pi_r$ are the internal forces occurring in the H–R formalism. When taken with the opposite sign, they are equivalent to the dissipative forces for states on the yield surface. Employing the elastic strain as a state parameter (which is the case in the *M–L* formalism, for instance), the fundamental differential equation of state takes the form

$$d\Phi' = p_\alpha\, dq^\alpha - \pi_\sigma\, dH^\sigma - s'\, d\theta, \qquad s' = s, \qquad \Phi' = \Phi \tag{2.40}$$

Thus, in all the theories employing the elastic strain as the 'geometric' state variable only a part $-\pi_\sigma$ of the *H–R* forces $-\pi_r$ is made explicit in (2.40). This implies different physical character of the thermostatic couplings—the notion of 'stress' is replaced by the energy-conjugate stress p_α, the notion of 'strain' is replaced by the elastic strain, isostatic refers to a constant value of p_α, isomeric—to a constant value of the elastic strain. The equality (2.39) means, however, that in all the theories the dissipated energy is the same, although the choice of thermodynamical rates (or 'fluxes') and dissipation-conjugate forces can, in general, be different.

It is a common interpretation employed while decomposing the total strain to identify dq^α and $-d^{(h)}q^\alpha$ with the elastic and plastic strain increments, respectively. However, such strain changes are different from those defined within the *H–R* formalism and, moreover, they depend upon the way the total strain is decomposed. Therefore, different definitions of the elastic strain imply the different decompositions of the total work into its elastic and plastic parts. Taking the additive decomposition (2.15) as proposed in ref. 30, we have $Q^\alpha_i - \delta^\alpha_i$ and $Q^\alpha_{ij} = 0$ so that the elastic strain increment conforming with the *H–R* formalism must be defined as

$$d^E q^\alpha = dq^\alpha - d^P q^\alpha, \quad d^E q^\alpha = \delta^\alpha_i (dq^i - d^P q^i) \equiv \delta^\alpha_i\, d^E q^i \tag{2.41}$$

In order to compare the *H–R* definition of the elastic strain increment with that of Mandel–Lee we need to consider the transformation $d^P q^i \rightarrow d^P q^\alpha$ which has the form

$$d^P q^\alpha = \Gamma^\alpha_i (d^P q^i + Q^i_\beta\, d^{(h)} q^\beta) - \rho'_o p_i m^{\alpha\beta}\, d^{(h)} Q^i_\beta \tag{2.42}$$

$$d^{(h)} Q^i_\beta = \frac{\partial q^i}{\partial q^\beta\, \partial H^r}$$

Because the *H–R* formalism employs a fixed reference configuration, it is here assumed that the instantaneous stress-free configuration $\overset{*}{\kappa}$, at a given instant, is treated as the new fixed reference configuration. Confining ourselves to the Green (or Almansi) strain measure, substituting $A = F^p(H^r)$, $E' = E^e$ into eqn. (2.11) and using eqn. (2.42) we get

$$\delta^P q^i \delta_i^\alpha = d^P q^\alpha - d^{(h)} q^\alpha + \rho_o' \delta_i^\gamma p_\gamma m^{\alpha\beta} d^{(h)} Q_\beta^i$$

because now $Q_\beta^i = \delta_\beta^i$, $Q_i^\alpha = \delta_i^\alpha$. Eliminating $d^{(h)}q^\alpha$ we arrive at

$$d^E q^\alpha = \delta_i^\alpha d^E q^i + \rho_o' \delta_i^\gamma p_\gamma m^{\alpha\beta} d^{(h)} Q_\beta^i$$

It is seen that $d^E q^\alpha$ as defined by eqn. (2.41) is only approximately equal to the H–R elastic strain increment. The correction accounting for the term $d^P q^\alpha$ in (2.41) has been introduced by Mandel in ref. 16.

Equation (2.25) implies the plastic change in entropy under the transformation (2.1) to be given by

$$d^P s'(\Upsilon^{\theta q}) + \frac{1}{\rho_o'} \beta_\alpha d^P q^\alpha = d^P s(\Upsilon^{\theta q}) \tag{2.43}$$

It can easily be proved that the temperature equation (1.23) remains valid if only the term $\pi_r \, dH^r$ is replaced by the rhs of eqn. (2.39). The partially integrated equations for free energy (1.31) and free enthalpy (1.36) hold as well, provided $\overset{*}{q}{}^i$ is replaced by $\overset{*}{q}{}^\alpha = 0$.

Another important conclusion can be drawn from the above considerations by superposing upon the transformation (2.1) the transformation (2.7) [60]: in the M-formalism, the change of elastic work defined by $p_\alpha \, dq^\alpha$, eqn. (2.40) is measure invariant, i.e. is not dependent upon the elastic strain measure in $\overset{*}{\kappa}$. Similarly, both the terms in the rhs of eqn. (2.39) are invariant under change of the strain measure in $\overset{*}{\kappa}$.

1.3. Mandel Formalism

1.3.1. $\overset{*}{\kappa}$-*description*

In what follows, matrices will be used representing tensors in the director triad $\mathbf{m}_k$ (cf. Section 1.2.1). Under the assumption $\overset{*}{R} = I$ (cf. (2.21)), such tensor representations coincide with the background components. Mandel [12–17], Lee [18–21] and the majority of other researchers [22, 23, 33, 40–42, 63, 64] have employed the Green measure for the description of the elastic strain

$$2E^e(1) = C^e - I, \qquad C^e = \overset{T}{F^e} F^e \tag{3.1}$$

According to the conclusion stated at the end of the previous section, any other strain measure E^e from the class (1.1) can be used instead of (3.1) provided U in (1.1) is replaced by the elastic stretch tensor

$U^{e}=(\overset{T}{F}{}^{e}F^{e})^{1/2}$. Free energy in the $\overset{*}{\kappa}$-description is given as

$$\Phi'=\Phi'(\theta, E^{e}, H^{\sigma}) \tag{3.2}$$

which is obtained from the function $\Phi(\theta, E, H^{\sigma})$ by eliminating E by means of (2.23). The fundamental equations in the M-formalism may be obtained directly by noting that the rate of total work in the actual state is given by $\operatorname{tr}(\sigma D)(1/\rho)$, where D is the deformation rate tensor

$$2D=V+V^{\mathrm{T}}, \qquad V=\dot{F}\overset{-1}{F} \tag{3.3}$$

and assuming the Green scale function in the definition of E^{e}. This simple presentation would not reveal all the relations of the theory to other formalisms, though.

Denote by T^{e} the energy-conjugate stress-tensor corresponding to E^{e} (which is a counterpart of p_{α} in (2.39)). Because $p_{\alpha}\,\mathrm{d}^{(h)}q^{\alpha}$ is invariant under choice of the elastic strain measure in $\overset{*}{\kappa}$, its tensorial counterpart may be found by taking $E^{e}(1)$ as the measure of the elastic strain. The relation between $E^{e}(1)$ and the Green measure in the Lagrangian description is given by eqn. (2.11) in which $A=F^{\mathrm{p}}$ and $E'=E^{e}(1)$. This formula can be treated as the co-ordinate transformation from the Lagrangian description to the $\overset{*}{\kappa}$ description so that

$$T^{e}(1)=\frac{1}{\rho}\overset{-1}{F}{}^{e}\sigma\overset{-T}{F}{}^{e} \tag{3.4}$$

and

$$-p_{\alpha}\,\mathrm{d}^{(h)}q^{\alpha}\Leftrightarrow\operatorname{tr}(P^{*}V^{*}) \tag{3.5}$$

where

$$P^{*}=\frac{1}{\rho}\overset{-1}{F}{}^{e}\sigma F^{e}=T^{e}(1)C^{e}, \qquad V^{*}=\dot{F}^{\mathrm{p}}\overset{-1}{F}{}^{\mathrm{p}} \tag{3.6}$$

The rate of dissipation of energy is given by

$$\Sigma_{D}=\operatorname{tr}P^{*}V^{*}+\pi_{\sigma}\dot{H}^{\sigma} \tag{3.7}$$

for all the states that lie on the yield surface. By introducing the free-enthalpy function

$$\Psi'=\Phi'-\operatorname{tr}(T^{e}E^{e}) \tag{3.8}$$

the state equations are expressed in terms of T^{e}, θ, H^{σ} as

$$E^{e}=-\frac{\partial\Psi'}{\partial T^{e}}, \qquad s=-\frac{\partial\Psi}{\partial\theta}, \qquad \pi_{\sigma}=-\frac{\partial\Psi}{\partial H^{\sigma}} \tag{3.9}$$

for the isoclinic configuration $\overset{*}{\kappa}$. Time differentiation of $(3.9)_1$ yields†

$$\dot{E}^{\mathrm{e}} = \overset{*}{\rho} M^* \dot{T}^{\mathrm{e}} + A^* \dot{\theta} + \frac{\mathrm{d}^{\mathrm{P}} E^{\mathrm{e}}}{\mathrm{d}t}; \qquad \frac{\mathrm{d}^{\mathrm{P}} E^{\mathrm{e}}}{\mathrm{d}t} = \frac{\partial E^{\mathrm{e}}}{\partial H^{\sigma}} \dot{H}^{\sigma} \tag{3.10}$$

where M^* and A^* are isothermal elastic compliances and thermal strain coefficients, respectively, both symmetric and defined in $\overset{*}{\kappa}$. The equation for temperature takes the form

$$C_{\mathrm{p}} \dot{\theta} = \Sigma_D - \theta \operatorname{tr}(A^* \dot{T}^{\mathrm{e}}) - \theta \frac{\mathrm{d}^{\mathrm{P}} s}{\mathrm{d}t} - \dot{Q}, \qquad \frac{\mathrm{d}^{\mathrm{P}} s}{\mathrm{d}t} = \frac{\partial s(\theta, T^{\mathrm{e}}, H^{\sigma})}{\partial H^{\sigma}} \dot{H}^{\sigma} \tag{3.11}$$

For $\dot{H}^{\sigma} = 0$ (thermoelastic straining) the first and the third terms in the rhs of eqn. (3.11) vanish.

Both $\Phi' = \Phi$ and Ψ' may always be presented in the form (1.31) and (1.36) provided we put $\overset{*}{q}_i = 0$. Replacing in (1.36) p_j by T^{e}, a^i by A^* and decomposing A^* according to (1.34) it is noted that

$$\frac{\mathrm{d}^{\mathrm{P}} E^{\mathrm{e}}}{\mathrm{d}t} = 0$$

if A^* and Ψ_C are independent of H^{σ}. This is normally the case for most metals and alloys in the significant range of straining.

One of the most controversial issues of the contemporary finite strain plasticity is the separation of 'elastic strain rate' from the total strain rate in the course of active plastic loading and associated decomposition of the total rate of mechanical work into its elastic and plastic parts (cf. refs. 19, 20, 50). Within the H–R formalism such a decomposition is not unique since rate of plastic work depends upon the choice of reference configuration and strain measure as shown in Section 1.2.2 and remarked by Hill in ref. 6. This problem appears in the $\overset{*}{\kappa}$-description as well unless $\mathrm{d}^{\mathrm{P}} E^{\mathrm{p}}/\mathrm{d}t = 0$. When $\mathrm{d}^{\mathrm{P}} E^{\mathrm{e}}/\mathrm{d}t \neq 0$ Mandel [16] defines the elastic strain rate as

$$\overset{*}{\varepsilon}{}^{\mathrm{e}} = \dot{E}^{\mathrm{e}} - \mathrm{d}^{\mathrm{P}} E^{\mathrm{e}}/\mathrm{d}t \tag{3.12}$$

even though the corresponding rate of work

$$\Sigma^{\mathrm{e}}_{\mathrm{w}} = \operatorname{tr}(T^{\mathrm{e}} \overset{*}{\varepsilon}{}^{\mathrm{e}}) \tag{3.13}$$

is not invariant under change of elastic strain E^{e} in $\overset{*}{\kappa}$ (since the term

† Here we use the following abbreviation: AB means either $A_{ik}B_{kj}$ or $A_{ijkl}B_{kl}$.

tr $(T^e\, d^P E^e/dt)$ is not invariant). Presumably, the Mandel formula (3.12) was influenced by the *H–R* definition of the elastic change of the total strain as discussed in Secton 1.2.3. The suggestion concerning the invariant definition of the rate of elastic work in $\overset{*}{\kappa}$ valid in the situation when elastic properties in $\overset{*}{\kappa}$ are influenced by prior plastic deformations is presented in Section 1.3.3.

1.3.2. *Eulerian Description*

All the equations corresponding to the $\overset{*}{\kappa}$-description can obviously be transformed to the actual configuration κ by considering the general transformation rule dependent upon F^p. We shall confine ourselves to the Green strain measure (3.1) as in such a case the transformation is the simplest. The basic relations are

$$V = V^e + V^p; \quad V^e = \dot{F}^e \overset{-1}{F^e}; \quad V^p = F^e V^* \overset{-1}{F^e}; \quad \dot{T}^e(1) = \frac{1}{\rho} \overset{-1}{F^e} \overset{\nabla}{\sigma} \overset{-T}{F^e},$$
$$\overset{-T}{F^e} \dot{E}^e(1) \overset{-1}{F^e} = -\tfrac{1}{2}(V^p + \overset{T}{V^p}); \quad \overset{\nabla}{\sigma} = \dot{\sigma} - V^e\sigma - \sigma \overset{T}{V^e} + \sigma \operatorname{tr} V. \tag{3.14}$$

Following Mandel [16] the elastic and plastic parts of D are defined by (cf. (3.12))

$$D^e = \overset{-T}{F^e} \dot{E}^e(1) \overset{-1}{F^e} - D = \tfrac{1}{2}(V^e + \overset{T}{V^e}) - D'$$
$$D^p = \tfrac{1}{2}(V^p + \overset{T}{V^p}) + D'; \quad D' = \overset{-T}{F^e} \frac{d^P E^e(1)}{dt} \overset{-1}{F^e} \tag{3.15}$$

so that

$$D = D^e + D^p \tag{3.16}$$

Substituting eqn. (3.10) into eqn. (3.15) and accounting for eqns. $(3.14)_2$ the basic rate equation of state in κ in the case of the Green strain measure is found as

$$D^e = M(1)\overset{\nabla}{\sigma} + A(1)\dot{\theta},$$
$$M_{ijkl}(1) = (\det F^e)\overset{-1}{F^e_{mi}}\overset{-1}{F^e_{nj}} M^*_{mnpr}(1) \overset{-1}{F^e_{pk}} \overset{-1}{F^e_{r1}}, \; A(1) = \overset{-T}{F^e} A^*(1) \overset{-1}{F^e} \tag{3.17}$$

Here, $M(1)$ and $A(1)$ are the Green-elastic compliances and thermal expansion in κ. The rate of energy dissipation (3.7) and the equation for temperature (3.11) can be transformed to κ yielding

$$\Sigma_D = \frac{1}{\rho}\,\mathrm{tr}\,[\delta(D^{\mathrm{p}} - D')] + \pi_\sigma \dot{H}^\sigma,$$
$$C_{\mathrm{p}}\dot{\theta} = \Sigma_D - \theta\,\mathrm{tr}\,(A\overset{\nabla}{\sigma}) - \theta\frac{\mathrm{d}^{\mathrm{P}} s'(Y^{\theta\mathrm{p}})}{\mathrm{d}t} - \dot{Q} \tag{3.18}$$

at the yield point. The decomposition of the rate of total work into its elastic and plastic parts reads [16]

$$\Sigma_{\mathrm{w}} = \frac{1}{\rho}\,\mathrm{tr}\,(\sigma D) = \Sigma_{\mathrm{w}}^{(e)} + \Sigma_{\mathrm{w}}^{(\mathrm{p})},$$
$$\Sigma_{\mathrm{w}}^{(e)} = \mathrm{tr}\,[T^{\mathrm{e}}(1)\overset{*}{\dot{\varepsilon}}{}^{\mathrm{e}}] = \frac{1}{\rho}\,\mathrm{tr}\,(\sigma D^{\mathrm{e}});\quad \Sigma_{\mathrm{w}}^{(\mathrm{p})} = \frac{1}{\rho}\,\mathrm{tr}\,(\sigma D^{\mathrm{p}}) \tag{3.19}$$

However, this decomposition is not invariant under the change of the elastic strain measure E^{e} in $\overset{*}{\kappa}$.

Equations (3.17) can be rearranged into the following two equivalent forms

$$\frac{\delta^{cc}\sigma}{\delta t} = L(1)D - B\dot{\theta} - L(1)D^{\mathrm{p}}(1)$$
$$\frac{\mathrm{D}\sigma}{\mathrm{D}t} = L(0)D - B\dot{\theta} - L(0)D^{\mathrm{p}}(0) \tag{3.20}$$

where

$$L(1) = \overset{-1}{M}(1),\quad B = L(1)A(1),$$
$$L_{ijkl}(0) = L_{ijkl}(1) + \tfrac{1}{2}(\delta_{il}\sigma_{kj} + \sigma_{ik}\,\delta_{jl} + \delta_{ik}\sigma_{ij} + \sigma_{il}\,\delta_{jk})$$
$$\frac{\delta^{cc}\sigma}{\delta t} = \dot{\sigma} - V\sigma - \sigma\overset{T}{V} + \sigma\,\mathrm{tr}\,V;\quad \frac{\mathrm{D}\sigma}{\mathrm{D}t} = \dot{\sigma} - \omega\sigma + \sigma\omega + \sigma\,\mathrm{tr}\,V,$$
$$2\omega = V - \overset{T}{V},\quad 2\omega^{\mathrm{e}} = V^{\mathrm{e}} - \overset{T}{V}{}^{\mathrm{e}},\quad 2\omega^{\mathrm{p}} = V^{\mathrm{p}} - \overset{T}{V}{}^{\mathrm{p}},\quad \omega = \omega^{\mathrm{e}} + \omega^{\mathrm{p}} \tag{3.21}$$

$$D^{\mathrm{p}}(1) = D^{\mathrm{p}} + M(1)[V^{\mathrm{p}}\sigma + \sigma\overset{T}{V}{}^{\mathrm{p}}]$$
$$D^{\mathrm{p}}(0) = D^{\mathrm{p}} + M(0)[(\omega^{\mathrm{p}} - D')\sigma - \sigma(\omega^{\mathrm{p}} + D')];\quad M(0) = \overset{-1}{L}(0) \tag{3.22}$$

Here, ω, ω^{e} and ω^{p} stand for the total spin in κ, 'elastic' spin resulting from the deformation with respect to $\overset{*}{\kappa}$ and 'plastic' spin resulting from the motion of $\overset{*}{\kappa}$ itself with respect to the director triad. The symbols $D^{\mathrm{p}}(1)$ and $D^{\mathrm{p}}(0)$ stand for tensors which could have been called the instantaneous plastic strain rate tensors had we accepted in the H–R formalism the configuration κ as the reference state and taken the Green and logarithmic strain measures, respectively. (The instantaneous tensor of thermal stresses B in κ is invariant under choice of the strain measure in κ.) Equations (3.22) illustrate the relations between plastic strain rate tensors in the H–R- and M-formalisms. At the same time, they are an example of the general relation (2.42).

Consider now the class of isotropic materials with respect to their elastic–plastic properties in $\overset{*}{\kappa}$ and assume that the elastic moduli in $\overset{*}{\kappa}$ are independent of the plastic strains. We have in such a case $D' = 0$, $\overset{T}{V}{}^* = V^*$ (cf. eqn. (5.6)) and $\omega^{\mathrm{p}} = 0$, the latter condition being a direct consequence of the equal principal directions of the symmetric matrices V^*, $T^{\mathrm{e}}(1)$ and $E^{\mathrm{e}}(1)$. Further

$$D^{\mathrm{p}} = D^{\mathrm{p}}(0) = R^{\mathrm{e}} D^* \overset{T}{R}{}^{\mathrm{e}},\ 2D^* = \overset{T}{V}{}^* + V^* = 2V^* \tag{3.23}$$

i.e. the plastic part $D^{\mathrm{p}}(0)$ of the strain rate tensor in κ is a rotated plastic strain rate tensor D^* in $\overset{*}{\kappa}$. Moreover, D^{p} is equal to the H–R plastic strain rate tensor in κ provided the latter is based on the logarithmic strain.

The equations for the particular case discussed can be shown to coincide with those developed by Lee; the proof of it would require considering a non-isoclinic reference configuration and is therefore omitted.

1.3.3. *Invariant Definition of the Incremental Elastic Mechanical Work* [60]

We shall confine ourselves in this section to isothermal processes at $\theta = \theta_o$. The free energy Φ' and free enthalpy Ψ' functions corresponding to a given elastic strain measure E^{e} can in accordance with eqns. (1.31) and (1.36) be presented as ($\overset{*}{q}{}^{\alpha} = 0$)

$$\begin{aligned} \Phi' &= \Phi_1'(q^{\alpha}, H^{\alpha}) + \overset{*}{\Phi}(H^{\sigma}) \\ \Psi' &= \Psi_1'(p_{\alpha}, H^{\sigma}) + \overset{*}{\Phi}(H^{\sigma}) \end{aligned} \tag{3.24}$$

Here, Φ' can be interpreted as a collective description of elastic strain potential energies of different elastic materials, induced by prior plastic straining of the element. The similar interpretation can be given to the dual function Ψ' because $\Phi'=0$, $\partial\Phi'/\partial q^\alpha=0$ for $q^\alpha=0$ and $\Psi'=0$, $\partial\Psi/\partial p_\alpha=0$ for $p_\alpha=0$. The function $\overset{*}{\Phi}=\overset{*}{u}(H^\sigma)-\theta_o\overset{*}{s}(H^\sigma)$ represents the stored free energy (cf. Section 1.1.3) whereas $\overset{*}{u}$ and $\overset{*}{s}$ are stored internal energy and entropy, respectively The symbol p_α stands for the energy-conjugate stresses corresponding to q^α and is thus the counterpart of the tensor T^e.

The function Φ_1' as well as the functions $d\Phi_1'=p_\alpha\, dq^\alpha+d^P\Phi_1'$ and $d^P\Phi_1'=d^P\Psi_1'$ are invariant with respect to the change of elastic strain measure. (The last equality results from duality of Ψ_1' and Φ_1' and the fact that H^σ is a passive variable in the Legendre transformation $\Psi_1'=\Phi_1'-p_\alpha q^\alpha$, cf. ref. 51.) Thus, at the yield point

$$d\Phi_1'=p_\alpha\, dq^\alpha+d^P\Psi_1' \tag{3.25}$$

represents the difference of potential energies of two different elastic materials produced by two neighbouring but different histories of plastic deformation of the element. Due to this clear physical interpretation and the invariance under change of elastic strain measure, eqn. (3.25) can serve as a basis for the decomposition of the total incremental work into the reversible and irreversible parts. First, we replace eqn. (3.13) by

$$d\Phi_1'=p_\alpha\, dq^\alpha-\pi_\sigma^{(2)}\, dH^\sigma \tag{3.26}$$

where

$$-\pi_\sigma^{(2)}(p_\alpha, H^\sigma)=\frac{\partial\Psi_1'}{\partial H^\sigma} \quad \text{and} \quad \pi_\sigma^{(2)}=0, \frac{\partial\pi_\sigma^{(2)}}{\partial p_\alpha}=0 \quad \text{for} \quad p_\alpha=0$$

so that $\pi_\sigma=\pi_\sigma^{(1)}+\pi_\sigma^{(2)}$, $-\pi_\sigma^{(1)}=\partial\overset{*}{\Phi}/\partial H^\sigma$. Now, if for a conjugate pair p_α, q^α the function Ψ_1' can be presented in the form

$$\Psi_1'=-p_\alpha e^\alpha(p_\alpha, H^\sigma); \qquad d^P\Psi_1'=-p_\alpha\, d^Pe^\alpha \tag{3.27}$$

then an 'elastic' increment of the elastic strain can be defined as $dq^\alpha-d^Pe^\alpha$ so that the work of p_α on such an increment equals

$$d\Phi_1'=p_\alpha(dq^\alpha-d^Pe^\alpha) \tag{3.28}$$

Because $d^P\Psi_1'$ is invariant the quantity d^Pe^α must transform in a covariant way with respect to dq^α under the change of the elastic strain measure. Dissipation of mechanical work given by (3.28) can be expressed as the sum

$$\pi_i\, dH^i=\pi_\sigma^{(1)}\, dH^\sigma-p_\alpha\, d^{(h)}q^\alpha+p_\alpha\, d^Pe^\alpha \tag{3.29}$$

The last term on the rhs represents the part of dissipation due to the change of elastic properties of the material induced by plastic straining.

The term $p_\alpha \, \mathrm{d}^{\mathrm{P}} e^\alpha$ may turn out to be negative if the element stiffens elastically with plastic deformation.

2. RATE EQUATIONS

2.1. Hill–Rice Formalism

We are now to shortly discuss the plastic flow rates as developed by Hill and Rice, [3, 8–10]. Both the associated and non-associated rules are considered and generalized to include the thermal effects. For rate independent materials, we write for regular yield points

$$\mathrm{d}H^r = k^r(\pi_s)\,\mathrm{d}\lambda, \qquad \mathrm{d}\lambda > 0 \tag{4.1}$$

where k^r may possibly depend on θ and H^r as well. Hence,

$$\mathrm{d}^{\mathrm{P}} p_i = -l_i(\Upsilon^{\theta q})\,\mathrm{d}\lambda\,; \qquad l_i = -\left[\frac{\partial \pi_r}{\partial q^i} k^r\right]_{\pi = \pi(\Upsilon^{\theta q})} \tag{4.2}$$

Let ν_i be the unique normal outward from the elastic domain in the q^i-space at a regular yield point $\Upsilon^{\theta q}$. Consider a segment $(\mathrm{d}q^i, \mathrm{d}\theta)$ in the q^i, θ space, associated with $\mathrm{d}H^r \neq 0$. We adopt the following generalization of the classical loading condition [47]:

$$\nu_i \, \mathrm{d}q^i + \eta \, \mathrm{d}\theta > 0 \quad \text{if} \quad \mathrm{d}H^r \neq 0 \tag{4.3}$$

Here η represents the sensitivity of the yield limit to temperature or just the thermal softening. The material element can be plastically deformed even if its overall physical dimensions are not altered provided the temperature varies and stresses correspond to a current yield point. With the assumption that the incremental response is piecewise linear and continuous we have

$$\mathrm{d}\lambda = \frac{1}{\rho_o H_o}(\nu_i \, \mathrm{d}q^i + \eta \, \mathrm{d}\theta) \tag{4.4}$$

since $\mathrm{d}\lambda$ must vanish for $\nu_i \, \mathrm{d}q^i + \eta \, \mathrm{d}\theta = 0$. Here, H_o is necessarily positive. Combining (4.2), (4.4), (1.12) and (1.23), we arrive at the general incremental form of the non-isothermal non-associated equations of thermoplasticity for an active plastic straining process ($\mathrm{d}\lambda > 0$):

$$\mathrm{d}p_i = \frac{1}{\rho_o} l_{ij} \, \mathrm{d}q^j - \frac{1}{\rho_o} \beta_i \, \mathrm{d}\theta - \frac{l_i}{\rho_o H_o}(\nu_j \, \mathrm{d}q^j + \eta \, \mathrm{d}\theta), \tag{4.5}$$

$$C_q \, \mathrm{d}\theta = (\Phi_o - s_o\theta)\,\mathrm{d}\lambda - \frac{\theta}{\rho_o} \beta_i \, \mathrm{d}q^i - \bar{\bar{\mathrm{d}}}Q \tag{4.6}$$

where

$$\Phi_o(\Upsilon^{\theta q}) = \pi_r(\Upsilon^{\theta q})k^r, \qquad s_o(\Upsilon^{\theta q}) = \frac{\partial \pi_r}{\partial \theta} k^r \tag{4.7}$$

Solving eqns. (4.5) and (4.6) with respect to $\mathrm{d}\theta$ one can express $\mathrm{d}p_i$ and $\mathrm{d}\lambda$ in terms of $\mathrm{d}q^i$ and $\bar{\bar{\mathrm{d}}}Q$ [60]

$$\begin{aligned} \mathrm{d}p_i &= \frac{1}{\rho_o} l_{ij}^{(a)}\, \mathrm{d}q^j + \frac{\beta_i}{\rho_o C_q} \bar{\bar{\mathrm{d}}}Q - l_i^{(a)}\, \mathrm{d}\lambda \\ \mathrm{d}\lambda &= \frac{1}{\rho_o H_o^{(a)}} \left[\nu_i^{(a)}\, \mathrm{d}q^i - \frac{\eta}{C_q} \bar{\bar{\mathrm{d}}}Q \right] > 0 \end{aligned} \tag{4.8}$$

provided

$$H_o^{(a)} = H_o - \frac{\eta}{\rho_o C_q} (\Phi_o - s_o \theta) > 0 \tag{4.9}$$

Here, $l_{ij}^{(a)}$, $l_i^{(a)}$ and $\nu_i^{(a)}$ are adiabatic counterparts of l_{ij}, l_i and ν_i

$$\begin{aligned} l_{ij}^{(a)} &= l_{ij} + \frac{\theta}{\rho_o C_q} \beta_i \beta_j; \qquad l_i^{(a)} = l_i + \beta_i \frac{\Phi_o - s_o \theta}{\rho_o C_q}; \\ \nu_i^{(a)} &= \nu_i - \frac{\eta \theta}{\rho_o C_q} \beta_i. \end{aligned} \tag{4.10}$$

The dual form of eqn. (4.5) can be obtained by using eqns. $(1.14)_1$, $(1.19)_3$ and (4.4) as

$$\begin{aligned} \mathrm{d}\lambda &= \frac{1}{\rho_o h_o} [\rho_o \mu^i\, \mathrm{d}p_i + \eta_1\, \mathrm{d}\theta] \\ \mathrm{d}q^i &= \rho_o m^{ij}\, \mathrm{d}p_j + a^i\, \mathrm{d}\theta + \frac{m^i}{h_o} [\rho_o \mu^j\, \mathrm{d}p_j + \eta_1\, \mathrm{d}\theta] \end{aligned} \tag{4.11}$$

where

$$\begin{aligned} m^i &= \frac{1}{\rho_o} \frac{\partial \pi_r(\Upsilon^{\theta p})}{\partial p_i} k^r = m^{ij} l_j; \qquad \mu^i(\Upsilon^{\theta p}) = \nu_j m^{ij} \\ \eta_1 &= \nu_i a^i + \eta = \mu^i \beta_i + \eta; \qquad h_o = H_o - l_{ij} \mu^i m^j = H_o - \nu_j m^j \end{aligned} \tag{4.12}$$

On the basis of eqns. (4.11) and (4.12) three types of the material behaviour can be distinguished:

$$\begin{aligned} &h_o > 0 \text{ (hardening)} \qquad \rho_o \mu^j\, \mathrm{d}p_j + \eta_1\, \mathrm{d}\theta > 0 \\ &h_o < 0 \text{ (softening)} \qquad \rho_o \mu^j\, \mathrm{d}p_j + \eta_1\, \mathrm{d}\theta < 0 \\ &h_o = 0 \text{ (neutral)} \qquad \rho_o \mu^j\, \mathrm{d}p_j + \eta_1\, \mathrm{d}\theta = 0 \end{aligned} \tag{4.13}$$

In the last two cases, dq^i is not uniquely determined by dp_i and $d\theta$. The similar classification can be carried out for adiabatic processes, $\bar{\bar{d}}Q=0$.

Associated plastic flow rules are obtained under the assumption that the dissipative forces are derivable from a potential $\bar{\bar{d}}\Sigma_D^{(1)}(dH^r)$, which is a non-linear differential form, positive definite and for rate-independent materials necessarily homogeneous of degree one with respect to dH^r, i.e.

$$\pi_r=\frac{\partial(\bar{\bar{d}}\Sigma_D^{(1)})}{\partial(dH^r)};\qquad \frac{\partial(\bar{\bar{d}}\Sigma_D^{(1)})}{\partial(dH^r)}\,dH^r=\bar{\bar{d}}\Sigma_D^{(1)}>0 \tag{4.14}$$

The matrix

$$\left\|\frac{\partial^2(\bar{\bar{d}}\Sigma_D^{(1)})}{\partial(dH^r)\,\partial(dH^s)}\right\|$$

is singular. Hence, the forces π_r are not independent. If it is additionally assumed that the rank of this matrix equals $n-1$, where n is the number of internal parameters, then there exists exactly one relation of the type

$$\mathscr{F}(\pi_r)=0;\qquad \mathscr{F}\,|_{\pi_r(\Upsilon^{\theta q})}-\overset{(q)}{\mathscr{F}}(\Upsilon^{\theta q});\qquad \mathscr{F}\,|_{\pi_r(\Upsilon^{\theta p})}-\overset{(p)}{\mathscr{F}}(\Upsilon^{\theta p}) \tag{4.15}$$

which is a generalized plastic flow surface in the π_r-space. For $dH^r=0$ the forces π_r are not uniquely determined by (4.14), i.e. at constrained equilibrium $dH^r=0$, $\pi_r\,dH^r=0$ but the dissipative forces have not to vanish. It can be proved, similarly as in ref. 62 that the inverse relation to (4.14) reads

$$dH^r=d\lambda\,\frac{\partial\mathscr{F}}{\partial\pi_r} \tag{4.16}$$

and we can identify k^r in (4.1) with the outward normal to the domain $\mathscr{F}<0$

$$k^r=\frac{\partial\mathscr{F}}{\partial\pi_r}. \tag{4.17}$$

The function $\mathscr{F}$ can additionally depend on θ and H^r. By substituting eqn. (4.17) into eqns. (4.2) and (4.12) the associated non-isothermic plastic flow laws are obtained as

$$l_i=-\frac{\partial\overset{(q)}{\mathscr{F}}}{\partial q^i}\equiv\nu_i;\qquad m^i=\frac{1}{\rho_o}\frac{\partial\overset{(p)}{\mathscr{F}}}{\partial p_i}\equiv\mu^i. \tag{4.18}$$

For the function $\mathscr{F}$ that is not dependent on θ we have

$$s_o = \frac{\partial \overset{(q)}{\mathscr{F}}(Y^{\theta q})}{\partial \theta}. \tag{4.19}$$

A different justification for the associated rule of the non-isothermic plastic flow can be based upon inequality (2.30) as mentioned in Section 1.2.2.

Under given H^r, the knowledge of dq^i and $d\theta$ (or $\bar{\bar{d}}Q$) allows by eqns. (4.5) (or (4.8)) to find the stress increments as the material reaction. This reaction may be estimated by considering a linear comparison solid with the constitutive equation

$$\rho_o \, dp_i = \mathscr{L}^{(c)}_{ij} \, dq^j; \qquad \mathscr{L}^{(c)}_{ij} = l_{ij} - \frac{1}{4rH_o}(l_i + r\nu_i)(l_j + r\nu_j) \tag{4.20}$$

where l_{ij}, l_i, ν_i and H_o are identical to the quantities appearing in eqn. (4.5) (or $l_{ij} = l^{(a)}_{ij}$, $l_i = l^{(a)}_i$, $H_o = H^{(a)}_o$—in the case of eqn. (4.8)) and r is an arbitrary positive scalar parameter. It was proved in refs. 57 and 59 that

$$(\Delta \, dq^i)\mathscr{L}^{(c)}_{ij}(\Delta \, dq^j) \leqslant \rho_o(\Delta \, dp_i)(\Delta \, dq^i) \tag{4.21}$$

for each finite difference

$$\Delta \, dq^i = (dq^i)^{(1)} - (dq^i)^{(2)} \tag{4.22}$$

and each $r > 0$. In the rhs of eqn. (4.21) we have a difference of the generalized stress increments calculated from the actual piecewise linear relation (4.5) for the same $d\theta$. The inequality (4.21) indicates that the hypothetical comparison solid (4.20) may serve as a tool to calculate lower bounds to the bifurcation states. For $\nu_i = l_i$, $r = 1$, the known Hill's inequality for isothermic processes can be rediscovered.

The incremental relations between dp_i, dq^i, $d\theta$ and $\bar{\bar{d}}Q$ can be presented in the form (4.5), (4.6), (4.8) and (4.11) under arbitrary choice of the reference state and strain measure. In this sense they are invariant, which was noted for the isothermal case by Hill [5]. Now it is noted that $d\lambda$ in (4.1) is invariant under (2.7) while l_i in (4.2) transforms as p_i. Assuming $\overset{(q)}{\mathscr{F}} < 0$ in the elastic domain it is always possible to adjust the magnitude of ν_i under the change of co-ordinates in such a way that ν_i transforms like p_i and the yield temperature sensitivity factor η is invariant. Hence $\rho_o H_o$ is also invariant. Both Φ_o and s_o are invariant by (2.25) and so is $\rho_o H^{(a)}_o$, whereas $l^{(a)}_i$ and $\nu^{(a)}_i$ transform like p_i in view of $(2.32)_3$ and $(2.33)_2$. However, none of the quantities $\rho_o h_o$, $\rho_o \mu^i\, dp_i$ and $\eta_1\, d\theta$ occurring in (4.11) is invariant, as remarked (for the isothermal case) by Hill [5]. Both $\rho_o m^i$ and $\rho_o \mu^i$ transform like $d^P q^i$ since $d\lambda$ is invariant. Thus, the inequalities (4.13) have no invariant character; the same material may 'harden' at one choice of the conjugate pair and 'soften' at another. The temperature sensitivity factor η_1 in (4.11) is also not invariant. This seems to be important as the values of the terms $\nu_i a^i$ and η in the expression for η_1 are comparable.

2.2. Mandel Formalism

2.2.1 $\overset{*}{\kappa}$-*description*

(i) In the M formalism dissipation in $\overset{*}{\kappa}$ has the form (3.7). Considering P^* and π_σ as the thermodynamic forces conjugate to the pair V^*, $\dot{H}^\sigma$ the rate equations at a yield point may be written as

$$\begin{aligned} V^* &= \dot{\lambda}\overset{*}{M}{}^{(p)}_{(1)}(P^*, \pi_\sigma, \theta, H^\sigma), \qquad \dot{\lambda} > 0 \\ \dot{H}^\sigma &= \dot{\lambda}\overset{*}{K}{}^{\sigma}_{(1)}(P^*, \pi_\sigma, \theta, H^\sigma). \end{aligned} \tag{5.1}$$

Using (3.6) and $(3.9)_1$ the functions $\overset{*}{M}{}^{(p)}_{(1)}$ and $\overset{*}{K}{}^{\sigma}_{(1)}$ can equivalently be expressed as [12–15]

$$\begin{aligned} V^* &= \dot{\lambda}\overset{*}{M}{}^{(p)}[T^e(1), \theta, H^\sigma] \\ \dot{H}^\sigma &= \dot{\lambda}\overset{*}{K}{}^{\sigma}[T^e(1), \theta, H^\sigma] \end{aligned} \tag{5.2}$$

The general equations (5.2) do not seem to have been specified so far for particular anisotropic materials.

The complete constitutive description requires the specification of 9 equations for the velocity gradient V^* in κ^*. The symmetric part $D^* = \frac{1}{2}(V^* + \overset{T}{V}{}^*)$ describes the rate of plastic strain in $\overset{*}{\kappa}$ whereas the antisymmetric part $\omega^* = \frac{1}{2}(V^* - \overset{T}{V}{}^*)$ describes the rate of material element rotation with respect to the director triad and thus, roughly speaking, it represents in some mean sense a relative mutual rotation of micro-elements in $\overset{*}{\kappa}$ associated with the overall plastic deformation. Distinguishing the director triad in working out a theory accounting for the microstructure seems natural—the example of the monocrystal illustrates this [53, 65]. However, no attempt has been made so far to indicate experimental action that should be undertaken in order to find at macrolevel the additional three equations occurring in $(5.1)_1$–$(5.2)_1$. According to Mandel [15], the director triad may be formed by three mutually orthogonal unit vectors $\mathbf{m}_k$. Vector $\mathbf{m}_1$ may be taken to represent a material line lying in the material plane with normal $\mathbf{m}_2$ and $\mathbf{m}_3 = \mathbf{m}_1 \times \mathbf{m}_2$ (of course in general neither $\mathbf{m}_2$ nor $\mathbf{m}_3$ represents material lines). Supposing that V^* is prescribed, one can derive the expression for the spin $\boldsymbol{\omega}^t$ representing the angular velocity of a triad in its conceptual motion in $\overset{*}{\kappa}$. Written in tensorial form $\boldsymbol{\omega}^t$ is given by [60]

$$\begin{aligned}\boldsymbol{\omega}^t = \boldsymbol{\omega}^* &+ D^*(m_1, m_2)[\mathbf{m}_2 \otimes \mathbf{m}_1 - \mathbf{m}_1 \otimes \mathbf{m}_2] \\ &+ D^*(m_1, m_3)[\mathbf{m}_3 \otimes \mathbf{m}_1 - \mathbf{m}_1 \otimes \mathbf{m}_3] \\ &+ D^*(\mathbf{m}_2, \mathbf{m}_3)[\mathbf{m}_2 \otimes \mathbf{m}_3 - \mathbf{m}_3 \otimes \mathbf{m}_2]\end{aligned}$$

where $\mathbf{a} \otimes \mathbf{b}$ denotes dyadic product of two vectors and

$$D^*(m_K, m_L) \equiv \mathbf{m}_K \,.\, (\mathbf{D}^* \mathbf{m}_L)$$

are the components of D^* in the co-ordinate axes that at a given instant coincide with the triad. In the case of isoclinic configuration $\mathbf{m}_K = \mathbf{e}_K = \text{const.}$ (in time) so that $\boldsymbol{\omega}^t$ must vanish and

$$\begin{aligned}\boldsymbol{\omega}^* = D^*(e_1, e_2)[\mathbf{e}_1 \otimes \mathbf{e}_2 - \mathbf{e}_2 \otimes \mathbf{e}_1] + D^*(e_1, e_3)[\mathbf{e}_1 \otimes \mathbf{e}_3 - \mathbf{e}_3 \otimes \mathbf{e}_1] \\ + D^*(e_2, e_3)[\mathbf{e}_3 \otimes \mathbf{e}_2 - \mathbf{e}_2 \otimes \mathbf{e}_3] \qquad (5.3)\end{aligned}$$

This equation may equivalently be rewritten in the form containing merely the couple (e_1, e_2)

$$\begin{aligned}\boldsymbol{\omega}^* = D^*(e_1, e_2)[\mathbf{e}_2 \otimes \mathbf{e}_1 - \mathbf{e}_1 \otimes \mathbf{e}_2] + [(\mathbf{D}^* \mathbf{e}_2) \otimes \mathbf{e}_2 - \mathbf{e}_2 \otimes (\mathbf{D}^* \mathbf{e}_2)] \\ + [\mathbf{e}_1 \otimes (\mathbf{D}^* \mathbf{e}_1) - (\mathbf{D}^* \mathbf{e}_1) \otimes \mathbf{e}_1] \qquad (5.4)\end{aligned}$$

Equation (5.3) or (5.4) specifies the antisymmetric part of $\overset{*}{M}{}^{(p)}$ which is expressed in terms of its symmetric part and the director triad chosen. The relation between ω^* and $T^e(1)$ resulting from (5.3) (or (5.4)) and from

$$D^* = \dot{\lambda} \overset{*}{M}{}^{(p)}_{(s)}, \qquad 2\overset{*}{M}{}^{(p)}_{(s)} = \overset{*}{M}{}^{(p)} + \overset{*}{M}{}^{(p)T} \qquad (5.5)$$

is not isotropic even for $\overset{*}{M}{}^{(p)}_{(s)}$ being an isotropic function of $T^e(1)$ for each fixed θ and H^σ (H^σ must all be then scalars). Anisotropy results here from the fact that different material fibres have different angular velocities with respect to the director triad chosen. In the case of materials which are fully isotropic with respect to their plastic properties in $\overset{*}{\kappa}$ this relative angular velocity difference is neglected within the framework of the M formalism by putting

$$\overset{*}{\omega} = 0 \qquad (5.6)$$

The above condition results directly from eqn. $(5.2)_1$ if we formally make use of the representation theorem for isotropic functions [66]. The condition (5.6) was also accepted by Wang [68] and Willis [69].

(ii) Equations (5.1) have been obtained by using a different approach by Halphen [33]. He neglected the possible influence of the previous plastic strains on elastic and thermal moduli in $\overset{*}{\kappa}$ (π_σ does not depend on $T^{(e.)}$) and assumed that thermodynamical rates V^* and $\dot{H}^\sigma$ have a potential so that

$$\overset{*}{M}{}^{(p)}_{(1)} = \frac{\partial \mathscr{F}}{\partial \overset{T}{P}{}^*}; \qquad K^\sigma_{(1)} = \frac{\partial \mathscr{F}}{\partial \pi_\sigma}, \qquad \sigma = 1, 2, \ldots, n \tag{5.7}$$

where

$$\mathscr{F}(P^*, \pi_\sigma, \theta, H^\sigma) = 0 \tag{5.8}$$

plays a role of the yield surface in the (P^*, π_σ)-space, whereas

$$\mathscr{F}_1(P^*, \theta, H^\sigma) = \mathscr{F}\,|_{\pi_\sigma(\theta, H^\sigma)} = 0 \tag{5.9}$$

is an equation of the corresponding yield surface in P^*-space. Writing the consistency relation in the form

$$\frac{1}{\rho} h_o \dot{\lambda} = \operatorname{tr}(N^* \dot{P}^*) + \frac{\partial \mathscr{F}_1}{\partial \theta} \dot{\theta} \tag{5.10}$$

where

$$N^* = \frac{\partial \mathscr{F}_1}{\partial \overset{T}{P}{}^*} \equiv \overset{*}{M}{}^{(p)}_{(1)}$$

in the case considered, and taking into account that $\dot{\lambda} > 0$, we have

$$\begin{aligned}
&h_o > 0; \qquad \operatorname{tr}(N^* \dot{P}^*) + \frac{\partial \mathscr{F}_1}{\partial \theta} \dot{\theta} > 0 \qquad \text{at hardening} \\
&h_o = 0; \qquad \operatorname{tr}(N^* \dot{P}^*) + \frac{\partial \mathscr{F}_1}{\partial \theta} \dot{\theta} = 0 \qquad \text{at neutral state} \\
&h_o < 0; \qquad \operatorname{tr}(N^* \dot{P}^*) + \frac{\partial \mathscr{F}_1}{\partial \theta} \dot{\theta} < 0 \qquad \text{at softening}
\end{aligned} \tag{5.11}$$

analogous to (4.13).

According to recently introduced terminology (Hill [6]) the neutral state at which $h_o = 0$ may be called an intrinsic eigenstate.

2.2.2. *Eulerian Description*

The transformation of eqns. (5.1)–(5.2) from $\overset{*}{\kappa}$ to κ can be carried out by means of eqns. $(3.14)_1$, $(3.15)_2$ and (3.20)–(3.22)

$$\begin{aligned} &V^{\mathrm{p}} = \dot{\lambda} M^{(\mathrm{p})}; \qquad D' = \dot{\lambda} K \\ &M^{(\mathrm{p})} = F^{\mathrm{e}} \overset{*}{M}{}^{(\mathrm{p})}_{(1)} \overset{-1}{F^{\mathrm{e}}}; \qquad K = \overset{-T}{F^{\mathrm{e}}} \left(\frac{\partial E^{\mathrm{e}}(1)}{\partial H^{\sigma}} K^{\sigma}_{(1)} \right) \end{aligned} \tag{5.12}$$

$$\begin{aligned} &\frac{\delta^{\mathrm{cc}}\sigma}{\delta t} = L(1)D - B\dot{\theta} - \dot{\lambda} L^{\mathrm{p}} \\ &L^{\mathrm{p}} = L(1)K + L' M^{(\mathrm{p})}; \qquad L'_{ijkl} = L_{ijkl}(1) + \delta_{ki}\sigma_{jl} + \delta_{jk}\sigma_{li} \end{aligned} \tag{5.13}$$

To find a relation between $\dot{\lambda}$ and D one needs to adopt a plastic loading condition. If we accept the assumptions of Halphen [33] discussed above, (ii) of Section 2.2.1, then eqn. (5.10) can be transformed to the form

$$h_{\mathrm{o}}\dot{\lambda} = \mathrm{tr}\,[M^{(\mathrm{p})}(\overset{\nabla}{\sigma} + 2\sigma D^{\mathrm{e}})] + \eta_1 \dot{\theta} \tag{5.14}$$

where

$$M^{(\mathrm{p})} = F^{\mathrm{e}} N^{*} \overset{-1}{F^{\mathrm{e}}}; \qquad \eta_1 = \rho \frac{\partial \mathscr{F}_1}{\partial \theta} \tag{5.15}$$

and $\overset{\nabla}{\sigma}$ is determined by $(3.14)_1$. Eliminating $\overset{\nabla}{\sigma}$ by $(3.17)_1$ and using (3.16), (5.7) and $(5.12)_1$ we arrive at

$$\begin{aligned} &\dot{\lambda} H_{\mathrm{o}} = \mathrm{tr}\,[D(L' M^{(\mathrm{p})})] - [\mathrm{tr}\,(M^{(\mathrm{p})} B)]\dot{\theta} + \eta_1 \dot{\theta} > 0 \\ &H_{\mathrm{o}} = h_{\mathrm{o}} + \mathrm{tr}\,[M^{(\mathrm{p})}(L' M^{(\mathrm{p})})] > 0 \end{aligned} \tag{5.16}$$

Note that eqns. (5.13) and (5.16) differ from those of the Eulerian H–R theory by having the matrix L' which is different from $L(1)$ (and, in general, non-symmetric). However, when $M^{(\mathrm{p})}$ is symmetric then L' in (5.13) and (5.16) can be replaced by $L(0)$ (cf. $(3.21)_2$) and $(5.13)_1$ becomes

$$\frac{\mathrm{D}\sigma}{\mathrm{D}t} = L(0)D - B\dot{\theta} - \dot{\lambda} L(0) M^{(\mathrm{p})} \tag{5.17}$$

so that the associated plastic flow laws for both M- and H–R-formalisms coincide. In particular this holds true for materials that are isotropic in $\overset{*}{\kappa}$ with respect to both the elastic and plastic properties. Further, $\mathrm{D}\sigma/\mathrm{D}t$ in (5.17) has a potential. Thus, the variational theorems and bifurcation conditions of Hill [34] can also be employed for the M-approach. Rewriting the plastic flow condition (5.9) as

$$\mathscr{F}_1(P^*, \theta, H^\sigma) = \mathscr{F}_o\left(\frac{1}{\rho}\sigma, F^e, \theta, H^\sigma\right)$$

it can be noticed that

$$\tfrac{1}{2}(M^{(p)} + \overset{T}{M}{}^{(p)}) = \rho\frac{\partial \mathscr{F}_o}{\partial \sigma}$$

The classical yield conditions for isotropic hardening can be derived by assuming that $\mathscr{F}_1$ is a function of three invariants $s_1 = \text{tr}\, P^* = 1/\rho\, \text{tr}\, \sigma$; $s_2 = \text{tr}\, P^{*2} = 1/\rho^2\, \text{tr}\, \sigma^2$; $s_3 = \text{tr}\, P^{*3} = 1/\rho^3\, \text{tr}\, \sigma^3$ while all H^σ parameters are scalars.

The simplification of the basic equations of M-formalism, resulting from the anticipated infinitesimal elastic strains, is discussed in refs. 16, 19, 21 and 64. In ref. 69 an isotropic, elastic-ideally plastic material model has been successfully combined with a hydrodynamical model describing the metal behaviour under large pressures.

We finally note that in both the approaches discussed the rate-equations contain partial derivatives of free energy (represented by π_i and π_σ). The relation between say, stored energy and hardening is visible at the so-called 'recovery' stage of annealing when the physical properties (e.g. the yield stress) that suffered changes as a result of prior plastic deformation tend to recover their original value. This is done at the expense of stored energy [61]. Obviously, the rate-independent theory can not describe the phenomena observed during recovery stage. It should, however, indicate the relation that exists between stored energy and hardening. In both approaches this can be effected by expressing rate-equations through internal thermodynamical forces.

2.3. Discussion and Conclusions

In this paper the macroscopic aspects of the theory of plasticity at unrestricted strains have been discussed. Starting from the concepts put forward by Hill and Rice and discussing thermostatic properties of elastic–plastic solids we have investigated the transformation rules and invariance properties for different quantities in the six-dimensional space. Equations for the increment of temperature as well as partially integrated forms of the free energy have been obtained.

The Mandel–Lee description has naturally resulted from the H–R approach as a transition from a fixed co-ordinate system to a moving co-ordinate system the changes of which correspond to the changes in the internal state of the material element. The Mandel formalism has been presented in the form which is independent of the choice of the strain measure in the conceptual stress-free configuration. It has been shown that the elastic work as defined by Mandel is not invariant; a different decomposition of the total work into the elastic and plastic parts has been suggested. It has further been shown that for fully isotropic materials the Hill–Rice and the Mandel approaches can be made equivalent. Relationships for plastic strains defined in the actual configuration have been compared for both the formalisms. A choice of three additional equations to determine the material elements orientation has been suggested; other possibilities exist. No evolution equations for internal parameters have been discussed. For large plastic strains this field does not yet seem to have been sufficiently explored.

REFERENCES

1. Hill, R. On constitutive inequalities for simple materials—I, *J. Mech. Phys. Solids*, **16** (1968), 229–242.
2. Hill, R. On constitutive inequalities for simple materials—II, *J. Mech. Phys. Solids*, **16** (1968), 315–322.
3. Hill, R. Aspects of invariance in solid mechanics. In: *Advances in Applied Mechanics*, Vol. 18 (Ed. Chia-Shien Yih), Academic Press, New York, 1978, 1–75.
4. Hill, R. On constitutive macro-variables for heterogeneous solids at finite strain, *Proc. R. Soc. Lond.*, Ser. **A 326** (1972), 131–147.
5. Hill, R. On the classical constitutive laws for elastic–plastic solids. In: *Recent Progress in Applied Mechanics*, F. K. G. Odgvist Volume, Wiley, New York, 1967, 241–249.
6. Hill, R. On intrinsic eigenstates in plasticity with generalized variables, *Math. Proc. Camb. Phil. Soc.*, **93** (1983), 177–189.
7. Hill, R. Invariance relations in thermoelasticity with generalized variables, *Math. Proc. Camb. Phil. Soc.*, **90** (1981) 373–384.
8. Hill, R. and J. R. Rice, Elastic potentials and the structure of inelastic constitutive laws, *SIAM J. appl. Math.*, **25** (1973), 448–461.
9. Rice, J. R. Inelastic constitutive relations for solids: an internal variable theory and its application to metal plasticity, *J. Mech. Phys. Solids* **19** (1971), 433–455.
10. Rice, J. R. Continuum mechanics and thermodynamics of plasticity in relation to microscale deformation mechanisms. In: *Constitutive Equations in Plasticity* (Ed. A. S. Argon), MIT Press, Cambridge, Mass., 1975, 23–80.
11. Rice, J. R. Continuum plasticity in relation to microscale deformation

mechanisms. In: *Metallurgical Effects at High Strain Rates* (Eds. R. W. Rohde, B. M. Butcher, J. R. Holland and C. H. Karners), Plenum Press, New York, 1973, 93–106.
12. MANDEL, J. Plasticité et viscoplasticité, *CISM Lecture Notes* No. 97, Udine, Springer, Wien, 1971.
13. MANDEL, J. Equations constitutives et directeurs dans les milieux plastiques et viscoplastiques, *Int. J. Solids Struct.*, **9** (1973), 725–740.
14. MANDEL, J. Director vectors and constitutive equations for plastic and viscoplastic media. In: *Problems of Plasticity* (Ed. A. Sawczuk), Nordhoff, Groningen, 1974, 135–143.
15. MANDEL, J. Thermodynamics and plasticity. In: *Foundations of Continuum Thermodynamics* (Eds. J. J. Delgado Domingers, N. R. Nina and J. H. Whitelaw), Macmillan, London, 1974, 283–304.
16. MANDEL, J. Sur la definition de la vitesse de deformation elastique et sa relation avec la vitesse de contrainte, *Int. J. Solids Struct.*, **17** (1981), 873–878.
17. MANDEL, J. Definition d'un repére privilégié pour l'etude des transformations anélastiques du polycristal, *J. Méc. Théor. Appl.*, **1** (1982), 1–23.
18. LEE, E. H. Elastic–plastic deformation at finite strains, *J. appl. Mech.*, **36** (1969), 1–6.
19. LEE, E. H. and R. M. MCMEEKING, Concerning elastic and plastic components of deformation, *Int. J. Solids Struct.* **16** (1980), 715–721.
20. LEE, E. H. Some comments on elastic–plastic analysis, *Int. J. Solids Struct.* **17** (1981), 859–872.
21. LUBARDA, V. A. and E. H. LEE, A correct definition of elastic and plastic deformation and its computational significance, *J. appl. Mech.*, **48** (1981), 35–40.
22. ARGYRIS, J. H. and M. KLEIBER, Incremental formulation in nonlinear mechanics and large strain elasto–plasticity, *Comp. Meth. appl. mech. Engng*, **11** (1977), 215–247.
23. ARGYRIS, J. H. and J. ST. DOLTSINIS, On the large strain inelastic analysis in natural formulation, Parts I, II, *Comp. Meth. appl. mech. Engng*, **20** (1979), 213–251; (1980), 91–128.
24. BEVER, M. B., D. L. HOLT, and A. L. TITCHENER, The stored energy of cold work. In: *Progress in Material Sciences*, Vol. 17, Pergamon Press, Oxford, 1973.
25. BUCHDAHL, H. A. *The Concepts of Classical Thermodynamics*. Cambridge University Press, Cambridge, 1966.
26. CALLEN, H. B. *Thermodynamics*, Wiley, New York, 1960.
27. CASEY, J. and P. M. NAGHDI, A remark on the use of the decomposition $F = F^e F^p$ in plasticity, *J. appl. Mech.*, **47** (1980), 672–675.
28. DAFALIAS, Y. F., Elasto–plastic coupling within a thermodynamic strain space formulation of plasticity, *Int. J. Non-Lin. Mech.*, **12** (1977), 327–337.
29. ECKART, G. Theory of elasticity and inelasticity, *Phys. Rev.*, **73** (1948), 373–380.
30. GREEN, A. E. and P. M. NAGHDI, General theory of an elastic–plastic continuum, *Arch. Rat. Mech. Anal.*, **18** (1965), 251–281.

31. GREEN, A. E. and P. M. NAGHDI, Thermodynamic development of elastic–plastic continua. In: *Irreversible Aspects of Continuum Mechanics and Transfer of Physical Characteristics in Moving Fluid*, Springer, Wien, 1966.
32. GREEN, A. E. and P. M. NAGHDI, Some remarks on elastic–plastic deformation at finite strain, *Int. J. Engng Sci.*, **9** (1971), 1219–1229.
33. HALPHEN, B. Sur le champ des vitesses en thermoplasticité finite, *Int. J. Solids Struct.*, **11** (1975), 947–960.
34. HILL, R. Uniqueness and extremum principles in self-adjoint boundary-value problems in continuum mechanics, *J. Mech. Phys. Solids*, **10** (1962), 185–194.
35. HOLSAPPLE, K. A. Elastic–plastic materials as simple materials, *ZAMM*, **53** (1973), 261–270.
36. HOLSAPPLE, K. A. A finite elastic–plastic theory and invariance requirements, *Acta Mech.*, **17** (1973), 277–290.
37. INOUE, T. and S. NAGAKI, A constitutive modelling of thermoviscoelastic–plastic materials, *J. Thermal Stresses*, **1** (1978), 53–61.
38. KESTIN, J. and J. R. RICE, Paradoxes in the application of thermodynamics to strained solids. In: *A Critical Review of Thermodynamics*, Mono Book Corp., 1970.
39. KESTIN, J. Thermodynamics in thermoplasticity. In: *Thermoplasticity*, Ossolineum, Wroclaw, 1973 (in Polish).
40. KLEIBER, M. Kinematics of deformation processes in materials subjected to finite elastic–plastic strains, *Int. J. Engng Sci.*, **13** (1975), 513–525.
41. KLEIBER, M. Large elasto–plastic deformations. Theory and numerical analysis of structures, Rept. No 13/78, Institute of Fundamental Technical Research, Warsaw, 1978 (in Polish).
42. KLEIBER, M. A note on finite strain elasto–plasticity. *Acta Mech.* (in press).
43. KLEIBER, M., J. A. KONIG and A. SAWCZUK, Studies on plastic structures, *Comp. Meth. appl. mech. Engng*, **33** (1982), 487–556.
44. LEHMANN, TH. Some aspects of non-isothermic large inelastic deformations, *Sol. Mech. Arch.*, **3** (1978), 261–317.
45. LEHMANN, TH. Some remarks on the decomposition of deformations and mechanical work, *Int. J. Engng Sci.*, **20** (1982), 281–288.
46. MROZ, Z. and B. RANIECKI, Variational principles in uncoupled thermoplasticity, *Int. J. Engng Sci.*, **11** (1973), 1133–1141.
47. MROZ, Z. and B. RANIECKI, A note on variational principles in coupled thermoplasticity. *Bull Acad. Polon. Sci., Ser. Sci. Techn.*, **XXIII** (1974), 133–139.
48. MROZ, Z. and B. RANIECKI, On the uniqueness problem in coupled thermoplasticity, *Int. J. Engng Sci.*, **14** (1976), 211–221.
49. NAGHDI, P. M. and J. A. TRAPP, The significance of formulating plasticity theory with reference to loading surfaces in strain space, *Int. J. Engng Sci.*, **13** (1975), 785–797.
50. NEMAT-NASSER, S. On finite deformation elasto–plasticity. *Int. J. Solids Struct.*, **18** (1982), 857–872.
51. NOBLE, B. and M. J. SEWELL, On dual extremum principles in applied mathematics, *J. Inst. Math. Appl.*, **9** (1972), 123–193.

52. NYE, J. F. *Physical Properties of Crystals*, Clarendon Press, Oxford, 1964.
53. PEIRCE, D., R. J. ASARO and A. NEEDLEMAN, An analysis of non-uniform and localized deformation in ductile single crystals, Overview 21, *Acta Metall.*, **30** (1982), 1087–1119.
54. PERZYNA, P. and W. WOJNO, On the constitutive equations of elastic/viscoplastic materials at finite strains, *Arch. Mech.*, **18** (1966), 85–100.
55. PERZYNA, P. Thermodynamic theory of viscoplasticity. In: *Advances in Applied Mechanics*, Vol. 11, Academic Press, New York, 1971, 313–354.
56. PERZYNA, P. and W. WOJNO, Thermodynamics of a rate-sensitive plastic material, *Arch. Mech.*, **20** (1968), 499–511.
57. RANIECKI, B. Uniqueness criteria in solids with non-associated plastic flow, *Bull. Acad. Polon. Sci., Ser. Sci. Techn.*, **XXVII** (1979), 391–399.
58. RANIECKI, B. and K. THERMANN, Infinitesimal thermoplasticity and kinematics of finite elastic–plastic deformations, Mitteilungen aus dem Institut fur Mechanik Universitat Bochum, 2, Juni 1978.
59. RANIECKI, B. and O. T. BRUHNS, Bounds to bifurcation stresses in solids with nonassociated plastic flow law at finite strain, *J. Mech. Phys. Solids*, **29** (1981), 153–172.
60. RANIECKI, B. Unpublished.
61. REED HILL, R. E., *Physical Metallurgy Principles*, Van Nostrand Co., New York, 1964.
62. SEDOV, L. I. *Mechanics of Continuous Medium*, Vol. 2, Nauka, Moscow, 1970 (in Russian).
63. SIDOROFF, F. The geometrical concept of intermediate configuration and elastic–plastic finite strain, *Arch. Mech.*, **25** (1973), 299–308.
64. SIDOROFF, F. Incremental constitutive equation for large strain elasto–plasticity, *Int. J. Engng Sci.*, **20** (1982), 19–26.
65. THEODOSIU, C. and F. SIDOROFF, A theory of finite elasto–viscoplasticity of single crystals, *Int. J. Engng Sci.*, **14** (1976), 165–176.
66. VAKULENKO, A. A. A thermodynamic investigation of stress–strain relations in isotropic elastic–plastic solids, *Dokl. A. N. SSSR*, **126** (1959), 736–739 (in Russian).
67. WANG, C. C. A new representation theorem for isotropic functions, *Arch. Rat. Mech. Anal.*, **36** (1970), 166–233.
68. WANG, Y. S. A simplified theory of the constitutive equations of metal plasticity at finite deformation, *J. appl. Mech.*, **40** (1973), 941–947.
69. WILLIS, J. R. Finite deformation solution of a dynamic problem of combined compressive and shear loading for elastic–plastic half-space, *J. Mech. Phys. Solids*, **23** (1975), 405–419.

APPENDIX

Consider the identities

$$p_i = \frac{\partial \Phi}{\partial q^i}\bigg|_{q^i = q^i(\Upsilon^{\theta p})}; \qquad s(\Upsilon^{\theta q})\Big|_{q^i = q^i(\Upsilon^{\theta p})} = s(\Upsilon^{\theta p}) \tag{A.1}$$

By differentiating the above expressions with respect to p_i and θ one can derive the thermodynamic identities of classical thermoelasticity as

$$(\beta_i - l_{ij}a^j)\big|_{q^i=q^i(Y^{\theta p})} = 0; \quad m^{ik}l_{kj}\big|_{q^i=q^i(Y^{\theta p})} = \delta^i_j;$$

$$\left(C_q + \frac{\theta}{\rho_o} l_{ij}a^i a^j\right)\bigg|_{q^i=q^i(Y^{\theta p})} = C_p \tag{A.2}$$

Since H^r is the passive variable in the Legendre transformation we have

$$\pi_r(Y^{\theta q})\big|_{q^i=q^i(Y^{\theta p})} = \pi_r(Y^{\theta p}) \tag{A.3}$$

By calculating partial differentials of both sides of (A.3) in turn with respect to θ, p_i and H^r we obtain the remaining identities

$$\left(\frac{\partial s(Y^{\theta q})}{\partial H^r} - \frac{\partial p_j}{\partial H^r} a^j\right)\bigg|_{q^i=q^i(Y^{\theta p})} = \frac{\partial s(Y^{\theta p})}{\partial H^r} \tag{A.4}$$

$$\frac{\partial q^i(Y^{\theta p})}{\partial H^r} + \rho_o \frac{\partial p_k}{\partial H^r}\bigg|_{q^i=q^i(Y^{\theta p})} m^{ki} = 0 \tag{A.5}$$

$$\frac{\partial \pi_r(Y^{\theta q})}{\partial H^s} - \frac{\partial p_i}{\partial H^s}\frac{\partial q^i}{\partial H^r}\bigg|_{q^i=q^i(Y^{\theta p})} = \frac{\partial \pi_r(Y^{\theta p})}{\partial H^s} \tag{A.6}$$

2

Appropriate Simple Idealizations for Finite Plasticity

D. C. DRUCKER

College of Engineering, University of Illinois, Urbana, USA

ABSTRACT

Finite elastoplasticity has become a very lively subject in recent years with much disagreement over proposed methods of taking finite deformations into account in constitutive relations. For good reasons worth recalling, the conventional incremental theory of plasticity does not explicitly include the continuum rotation associated with the shear deformation that is the dominant feature of many problems of practical importance. Yet this simple theory has been replaced in many computer programs and analyses by mathematical approaches and definitions of isotropic finite elasticity which do. As a consequence, incorrect inferences have been drawn for plastic stress–strain relations in general and for material instability in simple plastic shear in particular. When elastic strains and lattice rotations are negligible, rotation should be taken as zero in incremental plasticity theory no matter how large the continuum rotations computed from the shear displacement field.

1. INTRODUCTION

The concept of stress–strain relations or, more generally, constitutive relations developed slowly over the past centuries. Linear elasticity represented a great step forward as did the incremental plastic stress–strain relations of Levy based on the experiments and thoughts of Tresca. Small displacement theory, or at least small strain theory, was appropriate for the elastic response of all the materials then used for structures and machines. Plasticity theory, somewhat similarly, treated small incremental changes from an existing configuration.

Then curiosity and the engineering use of rubber and rubberlike materials with their large recoverable strains stimulated the development of modern finite elasticity. A remarkably simple yet powerful general theory for an initially isotropic material was developed for finite rotation as well as finite strain [5]. Calls arose for the corresponding treatment of plastic behavior. They went largely unheeded. Those proposals that were put forward were ignored by most researchers. However, attitudes began to change considerably as computing power grew and much interest centred on instability in shear. Rice [13] reminded us that the stress and the tangent modulus in the plastic range are of comparable magnitude. Therefore terms of stress σ multiplied by increments of plastic strain ($\sigma\, d\varepsilon^{p}$) could be and usually were of the same magnitude as the increment of stress $d\sigma$ associated with the increment of plastic strain $d\varepsilon^{p}$. This was well known in connection with necking in tension but he emphasized the possibility of instability in shear in a stable material.

The shear instability resulting from a small rotation, which accounts for the torsional buckling of an axially loaded cruciform column of work-hardening material [11], was not as well known. Concern over the possibility of a more general shear instability in two- and three-dimensional problems from landslides to shear bands in metal forming (Fig. 1), led to the inclusion in many computer programs of a mathematically permissible but physically inappropriate formulation of finite elastoplasticity with Jaumann rates of stress to eliminate unwanted effects of rigid body rotation. Great excitement was caused by the calculations by Nagtegaal and de Jong [9] of oscillating shear stress with increasing large shear strain for a kinematic hardening material. Lee and co-authors [7] pointed out that the sum of Jaumann spins kept growing without limit and did not represent the limited continuum rotation. Palgen and I suggested later [12, 4] that the continuum rotation itself was not a relevant quantity, that a theory of finite elastoplasticity which followed isotropic finite elasticity was misleading or incorrect in principle and physical relevance.

In mechanics, as in much of human activity, much that appears new and exciting turns out otherwise. On the other hand an important concept such as Shanley's approach to buckling in the plastic range may be ignored when first proposed. One of Professor Olszak's great

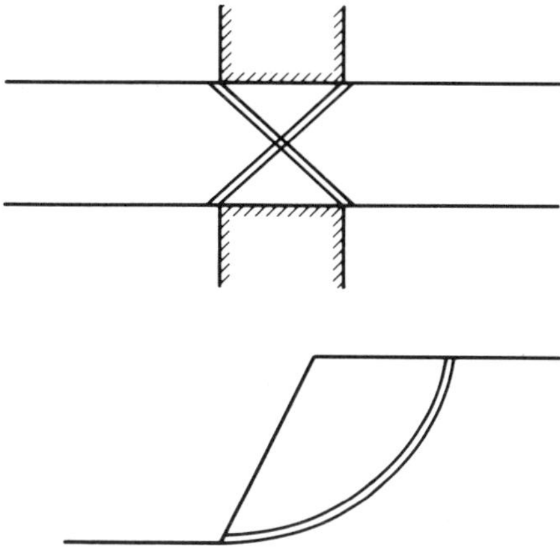

FIG. 1. Shear bands.

contributions over many years and in many forms [10] was to highlight those new developments worthy of attention by his colleagues in Poland and throughout the world.

2. SOME BASIC EXPERIMENTAL AND THEORETICAL FEATURES OF FINITE PLASTICITY

Much of what is known and understood about plasticity has come from observed behavior in simple tension and compression supplemented by tests of beams in bending and circular bars in torsion. Some of the salient features of such one-dimensional stress loadings, unloadings, and reloadings in the stable range are illustrated in Fig. 2 as a plot of shear stress S_T vs shear strain e_T for a highly ductile material. After several percent of plastic strain, the work hardening is moderate but the stress does continue to rise as the strain increases. The response to continued loading or to reversed loading is stable but is not strongly dependent on the magnitude of the plastic strain. The response to a path of forward loading into the plastic range following reversed loading of a few percent plastic strain is almost independent of the plastic strain at which reversed loading is begun. Just about identical steady state cyclic stress–strain curves [2] are reached for an unfractured ductile metal or alloy cycled plastically between $e_T^p + \Delta e_T$ and $e_T^p - \Delta e_T$ for a mean plastic strain of zero, 1%, 10%, or 1.

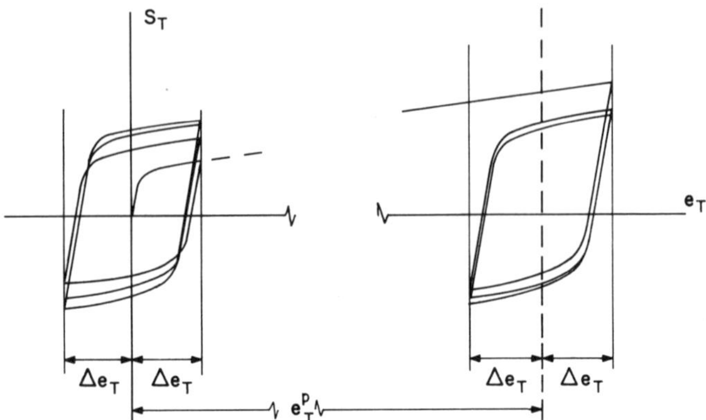

FIG. 2. Cyclic steady state is almost independent of e_T^p.

Concepts of dislocation pile-ups and cell structures are consistent with this one-dimensional picture of a kinematic type of hardening in an already work-hardened material at the widely different 'initial' plastic strains. Of course there is a limit. The pictures and their mathematical representation do alter at extremely large plastic strains. The microstructure then does change its character due to the opening of a significant number of voids or cracks or the appreciable disruption of the grain boundaries and grains.

The ability, in fact the necessity, to choose a highly plastically deformed configuration as the initial configuration for plastic stress–strain relations is implicit in the use of mathematical theories of plasticity. Metals and alloys often are subjected to very large global and local plastic deformations in their manufacture and fabrication. Rolling, drawing, extrusion, forging, forming, and welding are examples which illustrate the necessarily arbitrary nature of the choice of the initial state of zero plastic deformation in any appropriate set of incremental stress–strain relations.

Extension of physical understanding and mathematical representation in one dimension to multiaxial states of stress is neither unique nor obvious in principle. Different materials with similar responses to one-dimensional stress paths may and do behave very differently under multiaxial stress paths such as the addition of axial stress S_A to an existing shear stress S_T, or vice versa. Despite some excellent experimental investigations, only a little is known about the response of just

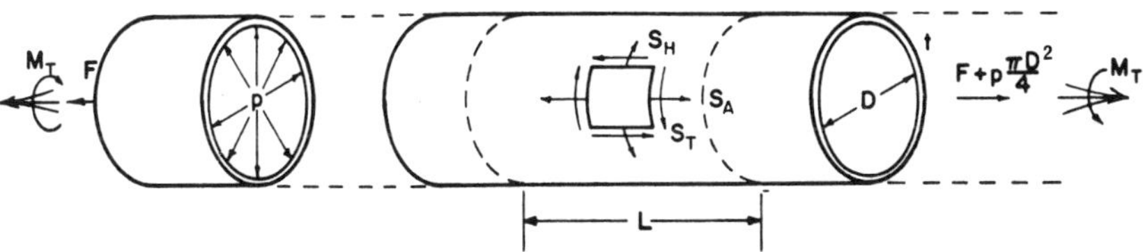

FIG. 3. Thin-walled circular tube specimen.

a few materials to such non-radial or non-proportional loading paths. The compelling reasons for the choice of a thin-walled circular tube test specimen (Fig. 3) under axial force F, twisting moment M_T and interior pressure p are familiar, as are the severe limitations [4]. Despite the paucity of data on most materials of practical importance, the general conceptual framework of a mathematical theory of plasticity is in existence [3].

By definition, the term plasticity denotes time-independence, a useful approximation to the physical behavior of structural metals at temperatures well below half their melting point and at low to moderate strain rates. Irreversibility requires that the theory be incremental in form. In principle, continual reconfiguration [6] is needed. At each instant of loading, at each point in a body, the current physical and geometric state of the material region serves as the new reference or initial state.

When simple stress–strain relations are sought, confusion and controversy do arise over the appropriate choice(s) of the increments of stress and strain, and about their mathematical and physical meaning. The issue does not disappear in a careful approach to the limit of zero increments because the tangent modulus in the plastic range is similar to the stress in magnitude [13]. The confusion is basically physical rather than mathematical. Each set of increments chosen is obtained from the experimental data by straightforward mathematical means and any one set of stresses or strains or their increments can be transformed to any other.

3. INCREMENTS OF STRESS

The incremental physical picture for the thin-walled tube is clear enough for the choice of either nominal or true stress in the conventional strength-of-materials sense. The conventional stresses S_T, S_A, S_H

are determined by the loads M_T, F, p and the geometry D, t where D is the tube diameter and t the tube thickness through such forms as

$$S_T = \frac{M_T}{\pi D^2 t/2} \tag{1}$$

which are entirely independent of material behavior provided axial symmetry is maintained. The increment in S_T is ΔS_T^{nom} if geometry change is ignored, ΔS_T^{true} if geometry change is taken into account.

$$\begin{aligned} \Delta S_T^{true} &= \Delta\left(\frac{M_T}{\pi D^2 t/2}\right) = \frac{\Delta M_T}{\pi D^2 t/2} - \frac{M_T}{\pi D^2 t/2}\frac{\Delta(\pi D^2 t/2)}{\pi D^2 t/2} \\ &= \Delta S_T^{nom} - S_T\left(\frac{2\Delta D}{D} + \frac{\Delta t}{t}\right) \\ &= \Delta S_T^{nom} - S_T(2\Delta e_H + \Delta e_R) \end{aligned} \tag{2}$$

Correspondingly,

$$\Delta S_A^{true} = \Delta S_A^{nom} - S_A(\Delta e_H + \Delta e_R) \tag{3}$$

Necking instability in simple tension occurs in a work-hardening material when $\Delta S_A^{nom} = \Delta S_A^{true} + S_A(\Delta e_H + \Delta e_R) < 0$ despite $\Delta S_A^{true} > 0$. The decrease in cross-sectional area ($\Delta e_H < 0, \Delta e_R < 0$) accompanying the axial extension as S_A increases and the plastic tangent modulus in tension decreases or remains constant must result at some stage in $\Delta S_A^{nom} < 0$. Similarly, if both increasing axial tension and twisting moment are applied to the tube, ΔS_T^{nom} will go negative at some stage. However twisting moment alone, unlike axial force alone, does not usually cause appreciable changes in the cross-sectional dimensions. In the absence of buckling, M_T as well as S_T^{true} increases as the angle of twist or conventional shear strain e_T increases.

Conventional stress, conventional strain, and their increments are obvious natural quantities to use for the axisymmetric circular tube. They were used effectively for a long time in the small strain range of elastic response and the large as well as small strain range of plastic response. However they do not fit comfortably into the unifying approach to finite elastic deformation properly based on a path independent strain energy density. In particular, for an initially isotropic elastic material the physically meaningful increments of stress are the changes of stress in a co-ordinate system that rotates in accord with the usual continuum or average rotation.

Somehow, these increments of stress in the rotated system came to be viewed as the best measure to use in incremental plastic stress–strain relations also. As an example, consider the changes $\Delta\tau$, $\Delta\sigma_a$, $\Delta\sigma_h$ in the Cauchy stresses τ, σ_a, σ_h (which coincide with the conventional true stresses S_T, S_A, S_H to start) due to infinitesimal changes in the true conventional stress ΔS_T, ΔS_A, ΔS_H and accompanying changes in the conventional strains Δe_T, Δe_A, Δe_H and the thickness. The Cauchy (true) stresses in the co-ordinate system rotated by $\Delta e_T/2$ (Fig. 4) are to first order

$$
\begin{aligned}
\tau + \Delta\tau &= S_T + \Delta S_T + (S_H - S_A)\Delta e_T/2 \\
\sigma_a + \Delta\sigma_a &= S_A + \Delta S_A + S_T \Delta e_T \\
\sigma_h + \Delta\sigma_h &= S_H + \Delta S_H - S_T \Delta e_T
\end{aligned}
\tag{4}
$$

where ΔS_T, ΔS_A, ΔS_H include the change in S_T, S_A, S_H due to the change in dimensions of the tube.

In particular, for $S_H = 0$,

$$\Delta\tau = \Delta S_T - S_A \Delta e_T/2 \tag{5}$$

If inappropriately it is $\Delta\tau$ rather than ΔS_T that is thought of as given by the usual elastic–plastic tangent modulus multiplied by $\Delta e_T/2$ (or Δe_T), then the normal stress term $-S_A\Delta e_T/2$ appears to have a worrisome destabilizing effect. As a compressive S_A increases in magnitude or the

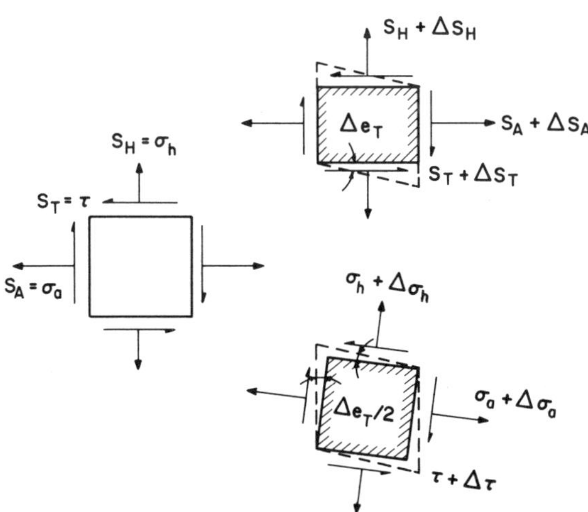

FIG. 4. Increments of stress in conventional and rotated co-ordinate system.

tangent modulus decreases or both, $\Delta\tau$ will not keep up with ΔS_T. However the tube test itself has the normal stresses acting on the shear planes and demonstrates the stability or instability in shear. No proper mathematical transformation can make any difference in that physical response.

It is true stress rather than nominal stress that should translate the experimental results on the tube in the plastic range to a continuum where all dimensions are of comparable size. However, the observed behavior is the real behavior no matter how the translation is handled. If there is stability for a given sequence of states of conventional true stress and strain in the tube there will be stability for the same sequence of states of stress and strain in the continuum. The continuum (average) rotation is irrelevant.

4. RELEVANT MATERIAL DIRECTIONS AND ROTATIONS

Consider again the combinations of finite shear, axial, and circumferential strain in a thin-walled circular tube of initially isotropic elastic material. At each stage for all paths of loading, the principal stress directions coincide with the principal strain directions and the principal stresses are the same for a given set of principal strains whether or not strain is accompanied by rotation due to twist. Separation of the displacement gradient into strain and rotation in the customary manner for the continuum not only is appropriate, it is an essential part of the proper physical description.

Addition of shear deformation to an already deformed configuration adds incremental principal strains and rotates the principal strain directions. The rotation does not affect the relation between the increments of principal stress and strain in an initially isotropic material, a relation which in general is anisotropic. Upon complete unloading the original isotropic picture of zero deformation is restored.

Deformation of an initially isotropic elastic–plastic material, on the contrary, produces an anisotropic material whose memory of its initial isotropy is lost for subsequent stress paths. All initial directions cease to be equal. Under general loading paths the principal strain picture is irrelevant. Separation of the deformation gradient into strain and rotation in the customary continuum manner is not physically meaningful. Separation of the increment of deformation gradient into incremental strain and incremental rotation requires the determination of the rotation of physically meaningful directions in the material, for example grain or lattice orientation. The usual continuum rotation increment, which implicitly treats all directions equally, is appropriate only for pure rigid body rotation.

Additional shearing deformation caused by additional twisting of the tube through a shear angle Δe_T does not to a first approximation rotate the cross-sectional planes, or the grains or material lattices. Consequently it adds a zero rotation increment, not the continuum rotation increment equal to the shear strain increment $\Delta e_T/2$. Schematically as pictured in Fig. 5, elastic strains are small so that the lattice directions remain fixed as dislocations or slip provide the necessary accommodation to large plastic deformation. This is in sharp contrast to finite

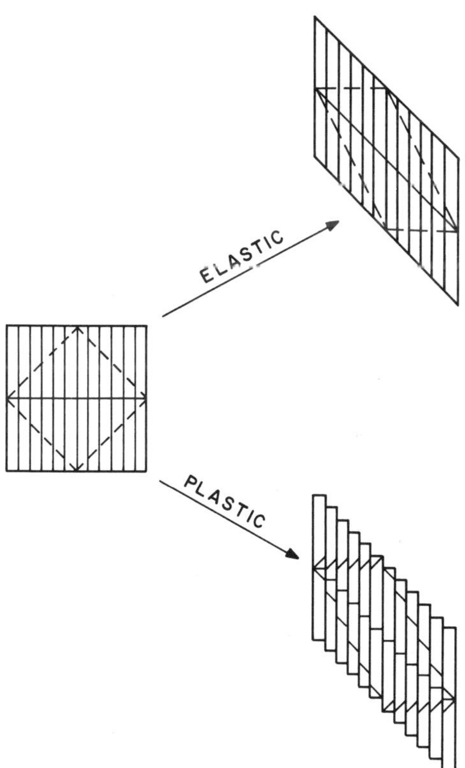

FIG. 5. Elastic vs plastic.

elastic deformation where the line of material originally in the axial direction remains continuous and rotates by Δe_T. Intermediate lines of material between the circumferential and the axial direction likewise remain continuous and rotate by intermediate amounts. On the other hand, if the elastic deformation accompanying finite plastic deformation is neglected, all local directions, like lattice directions, undergo zero rotation due to plastic deformation. All intermediate lines of material are discontinuous.

Of course, for an initially anisotropic elastic material it is equally necessary to follow the material directions that are analogous to lattice directions. If an axisymmetric thin-walled tube specimen can be made, the zero rotation of the cross-sectional planes now plays a key role as it does for plastic response. However the axial lines still would remain continuous and rotate appreciably. In general the local continuum or average rotation, which includes rigid body rotation, cannot by itself give the essence of the physical behavior associated with orientation. As Mandel [8] pointed out earlier, physically important material orientations must be included explicitly in constitutive relations.

Local lattice rotations or their equivalent in amorphous or polymeric materials are determined by the deformation occurring elsewhere in the body. In most instances there are directions of symmetry or other directions that can serve conveniently for reference to eliminate overall rigid body rotations. Examples are the axis of the thin-walled tube specimen, lines joining points of support of a body or structure, and (almost) rigid constraints on a continuum. Elastic deformation between these reference lines or surfaces and the region of interest can lead to significant local rotation of lattice directions.

Fortunately for many problems of finite plastic deformation, the elastic strains are negligible and the rotation associated with the elastic displacement field in the body can be ignored. Zero lattice rotation locally is the likely mode in most but by no means all continuum problems.

Large lattice rotations do take place along with large shear deformations in metal forming processes such as extrusion and drawing with large reductions. In problems of instability where rotations do enter, the lattice rotation that matters is not the local continuum rotation but is the rotation of the direction of the loads applied to the structure with respect to the local lattice directions in the material. In the absence of such rotation, a shear instability in a continuum cannot occur with a rising true stress–strain curve in shear.

Calculations of stress–strain relations for single crystals or polycrystals from a postulated response on slip planes should and do properly take rotation of the slip planes into account. Such a calculation is analogous to a stability calculation for a structure. Once done, the resulting stress–strain relations apply to a continuum with zero 'lattice' rotation.

5. PERMISSIBLE AND APPROPRIATE IDEALIZATIONS

The familiar incremental stress–strain relations of plasticity, employed prior to the worries about continuum rotation, did treat the rotation due to shear as zero and properly so. Experimental data had been obtained from thin-walled tube tests and other tests in which, for all practical purposes, lattice rotation was zero. Isotropic hardening forms and kinematic hardening forms separately and in combination serve as very useful idealizations of plastic behavior despite their inability to match plastic response in detail. Better representations are available at the cost of greatly increased complexity. For each idealization, crude or elaborate, local lattice rotations with respect to reference directions in the system must be taken into account as best that can be done. However, the local continuum or average rotation associated with shear deformation is irrelevant and should be ignored. It is already included in the stress–strain relations written in terms of the conventional stresses S_T, S_A, S_H and strains e_T, e_A, e_H, and their increments or rates.

6. CONCLUSION

Rotation of material with respect to the direction of load can lead to shear instability in the elastic or plastic range. However, the local (average) continuum rotation accompanying the torsional shear strain in tube tests is not an added rotation. Whatever effect large or small elastic or plastic shear deformation may have is included in the data. Terms such as $S_A \Delta e_T/2$, which appear in the increment $\Delta\tau$ of shear stress when reference axes are rotated with the continuum (average) rotation, are physically significant for initially isotropic elastic material. For such a material the directions of the principal stresses coincide with those of the principal strains. The principal stress–principal strain relation is independent of direction. Consequently the interpretation and understanding of the data from tube tests is clearest when the continuum rotation is followed. On the contrary, in the plastic domain it is the lattice or grain rotation that is physically significant. That rotation is zero on average in the tube tests. Consequently the $S_A \Delta e_T/2$ type of term associated with the 'continuum rotation' is irrelevant to the question of plastic instability due to normal stress on shear planes. In general, approaches suitable or even necessary for finite deformation of initially isotropic elastic material are not appropriate for finite plasticity. In particular the often zero and usually known lattice rotation which is central to finite plasticity is totally divorced from the local continuum rotation so central to the finite elastic response of initially isotropic materials. The old-fashioned incremental plasticity forms familiar from long and successful use are the proper forms in principle provided they follow any change in lattice orientation.

REFERENCES

1. Backofen, W. A. The torsion texture of copper, *J. Metals, Trans. AIME*, **188** (1950), 1454–1459; **197** (1953), 61–62.
2. Dolan, T. J. Nonlinear response under cyclic loading conditions, *Proc. the Ninth Midwestern Mechanics Conference*, Madison, Wis., (1965) 3–21. See also, Morrow, J. and G. M. Sinclair, Cycle-dependent stress relaxation, *Symposium on Basic Mechanisms of Fatigue*, ASTM STP 237, American Society for Testing and Materials (1958), 83–109.
3. Drucker, D. C. Plasticity. In: *Structural Mechanics, Proc. First Symposium on Naval Structural Mechanics* (Eds. J. N. Goodier and N. J. Hoff), Pergamon Press, Oxford, 1960, 407–455.
4. Drucker, D. C. From limited experimental information to appropriately idealized stress–strain relations. In: *Mechanics of Engineering Materials* (Eds. C. S. Desai and R. H. Gallagher), John Wiley, London (to appear).
5. Green, A. E. and W. Zerna. *Theoretical Elasticity*, Oxford University Press, Oxford, 1954.
6. Hill, R. Some basic principles in the mechanics of solids without a natural time, *J. Mech. Phys. Solids*, **7** (1959), 209–225.
7. Lee, E. H., R. L. Mallett and T. B. Wertheimer. Stress analysis for anisotropic hardening in finite deformation plasticity. *J. appl. Mech.*, **50** (1983), 544–60.
8. Mandel, J. Equations constitutives et directeurs dans les milieux plastiques et viscoplastiques, *Int. J. Solids Struct.*, **9** (1973), 725–740.
9. Nagtegaal, J. C. and J. E. de Jong. Some aspects of non-isotropic workhardening in finite strain plasticity, plasticity of metals at finite strain: theory, experiment and computation (Eds. E. H. Lee and R. L. Mallett), Rensselaer Polytechnic Inst., 1982, 65–102.

10. Olszak, W., Z. Mroz and P. Perzyna. *Recent Trends in the Development of the Theory of Plasticity*, Macmillan-Pergamon, New York, 1963.
11. Onat, E. T. and D. C. Drucker. Inelastic instability and incremental theories of plasticity, *J. Aero. Sci.*, **20** (1953), 181–186.
12. Palgen, L. and D. C. Drucker. The structure of stress–strain relations in finite elasto-plasticity, *Int. J. Solids Struct.*, **19** (1983), 519–531.
13. Rice, J. R. A note on the 'small strain' formulation for elastic–plastic problems, Tech. Rept. N00014-67-A-000318, Div. of Engr., Brown University, 1970. See also, McMeeking, R. M. and J. R. Rice, Finite-element formations for problems of large elastic–plastic deformation, *Int. J. Solids Struct.*, **11** (1975), 601–616.

NOTE ADDED IN PROOF

The absence of lattice rotation in torsion is to be thought of as holding in the very gross macroscopic sense. Lattice planes within grains and from grain to grain do rotate one way and another to accommodate the deformation pattern imposed. Without such rotations the stress–strain relations would be very different from those observed. As stated earlier, these rotations must be taken into account explicitly in going from slip behavior on slip planes to the response of single crystals and in predicting the behavior of a polycrystalline aggregate from the response of single crystals. However they are included fully in the experimental data on which proposed macroscopic constitutive relations are based. No additional lattice rotation is introduced by the finite shear deformation in torsion.

The microscopic lattice orientation adjustments are in fact most rapid at the earlier stages of plastic deformation. As Backofen [1] demonstrated in 1950, there is little additional rotational accommodation in going from a plastic shear strain of 3·95 in a torsion specimen to a strain of 5·25. There is essentially *no rotation at all* in reversed plastic torsion that returns the plastic strain from 5·25 back to zero.

It is in this sense that the continuum rotation associated with finite shear strain is irrelevant and that the associated lattice rotation should be taken as zero. (I am indebted to Professors Robert J. Asaro and Rodney J. Clifton for raising the question of the experimental evidence on lattice rotation and the need for this clarification of the physical meaning of zero lattice rotation in the continuum sense.)

3

Finite Deformation Effects in Plasticity Analysis

E. H. LEE

Department of Mechanical Engineering, Rensselaer Polytechnic Institute, Troy, New York, USA

ABSTRACT

The need to incorporate the physical properties of phenomena as faithfully as we are able in constructing constitutive relations is emphasized. In particular the importance of fully understanding the physical significance of all terms in a constitutive relation is stressed. As an example, the precise uncoupling of elastic and plastic contributions to total deformation is discussed in the development of the theory of isotropic hardening at finite strain. In the case of ductile metals exhibiting a Bauschinger effect it is shown that an anomaly arising from the choice of terms added to preserve objectivity in finite deformation analysis led to serious errors in stress evaluation.

1. INTRODUCTION

I decided to discuss this topic because there appear to be some misunderstandings in the literature concerning the relationship between various formulations of elastic–plastic theory appropriate for materials subjected to finite strain and rotation. I will restrict myself to the classical rate independent theory, albeit of flow or incremental type. However, many of the aspects considered will be relevant to rate-dependent analysis. It is almost a truism to state that a major objective in formulating constitutive relations is that they should express the physical properties of the material to be analyzed. This objective may be compromised somewhat in the case of materials which exhibit a complicated deformation response to applied stress, or vice versa, in which case a simplified approximation to the physical characteristics may provide an appropriate model. However, it is important to bear in mind the liberty thus taken with physical reality when utilizing the approximation in order to ensure that a property deduced from the analysis does not arise solely because of some aspect of the approximation. In the recent continuum mechanics literature there seems to have arisen a tendency to treat the formulation of constitutive relations in a more cavalier manner than is appropriate as, for example, expressed by the statement ‘constitutive assumption’ which appears to me to carry the implication of casualness in devising what may be amongst the most important ingredients in a theory. A constitutive relation should be inferred from an extensive study of experimental findings, a process hardly expressed by the word assumption. A related matter is the use of the word define in the context ‘we define this variable to be the plastic strain’. It seems to me that in this case, for example, plastic strain is well characterized physically and our responsibility is to express this physical concept in mathematical terms as faithfully as we are able to do and not simply to *define* some mathematical construct to embody this physical entity.

2. ELASTIC–PLASTIC THEORY WITH ISOTROPIC HARDENING

We seek a constitutive relation which expresses the total deformation produced in terms of the history of the stress that has been applied, that is the strain as a functional of the stress or vice versa. Since the total strain is the resultant of elastic and plastic deformations it is necessary to determine these contributions from the elasticity and plasticity laws and analyze how these components combine to yield the resultant deformation.

Subsequent to loading which generates both elastic and plastic strains, if unloading to zero stress involves change in the elastic component of strain only, the deformation prior to unloading can be simply decomposed into an elastic and plastic component in terms of the configurations shown in Fig. 1. Using the same Cartesian axes, $\mathbf{X}$ denotes material particle co-ordinates of an initially undisturbed body, after elastic–plastic deformation these particles occupy the positions

$$\mathbf{x} = \mathbf{x}(\mathbf{X}, t) \tag{1}$$

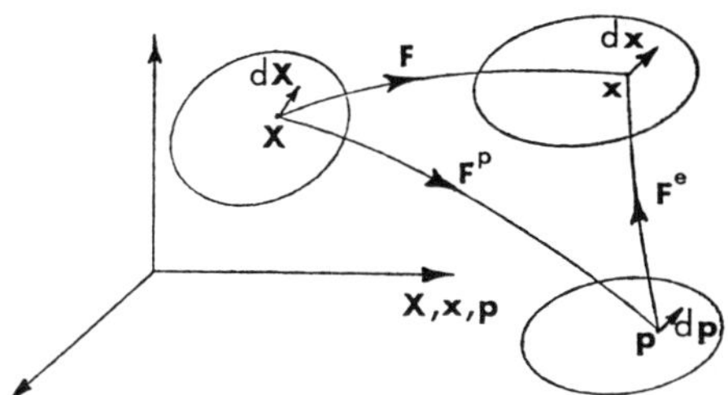

FIG. 1. Elastic–plastic deformation.

at time t, and after elastic unloading to zero stress, they occupy the positions

$$\mathbf{p} = \mathbf{p}(\mathbf{X}, t) \tag{2}$$

Since the macroscopic stress is zero in the configuration $\mathbf{p}$, the macroscopic elastic strain is zero although there will be microscopic elastic strains, for example, those associated with dislocations generated by the plastic flow. Thus in the configuration $\mathbf{p}$, the macroscopic deformation which can be expressed by the deformation gradient

$$F^{\text{p}}_{ij} = \partial p_i/\partial X_j\,;\ (\mathbf{F}^{\text{p}} = \partial \mathbf{p}/\partial \mathbf{X}) \tag{3}$$

constitutes purely plastic deformation. Since purely elastic deformation has occurred during the unloading $\mathbf{x} \rightarrow \mathbf{p}$, $\mathbf{F}^{\text{p}}$ also expresses the plastic deformation existing in the configuration $\mathbf{x}$. Since the elastic strain is zero in the configuration $\mathbf{p}$, the elastic deformation in the neighborhood of a particle in the configuration $\mathbf{x}$ can be expressed by the deformation gradient

$$\mathbf{F}^{\text{e}} = \partial \mathbf{x}/\partial \mathbf{p} \tag{4}$$

where $\mathbf{x}$ and $\mathbf{p}$ are associated with the same material particle $\mathbf{X}$. The chain rule then yields the kinematic relation

$$\mathbf{F} = \partial \mathbf{x}/\partial \mathbf{X} = \mathbf{F}^{\text{e}}\mathbf{F}^{\text{p}} \tag{5}$$

which decomposes the deformation at $\mathbf{x}$ into elastic and plastic components. This is termed the nonlinear coupling between elastic and plastic deformation in contrast to the summation relation for elastic and plastic strains

$$\boldsymbol{\varepsilon} = \boldsymbol{\varepsilon}^{\text{e}} + \boldsymbol{\varepsilon}^{\text{p}} \tag{6}$$

valid for infinitesimal deformation analysis.

Since unloading a body that has been subjected to nonhomogeneous plastic flow leaves it in a state of residual stress, unloading to zero stress will, in general, require a noncontinuous or multi-valued mapping of the particles in configuration **p** so that the deformation gradients $\mathbf{F}^e$ and $\mathbf{F}^p$ will not exist in a strict mathematical sense [5, 3]. This causes no difficulty in the analysis since $\mathbf{F}^e$ and $\mathbf{F}^p$ then become point functions of the particle co-ordinates **X** and play the same role as partial derivatives.

In order to obtain the elastic–plastic constitutive relation it is necessary to substitute into (5), or a more suitable modified version of (5), the elastic component in terms of the elastic law which involves the current stress acting and the plastic component according to the plasticity law which is of incremental or flow type, i.e. involves the rate of stress and rate of strain for the usually encountered strain-hardening material. Thus a differentiated version of (5) is needed and the most convenient form from the standpoint of incorporating the constitutive relation into the manipulation of stress-analysis boundary-value problems is to introduce the velocity gradient in the current configuration **x**. The particle velocity **v** is given by

$$\mathbf{v} = \partial \mathbf{x}/\partial t\,|_{\mathbf{X}} \tag{7}$$

and the velocity gradient

$$\mathbf{L} = \left.\frac{\partial \mathbf{v}}{\partial \mathbf{x}}\right|_t = \frac{\partial \mathbf{v}}{\partial \mathbf{X}}\frac{\partial \mathbf{X}}{\partial \mathbf{x}} = \dot{\mathbf{F}}\mathbf{F}^{-1} \tag{8}$$

where $\dot{\mathbf{F}}$ denotes $\partial \mathbf{F}/\partial t\,|_{\mathbf{X}}$. Substituting (5) for **F** in (8) yields

$$\mathbf{L} = \dot{\mathbf{F}}^e\mathbf{F}^{e^{-1}} + \mathbf{F}^e\dot{\mathbf{F}}^p\mathbf{F}^{p^{-1}}\mathbf{F}^{e^{-1}} \tag{9}$$

In elastic–plastic theory the physically achieved configuration $\mathbf{x}(\mathbf{X}, \mathbf{t})$ is our primary concern. When unloading occurs, the purely elastic deformation simply becomes a special case with the most general $\mathbf{F}^p$ corresponding to rigid-body rotation. The configuration **p** associated with **x** is needed to express the elastic and plastic components of the deformation existing in the state **x**. Thus the configuration **p** can be considered to constitute a mathematical entity which expresses convenient variables needed for the analysis. This state need never be achieved physically. We are therefore free to prescribe conditions on the state **p** for the convenience of the analysis as long as the property of being unstressed is maintained. Superimposed point-wise rotation is permissible since it is not necessary that the mapping be continuous, and within this framework it is convenient to select the configuration $\mathbf{p}$ so that the elastic unloading $\mathbf{x} \rightarrow \mathbf{p}$ takes place without rotation. Thus $\mathbf{F}^e$ expresses pure deformation and assumes the form $\mathbf{V}^e$, a symmetric matrix

$$\mathbf{F}^e = \mathbf{V}^e = \mathbf{V}^{eT} \tag{10}$$

where the superscript T denotes the transpose.

Thus (9) becomes

$$\mathbf{L} = \mathbf{D} + \mathbf{W} = \dot{\mathbf{V}}^e \mathbf{V}^{e^{-1}} + \mathbf{V}^e \dot{\mathbf{F}}^p \mathbf{F}^{p^{-1}} \mathbf{V}^{e^{-1}} \tag{11}$$

where $\mathbf{D}$, the symmetric part of $\mathbf{L}$, is the rate of deformation or stretching tensor and $\mathbf{W}$ the spin. Taking the symmetrical part of (11) gives

$$\mathbf{D} = \mathbf{D}^e + \mathbf{V}^e \mathbf{D}^p \mathbf{V}^{e^{-1}} \big|_S + \mathbf{V}^e \mathbf{W}^p \mathbf{V}^{e^{-1}} \big|_S \tag{12}$$

where $\mathbf{D}^e$ is the symmetric part of $\dot{\mathbf{V}}^e \mathbf{V}^{e^{-1}}$ and $\mathbf{D}^p$ and $\mathbf{W}^p$ the symmetric and anti-symmetric parts respectively of $\dot{\mathbf{F}}^p \mathbf{F}^{p^{-1}}$.

Relation (12) constitutes the convenient form of kinematic relation for use in elastic–plastic analysis in place of (5). Note that it introduced a coupling between elastic deformation and plastic rate of deformation and spin not present in the more commonly used kinematic relation

$$\dot{\boldsymbol{\varepsilon}} = \dot{\boldsymbol{\varepsilon}}^e + \dot{\boldsymbol{\varepsilon}}^p \tag{13}$$

the summation law for strain rates. Discussion of various aspects of (12) is to be found in ref. 2. Although the above discussion was based on elastic unloading to zero stress, the theory presented can be applied when a marked Bauschinger effect intrudes so that on unloading, reverse plastic flow occurs before the stress reaches zero [2]. This is likely to happen at large strains. By determining the strain-energy function in the elastic region which does not include zero stress, extrapolating this to zero stress would define a formal plastic strain associated with deformation within that elastic region such that the kinematic analysis presented would apply exactly wherever it had physical meaning, i.e. within that elastic region. Should the stress in fact be reduced to zero, renewed plastic flow would take place producing a new elastic region enclosing the stress origin. In this case the plastic flow could be measured and the original development applied directly. This plastic strain would differ from the formal one obtained by extrapolation, but the theory would apply exactly in both cases for stresses within their respective elastic regions.

As has been anticipated and in fact already led to the development of the kinematic rate relation (12) because the plasticity law is of incremental or flow type involving strain rate, it is now necessary to obtain a rate form of the elasticity law. Although in most cases elastic strains remain small because plastic flow limits stresses to the order of the yield stress, it is necessary to adopt a theory valid for finite deformation because, with large rotations occurring, it is necessary to use a correctly objective derivative of the elastic stress–strain relation in order to have the final constitutive relation valid under arbitrary superimposed time dependent rigid body rotation. Thus the overall objective is to express $\mathbf{D}^{e}$ and $\mathbf{D}^{p}$ in (12) in terms of a suitable stress rate which must be the same for both the elastic and plastic laws so that the combination in (12) will yield a total strain rate–stress rate relation. Both aspects—plasticity and elasticity—thus have a bearing on the choice of rate operator.

The plasticity law with isotropic hardening involved a yield condition

$$f = g(\boldsymbol{\tau}) - c < 0 \tag{14}$$

where $\boldsymbol{\tau}$ is the Kirchhoff stress, g an isotropic scalar function and c a scalar function of the history of plastic deformation. The concepts of plastic potential and strain hardening then provide a relation for the plastic deformation rate $\mathbf{D}^{p}$ of the form

$$D^{p}_{ij} = \frac{1}{h}\left(\frac{\partial f}{\partial \tau_{kl}}\dot{\tau}_{kl}\right)\frac{\partial f}{\partial \tau_{ij}} \tag{15}$$

where h is the scalar strain hardening modulus, a function of the history of plastic deformation, and $\dot{\tau}$ denotes the material derivative. The term in parentheses, $\dot{g}$ being a scalar invariant, is unaffected by superimposed rigid-body rotation. The Jaumann derivative of τ

$$\overset{\circ}{\boldsymbol{\tau}} = \dot{\boldsymbol{\tau}} - \mathbf{W}\boldsymbol{\tau} + \boldsymbol{\tau}\mathbf{W} \tag{16}$$

is a time derivative which includes the convection effects of translation with the material particle under study and also convection of the rotation of axes with spin $\mathbf{W}$. Since $\dot{g}$ is not influenced by rotation, the material derivative can be replaced by the Jaumann derivative without affecting the relation (15), and this, being objective, is appropriate for application to the elasticity law. Prager [8] has pointed out that of the objective derivatives in common use, the Jaumann derivative is the only one for which vanishing of the stress derivative implies that the invariants of stress are stationary. This is an important requirement for the validity of (15) for otherwise the terms in the parentheses could be zero, and hence also $\mathbf{D}^p$, while the yield condition f is changing—a physically unacceptable circumstance.

For elastic deformation under the deformation gradient $\mathbf{V}^e$ with no rotation, the elasticity law valid for finite deformation takes the form

$$\boldsymbol{\tau} = 2\mathbf{C}^e\, \partial\psi/\partial\mathbf{C}^e \tag{17}$$

where $\mathbf{C}^e = \mathbf{V}^e\mathbf{V}^e$ and ψ is the strain energy function. Operating on both sides with a Jaumann derivative it is shown in ref. 6 that (17) yields a linear relationship between $\overset{\circ}{\tau}$ and the Jaumann derivative of $\mathbf{C}^e$. This can then be combined with the plasticity law (15), the stress derivative there being changed to the Jaumann derivative, to yield a total strain rate–stress rate constitutive relation having the form

$$D_{ij} = \Lambda_{ijkl}\overset{\circ}{\tau}_{kl} \tag{18}$$

This procedure is purely deductive, the final elastic–plastic operator (18) being uniquely determined by the choice of derivative.

This process contrasts with the usual approach to formulating the elastic–plastic constitutive relation for finite strain applications which has been based on the concept of small elastic strains with Hooke's law being used in the well-known form

$$\varepsilon^e_{ij} = \frac{1+\nu}{E}\tau_{ij} - \frac{\nu}{E}\tau_{kk}\delta_{ij} \tag{19}$$

where $\boldsymbol{\varepsilon}^e$ is the infinitesimal strain. A rate form of (19) is taken to be

$$D^e_{ij} = \frac{1+\nu}{E}\check{\tau}_{ij} - \frac{\nu}{E}\check{\tau}_{kk}\delta_{ij} \tag{20}$$

where $\check{\boldsymbol{\tau}}$ is a time derivative of the stress tensor. Because the final constitutive relation must be valid for large plastic deformation, it must be so for large total strain and rotation and hence must be objective. The elastic law (20) and the plasticity law (15) are substituted into (13) to yield the elastic–plastic constitutive relation. For this to be valid, clearly the stress derivative $\check{\boldsymbol{\tau}}$ must be objective. As already discussed for the nonlinear kinematics approach, the Jaumann derivative is chosen to be compatible with the plasticity law. In the approach based on Hooke's law, however, (19) cannot be formally differentiated because it is not a valid equality in the context of finite deformation theory. It is evident that even for small rotations, $\boldsymbol{\varepsilon}^e$, the symmetric part of the displacement gradient is not changed whereas the stress components are. Thus in the absence of a formal derivative of (19), the Jaumann derivative was applied to the right-hand side and the velocity field was introduced with which to express the 'rate of strain' for the left hand side of (20). In contrast we have seen that the elasticity law valid for finite deformation can be formally differentiated to provide a strain rate term which can be substituted into the nonlinear kinematical relation (12) along with the plasticity law. In [6] these components were combined together in a purely deductive process to yield an elastic–plastic constitutive relation and a corresponding variational principle. A simplification of the final result valid for small elastic strain yielded the conventional structure.

Since the small-elastic-deformation theory based on the summability of elastic and plastic strain rates poses a simpler computational task and in most problems likely to arise in engineering practice is a close approximation to the more complex theory based on nonlinear kinematics, the former is usually the appropriate choice for stress evaluations. However, some problems, for example high pressure impact and explosively generated waves involve large elastic dilatational strains which require the finite-elastic-strain analysis. The more complete theory may also be needed for the study of instabilities and localization of deformation since such problems are particularly sensitive to minor variations in the constitutive relations. Also for basic studies concerning the structure of elastic–plastic constitutive relations the complete theory may be needed to investigate the coupling of elastic and plastic strains in such concepts as normality.

3. STRESS ANALYSIS FOR PLASTIC STRAIN-INDUCED ANISOTROPY

So far the discussion has been limited to isotropic hardening. However it is known that the anisotropic Bauschinger effect becomes more pronounced as the strain increases and can have a significant effect particularly if destressing occurs followed by reverse stressing. This often arises as the product emerges from a metal-forming process. It is therefore important to incorporate anisotropic hardening and the most common approach has been through kinematic hardening or a combination of this with isotropic hardening.

An unexpected outcome of a workshop: Plasticity of Metals at Finite

Strain introduced in a paper by J. C. Nagtegaal and J. E. de Jong [7], was the realization that the then state-of-the-art computer programs for stress analysis at finite strain of materials which exhibit plastic strain-induced anisotropy (e.g. the Bauschinger effect) predicted spurious oscillation of the shear stress generated by monotonically increasing shear strain. This involved huge errors in the calculated stress. The difficulty was not resolved during the workshop, but it later became clear [4] that erroneous mathematical representation of the influence of material rotation on the growth of the anisotropy was responsible for the anomaly.

The major component of the plastic anisotropy associated with the Bauschinger effect in a polycrystalline material is caused by a residual stress distribution, called the back stress $\boldsymbol{\alpha}$, generated by deformation of the polycrystalline material comprising a more or less random agglomeration of crystallites, each exhibiting the crystallographic anisotropy of a single crystal. Such residual stresses, which strengthen a structure for continued loading in the direction previously applied but reduce the yielding strength in reverse loading, commonly arise due to the development of nonhomogeneous plastic strain, as for example in autofrettage of a gun tube or pressure vessel. In the case of plastic yielding of a polycrystalline material, the effect of this residual stress distribution is expressed as a kinematic hardening law with a shift $\boldsymbol{\alpha}$ of the yield surface in stress space. The residual stresses vary appreciably from crystallite to crystallite in such a way that the macroscopic stress is zero. Those crystallites oriented so that they are amongst the first to flow plastically during the initial loading will have accumulated residual stresses of the opposite sign and thus on repeated loading in the original direction renewed plastic flow will be delayed. In contrast, reverse loading will experience a reduced yield stress. Averaged over a surface which intersects many crystallites, the magnitude of the residual stresses will approach zero since this average must approach the macroscopic stress. Thus the physical mechanism causing the anisotropy of the yield stress is embedded in the material on the crystallite scale.

In simple shearing deformation, material elements distort and rotate causing growth and rotation of the residual stresses and hence of the macroscopic plastic anisotropy which they cause. Because distortion and rotation are occurring simultaneously, different material lines through a material point rotate with different angular velocities, the average value of which is the spin $\mathbf{W}$.

A study of the micromechanics of the situation, either at the crystallite level, the dislocation level, or at both, may be needed to fully understand this question, but this has not yet been done. However, information can be gleaned from the macroscopic theory. A physically based model which provides a first approximation to the effect of material rotation on the evolution of the anisotropy and includes major mechanisms which govern the phenomenon was presented in ref. 4. It yielded promising results and eliminated the anomaly generated by the then state-of-the-art codes.

In order to appreciate the anomaly and its resolution in the stress analysis for monotonically increasing shear strain, it is important to examine the kinematics involved. Using rectangular Cartesian coordinates for the configuration at time t, simple shear strain, $\gamma = kt$, shown in Fig. 2, is generated by the steady-state velocity field

$$v_1 = kx_2, \qquad v_2 = v_3 = 0 \tag{21}$$

having the velocity gradient **L** with symmetric part **D**, the rate of deformation, and anti-symmetric part **W**, the spin.

$$\mathbf{L} = \frac{\partial v_i}{\partial x_j} = \begin{bmatrix} 0 & k & 0 \\ 0 & 0 & 0 \\ 0 & 0 & 0 \end{bmatrix}, \qquad \mathbf{D} = \begin{bmatrix} 0 & k/2 & 0 \\ k/2 & 0 & 0 \\ 0 & 0 & 0 \end{bmatrix},$$

$$\mathbf{W} = \begin{bmatrix} 0 & k/2 & 0 \\ -k/2 & 0 & 0 \\ 0 & 0 & 0 \end{bmatrix} \tag{22}$$

Since large strains are considered, elastic strains will be neglected and so the plastic strain rate about the current configuration is given by

$$\mathbf{D}^{\mathrm{p}} = \mathbf{D} \tag{23}$$

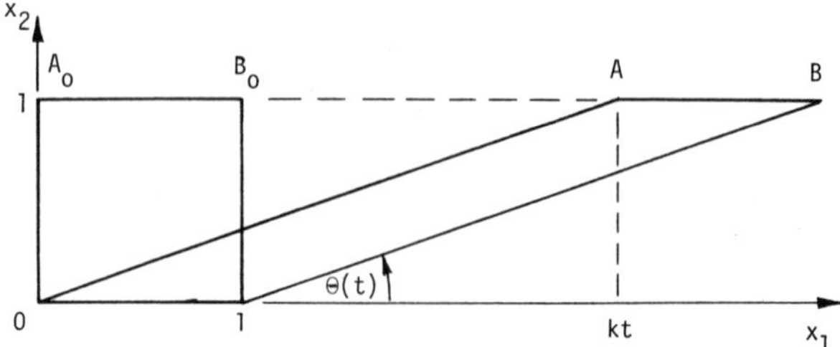

FIG. 2. Simple shear in the x_1 direction.

Because the velocity field is linear in $\mathbf{x}$, straight material lines remain straight as illustrated in Fig. 2. The velocity field is steady and uniform over the body so that the angular velocity of a line of elements depends only on its current orientation angle θ and is given by

$$\dot{\theta} = -k \sin^2 \theta \tag{24}$$

In spite of the constant spin tensor, (22), it is clear from Fig. 2 that the material elements initially along $0A_0$ only rotate through an angle $\pi/2$ as $t \to \infty$, and no material lines ever rotate by more than π.

In ref. 4 the anisotropic hardening was expressed as kinematic hardening, the constitutive relations for which were initially developed for infinitesimal displacement theory giving the evolution equation for the back stress $\boldsymbol{\alpha}$ in the form

$$\dot{\boldsymbol{\alpha}} = c(\bar{\varepsilon}^{\mathrm{p}})\mathbf{D}^{\mathrm{p}} \tag{25}$$

following the Prager–Ziegler model. This expresses the growth of $\boldsymbol{\alpha}$ due to the straining currently taking place but, because infinitesimal displacement theory neglects rotation effects, (25) ignores the change in $\boldsymbol{\alpha}$ caused by the rotation of the anisotropy already generated by previous plastic flow. If this rotation, caused by the rotation of the material in which the residual stress is embedded, corresponds to an angular velocity $\mathbf{W}^*$, then the resultant material derivative of $\boldsymbol{\alpha}$ relative to co-ordinates in fixed space is

$$\dot{\boldsymbol{\alpha}} = c(\bar{\varepsilon}^{\mathrm{p}})\mathbf{D}^{\mathrm{p}} + \mathbf{W}^*\boldsymbol{\alpha} - \boldsymbol{\alpha}\mathbf{W}^* \tag{26}$$

A study of the micromechanics at the crystallite level is needed to fully understand this question but since this has not yet been carried out an initial approach to the solution can be constructed from macroscopic considerations. The principal component of $\boldsymbol{\alpha}$ having the largest absolute magnitude produces the major influence on the yield surface, and hence on the stress field, and is carried in the lines of material elements oriented in the corresponding eigenvector direction. Thus the spin of these lines of material elements may be considered to incorporate the major rotational influence of the back stress generated by previous plastic flow.

The formerly accepted evolution law for kinematic hardening at finite deformation took the form

$$\mathring{\boldsymbol{\alpha}} = \dot{\boldsymbol{\alpha}} - \mathbf{W}\boldsymbol{\alpha} + \boldsymbol{\alpha}\mathbf{W} = c(\bar{\varepsilon}^{\mathrm{p}})\mathbf{D}^{\mathrm{p}} \tag{27}$$

which was generalized from (25) by use of the Jaumann derivative in order to make the equation objective, without further consideration of the physical significance of the terms added. Comparison with (26), and bearing in mind the physical model on the basis of which (26) was developed, implies that the material rotation effect on the anisotropy embedded in the material corresponds to material lines rotating at angular velocity $k/2$ which is inconsistent with the kinematics discussed. The nature of the connection between the material elements in an elastic–plastic continuum clearly rules out the validity of (27) for ductile metals. The structure of (26) suggests a modified interpretation by writing it in the form of a Jaumann type derivative with spin $\mathbf{W}^*$ which is shown in [4] to be objective.

Figures 3 and 4, taken from ref. 4 show the variations of shear stress σ_{12} and normal stress σ_{11} for increasing shear strain γ corresponding to the use of Jaumann type derivatives based on $\mathbf{W}$ and $\mathbf{W}^*$ as indicated. A linear hardening stress–strain curve in tension was assumed, and in Fig. 3 the curve deduced from the same tensile behavior assuming isotropic hardening is indicated (no normal stresses are generated in this case). It is clear that the incorrect formulation of the influence of material spin causes the oscillations associated with the conventional Jaumann derivative. The decreasing tangent modulus exhibited by the new solution arises because the direction of maximum tensile yield stress occurs at monotonically decreasing values of θ and hence rotates away from the direction of maximum tensile strain rate (which remains constant at $\theta = \pi/4$) and this causes a lessening of the

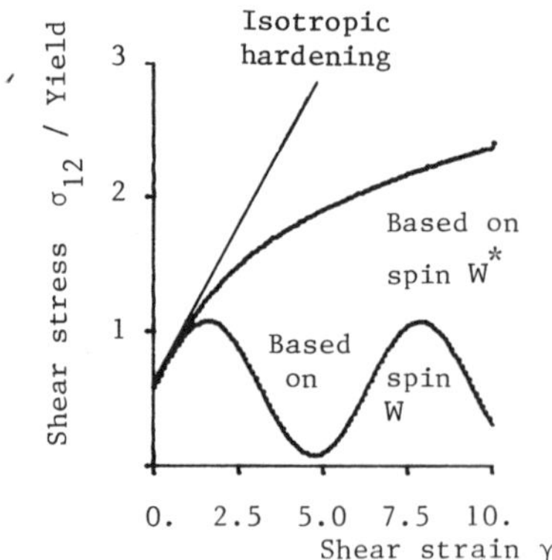

FIG. 3. Shear stress variation.

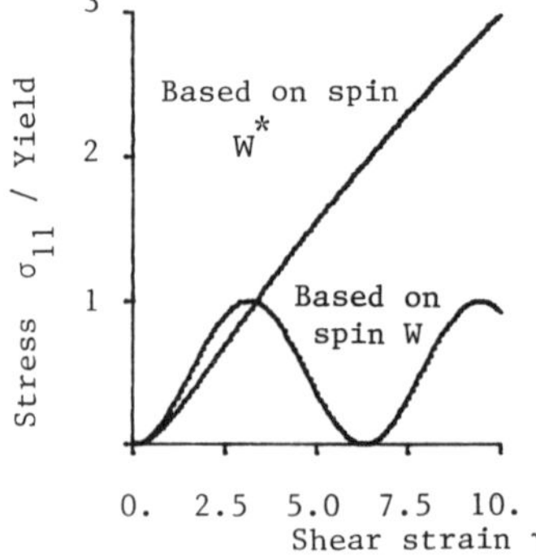

FIG. 4. Normal stress variation.

hardening. This phenomenon has been termed the rotational Bauschinger effect by J. J. Jonas [1].

This development emphasizes the importance of considering the physical significance of all terms occurring in a constitutive relation. It is clearly not sufficient to simply ensure that a relation is objective in generalizing one formulated for infinitesimal displacement theory since this can be done in an infinite number of ways; consideration must be given to the physical significance of the terms added.

Finite-element computer codes which incorporate kinematic hardening and are considered valid for finite strain are in active current use. In view of the research findings presented they can involve huge errors. There is thus an urgent need to clarify this question and to generate and demonstrate a reliable means of stress and deformation evaluation in this field of considerable technological importance. To date most forming analyses have been based on isotropic hardening theory but it is known that the Bauschinger effect, which is exhibited by many structural metals, can have an important influence on such technologically important phenomena as the generation of residual stresses due to forming. This will increase the demand for reliable analysis to incorporate anisotropic hardening into computer codes and hence to complete the research task considered in this paper.

ACKNOWLEDGEMENT

The results presented in this paper were obtained in the course of research sponsored by the U.S. Army Research Office under Contract No. DAAG 29-82-K-0016 with Rensselaer Polytechnic Institute. The author gratefully acknowledges this support.

REFERENCES

1. Jonas, J. J. *Plasticity of Metals at Finite Strain: Theory, Experiment and Computation* (Eds. E. H. Lee and R. L. Mallett), Div. Appl. Mech., Stanford University and Dept. of Mech. Eng, R. P. I., 1982, 727–734.
2. Lee, E. H. Some comments on elastic–plastic analysis, *Int. J. Solids Struct.*, **17** (1981), 859–872.
3. Lee, E. H. Finite deformation theory with nonlinear kinematics, *ibid.* [1], (1982), 107–120.
4. Lee, E. H., R. L. Mallett and T. B. Wertheimer, Stress analysis for

anisotropic hardening in finite-deformation plasticity, *J. appl. Mech.*, **50** (1983), 554–560.

5. Lee, E. H. and R. M. McMeeking, Concerning elastic and plastic components of deformation, *Int. J. Solids Struct.*, **16** (1980), 715–721.
6. Lubarda, V. A. and E. H. Lee, A correct definition of elastic and plastic deformation and its computational significance, *J. appl. Mech.*, **48** (1981), 35–40.
7. Nagtegaal, J. C. and J. E. de Jong, Some aspects of non-isotropic work-hardening in finite strain plasticity, *ibid.* [1], (1982), 65–102.
8. Prager, W. An elementary discussion of definitions of stress rate, *Q. appl. Math.*, **18** (1961), 403–407.

4

Recent Developments in Finite Deformation Plasticity

P. M. NAGHDI

Department of Mechanical Engineering, University of California, Berkeley, USA

ABSTRACT

An account is given here of some recent results in the rate-independent mechanical theory of finitely deformed elastic–plastic materials, in which the loading surface relative to strain space and the corresponding loading criteria are regarded as primary. The nature of constitutive restrictions is then discussed both in the context of the general theory and for special classes of materials. The predictive capability of the theory in relation to experimentally observed phenomena is briefly indicated.

1. PRELIMINARY REMARKS

Starting with the general constitutive equations of the purely mechanical theory of elastic–plastic materials contained in the work of Green and Naghdi [5, 6], the purpose of this paper is to present a rapid account of some aspects of more recent results in plasticity contained in the papers of Naghdi and Trapp [8–10] and Casey and Naghdi [1–4]. In the discussion that follows, we freely quote some of these works and for details refer the reader to the original papers.

The traditional formulation of plasticity theory employs yield surfaces in stress space, together with loading criteria which involve the time rate of stress. The nonlinear theory of Green and Naghdi [5, 6] was therefore formulated in stress (as well as temperature) space setting. However, as pointed out by Naghdi and Trapp [8] and further emphasized and elaborated upon by Naghdi [11], the stress space formulation of plasticity necessarily leads to unreliable results in any region such as that corresponding to the maximum point of the engineering stress vs engineering strain curve for uniaxial tension of a typical ductile metal. For additional remarks on this point, the reader is referred to the discussion following Fig. 1 of ref. 8. After also observing that the stress-space formulation does not reduce directly to the theory of elastic–perfectly plastic materials, and that a separate formulation for the latter is required, Naghdi and Trapp [8] proposed an alternative strain-space formulation of plasticity which (a) is valid for the full range of elastic–plastic deformation, and (b) includes, as a special case, the theory of elastic–perfectly plastic materials.

In [8], Naghdi and Trapp also noted that a correspondence between the two sets of loading criteria (in strain space and stress space) could be established for all conditions except that of loading from an elastic–plastic state. Subsequently, Casey and Naghdi [1] regarded the loading criteria of strain space as primary and determined the loading conditions in stress space by those of strain space through the constitutive equations of the theory. In this connection, it should be emphasized that the conditions induced in stress space during loading are not in general identical to those of the strain-space formulation, nor do they imply the loading conditions of the strain-space formulation. The different types of loading conditions that can exist in stress space in conjunction with loading criteria in strain space suggest a natural classification of strain-hardening behavior into three distinct types: hardening, softening and perfectly plastic [1, 2]. Geometrically, the yield surface in strain space is always moving outwards during loading, whereas the corresponding yield surface in stress space may be concurrently moving outwards, inwards, or may be stationary, depending on whether the material is hardening, softening, or exhibiting perfectly plastic behavior; in this connection, see Figs 1 and 2 of ref. 1. It is therefore clear that the stress space and strain space formulations are not equivalent as has been emphasized again more recently in ref. 3.

The discussion in the present paper is carried out within the scope of the purely mechanical theory, leaving aside relevant thermodynamical aspects of the subject as developed in refs. 5–7. Keeping the background information in the preceding two paragraphs in mind, the main constitutive results of the rate-independent theory contained in the work of Green and Naghdi [5, 6], Naghdi and Trapp [8] and Casey and Naghdi [1, 2] are summarized in Section 2. This is followed in Section

3 by a discussion of the consequences of the work assumption of Naghdi and Trapp [9, 10] and a brief indication of the predictive capabilities of the theory in relation to a number of experimentally observed phenomena described in ref. 4.

2. CONSTITUTIVE EQUATIONS OF THE MECHANICAL THEORY

Consider a body with material points X and identify the material point X with its position $\mathbf{X}$ in a fixed reference configuration in a Euclidean three-space. Let $\mathbf{x}$ denote the place occupied by X in the present configuration at time t. Then, a motion of the body is defined by a vector function $\boldsymbol{\chi}$ which assigns position $\mathbf{x}$ to each $\mathbf{X}$ at each instant of time, i.e. $\mathbf{x} = \boldsymbol{\chi}(\mathbf{X}, t)$. We assume that the vector function $\boldsymbol{\chi}$ is single-valued and continuously differentiable with respect to its arguments as many times as may be required in the subsequent analysis, except possibly at singular points, curves and surfaces. The deformation gradient $\mathbf{F}$ at $\mathbf{X}$ relative to the reference configuration is given by

$$\mathbf{F} = \mathbf{F}(\mathbf{X}, t) = \frac{\partial \boldsymbol{\chi}(\mathbf{X}, t)}{\partial \mathbf{X}}, \qquad \det \mathbf{F} \neq \mathbf{0} \tag{1}$$

We define a symmetric strain tensor $\mathbf{E}$ by

$$\mathbf{E} = \tfrac{1}{2}(\mathbf{F}^{\mathrm{T}}\mathbf{F} - \mathbf{I}) \tag{2}$$

where $\mathbf{F}^{\mathrm{T}}$ stands for the transpose of $\mathbf{F}$ and $\mathbf{I}$ is the unit tensor. Also, the velocity $\mathbf{v}$, the velocity gradient $\mathbf{L}$ and the rate of strain are defined by

$$\mathbf{v} = \dot{\mathbf{x}}, \qquad \mathbf{L} = \dot{\mathbf{F}}\mathbf{F}^{-1} \tag{3}$$

and

$$\dot{\mathbf{E}} = \mathbf{F}^{\mathrm{T}}\mathbf{D}\mathbf{F}, \qquad \mathbf{D} = \tfrac{1}{2}(\mathbf{L} + \mathbf{L}^{\mathrm{T}}) \tag{4}$$

where a superposed dot denotes the material time derivative with respect to the present time t holding $\mathbf{X}$ fixed, and $\mathbf{F}^{-1}$ is the inverse of $\mathbf{F}$.

We now summarize the main constitutive results governing the purely mechanical, rate-independent, theory of elastic–plastic continua contained in the work of Green and Naghdi [5, 6] and Naghdi and Trapp [8], along with the main ingredients of strain-hardening characterization given by Casey and Naghdi [1, 2]. Thus, in addition to the strain tensor $\mathbf{E}$ defined by (2), at each point of the continuum we assume the existence of a plastic strain specified by a second-order tensor $\mathbf{E}^{\mathrm{p}}$ and a measure of work hardening specified by the scalar function κ, and introduce the stress constitutive assumption

$$\mathbf{S} = \hat{\mathbf{S}}(\mathcal{U}), \qquad \mathcal{U} = (\mathbf{E}, \mathbf{E}^{\mathrm{p}}, \kappa) \tag{5}$$

We assume that, for fixed values of $\mathbf{E}^{\mathrm{p}}$ and κ, the expression $(5)_1$ is invertible in the form

$$\mathbf{E} = \hat{\mathbf{E}}(\mathcal{V}), \qquad \mathcal{V} = (\mathbf{S}, \mathbf{E}^{\mathrm{p}}, \kappa) \tag{6}$$

We use a strain-space formulation of plasticity and, as in the paper of Casey and Naghdi [1], also regard the loading criteria of the strain-space formulation as primary. Thus, we admit the existence of a continuously differentiable scalar-valued yield (or loading) function $g(\mathcal{U})$ such that

$$g(\mathcal{U}) = 0 \tag{7}$$

represents a closed orientable hypersurface $\partial\mathcal{E}$ of dimension five enclosing a region $\mathcal{E}$ of strain space. The function g is chosen so that $g(\mathcal{U}) < 0$ for all points in the interior of the region $\mathcal{E}$.

Corresponding to a motion $\boldsymbol{\chi}$, with each particle of the body we associate a smooth oriented curve C_{e} in strain space. The curve C_{e} is called a strain trajectory. Along the strain trajectory C_{e}, the time rates of $\mathbf{E}^{\mathrm{p}}$ and κ are assumed to have the forms [8]:

$$\dot{\mathbf{E}}^{\mathrm{p}} = \begin{cases} 0 \quad \text{if} \quad g < 0, & \text{(a)} \\ 0 \quad \text{if} \quad g = 0, \quad \hat{g} < 0 & \text{(b)} \\ 0 \quad \text{if} \quad g = 0, \quad \hat{g} = 0 & \text{(c)} \\ \mathcal{M}[\dot{\mathbf{E}}] = \lambda \hat{g} \boldsymbol{\rho}, \quad g = 0, \quad \hat{g} > 0 & \text{(d)} \end{cases} \tag{8}$$

and

$$\dot{\kappa} = \mathcal{C} \cdot \dot{\mathbf{E}}^{\mathrm{p}} \tag{9}$$

where $\mathcal{C}$ and $\mathcal{M}$ are, respectively, a symmetric second order and a fourth order tensor function of the variables $(5)_2$,

$$\hat{g} = \frac{\partial g}{\partial \mathbf{E}} \cdot \dot{\mathbf{E}} \tag{10}$$

$$\mathcal{M} = \lambda \boldsymbol{\rho} \otimes \frac{\partial g}{\partial \mathbf{E}} \tag{11}$$

and where λ and $\boldsymbol{\rho}$ are, respectively, a scalar valued and a symmetric second order tensor-valued function of the variables $\mathcal{U}$ and the symbol $\otimes$ denotes tensor product. The conditions involving g and $\hat{g}$ in (8) are the loading criteria of the strain-space formulation. Using conventional terminology, these four conditions in the order listed correspond to (a) an elastic state (or point in strain space for which $g<0$); (b) unloading from an elastic–plastic state, i.e. a point in strain space for which $g=0$; (c) neutral loading from an elastic–plastic state; and (d) loading from an elastic–plastic state. We assume that the coefficient of $\hat{g}$ in (8d) is nonzero on the yield surface; and, without loss in generality, we set $\boldsymbol{\rho}\neq 0$, $\lambda>0$. The stipulation that during loading the strain trajectory remains on the yield surface $\partial\mathscr{E}$ so that $\dot{g}=0$ (with $g=0$, $g>0$) yields the so-called 'consistency' condition given by

$$1+\lambda\boldsymbol{\rho}\cdot\left(\frac{\partial g}{\partial \mathbf{E}^{\mathrm{p}}}+\frac{\partial g}{\partial \kappa}\mathscr{C}\right)=0 \tag{12}$$

where (8d), (11) and (9) have been used.

For a given motion $\boldsymbol{\chi}$ and associated strain trajectory C_e, by $(5)_1$, (8), (9) and appropriate initial conditions for $\mathbf{E}^{\mathrm{p}}$, κ we obtain the corresponding stress trajectory C_s. Moreover, for a given loading function $g(\mathcal{U})$, with the help of $(6)_1$, we can obtain a corresponding function $f(\mathcal{V})$ by means of the formula

$$g(\mathcal{U})=g(\mathbf{E}(\mathcal{V}),\mathbf{E}^{\mathrm{p}},\kappa)=f(\mathcal{V}) \tag{13}$$

Because of the smoothness of $(6)_1$, for fixed values of $\mathbf{E}^{\mathrm{p}}$ and κ, the equation

$$f(\mathcal{V})=0 \tag{14}$$

represents a hypersurface $\partial\mathscr{S}$ of dimension five in stress space having the same geometrical properties as the hypersurface $\partial\mathscr{E}$ in strain space. The open region enclosed by $\partial\mathscr{S}$ is denoted by $\mathscr{S}$. It follows from (13) that a point in strain space belongs to the region $\mathscr{E}$ (i.e. $g(\mathcal{U})<0$) if and only if the corresponding point in stress space satisfies $f(\mathcal{V})<0$ and hence belongs to $\mathscr{S}$. Likewise, a point in strain space lies on the yield surface $\partial\mathscr{E}$ if and only if the corresponding point in stress space lies on $\partial\mathscr{S}$. In view of these properties, we refer to f as the yield (or loading) function in stress space, to $\partial\mathscr{S}$ as the yield (or loading) surface in stress space, and to $\mathscr{S}$ as the elastic region in stress space.

From the material derivative of (13), we have

$$\dot{g} = \dot{f} = \hat{f} + \frac{\partial f}{\partial \mathbf{E}^{\mathrm{p}}} \cdot \dot{\mathbf{E}}^{\mathrm{p}} + \frac{\partial f}{\partial \kappa} \dot{\kappa} \tag{15}$$

where

$$\hat{f} = \frac{\partial f}{\partial \mathbf{S}} \dot{\mathbf{S}} \tag{16}$$

is the inner product of $\partial f/\partial \mathbf{S}$ and the tangent vector to the stress trajectory C_s. Now certain conditions are implied in stress space by the loading criteria of strain space for the three cases (8a, b, c). In fact, with the use of (8), (9), (13) and (15), the criteria for an elastic state, unloading and neutral loading in (8a, b, c) imply, respectively, the conditions

$$\text{(a) } f<0; \qquad \text{(b) } f=0, \hat{f}<0; \qquad \text{(c) } f=0, \hat{f}=0 \tag{17}$$

in stress space. The condition for loading in stress space requires an additional consideration and is intimately related to strain-hardening characterization as discussed in refs. 1 and 2 and further elaborated upon in ref. 4. We do not present here the details of the strain-hardening characterization, but display below a table which indicates the relationships between the loading criteria in strain space and associated conditions in stress space.

As is evident from Table I, during hardening behavior, the loading conditions of the stress-space and strain-space formulations imply one another. However, for softening and perfectly plastic behavior, no such equivalence exists. This non-equivalence has been emphasized in ref. 3.

TABLE I
Relationships between loading criteria in strain space and associated conditions in stress space (with $g = f = 0$)

Strain-space loading criterion	*Hardening*	*Softening*	*Perfectly plastic*
Unloading	$\hat{g}<0 \Leftrightarrow \hat{f}<0$	$\hat{g}<0 \Rightarrow \hat{f}<0$	$\hat{g}<0 \Leftrightarrow \hat{f}<0$
Neutral loading	$\hat{g}=0 \Leftrightarrow \hat{f}=0$	$\hat{g}=0 \Leftrightarrow \hat{f}=0$	$\hat{g}=0 \Rightarrow \hat{f}=0$
Loading	$\hat{g}>0 \Leftrightarrow \hat{f}>0$	$\hat{g}>0 \Rightarrow \hat{f}<0$	$\hat{g}>0 \Rightarrow \hat{f}=0$

3. CONSEQUENCES OF A WORK ASSUMPTION. ADDITIONAL REMARKS

A physically plausible work assumption was introduced by Naghdi and Trapp [9] and was used by them to impose restrictions on constitutive equations of finitely deformed elastic–plastic materials [9, 10]. The relationship of this work assumption to the postulates of Drucker and Il'iushin is discussed in refs. 9 and 10 and need not be repeated here. For our present purpose, the work assumption of ref. 9 may be stated as: The external work done on an elastic–plastic body in any smooth homogeneous cycle of deformation is nonnegative. From this assumption, it follows that a symmetric second order tensor $\boldsymbol{\sigma}$ defined by

$$\boldsymbol{\sigma}=\left\{\frac{\partial\hat{\mathbf{S}}}{\partial\mathbf{E}^{\mathrm{p}}}+\frac{\partial\hat{\mathbf{S}}}{\partial\kappa}\otimes\mathscr{C}\right\}[\boldsymbol{\rho}] \tag{18}$$

is parallel to the normal to the yield surface $\partial\mathscr{E}$ in strain space and points into the elastic region $\mathscr{E}$:

$$\boldsymbol{\sigma}=-\gamma^{*}\frac{\partial g}{\partial\mathbf{E}},\qquad \gamma^{*}\geqslant 0,\qquad (g=0) \tag{19}$$

where the scalar function γ^{*} depends only on the variables $\mathscr{U}$ (or $\mathscr{V}$) and the various response functions in (18) were introduced in (5), (8) and (9).

For certain purposes, it is convenient to express the constitutive equation $(5)_1$ in terms of an equivalent set of kinematic variables in the form

$$\mathbf{S}=\bar{\mathbf{S}}(\bar{\mathscr{U}}),\qquad \bar{\mathscr{U}}=(\mathbf{E}-\mathbf{E}^{\mathrm{p}},\mathbf{E}^{\mathrm{p}},\kappa) \tag{20}$$

It then follows from (19) that [9, 4]:

$$-\boldsymbol{\rho}+\frac{\partial\hat{\mathbf{E}}}{\partial\mathbf{S}}\left[\left(\frac{\partial\bar{\mathbf{S}}}{\partial\mathbf{E}^{\mathrm{p}}}+\frac{\partial\bar{\mathbf{S}}}{\partial\kappa}\otimes\mathscr{C}\right)[\boldsymbol{\rho}]\right]=-\gamma^{*}\frac{\partial f}{\partial\mathbf{S}} \tag{21}$$

with $g=f=0$. In the special case that the response function $\bar{\mathbf{S}}$ is independent of its second and third arguments, i.e. if

$$\frac{\partial\bar{\mathbf{S}}}{\partial\mathbf{E}^{\mathrm{p}}}=\mathbf{0},\qquad \frac{\partial\bar{\mathbf{S}}}{\partial\kappa}=0 \tag{22}$$

then (21) yields

$$\boldsymbol{\rho}=\gamma^{*}\frac{\partial f}{\partial\mathbf{S}}\neq 0,\qquad \gamma^{*}>0\qquad (g=f=0) \tag{23}$$

and hence by (8d) during loading we have

$$\dot{\mathbf{E}}^{\mathrm{p}} = \lambda\gamma^{*}\hat{g}\frac{\partial f}{\partial \mathbf{S}} \neq 0 \tag{24}$$

Consider next the case in which the conditions (22) are satisfied and the response function $\bar{\mathbf{S}}$ is linear in the variable $\mathbf{E}-\mathbf{E}^{\mathrm{p}}$ with constant coefficients. In this case, in addition to the condition (19) or (23), it is also possible to derive a second inequality from which convexity of the yield surface follows in both strain space and stress space [10].

The condition (19) has a further implication in regard to the form of a function which defines strain-hardening behavior. We do not elaborate on this here and refer the reader to refs. 1 and 4 for details.

Before closing we call attention to the predictive capabilities of the theory presented here in relation to a number of experimentally observed phenomena. These are demonstrated or discussed in some detail in ref. 4, Section 5, where references to the original papers are also cited. The predictions are effected with the use of special constitutive equations and pertain to such strain-hardening behavior as isotropic and kinematic hardening, strength-differential effect, ratcheting of strain by stress cycling between fixed values of stress, and saturation hardening by strain cycling between fixed values of strain.

ACKNOWLEDGEMENT

The results reported here were obtained in the course of research supported by the U.S. Office of Naval Research under Contract N00014-75-C-0148, Project NR 064-436 with the University of California, Berkeley.

REFERENCES

1. Casey, J. and P. M. Naghdi. On the characterization of strain-hardening in plasticity, *J. appl. Mech.*, **48** (1981), 285–296.
2. Casey, J. and P. M. Naghdi. A remark on the definition of hardening, softening and perfectly plastic behavior, *Acta Mech.*, **48** (1983), 91–94.
3. Casey, J. and P. M. Naghdi. On the nonequivalence of the stress space and strain space formulations of plasticity theory, *J. appl. Mech.*, **50** (1983), 350–354.
4. Casey, J. and P. M. Naghdi. Strain-hardening response of elastic–plastic

materials. In: *Mechanics of Engineering Materials* (Eds. C. S. Desai and R. H. Gallagher), John Wiley, Chichester, U.K. 1984 (in press).

5. Green, A. E. and P. M. Naghdi. A general theory of an elastic–plastic continuum, *Arch. Rat. Mech. Anal.*, **18** (1965), 251–281.
6. Green, A. E. and P. M. Naghdi. A thermodynamic development of elastic–plastic continua, *Proc. IUTAM Symp. on Irreversible Aspects of Continuum Mechanics and Transfer of Physical Characteristics in Moving Fluids* (Eds. H. Parkus and L. I. Sedov), Springer-Verlag, Berlin, 1966, 117–131.
7. Green, A. E. and P. M. Naghdi. On thermodynamical restrictions in the theory of elastic–plastic materials, *Acta Mech.*, **30** (1978), 157–162.
8. Naghdi, P. M. and J. A. Trapp. The significance of formulating plasticity with reference to loading surfaces in strain space, *Int. J. Engng Sci.*, **13** (1975), 785–797.
9. Naghdi, P. M. and J. A. Trapp. Restrictions on constitutive equations of finitely deformed elastic–plastic materials, *Q. J. Mech. appl. Math.*, **28** (1975), 25–46.
10. Naghdi, P. M. and J. A. Trapp. On the nature of normality of plastic strain rate and convexity of yield surfaces in plasticity, *J. appl. Mech.*, **42** (1975), 61–66.
11. Naghdi, P. M. Some constitutive restrictions in plasticity, *Proc. Symp. on Constitutive Equations in Viscoplasticity: Computational and Engineering Aspects*, AMD, Vol. 20, American Society of Mechanical Engineers, 1976, 79–93.

5

Micromechanically Based Finite Plasticity

S. NEMAT-NASSER

Department of Civil Engineering, The Technological Institute, Northwestern University, Evanston, Illinois, USA

ABSTRACT

For arbitrarily large strains and rotations, rate-independent and rate-dependent constitutive relations of single crystals are developed and used to calculate the overall response of a polycrystal.

1. INTRODUCTION

In this paper the overall behavior of polycrystalline solids at finite strains and rotations is characterized in terms of the single-crystal response and Hill's [5] fundamental averaging technique. The basic averaging theorems are briefly examined in Section 2. In Section 3, general rate-independent and rate-dependent elasto–plastic constitutive relations of single crystals are developed. Finite rotations and strains are considered, and it is assumed that the inelasticity of the crystal stems from rate-independent or rate-dependent slip on crystallographic planes of the crystal. In Section 4 the overall instantaneous moduli of the polycrystal are calculated for arbitrarily large strains and rotations, in terms of those of the single crystal constituents.

2. PRELIMINARIES

For simplicity a fixed rectangular Cartesian co-ordinate system is used. The co-ordinate axes are denoted by x_i or X_A, i, $A = 1, 2, 3$, or collectively by $\mathbf{x}$ or $\mathbf{X}$.

The considered crystalline solid may have undergone a finite, generally inelastic, deformation, and in its current deformed state, residual stresses of different magnitude may exist within its various micro-elements which, in this work, will be assumed to be single crystals. Consider a rate of deformation from the current state, and denote the overall macroscopic velocity gradient by $\mathbf{L}$ with components L_{ij}. This overall velocity gradient, in general, consists of two accompanying contributions, $\mathbf{L}^{\text{in}}$ and $\mathbf{L}^{\text{e}}$, which will be referred to as the 'inelastic' and 'elastic' parts, respectively. The inelastic contribution, $\mathbf{L}^{\text{in}}$, in crystalline solids is often regarded to stem from slip over crystallographic planes, which may be rate-dependent or possibly rate-independent, and from grain-boundary sliding, void formation and growth, and other microstructural events. The elastic contribution, $\mathbf{L}^{\text{e}}$, on the other hand, arises from the crystal lattice distortion, and serves to accommodate the inelastic deformation rate in such a manner as to produce an overall compatible velocity gradient. Within each micro-element or crystal, the *local* inelastic, $\mathbf{l}^{\text{in}}$, the elastic, $\mathbf{l}^{\text{e}}$, and the total, $\mathbf{l}$, velocity gradients are regarded as *uniform*, but these, in general, vary from crystal to crystal. When all microstructural changes are 'locked' (a thought experiment), the response at the local level, as well as globally, would be 'elastic'. The inelastic response, $\mathbf{l}^{\text{in}}$, as well as the accommodating elastic response, $\mathbf{l}^{\text{e}}$, of a typical crystal is affected by the overall constraint that the remaining crystals impose on this element. These quantities would be different if the single crystal were to undergo the inelastic and elastic velocity gradients in isolation and without the constraint of the surrounding crystals. Thus, in general, the actual local quantities $\mathbf{l}^{\text{e}}$ and $\mathbf{l}^{\text{in}}$ depend on the corresponding elastic and inelastic deformation rates and spins of all other micro-elements, especially on those of their immediate neighboring crystals.

In view of the above comments, one basic requirement for obtaining the overall constitutive relations of polycrystals in terms of the single crystal response, is to express the local velocity gradient, $\mathbf{l}$, in terms of the overall velocity gradient, $\mathbf{L}$. For infinitesimal deformations, this may be accomplished with the aid of a 'self-consistent scheme' and Eshelby's [2] solution of the stress and strain fields in an ellipsoidal region which undergoes a uniform transformation strain while embedded in an infinitely extended linearly elastic homogeneous solid. For finite deformations and rotations, on the other hand, a suitable averaging scheme in terms of *appropriate* dynamical and kinematical variables must first be carefully developed, and then these appropriate local quantities must be expressed in terms of the corresponding overall data.

2.1. Averaging Theorems (Hill [5])

We choose the current configuration of the deformed and stress solid as the reference ground state. Then the Cauchy, the Kirchhoff, and the nominal stresses coincide but, of course, not their rates. We use lower case letters for the local (micro) variables, and the corresponding upper case letters for the associated overall (macro) quantities. Let **n** with components n_{ij} denote the local nominal stress rate, the overall nominal stress rate being $\dot{\mathbf{N}}$ with components $\dot{N}_{ij}$. In the absence of body forces, the continued equilibrium requires

$$\begin{aligned} \dot{n}_{ij,i} &= 0 \quad \text{in } V \\ \dot{n}_{ij}\nu_i &= \dot{t}_j \quad \text{on } S \end{aligned} \tag{2.1}$$

where ν_i is the outward unit vector on the surface S of the solid with volume V, and $\dot{t}_j$ stands for the prescribed traction rate; in $(2.1)_1$, comma followed by an index denotes partial differentiation with respect to the corresponding co-ordinate, and superimposed dot denotes material time differentiation.

From the identity

$$\frac{1}{V}\int_S x_i \dot{t}_j \, \mathrm{d}S = \frac{1}{V}\int_V \dot{n}_{ij} \, \mathrm{d}V \tag{2.2}$$

which is deduced from the divergence theorem and (2.1), it follows that the volume average

$$\langle \dot{n}_{ij} \rangle \equiv \frac{1}{V}\int_V \dot{n}_{ij} \, \mathrm{d}V \tag{2.3}$$

is uniquely defined by the surface data alone. In a similar manner, it follows from

$$\langle l_{ij} \rangle \equiv \frac{1}{V}\int_V l_{ij} \, \mathrm{d}V = \frac{1}{V}\int_S v_i \nu_j \, \mathrm{d}S \tag{2.4}$$

where v_i is the velocity field and $l_{ij} \equiv v_{i,j}$, that the volume average of the velocity gradient is uniquely defined by the surface velocity data. Thus the nominal stress rate, $\dot{\mathbf{N}}$, and the velocity gradient, $\mathbf{L}$, of a macro-element (a continuum material particle) may reasonably be defined by

$$\dot{N}_{ij} \equiv \langle \dot{n}_{ij} \rangle \quad \text{and} \quad L_{ij} \equiv \langle l_{ij} \rangle \tag{2.5}$$

which are the local dynamical and kinematical quantities of the corresponding continuum with microstructure. For this continuum, the local (in the macroscopic sense) velocity field is

$$V_i = L_{ij} x_j \tag{2.6}$$

with L_{ij} defined by $(2.5)_2$.

2.2. Other Measures

Let $\mathbf{T}$ and $\boldsymbol{\Sigma}$ denote the overall (macro) Kirchhoff and Cauchy stresses, respectively, and denote the corresponding local (micro) quantities by $\boldsymbol{\tau}$ and $\boldsymbol{\sigma}$. With the current configuration as the reference ground state, we have $\mathbf{N} = \mathbf{T} = \boldsymbol{\Sigma}$ and $\mathbf{n} = \boldsymbol{\tau} = \boldsymbol{\sigma}$, but, of course, $\dot{\mathbf{N}} \neq \dot{\mathbf{T}} \neq \dot{\boldsymbol{\Sigma}}$ and $\dot{\mathbf{n}} \neq \dot{\boldsymbol{\tau}} \neq \dot{\boldsymbol{\sigma}}$. Indeed,

$$\begin{aligned} \dot{\tau}_{ij} &= \dot{\sigma}_{ij} + d_{kk}\sigma_{ij} = \dot{n}_{ij} + l_{ik}\sigma_{kj} \\ \dot{T}_{ij} &\equiv \dot{\Sigma}_{ij} + D_{kk}\Sigma_{ij} \equiv \dot{N}_{ij} + L_{ik}\Sigma_{kj} \end{aligned} \tag{2.7}$$

where

$$d_{ij} \equiv \tfrac{1}{2}(l_{ij} + l_{ji}) \quad \text{and} \quad D_{ij} \equiv \tfrac{1}{2}(L_{ij} + L_{ji}) \tag{2.8}$$

are the micro- and the macrodeformation rates, respectively; the corresponding spins are

$$w_{ij} \equiv \tfrac{1}{2}(l_{ij} - l_{ji}) \quad \text{and} \quad W_{ij} \equiv \tfrac{1}{2}(L_{ij} - L_{ji}) \tag{2.9}$$

From (2.7), it is clear that the volume average of the Kirchhoff and Cauchy stress rates is not, in general, equal to the overall Kirchhoff and Cauchy stress rates. These overall rate quantities must be *defined* in terms of the overall nominal stress rate. For example, we set

$$\dot{T}_{ij} \equiv \dot{\Sigma}_{ij} + D_{kk}\Sigma_{ij} \equiv \tfrac{1}{2}(\dot{N}_{ij} + \dot{N}_{ji} + L_{ik}\Sigma_{kj} + L_{jk}\Sigma_{ki}) \tag{2.10}$$

This then *defines* $\dot{T}_{ij}$ and $\dot{\Sigma}_{ij}$ once the nominal stress, its rate, and the velocity gradient are known. Note that, in an actual calculation, the current stress state is known from possibly a sequence of incremental calculations. Note also that, from (2.1) and (2.2) written for the nominal stress with the current state as the reference one, it follows that

$$\mathbf{N} = \boldsymbol{\Sigma} = \mathbf{T} = \langle \mathbf{n} \rangle = \langle \boldsymbol{\sigma} \rangle = \langle \boldsymbol{\tau} \rangle \tag{2.11}$$

3. RATE-DEPENDENT PLASTICITY OF SINGLE CRYSTALS

We consider finite inelastic deformation of a typical single crystal, and assume that the rate-dependent (or rate-independent) inelastic flow stems from slip on crystallographic planes. Figure 1(a) and (b) are micrographs of the slip pattern in a Waspaloy polycrystal sample tested in uniaxial tension, which show that the actual microdeformation may be more complicated than the simple slip-model just mentioned, but the model does capture the main physics of the process, and has been successfully used in other contexts since the pioneering work of Taylor [11]; see Nemat-Nasser [8] for references and an overview.

For the single crystal, we set

$$l_{ij} = d_{ij} + w_{ij} \tag{3.1}$$

$$d_{ij} = d^{\mathrm{e}}_{ij} + d^{\mathrm{in}}_{ij}, \qquad w_{ij} = w^{\mathrm{e}}_{ij} + w^{\mathrm{in}}_{ij} \tag{3.2}$$

where we view d^{in}_{ij} and w^{in}_{ij} as the inelastic deformation rate and spin that are produced by slip on suitably oriented slip systems, and regard d^{e}_{ij} and w^{e}_{ij} as the accommodating deformation rate and spin which are produced by elastic lattice distortion. The inelastic quantities are defined by

$$d^{\mathrm{in}}_{ij} = r^{\alpha}_{ij}\dot{\gamma}^{\alpha}, \qquad w^{\mathrm{in}}_{ij} = \omega^{\alpha}_{ij}\dot{\gamma}^{\alpha} \tag{3.3}$$

where α is summed over all active slip systems, and the notation

$$r^{\alpha}_{ij} \equiv \tfrac{1}{2}(s^{\alpha}_i n^{\alpha}_j + s^{\alpha}_j n^{\alpha}_i), \qquad \omega^{\alpha}_{ij} \equiv \tfrac{1}{2}(s^{\alpha}_i n^{\alpha}_j - s^{\alpha}_j n^{\alpha}_i) \qquad (\alpha \text{ not summed}) \tag{3.4}$$

is used; here $\mathbf{n}^{\alpha}$ is the unit vector normal to the αth slip plane, and $\mathbf{s}^{\alpha}$ is the unit vector in the corresponding slip direction.

To obtain a rather general elastic–viscoplastic model, we assume that the shear rate, $\dot{\gamma}^{\alpha}$, of the αth slip system is governed by the following rate relation:

$$\dot{\tau}^{\alpha} + k^{\alpha\beta}\tau^{\beta} + \tan\eta_1\dot{\sigma}^{\alpha} + \tan\eta_2\dot{p} = h^{\alpha\beta}\dot{\gamma}^{\beta}, \qquad \alpha, \beta = 1, 2, \ldots, n \tag{3.5}$$

where β is summed on all active slip systems; $h^{\alpha\beta}$ is the hardening matrix which, in general, depends on the total slips, γ^{α}; τ^{α} is the resolved shear stress; σ^{α} is the normal traction on the αth system; p is the mean stress; $\tan\eta_1$ and $\tan\eta_2$ represent pressure sensitivity

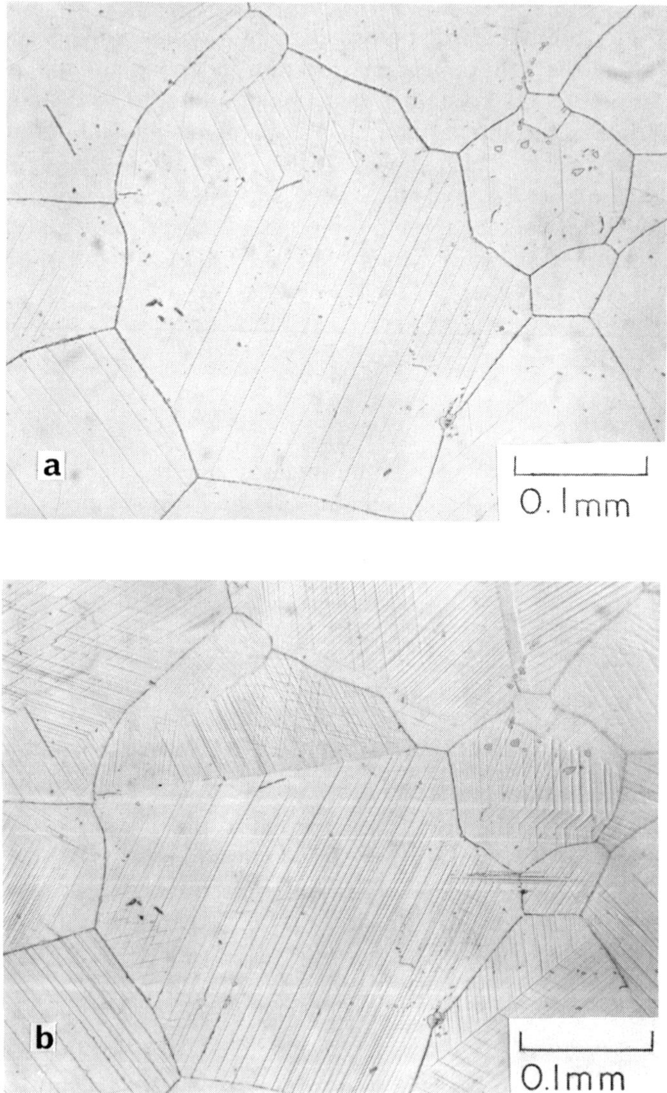

FIG. 1. Micrographs of slip systems in a Waspaloy tensile specimen at (a) 1% extension and (b) 3·5% extension.

parameters (Nemat-Nasser, Mehrabadi and Iwakuma [10]; Nemat-Nasser [8]); and $k^{\alpha\beta}$ is the rate-dependency matrix which, in general, depends on the stress-state, as well as possibly on the total slips. When no coupling between slip systems is assumed, matrices $k^{\alpha\beta}$ and $h^{\alpha\beta}$ are diagonal. In general, however, slips are highly coupled (Kocks [7]).

We have

$$\tau^{\alpha} = r^{\alpha}_{ij}\sigma_{ij}, \qquad \sigma^{\alpha} = \eta^{\alpha}_{ij}\sigma_{ij} \tag{3.6}$$

where

$$\eta^{\alpha}_{ij} \equiv n^{\alpha}_{i} n^{\alpha}_{j} \qquad (\alpha \text{ not summed}) \tag*{(3.4)$_3$}$$

We shall assume that inelastic flow does not affect the lattice structure,† and set

$$\dot{n}^{\alpha}_{i} = w^{e}_{ij} n^{\alpha}_{j}, \qquad \dot{s}^{\alpha}_{i} = w^{e}_{ij} s^{\alpha}_{j} \tag{3.7}$$

The relation (3.5) now becomes

$$q^{\alpha}_{ij}\overset{\circ}{\sigma}{}^{*}_{ij} + k^{\alpha\beta} r^{\beta}_{ij}\sigma_{ij} = h^{\alpha\beta}\dot{\gamma}^{\beta} \tag{3.8}$$

where β is summed over all active slip systems. In (3.8), the following notation is used:

$$q^{\alpha}_{ij} \equiv r^{\alpha}_{ij} + \eta^{\alpha}_{ij} \tan \eta_1 + \tfrac{1}{3}\delta_{ij} \tan \eta_2 \tag{3.9}$$

where δ_{ij} is the Kronecker delta and other terms are defined in (3.4) and (3.5); moreover,

$$\overset{\circ}{\sigma}{}^{*}_{ij} \equiv \dot{\sigma}_{ij} - w^{e}_{ik}\sigma_{kj} - w^{e}_{jk}\sigma_{ki} \tag{3.10}$$

where the superscript asterisk emphasizes that the Jaumann rate (denoted by superimposed $\circ$) is taken co-rotational with only the elastic spin.

For crystalline solids the elastic lattice distortion is often very small compared with the inelastic strains and rotations, so that one may assume the simple relation

$$\overset{\circ}{\tau}{}^{*}_{ij} = \mathscr{L}^{\Omega}_{ijkl} d^{e}_{kl} \tag{3.11}$$

which connects the Jaumann rate of the Kirchhoff stress—measured co-rotational with lattice—and the lattice elastic deformation rate, through the crystal's elasticity tensor $\mathscr{L}^{\Omega}$, where the following notation is used:

$$\overset{\circ}{\tau}{}^{*}_{ij} \equiv \overset{\circ}{\sigma}{}^{*}_{ij} + d^{e}_{kk}\sigma_{ij} = \dot{\tau}_{ij} - w^{e}_{ik}\tau_{kj} - w^{e}_{jk}\tau_{ki} \tag{3.12}$$

† This is a reasonable assumption, but it can be relaxed if desired.

From this and (3.6)–(3.11), it follows that

$$\overset{\circ}{\tau}_{ij} \equiv \dot{\tau}_{ij} - w_{ik}\tau_{kj} - w_{jk}\tau_{ki} = \mathscr{L}^{\Omega}_{ijkl}d_{kl} - \{\mathscr{L}^{\Omega}_{ijkl}r^{\alpha}_{kl} + \omega^{\alpha}_{ik}\sigma_{kj} + \omega^{\alpha}_{jk}\sigma_{ki}\}\dot{\gamma}^{\alpha} \tag{3.13}$$

It is simpler to use direct notation and write (3.13) as

$$\overset{\circ}{\boldsymbol{\tau}} = \mathscr{L}^{\Omega} : \mathbf{d} - \{\mathscr{L}^{\Omega} : \mathbf{r}^{\alpha} + \boldsymbol{\omega}^{\alpha}\boldsymbol{\sigma} - \boldsymbol{\sigma}\boldsymbol{\omega}^{\alpha}\}\dot{\gamma}^{\alpha} \tag{3.14}$$

where the meaning of each expression is obtained by comparison with (3.13). In the sequel both the index and the direction notation will be used.

It should be pointed out that inelastic volume changes are excluded because of the assumptions (3.3) and (3.4). Inelastic dilatancy (or densification) can easily be included if desired, in the manner first discussed by Nemat-Nasser, Mehrabadi and Iwakuma [10]; see also Nemat-Nasser [8]. This is done by replacing r^{α}_{ij} in (3.3) by

$$p^{\alpha}_{ij} = r^{\alpha}_{ij} + \tan \nu_1 \eta^{\alpha}_{ij} + \tan \nu_2 \delta_{ij} \tag{3.15}$$

Then (3.14) becomes

$$\overset{\circ}{\boldsymbol{\tau}} = \mathscr{L}^{\Omega} : \mathbf{d} - \{\mathscr{L}^{\Omega} : \mathbf{p}^{\alpha} + \boldsymbol{\omega}^{\alpha}\boldsymbol{\sigma} - \boldsymbol{\sigma}\boldsymbol{\omega} - p^{\alpha}_{kk}\boldsymbol{\sigma}\}\dot{\gamma}^{\alpha} \tag{3.16}$$

For the sake of completeness we shall use this version of the theory in the sequel.

The slip rate $\dot{\gamma}^{\alpha}$ can be obtained from (3.8),

$$\dot{\gamma}^{\alpha} = M^{\alpha\beta}\{\mathbf{q}^{\beta} : (\mathscr{L}^{\Omega} - \boldsymbol{\sigma}\otimes\boldsymbol{\delta}) : \mathbf{d} + k^{\beta\theta}\mathbf{r}^{\theta} : \boldsymbol{\sigma}\} \tag{3.17}$$

where α, β, $\theta = 1, 2, \ldots, n$, when there are n active slip systems, and repeated superscripts are summed. In (3.17), $\boldsymbol{\delta}$ is the Kronecker delta (with components δ_{ij}), and the components of $\boldsymbol{\sigma}\otimes\boldsymbol{\delta}$ are $\sigma_{ij}\delta_{kl}$; the colon stands for double contraction. The matrix $M^{\alpha\beta}$ is given by

$$M^{\alpha\beta} = \{h^{\alpha\beta} + \mathbf{q}^{\alpha} : (\mathscr{L}^{\Omega} - \boldsymbol{\sigma}\otimes\boldsymbol{\delta}) : \mathbf{p}^{\beta}\}^{-1} \tag{3.18}$$

In certain cases the inverse may not exist. However, one may always choose a set of active slip systems and solve (3.17). This has been discussed by Hill [4], Havner [3], and others.

Equations (3.16) and (3.17) complete the constitutive relations of a single crystal which deforms inelastically by rate-dependent (viscoplastic) slip, and elastically by lattice distortion, both at finite strains and rotations. To obtain the rate-independent theory, we set $k^{\beta\theta}$ equal to zero (Iwakuma and Nemat-Nasser [6]). On the other hand, when steady-state creep is involved, we may greatly simplify the above results, and write, in place of (3.3) and (3.5),

$$\mathbf{d}^{\text{in}} = \mathbf{p}^{\alpha}\dot{\gamma}^{\alpha} \tag{3.19}$$

$$\dot{\gamma}^{\alpha} = k^{\alpha\beta}\tau^{\beta} = k^{\alpha\beta}\mathbf{r}^{\beta} : \boldsymbol{\sigma} \tag{3.20}$$

This then leads to

$$\overset{\circ}{\boldsymbol{\tau}} = \mathscr{L}^{\Omega} : \mathbf{d} - \{\mathscr{L}^{\Omega} : \mathbf{p}^{\alpha} + \boldsymbol{\omega}^{\alpha}\boldsymbol{\sigma} - \boldsymbol{\sigma}\boldsymbol{\omega}^{\alpha} - p^{\alpha}_{kk}\boldsymbol{\sigma}\}k^{\alpha\beta}\mathbf{r}^{\beta} : \boldsymbol{\sigma} \tag{3.21}$$

As mentioned before, matrix $k^{\alpha\beta}$, in general, depends on the current stresses. For creep arising from transgranular mass diffusion, a power-law seems to apply over a wide range of stresses and deformation rates (Ashby and Frost [1]). Suppose $k^{\alpha\beta}$ is diagonal. Then one may write

$$k^{\alpha\alpha} \equiv k^{\alpha} = k^{\alpha}_{0}\,|\tau^{\alpha}/\tau_{0}|^{m} \tag{3.22}$$

where k^{α}_{0}, in general, is temperature dependent,

$$k^{\alpha}_{0} = k^{\alpha}_{1}\exp\{-Q/kT\} \tag{3.23}$$

here, Q is the activation energy, k is Boltzmann's constant, and T is the absolute temperature. Instead of the resolved shear stress, τ^{α}, one may use in (3.22) the second invariant of the stress deviator,

$$\bar{\tau} = (\tfrac{1}{2}\tau'_{ij}\tau'_{ij})^{1/2} \tag{3.24}$$

where prime denotes the deviatoric part.

4. FUNDAMENTAL RESULTS FOR POLYCRYSTALS

One basic requirement for obtaining the macroscopic material response from micro-modeling, is to characterize the deformation rate and the stress rate at the micro-level in terms of the far-field data. For small deformations (infinitesimal theory), this is accomplished by using Eshelby's [2] solution of the stress and strain fields in an ellipsoidal region which undergoes a uniform transformation strain while embedded in an infinitely extended linearly elastic homogeneous solid. For finite deformations and rotations, Nemat-Nasser and Iwakuma [9] and Iwakuma and Nemat-Nasser [6] have obtained a similar result in terms of the velocity gradient and the nominal stress rate.

To discuss this fundamental result, we confine attention to rate-independent relations for the single crystal and write, from (2.1), (3.16), and (3.17),

$$\dot{n}_{ij} = \mathscr{F}^{\Omega}_{ijkl}l_{kl} \tag{4.1}$$

$$\mathcal{F}^{\Omega}_{ijkl}=\mathcal{L}^{\Omega}_{ijkl}-\tfrac{1}{2}(\delta_{ik}\delta_{lm}+\delta_{il}\delta_{km})\sigma_{mj}+\tfrac{1}{2}(\delta_{jk}\delta_{lm}-\delta_{jl}\delta_{km})\sigma_{mi}$$
$$-(\mathcal{L}^{\Omega}_{ijrs}p^{\alpha}_{rs}-\sigma_{ij}p^{\alpha}_{rr}+\omega^{\alpha}_{ir}\sigma_{rj}+\omega^{\alpha}_{jr}\sigma_{ri})$$
$$\times M^{\alpha\beta}p^{\beta}_{mn}(\mathcal{L}^{\Omega}_{mnkl}-\sigma_{mn}\delta_{kl}) \tag{4.2}$$

Now, we consider an infinitely extended homogeneous elastic–plastic solid with constitutive relations

$$\dot{N}_{ij}=\mathcal{F}_{ijkl}L_{kl} \tag{4.3}$$

containing an ellipsoidal crystal, Ω, with constitutive relations defined by (4.1). When the far-field uniform velocity gradient L_{ij} is prescribed, the velocity gradient in Ω becomes

$$l_{ij}=\mathcal{A}^{\Omega}_{ijkl}L_{kl} \tag{4.4}$$

where the concentration tensor $\mathcal{A}^{\Omega}$ is defined in terms of $\mathcal{F}$ and the geometry of Ω; see Iwakuma and Nemat-Nasser [6].

Since $\langle l_{ij}\rangle=L_{ij}$, the self-consistency requires that $\langle\mathcal{A}^{\Omega}\rangle=\mathbf{I}$, the identity tensor. To ensure this here we follow Walpole [12] and normalize $\mathcal{A}^{\Omega}$ as

$$\bar{\mathcal{A}}^{\Omega}=\mathcal{A}^{\Omega}\langle\mathcal{A}^{\Omega}\rangle^{-1} \tag{4.5}$$

and use this normalized tensor in place of $\mathcal{A}^{\Omega}$ in (4.4). Then it follows that

$$\dot{N}_{ij}=\langle\dot{n}_{ij}\rangle=\langle\mathcal{F}^{\Omega}_{ijmn}\bar{\mathcal{A}}^{\Omega}_{mnkl}\rangle L_{kl} \tag{4.6}$$

Thus, the overall instantaneous moduli are given by

$$\mathcal{F}=\langle\mathcal{F}^{\Omega}:\bar{\mathcal{A}}^{\Omega}\rangle \tag{4.7}$$

It has been shown by Iwakuma and Nemat-Nasser [6] that $\mathcal{A}^{\Omega}$ and hence $\mathcal{F}$ can be calculated as long as the operator $\mathcal{F}_{ijkl}\,\partial^2/\partial x_i\,\partial x_l$ is elliptic. Once this operator ceases to be elliptic, the overall uniform deformation ceases to be stable, leading to possible localized deformations. Iwakuma and Nemat-Nasser [6] present specific examples and give detailed illustrations of this and related points for two-dimensional problems.

ACKNOWLEDGEMENT

This work has been supported by the Army Research Office under Contract DAAG29-82-K-0147 to Northwestern University.

REFERENCES

1. Ashby, M. F. and H. J. Frost. The kinetics of inelastic deformation above 0°K. In: *Constitutive Equations in Plasticity* (Ed. A. S. Argon), The MIT Press, Cambridge, Mass., 1975, 117–147.
2. Eshelby, J. D. The determination of the elastic field of an ellipsoidal inclusion, and related problems, *Proc. R. Soc. Lond.*, **A241** (1957), 376–396.
3. Havner, K. S. The theory of finite plastic deformation of crystalline solids. In: *Mechanics of Solids: The Rodney Hill 60th Anniversary Volume* (Eds H. G. Hopkins and M. J. Sewell), Pergamon Press, Oxford, 1982, 265–302.
4. Hill, R. Generalized constitutive relations for incremental deformation of metal crystals by multislip, *J. Mech. Phys. Solids*, **14** (1966), 95–102.
5. Hill, R. On constitutive macro-variables for heterogeneous solids at finite strain, *Proc. R. Soc. Lond.*, **A326** (1972), 131–147.
6. Iwakuma, T. and S. Nemat-Nasser. Finite elastic–plastic deformation of polycrystalline metals, *Proc. R. Soc. Lond.*, **A394** (1984), 87–119.
7. Kocks, U. F. The relation between polycrystal deformation and single-crystal deformation, *Met. Trans.*, **1** (1970), 1121–1143.
8. Nemat-Nasser, S. On finite plastic flow of crystalline solids and geomaterials, *J. appl. Mech.* (50th Anniversary Issue), **40** (46) (1983), 1114–26.
9. Nemat-Nasser, S. and T. Iwakuma. Finite elastic–plastic deformation of composites. In: *Mechanics of Composite Materials: Recent Advances* (Eds Z. Hashin and C. T. Herakovich), Pergamon Press, Oxford, 1983, 47–55.
10. Nemat-Nasser, S., M. M. Mehrabadi and T. Iwakuma. On certain macroscopic and microscopic aspects of plastic flow of ductile materials. In: *Three-Dimensional Constitutive Relations and Ductile Fracture, Proc. IUTAM Symposium*, Dourdan, France (Ed. S. Nemat-Nasser), North-Holland, Amsterdam, 1980, 157–172.
11. Taylor, G. I. Plastic strain in metals, *J. Inst. Metals*, **62** (1), (1938), 307–324.
12. Walpole, L. J. On the overall elastic moduli of composite materials, *J. Mech. Phys. Solids*, **17** (1969), 235–251.

6

Models of Metal Plasticity: Theory and Experiment

J. F. BESSELING

Laboratory for Engineering Mechanics, Delft University of Technology, Delft, The Netherlands

ABSTRACT

The effects of inhomogeneous inelastic deformation on a microscale are represented by a fraction (or overlay) model in continuum theory. The model is a special case of thermodynamic internal state-variable theory and is applicable to small and to arbitrarily large deformations. A considerable number of records of uniaxial and biaxial stress and strain histories for an austenitic and a ferritic steel are available from experiments on thin-walled tubes. The evaluation of plasticity and creep experiments carried out so far has shown the fraction model to be a well-defined concept that compares favourably with a number of other commonly used models. For detailed comparisons references are given.

1. INTRODUCTION

A sharp distinction can be made between constitutive theories for the description of inelastic material behaviour in which the material is represented by a model, that constitutes a completely defined thermodynamic system, and theories in which the material is characterized by response functionals. In theories of the first kind the state of the material is at any moment supposed to be fully determined by the value of state variables, while in the theories of the second kind the state is determined by the history of the external action as represented in the response functional. Though the latter theories start out from much broader concepts than the state variable approach, for arriving at a well-defined, experimentally verifiable description, the simpler approach has proved to be much more practical.

The author translated a well-known one-dimensional model, consisting of a parallel arrangement of physically different elements for the representation of the Bauschinger effect and of other hereditary effects in inelastic deformation, in terms of a three-dimensional theory of plasticity [1]. This so-called fraction model was later on also referred to as an overlay model. The presentation of the model in [2] ended with the remark that no attempt had as yet been made to obtain a comprehensive comparison between the predictions of the theory and available experimental data. The first efforts to make this comparison were stranded because of the insufficiency of the information given in the publications dealing with experiments in creep and plasticity. For a detailed check on the value of a particular model for the inelastic behaviour of solids uninterpreted records of independent measurements of strain and stress histories are needed.

After a number of preliminary plasticity experiments on thin-walled tubes of an aluminium alloy, and creep experiments on tensile specimens of a magnesium alloy [12], the research in connection with the joint German-Benelux fast breeder project made it possible to carry out an extensive program for the verification of plasticity and creep models. The program has produced a considerable number of records of uniaxial and biaxial stress and strain histories, that are now available for future reference. The materials tested were a 304 stainless steel (WN 1.4948) and a $2\frac{1}{4}$Cr-1Mo ferritic steel stabilized with niobium (WN 1.6770).

This paper cannot cover the work of a large group of people, extending over a number of years. References will be given for more detailed information on the experimental and theoretical program, that unfortunately came abruptly to an end, when under political pressure the research connected with the fast breeder project was reduced to a minimum.

2. CONSTITUTIVE RELATIONS FOR CREEP AND PLASTICITY

The industrially produced metals are conglomerations of differently orientated, anisotropic microbodies of a more or less well defined structure. But even if an accurate mathematical description of the processes on a microlevel would be possible, the complicated interaction of the microbodies would rule out a rigorous derivation of the macroscopic behaviour. Therefore the actual material is replaced by a continuum with at all times the same macroscopic mass density as the actual material. Thus a mathematical description of the mass flow can be given in terms of this continuum model, but we should realize that the so-called material points of the continuum define by no means the place of the actual material particles.

While it would be obviously inappropriate for fluids, for solids it is tempting to relate the geometric configuration of the material points of the continuum to the geometric configuration of the particles of the actual material. Then the thermodynamic state, insofar as it is geometry dependent, will be determined by the deformation of the continuum with respect to a reference geometry of this continuum. For the reversible process of elastic deformation this approach is successful, but for plastic deformation it must fail, even as a first approximation, if the deformations become very large. The material of a plate will have lost its remembrance of the shape of the block from which the plate has been produced by a hot-rolling process.

Nevertheless subsequent elastic properties can be predicted fairly accurately to be unchanged, which in ref. 3 prompted the idea of a generalization of Eckart's natural reference state [8] for a thermodynamic definition of stress. Stress is then by a thermodynamic equation of state uniquely determined by a difference in entropy and by a deformation with respect to a natural reference state, attributed to each material point of the continuum. The change of the natural reference state, in particular the change in geometrical configuration between the actual state and the natural reference state, is not kinematically defined by the motion of the continuum, but constitutes the essence of the theory of creep and plasticity. The glide mechanisms of inelastic deformation of metals will cause a change of neighbours for actual material particles; this is in contrast to the material points of the continuum, where a priori neighbours remain neighbours. The theory based upon the existence of a natural reference state may also be looked upon as a special case of the internal state variable theory.

Let us recapitulate the theory of the natural reference state in terms of cartesian co-ordinates and cartesian components of vector and tensor quantities.

With x_i as the cartesian co-ordinates of a material point at time t, ρ

as mass density, $\dot{u}_i$ as velocity components, t_{ij} as stress tensor components, ρf_i as force density components, $\ddot{u}_i$ as material acceleration components, t_i as stress vector components, T as absolute temperature, $T\dot{h}_i$ as heat-flux components, n_i as unit normal components, s as entropy per unit mass, σ as entropy production per unit mass, and e as internal energy per unit mass, we list the following basic equations of continuum thermomechanics:
conservation of mass:

$$\frac{\partial \rho}{\partial t} + \dot{u}_i \frac{\partial \rho}{\partial x_i} + \rho \frac{\partial \dot{u}_i}{\partial x_i} = 0 \tag{1}$$

balance of momentum:

$$\frac{\partial t_{ij}}{\partial x_j} + \rho f_i = \rho \ddot{u}_i = \rho\left(\frac{\partial \dot{u}_i}{\partial t} + \dot{u}_k \frac{\partial \dot{u}_i}{\partial x_k}\right) \tag{2}$$

balance of moment of momentum:

$$t_{ij} = t_{ji} \tag{3}$$

conservation of energy:

$$\int_{\partial B} (t_i \dot{u}_i - T\dot{h}_i n_i)\,\mathrm{d}A + \int_B \rho f_i \dot{u}_i\,\mathrm{d}V = \int_B \rho(\dot{e} + \ddot{u}_i \dot{u}_i)\,\mathrm{d}V \tag{4}$$

entropy production:

$$\rho\sigma = \rho\dot{s} + \frac{\partial \dot{h}_i}{\partial x_i} \geqslant 0 \tag{5}$$

By stating the law of heat conduction with k_{ij} as elements of the positive definite heat conductivity matrix,

$$k_{ij}\frac{\partial T}{\partial x_j} = -T\dot{h}_i \tag{6}$$

we may combine the basic equations into one inequality, to be fulfilled by the constitutive theory for creep and plasticity

$$T\rho\sigma - \frac{k_{ij}}{T}\frac{\partial T}{\partial x_i}\frac{\partial T}{\partial x_j} = t_{ij}d_{ij} - \rho(\dot{e} - T\dot{s}) \geqslant 0 \tag{7}$$

where d_{ij} denotes the rate of deformation tensor of the continuum.

$$d_{ij} = \frac{1}{2}\left(\frac{\partial \dot{u}_i}{\partial x_j} + \frac{\partial \dot{u}_j}{\partial x_i}\right) \tag{8}$$

The internal energy per unit mass, e, that we have introduced, is without physical content as long as we do not specify the variables, that determine its value. The variables are so-called state variables, which in the inertial system that we are considering are independent of the velocity. Besides on the entropy s the internal energy will depend on state variables of a geometric nature.

The theory that we shall consider rests upon the assertion that *at any moment* the geometric configuration of the particles of the actual material, imbedded in a small neighbourhood of a material point of the continuum, differs from the configuration in a *fixed* natural reference state of the actual material particles by an invertible transformation of line-elements.

$$\mathrm{d}a_\alpha = b_{\alpha i}\,\mathrm{d}x_i, \quad \det(b_{\alpha i}) > 0 \tag{9}$$

With

$$b_{\alpha i} b_{\beta i} = C_{\alpha\beta}$$

we have by the polar decomposition theorem

$$b_{\alpha i} = C^{1/2}_{\alpha\beta} R_{\beta i} \tag{10}$$

where

$$C^{1/2}_{\alpha\beta} = \sum_{p=1}^{3} \sqrt{C^{(p)}}\, n^{(p)}_\alpha n^{(p)}_\beta, \; C^{(p)} > 0$$

$$R_{\alpha i} R_{\beta i} = \delta_{\alpha\beta}$$

The tensor components $C_{\alpha\beta}$ define a deformation, the components $R_{\alpha i}$ a rigid rotation. The latter determine the orientation of the material with respect to orthogonal triads, with fixed directions in space, attached to the material points of the continuum. Together with the entropy per unit mass, s, the tensor components $C_{\alpha\beta}$, independent of rigid rotations, are proper state variables in an internal energy function

$$e = e(C_{\alpha\beta}, s) \tag{11}$$

The change of the natural reference state, characterized by $\dot{\overline{\mathrm{d}a_\alpha}}$, may be expressed in terms of quantities, determined by the local creep and plasticity process, as well as in terms of the rate of change of the tensor $b_{\alpha i}$ and the velocity field of the continuum in motion.

$$\dot{\overline{\mathrm{d}a_\alpha}} = p_{\alpha i}\,\mathrm{d}x_i = \left(\dot{b}_{\alpha i} + b_{\alpha j}\frac{\partial \dot{x}_j}{\partial x_i}\right)\mathrm{d}x_i$$

or

$$\dot{b}_{\alpha i} = p_{\alpha i} - b_{\alpha j}\frac{\partial \dot{x}_j}{\partial x_i} \tag{12}$$

With the aid of (12) the rate of change of the state variables $C_{\alpha\beta}$ can be expressed by

$$\dot{C}_{\alpha\beta} = \dot{b}_{\alpha i}b_{\beta i} + b_{\alpha i}\dot{b}_{\beta i} = p_{\alpha i}b_{\beta i} + b_{\alpha i}p_{\beta i} - 2b_{\alpha i}b_{\beta j}d_{ij} \tag{13}$$

On the basis of the internal energy function (11) the expression (13) transforms the inequality (7) into

$$\left(t_{ij} + 2\rho\frac{\partial e}{\partial C_{\alpha\beta}}\, b_{\alpha i}b_{\beta j}\right)d_{ij} + \rho\left(T - \frac{\partial e}{\partial s}\right)\dot{s} - 2\rho\frac{\partial e}{\partial C_{\alpha\beta}}\, b_{\alpha i}p_{\beta i} \geqslant 0 \tag{14}$$

With the thermodynamic definition of stress and temperature,

$$\begin{aligned} t_{ij} &= -2\rho\frac{\partial e}{\partial C_{\alpha\beta}}\, b_{\alpha i}b_{\beta j} \\ T &= \frac{\partial e}{\partial s} \end{aligned} \tag{15}$$

the energy dissipation due to creep and plasticity is given by

$$-2\rho\frac{\partial e}{\partial C_{\alpha\beta}}\, b_{\alpha i}p_{\beta i} = t_{ij}b_{j\alpha}^{-1}p_{\alpha i} \geqslant 0 \tag{16}$$

The theory is completed by constitutive equations for the tensor components $p_{\alpha i}$, which are equal to zero in the case of thermoelastic deformation.

First we make the obvious assumption that the natural reference state is characterized by a constant mass density ρ_o. In that case, as is shown in detail in ref. 6, only the deviator components s_{ij} of the stress tensor contribute to the energy dissipation. With

$$s_{ij} = t_{ij} + p\delta_{ij}, \qquad s_{ii} = 0 \tag{17}$$

we then have the inequality

$$s_{ij}b_{j\alpha}^{-1}p_{\alpha i} \geqslant 0 \tag{18}$$

For creep we introduce a stress and temperature dependent energy dissipation function π,

$$s_{ij}b_{j\alpha}^{-1}p_{\alpha i} = \pi\left(p,\, T,\, \frac{\rho_o}{\rho}\, s_{ij}b_{i\alpha}^{-1}b_{\beta j}\right) \geqslant 0 \tag{19}$$

Notice that $s_{ij}b_{i\alpha}^{-1}b_{j\beta}$ are proper state variables [15], since also the rate of energy dissipation must be independent of rigid rotations of the material.

Time independent plasticity may be regarded as a limiting case of highly nonlinear creep. The yield surface then represents a surface of indeterminate rate of energy dissipation. This surface can be defined by

$$\phi = \phi\left(p, T, \frac{\rho_o}{\rho} s_{ij}b_{i\alpha}^{-1}b_{j\beta}\right) = 0 \tag{20}$$

It was recently noted [9] that the factor ρ_o/ρ is needed in order to obtain the desired symmetry properties in the plasticity relations if the constitutive equations for $p_{\alpha i}$ are based upon the normality condition for the tensor dual to s_{ij} in the expression for the rate of energy dissipation (18).

$$b_{j\alpha}^{-1}p_{\alpha i} = \lambda \frac{\partial \pi}{\partial s_{ij}} + \mu \frac{\partial \phi}{\partial s_{ij}} \tag{21}$$

where $\mu \geqslant 0$ if $\phi = 0$ and $\dot{\phi} \geqslant 0$

$\mu = 0$ if $\phi < 0$ or $\phi = 0, \quad \dot{\phi} < 0.$

For (12) and (13) we have now

$$\begin{aligned} \dot{b}_{\alpha i} &= b_{\alpha k}\left(\lambda \frac{\partial \pi}{\partial s_{ik}} + \mu \frac{\partial \phi}{\partial s_{ik}} - \frac{\partial \dot{x}_k}{\partial x_i}\right) \\ \dot{C}_{\alpha\beta} &= 2b_{\alpha i}b_{\beta j}\left(\lambda \frac{\partial \pi}{\partial s_{ij}} + \mu \frac{\partial \phi}{\partial s_{ij}} - d_{ij}\right) \end{aligned} \tag{22}$$

With the aid of (22) and with the spin tensor components

$$\omega_{ij} = \frac{1}{2}\left(\frac{\partial \dot{u}_i}{\partial x_j} - \frac{\partial \dot{u}_j}{\partial x_i}\right) \tag{23}$$

the material derivatives of the stress tensor components can be written as follows

$$\dot{t}_{ij} - t_{kj}\omega_{ik} + t_{ik}\omega_{kj} = \frac{\dot{\rho}}{\rho} t_{ij} + L_{ijmn}\left(d_{mn} - \lambda \frac{\partial \pi}{\partial s_{mn}} - \mu \frac{\partial \phi}{\partial s_{mn}}\right) + \frac{\partial t_{ij}}{\partial s} \dot{s}$$

where

$$L_{ijmn} = 4\rho \frac{\partial^2 e}{\partial C_{\alpha\beta}\, \partial C_{\gamma\delta}} b_{\alpha i} b_{\beta j} b_{\gamma m} b_{\delta n} - \tfrac{1}{2}(t_{jn}\delta_{im} + t_{in}\delta_{jm} + t_{jm}\delta_{in} + t_{im}\delta_{jn}) \tag{24}$$

The scalar factor λ is determined by the creep function.

$$\lambda s_{ij} \frac{\partial \pi}{\partial s_{ij}} = \pi \Rightarrow \lambda = \left(s_{ij} \frac{\partial \pi}{\partial s_{ij}} \right)^{-1} \pi \tag{25}$$

The calculation of the factor μ is more complicated. After putting

$$s^*_{\alpha\beta} = \frac{\rho_o}{\rho} s_{ij} b^{-1}_{i\alpha} b_{j\beta} \tag{26}$$

we must require

$$\dot{\phi} = \frac{\partial \phi}{\partial s^*_{\alpha\beta}} \dot{s}^*_{\alpha\beta} + \frac{\partial \phi}{\partial p} \dot{p} + \frac{\partial \phi}{\partial T} \dot{T} \geqslant 0 \tag{27}$$

By observing that [15]

$$\begin{aligned} b^{-1}_{j\alpha} \dot{b}_{\alpha i} - b^{-1}_{i\alpha} \dot{b}_{\alpha j} &= 2\omega_{ij}, \\ b^{-1}_{j\alpha} \dot{b}_{\alpha i} + b^{-1}_{i\alpha} \dot{b}_{\alpha j} &= 2\left(\lambda \frac{\partial \pi}{\partial s_{ij}} + \mu \frac{\partial \phi}{\partial s_{ij}} - d_{ij} \right), \end{aligned} \tag{28}$$

we can write

$$\begin{aligned} \frac{\partial \phi}{\partial s^*_{\alpha\beta}} \dot{s}^*_{\alpha\beta} = \frac{\partial \phi}{\partial s_{ij}} \Bigg(& \frac{\dot{\rho}}{\rho} s_{ij} + s_{kj}\omega_{ik} - s_{ik}\omega_{kj} + \frac{\partial s_{ij}}{\partial s} \dot{s} \\ & + L_{ijmn} \left(d_{mn} - \lambda \frac{\partial \pi}{\partial s_{mn}} - \mu \frac{\partial \phi}{\partial s_{mn}} \right) - \frac{\dot{\rho}}{\rho} s_{ij} - s_{kj}\omega_{ik} + s_{ik}\omega_{kj} \Bigg) \end{aligned} \tag{29}$$

Here we see that because of the factor ρ_o/ρ the terms with $(\dot{\rho}/\rho)s_{ij}$ cancel and, as could be expected, the rigid rotations drop out since proper state variables were used in the yield function ϕ. The factor μ can now be written as $(0 \leqslant h < 1)$

$$\mu = (1-h) \frac{\dfrac{\partial \phi}{\partial s_{ij}} L_{ijmn} \left(d_{mn} - \lambda \dfrac{\partial \pi}{\partial s_{mn}} \right) + \dfrac{\partial \phi}{\partial s_{ij}} \dfrac{\partial s_{ij}}{\partial s} \dot{s} + \dfrac{\partial \phi}{\partial p} \dot{p} + \dfrac{\partial \phi}{\partial T} \dot{T}}{\dfrac{\partial \phi}{\partial s_{kl}} L_{klab} \dfrac{\partial \phi}{\partial s_{ab}}} \tag{30}$$

with the restriction $\mu \geqslant 0$.

For metals the stress deviator components s_{ij} are independent of the entropy s. Usually the pressure dependence of the yield function is neglected. We then have

$$\dot{t}_{ij} - t_{kj}\omega_{ik} + t_{ik}\omega_{kj} = [L_{ijmn} - (1-h)Y_{ijmn} - t_{ij}\delta_{mn}]$$

$$\times\left(d_{mn} - \lambda\frac{\partial\pi}{\partial s_{mn}}\right) - (1-h)\frac{L_{ijmn}\dfrac{\partial\phi}{\partial s_{mn}}\dfrac{\partial\phi}{\partial T}\dot{T}}{\dfrac{\partial\phi}{\partial s_{kl}}L_{klab}\dfrac{\partial\phi}{\partial s_{ab}}} - \frac{\partial p}{\partial s}\dot{s} \quad (31)$$

with

$$Y_{ijmn} = \frac{L_{ijpq}\dfrac{\partial\phi}{\partial s_{pq}}\dfrac{\partial\phi}{\partial s_{rs}}L_{rsmn}}{\dfrac{\partial\phi}{\partial s_{kl}}L_{klab}\dfrac{\partial\phi}{\partial s_{ab}}} \quad (32)$$

Note that no restrictions have been placed upon the amount of creep or plastic deformation, nor have the elastic deformations been assumed to be small.

But for metals the elastic deformations can only be small and then it is advantageous to replace the state variables $C_{\alpha\beta}$ by the elastic strain tensor $E'_{\alpha\beta}$.

$$E'_{\alpha\beta} = \tfrac{1}{2}(C^{-1}_{\alpha\beta} - \delta_{\alpha\beta}), \qquad e'_{\alpha\beta} = E'_{\alpha\beta} - \tfrac{1}{3}E'_{\gamma\gamma}\delta_{\alpha\beta} \quad (33)$$

For an elastically isotropic metal the internal energy function depends on the deformation through the invariants

$$I_1 = E'_{\alpha\alpha}, \qquad I_2 = \tfrac{1}{2}e'_{\alpha\beta}e'_{\alpha\beta} \quad (34)$$

In the case of small changes of entropy the internal energy can be given in the form

$$\rho_o e = \rho_o T_0 s + \tfrac{1}{2}C\left(1 + \frac{C\alpha^2 T_0}{\rho_o c_v}\right)I_1^2 + 2GI_2 - \frac{C\alpha T_0}{c_v}I_1 s + \frac{\rho_o T_0}{2c_v}s^2 \quad (35)$$

T_0 represents the temperature of the natural reference state. The remaining four constants have been expressed in terms of the commonly used physical constants: the coefficient of thermal expansion, α, the specific heat at constant volume, c_v, the isothermal bulk modulus of elasticity, C, and the shear modulus, G.

If we take into account that the values of the moduli C and G are several orders of magnitude larger than the values the stresses may assume under normal conditions, while further for small values of $E'_{\alpha\beta}$ $(b_{\alpha i}-R_{\alpha i})\ll 1$ holds, then we derive for L_{ijmn} according to (24) and (35):

$$L_{ijmn}=C\left(1+\frac{C\alpha^2T_0}{\rho_o c_v}\right)\delta_{ij}\delta_{mn}+2G(\delta_{im}\delta_{jn}-\tfrac{1}{3}\delta_{ij}\delta_{mn}) \quad (36)$$

In order to eliminate the entropy s from our equations we consider the results from (15) and (35)

$$T-T_0=-\frac{C\alpha T_0}{\rho_o c_v}I_1+\frac{T_0 s}{c_v}, \qquad \frac{\partial p}{\partial s}=\frac{C\alpha T_0}{c_v} \quad (37)$$

and obtain

$$\frac{\partial p}{\partial s}\dot{s}=C\alpha\dot{T}+\frac{C^2\alpha^2T_0}{\rho_o c_v}d_{kk} \quad (38)$$

Next we introduce for creep an energy dissipation function of the form

$$\pi=\gamma(T)\left[\frac{3J_2-(1-\beta)(\sigma^y)^2}{\beta(\sigma^y)^2}\right]^n, \qquad J_2=\tfrac{1}{2}s_{ij}s_{ij}, \qquad 0<\beta\leqslant 1, \quad n\geqslant 1 \quad (39)$$

provided

$$3J_2\geqslant(1-\beta)(\sigma^y)^2, \quad \text{else} \quad \pi=0. \quad (40)$$

This function gives in accordance with (25)

$$\lambda\frac{\partial\pi}{\partial s_{ij}}=c_{ij}=\tfrac{1}{2}\gamma(T)\left[\frac{3J_2-(1-\beta)(\sigma^y)^2}{\beta(\sigma^y)^2}\right]^n\frac{s_{ij}}{J_2} \quad (41)$$

We plot the rate of energy dissipation according to (39) as a function of J_2 and we note that for $\beta=1$ and $n=1$ we have linearly visco-elastic behaviour. On the other hand for $\beta\to 0$ the behaviour becomes highly non-linear and in the limit it is represented by time independent plasticity with a yield surface

$$\phi=3J_2-(\sigma^y)^2=0 \quad (42)$$

According to (32) the yield coefficients are then given by

$$Y_{ijmn}=GJ_2^{-1}s_{ij}s_{mn} \quad (43)$$

In reality material behaviour is neither linearly visco-elastic nor time independent elastic–plastic. Furthermore inelastic deformation turns

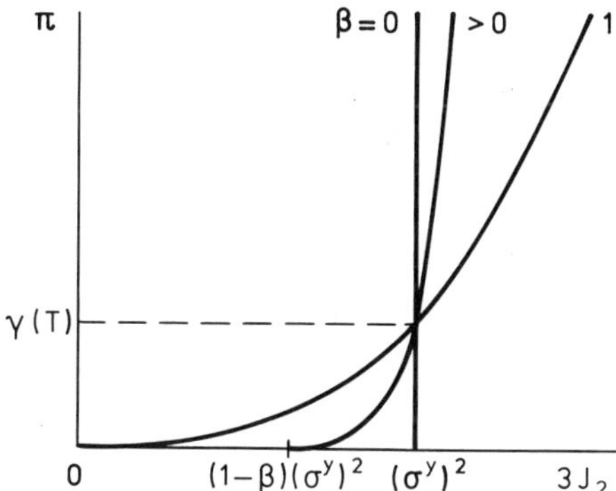

FIG. 1. Rate of energy dissipation.

an initially isotropic material always into a material that is anisotropic with respect to its inelastic properties, even though its elastic properties in the gross remain isotropic.

3. THE FRACTION MODEL

In elastic deformation the distribution of the internal energy on a microscopic scale is irrelevant to the macroscopic deformation problem, since this distribution is independent of the history of deformation. The internal energy can be replaced by the internal energy of a fictitious homogeneous continuum with respect to a fixed natural reference state imbedded in this continuum. In inelastic deformation, however, the dissipation of energy cannot in general be replaced by the dissipation in a homogeneous continuum. Inelastic deformation of metals, i.e. glide in crystal lattices, will be initiated at certain points and will spread in subsequent loading over the whole volume of the material in the neighbourhood of a material point of the continuum. This process can be modelled by conceiving the material to be composed of a limited number of portions, which can be represented by subelements of an element of volume dV, each with different parameters in the dissipation function π or in the yield function ϕ. This we call the fraction model. Others have introduced it under the name overlay model [14].

As a consequence of the difference in inelastic behaviour of the various fractions, which model together the whole of the material, each of the fractions must be attributed its own natural reference state in the continuum. For a number N of fractions we have as an expression for the internal energy per unit mass

$$e = \sum_{k=1}^{N} \psi^k e^k(C^k_{\alpha\beta}, s), \qquad \sum_{k=1}^{N} \psi^k = 1 \tag{44}$$

Note that we make no difference in change of entropy between actual state and natural reference state for the various fractions. Clearly the fraction model is a case of internal state variable theory, where the set of geometrical state variables has been enlarged from $C_{\alpha\beta}$ to $C^k_{\alpha\beta}$, $k = 1, \ldots, N$.

So far the fraction model has not been extended beyond the version presented in refs. 1 and 2. The various fractions are treated as independent thermodynamic systems, be it that they are imbedded in the same continuum and are therefore subjected to the same macroscopic rate of deformation.

The behaviour of each fraction is being described by the same internal energy function $e^k(C^k_{\alpha\beta}, s)$, but by dissipation functions and yield surfaces with different material parameters. For instance for the fraction k we define

$$\rho_o e^k = \rho_o T_0 s + \tfrac{1}{2}C\left(1 + \frac{C\alpha^2 T_0}{\rho_o c_v}\right)I_1^2 + 2GI_2^k - \frac{C\alpha T_0}{c_v} I_1 s + \frac{\rho_o T_0}{2c_v} s^2 \tag{45}$$

$$\pi^k = \gamma(T)\left[\frac{3J_2^k - (1-\beta^k)(\sigma^{yk})^2}{\beta^k(\sigma^{yk})^2}\right]^n, \quad J_2^k = \tfrac{1}{2}s^k_{ij}s^k_{ij}, \quad 0 \leqslant \beta^k < 1, \quad n \geqslant 1 \tag{46}$$

for $3J_2^k > (1-\beta^k)(\sigma^{yk})^2$, else $\pi^k = 0$

$$\phi^k = 3J_2^k - (\sigma^{yk})^2 = 0 \tag{47}$$

Here we have used $I_1^k = I_1$, which expresses the fact that for metals the small volume changes are an elastic phenomenon. After inelastic deformation the internal energies per unit mass of the various fractions differ only by the values of the elastic deformation in the term $2GI_2^k$.

In the dissipation function π^k we could give each fraction its own value for $\gamma(\tau)$ and for n, in addition to the fraction parameters β^k and σ^{yk}, but as long as comparison with experimental results does not compel us to accept this complication we should refrain from it. The more parameters we introduce, the less the predictive strength of the model will be. In fact the attraction of the fraction model stems mainly from the possibility of a rather straightforward determination of the main material parameters from simple tests in the laboratory in the case of plastic deformation.

Because of (44) we may consider the macroscopic stress as the weighted sum of the fractional stresses

$$t_{ij} = \sum_{k=1}^{N} \psi^k t_{ij}^k, \qquad t_{ij}^k = -2\rho \frac{\partial e^k}{\partial C_{\alpha\beta}^k} b_{\alpha i}^k b_{\beta j}^k \tag{48}$$

With respect to creep and plasticity the fractional stresses obey constitutive relations of the type (31). Specializing to isothermal tests in uniaxial tension and compression we obtain from (45) and (47) for time independent plasticity

$$\dot{t}_{xx} = 3G\left[1 - \sum_{k=1}^{N} \psi^k (1-h^k)\right](d_{xx} - \tfrac{1}{3}d_{ii})$$

$$d_{ii} = \frac{1}{3C}\dot{t}_{xx} \tag{49}$$

where for the hardening parameters h^k holds

$$h^k \ll 1 \quad \text{for} \quad t_{xx}^k = \sigma^{yk}, \qquad t_{xx}^k d_{xx} > 0$$

$$h^k = 1 \quad \text{for} \quad t_{xx}^k < \sigma^{yk} \quad \text{or} \quad t_{xx}^k = \sigma^{yk}, \qquad t_{xx}^k d_{xx} \leqslant 0 \tag{50}$$

A value $h^k = 0$ implies plastic deformation of fraction k at constant stress. The parameter h^k was introduced in ref. 1 for the description of isotropic work hardening. A combination of elastic–ideally plastic elements not only leads to an overestimation of the anisotropy of hardening, but it would also imply that a small fraction of the material would behave elastically at large deformations if, as is usually the case, the stress–strain curve at these deformations still shows some hardening. For metals this is to be rejected on physical grounds. The so-called kinematic hardening model, which in essence is a two-fraction model with one fraction remaining elastic, also suffers from this defect.

A better description is obtained if the stress–strain curve of Fig. 2 is approximated by a broken line, the breakpoints defining the elastic limits of the various fractions, such that from a last breakpoint (say point 4 in Fig. 2) the slope of the curve gives a first estimate of $h^k = h$.

$$3Gh = \frac{dt_{xx}}{de_{xx}} \tag{51}$$

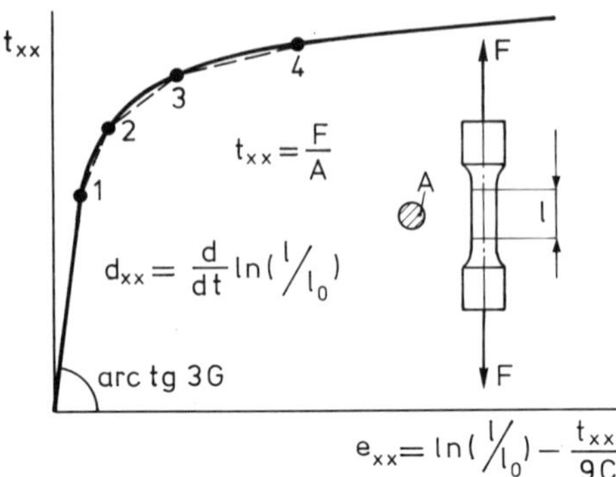

FIG. 2. Experimental stress–strain curve.

Next on the basis of (49) the values of ψ^k and σ^{yk} can be derived successively from the breakpoints and the slopes of the broken line approximating the stress–strain curve.

The values of the isotropic hardening parameters do not greatly affect the values of ψ^k and σ^{yk}, but they are essential because of the cumulative effect of the isotropic hardening. For more details in the determination of the hardening parameters h^k we must refer to refs. 10, 6, 11, and 14.

Another correction of the basic model, which for instance was found to be essential for a 304 stainless steel at room temperature pertains to the strain-rate effects. As indicated in Fig. 3 a considerable drop in stress occurs if the loading of a specimen is interrupted for several

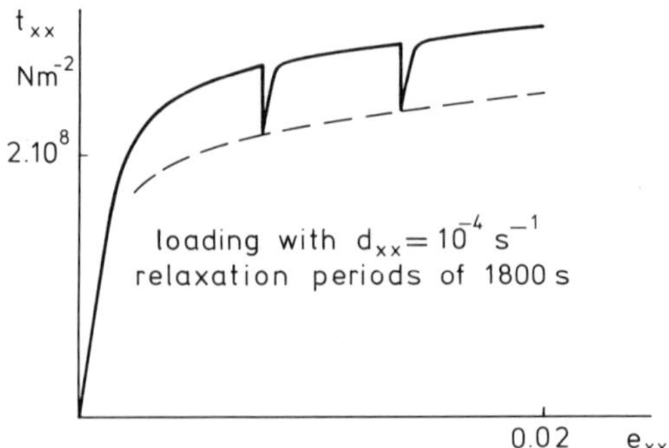

FIG. 3. Time dependent stress–strain data.

minutes. It has been shown in ref. 11 that an accurate description of these phenomena can be obtained if the first fraction is given the properties of a Bingham model which are equivalent to the properties connected with a dissipation function of the type (46). Stress–strain curves, obtained for three different, suitably chosen strain rates, suffice for the determination of the material parameters γ, β^1 and n. The value of ψ^1 is obtained, either from the initial slope after a relaxation period [11], or from an initial, time independent approximation of the stress–strain curve.

While the application of the fraction model to plastic deformation of steels and aluminium alloys at room temperatures has shown its great merits as compared to other models, for the description of the creep behaviour at higher temperatures much less progress has been made with the fraction model. The determination of the material parameters is not as straightforward as in the case of plastic deformation. Hardening and softening effects of creep-deformation play a predominant role. A recent study [13] of the creep data obtained for the ferritic steel has clarified these phenomena with the prospect of further progress in the application of the fraction model, also in the creep range.

4. EXPERIMENTS

In a period of approximately ten years a large number of experiments has been performed on tubular specimens, which were loaded in combinations of tension and compression, torsion, and internal pressure along carefully selected stress and strain paths. During this loading the strain rate, ranging from 10^{-4} to $10^{-6}\,s^{-1}$, was kept approximately constant and the forces and deformations were measured every 2–5 s. The experiments were restricted to a strain rate of a few per cent. The accuracy of the experimental data is typically 1%.

A data logging system was used to register the voltages coming from the amplifiers onto punch tape. The monitoring was done with x–y and x–t recorders. After correcting obvious punching errors the registered signals were recorded onto magnetic tape along with a record of the conditions under which the experiment was carried out. This includes a record of the mechanical measurements on the tube before and after the test. Detailed information on the experimental set-up at the time of the tests on the austenitic tubes is given in ref. 5, while ref. 10 supplies the data on the tests on the ferritic specimens.

The magnetic tapes containing the complete information from the tests carried out so far in the Laboratory of Engineering Mechanics at Delft can be made available for all research programs that aim at developing and evaluating plasticity models.

The conclusions reached in the course of the program connected with the fast breeder project imply that the fraction model gives a quite accurate description for almost all the loading histories considered. In particular no evidence was found in any of the tests that the yield surface may develop corners or vertices in the course of loading. However the main source of the discrepancies still occurring between model and experiment can be found in the description of the gradual transition from elastic to fully plastic behaviour, especially after load reversals. The experimental data show that the linear elastic range after a load reversal becomes smaller after successive load cycles. Hence the diameter of the yield surface of the first fraction, which is already much smaller to begin with than the diameter of the yield surface in any one-yield surface theory, tends to become even smaller, thereby increasing the curvature of the inner yield surface at the loading point even further.

REFERENCES

1. Besseling, J. F. A theory of plastic flow for anisotropic hardening in plastic deformation of an initially isotropic material, Nat. Aero. Res. Inst., Amsterdam, 1953, Report S-410.
2. Besseling, J. F. A theory of elastic, plastic and creep deformations of an initially isotropic material, *J. appl. Mech.*, **25** (1958), 529–36.
3. Besseling, J. F. A thermodynamic approach to rheology. In: *Proc. IUTAM-Symp. On Irreversible Aspects of Continuum Mechanics*, Springer, Wien, 1968, 16–53.
4. Besseling, J. F. Plasticity and creep theory in engineering mechanics. In: *Topics in Continuum Mechanics*, (Eds J. L. Zeman and F. Ziegler), Springer, Wien, 1974, 115–136.
5. Booij, J. and K. van der Werff. A description of plasticity experiments on austenitic stainless steel tubes, Delft University of Technology, Lab. Eng. Mech., 1975, WTHD 77.
6. Booij, J. Computer program for testing constitutive relations in creep and plasticity, Delft University of Technology, Lab. Eng. Mech., 1977, WTHD 104.
7. Booij, J. A study of the constitutive relations of aluminium alloy specimens loaded in plane stress (to be published in: *Plastic Behavior of Anisotropic Solids*, Ed. J. P. Boehler).

8. ECKART, C. The thermodynamics of irreversible processes IV: The theory of elasticity and anelasticity, *Phys. Rev.*, **73,** 2 (1948), 373–382.
9. HEIJDEN, A. M. A. VAN DER and J. F. BESSELING. A large strain plasticity theory and the symmetry properties of its constitutive equations. To be published in *Comp. and Struct.*
10. HUETINK, J. Evaluation of biaxial plasticity experiments on the ferritic steel WN 1.6770, TNO-Institute for Mechanical Constructions, Delft, 1979, Report nr. 5031021-79-4.
11. JANSSEN, G. T. M. Evaluation of biaxial plasticity experiments on the austenitic steel WN 1.4948, TNO-Institute for Mechanical Constructions, Delft, 1980, Report nr. 5031201-80-1.
12. LAMBERMONT, J. H. and J. F. BESSELING. An experimental and theoretical investigation of creep under uniaxial stress of a Mg alloy. In: *Proc. IUTAM-Symp. On Creep in Structures*, Springer, Berlin, 1972, 38–63.
13. LAMBERMONT, J. H. The creep equations of the niobium stabilized $2\frac{1}{4}$Cr-1Mo ferritic steel WN 1.6770, Part I, TNO-Institute for Mechanical Constructions, Delft, 1982, Report nr. 5011041-82-1.
14. MEIJERS, P. and F. ROODE. Experimental verification of constitutive equations for creep and plasticity based on overlay models, *J. Pressure Vessel Technology*, **105** (1983), 277–84.
15. GIESSEN, E. V.D., Private Communication, 1984.

7

On a Generalized Constitutive Law in Thermoplasticity

TH. LEHMANN

Lehrstuhl für Mechanik I, Ruhr-Universität Bochum, Federal Republic of Germany

ABSTRACT

A phenomenological theory of large, non-isothermic inelastic deformations of polycrystalline material can be based on a material model containing four different mechanisms. Adopting the assumption of local thermodynamical equilibrium the thermodynamical frame for the definition of the constitutive law is represented by the balance equations for the free enthalpy and the entropy. The constitutive law consists of:

(a) *state function for the specific free enthalpy governing at the same time the reversible part of the thermomechanical process;*
(b) *evolution laws for the remaining deformations;*
(c) *evolution laws for the internal variables;*
(d) *laws ruling the energy fluxes;*
(e) *laws of entropy production.*

One possible definition of a consistent generalized constitutive law is given and discussed with respect to some consequences.

1. INTRODUCTION

This paper deals with phenomenological considerations of large, non-isothermic deformations of solid polycrystalline solids within the frame of classical continuum mechanics and thermodynamics.

Such a phenomenological theory can be based on a material model

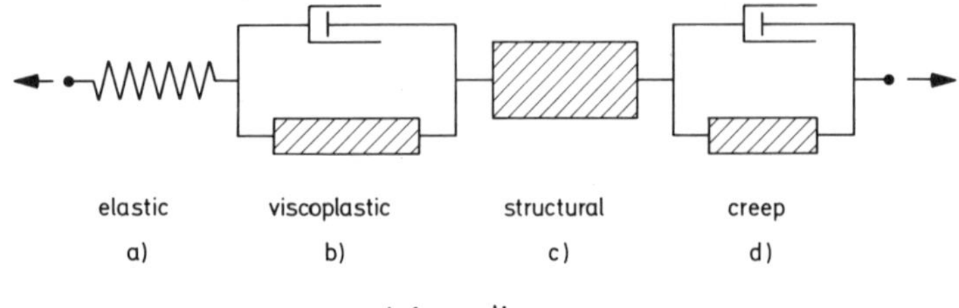

FIG. 1. Material model.

as shown in Fig. 1. It contains four different mechanisms in one row:

(a) an elastic element rendering reversible thermomechanical processes, which are governed by thermodynamical state equations;
(b) a complex viscoplastic element consisting of a rate insensitive plastic element (responsible also for certain changes of the accompanying constrained equilibrium state) and a parallel arranged rate sensitive viscous element;
(c) an element reflecting certain internal processes leading to changes of the material structure independent of viscoplastic flow and creep (structural deformations);
(d) a complex element representing unlimited long time creep and relaxation phenomena, particularly at high temperature, connected likewise with changes of internal material structure.

Of course, this material model can be refined. The elastic element, for instance, can be replaced by a viscoelastic one. Furthermore we can introduce more complex viscoplastic and creep elements. For many technical applications, however, a theory based on this model may be sufficient with respect to accuracy on the one hand and facility of handling on the other hand. Concerning the description of a thermomechanical process we have to distinguish between the description in the space of process variables and the accompanying description of such processes in the space of thermodynamical state. Within the frame of our phenomenological theory we suppose that the thermodynamical state of each material element is uniquely defined by the values of a finite set of well defined external and internal state variables. This means we adopt the assumption of local thermodynamical equilibrium even in irreversible processes due to non-equilibrium states of the body as the whole. A theory based on such an assumption, is limited to processes running not too far from thermodynamical equilibrium.

A thermomechanical process starts in the initial state $\mathring{\mathscr{L}}$ of the body defined by the initial geometrical configuration of the body and by the initial values of the thermodynamical state variables of each material element. It is governed by the history of the independent process variables. These are

(a) the prescribed thermomechanical boundary conditions,
(b) the prescribed specific body forces and energy sources.

The process description aspires to delivering (explicitly or implicitly) the values of the dependent process variables in the actual state $\mathscr{L}$ of the body. These dependent process variables are

(a) the non-prescribed thermomechanical quantities at the boundary,
(b) the displacements (strains), the stresses, and the temperature as well as the non-prescribed specific body forces and energy sources inside the body,
(c) all other interesting quantities, such as, for instance, the values of the internal state variables, the interchanged energies, etc.

The link between the initial state and the history of the independent process variables on the one hand and the resulting actual state on the other hand is formed by

1. the general field equations (material independent balance and compatibility equations),
2. the constitutive law.

We shall focus our considerations on the constitutive law.

The real thermomechanical process carries the body from the initial state $\mathring{\mathscr{L}}$ into the actual state $\mathscr{L}$. All physical quantities are acting in the respective current configuration of the body. Therefore we relate all quantities to a body-fixed co-moving co-ordinate system in its actual configuration. The real thermomechanical process embraces reversible parts governed by state equations and other parts which are not necessarily dissipative processes. We attach to the actual state $\mathscr{L}$ of the body an accompanying fictitious reference state $\overset{*}{\mathscr{L}}$ by a fictitious reversible process which carries each material element from its actual

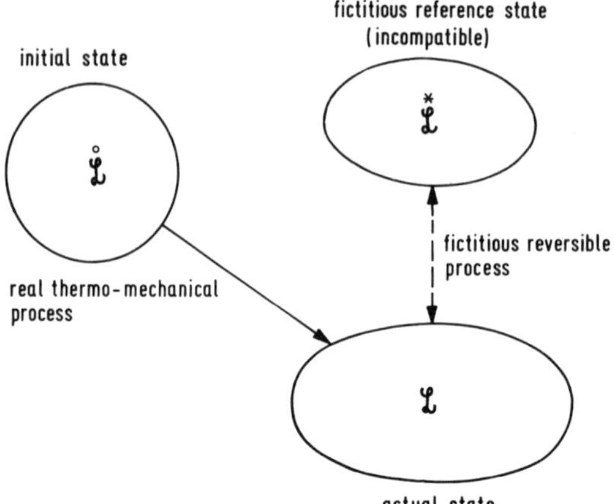

FIG. 2. Thermomechanical process.

thermodynamical state into an unstressed state at reference temperature $\overset{*}{T}$. During this fictitious process the internal variables are kept constant (Fig. 2).

A realization of this fictitious reversible process is, in general, impossible for two reasons:

1. The resulting reference state $\overset{*}{\mathscr{L}}$ represents in general an incompatible configuration.
2. A real process carrying a material element from the actual state into the reference state of stress and temperature may lead to changes of the internal variables which are excluded by our definition of the fictitious reversible process.

The defined fictitious process ensures a unique definition of what we call reversible energy. The energy really gained back, however, depends on the further history of the thermomechanical process.

From the preceding remarks it follows immediately that we can not define a (real or fictitious) process leading from the initial state $\overset{\circ}{\mathscr{L}}$ to the reference state $\overset{*}{\mathscr{L}}$. Therefore it becomes unnecessary to introduce any strain tensor defining the non-reversible deformations. We need, however, a unique decomposition of mechanical work into its reversible part and its remaining parts. Furthermore we require a unique measure for the reversible strains serving as thermodynamical state variables. In the following we shall at first concretize the mechanical and thermodynamical frame of our phenomenological theory. Then we shall discuss a possible form of the constitutive law followed by a simplified example for an elastic–plastic and a corresponding elasto–viscoplastic material. It will be shown from which fundamental experiments the parameters and functionals entering this constitutive law can be determined including some results. Finally some problems will be discussed briefly which are very sensitive with respect to the constitutive law.

2. MECHANICAL AND THERMODYNAMICAL FRAME

We introduce a body-fixed co-ordinate system ξ^i. The corresponding base vectors and metric are

in the initial configuration $\mathring{\mathscr{L}}$ at time $\mathring{t}$:

$$\mathring{\mathbf{g}}_i = \mathbf{g}_i(\xi^r, \mathring{t}), \qquad \mathring{g}_{ik} = g_{ik}(\xi^r, \mathring{t}) \tag{1}$$

in the actual configuration $\mathscr{L}$ at time t:

$$\mathbf{g}_i = \mathbf{g}_i(\xi^r, t), \qquad g_{ik} = g_{ik}(\xi^r, t) \tag{2}$$

The change of the metric can be measured by the quantities [7, 17]

$$q^i_k = \mathring{g}^{ir} g_{rk}, \qquad (q^{-1})^i_k = g^{ir}\mathring{g}_{rk} \tag{3}$$

which transform the metric according to the relations

$$g_{ik} = q^r_i \mathring{g}_{rk} = \mathring{g}_{ir} q^r_k, \qquad \mathring{g}_{ik} = (q^{-1})^r_i g_{rk} = \mathring{g}_{ir}(q^{-1})^r_k \tag{4}$$

The tensor

$$\mathbf{q} = q^i_k \mathbf{g}_i \otimes \mathbf{g}^k \tag{5}$$

corresponds to the left Cauchy tensor in mixed-variant notation.

Instead of the tensor $\mathbf{q}$ we may also introduce any other strain tensor by means of isotropic tensor functions, for instance,

$$\boldsymbol{\varepsilon} = \varepsilon^i_k \mathbf{g}_i \otimes \mathbf{g}^k = \tfrac{1}{2}(\ln q)^i_k \mathbf{g}_i \otimes \mathbf{g}^k \tag{6}$$

The deformation rate is defined by

$$\mathbf{d} = d^i_k \mathbf{g}_i \otimes \mathbf{g}^k = \tfrac{1}{2}(q^{-1})^i_r (\dot{q})^r_{\circ k} \mathbf{g}_i \otimes \mathbf{g}^k \tag{7}$$

where $(\cdot)$ means the partial derivative with respect to time, ξ^i held constant [7, 8].

The stress acting in the actual configuration is described by the Cauchy stress tensor

$$\boldsymbol{\sigma} = \sigma^i_k \mathbf{g}_i \otimes \mathbf{g}^k \tag{8}$$

With respect to energy considerations we introduce the weighted Cauchy stress tensor

$$\mathbf{S} = \frac{\overset{\circ}{\rho}}{\rho}\,\sigma^{i}_{\cdot k}\mathbf{g}_i \otimes \mathbf{g}^k = S^{i}_{\cdot k}\mathbf{g}_i \otimes \mathbf{g}^k \tag{9}$$

where ρ and $\overset{\circ}{\rho}$ denote the mass density in the actual state and in the initial state, respectively.

The total work rate is given by

$$\dot{w} = \frac{1}{\overset{\circ}{\rho}}\,S^{i}_{\cdot k}d^{k}_{\cdot i} = \frac{1}{\overset{\circ}{\rho}}\,S^{i}_{\cdot r}(q^{-1})^{r}_{\cdot k}\tfrac{1}{2}(\dot{q})^{k}_{\cdot i} \tag{10}$$

$S^{i}_{\cdot r}(q^{-1})^{r}_{\cdot k}$ and $\frac{1}{2}q^{i}_{\cdot k}$ represent a conjugate pair of stress and strain. In the case of isotropy also $S^{i}_{\cdot k}$ itself and $\varepsilon^{i}_{\cdot k}$ according to (6) can be introduced as a conjugate pair [17, 8].

The total work rate can be decomposed additively into the reversible part $\underset{(r)}{\dot{w}}$ and the remaining part $\underset{(i)}{\dot{w}}$ according to

$$\dot{w} = \underset{(r)}{\dot{w}} + \underset{(i)}{\dot{w}} \tag{11}$$

From (11) it follows that also the total deformation rate can be split

$$\dot{w} = \frac{1}{\overset{\circ}{\rho}}\,S^{i}_{\cdot k}d^{k}_{\cdot i} = \frac{1}{\overset{\circ}{\rho}}\,S^{i}_{\cdot k}\underset{(r)}{d^{k}_{\cdot i}} + \frac{1}{\overset{\circ}{\rho}}\,S^{i}_{\cdot k}\underset{(i)}{d^{k}_{\cdot i}} \tag{12}$$

with

$$d^{i}_{\cdot k} = \underset{(r)}{d^{i}_{\cdot k}} + \underset{(i)}{d^{i}_{\cdot k}}$$

Since the reversible mechanical work is governed by thermodynamical state equations $\underset{(r)}{\dot{w}}$ must be expressible in the form

$$\underset{(r)}{\dot{w}} = \frac{1}{\overset{\circ}{\rho}}\,S^{i}_{\cdot k}\underset{(r)}{d^{k}_{\cdot i}} = \frac{1}{\overset{\circ}{\rho}}\,\tilde{S}^{i}_{\cdot k}(\underset{(r)}{\dot{\tilde{\varepsilon}}})^{k}_{\cdot i} \tag{13}$$

where $\tilde{S}^{i}_{\cdot k}$ and $\underset{(r)}{\tilde{\varepsilon}^{i}_{\cdot k}}$ represent a suitable conjugate pair of stress and reversible strain.

The requirements which have to be fulfilled by the definition of $\underset{(r)}{\tilde{\varepsilon}^{i}_{\cdot k}}$ are discussed in refs. 9 and 8. It comes out that a multiplicative decomposition of the quantity $q^{i}_{\cdot k}$ introduced by (3) leads to a satisfying definition based on the following relations

$$q^i_k = \overset{\circ}{g}{}^{ir} g_{rk} = \overset{\circ}{g}{}^{im} \overset{*}{g}_{ms} \overset{*}{g}{}^{sr} g_{rk}$$

$$= \underset{(i)}{q}{}^i_{\cdot s} \underset{(r)}{q}{}^s_k \tag{14}$$

where the superposed * relates to the fictitious reference configuration $\overset{*}{\mathscr{L}}$. Restricting ourselves to isotropic reversible behaviour from (14) we derive

$$\underset{(r)}{q}{}^i_k = \overset{*}{g}{}^{im} g_{mk}, \quad \underset{(r)}{d}{}^i_{\cdot k} = (\underset{(r)}{q}{}^{-1})^i_r \tfrac{1}{2} (\underset{(r)}{\dot{q}})^r_{\cdot k} \tag{15}$$

$$\left.\begin{aligned} \underset{(r)}{\dot{w}} &= \frac{1}{\overset{\circ}{\rho}} S^i_k \underset{(r)}{d}{}^k_i = \frac{1}{\overset{\circ}{\rho}} S^i_r (\underset{(r)}{q}{}^{-1})^r_k \underset{(r)}{\overset{\nabla}{q}}{}^k_i = \frac{1}{\overset{\circ}{\rho}} S^i_k \underset{(r)}{\overset{\nabla}{\varepsilon}}{}^k_i \\ \text{with} \qquad & \\ \underset{(r)}{\varepsilon}{}^i_k &= \tfrac{1}{2} (\ln \underset{(r)}{q})^i_k \end{aligned}\right\} \tag{16}$$

where the superscript ∇ denotes the covariant time derivative which is defined by

$$\overset{\nabla}{q}{}^i_k = (\dot{q})^i_{\cdot k} + d^i_r q^r_k - d^r_k q^i_r \tag{17}$$

With the notation
u: specific internal energy
q^i: energy flux
r: specific energy sources
the first law of thermodynamics states

$$\dot{u} = \dot{w} - \frac{1}{\rho} q^i \bigg|_i + r = \underset{(r)}{\dot{w}} + \underset{(i)}{\dot{w}} - \frac{1}{\rho} q^i \bigg|_i + r \tag{18}$$

The energy flux includes heat flux and other energy fluxes, which may, for instance, be due to diffusion of self-equilibrated microstress fields. Such energy fluxes different from heat may be mostly small in solid bodies and therefore negligible in many cases. However, they have to be taken into account in general considerations.

Within the frame of our assumptions u must be expressible as a unique function of a finite set of thermodynamical state variables. This set may consist of

$\underset{(r)}{\varepsilon}{}^i_k$ reversible strain,

s specific entropy,

b, β^i_k internal variables representing an arbitrary set of such variables.

Therefore we can write

$$u = u(\underset{(r)}{\varepsilon^i_k}, s, b, \beta^i_k) \tag{19}$$

With respect to the following considerations it is advantageous to replace $\underset{(r)}{\varepsilon^i_k}$ and s by their conjugated state variables, i.e. the stress S^i_k according to relation (16) and the absolute temperature T. This can be achieved by a double Legendre transformation leading to the definition of specific free enthalpy

$$\psi = u - \frac{1}{\mathring{\rho}} S^i_k \underset{(r)}{\varepsilon^k_i} - Ts = \psi(S^i_k, T, b, \beta^i_k) \tag{20}$$

From the properties of the Legendre transformation we derive immediately:

thermic state equation:

$$\underset{(r)}{\varepsilon^i_k} = -\mathring{\rho} \frac{\partial \psi}{\partial S^k_i} = \underset{(r)}{\varepsilon^i_k}(S^i_k, T, b, \beta^i_k) \tag{21}$$

caloric state equation:

$$s = -\frac{\partial \psi}{\partial T} = s(S^i_k, T, b, \beta^i_k) \tag{22}$$

Using these relations furthermore we obtain from (18)–(22) (for details see refs. 8 and 10) the balance equation for reversible specific work:

$$\underset{(r)}{\dot{w}} = -S^i_k \left\{ \frac{\partial^2 \psi}{\partial S^r_s \, \partial S^i_k} \overset{\nabla}{S}{}^r_s + \frac{\partial^2 \psi}{\partial T \, \partial S^i_k} \dot{T} + \frac{\partial^2 \psi}{\partial b \, \partial S^i_k} \dot{b} + \frac{2}{\partial \beta^r_s \, \partial S^i_k} \overset{\nabla}{\beta}{}^r_s \right\} \tag{23}$$

the balance equation for residual specific energy supply:

$$\begin{aligned} \underset{(i)}{\dot{w}} - \frac{1}{\rho} q^i|_i + r = {} & -T \left\{ \frac{\partial^2 \psi}{\partial S^i_k \, \partial T} \overset{\nabla}{S}{}^i_k + \frac{\partial^2 \psi}{\partial T^2} \dot{T} \right\} \\ & + \frac{\partial}{\partial b} \left\{ \psi - T \frac{\partial \psi}{\partial T} \right\} \dot{b} + \frac{\partial}{\partial \beta^i_k} \left\{ \psi - T \frac{\partial \psi}{\partial T} \right\} \overset{\nabla}{\beta}{}^i_k \end{aligned} \tag{24}$$

and finally the balance equation for specific entropy (Gibbs equation):

$$T\dot{s} = \underset{(i)}{\dot{w}} - \frac{1}{\rho} q^i|_i + r - \frac{\partial \psi}{\partial b} \dot{b} - \frac{\partial \psi}{\partial \beta^i_k} \overset{\nabla}{\beta}{}^i_k \tag{25}$$

In the balance equation (25) for specific entropy we have to decompose $T\dot{s}$ into its reversible part $T\underset{(r)}{\dot{s}}$ and its irreversible (dissipative) part $T\underset{(d)}{\dot{s}}$. The irreversible part $T\underset{(d)}{\dot{s}}$ consists of

1. the rate of immediately dissipated specific work

$$\underset{(d)}{\dot{w}} = \underset{(i)}{\dot{w}} - \underset{(h)}{\dot{w}} \tag{26}$$

where $\underset{(h)}{w}$ denotes the specific mechanical work stored in the internal structure of the material;

2. the irreversible part of heat flux $\underset{(T)}{q^i}$, i.e.

$$-\frac{1}{\rho T}\underset{(T)}{q^i}\, T|_i = -\frac{1}{\rho T}(q^i - \underset{(h)}{q^i})T|_i \tag{27}$$

where $\underset{(h)}{q^i}$ represents the energy fluxes apart from heat;

3. the entropy production (summarized in $T\dot{\eta}$) due to dissipative processes involved in remaining energy fluxes $\underset{(h)}{q^i}$, in energy supply by sources r, and in changes of the internal material structure.

With these notations the second law of thermodynamics reads

$$T\underset{(d)}{\dot{s}} = \underset{(d)}{\dot{w}} - \frac{1}{\rho T}\underset{(T)}{q^i}\, T|_i + T\dot{\eta} \geqslant 0 \tag{28}$$

3. A POSSIBLE GENERALIZED FORM FOR THE CONSTITUTIVE LAW IN THERMOPLASTICITY

Within the frame given by the relations (26)–(28) the constitutive law has to be defined. It consists of

(a) state function for the specific free enthalpy determining also immediately the reversible processes;
(b) evolution laws for the different components of the remaining (inelastic) deformations;
(c) evolution laws for the internal variables;
(d) flux laws for the different kinds of energy (heat flux and others);
(e) laws of entropy production, specifying $\underset{(d)}{\dot{w}}$ and $\dot{\eta}$.

In the following we discuss a possible form for the constitutive law of elastic–plastic and elastic–viscoplastic materials. This constitutive law is generalized in some respects in comparison with usual formulation (see, for instance, refs. 1, 2, 4, 6, 13–16 and 18).

For simplicity we disregard long time creep or relaxation processes reflected by part (d) of the material model. Likewise solid phase transformations are not taken into account. Such phenomena, however, can be inserted into the constitutive law supplementarily without changes in its general structure.

We suppose the specific enthalpy can be written in the form

$$\psi(S^i_k, T, b, \beta^i_k) = \psi^*(S^i_k, T) + \psi^{**}(T, b, \beta^i_k) \tag{29}$$

This means that the thermoelastic behaviour is not influenced by changes of the internal variables. This implies that the thermoelastic behaviour is isotropic. For polycrystalline materials this assumption seems to be justified in many cases.

From (29) we derive the thermic state equation (21) which can be transformed into an incremental law of the form

$$\underset{(r)}{d^i_k} = \underset{(r)}{d^i_k}(S^i_k, T, \overset{\nabla}{S}{}^i_k, \dot{T}) \tag{30}$$

In many cases we can approximate this law by a simpler hypoelastic law [8, 11]

$$\underset{(r)}{d^i_k} = \frac{1}{2G}\overset{\nabla}{t}{}^i_k + \frac{1}{9K}(\dot{S})^r_r + \alpha\dot{T}\,\delta^i_k \tag{31}$$

In this relation G is shear modulus, K is bulk modulus, t^i_k is weighted stress deviator, α is coefficient of thermal expansion.

The inelastic work and deformation rate, which are not governed by state equations, can be decomposed according to the introduced material model into

$$\underset{(i)}{\dot{w}} = \underset{(p)}{\dot{w}} + \underset{(s)}{\dot{w}} + \underset{(c)}{\dot{w}} \tag{32}$$

$$\underset{(i)}{d^i_k} = \underset{(p)}{d^i_k} + \underset{(s)}{d^i_k} + \underset{(c)}{d^i_k} \tag{33}$$

The indices in parentheses relate to the following processes:

(p) plastic or viscoplastic processes,
(s) additional processes leading to certain changes of the internal structure,
(c) long time creep or relaxation processes.

As already mentioned we disregard the last component. Concerning the two other components we assume that their occurrence is subjected to certain constraints (yield conditions):

$$\underset{(p)}{F}(S^i_k, T, b, \beta^i_k)\begin{cases} =0 & \text{(plastic)} \quad (34a) \\ >0 & \text{(viscoplastic)} \quad (34b) \end{cases}$$

$$\underset{(s)}{F}(S^i_k, T, b, \beta^i_k)=0 \tag{35}$$

The corresponding evolution laws for the respective deformations can be written in the general form

$$\underset{(p)}{d^i_k} = \begin{cases} \lambda \dfrac{\partial \underset{(p)}{F}}{\partial S^k_i} & \text{(plastic)} \quad (36a) \\[2ex] \Phi(\underset{(p)}{F}) \dfrac{\partial \underset{(p)}{F}}{\partial S^k_i} & \text{(viscoplastic)} \quad (36b) \end{cases}$$

$$\underset{(s)}{d^i_k} = \underset{(s)}{d^i_k}(S^i_k, T, b, \beta^i_k, \overset{\nabla}{S}{}^i_k, \dot{T}) \tag{37}$$

It must be emphasized that the yield conditions $\underset{(p)}{F}$ and $\underset{(s)}{F}$ and the respective evolution laws of the inelastic deformations represent different mechanisms. Plastic and viscoplastic deformations are characterized by slip processes; their direction is governed by the actual stresses. The separate deformations represented by $\underset{(s)}{d^i_k}$, however, correspond essentially to the production and redistribution of lattice defects; thus their direction is mainly governed by the stress increments. The magnitude of the deformation rate $\underset{(s)}{d^i_k}$ is in the order of the reversible deformation rate $\underset{(r)}{d^i_k}$ or slightly larger. Therefore in proportional loading histories the plastic or viscoplastic deformations prevail. On the other hand the sensitivity to yield according to the mechanism producing $\underset{(s)}{d^i_k}$ is higher in comparison with plastic or viscoplastic deformations. The situation is sketched in Fig. 3. The yield surface $\underset{(s)}{F}$ can be detected experimentally only after partial unloading; it requires a very high accuracy of measurement [13, 5]. After total unloading something between $\underset{(s)}{F}$ (in its actual configuration) and $\underset{(p)}{F}$ is determined

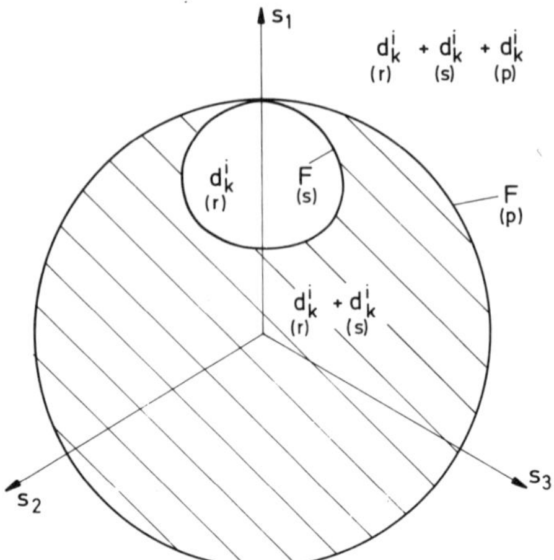

FIG. 3. Yield surfaces $\underset{(s)}{F}$ and $\underset{(p)}{F}$.

as yield surface depending on the definition of yield [5, 3] (see Figs. 4 and 5). The existence of two independent yield mechanisms explains also the observable hysteresis in unloading and reloading [8] and some phenomena in other cyclic processes. Of course the real material behaviour is still much more complex. We may mention the influence of time on the determination of yield surfaces [13, 5]. However, we can neglect such influences in many cases.

The evolution laws for the internal variables represent balance equations. This means they are first-order partial differential equations containing also spatial gradients of the state variables. If, however, the energy fluxes apart from heat can be disregarded the balance equations reduce to simple evolution laws, i.e. to first order ordinary differential equations in time of the form

$$\dot{b} = \dot{b}(S^i_{\ k}, T, b, \beta^i_{\ k}; \overset{\nabla}{S}{}^i_k, \dot{T}) \tag{38a}$$

$$\overset{\nabla}{\beta}{}^i_k = \overset{\nabla}{\beta}{}^i_k(S^i_{\ k}, T, b, \beta^i_{\ k}; \overset{\nabla}{S}{}^i_k, \dot{T}) \tag{38b}$$

This simplification may be supposed in the following. The evolution laws can be classified according to whether or not the increments $\overset{\nabla}{S}{}^i_k$ and $\dot{T}$ enter the right hand side [8].

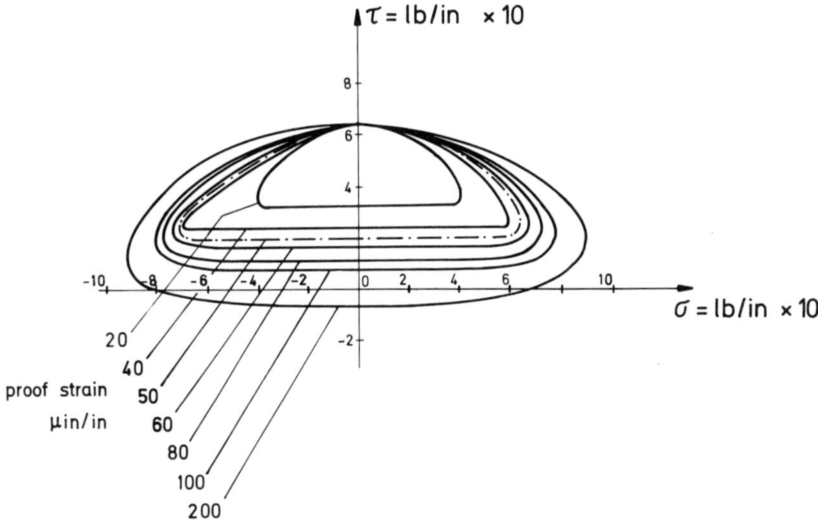

FIG. 4. Effect of proof strain on subsequent yield surface after partial unloading (Ikegami [5]).

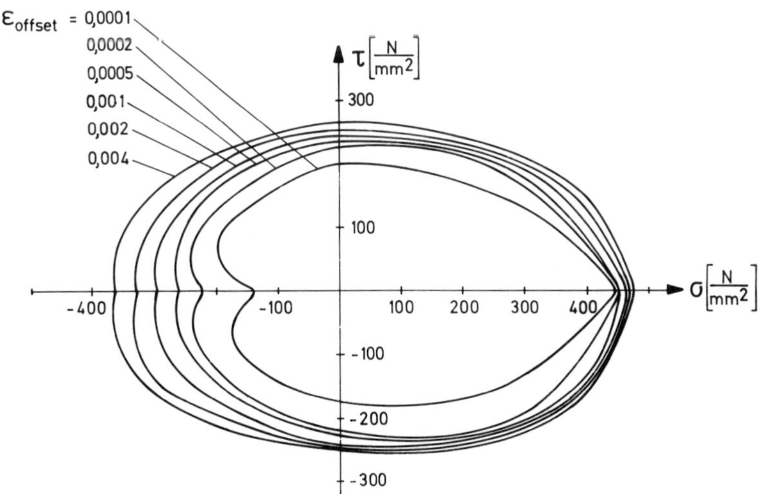

FIG. 5. Effect of proof strain on subsequent yield surface, after total unloading (Gupta and Lauert [3]).

It remains to define the flux laws of energy and the laws of entropy production. Disregarding other fluxes only the flux law for heat is left. It may be assumed to be in the usual form (omitting the subscript T)

$$q^i = -\bar{\lambda} T|_i \tag{39}$$

where $\bar{\lambda}$ represents the coefficient of heat conduction. The definitions of $\underset{(d)}{\dot{w}}$ and $\dot{\eta}$ will be discussed in connection with the following example.

4. A SIMPLIFIED EXAMPLE

We consider an elastic–plastic material. The extension to elastic–viscoplastic materials is achieved easily [8]. When we are interested only in large deformations we may neglect the small deformations occurring between the yield surfaces $\underset{(s)}{F}$ and $\underset{(p)}{F}$. This does not mean, however, that we can disregard the deformation rate $\underset{(s)}{d^i_k}$ totally. This point will be discussed later.

We assume the specific enthalpy (29) given in the form

$$\Psi(S^i_k, T, B, \beta^i_k) = \Psi^*(S^i_k, T) + \underset{(b)}{\Psi^{**}}(T, b) + \underset{(\beta)}{\Psi^{**}}(T, B) \tag{40}$$

with

$$B = \beta^i_k \beta^k_i$$

The evolution law for the reversible deformation can be derived from (40) or may be assumed in the hypoelastic approximation (31). The balance equation (24) now reads

$$\underset{(h)}{\dot{w}} + \underset{(d)}{\dot{w}} - \frac{1}{\rho} q^i|_i + r = c_p \dot{T} + B^k_i \overset{\nabla}{S^i_k} + h\dot{b} + g 2\beta^i_k \overset{\nabla}{\beta^k_i} \tag{41}$$

with

$$c_p = -T\frac{\partial^2 \Psi}{\partial T^2}, \; B^k_i = -T\frac{\partial^2 \Psi^*}{\partial S^k_i \, \partial T}$$

$$h = \frac{\partial}{\partial b}\left\{\underset{(b)}{\Psi^{**}} - T\frac{\partial \underset{(b)}{\Psi^{**}}}{\partial T}\right\}, \; g = \frac{\partial}{\partial B}\left\{\underset{(\beta)}{\Psi^{**}} - T\frac{\partial \underset{(\beta)}{\Psi^{**}}}{\partial T}\right\}$$

Neglecting the deformations between the yield surfaces $\underset{(s)}{F}$ and $\underset{(p)}{F}$ we suppose a unified yield condition of the form

$$\underset{(p)}{F} = \underset{(s)}{F} = F = (t^i_k - \beta^i_k)(t^k_i - \beta^k_i) - k^2(b, T) = f(t^i_k, \beta^i_k) - k^2(b, T) = 0 \tag{42}$$

The continuation of inelastic deformation requires

$$\begin{aligned}\dot{F} = 0 &= \frac{\partial F}{\partial S^i_k} \overset{\nabla}{S}{}^i_k + \frac{\partial F}{\partial T} \dot{T} + \frac{\partial F}{\partial b} \dot{b} + \frac{\partial F}{\partial \beta^i_k} \overset{\nabla}{\beta}{}^i_k \\ &= 2(t^i_k - \beta^i_k)(\overset{\nabla}{t}{}^k_i - \overset{\nabla}{\beta}{}^k_i) - \frac{\partial k^2}{\partial T} \dot{T} - \frac{\partial k^2}{\partial b} \dot{b}\end{aligned} \tag{43}$$

Concerning the evolution for plastic deformations we obtain according to (36a)

$$\underset{(p)}{d^i_k} = \lambda \frac{\partial F}{\partial S^k_i} = \lambda 2(t^i_k - \beta^i_k) \tag{44}$$

With respect to $\underset{(s)}{d^i_k}$ we assume

$$\underset{(s)}{d^i_k} = \kappa(\overset{\nabla}{t}{}^i_k - \overset{\nabla}{\beta}{}^i_k) \tag{45}$$

This means we consider these deformations to be governed by the changes of the effective stress $\tilde{t}^i_k = t^i_k - \beta^i_k$.

Concerning the evolution laws for the internal variables we suppose

$$\dot{b} = \frac{\delta}{\overset{\circ}{\rho}} \tilde{t}^i_k \underset{(s)}{d^k_i} + \frac{\mu}{\overset{\circ}{\rho}} \tilde{t}^i_k \underset{(p)}{d^k_i} - b \underset{(b)}{\vartheta}$$

$$\text{with side condition } \underset{(b)}{\vartheta} \begin{cases} >0 \text{ for } T > T_R(S^i_k, b, \beta^i_k) \\ =0 \text{ for } T \leq T_R(S^i_k, b, \beta^i_k) \end{cases} \tag{46}$$

$$\overset{\nabla}{\beta}{}^i_k = \zeta \underset{(s)}{d^i_k} + \nu \underset{(p)}{d^i_k} - \beta^i_k \underset{(\beta)}{\vartheta}$$

$$\text{with side condition } \underset{(\beta)}{\vartheta} \begin{cases} >0 \text{ for } T > T_R(S^i_k, b, \beta^i_k) \\ =0 \text{ for } T \leq T_R(S^i_k, b, \beta^i_k) \end{cases} \tag{47}$$

This means an interaction between hardening or softening due to inelastic deformations on the one hand and annealing processes like recrystallization on the other hand which represent internal dissipative processes.

Moreover, the larger part of plastic work is dissipated. Therefore we have to put for the entropy production (with $r=0$)

$$T\dot{\eta} = h\underset{(b)}{b\vartheta} + g2B\underset{(\beta)}{\vartheta} \tag{48}$$

$$\underset{(d)}{\dot{w}} = \frac{1-\xi}{\mathring{\rho}} t^i_k \underset{(p)}{d^k_i} \quad \text{with} \quad 0 \leqslant \xi \ll 1 \tag{49}$$

Within the balance equation (41) now we have to balance $\underset{(h)}{\dot{w}}$ with the non-dissipative terms of $h\dot{b} + g2\beta^k_i \overset{\nabla}{\beta^i_k}$.

This yields

$$\begin{aligned}\underset{(h)}{\dot{w}} = \underset{(i)}{\dot{w}} - \underset{(d)}{\dot{w}} &= \frac{1}{\mathring{\rho}} t^i_k \underset{(s)}{d^k_i} + \frac{\xi}{\mathring{\rho}} t^i_k \underset{(p)}{d^k_i} \\ &= \left\{\frac{\delta h}{\mathring{\rho}}(t^i_k - \beta^i_k) + \zeta g 2\beta^i_k\right\} \underset{(s)}{d^k_i} \\ &\quad + \left\{\frac{\mu h}{\mathring{\rho}}(t^i_k - \beta^i_k) + \nu g 2\beta^i_k\right\} \underset{(p)}{d^k_i}\end{aligned} \tag{50}$$

From (57) we conclude

$$\begin{aligned}\delta h = 1, \ \zeta &= \frac{1}{2\mathring{\rho} g} \\ 0 \leqslant \mu h = \xi \ll 1, \ \nu &= \frac{\xi}{2\mathring{\rho} g}\end{aligned} \tag{51}$$

With these results following from energy balance and using eqns. (44) and (45) the evolution laws for the internal variables read

$$\begin{aligned}\dot{b} = \frac{1}{1+\kappa\zeta}\frac{\kappa}{\mathring{\rho} h}\{(t^i_k - \beta^i_k)\overset{\nabla}{t^k_i} - \xi\zeta 2k^2\lambda + (t^i_k - \beta^i_k)\beta^k_i \underset{(\beta)}{\vartheta}\} \\ + \frac{\xi}{\mathring{\rho} h} 2k^2\lambda - b\underset{(b)}{\vartheta}\end{aligned} \tag{52}$$

$$\overset{\nabla}{\beta^i_k} = \frac{1}{1+\kappa\zeta}\{\kappa\zeta \overset{\nabla}{t^i_k} + \xi\zeta 2(t^i_k - \beta^i_k)\lambda - \beta^i_k \underset{(\beta)}{\vartheta}\} \tag{53}$$

Inserting the expressions (52) and (53) into the consistency condition (43) we obtain an equation for λ. Furthermore we can derive from the consistency condition (43) the so-called loading condition on account of the requirement $\lambda > 0$. With these results for λ finally we get the explicit forms of the evolution laws (52) and (53) for the internal variables and of the evolution laws (44) and (45) for the inelastic deformations. Adding the law (39) for heat flux the constitutive law is defined totally.

We recognize that the following functions enter the set of evolution laws for the internal variables and the inelastic deformations:

$$h = \frac{\partial}{\partial b}\left\{ \underset{(b)}{\psi^{**}} - T\frac{\partial \underset{(b)}{\psi^{**}}}{\partial T}\right\} = h(b, T) \tag{54a}$$

$$g = \frac{\partial}{\partial B}\left\{ \underset{(\beta)}{\psi^{**}} - T\frac{\partial \underset{(\beta)}{\psi^{**}}}{\partial T}\right\} = g(B, T) = \frac{1}{2\mathring{\rho}\zeta} \tag{54b}$$

$$k^2 = k^2(b, T) \tag{54c}$$

$$\kappa = \kappa(S^i_k, T, b, \beta^i_k) \tag{54d}$$

$$\xi = \xi(S^i_k, T, b, \beta^i_k) \tag{54e}$$

$$\underset{(b)}{\vartheta} = \underset{(b)}{\vartheta}(S^i_k, T, b, \beta^i_k), \qquad \underset{(\beta)}{\vartheta} = \underset{(\beta)}{\vartheta}(S^i_k, T, b, \beta^i_k) \tag{54f}$$

These functions (or parameters, respectively) have to be determined experimentally. It may be mentioned that in such cases where $\underset{(b)}{\psi^{**}}(b, T)$ can be decomposed into two terms depending only on b or T, respectively, it can be taken that $h = 1$ by a suitable choice of the internal variable b.

The preceding considerations can be easily extended to elastic–viscoplastic materials. We have only to replace the evolution law (44) for the plastic deformations by the corresponding evolution law for viscoplastic deformations based on (36b). This is done in ref. 8. The set of evolution equations becomes simpler in this case. The fundamental difference, however, is that the evolution law for the viscoplastic deformations is rate dependent in any case and not only in the case of rate-dependent internal processes like recrystallization.

5. FUNDAMENTAL EXPERIMENTS CONCERNING THE CONSTITUTIVE LAW

We focus our considerations on the determination of the functions (54a)–(54e). Recrystallization processes and other internal processes can be investigated separately. Therefore we disregard such processes in the following. The quantities (54a)–(54c) can be calculated from the investigation of subsequent yield surfaces after different preloading histories at different temperature levels.

The quantity ξ (54e) (characterizing the non-dissipative part of plastic work) plays a subordinate role in the hardening laws. We may assume ξ as a constant of magnitude $\approx 0{\cdot}1$. The quantity κ has a very small influence on processes with proportional loading like, for instance, simple tension tests. The influence becomes important, however, in processes with rotating principal stress axes (as in shear processes) and in processes with abrupt changes of the loading path. Therefore experiments with such processes at different temperature levels can deliver suitable information about κ (S^i_k, T, b, β^i_k).

We have started an extended experimental research program sponsored by Deutsche Forschungsgemeinschaft and by Stiftung Volkswagenwerk, to identify the material functions and parameters entering the preceding formulated constitutive law and the corresponding law for elastic–viscoplastic material, particularly, for mild steel and one aluminium alloy in the first step. First results seem to indicate:

1. In proportional loading histories the anisotropic hardening rate decreases faster than the isotropic hardening rate. In cyclic loading processes the isotropic hardening rate tends to zero whereas the anisotropic hardening tends to cyclic behaviour.
2. The function κ (54d) seems to be restricted roughly to the range

$$0 < \kappa 2G < 10 \tag{55}$$

   This means the occurrence of the deformation rate $\underset{(s)}{d^i_k}$, governed in its magnitude by κ, has the effect that the shear modulus G seems to be diminished during plastic deformations.
3. The isotropic and the anisotropic hardening rate as well as κ decrease with increasing temperature.

The proposed phenomenological theory seems to prove a suitable frame for the description of the thermomechanical processes under consideration. It requires, however, much more experimental work to obtain reliable quantitative information about the material functions entering the constitutive law.

6. CONCLUDING REMARKS

The proposed generalized phenomenological theory of thermoplasticity tries to take into account some phenomena not covered by the classical theory, such as, for instance, the deviations from the so-called normality rule. On the other hand this theory intends to be applicable to technical problems without a tremendous increase of expenditure. Therefore the concept of yield condition is retained. The flexibility of the theoretical frame, however, is much enlarged by introducing an additional yield mechanism (even without a separate yield condition) allowing for (small) deviations from the normality rule.

These effects do not play an important role in proportional loading processes. They become, however, essential in non-proportional loading processes and in processes with rotating principal stress axes against the material elements. The introduction of a separate yield condition $\underset{(s)}{F}$ different from $\underset{(p)}{F}$ is only necessary when we want to investigate the transition from elastic to plastic deformations after unloading processes. The occurrence of a second yield mechanism governed by the stress increments, however, remains in all mentioned cascs, csscntial. It can bc shown [12] that, for instancc, sccond ordcr effects (Poynting effects) in shear processes, the behaviour after abrupt changes of the loading path, and bifurcation problems can be described with better agreement with experimental results than the classical theory of plasticity does.

REFERENCES

1. BEEVERS, C. E. and J. BREE, A thermodynamic theory of isotropic elastic–plastic materials, *Arch. Mech.*, **31** (1979), 385–395.
2. GREEN, A. E. and P. M. NAGHDI, A thermodynamic development of elastic–plastic continua. In: *Proc. IUTAM Symp. on Irreversible Aspects of Continuum Mechanics on Transfer of Physical Characteristics in Moving Fluids* (Ed. H. Parkus and L. Sedov), Springer, Berlin, 1968, 117–131.
3. GUPTA, N. K. and H.-A. LAUERT, A study of yield surface upon reversal of loading under biaxial stress, *ZAMM*, **63** (1983), 497–504.
4. HART, E. W. Constitutive relations for the nonelastic deformation of metals, *Trans. ASME, J. Engng Mater. Technol.*, **98** (1976), 193–202.
5. IKEGAMI, K. Experimental plasticity on the anisotropy of metals. In: Mechanical behaviour of anisotropic solids, *Proc. Euromech. Coll. 115* (Ed. J. P. Boehler), Martinus Nijhoff, The Hague, 1982, 201–242.

6. LEE, E. H. Elastic–plastic deformations at finite strain, *J. appl. Mech.*, **36** (1969), 1–6.
7. LEHMANN, TH. Some remarks on kinematics and constitutive equations of inelastic solids. In: *Mechanics of Inelastic Media and Structures* (Ed. O. Mahrenholtz and A. Sawczuk), Pol. Sci. Publ. Warszawa, 1982, 161–178.
8. LEHMANN, TH. General frame for the definition of constitutive laws for large non-isothermic elastic–plastic and elastic–viscoplastic deformations. In: *The Constitutive Law in Thermoplasticity* (Ed. Th. Lehmann), CISM-Courses and Lecture No. 281, Springer, Wien, 1984.
9. LEHMANN, TH. Some remarks on the decomposition of deformations and mechanical work, *Int. J. Engng Sci.*, **20** (1982), 281–288.
10. LEHMANN, TH. On the concept of stress–strain relations in plasticity, *Acta Mech.*, **42** (1982), 263–275.
11. LEHMANN, TH. On large elastic–plastic deformations. In: *Foundation of Plasticity* (Ed. A. Sawczuk), Noordhoff, Leyden, 1973, 571–585.
12. LEHMANN, TH. Some theoretical considerations and experimental results concerning elastic–plastic stress–strain relations, *Ing. Arch.*, **52** (1982), 391–403.
13. PHILLIPS, A. The foundation of thermoplasticity—experiments and theory. In: *Topics in Applied Continuum Mechanics* (Ed. J. L. Zeman and F. Ziegler), Springer, Wien, 1974.
14. PERZYNA, P. Thermodynamic theory of viscoplasticity. In: *Advances in Applied Mechanics*, Vol. 11, Academic Press, New York, 1971, 313–354.
15. RICE, J. R. Inelastic constitutive relations for solids: An internal-variable theory and its application to metal plasticity, *J. Mech. Phys. Solids*, **19** (1971), 433–455.
16. SEDOV, L. I. *Introduction to the Mechanics of a Continuous Medium* Addison-Wesley, London, 1965.
17. THERMANN, K. Foundations of large deformations. In: *The Constitutive Law in Thermoplasticity* (Ed. Th. Lehmann), CISM Courses and Lecture No. 281, Springer, Wien, 1984.
18. TING, E. C. A thermodynamical theory for finite elastic–plastic deformations, ZAMP, **22** (1971), 702–713.

8

A Missing Link in the Macroscopic Constitutive Formulation of Large Plastic Deformations

YANNIS F. DAFALIAS

Department of Civil Engineering, University of California, Davis, USA

ABSTRACT

On the basis of the tensorial representation of isotropic functions, the general form of macroscopic constitutive relations for plastic spin is discussed, offering a missing link between Mandel's theory of director vectors and its possible application. Based on the foregoing, the overall constitutive formulation is presented in general terms for small elastic but large plastic deformations extending some of Mandel's results, introducing damage or elastoplastic coupling and unifying the approach for initially isotropic and anisotropic materials. The preceding are illustrated and compared with other approaches by performing a detailed analysis of large simple shear for a material which hardens kinematically.

1. INTRODUCTION

The definition of the proper co-rotational rates for the stress and internal structure tensorial variables is still being debated in the macroscopic constitutive formulation of large deformation elastoplasticity. In his elucidative work Mandel [14] supplemented the notion of the multiplicative decomposition of the deformation gradient into an elastic and a plastic part, introduced by Lee and Liu [11], with the notion of director vectors attached to the material substructure in the relaxed configuration. Motivated by well understood concepts of the micromechanics of crystalline structures, Mandel defined as proper co-rotational rate the one associated with the spin of the triad of the director vectors. However, he never tried to present a systematic macroscopic derivation of analytical expressions for this spin and, instead, he attempted to approach the problem from microstructural considerations, a formidable task not yet fully developed. The principal objective of this work is to provide this missing link for the macroscopic application of Mandel's theory, to extend and generalize certain aspects of it and illustrate it by examples.

An immediate consequence of the notion of the preceding spin, is that constitutive relations are required not only for the plastic rate of deformation but also for the plastic spin which when added to the spin of the director vectors yields the material spin. Thus, our focus will be the macroscopic formulation of constitutive relations for the plastic spin which becomes the key to our initial objective. This is achieved by using the representation theorems for isotropic functions, particularly as presented in ref. 16. Some recent advances in these theorems allow us to unify under certain conditions the constitutive formulation for initially isotropic and anisotropic materials and reach certain novel conclusions not foreseen by Mandel. The concept of damage and/or elasoplastic coupling offers a new element in this formulation which is presented in direct and inverse form for small elastic but large plastic deformations. The foregoing are illustrated by an example on a material which hardens kinematically. For such a material the detailed analysis of large simple shear is performed for three other co-rotational rates and the results are compared with those obtained for a co-rotational rate defined according to the present general development. In most cases it was possible to obtain in closed form the analytical expressions of the developing stresses for rigid-plastic material response. The relative merits of the different rates are discussed.

Tensors (first, second and fourth order) will be denoted by bold-face characters in direct notation. Their juxtaposition implies the usual summation operation over two adjacent indices, while dots and double dots indicate the summation products $\mathbf{a} \cdot \boldsymbol{\sigma} = a_{ij}\sigma_{ij}$, $\mathbf{a} : \boldsymbol{\sigma} = a_{ij}\sigma_{ji}$, the former applied to as many indices as desired for different order tensors. An operation over a representative element of a set is understood as being repeated over all pertinent elements of the set. A superposed dot indicates the rate, a superscript T the transpose and the symbol $\otimes$ the tensor product.

2. BASIC KINEMATICS

All quantities will be referred to a fixed cartesian co-ordinate system $\mathbf{x}$ unless stated otherwise. The co-rotational rate of a second order tensor $\mathbf{a}$ and a vector $\mathbf{m}$ with respect to an antisymmetric tensor $\boldsymbol{\Omega}$ (usually representing a spin) is denoted by a superposed symbol, e.g. $\circ$, and defined by

$$\overset{\circ}{\mathbf{a}} = \dot{\mathbf{a}} + \mathbf{a}\boldsymbol{\Omega} - \boldsymbol{\Omega}\mathbf{a} \tag{1a}$$

$$\overset{\circ}{\mathbf{m}} = \dot{\mathbf{m}} - \boldsymbol{\Omega}\mathbf{m} \tag{1b}$$

The multiplicative decomposition of the deformation gradient into an elastic part $\mathbf{V}$ and a plastic part $\mathbf{P}$ [11] with respect to the particular relaxed configuration κ_0 for which $\mathbf{V} = \mathbf{V}^{\mathrm{T}}$ [14] is given by

$$\mathbf{F} = \mathbf{VP} \tag{2}$$

Mandel [14] supplemented the preceding notion with the concept of a triad of director vectors attached to the material substructure at κ_0, and defined by the orthogonal transformation $\mathbf{x} = \boldsymbol{\beta}\hat{\mathbf{x}}$ with $\boldsymbol{\beta}\boldsymbol{\beta}^{\mathrm{T}} = \mathbf{I}$. The triad rotates with a spin $\boldsymbol{\omega} = \dot{\boldsymbol{\beta}}\boldsymbol{\beta}^{\mathrm{T}}$. In [14] this spin was symbolized by $\boldsymbol{\omega}_{\mathrm{D}}^{1}$. Denoting by a superposed $\circ$ the co-rotational rate with respect to $\boldsymbol{\omega}$ (if so stated, the $\circ$ will denote sometimes a rate with respect to any $\boldsymbol{\Omega}$) we obtain from eqn. (2)

$$\dot{\mathbf{F}}\mathbf{F}^{-1} = \mathbf{D} + \mathbf{W} = \dot{\mathbf{V}}\mathbf{V}^{-1} + \mathbf{V}\dot{\mathbf{P}}\mathbf{P}^{-1}\mathbf{V}^{-1} = \boldsymbol{\omega} + \overset{\circ}{\mathbf{V}}\mathbf{V}^{-1} + \mathbf{V}\overset{\circ}{\mathbf{P}}\mathbf{P}^{-1}\mathbf{V}^{-1} \tag{3}$$

where $\overset{\circ}{\mathbf{V}}$ and $\overset{\circ}{\mathbf{P}}$ are defined along the lines of eqns. (1a) and (1b), respectively, because the $\mathbf{P}$ is attached to κ_0 by one of its indices only, and $\mathbf{D}$ and $\mathbf{W}$ represent the material rate of deformation and spin, respectively. The last part of eqn. (3) was explicitly given by Mandel in ref. 1 only for initially isotropic materials (Mandel preferred to work with the so-called isoclinic relaxed configuration). Restricting further attention to small elastic deformations for which $\mathbf{V} \simeq \mathbf{I}$ and κ_0 coincides with the current configuration, we obtain from eqn. (3)

$$\mathbf{D} = \overset{\circ}{\mathbf{V}} + (\overset{\circ}{\mathbf{P}}\mathbf{P}^{-1})_{\mathrm{s}} = \mathbf{D}^{\mathrm{e}} + \mathbf{D}^{\mathrm{p}} \tag{4a}$$

$$\mathbf{W} = \boldsymbol{\omega} + (\overset{\circ}{\mathbf{P}}\mathbf{P}^{-1})_{\mathrm{a}} = \boldsymbol{\omega} + \mathbf{W}^{\mathrm{p}} \tag{4b}$$

with $\mathbf{D}^{\mathrm{e}} = \overset{\circ}{\mathbf{V}}$ (elastic), $\mathbf{D}^{\mathrm{p}}$ (plastic) and $\mathbf{W}^{\mathrm{p}}$ the rates of deformation and plastic spin, respectively, and with the subscripts s and a denoting the symmetric and antisymmetric parts. The basic novelty of Mandel's theory is that constitutive relations are required not only for $\mathbf{D}^p$ but also for $\mathbf{W}^p$. This has also been proposed by Kratochvil [9].

3. GENERAL FORMULATION OF CONSTITUTIVE RELATIONS

The state variables will be the Cauchy stress $\boldsymbol{\sigma}$ and a set $\mathbf{s}$ of structure variables consisting of second-order tensors $\mathbf{a}$ (not necessarily symmetric), vectors $\mathbf{m}$ and scalars k. Temperature is omitted for simplicity.

Some preliminary statements on isotropic functions will be made. Given any orthogonal transformation $\mathbf{Q}$, a function of time in general, let $\mathbf{Q}[\mathbf{s}]$ denote the $\mathbf{QaQ}^T$, $\mathbf{Qm}$ or k according to the tensorial character of $\mathbf{s}$. For an isotropic function $\mathbf{f}(\mathbf{s}) = \mathbf{Q}[\mathbf{f}(\mathbf{Q}^T[\mathbf{s}])]$ it can be shown that

$$\mathring{\mathbf{f}} = \frac{\partial \mathbf{f}}{\partial \mathbf{s}} \cdot \mathring{\mathbf{s}} \tag{5}$$

for co-rotational rates with respect to $\boldsymbol{\Omega} = \dot{\mathbf{Q}}\mathbf{Q}^T$, and with the understanding that $\circ$ implies the material time derivative for scalar valued quantities.

The essence of the following statement has already been employed [2] and recently iterated as a theorem in a general way by Liu [13], a version of which is: a scalar, vector or tensor (second order) valued function $\mathbf{f}$ of a set $\mathbf{s}_1$ is invariant under a group of orthogonal (proper orthogonal) transformations $\mathbf{Q}$ defined by $\mathbf{Q}[\mathbf{s}_2] = \mathbf{s}_2$ for another given set $\mathbf{s}_2$ of tensors and unit vectors, if and only if $\mathbf{f}$ can be represented as an isotropic (hemitropic) function of the variables of both sets $\mathbf{s}_1$ and $\mathbf{s}_2$.

For an observer who rotates with spin $\boldsymbol{\omega} = \dot{\boldsymbol{\beta}}\boldsymbol{\beta}^T$ on the system $\hat{\mathbf{x}} = \boldsymbol{\beta}^T\mathbf{x}$, there exist symmetric and antisymmetric tensor valued functions $\mathbf{N}^p$ and $\boldsymbol{\Omega}^p$, respectively, of $\hat{\boldsymbol{\sigma}} = \boldsymbol{\beta}^T\boldsymbol{\sigma}\boldsymbol{\beta}$ and $\hat{\mathbf{s}} = \boldsymbol{\beta}^T[\mathbf{s}]$ such that with reference to $\hat{\mathbf{x}}$

$$\hat{\mathbf{D}}^p = \boldsymbol{\beta}^T\mathbf{D}^p\boldsymbol{\beta} = \langle\lambda\rangle\mathbf{N}^p(\hat{\boldsymbol{\sigma}}, \hat{\mathbf{s}}), \qquad \hat{\mathbf{W}}^p = \boldsymbol{\beta}^T\mathbf{W}^p\boldsymbol{\beta} = \langle\lambda\rangle\boldsymbol{\Omega}^p(\hat{\boldsymbol{\sigma}}, \hat{\mathbf{s}}) \tag{6}$$

with λ a properly defined scalar loading index (to be introduced later in relation to a yield/damage criterion in stress space) and $\langle\ \rangle$ are the Macauley brackets giving $\langle\lambda\rangle = \lambda$ if $\lambda > 0$ and $\langle\lambda\rangle = 0$ if $\lambda \leq 0$. Assume henceforth that initial symmetries, if any, with respect to the initial relaxed configuration in $\hat{\mathbf{x}}$ are defined by a set $\hat{\mathbf{s}}_2$ of structure variables and the associated $\mathbf{Q}$ such that $\mathbf{Q}[\hat{\mathbf{s}}_2] = \hat{\mathbf{s}}_2$. The $\hat{\mathbf{s}}_2$ is a subset of $\hat{\mathbf{s}}$, possibly **I** (identity) for initial isotropy. According to the previous theorem [13] the $\mathbf{N}^p$ and $\mathbf{\Omega}^p$ must be isotropic functions of their arguments, thus eqn. (6) becomes

$$\mathbf{D}^p = \langle\lambda\rangle \mathbf{N}^p(\boldsymbol{\sigma}, \mathbf{s}), \qquad \mathbf{W}^p = \langle\lambda\rangle \mathbf{\Omega}^p(\boldsymbol{\sigma}, \mathbf{s}) \tag{7}$$

The importance of eqn. (7) cannot be overestimated. It applies to both initially isotropic and anisotropic materials unifying the development, and notice that the isotropy of $\mathbf{N}^p$ and $\mathbf{\Omega}^p$ was not derived from the requirement of invariance under superposed rigid body motion, which is satisfied anyway at the outset by stating eqn. (6). In other approaches eqn. $(7)_1$ is stated directly in a Eulerian formulation and the aforementioned invariance renders $\mathbf{N}^p$ an isotropic function, but one is then left to wonder what sort of co-rotational rate is proper to use since no concept of substructure is introduced. The remarkably simple fact of eliminating $\boldsymbol{\beta}$ in going from eqn. (6) to eqn. (7) is a novel proposition based on the previous theorem. Mandel has obtained the same result only for initially isotropic materials. The notion of $\boldsymbol{\beta}$ may exist implicitly in eqn. (7) by means of the orientation of those $\hat{\mathbf{s}}_2 = \boldsymbol{\beta}^T[\mathbf{s}_2]$, if any, for which $\mathring{\mathbf{s}}_2 = \boldsymbol{\beta}[\dot{\hat{\mathbf{s}}}_2] = \mathbf{0}$. The general functional form of $\mathbf{N}^p$ and $\mathbf{\Omega}^p$ can be found using the representation theorems for isotropic functions [16]. The fact that such isotropic functions are used does not imply that the material is isotropic in κ_0, due to the tensorial character of $\mathbf{s}$.

Since the $\mathbf{s}$ are attached to the material substructure (crystal lattice, network of cracks, etc.) their proper co-rotational rate is with respect to $\boldsymbol{\omega}$. Thus, using the relation $\dot{\hat{\mathbf{s}}} = \boldsymbol{\beta}^T[\mathring{\mathbf{s}}] = \langle\lambda\rangle \bar{\mathbf{s}}(\hat{\boldsymbol{\sigma}}, \hat{\mathbf{s}})$, following arguments similar to the ones which led us from eqn. (6) to eqn. (7), and denoting by a superposed ∇ the Jaumann co-rotational rate defined by eqns. (1) when $\mathbf{\Omega} = \mathbf{W}$, we have the equivalent relations

$$\mathring{\mathbf{s}} = \langle\lambda\rangle \bar{\mathbf{s}}(\boldsymbol{\sigma}, \mathbf{s}) \tag{8a}$$

$$\overset{\nabla}{\mathbf{a}} = \langle\lambda\rangle(\bar{\mathbf{a}} + \mathbf{a}\mathbf{\Omega}^p - \mathbf{\Omega}^p\mathbf{a}), \qquad \overset{\nabla}{\mathbf{m}} = \langle\lambda\rangle(\bar{\mathbf{m}} - \mathbf{\Omega}^p\mathbf{m}), \qquad \dot{k} = \langle\lambda\rangle\bar{k} \tag{8b}$$

with $\bar{\mathbf{s}}$ (i.e. $\bar{\mathbf{a}}, \bar{\mathbf{m}}, \bar{k}$) isotropic functions of $\boldsymbol{\sigma}$, $\mathbf{s}$ and where use of eqns. (4b) and $(7)_2$ was made. If a structure variable does not contribute to hardening or damage, such as a vector indicating a preferred direction, one sets $\bar{\mathbf{s}} \equiv \mathbf{0}$.

For the elastic-damage relations let $\psi(\boldsymbol{\sigma}, \mathbf{s})$ represent the complementary free energy (negative enthalpy) per unit volume of κ_0, isotropic function of $\boldsymbol{\sigma}$, $\mathbf{s}$ for the same reasons as applied to $\mathbf{N}^p$, $\mathbf{\Omega}^p$ and

$\bar{\mathbf{s}}$. The small elastic strain $\boldsymbol{\varepsilon}^e$ is obtained by $\boldsymbol{\varepsilon}^e = (\partial\psi/\partial\boldsymbol{\sigma})$. Recalling eqn. (5), we obtain

$$\mathring{\boldsymbol{\varepsilon}}^e = \mathbf{D}^e = \frac{\partial^2\psi}{\partial\boldsymbol{\sigma}\otimes\partial\boldsymbol{\sigma}} : \mathring{\boldsymbol{\sigma}} + \frac{\partial^2\psi}{\partial\boldsymbol{\sigma}\otimes\partial\mathbf{s}} \boldsymbol{.}\, \mathring{\mathbf{s}} = \mathbf{D}^r + \mathbf{D}^c = \mathscr{L}^{-1} : \mathring{\boldsymbol{\sigma}} + \langle\lambda\rangle\mathbf{N}^c \quad (9a)$$

$$\mathscr{L}^{-1} = \frac{\partial^2\psi}{\partial\boldsymbol{\sigma}\otimes\partial\boldsymbol{\sigma}}, \qquad \mathbf{N}^c = \frac{\partial^2\psi}{\partial\boldsymbol{\sigma}\otimes\partial\mathbf{s}} \boldsymbol{.}\, \bar{\mathbf{s}} \quad (9b)$$

where $\mathbf{D}^r$ represents the reversible rate of deformation and $\mathbf{D}^c$ the damage induced or elastoplastic coupling rate of deformation.

The yield/damage criterion is expressed by an isotropic function (same reasons for isotropy) as

$$f(\boldsymbol{\sigma}, \mathbf{s}) = 0 \quad (10)$$

Defining $\lambda = (1/H)\mathbf{N}^n : \mathring{\boldsymbol{\sigma}}$ for a state on $f = 0$ with $\mathbf{N}^n = (\partial f/\partial\boldsymbol{\sigma})_s$, the plastic/damage modulus H is obtained from the consistency condition $\dot{f} = 0$ and use of eqns. (5) and (8a) as $H = -(\partial f/\partial\mathbf{s}) \boldsymbol{.}\, \bar{\mathbf{s}}$. Using all the preceding we can finally write the following complete set of equations:

$$\mathbf{D} = \mathbf{D}^r + \mathbf{D}^c + \mathbf{D}^p = \mathscr{L}^{-1} : \mathring{\boldsymbol{\sigma}} + \langle\lambda\rangle(\mathbf{N}^p + \mathbf{N}^c)$$

$$= \mathscr{L}^{-1} : \overset{\nabla}{\boldsymbol{\sigma}} + \langle\lambda\rangle[\mathbf{N}^p + \mathbf{N}^c - \mathscr{L}^{-1} : (\boldsymbol{\sigma}\boldsymbol{\Omega}^p - \boldsymbol{\Omega}^p\boldsymbol{\sigma})] = \boldsymbol{\Lambda}^{-1} : \overset{\nabla}{\boldsymbol{\sigma}} \quad (11a)$$

$$\lambda = \frac{1}{H}\mathbf{N}^n : \mathring{\boldsymbol{\sigma}} = \frac{\mathbf{N}^n : \overset{\nabla}{\boldsymbol{\sigma}}}{H + \mathbf{N}^n : (\boldsymbol{\sigma}\boldsymbol{\Omega}^p - \boldsymbol{\Omega}^p\boldsymbol{\sigma})} = \frac{\mathbf{N}^n : \mathscr{L} : \mathbf{D}}{H + \mathbf{N}^n : \mathscr{L} : (\mathbf{N}^p + \mathbf{N}^c)} \quad (11b)$$

$$\boldsymbol{\Lambda} = \mathscr{L} - h(\lambda)\frac{[\mathscr{L} : (\mathbf{N}^p + \mathbf{N}^c - \mathscr{L}^{-1} : (\boldsymbol{\sigma}\boldsymbol{\Omega}^p - \boldsymbol{\Omega}^p\boldsymbol{\sigma}))] \otimes (\mathbf{N}^n : \mathscr{L})}{H + \mathbf{N}^n : \mathscr{L} : (\mathbf{N}^p + \mathbf{N}^c)} \quad (11c)$$

with $h(\lambda)$ the Heaviside step function. It is possible to have plasticity without damage, in which case $\mathbf{N}^c = \mathbf{0}$, and damage without plasticity in which case $\mathbf{N}^p = \mathbf{0}$ and $\boldsymbol{\Omega}^p = \mathbf{0}$. In the latter case $f = 0$ acts only as a damage criterion and $\mathbf{W} = \boldsymbol{\omega}$. The assumption implied here is that when damage and plasticity appear simultaneously, they are governed by the same $f = 0$ and associated λ. The normality structure is obtained if $\mathbf{N}^n$ is proportional to $\mathbf{N}^p + \mathbf{N}^c$ and the terms involving $\boldsymbol{\Omega}^p$ in eqn. (11c) are zero or negligible if terms of the order stress/elastic moduli are neglected compared to one. This renders the moduli $\boldsymbol{\Lambda}$ symmetric with respect to adjacent pairs of indices. Also due to assumed small elastic deformations, $\mathbf{D}^c$ can be neglected compared to large $\mathbf{D}^p$. The denominator of the second expression for λ in eqn. (11b) represents a quantity equivalent to what was called the effective modulus in crystalline plasticity [7]. It is remarkable that the only, but important, trace left by the notion of director vectors is the existence of $\boldsymbol{\Omega}^{\mathrm{p}}$, especially in eqns. (8b) and the third member of eqn. (11b). Mandel had to carry explicitly $\boldsymbol{\beta}$ until the end since he did not employ the theorems in refs. 13 and 16.

4. THE MISSING LINK

It is clear that the missing link between the theoretical development by Mandel and its practical application is the constitutive relations for $\mathbf{W}^{\mathrm{p}}$, and for that purpose for $\boldsymbol{\Omega}^{\mathrm{p}}$. This can also be considered a link between microscopic theories [7], where such plastic spin is explicitly considered, and macroscopic theories where as a rule it is either ignored or improperly considered. As it was already suggested the representation theorems can provide the basic mathematical tool. Kratochvil [10] appears to be the first to make such a suggestion, but only in relation to continuously isotropic materials for which $\boldsymbol{\Omega}^{\mathrm{p}} = \mathbf{0}$, as Mandel has already pointed out without the need to refer to ref. 16. We shall see that much can be gained by such a reference.

Considering a second order symmetric tensor $\mathbf{a}$ as the only structure variable for simplicity, we have according to ref. 16

$$\mathbf{N}^{\mathrm{p}} = \phi_1 \mathbf{I} + \phi_2 \boldsymbol{\sigma} + \phi_3 \mathbf{a} + \phi_4 \boldsymbol{\sigma}^2 + \phi_5 \mathbf{a}^2 + \phi_6(\mathbf{a}\boldsymbol{\sigma} + \boldsymbol{\sigma}\mathbf{a}) + \phi_7(\mathbf{a}^2\boldsymbol{\sigma} + \boldsymbol{\sigma}\mathbf{a}^2) + \phi_8(\mathbf{a}\boldsymbol{\sigma}^2 + \boldsymbol{\sigma}^2\mathbf{a}) \tag{12a}$$

$$\boldsymbol{\Omega}^{\mathrm{p}} = \eta_1(\mathbf{a}\boldsymbol{\sigma} - \boldsymbol{\sigma}\mathbf{a}) + \eta_2(\mathbf{a}^2\boldsymbol{\sigma} - \boldsymbol{\sigma}\mathbf{a}^2) + \eta_3(\mathbf{a}\boldsymbol{\sigma}^2 - \boldsymbol{\sigma}^2\mathbf{a}) + \eta_4(\mathbf{a}\boldsymbol{\sigma}\mathbf{a}^2 - \mathbf{a}^2\boldsymbol{\sigma}\mathbf{a}) + \eta_5(\boldsymbol{\sigma}\mathbf{a}\boldsymbol{\sigma}^2 - \boldsymbol{\sigma}^2\mathbf{a}\boldsymbol{\sigma}) \tag{12b}$$

the ϕ_{i}'s and η_{i}'s being scalar-valued isotropic functions of the invariants of $\boldsymbol{\sigma}$ and $\mathbf{a}$. It follows from eqn. (12b) that a sufficient condition for $\boldsymbol{\Omega}^{\mathrm{p}} = \mathbf{0}$ is that $\boldsymbol{\sigma}$ and $\mathbf{a}$ commute, for which a sufficient and necessary (due to the symmetry of $\boldsymbol{\sigma}$, $\boldsymbol{\alpha}$) condition is that there is a set of common eigenvectors for $\boldsymbol{\sigma}$ and $\mathbf{a}$ (coaxiality). Thus, it follows from eqns. (4b) and $(7)_2$ that it is possible to have $\mathbf{W} = \boldsymbol{\omega}$ during plastic loading even for a material which is anisotropic in κ_0. The foregoing can be extended to the case where the elastic deformations are finite (eqn. (3)), as follows. Without changing notation for simplicity, assume that $\boldsymbol{\sigma}$ and $\mathbf{a}$ refer to the κ_0 configuration (in this case $\boldsymbol{\sigma}$ can represent a symmetric Piola–Kirchhoff stress tensor). With $\mathbf{a}\boldsymbol{\sigma} = \boldsymbol{\sigma}\mathbf{a}$ it follows that $(\mathring{\mathbf{P}}\mathbf{P}^{-1})_{\mathrm{a}} = \mathbf{0}$. Thus, $\mathring{\mathbf{P}}\mathbf{P}^{-1} = (\mathring{\mathbf{P}}\mathbf{P}^{-1})_{\mathrm{s}}$ and commutes with $\mathbf{V}$ because both have a representation of the form (12a) being symmetric and isotropic functions of $\boldsymbol{\sigma}$ and $\mathbf{a}$. Hence, $\mathbf{V}$ cancels $\mathbf{V}^{-1}$ in eqn. (3) and, consequently $(\mathbf{V}\mathring{\mathbf{P}}\mathbf{P}^{-1}\mathbf{V}^{-1})_{\mathrm{a}}=\mathbf{0}$ and $\mathbf{W}=\boldsymbol{\omega}+(\mathring{\mathbf{V}}\mathbf{V}^{-1})_{\mathrm{a}}$. The preceding observations, based on the representation theorems [2, 13, 16], are novel in the context of finite deformation plasticity. Generalization to many structure variables appears to be straightforward.

With $\mathbf{a}^{\mathrm{d}}$ denoting the deviatoric part of $\mathbf{a}$, the requirement of continuous transition from anisotropy to isotropy and vice versa is satisfied if the sufficient conditions $\lim \phi_i(\mathbf{a}^{\mathrm{d}})^{x_i}=\mathbf{0}$ and $\lim \eta_i(\mathbf{a}^{\mathrm{d}})^{y_i}=\mathbf{0}$ as $\mathbf{a}^{\mathrm{d}}\rightarrow\mathbf{0}$ are imposed, with x_i, y_i the corresponding proper exponents in the expressions (12). For a single generator involving $\mathbf{a}$, this is also a necessary condition. The values of ϕ_i's and η_i's can be obtained in principle by proper experiments, a formidable task especially for the η_i's unless one restricts the description to fewer generators. We will propose an alternative approach here, which is to a certain extent heuristic, in order to reduce the number of η_i's and still be general enough. The spin of a material line element in κ_0 parallel to a unit vector $\mathbf{n}_i$ is obtained by standard methods of continuum mechanics to within an arbitrary spin about the segment as

$$\mathbf{W}_i=\boldsymbol{W}-[(\mathbf{n}_i\otimes\mathbf{n}_i)\mathbf{D}^{\mathrm{p}}-\mathbf{D}^{\mathrm{p}}(\mathbf{n}_i\otimes\mathbf{n}_i)] \tag{13}$$

If $\mathbf{n}_i$ lies along one of the eigenvectors of $\mathbf{a}$ with λ_i the corresponding eigenvalue, a modified 'weighted' spin without a kinematical interpretation can be proposed as

$$\bar{\mathbf{W}}_i=\mathbf{W}-3\eta_a\lambda_i[(\mathbf{n}_i\otimes\mathbf{n}_i)\mathbf{D}^{\mathrm{p}}-\mathbf{D}^{\mathrm{p}}(\mathbf{n}_i\otimes\mathbf{n}_i)] \tag{14}$$

where η_a is a material function of the invariants of $\mathbf{a}$. Summing up the three $\bar{\mathbf{W}}_i$, as given by eqn. (14) for each one of the three $\mathbf{n}_i$ and corresponding λ_i, employing the spectral representation for $\mathbf{a}$ and taking the mean value, it follows

$$\mathbf{W}_a=\tfrac{1}{3}\sum_{i=1}^{3}\bar{\mathbf{W}}_i=\mathbf{W}-\eta_a(\mathbf{a}\mathbf{D}^{\mathrm{p}}-\mathbf{D}^{\mathrm{p}}\mathbf{a}) \tag{15}$$

Recalling eqn. (4b), the last term in eqn. (15) can be interpreted as the contribution of $\mathbf{a}$ to $\mathbf{W}^{\mathrm{p}}$, and $\mathbf{W}_a$ as its contribution to $\boldsymbol{\omega}$. Based on this last observation and assuming there are many $\mathbf{a}$'s denoted by $\mathbf{a}_i$, $i=1,2,\ldots,n$, summing up the corresponding $\mathbf{W}_{\mathrm{a}}$'s for all $\mathbf{a}$'s, taking the mean value and considering eqn. (7), we obtain

$$\boldsymbol{\Omega}^{\mathrm{p}}=\mathbf{a}_{\mathrm{eq}}\mathbf{N}^{\mathrm{p}}-\mathbf{N}^{\mathrm{p}}\mathbf{a}_{\mathrm{eq}},\qquad \mathbf{a}_{\mathrm{eq}}=\frac{1}{n}\sum_{i=1}^{i=n}\eta_{a_i}\mathbf{a}_i \tag{16}$$

with $\mathbf{a}_{eq}$ an 'equivalent' structure tensor. Thus, eqn. (16) defines $\boldsymbol{\Omega}^p$ once $\mathbf{N}^p$ is known from eqn. (12a) by means of n functions η_{a_i}, one for each $\mathbf{a}_i$. The so defined $\boldsymbol{\Omega}^p$ will in general have fewer generators. For example for $\boldsymbol{\sigma}$ and $\mathbf{a}$, the $\boldsymbol{\sigma}\mathbf{a}\boldsymbol{\sigma}^2-\boldsymbol{\sigma}^2\mathbf{a}\boldsymbol{\sigma}$ will be missing. If one assumes also an equation of the form (15) for $\boldsymbol{\sigma}$, then all generators of eqn. (12b) appear as well as some redundant elements. As a starting point we propose here to use only the first generator in eqn. (12b) and write

$$\boldsymbol{\Omega}^p = \eta(\mathbf{a}\boldsymbol{\sigma} - \boldsymbol{\sigma}\mathbf{a}) \tag{17}$$

The expression (17) can also be obtained from eqn. (16) for one $\mathbf{a}$, using only the ϕ_1, ϕ_2 and ϕ_3 terms of $\mathbf{N}^p$ (eqn. (12a)). The form of eqn. (17) was recently proposed by Dafalias [3] in relation to kinematic hardening. The foregoing arguments can be extended to the case where the set $\mathbf{s}$ includes vectors and antisymmetric tensors.

5. EXAMPLE ON KINEMATIC HARDENING AT LARGE SIMPLE SHEAR

The preceding can be applied to different kinds of initial symmetries (orthotropy, transverse isotropy, etc.) as shown by Dafalias [4], where the anisotropic structure variables $\mathbf{s}$ can usually be represented by $\mathbf{n}_i \otimes \mathbf{n}_i$ with $\mathbf{n}_i$ the unit vectors along preferred directions [2]. In such cases recall that $\mathring{\mathbf{n}}_i = \mathbf{0}$. Here attention will be focused on an initially isotropic material with a Mises type yield criterion which hardens kinematically. Different approaches in defining the co-rotational rates, including the one emerging from the present general formulation according to eqn. (17), will be compared in relation to their prediction for large simple shear.

Summarizing the basic relations, the yield surface is given by $f = (3/2)(\boldsymbol{\sigma}^d - \boldsymbol{\alpha}):(\boldsymbol{\sigma}^d - \boldsymbol{\alpha}) - k_0^2 = 0$ with $\boldsymbol{\sigma}^d$ the stress deviator and $\boldsymbol{\alpha}$ the deviatoric shift or back-stress tensor, the only structure variable for constant k_0. Subsequently the numerical values of all stress quantities in the text and the figures, including plastic moduli, will be considered normalized by k_0. The rate equations are

$$\lambda = \frac{3}{2}\frac{1}{h_\alpha}\mathring{\boldsymbol{\sigma}}:\mathbf{n}, \qquad \mathbf{D}^p = \langle\lambda\rangle\mathbf{n}, \qquad \mathring{\boldsymbol{\alpha}} = \tfrac{2}{3}h_\alpha \mathbf{D}^p \tag{18}$$

with h_α the constant plastic modulus in uniaxial tension–compression, $\mathbf{n} = (3/2)^{1/2}(\boldsymbol{\sigma}^d - \boldsymbol{\alpha})/k_0$ the outward unit normal to $f = 0$ and where the expression for $\mathring{\boldsymbol{\alpha}}$ is of the form (8a). For the case of large simple shear

γ along the x_1 direction in the x_1–x_2 plane and under the assumption of rigid-plastic response, we obtain $D^p_{12} = D^p_{21} = W_{12} = -W_{21} = \dot{\gamma}/2$, all other components being zero. In addition, for $\sigma_{33} = 0$, zero initial value for $\boldsymbol{\alpha}$, $\mathring{\boldsymbol{\alpha}}$ defined according to eqn. (1a) for a chosen $\boldsymbol{\Omega}$, and denoting by a prime the derivative with respect to γ, a procedure similar to the one in ref. 3 yields the following set of relations

$$\alpha'_{11} = 2z\alpha_{12} \tag{19a}$$

$$\alpha'_{12} = (1/3)h_\alpha - 2z\alpha_{11} \tag{19b}$$

and $\sigma_{11} = \sigma^d_{11} = \alpha_{11} = -\sigma_{22} = -\sigma^d_{22} = -\alpha_{22}$, $\sigma_{12} = \pm(k_0/\sqrt{3}) + \alpha_{12}$ with all other components being zero. The $\pm(k_0/\sqrt{3})$ for σ_{12} correspond to $\dot{\gamma} > 0$ and $\dot{\gamma} < 0$, respectively. The quantity z is defined by $z = \omega/\dot{\gamma}$, where $\omega = \Omega_{12} = -\Omega_{21}$ and all other components of $\boldsymbol{\Omega}$ are equal to zero because of the plane deformation. It is precisely the definition of $\boldsymbol{\Omega}$ that differentiates one approach from another and yields different solutions for the system of eqns. (19). In passing, notice that one could substitute $(\alpha_{11} - \alpha_{22})/2$ for α_{11} in the system of eqns. (19), if the simplifying assumption $\alpha_{11} + \alpha_{22} = 0$ (as a result of the assumed $\alpha_{11} = \alpha_{22} = 0$ at $\gamma = 0$) was not made.

5.1. Solution for $\boldsymbol{\Omega} = \mathbf{W}$ (Jaumann rate)

In this case $z = 1/2$ and as shown in ref. 3 the solution of eqns. (19) is

$$\alpha_{11} = \tfrac{1}{3}h_\alpha(1 - \cos\gamma), \qquad \alpha_{12} = \tfrac{1}{3}h_\alpha \sin\gamma, \qquad (\alpha_{11} - \tfrac{1}{3}h_\alpha)^2 + \alpha_{12}^2 = (\tfrac{1}{3}h_\alpha)^2 \tag{20}$$

The α_{11}–α_{12} trajectory is a circular path, as seen from eqn. $(20)_3$. The resulting stress oscillations were shown in ref. 15 by a numerical solution of eqns. (19).

5.2. Solution for $\boldsymbol{\Omega} = \dot{\mathbf{R}}\mathbf{R}^{\mathrm{T}}$

With $\mathbf{R}$ being the orthogonal part of the polar decomposition of $\mathbf{F}$, it is shown in ref. 3 that $z = 2/(\gamma^2 + 4)$ and with $\phi = \tan^{-1}(\gamma/2)$ the closed form solution of the corresponding non-autonomous system of eqns. (19) is given by

$$\alpha_{11} = \tfrac{1}{3}h_\alpha[4\cos 2\phi \ln(\cos\phi) - 2\sin 2\phi(\tan\phi - 2\phi)] \tag{21a}$$

$$\alpha_{12} = \tfrac{1}{3}h_\alpha[2\sin^2\phi \tan\phi + 4\phi\cos 2\phi - (1 + 4\ln(\cos\phi))\sin 2\phi] \tag{21b}$$

The α_{11}, α_{12} increase monotonically with γ. A co-rotational stress rate associated with $\dot{\mathbf{R}}\mathbf{R}^{\mathrm{T}}$ was proposed within a different framework by

Green and Naghdi for plasticity [6] and Dienes for hypoelasticity [5]. As pointed out in ref. 3 the $\dot{\mathbf{R}}\mathbf{R}^T$ cannot be determined by the current state of $\boldsymbol{\sigma}$ and $\boldsymbol{\alpha}$.

5.3. Solution for $\boldsymbol{\Omega} = \mathbf{W}_i$

This was proposed by Lee *et al.* [12] with $\mathbf{W}_i$ given by eqn. (13) where $\mathbf{n}_i$ is the unit eigenvector of $\boldsymbol{\alpha}$ associated with the absolutely largest eigenvalue (the symbol $\mathbf{W}^*$ was used in ref. 12 instead of $\mathbf{W}_i$). One now can easily show that

$$z = \sin^2\theta = \frac{1}{2}\left(1 - \frac{\alpha_{11}}{(\alpha_{11}^2 + \alpha_{12}^2)^{1/2}}\right) \tag{22}$$

where θ is the angle between the x_1 axis and the eigenvector of $\boldsymbol{\alpha}$ with the positive eigenvalue, recalling that $\boldsymbol{\alpha}$ is deviatoric with $\alpha_{33} = 0$. For such a z there is no equilibrium point and α_{11}, α_{12} increase monotonically with γ. Numerical solutions of eqns. (19) were provided in [12] for $h_\alpha = 1{\cdot}5$. Observe that if $\boldsymbol{\alpha}$ goes through $\mathbf{0}$ under simple shear loading conditions, the angle θ changes discontinuously by $\pi/2$ and unless $\theta = \pi/4$ it will cause a discontinuous change of z, thus of the spin. This will not affect $\mathring{\boldsymbol{\alpha}}$ since $\boldsymbol{\alpha} = \mathbf{0}$, but it will cause a discontinuity of $\mathring{\boldsymbol{\sigma}}$ (see also discussion in ref. 3).

The foregoing are illustrated in Figs. 1(a), 2(a) and 3(a) where the solutions of eqns. (19) for the three spins $\mathbf{W}$, $\dot{\mathbf{R}}\mathbf{R}^T$ and $\mathbf{W}_i$ with $h_\alpha = 1{\cdot}5$ are presented in the σ_{11}–γ, σ_{12}–γ and α_{11}–α_{12} plots for zero initial value of $\boldsymbol{\alpha}$. Recalling that $\sigma_{12} = (k_0/\sqrt{3}) + \alpha_{12}$ for $\dot{\gamma} > 0$ and the normalization by k_0, the values of α_{12} versus γ can be measured from the dashed line $\sigma_{12} = (1/\sqrt{3})$ in the σ_{12}–γ plots. For the spin $\mathbf{W}_i$ the solution was taken from ref. 12.

5.4. Solution for $\boldsymbol{\Omega} = \boldsymbol{\omega}$

Within the general framework presented here and according to eqns. (4b), $(7)_2$, (17) and (18) one has

$$\boldsymbol{\omega} = \mathbf{W} - \langle\lambda\rangle\eta(\boldsymbol{\alpha}\boldsymbol{\sigma}^{\mathrm{d}} - \boldsymbol{\sigma}^{\mathrm{d}}\boldsymbol{\alpha}) = W - \eta k_0\sqrt{\tfrac{2}{3}}(\boldsymbol{\alpha}\mathbf{D}^{\mathrm{p}} - \mathbf{D}^{\mathrm{p}}\boldsymbol{\alpha}) \tag{23}$$

Equation (23) was proposed in ref. 3 but its implication not examined in detail. Quite interestingly if one sets $\eta = \sqrt{3}/(2k_0(\mathrm{tr}\,\boldsymbol{\alpha}^2)^{1/2})$ in eqn. (23) and applies it to the case of simple shear, the expression (22) is retrieved. Recall, however, that this is an improper choice of η since then $\lim(\eta\boldsymbol{\alpha}) \neq \mathbf{0}$ as $\boldsymbol{\alpha} \to \mathbf{0}$. Instead, we propose to set $\eta = (3/2)^{1/2}\rho/2k_0$

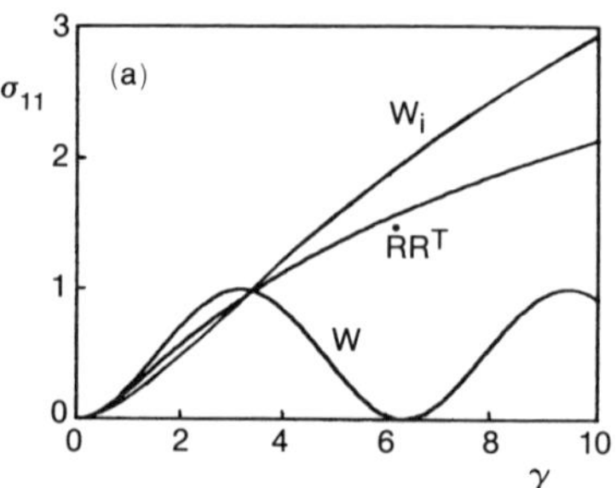

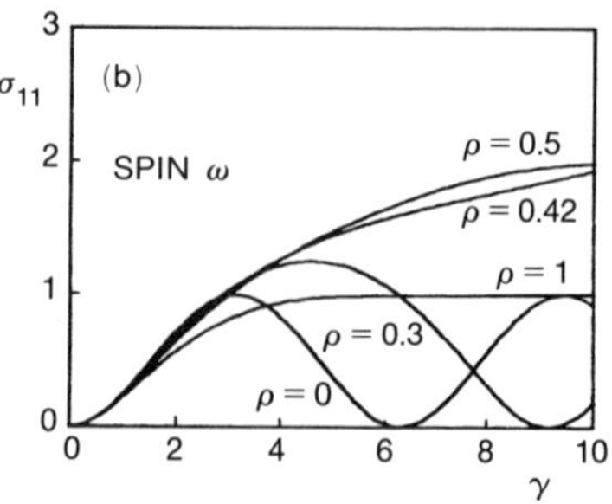

FIG. 1. Normal stress versus shear strain for co-rotational rates associated with the spins (a) $\mathbf{W}$, $\dot{\mathbf{R}}\mathbf{R}^{\mathrm{T}}$, $\mathbf{W}_i$ and (b) $\boldsymbol{\omega}$ for different values of ρ. Stress quantities are normalized by k_0.

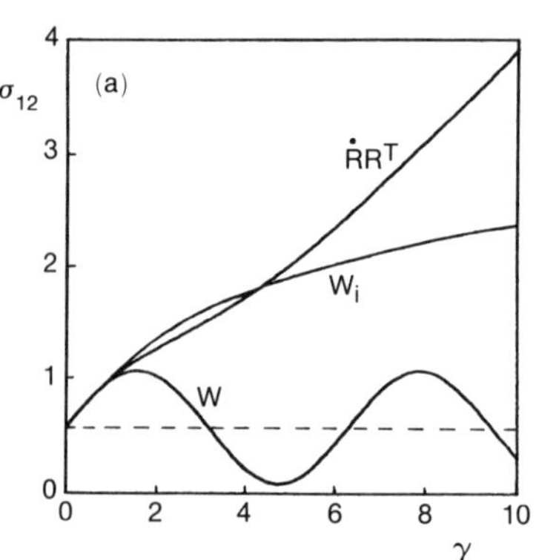

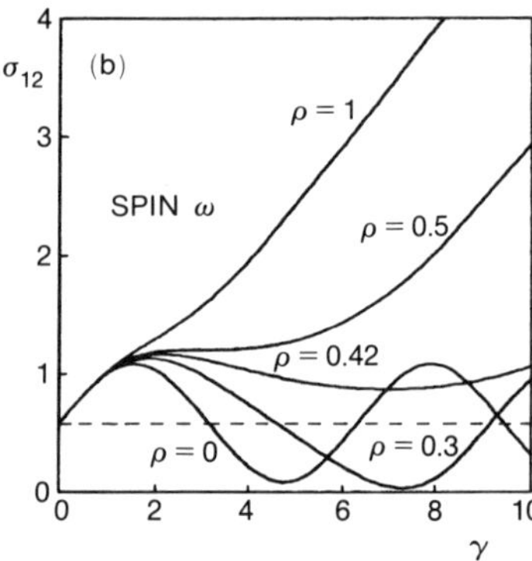

FIG. 2. Shear stress versus shear strain for co-rotational rates associated with the spins (a) $\mathbf{W}$, $\dot{\mathbf{R}}\mathbf{R}^{\mathrm{T}}$, $\mathbf{W}_i$ and (b) $\boldsymbol{\omega}$ for different values of ρ. Stress quantities are normalized by k_0.

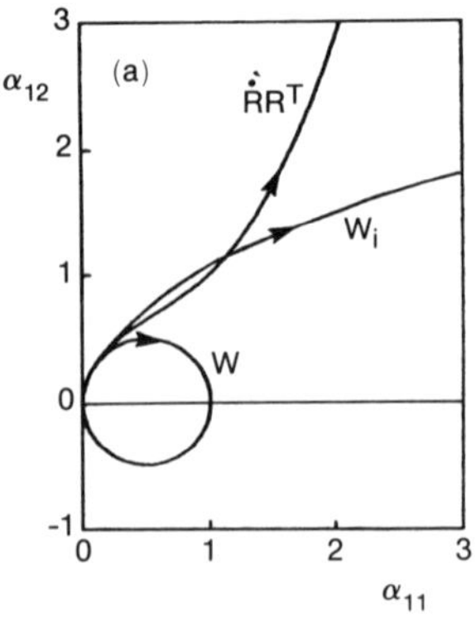

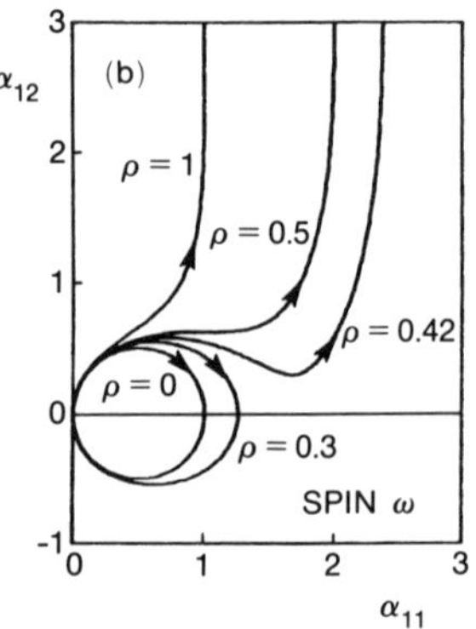

FIG. 3. Back-stress trajectories at simple shear for co-rotational rates associated with the spins (a) $\mathbf{W}$, $\dot{\mathbf{R}}\mathbf{R}^{\mathrm{T}}$, $\mathbf{W}_i$ and (b) $\boldsymbol{\omega}$ for different values of ρ. Stress quantities are normalized by k_0.

with ρ a constant (it could be a function of invariants), which yields for simple shear (recall $\alpha_{11}+\alpha_{22}=0$)

$$z=\tfrac{1}{2}(1-\rho\alpha_{11}) \tag{24}$$

The ρ has the dimensions of $(\text{stress})^{-1}$, thus its numerical values will be considered normalized by k_0^{-1}. One must have $\rho \geqslant 0$ on physical grounds. Otherwise, the spin of the director vectors would be greater than the material spin at the first stages of simple shear. Clearly for $\rho=0$ one obtains the Jaumann rate. By standard methods of analysis of systems of non-linear ordinary differential equations and phase diagram representations, the solution of eqns (19) for z given by eqn. (24) yields the family of trajectories

$$\alpha_{11}^2+\alpha_{12}^2+\frac{2}{3}\frac{h_\alpha}{\rho}\ln\left[\pm(1-\rho\alpha_{11})\right]=C_0 \tag{25}$$

for different values of the constant C_0. The $\pm$ correspond to $\alpha_{11}<1/\rho$ and $\alpha_{11}>1/\rho$, respectively, and there is a singular line at $\alpha_{11}=1/\rho$.

In order to proceed in a general way it is necessary to amend the initial condition $\boldsymbol{\alpha}=\mathbf{0}$ at $\gamma=0$ assumed so far, and allow for non-zero initial values of α_{11}, α_{22} and α_{12} but always with $\alpha_{11}+\alpha_{22}=0$. Further investigation depends on the sign of the quantity $\Delta=1-(4/3)\rho h_\alpha$. For $\Delta>0$, Fig. 4 shows a typical set of trajectories for $h_\alpha=1{\cdot}5$ and $\rho=0{\cdot}3$ on the half plane $\alpha_{12}>0$, symmetric with respect to the α_{11} axis. There are two equilibrium points of the system of eqns. (19) on the α_{11} axis, a centre $c=(1-\sqrt{\Delta})/2\rho=0{\cdot}61$ (for the assumed values of h_α, ρ) and a saddle point $s=(1+\sqrt{\Delta})/2\rho=2{\cdot}72$. The singular line (dashed line in

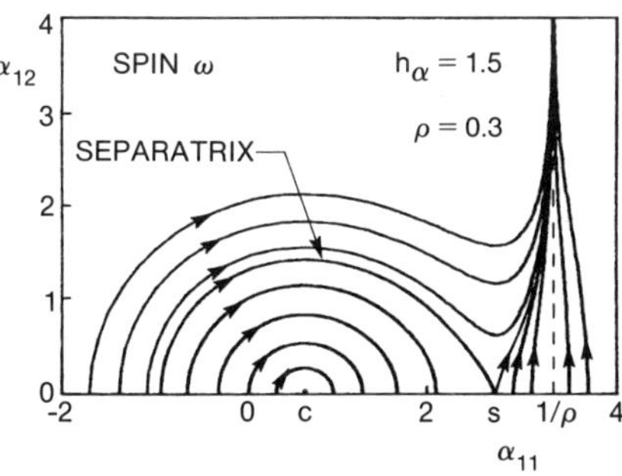

FIG. 4. A family of back-stress trajectories for different initial values at simple shear and for co-rotational rates associated with the spin $\boldsymbol{\omega}$. Stress quantities are normalized by k_0.

Fig. 4) goes through $\alpha_{11} = 1/\rho = 3{\cdot}33$. Using eqn. (25) the analytical expression for the separatrix (the trajectory which goes through the saddle point s) is given by

$$y(\alpha_{11}, \alpha_{12}) = \alpha_{11}^2 + \alpha_{12}^2 - \left(\frac{1+\sqrt{\Delta}}{2\rho}\right)^2 + \frac{2}{3}\frac{h_\alpha}{\rho} \ln\left[\frac{2(1-\rho\alpha_{11})}{1-\sqrt{\Delta}}\right] = 0 \quad (26)$$

For any initial point α_{11}, α_{12}, the subsequent 'motion' will follow the corresponding trajectory in the sense shown by arrows for $\dot{\gamma} > 0$, and in the opposite sense for $\dot{\gamma} < 0$. If such a point lies inside the branch of the separatrix shown by an arrow in Fig. 4, i.e. for which $y < 0$ and $\alpha_{11} < s$, the subsequent 'motion' is bounded (stable) and steady-state oscillations with respect to changing γ occur for both α_{11} and α_{12} with the corresponding trajectory circling the centre. If either $y > 0$ or $\alpha_{11} > s$ the 'motion' is unbounded (unstable) with $\alpha_{11} \to 1/\rho$ and $\alpha_{12} \to \infty$ as $\gamma \to \infty$ and there is the possibility to have $(d\sigma_{12}/d\gamma) \leqslant 0$ for some values of γ depending on the initial values of α_{11}, α_{12}. For $\Delta < 0$ there are no equilibrium points and as $\gamma \to \infty$ we have $\alpha_{11} \to 1/\rho$ and $\alpha_{12} \to \infty$ with $(d\sigma_{12}/d\gamma) > 0$ always. For $\Delta = 0$ the centre and saddle point coincide forming a cusp at $\alpha_{11} = (1/2\rho)$ and as $\gamma \to \infty$ we obtain again $\alpha_{11} \to 1/\rho$ and $\alpha_{12} \to \infty$ with $(d\sigma_{12}/d\gamma) = 0$ at $\alpha_{11} = (1/2\rho)$.

Using the results of the preceding investigation, we can now study the response with respect to the values of h_α and ρ for a loading which initiates with $\boldsymbol{\alpha} = \mathbf{0}$ at $\gamma = 0$. For $\Delta \leqslant 0$ the conclusion of the last paragraph applies. For $\Delta > 0$ we define first the quantity

$$\bar{y}(h_\alpha, \rho) = (1+\sqrt{\Delta})^2 + \tfrac{8}{3}\rho h_\alpha \ln\left[\tfrac{1}{2}(1-\sqrt{\Delta})\right] \quad (27)$$

from $y(0, 0) = 0$, i.e. by forcing the separatrix to go through the origin. It can now be concluded that for $\bar{y} > 0$ we have steady-state oscillations, for $\bar{y} = 0$ convergence towards the saddle point and for $\bar{y} < 0$ unbounded response with $\alpha_{11} \to 1/\rho$ and $\alpha_{12} \to \infty$ as $\gamma \to \infty$, but with $(d\sigma_{12}/d\gamma) \leqslant 0$ for some values of γ (the latter case could be misinterpreted as oscillations from a numerical solution only). An illustration of the foregoing, which provides also a direct comparison with the corresponding graphical representation of the solution for the previous co-rotational rates, is shown in Figs. 1(b), 2(b) and 3(b) for $h_\alpha = 1{\cdot}5$ again and 5 different values of $\rho = 0$ (Jaumann rate, $\Delta > 0, \bar{y} > 0$), $\rho = 0{\cdot}3$ ($\Delta > 0, \bar{y} > 0$), $\rho = 0{\cdot}42$ ($\Delta > 0, \bar{y} < 0$), $\rho = 0{\cdot}5$ ($\Delta = 0, \bar{y} < 0$) and $\rho = 1$ ($\Delta < 0$). The $\bar{y} = 0$ for $\rho = 0{\cdot}40725$, not shown in the figures. While the α_{11}–α_{12} plots can be obtained from the analytical expression (25), the plots versus γ are obtained by numerical integration of eqns.

(19). The maximum value of γ implicit in some of the α_{11}–α_{12} plots may exceed the value $\gamma=10$ used for the rest. It is clear that the co-rotational rate with respect to $\boldsymbol{\omega}$ offers a variety of responses depending on the constitutive parameter ρ, an obvious advantage over the other rates in addition to its physical interpretation.

The preceding rigorous analysis has more theoretical than applied interest, but it is necessary in order to understand the effect of the constitutive assumptions on the structure of the governing equations, a goal which cannot be achieved by numerical solutions alone. From a more applied perspective let us observe that the linear purely kinematic hardening used so far is an unrealistic constitutive assumption even for simple uniaxial loading. The induced oscillations and/or the very large stress response in simple shear are related to this assumption, if not caused exclusively by it as already shown analytically in the foregoing. To be sure, fading (but not steady-state) oscillations have been observed [8] under certain loading conditions, but large stress increase appears unlikely to occur before fracture. In an effort towards a more realistic approach, a preliminary investigation was conducted using the evanescent memory kinematic hardening model [1] for which

$$\mathring{\boldsymbol{\alpha}}=\tfrac{2}{3}h_\alpha\mathbf{D}^{\mathrm{p}}-c_{\mathrm{r}}(\tfrac{2}{3}\mathbf{D}^{\mathrm{p}}:\mathbf{D}^{\mathrm{p}})^{1/2}\boldsymbol{\alpha} \tag{28}$$

instead of eqn. $(18)_3$, with c_{r} a dimensionless material parameter and where again the expression for $\mathring{\boldsymbol{\alpha}}$ is of the form (8a). The system of eqns. (19) is now modified by subtracting the terms $\operatorname{sgn}(\dot{\gamma})(c_{\mathrm{r}}/\sqrt{3})\alpha_{11}$ and $\operatorname{sgn}(\dot{\gamma})(c_{\mathrm{r}}/\sqrt{3})\alpha_{12}$ from the right hand side of eqns. (19a) and (19b), respectively, with $\operatorname{sgn}(\dot{\gamma})$ denoting the sign of $\dot{\gamma}$ which is the same as that of γ for monotonic change. For $\boldsymbol{\alpha}=\mathbf{0}$ at $\gamma=0$ and with z given by eqn. (24), two σ_{12}–γ plots are shown in Fig. 5 for $h_\alpha=3/2$, $c_{\mathrm{r}}=\sqrt{3}/2$ and two values of $\rho=0$ (Jaumann rate) and $\rho=0{\cdot}5$. For $\rho=0$ the solution can be obtained in closed form by

$$\operatorname{sgn}(\gamma)\sigma_{12}=\frac{k_0}{\sqrt{3}}+\frac{h_\alpha}{3+c_r^2}\left[\frac{c_{\mathrm{r}}}{\sqrt{3}}+\exp\left(-\frac{c_{\mathrm{r}}}{\sqrt{3}}|\gamma|\right)\left[\sin|\gamma|-\frac{c_{\mathrm{r}}}{\sqrt{3}}\cos\gamma\right]\right] \tag{29}$$

The stress response shows an asymptotic convergence towards an equilibrium point (a stable spiral in stress space for the given numerical values), and the induced stress oscillations fade away as the equilibrium point is approached with increasing γ. Notice the much different scale for the stress compared to the one used in the other figures.

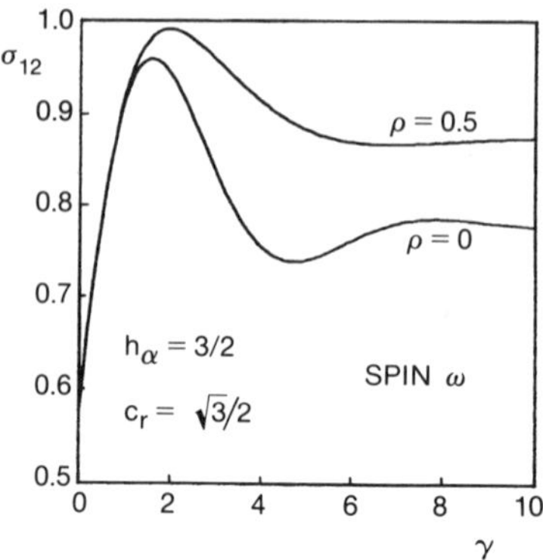

FIG. 5. The effect of the evanescent memory kinematic hardening rule on the shear stress–shear strain response for spin $\boldsymbol{\omega}$. Stress quantities are normalized by k_0.

Further investigation will be reported elsewhere for values of the parameters representing a more realistic material response.

6. CONCLUSIONS

In conclusion it may be said that by providing a means to construct macroscopically proper expressions for the plastic spin and, therefore, the spin of the director vectors, the present work brings the macroscopic formulation of large deformation plasticity in parallel line to the one of the microscopic approach employed for certain media (crystalline metals, etc.), but from a different scale and perspective. In this sense it may provide on one hand a working macroscopic framework for the present, and on the other hand a final goal for the microscopic considerations when in the future the gap from micro to macrolevel is covered by proper averaging processes which result in the emergence of macroscopic tensorial structure variables.

ACKNOWLEDGEMENT

Part of this work was accomplished during the sabbatical leave of absence of the author during Fall 1982 at the Laboratoire de Mécanique et Technologie, Cachan, ENSET, Université de Paris VI, partial support of which is gratefully acknowledged. Dr René Billardon from the Laboratoire and Mr Masoud S. Ranjbari from U. C. Davis provided numerical solutions when necessary.

REFERENCES

1. Armstrong, P. J. and C. O. Frederick, A mathematical representation of the multiaxial Bauschinger effect, CEGB report n° RD/B/N731, 1966.
2. Boehler, J. P. Lois de comportement anisotrope des milieux continus, *J. de Mecanique,* **17** (1978), 153–190.
3. Dafalias, Y. F. Corotational rates for kinematic hardening at large plastic deformations, *J. appl. Mech.,* **50** (1983), 561–565.
4. Dafalias, Y. F. On the evolution of structure variables in anisotropic yield criteria at large plastic transformations. In: *Failure Criteria of Structured Media* (Ed. J. P. Boehler), Colloque International du C.N.R.S. no. 351, Villard-de-Lans, June 1983, in press.
5. Dienes, J. K. On the analysis of rotation and stress rate in deforming bodies, *Acta Mech.,* **32** (1979), 217–232.
6. Green, A. E. and P. M. Naghdi. A general theory of an elastic–plastic continuum, *Arch. Ratl Mech. Anal.,* **18** (1965), 251–281.
7. Havner, K. S. and A. H. Shalaby. Further investigation of a new hardening law in crystal plasticity, *J. appl. Mech.,* **45** (1978), 500–506.
8. Jaoul, B. *Etude de la Plasticite et Application aux Metaux,* Dunod, Paris, 1965.
9. Kratochvil, J. Finite-strain theory of crystalline elastic–inelastic materials, *J. appl. Phys.,* **42** (1971), 1104–1108.
10. Kratochvil, J. Comment on elastic and plastic rotations. In: *Problems of Plasticity* (Ed. A. Sawczuk), Noordhoff, Leyden, 1972, 413–416.
11. Lee, E. H. and D. T. Liu. Finite-strain elastic–plastic theory with application to plane-wave analysis, *J. appl. Phys.,* **38** (1967), 19–27.
12. Lee, E. H., R. L. Mallett, and T. B. Wertheimer. Stress analysis for kinematic hardening in finite-deformation plasticity, *SUDAM* 81-11, Stanford University, 1981, revised as Metal Forming Report 4/82, Rensselaer Polytechnic Inst., 1982.
13. Liu, I. S. On representations of anisotropic invariants, *Int. J. Engng Sci.,* **20** (1982), 1099–1109.
14. Mandel, J. *Plasticité classique et viscoplasticité,* Courses and Lectures, No. 97, International Centre for Mechanical Sciences, Udine, Springer, New York, 1971.
15. Nagtegaal, J. C. and J. E. de Jong. Some aspects of non-isotropic workhardening in finite strain plasticity. In: *Plasticity of Metals at Finite Strain: Theory, Experiment and Computation* (Ed. E. H. Lee and R. L. Mallett), Div. Appl. Mech. Stanford University and Dept. Mech. Engng Rensselaer Poly. Inst., 1982, 65–102.
16. Wang, C. C. A new representation theorem for isotropic functions: an answer to Professor G. F. Smith's criticism of my paper on representations for isotropic functions, *Arch. Ratl Mech. Anal.,* **36** (1970), 166–223.

9

Experimental Verification of Endochronic Plasticity in Spatially Varying Strain Fields

K. C. VALANIS and JINGHONG FAN

Department of Electrical and Computer Engineering, University of Cincinnati, Ohio, USA

ABSTRACT

The validity and predictive capability of endochronic plasticity are tested experimentally in the case of a complex boundary value problem involving a highly heterogeneous elastoplastic strain field in a rectangular plate, with two notches symmetrically placed with respect to the longitudinal axis of symmetry. The plate is loaded cyclically in the direction of this axis. Excellent agreement between experimental and calculated results is shown along two perpendicular directions in the vicinity of the notch tip for a cyclic history up to three reversals.

1. INTRODUCTION

In the past few years we have seen an increasing incidence of the use of endochronic plasticity as a vehicle for the description of the mechanical response of materials, both under static and dynamic conditions. See refs. 1, 2, 5, 7, 8, 10–12 as typical examples. Hitherto, however, these applications fall basically into two categories, that in which the history of strain is complex but only one dimension is involved, or that where more dimensions are considered but the strain and stress fields are homogeneous.

In a recent reference [9], Valanis and Fan used the constitutive equations of endochronic plasticity, where the intrinsic time measure was defined in terms of the plastic strain increment [7, 8, 10, 13], to analyse in conjunction with a finite element code the stress and plastic strain distribution in a plate with two symmetrically placed notches, subject to a cyclic loading history. While the results obtained are of great value to crack propagation problems they were still subject to experimental verification.

In the present paper endochronic plasticity is put to a much more rigorous test. First an incremental form of the theory is developed and a finite element code for the constitutive equation is proposed. This sets the stage for testing the predictive capability of the theory in more complex multidimensional strain fields in the presence of reversed loading histories. This is achieved by designing a special experiment involving multidimensional spatially varying plastic strain fields in a notched plate and comparing the experimental results with the calculated values thereby testing the predictive capability of the endochronic theory.

2. AN INCREMENTAL FORM OF THE ENDOCHRONIC ELASTOPLASTIC CONSTITUTIVE EQUATION IN TERMS OF $\{d\sigma\}$ AND $\{d\varepsilon\}$

The following are the endochronic constitutive equations for plastically incompressible isotropic materials and small deformation [7, 10, 11, 13].

$$\mathbf{s} = \int_0^z \rho(z-z')\frac{\partial \mathbf{e}^{\mathrm{p}}}{\mathrm{d}z'}\,\mathrm{d}z' \tag{2.1}$$

$$\mathrm{d}\zeta = \|\mathrm{d}\mathbf{e}^{\mathrm{p}}\| \tag{2.1a}$$

$$\mathrm{d}z = \mathrm{d}\zeta / f(\zeta) \tag{2.1b}$$

where $f(\zeta)$ is a positive function and is monotonically increasing for hardening materials,

$$\sigma_{kk} = 3K\varepsilon_{kk} \tag{2.2}$$

$$\mathrm{d}\mathbf{e}^{\mathrm{p}} = \mathrm{d}\mathbf{e} - \frac{1}{2\mu}\,\mathrm{d}\mathbf{s} \tag{2.3}$$

where by definition

$$\mathrm{d}e_{ij} = \mathrm{d}\varepsilon_{ij} - \tfrac{1}{3}\,\mathrm{d}\varepsilon_{\alpha\alpha}\delta_{ij} \tag{2.4a}$$

$$\mathrm{d}s_{ij} = \mathrm{d}\sigma_{ij} - \tfrac{1}{3}\,\mathrm{d}\sigma_{\alpha\alpha}\delta_{ij} \tag{2.4b}$$

In this paper the form of $\rho(z)$ given by eqn. (2.5) was used in eqn. (2.1)

$$\rho(z)=\sum_{r=1}^{\infty} R_r e^{-\beta_r z} \tag{2.5}$$

with the conditions that β_r and R_r are positive for all r and

$$\sum_{r=1}^{\infty} R_r=\infty \tag{2.6a}$$

$$\sum_{r=1}^{\infty} \frac{R_r}{\beta_r}<\infty \tag{2.6b}$$

This form of $\rho(z)$ is continuous and differentiable in the open interval $(0, \infty)$. Specifically in the case where the infinitely large value of $\rho(0)$ is approximated by a suitably large value, as is done in this paper, the incremental form of eqn. (2.1) specified below can be used in conjunction with a finite element code. One may differentiate eqn. (2.1) with respect to z to obtain the following differential form of the endochronic constitutive equation:

$$d\mathbf{s}=\rho(0)\, d\mathbf{e}^{p}+\mathbf{h}(z)\, dz \tag{2.7}$$

where

$$\mathbf{h}(z)=\int_0^z \hat{\rho}(z-z')\frac{\partial \mathbf{e}^{p}}{\partial z'}\, dz' \tag{2.8}$$

and

$$\hat{\rho}(z)=\frac{d\rho}{dz} \tag{2.8a}$$

The elastoplastic constitutive equations (2.3) and (2.7) can then be combined and expressed in the differential form

$$ds_{ij}=2\hat{\mu}\left\{de_{ij}+\frac{1}{\rho(0)}h_{ij}(z)\, dz\right\} \tag{2.9}$$

where

$$\hat{\mu}=\rho(0)\left\{1+\frac{\rho(0)}{2\mu}\right\}^{-1} \tag{2.9a}$$

Alternatively, for computational purposes the incremental form given by eqn. (2.10) may be used, i.e.

$$\Delta s_{ij}=2\hat{\mu}\left\{\Delta e_{ij}+\frac{1}{\rho(0)}h_{ij}(z)\,\Delta z\right\} \tag{2.10}$$

Substituting (2.4a, b) into (2.9) and using (2.2) one obtains the operational incremental form of the elastoplastic constitutive equation in matrix notation as follows:

$$\{d\sigma\}=\{D\}\{d\varepsilon\}+\{dH_p\} \tag{2.11}$$

where

$$\{D\}=\begin{Bmatrix} c_1 & c_2 & 0 \\ c_2 & c_1 & 0 \\ 0 & 0 & \hat{\mu} \end{Bmatrix} \tag{2.12}$$

and

$$\{dH_p\}=\begin{Bmatrix} dH_{px} \\ dH_{py} \\ dH_{pxy} \end{Bmatrix} \tag{2.13}$$

In plane stress

$$c_1=\frac{12K\hat{\mu}+4\hat{\mu}^2}{3K+4\hat{\mu}} \tag{2.14}$$

$$c_2=\frac{6K\hat{\mu}-4\hat{\mu}^2}{3K+4\hat{\mu}} \tag{2.15}$$

$$D_1=\frac{2\hat{\mu}(3k-2\hat{\mu})}{3k+4\hat{\mu}} \tag{2.16}$$

$$dH_{px}=\{2\hat{\mu}h_x(z)-D_1h_Z(z)\}\,dz/\rho(0) \tag{2.17}$$

$$dH_{py}=\{2\hat{\mu}h_y(z)-D_1h_Z(z)\}\,dz/\rho(0) \tag{2.18}$$

$$dH_{pxy}=2\hat{\mu}h_{xy}(z)\,dz/\rho(0) \tag{2.19}$$

In plane strain

$$c_1=\frac{3K+4\hat{\mu}}{3} \tag{2.20}$$

$$c_2=\frac{3K-2\hat{\mu}}{3} \tag{2.21}$$

$$dH_{px}=2\hat{\mu}h_x(z)\,dz/\rho(0) \tag{2.22}$$

$$dH_{py}=2\hat{\mu}h_y(z)\,dz/\rho(0) \tag{2.23}$$

$$dH_{xy}=2\hat{\mu}h_{xy}(z)\,dz/\rho(0) \tag{2.24}$$

We note that $\{D\}$ is an adequate approximation to the elastic matrix

$\{E\}$. It is evident from eqn. (2.9a) that when $\rho(0) \rightarrow \infty$, $\{D\}$ becomes the elastic matrix $\{E\}$. Take plane stress as an example:

$$\lim_{\rho(0)\rightarrow\infty} \{D\} = \{E\} = \frac{E}{2(1+\nu)(1-\nu)} \begin{Bmatrix} 2 & 2\nu & 0 \\ 2\nu & 2 & 0 \\ 0 & 0 & (1-\nu) \end{Bmatrix} \tag{2.25}$$

We use axial tension to show the geometric meaning of eqn. (2.11). From a point A on the simple tension curve (Fig. 1) draw a straight line AB, the slope of which is Young's modulus E and its horizontal projection is $\mathrm{d}\varepsilon$. For simple tension

$$\{D\}\{\mathrm{d}\varepsilon\} = E\,\mathrm{d}\varepsilon \tag{2.26}$$

and

$$BD = E\,\mathrm{d}\varepsilon$$

So, BD can be considered as the first term of the right-hand side of (2.11). Since CD is equal to $\mathrm{d}\sigma$, the geometric meaning of $\mathrm{d}H_p$ is represented by the segment BC the value of which is negative for simple tension.

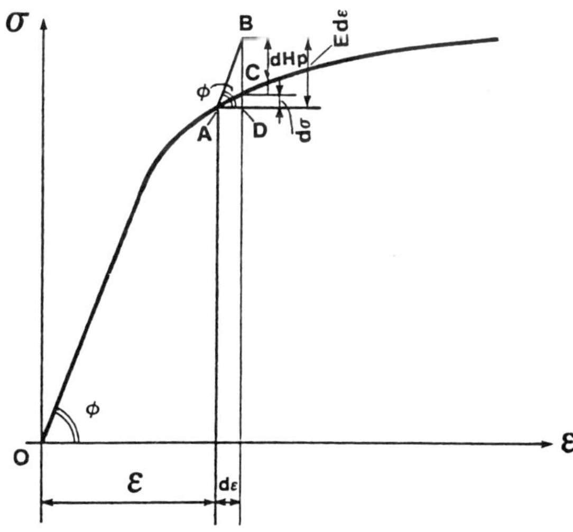

FIG. 1. The geometric meaning of incremental endochronic elasto–plastic constitutive equation.

3. A FINITE ELEMENT CODE FOR THE ENDOCHRONIC THEORY OF PLASTICITY

Using (2.11) and the principle of virtual work, one may formulate an initial stress finite element computational algorithm of the endochronic theory. In fact, with reference to a generic element we have

$$\iiint_v \{\sigma\}^{\mathrm{T}}\{\delta\varepsilon\}\,\mathrm{d}v = \{p_{\mathrm{ex}}\}^{\mathrm{T}}\delta\{q\} \tag{3.1}$$

and $\{p_{\mathrm{ex}}\}$ and $\{q\}$ are respectively the vectors of nodal external forces and displacements of the element. Substituting (2.11) into (3.1) one finds that

$$\{K\}\{\Delta q\} = \{\Delta p_{\mathrm{ex}}\} + \{\Delta p_{\mathrm{p}}\} \tag{3.2}$$

where $\{K\}$ is the stiffness matrix of the element and is the same as the stiffness matrix of an element in the usual elastic analysis but the constants C_1, C_2 are obtained from eqns. (2.14–2.16) or (2.20–2.21).

The quantity $\{\Delta P_{\mathrm{p}}\}$ is the incremental plastic pseudo-force vector for a typical triangular element used in the analysis and has the form

$$\begin{aligned}(\Delta P_{\mathrm{px}})_i &= -\frac{t}{2}(\alpha_i\,\Delta H_{\mathrm{px}} + \beta_i\,\Delta H_{\mathrm{pxy}})\\ (\Delta P_{\mathrm{py}})_i &= -\frac{t}{2}(\beta_i\,\Delta H_{\mathrm{py}} + \alpha_i\,\Delta H_{\mathrm{pxy}})\end{aligned} \qquad i = 1, 2, 3 \tag{3.3}$$

Where the components of $\{\Delta H_{\mathrm{p}}\}$ are given in eqns. (2.17–2.19) or (2.22–2.24) by changing operator 'd' to 'Δ', while α_i and β_i are related to the differences of nodal co-ordinates, i.e.

$$\alpha_i = \tfrac{1}{2}e_{ijk}\,\Delta y_{jk}, \qquad \beta_i = \tfrac{1}{2}e_{ijk}\,\Delta x_{jk} \qquad (i = 1, 2, 3) \tag{3.4}$$

where $\Delta y_{jk} = y_j - y_k$, $\Delta x_{jk} = x_k - x_j$ and e_{ijk} is the permutation symbol.

From eqns. (3.2) and (3.3) one obtains the total stiffness matrix $\{k\}$, total plastic pseudo-force matrix $\sum\{\Delta P_{\mathrm{p}}\}$ and the linear simultaneous equations for the structure.

On the basis of the above formula, a computer program called Endo-2 was developed to analyse the material response to monotonic and cyclic loading. This program is written for cases of both plane stress and plane strain, especially for a plate with two notches in which the finite element mesh of one quarter consists of 230 nodes and 413 elements. Figure 2(a) shows a finite element mesh of the region which

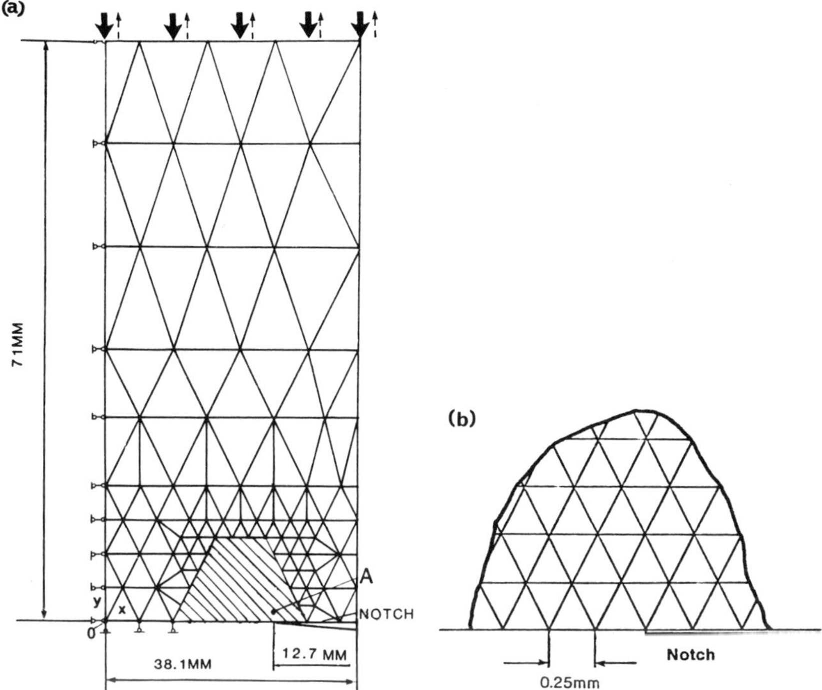

FIG. 2. Plate geometry and grid arrangement.

is far from the tip of the notch. In the shaded region of Fig. 2(a) close to the notch a much finer mesh is used (see Fig. 2(b)), the smallest size of which is 0·25 mm.

The incremental elasto–plastic solution is obtained by iteratively revising the right-hand side of the system of finite element equations by the addition of a vector of plastic pseudo-forces. The iteration process is continued until the maximum difference in the values of Δz between any two consecutive iterations is less than some defined tolerance. A new loading or unloading process is then initiated.

In the initial stress method of classical plasticity one [14] usually stops the iteration process if the difference of plastic pseudo-force vector of two consecutive iterations is sufficiently small. We use scale Δz instead of the pseudo-force vector as a convergence criterion not only because of its simplicity but because of its crucial role in endochronic plasticity.

4. SPECIMENS AND EQUIPMENT

Specimens were machined from 25·4 mm thick of oxygen free high conductivity (OFHC) copper (812 mm × 914 mm).

To determine the hardening function $f(\zeta)$ and kernel function $\rho(z)$ a cyclic tension–compression test was conducted under strain controlled conditions ($\pm 3000 \times 10^{-6}$) on a round specimen. The ratio of nominal length to diameter is 2·5 in order to prevent the incidence of buckling. The test was performed on a servohydraulic universal testing system ($4{\cdot}45 \times 10^{4}$ N capacity). The axial strain was measured by a low pressure point contact clamp-on extensometer mounted on the gauge section. Plotting was done by a Hewlett–Packard 70156 X-Y Plotter.

How the functions $\rho(z)$ and $f(\zeta)$ were determined is the subject of discussion in Section 5. Figure 3 shows an overall view of the notched specimen. Two specimens of this kind including strain gauge arrangement are described below. Specimen 1 is designed to determine the distribution of strain ε_y along the notch line oo′ (Fig. 4). To measure the strain as close to the tip of the notch as possible the smallest strain gauges MA-13-008CL-120 were used. The nominal length of this type of gauge is 0·2 mm, which is slightly smaller than the fine element near the tip (the element side length is 0·25 mm). Further from the tip an assembled gauge group EA-13-031ME-120 was arranged along oo′.

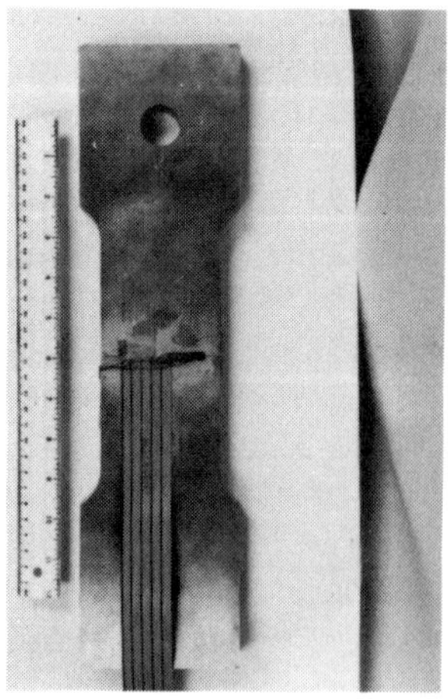

FIG. 3. A view of notched specimen.

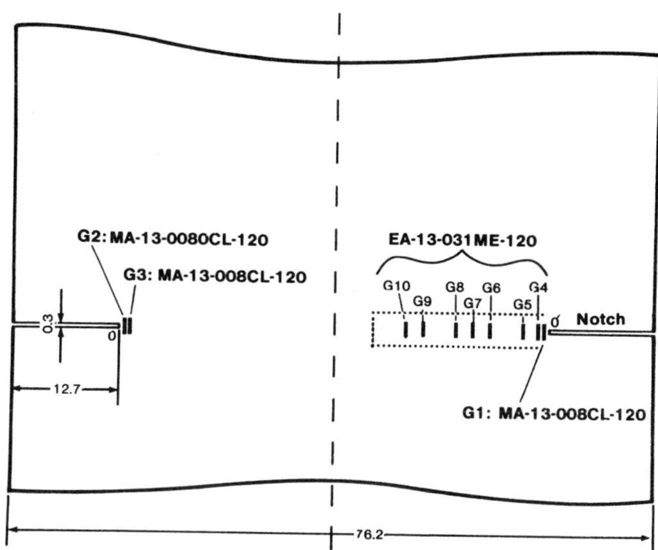

FIG. 4. Arrangements of strain gauges for determining the distribution of strain ε_y along the notch line oo′.

To determine the position of strain gauges more accurately, pictures of the specimen were taken and then magnified 3·6 times. Using these pictures strain gauges were easily located. After much effort the shortest distance (about 0·28 mm) from the gauge centre to the tip was obtained. The amplitude of the applied maximum stress was $3{\cdot}68\times 10^7$ Pa.

Specimen 2 is designed to determine the distribution of strain ε_y along the vertical line ob above the top of notch (Fig. 5). An assembled gauge group EA-13-020MT-120 was attached to the specimen. The amplitude of the applied maximum stress is $2{\cdot}3\times 10^7$ Pa, a value that is smaller than the previous one, so as to probe the non-linearity of the strain distribution.

To change the connection between strain gauges and indicator easily and avoid the need for a large number of cables, a ribbon cable (25 conductor) which is used in microcomputers was adopted. One end of the ribbon cable is attached to the specimen and soldered with the strain gauges (Fig. 3). The other end is connected to a DB 25 connector (female and male), and through it connected to the bridge switch gain (Model BSG-20), by which the channel of input to SR-4 indicator

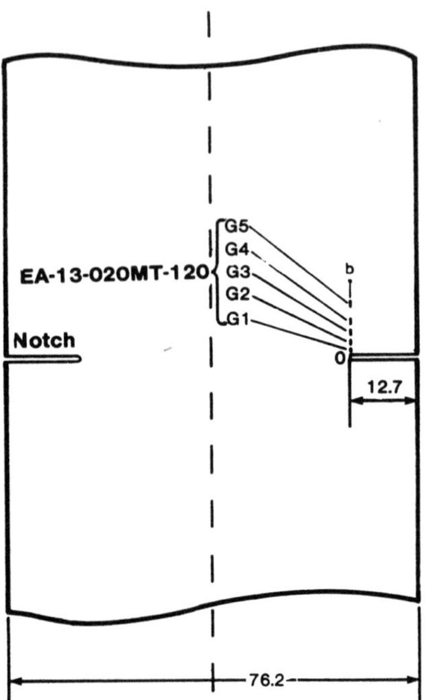

FIG. 5. Arrangements of strain gauges for determining the distribution of strain ε_y along the vertical line ob at the top of notch tip.

can be changed (Fig. 6). For each loading increment the time of the strain reading and recording of 12 gauges is about 2 min. During this period the loading was precisely kept to a constant value. In the test the strain gauge factor switch was put at 2·0. The recorded value ε_r of the strain was modified by its own strain gauge factor k_ε to get the actual strain ε_a given by the relation

$$\varepsilon_a = \frac{2}{k_\varepsilon}\,\varepsilon_r$$

Two symmetrically placed gauges were attached in the uniform deformation region of the specimen, the recorded reading of which shows good symmetry at the clamp. All equipment including fixtures worked very well. The indicator was stable during the whole recording process. The experimental data, therefore, were obtained with good accuracy.

FIG. 6. An overall view of testing system.

5. COMPARISON BETWEEN EXPERIMENTAL DATA AND CALCULATED RESULTS

In isotropic endochronic plasticity the material behaviour is determined by two elastic constants μ and K and two material functions $\rho(z)$ and $f(\zeta)$. We felt that these functions should be determined independently in a homogeneous strain field using a uniform specimen and their forms so found should then be used in the finite element code to calculate the heterogeneous strain field in the notched plate. This approach ensures objectivity and eliminates a posteriori manipulation of the material functions to fit the data.

5.1. Determination of the Functions $\rho(z)$ and $f(\zeta)$

The function $\rho(z)$ was found from a monotonic stress–strain test on a circular specimen discussed in ref. 3 and $f(\zeta)$ was determined from a strain controlled cyclic tension compression test on a similar bar, by a procedure which we proceed to discuss.

Our observations in experiments on copper show that $f(z)$ is a well-behaved slowly varying function, while $\rho(z)$ has a singularity at $z=0$, and decays at a high rate. These observations provide the mathematical foundation for the determination of these two functions.† Specifically, we stipulate that $f(z)$ remains sensibly constant in the interval $0 \leqslant z \leqslant z^*$, over which $\rho(z)$ decays to a negligible value. Thus in an idealized sense

$$f(\zeta)=1, \qquad 0 \leqslant \zeta \leqslant \zeta^* \tag{5.1}$$

$$\rho(z)=0, \qquad z^* \leqslant z \tag{5.2}$$

where $z=\zeta$ in the interval $0 \leqslant \zeta \leqslant \zeta^*$ as a result of eqn. (2.1b). This is a material with a fading and finite memory with respect to z.

In a uniaxial test, eqn. (2.1) in conjunction with the assumption of plastic incompressibility gives the relation

$$\sigma = \int_0^z E(z-z') \frac{\partial \varepsilon^{\mathrm{p}}}{\partial z'} \, \mathrm{d}z' \tag{5.3}$$

where $E(z)=\frac{3}{2}\rho(z)$ and σ and ε^{p} are the axial stress and axial plastic strain respectively. In view of eqns. (2.1a) and (2.1b) and under monotonic conditions

$$\sigma = \sqrt{\frac{2}{3}} \int_0^z E(z-z') f^*(z') \, \mathrm{d}z' \tag{5.4}$$

where $f^*(z)=f(z(\zeta))$. In view of conditions (5.1) and (5.2), eqn. (5.4) becomes

$$\sigma = \sqrt{\frac{2}{3}} \int_0^z E(z-z') \, \mathrm{d}z' = \sqrt{\frac{2}{3}} \int_0^z E(z') \, \mathrm{d}z' \tag{5.5}$$

where

$$z = \sqrt{\tfrac{3}{2}} \, \varepsilon^{\mathrm{p}} \tag{5.6}$$

Thus when σ is expressed as a function of the variable z it follows from eqn. (5.5) that

$$E(z) = \sqrt{\frac{3}{2}} \frac{\mathrm{d}\sigma}{\mathrm{d}z} \tag{5.7}$$

alternatively, it follows from eqn. (5.6) that

$$E(\sqrt{\tfrac{3}{2}} \, \varepsilon^{\mathrm{p}}) = \frac{\mathrm{d}\sigma}{\mathrm{d}\varepsilon^{\mathrm{p}}} \tag{5.8}$$

† When such conditions do not prevail, other methods exist for the determination of these two functions. See ref. 10.

and since $\rho = \frac{2}{3}E$ it follows also that,

$$\rho(\sqrt{\tfrac{3}{2}}\,\varepsilon^{\mathrm{p}}) = \frac{2}{3}\frac{\mathrm{d}\sigma}{\mathrm{d}\varepsilon^{\mathrm{p}}} \tag{5.9}$$

Hence, $\rho(z)$ is obtained from the slope of the uniaxial stress–plastic strain curve under monotonic loading conditions.

However, the reader is warned that this is true only if the reference state of the metal is a fully annealed state. If a prior history exists, the relation given by eqn. (5.8) is no longer valid. Reference states with prestrain will not be discussed in this paper.

5.2. Determination of $f(z)$

Consider a 'constant plastic strain amplitude' experiment which is depicted in Fig. 7. Further let the amplitude Δ be larger than z^*. Let σ_n be the magnitude of the stress at the completion of nth reversal. Then in view of eqn. (5.3)

$$\sigma_n = \int_0^{z_n} E(z - z')\frac{\partial \varepsilon^{\mathrm{p}}}{\mathrm{d}z'}\,\mathrm{d}z' \tag{5.10}$$

Application of the finite memory principle reduces the above equation to the form

$$|\sigma_n| = \sqrt{\tfrac{2}{3}}\,f^*(\bar{z})\int_{z_{n-\Delta}}^{z_n} E(z - z')\,\mathrm{d}z' \tag{5.11}$$

where $z_{n-1} \leqslant \bar{z}_n \leqslant z_n$.

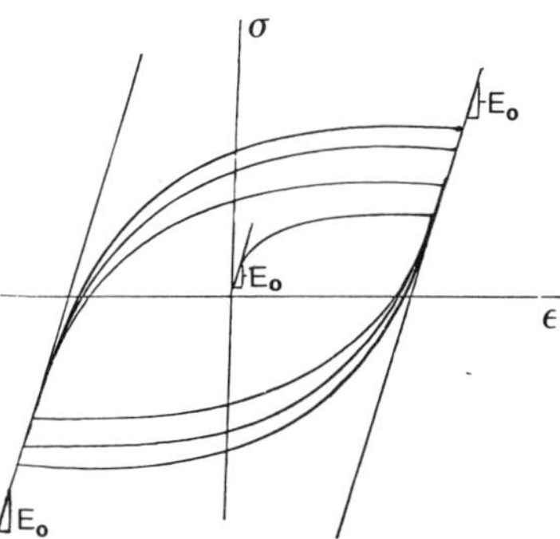

FIG. 7. Schematic representation of an experiment with constant plastic strain amplitude.

We recall that since $\Delta > z^*$, then in the interval $0 \leqslant z' \leqslant \Delta$, $z = \zeta$ and hence the expression on the right hand side of eqn. (5.11) is simply $f^*(\bar{z}_n)$ times σ_1 which is the magnitude of the stress and the completion of the first quarter-cycle. Thus

$$|\sigma_n| = f^*(\bar{z}_n)\sigma_1 \tag{5.12}$$

and

$$f(\bar{\zeta}_n) = \frac{|\sigma_n|}{\sigma_1} \tag{5.12a}$$

where $\zeta_{n-1} \leqslant \bar{\zeta}_n \leqslant \zeta_n$. Furthermore since f is a slowly varying function we may get

$$f(\bar{\zeta}_n) \doteq f(\zeta_n) = \frac{|\sigma_n|}{\sigma_1} \tag{5.13}$$

Equation (5.13) affords an easy experimental determination of the hardening function for all ζ.

If a laboratory is not equipped with sophisticated strain amplitude control then a fixed total strain amplitude experiment may be performed. In this case Δ is not fixed and decreases with z if the material hardens during cycling. However, if $\Delta_{\min} > z^*$, the above analysis still applies and eqn. (5.13) is still valid. The function $\rho(z)$ was approximated by a series of three exponential terms and was found to be as follows

$$\rho(z) = \sum_{r=1}^{3} A_r e^{-\alpha_r z} \quad \text{(GPA)}$$

where $A_{1,2,3} = (592, 220, 46)$ and $\alpha_{1,2,3} = (27{\cdot}5, 11{\cdot}5, 7{\cdot}67) \times 10^3$ and the function $f(\zeta)$ was found to be of the form

$$f(\zeta) = 1 + 0{\cdot}53\zeta^{0{\cdot}72}$$

5.3. Experimental Results

The measured distribution of strain ε_y along the notch line oo′ of specimen 1 (Fig. 4) when the amplitude of applied stress was $3{\cdot}68 \times 10^7$ Pa is shown in Table I.

The experimentally determined strain distribution ε_y along vertical line ob of specimen 2 (Fig. 5) for the amplitude of applied maximum stress $2{\cdot}3 \times 10^7$ Pa is shown in Table II.

TABLE I
Experimental results of strain distribution along notch line oo′ of specimen 1
(Applied Maximum Stress: $3{\cdot}68 \times 10^7$ Pa)

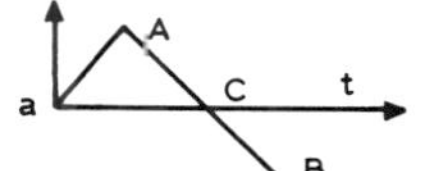

No.	†*Distance to notch tip (mm)*	*Gauge designation*	*Gauge factor* K_ε	*Positive peak* A		*Unloading* C		*Negative peak* B	
				Strain record $\varepsilon_r \times 10^6$	*Actual strain* $\varepsilon_a \times 10^6$	*Strain record* $\varepsilon_r \times 10^6$	*Actual strain* $\varepsilon_a \times 10^6$	*Strain record* $\varepsilon_r \times 10^6$	*Actual strain* $\varepsilon_a \times 10^6$
1	0·3	MA-13-008CL-120	2·32	3305	2850	1820	1570	−3200	−2760
2	0·35	MA-13-008CL-120	2·32	2300	1980	880	760	−2395	−2065
3	0·6	MA-13-008CL-120	2·32	1275	1100	600	520	−1130	−975
4	0·85	EA-13-031ME-120	2·17	1975	1820	750	690	−910	−840
5	2·9	EA-13-031ME-120	2·17	705	650	340	310	−290	−270
6	6·95	EA-13-031ME-120	2·17	520	480	195	180	−245	−225
7	9	EA-13-031ME-120	2·17	470	430	145	135	−280	−260
8	11	EA-13-031ME-120	2·17	470	430	160	150	−315	−290
9	15	EA-13-031ME-120	2·17	435	400	135	125	−310	−285
10	17	EA-13-031ME-120	2·17	420	390	120	110	−295	−270

† The accuracy of this distance is achieved with the adopted method using magnified pictures and the spacing between gauges for EA-13-031ME-120 is ensured by strain gauge itself.

TABLE II
The experimental strain distribution ε_y along vertical line ob of specimen 2
(Applied Maximum Stress: $2\cdot3 \times 10^7$ Pa)

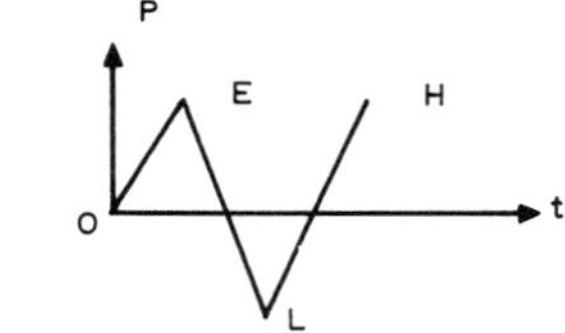

No.	*Distance to notch line oo′ (mm)*	*Gauge designation*	*Gauge factor* K_E	*First positive peak E*		*Negative peak L*		*Second positive peak H*	
				Strain record $\varepsilon_r \times 10^6$	*Actual strain* $\varepsilon_a \times 10^6$	*Strain record* $\varepsilon_r \times 10^6$	*Actual strain* $\varepsilon_a \times 10^6$	*Strain record* $\varepsilon_r \times 10^6$	*Actual strain* $\varepsilon_a \times 10^6$
1	0·75	MA-13-020MT-120	2·03	775	760	−515	−510	755	745
2	2·5	MA-13-020MT-120	2·03	335	330	−355	−350	330	325
3	5·2	MA-13-020MT-120	2·03	255	250	−265	−265	255	250
4	6·1	MA-13-020MT-120	2·03	240	235	−240	−235	235	230
5	8·7	MA-13-020MT-120	2·03	220	215	−255	−250	195	190

5.4. Calculated Results

The calculations were conducted by an electronic computer (AMDAHL 470 V/7A, close to IBM 370) in the computer centre of the University of Cincinnati. There are 413 elements and 230 nodes in both specimens 1 and 2. By 'varying band storage' the amount of storage for total stiffness matrix is 17698. The details of computation, for specimen 2 say, are described below. The incremental loading for each step is 4% of maximum loading (i.e. 890 N). The average number of iterations for each incremental loading was about 10, varying from 3 to 20. The computer time for each iteration was about 3·36 s, most of which is used to solve the 460 simultaneous equations. Since the locations of the elements and the strain gauges did not coincide exactly, we compared the calculated results with experimental data in terms of plotted curves. Comparisons were made over a wide range of magnitude of applied maximum stress, location and type of histories.

Measured and calculated strain distributions ε_y along the notch centre line oo′ are shown, for applied stress amplitude $3{\cdot}7\times10^7$ Pa

(i) at first tensile peak A (Fig. 8)

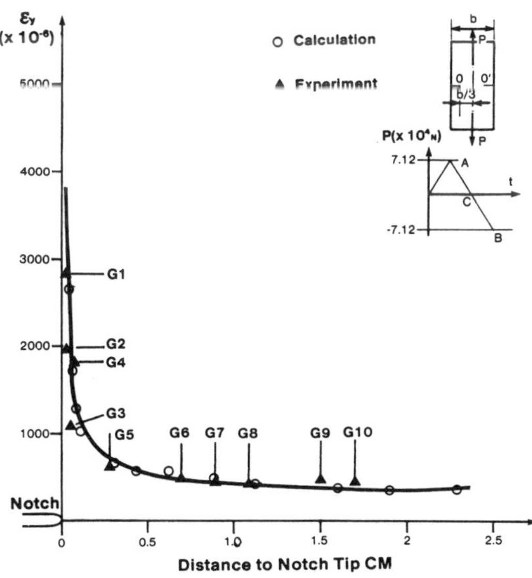

FIG. 8. Comparison between experimental and calculated distribution of strain ε_y along notch line oo′ at positive peak A.

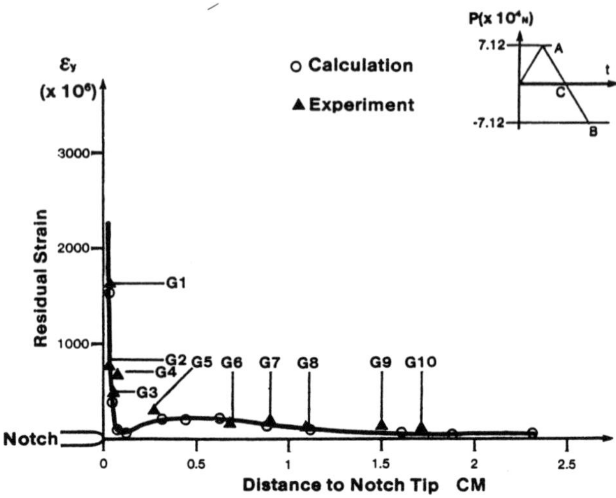

FIG. 9. Comparison between experimental and calculated residual strain distribution ε_y along notch line oo′ at point C.

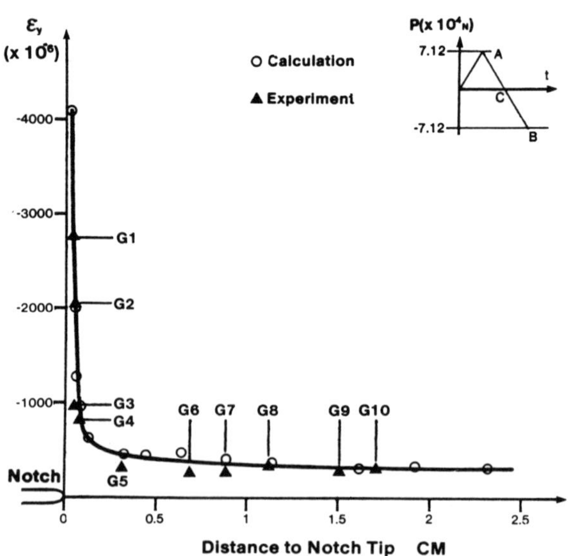

FIG. 10. Comparison between experimental and calculated strain distribution ε_y along notch line oo′ at negative peak B.

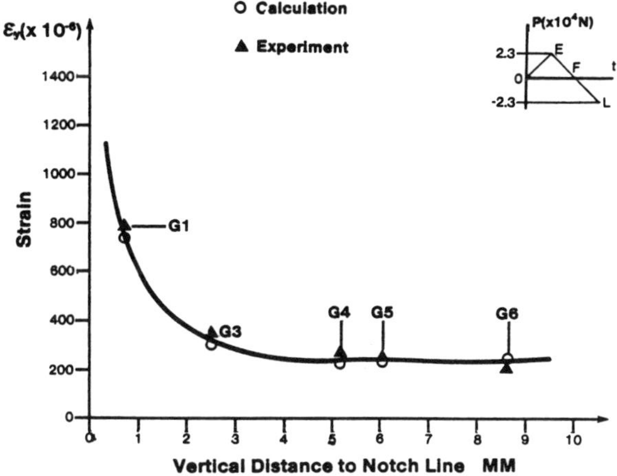

FIG. 11. Comparison between experimental and calculated strain distribution ε_y at the positive peak E along vertical line ob at the top of notch tip.

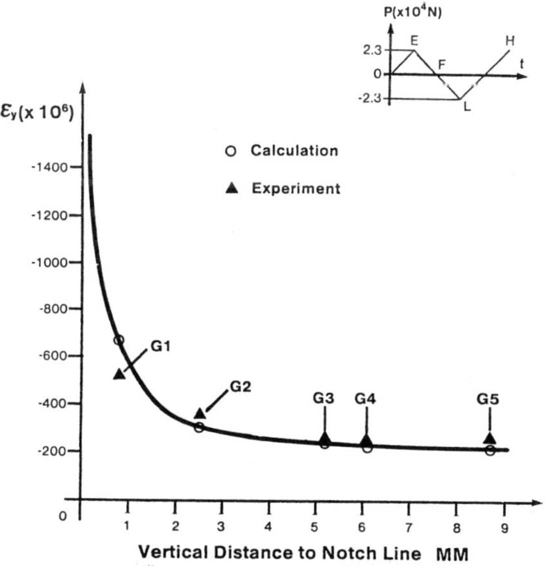

FIG. 12. Comparison between experimental and calculated strain distribution ε_y at the negative peak L along the vertical line ob at the top of notch tip.

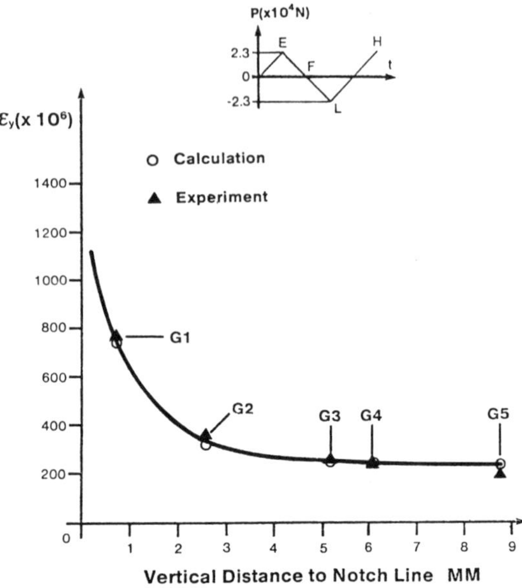

FIG. 13. Comparison between experimental and calculated strain distribution ε_y at the positive peak H along the vertical line ob at the top of notch tip.

(ii) at first unloading point C (Fig. 9)
(iii) at first compressive loading peak B (Fig. 10).

Letter designations as shown in Table I.

Measured and calculated strain distributions ε_y along the vertical line ob are shown for applied stress amplitude $2{\cdot}3 \times 10^7$ Pa

(i) at first tensile peak E (Fig. 11)
(ii) at first compressive peak L (Fig. 12)
(iii) at second loading peak H (Fig. 13).

Letter designations as shown in Table II.

Despite the complexity of the boundary value problem and the inherent experimental difficulties the agreement between experimental and calculated results is excellent both from the aspect of tendency and magnitude.

6. CONCLUSIONS

Recently more attention has been paid to the computational capability, validity and predictive power of the endochronic theory.† Attention has also been given to how the revolutionary concept of plasticity without a yield surface can be used in the stress and strain elasto–plastic analysis of complicated stress fields. By designing a special experiment, the validity and predictive ability of the endochronic theory was shown in this work, using the method developed in the case of a complex boundary value problem involving a highly heterogeneous elasto–plastic strain field. The agreement between calculated and experimental results in the vicinity of the notch tip for a notched specimen cyclic loaded in its own plane is excellent. This is not only evidence and a good example of how this concept can be adopted and applied, but also, to our understanding, a signal to take notice that endochronic plasticity has begun to enter a fully developed period.

ACKNOWLEDGEMENT

The authors wish to thank the Metcut Research Associate Corporation and specifically Dr John Cammett for their assistance and gracious cooperation without which the experimental part of this work would not have been possible.

REFERENCES

1. Bazant, Z. P. Endochronic inelasticity and incremental plasticity, *Int. J. Solids Struct.*, **14** (1978), 697.
2. Bazant, Z. P. and P. D. Bhat. Endochronic theory of inelasticity and failure of concrete, *J. Eng. Mech. Div., Proc. ASCE*, **102** (1976), 701.

† For instance, the criticism from Michno, Findley and Lamba is as follows: 'While the concept of intrinsic time seems to be a good one, the following mathematical complexity is much greater than the problem warrants. Such theories would be of a greater significance if they could predict experimental phenomena that have not been observed, instead of modeling previously observed phenomena that are easy to reproduce in the laboratory'. (Lamba [4]), and 'Additional experimental data involving more general loading histories are needed to shed additional light on the applicability of these theories'. (Michno and Findley [6]).

3. Fan, Jinghong. A Comprehensive Numerical Study and Experimental Verification of Endochronic Plasticity, Ph.D. Dissertation, Department of Aerospace Engineering and Applied Mechanics, University of Cincinnati, 1983.
4. Lamba, H. S. Nonproportional Cyclic Plasticity, T.&A.M. Report No. 413, UILU-Eng. 76-6008.
5. Liu, H. C. and H. C. Wu. On the improved endochronic theory of viscoplasticity and its application to plastic wave propagation, *Int. J. Solids Struct.* (in press).
6. Michno, M. J., Jr. and W. N. Findley. A Historical Perspective of Yield Surface Investigations for Metals, Report EMRL 47, Brown University, 1972.
7. Valanis, K. C. Endochronic Theory with Proper Hysteresis Loop Closure Properties, S-CUBED Report SSS-R-80-4182, 1979.
8. Valanis, K. C. Fundamental Consequences of a New Intrinsic Time Measure Plasticity as a Limit of the Endochronic Theory, Report G224-DME-78-001, The University of Iowa, 1978, or *Arch. Mechan.*, 1980, 171–191.
9. Valanis, K. C. and Jinghong Fan. Endochronic analysis of cyclic elastoplastic strain fields in a notched plate, *J. appl. Mech.* (in press).
10. Valanis, K. C. and C. F. Lee. Some recent developments in the endochronic theory with applications, *Nucl. Engng Design*, **69** (1982), 327.
11. Valanis, K. C. and C. F. Lee. Endochronic theory of cyclic plasticity with applications, *J. appl. Mech.* (submitted), March, 1983.
12. Valanis, K. C. and T. Nashiro. On the Tension–Torsion Problem in Brass Tubes (unpublished).
13. Valanis, K. C. and H. E. Read. A new endochronic plasticity model for soils. In: *Soil Mechanics—Transient and Cyclic Loads* (Ed. G. N. Paude and O. C. Zienkiewicz), John Wiley, New York, 1982.
14. Zienkiewicz, O. C., S. Valliappan, and I. P. King. Elasto–plastic solutions of engineering problems 'Initial stress' finite element approach, *Int. J. Num. Methods Engng*, **1** (1969), 75–100.

Discussions

THE USE OF PLASTIC STRAIN AS A STATE VARIABLE

E. H. Lee (Rensselaer Polytechnic Institute, Troy, NY, USA): Whenever plastic strain has been used as a state variable in constitutive relations to express the laws of plasticity, it has proved to be inappropriate. This was clearly evident in the deformation theory of plasticity in which the stress was expressed as a function of the plastic strain (as for elasticity) and the anomaly arose that, following plastic flow, if the stress point moved inside the yield surface and then intersected it at a different stress on reloading, the plastic strain would have to jump instantaneously to the plastic strain associated with the newly applied stress. That the use of the variable, plastic strain, is likely to be unsatisfactory in kinetic considerations is evident by considering, for example, the case of a body subjected to tensile strain followed by compressive strain so that the resultant deformation is zero. Clearly strain hardening and strain induced anisotropy could well have been generated even though the resulting plastic strain is zero.

Of course for an elastic–plastic medium, the stress can be expressed in terms of the elastic strain and hence (for infinitesimal strain theory) in terms of $(\boldsymbol{\varepsilon}-\boldsymbol{\varepsilon}^{\mathrm{p}})$, with an appropriate modification for finite strain. In this formulation, plastic strain must occur in conjunction with the total strain and not as an independent variable. *Elastic* strain is the effective independent variable.

In the case of isotropic hardening using the Mises yield condition, hardening is expressed in terms of the generalized plastic strain invariant $\bar{\varepsilon}^{\mathrm{p}}$ given by the time integral of $\sqrt{\dot{\varepsilon}^{\mathrm{p}}_{ij}\dot{\varepsilon}^{\mathrm{p}}_{ij}}$, a *functional* of the plastic strain, and this suggests the pattern in which plastic deformation must appear in the kinetic laws governing plastic flow.

In the case of strain-induced anisotropy, as exhibited by the Bauschinger effect, application of continuum thermodynamics of dissipative processes and homogenization presented at this symposium has led to kinematic hardening in which the back stress $\boldsymbol{\alpha}$ (the shift in the centre of the yield surface in stress space) takes the form: a function of $\boldsymbol{\varepsilon}^{\mathrm{p}}$, thus introducing $\boldsymbol{\varepsilon}^{\mathrm{p}}$ as a state variable. This result is in basic disagreement with measurements of the Bauschinger effect, since, for strains large compared with the elastic yield-point strain, the *range* of additional strain which generates an increase in the yield stress due to the associated strain hardening, and softening for reverse stressing, includes a much smaller strain range than the total plastic strain $\boldsymbol{\varepsilon}^{\mathrm{p}}$. The sign of the Bauschinger effect can be changed for smaller additional strain increments of either sign, otherwise the normal type of hysteresis loop would only be obtained for strain oscillations about the origin, for which the sign of the Bauschinger effect for each leg of the loop would change because of the change in the sign of the plastic strain (e.g. tensile or compressive). In fact the usual form of hysteresis loop is obtained if both legs of the loop occur at strains of the same sign. The back stress $\boldsymbol{\alpha}$ must thus be a functional of $\boldsymbol{\varepsilon}^{\mathrm{p}}$. Were the back stress a monotonically changing function of $\boldsymbol{\varepsilon}^{\mathrm{p}}$, a loop traversed in the region of tensile strain would have the stress range in the tensile part of the loop larger than that in the compressive part and this does not happen. A loop associated with a fixed range of strain becomes symmetric in stress range whatever the sign of the average strain.

The functional nature of the influence of plastic strain is expressed by the terms 'flow or incremental' laws in contrast to the 'deformation' theory of plasticity. The classical theories involve rates or increments of strain which are integrated over the straining history to determine the stress variation. It is fortunate that this is so since in many technological applications stress analysis must be carried out on a component which has been previously deformed plastically during manufacture, such as an extruded rod, and the plastic strain measured from its virgin state is not known. A usable theory must be such that measurements on a specimen yield the necessary characteristics to predict the stress history due to subsequent deformation. For example, for combined isotropic–kinematic hardening, measurement of the yield surface will determine σ_0 the isotropic component of the yield stress and $\boldsymbol{\alpha}$ the back stress or anisotropic component. Use of these, and previous studies of strain-induced hardening on the same material, would permit stress analysis of the continuing deformation. Since the plastic strain cannot be determined without knowledge of the previous history of deformation from the virgin state of the material, a law involving total plastic strain cannot be implemented directly.

It seems to me that the ability to make measurements of material characteristics at any stage in a deformation process, which will permit analysis of the subsequent deformation and utilization of variables in conformity with this requirement is crucial for application and is in conformity with the flow or incremental structure of plasticity theory. It also expresses aspects of the physics of plastic deformation in that the plastic strain does not determine a prescribed dislocation distribution, the history of straining is needed. Thus in contrast to some comments at the symposium, I feel that in kinematic hardening, for example, use of $\boldsymbol{\alpha}$ as a state variable with an evolution differential equation for the growth of $\boldsymbol{\alpha}$ in terms of plastic strain rate, as described in my paper for example, cannot be arbitrarily replaced by use of the variable $\boldsymbol{\varepsilon}^{\mathrm{p}}$ in place of $\boldsymbol{\alpha}$. It seems to me that the important difference is that the nature of plastic flow demands that the stress at any time depends on the history of deformation and that specific physical characteristics, other than purely kinematic shape changes, generated by plastic flow are functionals and not functions of the plastic strain.

It has been incorrectly observed that use of the plastic deformation gradient $\mathbf{F}^{\mathrm{p}}$ in the finite deformation theory of plasticity [1] involves utilization of the plastic strain variable, but in the kinetic part of the analysis it appears only in the rate of plastic strain term $\mathbf{D}^{\mathrm{p}}$, the symmetric part of the plastic velocity gradient:

$$\frac{\partial \mathbf{v}^{\mathrm{p}}}{\partial \mathbf{p}} = \frac{\partial \mathbf{v}^{\mathrm{p}}}{\partial \mathbf{X}} \frac{\partial \mathbf{X}}{\partial \mathbf{p}} = \dot{\mathbf{F}}^{\mathrm{p}} \mathbf{F}^{\mathrm{p}-1} \tag{1}$$

where $\mathbf{p}$ is the position variable in the plastically deformed state and $\mathbf{X}$ that in the reference state, usually the undeformed state. However, since $\partial \mathbf{v}^{\mathrm{p}}/\partial \mathbf{p}$ is defined entirely in the current plastically deformed state, the expression would not be influenced by whatever reference state $\mathbf{X}$ is used, so that the aspect of $\mathbf{F}^{\mathrm{p}}$ which involves $\mathbf{X}$ does not in fact have a specific influence on the kinetic parts of elastic–plastic analysis. The resulting functional equations, the flow law and the evolution equation for $\boldsymbol{\alpha}$, are integrated to determine the stress and velocity fields from which the plastic strain and other kinematic quantities can then be determined by integration after the elastic–plastic analysis has been completed.

Reference

1. Lee, E. H. Elastic–plastic deformation at finite strains, *J. appl. Mech.*, **36** (1969), 1–6.

ELASTIC–PLASTIC MATERIALS AT FINITE STRAINS

S. Nemat-Nasser (*Northwestern University, Evanston, Illinois, USA*): The authors (Kleiber and Raniecki) have attempted to cover a large number of topics in finite plasticity. While such an ambitious effort is admirable, it is bound to lead to superficial discussions of important issues and sometimes incorrect statements. One glaring mistake relates to the rate of plastic work. The authors state after their eqn. (2.26) and again in their Subsection 2.3 that the rate of plastic work depends on the strain measure and the reference state used. Obviously this cannot be true; otherwise by simply changing the strain measure, i.e. the *scale* for measuring the shape change of a material element, one would obtain a different amount of plastic dissipation for exactly the *same physical history of deformation.*

What the authors do not seem to realize is that they define the increment of plastic work in terms of a cycle of *stress.*

Since the stress rate in finite deformation is measure-dependent, as has been extensively discussed by Hill [1], the increment of work in a cycle of loading, whether plastic or otherwise, naturally would be strain-measure-dependent, because it will depend on *which stress is being cycled.* On the other hand, if one works in the strain space, then the increment of plastic work in a *strain cycle* will be measure-independent.

The distinction between these two ways of defining the rate of plastic work has been examined in the literature in connection with Drucker's postulate which involves a stress cycle, and Il'yushin's postulate which involves a strain cycle. As pointed out by Hill [1] and Hill and Rice [2], Il'yushin's postulate, being based on work in the strain space, is measure-independent, and therefore, the corresponding rate of plastic work will be measure-independent; for a discussion, see Subsection 3.2, p. 454, of Hill and Rice [2], and Subsection 4.2, p. 1122, of Nemat-Nasser [4].

The authors state, in connection with their eqn. (2.37), that their observation of the rate of plastic work being measure-dependent 'contradicts the basic Nemat-Nasser assumption in his controversial paper · · ·' (Nemat-Nasser [3]). First, since Nemat-Nasser's [3] development is in the strain space and since Nemat-Nasser defines the rate of plastic strain *directly in terms of the micro-structural changes* (i.e. in terms of the change in the internal variables, $d\mathbf{H}$ (see Nemat-Nasser's eqns. $(2.7)_1$ and (2.13)), and *not* by considering a cycle of *stress*), the corresponding rate of plastic work (equal to the rate of dissipation for the case considered by Nemat-Nasser) is, in fact, measure-independent without any assumptions. Second, the basic development of Nemat-Nasser is *kinematical*, and therefore, the kinetic results relating to the rate of plastic dissipation (at the end of Subsection 2.1) are consequences that are pointed out in passing. Third, the issue of the journal containing Nemat-Nasser's paper was available when the authors presumably were finishing their paper. Their use of the word 'controversial' shortly after the article had appeared, therefore, is puzzling.

I am sorry that I have to submit a negative discussion of this paper with such an impressive scope, but I feel obliged to caution unsuspecting readers.

References

1. HILL, R. On constitutive inequalities for simple materials, I and II, *J. Mech. Phys. Solids*, **16** (1968), 229–242 and 315–322.
2. HILL, R. and J. R. RICE. Elastic potentials and the structure of inelastic constitutive laws, *SIAM J. appl. Math.*, **25** (3), (1973), 448–461.
3. NEMAT-NASSER, S. On finite deformation elasto-plasticity, *Int. J. Solids Struct.*, **18** (10), (1982), 857–872.
4. NEMAT-NASSER, S. On finite plastic flow of crystalline solids and geomaterials, *J. appl. mech.*, **50** (4b), (1983), 1114–1126.

Authors' Reply (B. Raniecki)

The estimation of the true value of our paper we leave to readers. I shall confine myself to the merits.

(i) In order that the discussion may be helpful in understanding the subtle aspects of the foundation of the plasticity theory, one has to agree with a number of essential observations upon which the Hill–Rice general framework is based and also with a number of the important issues of the theory. Consider isothermal processes at the reference temperature $\theta = \theta_0$.†

1. The notion of a q^i-cycle (strain-cycle) is the invariant concept.
2. In the case of elastic–plastic solids the notion (the notion and not the work itself) of a p_i-cycle (generalized stress-cycle) is not the invariant concept: physically cycling one set of p_i does not

† Here the notation employed and reference numbers given follow those of the paper under discussion.

automatically cycle any other, even with the current configuration as a common reference (Hill, ref. 3, p. 39). Exceptional cases are cycles that start and end at stress-free states. The notion of a group of such cycles is an invariant concept.

3. For rate-independent elastic–plastic solids, the plastic part of the increment of any constitutive function of p_i and H^r, say $\mathrm{d}^{\mathrm{p}}\kappa(p_i, H^r)$, can generally be affected physically only by a *cycle of stressing* (Hill and Rice, ref. 8, p. 451; Hill, ref. 3, pp. 36–7).
4. The general structure of incremental constitutive equations in H–R theory adheres to the classical view, i.e.

$$\mathrm{d}p_i = \frac{1}{\rho_o} l_{ij}\, \mathrm{d}q^j + \mathrm{d}^{\mathrm{p}}p_i = \frac{1}{\rho_o} l_{ij}(\mathrm{d}q^j - \mathrm{d}^{\mathrm{P}}q^j) \tag{D.1}$$

$$\frac{1}{\rho_o} l_{ij}\, \mathrm{d}^{\mathrm{p}}q^j = -\mathrm{d}^{\mathrm{P}}p_i \tag{D.2}$$

and it is invariant.

5. The work expanded in the plastic part of $\mathrm{d}q^i$, $(\mathrm{d}W)^{\mathrm{P}} \equiv p_i\, \mathrm{d}^{\mathrm{p}}q^i$, is commonly identified as the incremental 'work of plastic deformation'. Here, $\mathrm{d}^{\mathrm{p}}q^i$ is the same as occurring in (D.1), it is operationally defined as indicated in point 3 above, and it is so expressed in terms of $\mathrm{d}H^r$ that the plastic part of the free enthalpy $\Psi(p_i, H^r)$ increment is its potential,

$$-\mathrm{d}^{\mathrm{P}}q^i = \frac{\partial[\mathrm{d}^{\mathrm{p}}\Psi(p_j, H^r)]}{\partial p_i} \tag{D.3}$$

6. The immediate consequence of point 2 is that the physical dimension change of the material element corresponding to $\mathrm{d}^{\mathrm{P}}q^i$ is not independent of the particular generalized co-ordinate (e.g. strain measure) used (Rice [9], p. 438); that is, the partitioning of incremental strain into elastic and plastic parts is itself not an invariant concept (Hill [6], p. 183). In particular, the additive decomposition of Eulerian strain rate tensor $\mathbf{D}$, $\mathbf{D} = \mathbf{D}^{\mathrm{e}} + \mathbf{D}^{\mathrm{p}}$, *is not invariant.* Even if the actual configuration is regarded as common reference, $\mathbf{D}^{\mathrm{p}}$, likewise stress rate, depends on the choice of strain measure.
7. The other immediate and intuitive consequence of points 2 and 6 above is that $(\mathrm{d}W)^{\mathrm{P}}$ *is not co-ordinate-invariant.* In particular, in the Eulerian description, $\boldsymbol{\sigma} \cdot \mathbf{D}^{\mathrm{p}}/\rho$ is not invariant since all conjugate stresses coincide with $\boldsymbol{\sigma}/\rho$ when the actual configuration is regarded as common reference.

Point 7 proves (in terms of manipulative arguments) the statement which the discussant has categorized as 'glaring mistake'! It is surprising that such an elementary observation, which can easily be illustrated on simple tension, can be questioned. The rigorous analytical proof of points 6 and 7 in terms of concise notation, following from the use of generalized coordinates [3, 6], is given in Subsection 1.2.3 of our paper. The other 'incorrect statements' in our paper are known only to the discussant.

The network expanded in any *actual* process of elastic–plastic straining, whether cyclic or not, *is always the total work*. Its partitioning into elastic and plastic parts must, therefore, entail certain concepts. When $(\mathrm{d}W)^{\mathrm{P}} = p_i\, \mathrm{d}^{\mathrm{P}}q^i$ occurring in *H–R* theory is identified as the incremental plastic work, the concept is clear—unfortunately the plastic work is not invariant. This is why the search for a definition of invariant incremental plastic work is desirable. The closest to an experimental insight is the incremental plastic work, $\bar{\bar{\mathrm{d}}}W^{\mathrm{P}}$ say, which is understood as the incremental total work of straining from an elastic–plastic state $P_1(q^i, H^r)$ to neighbouring elastic–plastic state $P_2(q^i + \mathrm{d}q^i, H^r + \mathrm{d}H^r)$ reduced by the difference between the elastic potential energies at states P_1 and P_2. The elastic potential energy at any state P is identified with the work that could be recovered in the conceptual purely elastic process ($H^r = \text{const.}$) of destressing from P to the corresponding stress-free state. Being expressed in terms of work, $\bar{\bar{\mathrm{d}}}W^{\mathrm{P}}$ is automatically invariant, moreover $\bar{\bar{\mathrm{d}}}W^{\mathrm{P}} = -\mathrm{d}^{\Gamma}\Phi_N(q^i, H^r)$ (substitute $\theta = \theta_0$ into eqn. (1.31) in our paper). The essence of this idea is presented in Subsection 1.3.3 of our paper in the context of *M* theory. *In general* $\bar{\bar{\mathrm{d}}}W^{\mathrm{P}} \neq (\mathrm{d}W)^{\mathrm{P}}$ *so that 'the conjugate plastic strain increment' cannot be expressed through eqn.* (*D*.3).

(ii) I am obliged to explain why we consider Nemat-Nasser's paper [50] to be controversial. There are several reasons. The most relevant to the matter in hand is the fact that in ref. 50 the subtle difference between $(\mathrm{d}W)^{\mathrm{P}}$ and invariant incremental plastic work is overlooked. This oversight is the source of a number of unprecise statements. The glaring mistake is reiterated in the discussion. What the discussant does not seem to realize is the following fact: whenever plastic strain rate is so defined that its product with the work-conjugate stress results in an invariant rate of plastic work then, in general, the plastic strain rate cannot be expressed in terms of $\dot{H}^r$ by eqn. (2.7) given in ref. 50 (eqn. D.3, herein); moreover, the rate constitutive equations cannot have the form presented in Section 2 of ref. 50 (D.1 and D.2, herein). The pitfall which the author has not avoided is similar in type to the one mentioned by Hill (ref. 3, p. 29).

Note: As I have only had a few days at my disposal to write this reply it has not been possible to contact my coauthor Professor Kleiber who at this time is in the USA. However, the shortage of time has in no way influenced the essential nature of my reply.

Editor's Remark (A. Sawczuk)

Reflecting on finite deformation elastoplasticity some time ago, when the idea and the need for the Plasticity Today Symposium appeared, it was clear to me that a special session should be devoted to the problem, where various concepts and results could be presented. The reason for this was not because the essentials could not be seized and transmitted to us through individual minds—on the contrary, we need only mention the reflections by Hill, Mandel and Rice. It was felt rather that various approaches and results had to be regarded without passion and in a comparative manner in order to make the essentials fully accessible to application-oriented minds like my own.

I feel no necessity to specify the reason for my conviction that, belonging to a younger generation of mechanicians, Professors Kleiber and Raniecki were the proper persons for the task. But every intellectual effort is marked by the personality of a thinker, by his competence and his own manner of presenting the problem. All this may deform, by no means purposely, the Cartesian frame of rational thinking: namely, specification of the problem, its separation into elements accessible to an analysis, scrutiny of the elements, and reflection as to whether any element has been forgotten when attempting a conclusion.

As the General Lecture on finite elastoplasticity provoked a discussion, an obviously natural and necessary thing in research, it was felt pertinent to have the discussions printed in the Proceedings, as they concerned essential points and appeared instructive.

I do not feel it either appropriate or necessary to conclude the discussion by presenting my own judgement, as it is obviously biased by my only partial competence in the problem and by my personal attitude. Leaving the contributions to discussions in their original forms allows the reader to make his own reflections as to the questions raised and the need for scientific discussions on elastoplasticity in order to arrive at the proper conceptual setting of the domain. Emotions, although not essential in any search for the truth, are a normal part of human behaviour even when presenting results of an intellectual kind.

The 'Discussions' and the 'Reply' are left in the forms submitted. The sincerity of the discussants and their involvement in our search for disclosing and making understandable the laws of nature in this specific domain of our intellectual inquiry have to be appreciated.

Section 2

EXPERIMENTAL PLASTICITY

10

Experiments and the Micromechanics of Viscoplasticity

R. J. CLIFTON

Division of Engineering, Brown University, Providence, Rhode Island, USA

ABSTRACT

Dislocation motion is viewed as consisting of the thermally activated motion of dislocations past obstacles and the viscous glide of dislocations between obstacles. This view is supported by so-called 'direct dislocation velocity measurements' in which crystals are subjected to a known applied shear stress for a prescribed duration. Such experiments indicate that dislocation velocity increases strongly with increasing resolved shear stress at low stress levels and becomes essentially proportional to the resolved shear stress at high stress levels. Furthermore, when the resolved shear stress is proportional to the dislocation velocity the drag coefficient that appears as a proportionality constant increases with temperature over wide ranges of temperature—consistent with the drag force being due to phonon viscosity.

An overview is given of experimental evidence that contributes to improved understanding of mechanisms of viscoplastic deformation at high strain rates. Evidence of a transition in the rate controlling mechanism for viscoplastic response of polycrystalline metals at high strain rates is presented.

1. INTRODUCTION

In this paper, quite different aspects of plasticity are considered than those reported on by others at this meeting. The emphasis herein is on improved understanding of physical models of plastic deformation, especially at high strain rates. Recent experiments for examining dynamic plastic response are considered. Experimental results are interpreted within the framework of rate dependent (i.e. visco-plastic) models of plastic flow. Emphasis is placed on physical models of plastic flow based on the mechanics of mobile dislocations. Infinitesimal deformations are considered in order to focus attention on the rate controlling processes of plastic deformation, without the complicating effects of large deformations.

The mobility of individual dislocations has been studied by means of experiments introduced by Johnston and Gilman [7] to make so-called 'direct dislocation velocity measurements'. In these experiments measurements are made of the distances moved by dislocations during the loading of single crystals by known stress pulses. Johnston and Gilman, whose experiments were on LiF crystals, determined the initial and final positions of dislocations by etching the crystals to reveal etch pits where the dislocation lines pierce the surfaces. The velocity of the dislocations is taken to be the distance moved by the dislocations divided by the pulse duration. The corresponding resolved shear stress is taken to be the value corresponding to the plateau of the stress pulse. Examples of the resulting curves of dislocation velocity versus resolved shear stress are shown in Fig. 1.

For the highest dislocation velocities shown on the curves for LiF obtained by Johnston and Gilman [7] the stress pulses were longitudinal stress waves with pulse durations of the order of 20 μs. Pulses of such short duration are necessary to prevent the moving dislocations from leaving the crystal during the pulse. In the high dislocation velocity regime the curves appear to take a slope corresponding to a proportional relation between dislocation velocity and resolved shear stress. Such a linear viscous relation is usually written in the form

$$\tau b = B v_{\mathrm{d}} \tag{1}$$

where τ is the resolved shear stress acting on the dislocation, b is the Burgers vector of the dislocation, B is the drag coefficient and v_{d} is the dislocation velocity. The drag force τb in (1) is, at temperatures ranging from a few degrees absolute to one-half the melting temperature, generally attributed to the interaction of the moving dislocations with thermal phonons. Vreeland and co-workers [4, 6, 14] investigated such phonon drag mechanisms by means of torsional stress pulse experiments on a number of high purity, annealed metal crystals (Zn, Al, Cu). In these experiments a line was scribed across a diameter of

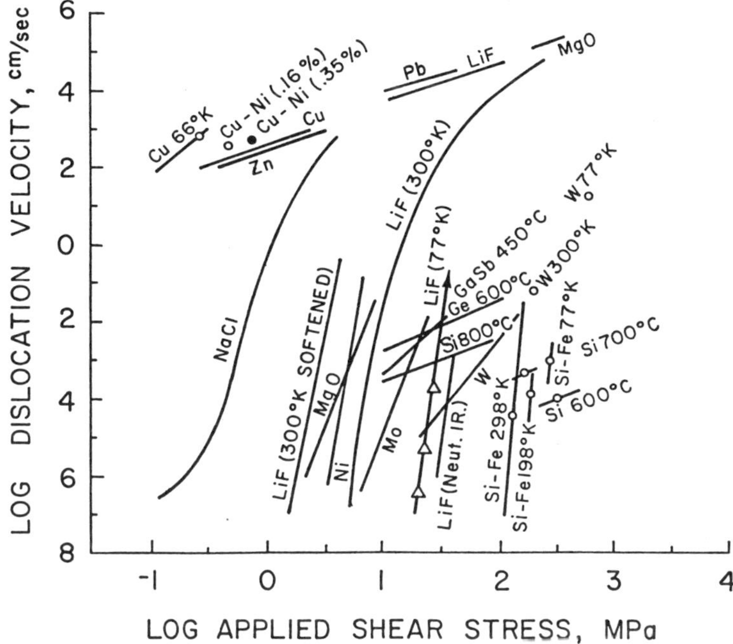

FIG. 1. Dislocation mobility in various materials (from ref. 8).

the circular cylindrical specimen to produce a source of fresh dislocation loops. After the specimen is subjected to the stress pulse the positions of these loops are observed by means of X-ray topography. The diameter of the expanded loops is observed to increase with increasing distance from the axis of the specimen. This observation, which corresponds to v_d in (1) increasing in proportion to the radial co-ordinate, coupled with the corresponding proportional change of shear stress with radius according to elastic torsion theory, supports the validity of (1). Use of elastic torsion theory is applicable because the dislocation density is sufficiently low that the straining is essentially elastic even though a few dislocations move. An attractive feature of these experiments is the exact one-dimensionality of the torsional waves, in contrast to the case of longitudinal bar waves for which the interpretation of experimental results is complicated by lateral inertia effects. An important contribution of the research by Vreeland and co-workers [4, 6, 14] is the systematic study of the influence of temperature which showed that the drag coefficient increases essentially linearly with increasing temperature in agreement with theoretical predictions of dislocation drag due to phonon damping. One drawback of these experiments is the limitation of the pulse amplitude to relatively low levels for which the dislocation velocities are not more than approximately 1% of an elastic shear wave speed for the material. This limitation arises because of the limited torque carrying capacity of a thin disk that is vaporized to release a stored torque and generate the stress pulse used for the experiment.

Extensions of dislocation mobility measurements to higher dislocation velocities by means of plate impact experiments will be described subsequently. First, however, a brief analysis of plastic deformation due to dislocation motion will be given in order to motivate the choice of experiments to be considered. As a result, measurements of both dislocation velocity and mobile dislocation density for known loading histories are reported. Finally, the combined influences of dislocation velocity and mobile dislocation density at very high strain rates are presented for pressure-shear plate impact experiments on polycrystalline aluminum specimens.

2. CRYSTALLINE PLASTICITY

Plastic deformation of crystalline materials occurs primarily through the glide motion of dislocations. For single crystals being deformed homogeneously by slip on slip systems $k = 1, \ldots, n$ the plastic strain rate tensor is

$$\dot{\mathbf{E}}^{\mathrm{p}} = \sum_{k=1}^{n} \dot{\gamma}_k^{\mathrm{p}} (\mathbf{m}^k \otimes \mathbf{n}^k)_s \tag{2}$$

where $\mathbf{m}^k, \mathbf{n}^k$ are unit vectors in, respectively, the slip direction and the direction of the normal to the slip plane for the kth slip system; the symbol $(\mathbf{m}^k \otimes \mathbf{n}^k)_s$ denotes the symmetric part of the tensor that is the tensor product of $\mathbf{m}^k$ and $\mathbf{n}^k$. The quantity $\dot{\gamma}_k^{\mathrm{p}}$ in (2) is the plastic shearing rate defined by

$$\dot{\gamma}_k^{\mathrm{p}} = \frac{1}{V} \int_{C^k} b^k v_{\mathrm{d}}^k \, \mathrm{d}l^k \tag{3}$$

where b^k is the Burgers vector, v_{d}^k is the dislocation velocity, and $\mathrm{d}l^k$ is the length of an element of the mobile dislocation line length l^k along the curve C^k in an elemental volume V. The usual expression

$$\dot{\gamma}^{\mathrm{p}}_{\mathrm{k}} = b^k N^k_{\mathrm{m}} \bar{v}^k_{\mathrm{d}} \tag{4}$$

for the plastic shearing rate is obtained by introducing the average dislocation velocity

$$\bar{v}^k_{\mathrm{d}} = \frac{1}{l^k} \int_{C^k} v^k_{\mathrm{d}} \, \mathrm{d}l^k \tag{5}$$

and the mobile dislocation density

$$N^k_{\mathrm{m}} = \frac{l^k}{V} \tag{6}$$

From (2) and (4) the plastic strain rate tensor $\dot{\mathbf{E}}^{\mathrm{p}}$ can be evaluated for a given loading history if the product $N^k_{\mathrm{m}} \bar{v}^k_{\mathrm{d}}$ is known throughout the loading history; for finite deformations it is also necessary to know the rotations of the vectors $\mathbf{m}^k$, $\mathbf{n}^k$ as the deformation proceeds. The tensor $\dot{\mathbf{W}}^{\mathrm{p}}$ which characterizes the rate of rotation of material elements due to the plastic shearing (4) is given by

$$\dot{\mathbf{W}}^{\mathrm{p}} = \sum_{k=1}^{n} \dot{\gamma}^{\mathrm{p}}_{\mathrm{k}} (\mathbf{m}^k \otimes \mathbf{n}^k)_{\mathrm{A}} \tag{7}$$

where the subscript A denotes the antisymmetric part of the associated tensor. Thus, the determination of the rate of rotation tensor $\dot{\mathbf{W}}^{\mathrm{p}}$ requires the same information as that needed for the plastic strain rate tensor $\dot{\mathbf{E}}^{\mathrm{p}}$. Since these two tensors characterize completely the kinematics of plastic deformation of crystals it is evident that experiments which provide measurements of $N^k_{\mathrm{m}} v^k_{\mathrm{d}}$ are fundamental to the development of physical models of plastic flow. Furthermore, for polycrystalline materials, constitutive equations that describe the evolution of $N^k_{\mathrm{m}} v^k_{\mathrm{d}}$ for a given loading history are required for modelling the plastic flow, including the effects of the development of texture due to the rotation of grains.

3. DISLOCATION VELOCITY MEASUREMENTS

Insight into the form of the dependence of dislocation velocity on the current state of stress can be obtained from the Second Law of Thermodynamics. Following Rice [15] one can show that the Second

Law gives the restriction

$$\tau^k b^k v_d^k \geqslant 0 \tag{8}$$

where τ^k, the resolved shear stress for the kth slip system, is defined by

$$\tau^k = \mathbf{m}^k \cdot \mathbf{T}\mathbf{n}^k \tag{9}$$

in which **T** is the non-singular part of the stress tensor. The inequality (8) is a requirement that dislocation glide be a dissipative process. In order for this requirement to be satisfied at sufficiently small values of the resolved shear stress τ^k, the sign of the dislocation velocity v_d^k must be determined by the resolved shear stress τ^k. Thus, for sufficiently small values of resolved shear stress, the dependence of v_d^k on τ^k must dominate its dependence on other resolved shear stresses. Because this primary dependence of v_d^k on τ^k is expected to continue to hold at large values of resolved shear stress, the average dislocation velocity $\bar{v}_d^k$ is generally regarded as depending only on the corresponding resolved shear stress (see Fig. 1).

Data for the highest dislocation velocities shown in Fig. 1 for LiF [12] and MgO [8] were obtained using the plate impact configuration shown in Fig. 2. The single crystal specimen is impacted by a thin flyer plate to produce a square stress pulse with a duration equal to the round-trip transit of longitudinal waves propagating in the thickness direction of the flyer. This compressive pulse propagates into the impedance-matching, momentum trap and reflects from the traction-free rear surface as a tensile pulse. Because the momentum trap is not bonded to the specimen the tensile pulse causes the momentum trap to fly off, leaving the specimen at rest and unstressed except for unloading waves from lateral boundaries. These unloading waves are directed into corridors around the periphery of the crystal by means of the eight-pointed star shape used for the flyer plate [11]. Thus, the central region of the crystal is subjected to essentially a single plane pulse, as required for a straightforward interpretation of the experiment. Details of the pulse shape are obtained by means of a laser displacement interferometer. Subsequent impact of the specimen is prevented by stopping the projectile with the impact of the projectile plate on the anvil.

Etch-pit techniques are used to observe dislocation configurations caused by the known stress pulse. In the bulk of the central region of the crystal these configurations are observed to correspond to circular-shaped regions of glide on the active slip planes. The radius of these

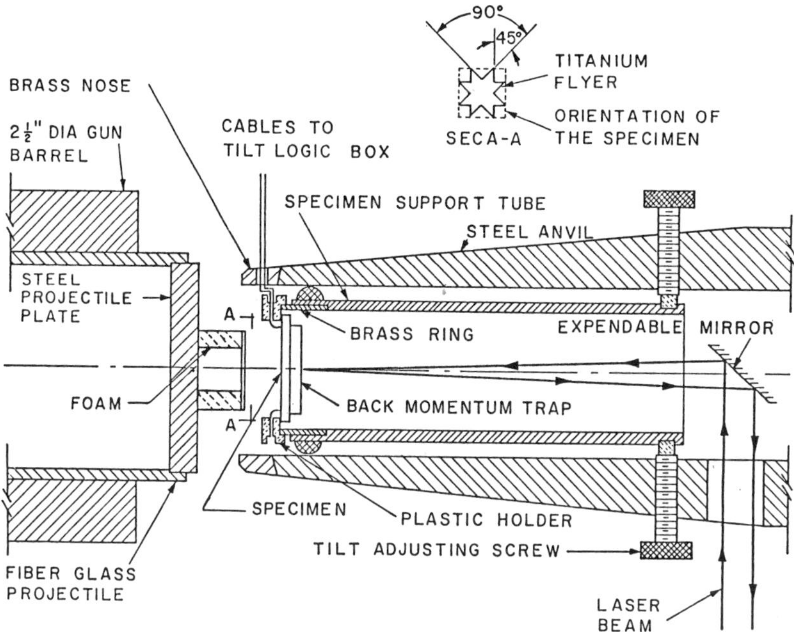

FIG. 2. Schematic of a plate impact experiment for determining dislocation mobility (from ref. 12).

regions is taken to be the distance moved by a dislocation during the pulse. This distance divided by the pulse duration gives the average dislocation velocity.

For high-purity, annealed LiF, which has been investigated most thoroughly, the dislocation velocity is observed to increase in proportion to the resolved shear stress, in agreement with the linear viscous drag relation (1) with B in the range $3{\cdot}0\text{–}4{\cdot}0 \times 10^{-5}$ N s/m^2 [12]. Dislocation velocities obtained in these experiments were in the range $0{\cdot}1c_2$–$0{\cdot}25c_2$, where c_2 is the relevant elastic shear wave speed. Similar values of B at lower dislocation velocities have been obtained for similar crystals in low velocity plate impact experiments [2] and ultrasonic attenuation experiments [1]. Thus, it appears that the intrinsic resistance of the lattice to the motion of dislocations is essentially proportioned to the dislocation velocity over a wide range of dislocation velocities below $0{\cdot}25c_2$. At higher velocities, as relativistic effects become important, it is expected that the drag force will increase more rapidly with increasing dislocation velocity; however, it is not expected that these effects will play much of a role until dislocation velocities exceed $0{\cdot}5c_2$.

4. MOBILE DISLOCATION DENSITY MEASUREMENTS

The observation of dislocation configurations caused by a known stress pulse can also be used to gain insight into the generation of dislocations during such a pulse. However, such observations do not allow one to infer directly the mobile dislocation density at any time during the pulse. Direct inference of mobile dislocation density before, during, and after the passage of a stress pulse has been made by Shioiri *et al.* [16] using an ultrasonic wave technique. The experimental setup is shown in Fig. 3. The specimen is sandwiched between two elastic bars as in a Kolsky bar experiment [10]. A compressional pulse with a duration of approximately 300 μs is generated in one of the pressure bars by impacting the remote end with another bar. As the compressional pulse propagates through the specimen, the specimen is deformed plastically. To investigate the mobile dislocation density that exists during this plastic deformation an ultrasonic wave (11 MHz) is propagated through the specimen in a direction perpendicular to the axis of the pressure bars. This ultrasonic wave is monitored at 18 μs intervals to determine the change in attenuation $\Delta\lambda$ and the change in wave velocity $\Delta V/V$. Results are shown in Fig. 4 for polycrystalline,

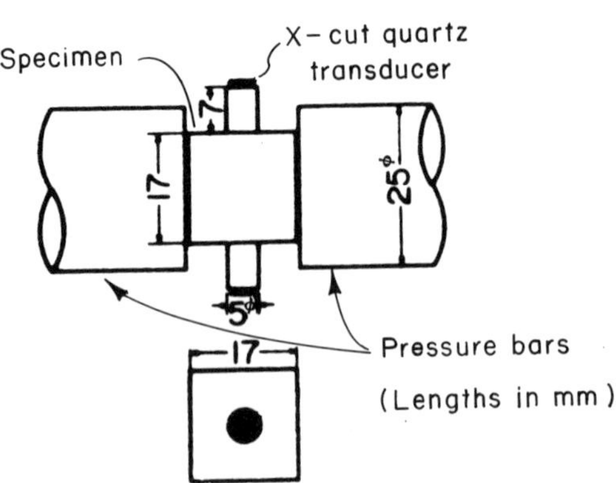

FIG. 3. Schematic of experiment for ultrasonic probing of specimen being deformed at high strain rates (from ref. 16).

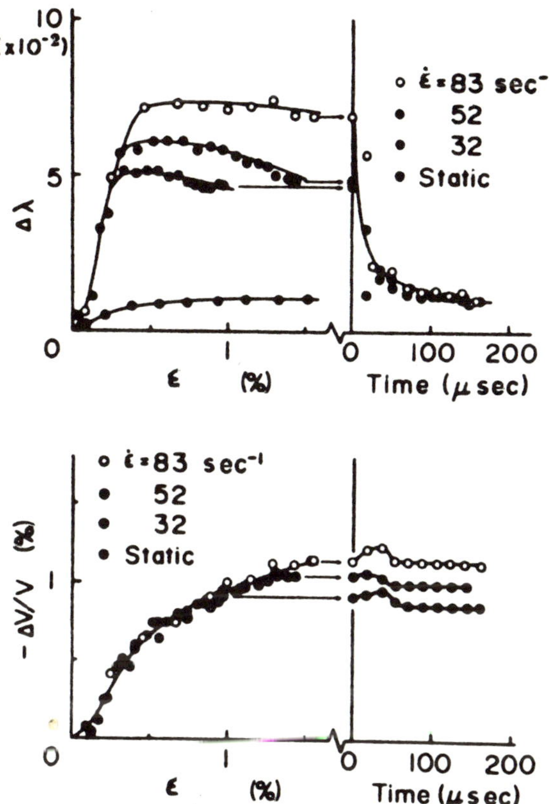

FIG. 4. Ultrasonic attenuation and wave speed changes during and after dynamic plastic deformation (from ref. 16).

annealed OFHC copper [16]. The attenuation during plastic deformation increases with increasing strain rate whereas the corresponding wave velocity is independent of the strain rate. After the specimen is unloaded, the attenuation decreases to approximately the values that are obtained for quasi-static deformation to the same total plastic strain. The velocity change $\Delta V/V$ is essentially unchanged by unloading.

These results are interpreted by Shioiri *et al.* [16] within the framework of the Granato–Lücke theory [5] for the interaction of pinned dislocations with a harmonic stress wave. Dislocations are viewed as consisting of tightly pinned dislocations and relatively free mobile dislocations. Shioiri *et al.* [16] view the tightly pinned dislocations as tangled cell wall dislocations although it appears doubtful that cell walls are established for the small plastic strains obtained in the experiments reported. However, the identification of the tightly pinned dislocations with cell wall dislocations is not essential for the interpretation of the experimental results.

According to the Granato–Lücke theory [5], the attenuation, expressed in terms of the decrement $\Delta\lambda$ (one-half the fractional energy loss per cycle), is approximately

$$\Delta\lambda = \frac{8N_m\omega\, dv^2}{\pi^2[(\omega_o^2-\omega^2)^2+\omega^2d^2]} \tag{10}$$

where ω is the frequency of the ultrasonic wave, ω_o is the natural frequency of the vibrating dislocation segments, N_m is the density of vibrating dislocations, v is the velocity of wave propagation and d is the drag coefficient B divided by the mass per unit length of the dislocation line. The corresponding change in the velocity of propagation is approximately

$$\frac{\Delta v}{v} = \frac{-4N_m v^2(\omega_o^2-\omega^2)}{\pi^3[(\omega_o^2-\omega^2)^2+\omega^2d^2]} \tag{11}$$

Equations (10) and (11) are derived for dislocation segments of uniform length, oriented perpendicular to the direction of propagation of the ultrasonic wave. Shioiri *et al.* [16] incorporated the effects of non-uniform dislocation lengths and non-uniform dislocation segment orientations. The increased attenuation at high strain rates was attributed to increased numbers of mobile dislocations. The density of relatively free, mobile dislocations was calculated using the high frequency limit (i.e. $\omega \gg \omega_o$) of eqn. (10). These mobile dislocation densities varied from $2{\cdot}3\times10^7\,\text{cm}^{-2}$ at a strain rate of $32\,\text{s}^{-1}$ to $4{\cdot}0\times10^7\,\text{cm}^{-2}$ at a strain rate of $83\,\text{s}^{-1}$. The corresponding density of tightly pinned dislocations, calculated from the low frequency limit of (10) and (11) with the measured values for the rate insensitive attenuation and velocity changes, is $1{\cdot}2\times10^9\,\text{cm}^{-2}$. Thus, less than 2% of the total dislocation density is viewed as mobile at the reported strain rate. Although further research is required to determine the validity of such estimates of mobile dislocation density, the results appear reasonable and the method appears attractive.

5. PLASTIC RESPONSE AT HIGH SHEAR STRAIN RATES

One means for studying the plastic response of materials at strain rates greater than those attained in Kolsky bar experiments is to use the pressure-shear impact configuration shown in Fig. 5 [13]. A thin specimen (e.g. 0·2–0·4 mm thick) is sandwiched between two hard plates that remain elastic throughout the experiment. The flyer, specimen, and anvil plates are aligned to be parallel, but inclined relative to the direction of approach. Details of the experiment are reported in ref. 13.

The stress wave transmitted through the specimen is monitored at the rear surface of the anvil plate by means of a normal velocity interferometer (NVI) and a transverse displacement interferometer (TDI) [9]. Only the motion during the time before unloading waves arrive from lateral boundaries is of interest. During this time the waves propagating in the three plates can be analyzed as plane waves propagating in the direction perpendicular to the impact face. Linear elastic wave theory is used to describe the wave propagation in the anvil and flyer plates. As in Kolsky bar experiments the time of interest in the experiment begins after the waves in the specimen have reflected from its faces several times so that a nominally homogeneous state of stress is established in the specimen. Then, the nominal

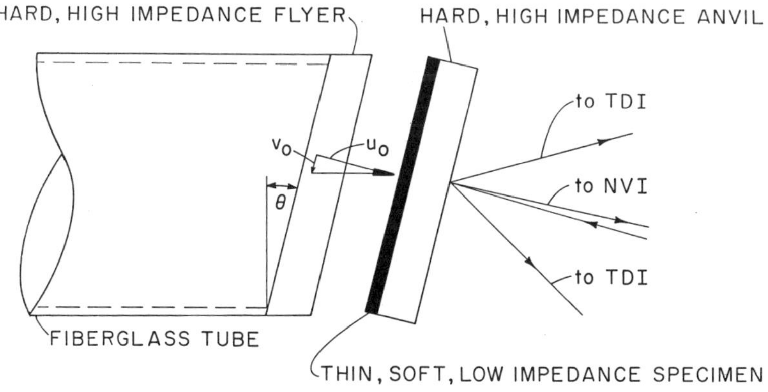

FIG. 5. Schematic of high strain rate pressure–shear experiment (from ref. 13).

longitudinal and shear strain rates in the specimen are

$$\dot{\varepsilon} = \frac{u_o - u_{fs}}{h} \tag{12a}$$

$$\dot{\gamma} = \frac{v_o - v_{fs}}{h} \tag{12b}$$

where u_o, v_o are the normal and in-plane components of the flyer velocity, and u_{fs}, v_{fs} are the normal and in-plane components of the velocity of the rear surface of the anvil plate; h is the specimen thickness. The normal and shear tractions at the rear surface of the specimen are

$$\sigma = \tfrac{1}{2}\rho c_1 u_{fs} \tag{13a}$$

$$\tau = \tfrac{1}{2}\rho c_2 v_{fs} \tag{13b}$$

where ρ, c_1, c_2 are, respectively, the mass density, the elastic longitudinal wave speed, and the elastic shear wave speed for the anvil. Because of the specimen's elastic stiffness to volume changes, the longitudinal stress increases with each wave reflection until, after a few reflections, the difference between the longitudinal components of particle velocity at the two faces of the specimen becomes negligible so that $u_{fs} \to u_o$ and $\dot{\varepsilon} \to 0$. Furthermore, since the longitudinal wave speed is greater than the shear wave speed this state of constant volume, which corresponds to a state of constant hydrostatic pressure, is reached before a nominally homogeneous state of shear stress is established in the specimen. Thus, during the time of interest for studying the plastic response in shear the hydrostatic pressure is essentially constant. Numerical simulation of the experiment indicates that the value of the hydrostatic pressure is approximately equal to the value of the normal stress σ obtained from (13a) as $u_{fs} \to u_o$.

A dynamic stress–strain curve for shearing deformation is obtained by plotting the shear stress τ from (13b) versus the shear strain γ obtained from the integration of (12b). Representative dynamic stress–strain curves for annealed 1100–0 aluminum are shown in Fig. 6. The steeply rising parts of the curves should be disregarded because the corresponding deformation occurs before nominally homogeneous states of stress are established. After the risetime the strain rate and the flow stress become essentially constant. The shear strain rates for the experiments shown are approximately two orders of magnitude larger than are obtained in conventional Kolsky bar experiments. The flow

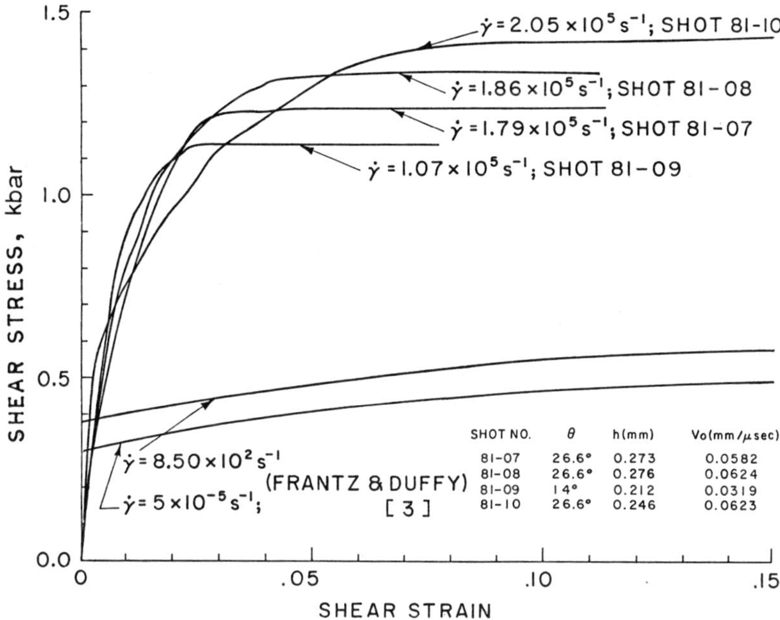

FIG. 6. Dynamic stress–strain curves for 1100-0 aluminum (from ref. 13) (hydrostatic pressure is approximately 2·8 GPa for the pressure–shear experiments).

stress increases strongly with increasing strain rate at strain rates of $10^5\ s^{-1}$. Stress levels are much higher than those shown for dynamic torsion tests at strain rates of $0{\cdot}85 \times 10^3\ s^{-1}$. The stress–strain curves at the lower strain rates in Fig. 6 have been shifted to the left by approximately a 1·5% shear strain to provide more direct comparison with the curves obtained from pressure–shear experiments in which the longitudinal compressive wave causes an additional effective shear strain of approximately 1·5%.

The increase in flow stress between the torsion experiments and the pressure–shear experiments appears to be due primarily to the increase in strain rate, not to the hydrostatic pressure which is present in the pressure–shear experiments and not in the torsion experiments. The relative importance of the effects of strain rate and pressure are examined in pressure–shear experiments by varying the angle θ and the projectile velocity V_o; constant pressure (shear strain rate) and varying shear strain rate (pressure) is obtained by keeping the normal component (in-plane component) of the projectile velocity constant while the other component is varied. Systematic variation of pressure and shear strain rate in a series of 27 experiments on 1100–0 aluminum indicated that the effect of hydrostatic pressure was statistically insignificant.

At the highest stress levels of the pressure–shear experiments, it is expected that thermal activation is no longer required for dislocation motion. Then the average dislocation velocity is limited by the intrinsic resistance of the clear lattice to the motion of dislocations. From (1)–(6) the form of the relationship between the plastic shear strain rate $\dot{\gamma}^{\mathrm{p}}$ and the shear stress becomes

$$\dot{\gamma}^{\mathrm{p}} = \frac{b^2 N_{\mathrm{m}}}{\beta B}\tau \tag{14}$$

where N_{m} is the total mobile dislocation density on all slip systems. The effects of multiple slip systems in each crystal and different orientations of the crystals that comprise the polycrystalline aggregate are accounted for by the dimensionless parameter β. For randomly oriented, face-centered cubic crystals with the same mobile dislocation density on all slip systems the calculated value of β is 2·97 [18], based on Taylor's assumption [17] of uniform plastic straining in all crystals. From (14), the mobile dislocation density N_{m} increases from a value of $2{\cdot}4\times10^8\,\mathrm{cm}^{-2}$ at a strain rate $\dot{\gamma}^{\mathrm{P}} = 1\times10^5\,\mathrm{s}^{-1}$ to $3{\cdot}7\times10^8\,\mathrm{cm}^{-2}$ at a strain rate $\dot{\gamma}^{\mathrm{P}} = 3{\cdot}1\times10^5\,\mathrm{s}^{-1}$. This trend is analogous to that reported by Shioiri *et al.* [16] from measurements of the attenuation of ultrasonic waves.

ACKNOWLEDGEMENTS

Research of the author and his collaborators that is described here has been supported by NSF, ARO and the Brown University NSF/MRL.

REFERENCES

1. Fanti, F., J. Holder and A. V. Granato. Viscous drag on dislocations in LiF and NaCl, *J. Acoust. Soc. Amer.*, **45** (1969), 1356–1366.
2. Flinn, J. E. and R. F. Tinder. Dislocation mobility observations in LiF monocrystals after impact loading, *Scripta Metallurgica*, **8** (1974), 689–694.

3. FRANTZ, R. A., JR. and J. DUFFY. The dynamic stress–strain behavior in torsion of 1100–0 aluminum subjected to a sharp increase in strain rate, *ASME J. appl. Mech.*, **39** (1972), 939–945.
4. GORMAN, J. A., D. S. WOOD and T. VREELAND, JR. Mobility of dislocations in aluminum, *J. appl. Phys.*, **40** (1969), 833–841.
5. GRANATO, A. and K. LÜCKE. Theory of mechanical damping due to dislocations, *J. appl. Phys.*, **27** (1956), 583–593.
6. GREENMAN, W. F., T. VREELAND, JR. and D. S. WOOD. Dislocation mobility in copper, *J. appl. Phys.*, **38** (1967), 3595–3603.
7. JOHNSTON, W. G. and J. J. GILMAN. Dislocation velocities, dislocation densities, and plastic flow in lithium fluoride crystals, *J. appl. Phys.*, **30** (1959), 129–144.
8. KIM, K. S. and R. J. CLIFTON. Dislocation motion in MgO crystals under plate impact, *J. Materials Sci.* **19** (1984), 1428–1438.
9. KIM, K. S., R. J. CLIFTON, and P. KUMAR. A combined normal and transverse displacement interferometer with an application to impact of Y-cut quartz, *J. appl. Phys.*, **48** (1977), 4132–4139.
10. KOLSKY, H. An investigation of the mechanical properties of materials at very high rates of loading, *Proc. Phys. Soc. Lond.*, Series B, **62** (1949), 676–700.
11. KUMAR, P. and R. J. CLIFTON. A star-shaped flyer for plate-impact recovery experiments, *J. appl. Phys.*, **48** (1977) 4850–4852.
12. KUMAR, P. and R. J. CLIFTON. Dislocation motion and generation in LiF single crystals subjected to plate impact, *J. appl. Phys.*, **50** (1979), 4747–4762.
13. LI, C. H. A pressure-shear experiment for studying the dynamic plastic response of metals at shear strain rates of $10^5\,s^{-1}$, Ph.D. Thesis, Brown University, Providence, RI, 1982.
14. POPE, D. P., T. VREELAND, JR. and D. S. WOOD. Mobility of edge dislocations in the basal slip system of zinc, *J. appl. Phys.*, **38** (1967), 4011–4018.
15. RICE, J. R. Continuum mechanics and thermodynamics of plasticity in relation of microscale deformation mechanisms. In: *Constitutive Equations in Plasticity* (Ed. A. S. Argon), M.I.T. Press, Cambridge, Mass., 1975, 23–79.
16. SHIOIRI, J., K. SATOH and K. NISHIMURA. Experimental studies on the behavior of dislocations in copper at high rates of strain. In: *High Velocity Deformation of Solids* (Eds. K. Kawata and J. Shioiri), Springer, New York, 1978, 50–66.
17. TAYLOR, G. I. Plastic strain in metals, *J. Inst. Metals*, **62** (1938), 307–327.
18. YANG, J. P. Calculations of visco-plastic response of polycrystals from slip theory for FCC crystals, Sc.M. Thesis, Brown University, Providence, RI, 1983.

11

On the Evolution, from Experiment, of General Constitutive Equations in a Continuum Theory for Finite Plastic Strain

JAMES F. BELL

Johns Hopkins University, Baltimore, Maryland, USA

ABSTRACT

Described is the culmination of several years of effort to proceed from an observed generalization in experiment to an internally consistent continuum theory. New, symmetric stress and strain tensors are introduced, which include large rotations in their formulation. With these, an application of the theory of constraints to an internal constraint found in experiment provides the desired close correlation between experiment and theory for finite strain. Experiments are described which consider statements on invariants, principal values, and reference configurations.

Both annealed metals and 'as received' structural metal alloys of aluminum, copper, and steel fall naturally in the framework of the same theory in terms of relative reference states. The same framework includes the material stability structure earlier observed in experiment. The result thus is that an incremental continuum theory of finite strain plasticity correlates in detail with observation for arbitrary stress path, strain rates, and composition of ordered solids.

1. ON THE INCREMENTAL CONTINUUM THEORY OF FINITE STRAIN PLASTICITY

As in other areas of continuum mechanics, the study of finite strain plasticity may be approached from at least two perspectives. On the basis of reasonable conjecture one may postulate a continuum theory, or perhaps merely a constitutive premise, and then face the problem of finding corroborating support in experiment. In contrast, still on the basis of reasonable conjecture but without reference to any particular continuum theory or constitutive premise, one may obtain and review a variety of experimental evidence to determine what order and unity exist, and, if order and unity are found, then face the often mathematically difficult problem of finding an internally consistent continuum theory which will correlate with the generalization derived from experiment.

The present paper will describe a tractable final stage of this second approach. New, symmetric stress and strain tensors and an internal constraint generated from experiment, provide a continuum theory of finite strain plasticity closely compatible with experiment at very large strain.

Order and unity are achieved when one introduces stress components, σ_{ij} and strain components, ε_{ij}, referred to a 'genesis' or original, undeformed reference configuration. The statement for total stress, $S_{ij} = \sigma_{ij} - \frac{1}{3}\sigma_{ii}\delta_{ij}$, in terms of which the generalization is formulated, departs from convention: at large strain, $(1/3)\sigma_{ii}$ as defined, is not a hydrostatic stress component!

Persistence in probing an experimental generalization based upon asserting such a total stress, S_{ij}, has been fruitful, to wit, this form is found to be implicit in the subsequent continuum theory as a consequence of the theory of constraints when applied to the second key element of the experimental generalization, $\varepsilon_{ii} = 0$, or $\operatorname{tr}\mathbf{E} = 0$. This constraint, $\operatorname{tr}\mathbf{E} = 0$, was delineated in results from my experiments on finite strain over a decade ago (see ref. 4, Section 4.35).

From experiment we know that the forms, $\boldsymbol{\sigma}$, $\mathbf{S}$, and $\mathbf{E}$, are symmetric, i.e. $\boldsymbol{\sigma} = \boldsymbol{\sigma}^{\mathrm{T}}$, $\mathbf{S} = \mathbf{S}^{\mathrm{T}}$, $\mathbf{E} = \mathbf{E}^{\mathrm{T}}$. Introducing a generalized scalar measure of stress, $T = \sqrt{\operatorname{tr}\mathbf{S}^2}$, and a generalized incremental strain, $\mathrm{d}\Gamma = \sqrt{\operatorname{tr}(\mathrm{d}\mathbf{E})^2}$, the order and unity found from two decades of systematic research comprising experiments on fully annealed solids is summarized in the three statements of eqns. (1), (2) and (3), where W is the laboratory measured work.

$$\mathrm{d}W = S_{ij}\,\mathrm{d}\varepsilon_{ij} = \sigma_{ij}\,\mathrm{d}\varepsilon_{ij} = T\,\mathrm{d}\Gamma \tag{1}$$

$$\frac{\mathrm{d}\varepsilon_x}{S_x} = \frac{\mathrm{d}\varepsilon_y}{S_y} = \frac{\mathrm{d}\varepsilon_z}{S_z} = \frac{\mathrm{d}\varepsilon_{xy}}{S_{xy}} = \frac{\mathrm{d}\Gamma}{T} = \frac{2\mathrm{d}T}{\beta^2} \tag{2}$$

$$\operatorname{tr}\mathbf{E} = 0 \tag{3}$$

where

$$dT = \frac{S_{ij}\, dS_{ij}}{T} \tag{4}$$

Since a detailed analysis of the continuum theory given here is contained in ref. 9, the present paper includes only the results obtained.

For x^i and X^k, with X^k the co-ordinate of particles in a dead annealed or 'genesis' reference state, and **F** the deformation gradient with components dx^i/dX^k, decompose **F** into a rigid rotation **R** followed by a pure homogeneous deformation, **V**.

$$\mathbf{F} = \mathbf{V}\mathbf{R} \tag{5}$$

$$\mathbf{R}^{-1} = \mathbf{R}^{\mathrm{T}} \tag{6}$$

$$\mathbf{V} = \mathbf{V}^{\mathrm{T}} \tag{7}$$

Then **V** has as principal values the principal stretches, λ_1, λ_2, and λ_3, or $\lambda_i = 1 + \varepsilon_i$. The appropriate strain tensor is then

$$\mathbf{E} = \mathbf{V} - \mathbf{1} \tag{8}$$

with the constraint of eqn. (3), $\mathrm{tr}\, \mathbf{E} = 0$.

Let W denote, as above, the 'strain energy' or work done per unit undeformed volume. Let $\mathbf{T}_{\mathrm{R}}$ denote the first Piola Kirchhoff stress, then

$$\dot{W} = \mathrm{tr}\, \mathbf{T}_{\mathrm{R}}^{\mathrm{T}} \dot{F} \tag{9}$$

or

$$\dot{W} = \mathrm{tr}\, \mathbf{T}_{\mathrm{R}}^{\mathrm{T}} (\dot{\mathbf{V}}\mathbf{R} + \mathbf{V}\dot{\mathbf{R}}) \tag{10}$$

or

$$\dot{W} = \mathrm{tr}\, \mathbf{R}\mathbf{T}_{\mathrm{R}}^{\mathrm{T}} \dot{\mathbf{E}} + \mathrm{tr}\, \mathbf{T}_{\mathrm{R}}^{\mathrm{T}} \mathbf{V}\dot{\mathbf{R}} \tag{11}$$

In discussion of the theory contained in eqns. (1), (2) and (3), J. L. Ericksen has pointed out (see ref. 9) that for isotropic solids, with W depending only upon the invariants of **E**, the second term on the right side of eqn. (11) vanishes. Designating σ_{ij} as $\boldsymbol{\sigma}$, and referring to eqn. (1), we have

$$\boldsymbol{\sigma} = \mathbf{R}\mathbf{T}_{\mathrm{R}}^{\mathrm{T}} \tag{12}$$

Let **N** be a stress which does no work in any motion which satisfies the constraint, $\mathrm{tr}\, \mathbf{E} = 0$ (eqn. (3)). Hence we have $\mathrm{tr}\, \mathbf{N}\dot{\mathbf{F}} = 0$. As with eqn. (11), this reduces to the single term

$$\mathrm{tr}\, \mathbf{R}\mathbf{N}\dot{\mathbf{E}} = 0 \tag{13}$$

If $\mathbf{RN}=p\mathbf{1}$, then we have p tr $\dot{\mathbf{E}}=0$. Since tr $\mathbf{E}=0$ and hence tr $\dot{\mathbf{E}}=0$, $\mathbf{N}=p\mathbf{R}^{\mathrm{T}}$ is admissible. With $p=-(1/3)\mathrm{tr}\,\boldsymbol{\sigma}$,

$$\mathbf{RN}=-(\tfrac{1}{3}\,\mathrm{tr}\,\boldsymbol{\sigma})\mathbf{1} \tag{14}$$

Designating $\mathbf{S}$ as the total stress, we thus have for $\mathbf{S}=\mathbf{RT}_{\mathrm{R}}^{\mathrm{T}}+\mathbf{RN}$

$$\mathbf{S}=\boldsymbol{\sigma}-(\tfrac{1}{3}\,\mathrm{tr}\,\boldsymbol{\sigma})\mathbf{1} \tag{15}$$

or

$$S_{ij}=\sigma_{ij}-\tfrac{1}{3}\sigma_{ii}\delta_{ij} \tag{16}$$

Equation (16) is the form introduced above from experiment. Hence, compatible with continuum theory and experiment, we have from eqns. (1) and (2)

$$\mathrm{d}\Gamma=\frac{2T\,\mathrm{d}T}{\beta^2} \tag{17}$$

as a fundamental parabolic property of the finite strain of all of the 50 ordered solids that have been studied in my laboratory.

Of equal import, from eqn. (2) we obtain the incremental constitutive equations for finite plastic strain (eqns. (18)).

$$\mathrm{d}\mathbf{E}=\frac{2\mathbf{S}\,\mathrm{d}T}{\beta^2} \tag{18}$$

or

$$\mathrm{d}\varepsilon_{ij}=\frac{2S_{ij}\,\mathrm{d}T}{\beta^2} \tag{19}$$

where β is a measured material constant. Contrary to assumption in theories of the Prandtl–Reuss type, there are no linear elastic components in eqn. (19). Direct experimental study (ref. 7, Section 7) demonstrates that for a loading stress path at finite strain such terms should be excluded.

Turning to the experiments which preceded the formulation of the theory, we examine, first, what experiment discloses regarding the invariant properties and principal values of $\boldsymbol{\sigma}$, $\mathbf{S}$ and $\mathbf{E}$. Next we consider the measurement of rotation during large deformation and the conclusions we may draw from experiment regarding an appropriate reference configuration at large strain for arbitrary cyclical loading stress paths.

To study invariant properties, we consider proportional loading in plane stress, i.e. the ratio of principal stresses, σ_2/σ_1 is constant. For the simultaneous tension–torsion of thin-walled tubes, σ_2/σ_1 ranges from −1 to 0. For the simultaneous tension vs internal pressure, σ_2/σ_1 ranges from 0 to +1. Data from thin-walled tubes are cross-checked by single compression and two-dimensional compression tests on solid cubical blocks, where $\sigma_2/\sigma_1 = 0$ and $\sigma_2/\sigma_1 = 1/2$, respectively. The ratio of principal stresses for plane stress is given by eqn. (20), where $\boldsymbol{\sigma}$ is defined by eqn. (12).

$$\frac{\sigma_2}{\sigma_1} = \frac{\sigma_x + \sigma_y - \sqrt{(\sigma_x - \sigma_y)^2 + 4S_{xy}^2}}{\sigma_x + \sigma_y + \sqrt{(\sigma_x - \sigma_y)^2 + 4S_{xy}^2}} \tag{20}$$

The first invariants of $\mathbf{S}$ and $\mathbf{E}$ equal zero, i.e. $I_{\mathbf{S}} = 0$ and $I_{d\mathbf{E}} = 0$. The second invariants $II_{\mathbf{S}}$ and $II_{d\mathbf{E}}$ provide $T = \sqrt{(2/3)\sigma_x^2 + 2S_{xy}^2}$ and $d\Gamma = \sqrt{(3/2)(d\varepsilon_x)^2 + (1/2)(ds)^2}$ for the generalized stress and strain in simultaneous axial tension and torsion. Note that $S_{xy} = (\text{applied torque})/r_{m0}A_0$ and $ds = 2d\varepsilon_{xy} = r_{m0}\, d\theta/l_0$ where r_{m0}, A_0, and l_0 are the initial values of the mean radius, the area, and the length of the thin-walled tube; and $d\theta$ is the angle of twist in radians.

For proportional loading along principal axes, with $S_{xy} = 0$, the second invariants lead to $T = \sqrt{S_1^2 + S_2^2 + S_3^2}$ and $d\Gamma = \sqrt{(d\varepsilon_1)^2 + (d\varepsilon_2)^2 + (d\varepsilon_3)^2}$, with the incremental constitutive eqns. (19) becoming eqns. (21).

$$d\varepsilon_i = \frac{2S_i\, dT}{\beta^2} \tag{21}$$

For simultaneous tension and torsion, on the other hand, eqns. (19) become eqns. (22) and (23).

$$d\varepsilon_x = \frac{4\sigma_x\, dT}{3\beta^2} \tag{22}$$

$$ds = \frac{4S_{xy}\, dT}{\beta^2} \tag{23}$$

For simple torsion, eqn. (20) provides $\sigma_2/\sigma_1 = -1$, with interesting kinematical implications at large strain, which are supported by the data from experiment. With $\sigma_z = \sigma_y = 0$, many values of σ_x were chosen for study in simultaneous tension–torsion, with $\sigma_2/\sigma_1 =$ constant, i.e. proportional loading. Three of these are shown as T^2 vs Γ plots in Fig. 1 for $\sigma_2/\sigma_1 = -1$, $\sigma_2/\sigma_1 = -1/4$, and $\sigma_2/\sigma_1 = -0{\cdot}45$. Also included in Fig. 1 are the results for a thin-walled tube in simple tension for which $\sigma_2/\sigma_1 = 0$, and the results for a single axial *compression* test on a radically different geometric form, namely a solid cubical

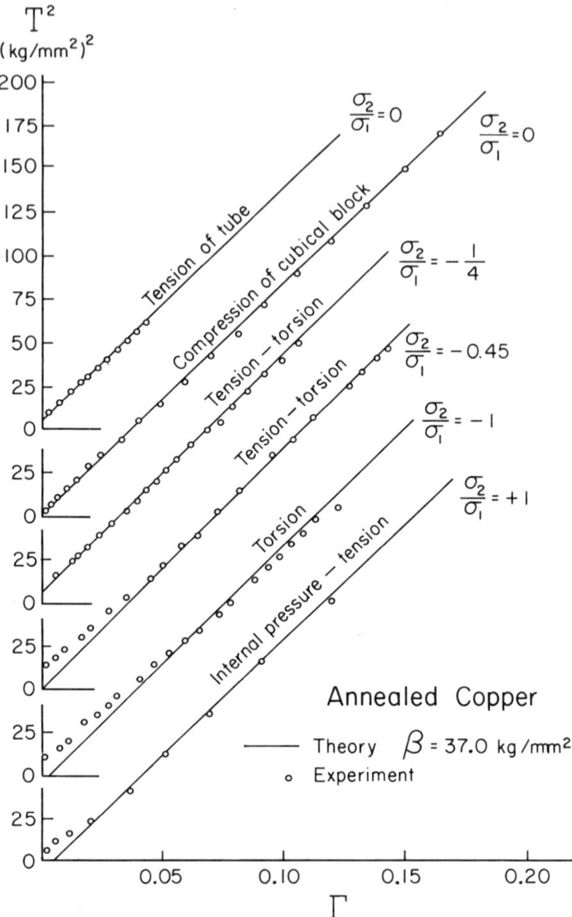

FIG. 1. Experimental data demonstrating that the second invariants, II_S and II_E are independent of principal stress ratio σ_2/σ_1.

block, for which the ratio also is $\sigma_2/\sigma_1 = 0$. All of these data are for annealed copper.

Elsewhere [9], I have described the correlation between the current theory and measurement for data from internal pressure vs tension tests of Evan A. Davis on thin-walled tubes of annealed copper, for eight σ_2/σ_1 ratios in the range from 0 to +1. From this group, one test for $\sigma_2/\sigma_1 = +1$ is included in Fig. 1.

The two-dimensional compression experiment was introduced by

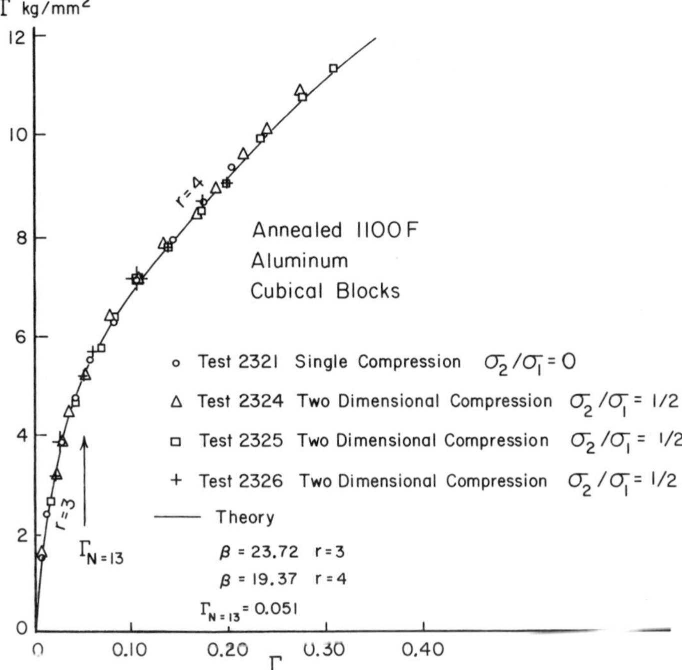

FIG. 2. A comparison of one- and two-dimensional compression in cubical blocks of annealed aluminum, showing that the second invariants are independent of stress path to 30% generalized strain.

Bridgman [11] nearly forty years ago. With σ_1 arbitrary, σ_2 is chosen so that $\varepsilon_2 = 0$ throughout the test. For $d\varepsilon_2 = 0$, from eqns. (21) and eqns. (16), $S_2 = 0$, or $\sigma_2/\sigma_1 = +1/2$. In Fig. 2, T vs Γ plots of the data for a single compression test for which $\sigma_2/\sigma_1 = 0$ are compared with the data from three tests in two-dimensional compression. The specimens are all identical cubical blocks of annealed aluminum.

The invariants and principal values of $\boldsymbol{\sigma}$, $\mathbf{S}$, and $\mathbf{E}$ are indeed proper statements, as is made evident by the fact that T and Γ provide the common response function seen in Figs. 1 and 2.

In non-proportional loading, finite strains extending to 30% or 40% are accompanied by an equally large rotation of principal axes. Viewed from the present perspective, one sees that in the data on non-proportional loading for annealed copper [6] and for annealed mild steel [10] for which experiment and theory, eqns. (1)–(19), are in close accord to large finite strain, rotations of principal axes as high as 40° or 50° also are present.

Early in the present study, measurements for loading into the finite strain domain were followed by unloading and reloading. In a reloading along the same or some alternative stress path, the question of reference configuration arises. Should stress and strain components be referred to the last unloaded state or to the original, undeformed, 'genesis' state? The results were definitive.

When the reloading reached and then exceeded the previous maximum generalized stress, T(OYS), there was disorder among the data if the stress and strain components were defined in terms of the last unloaded reference configuration. The character of the observed strain path altered from one measurement to another, depending upon the history of the stress path which preceded the last unloading. On the other hand, with a restatement of stress and strain components in terms of the 'genesis' or original undeformed reference configuration, a single theory describes all re-entry to finite strain beyond the maximum stress, T(OYS), of the previous loading. This genesis, undeformed reference configuration of the fully annealed solid, in terms of which $\boldsymbol{\sigma}$ and $\mathbf{E}$ are defined, is the same as that which defines the stage III finite strain deformation of the free single crystal. It is the reference configuration implicit in aggregate theories.

The loading surface corresponding to the last maximum generalized stress, T, is referred to as an *outer yield surface*, T(OYS). The significance of the different roles of two yield surfaces bounding an intermediate yield region was first recognized and studied by Moon in 1973 [13]. The inner and outer yield surfaces have very different properties, particularly with respect to the memory of previous loading. For large finite strain plasticity, one is concerned only with the characteristics of the outer yield surface. Therefore, in a new series of experiments, I recently have reviewed and restudied [9] Taylor and Quinney's [14] widely quoted experiment on the determination of yield surfaces. In terms of $\boldsymbol{\sigma}$ of eqn. (12), the yield surface is of the type introduced by von Mises. Furthermore, when re-entry of the yield surface is followed by further loading into the finite strain domain, the strain history is given by the continuum theory of eqns. (1)–(19). The work hardening is isotropic; each successive loading surface is in accord with the von Mises description. For annealed aluminum, the dynamic properties of the outer yield surface have been described elsewhere [8].

A more severe test of theory by experiment than that proposed by Taylor and Quinney is shown in an abbreviated form in Fig. 3. It is described in detail in ref. 9. To simplify the presentation, yield surfaces are represented as circles rather than as ellipses, hence the following change of variables was made:

$$\Sigma = \sqrt{\frac{2}{3}}\,\sigma_x\,;\ \eta = \sqrt{\frac{3}{2}}\,\varepsilon_x\,;\ Q = \sqrt{2}S_{xy}\,;\quad \text{and}\quad q = \frac{s}{\sqrt{2}} = \sqrt{2}\varepsilon_{xy}$$

In the totally plastic region a linear reloading path, making an angle of 57° with the shear stress abscissa, $Q = \sqrt{2}S$, is located in different sections of Σ vs Q stress space. The origins for two such reloadings are shown in Fig. 3. The response in η vs q strain space also is shown in Fig. 3.

The present theory predicts that this reloading stress path may be located so that a torque applied in one direction will produce a rotation in the opposite direction!

In each instance, the theory (arrows) and the measurement (circles) are in accord, including test 2314 in which, as predicted, we see that while the torque component is counterclockwise, the initial rotation is indeed clockwise; and test 2312 in which, on the other hand, the torque is clockwise and the initial rotation is counterclockwise.

In an effort to inhibit the retention of a memory of the last outer yield surface and of the original, genesis reference configuration, we seek a drastically different experiment. A thin-walled tube is loaded cyclically in alternating clockwise and counterclockwise torque while at the same time it is subjected to alternating tension and compression. The compression stress rate is lower than the tension stress rate so that the net tensile load will increase at some chosen rate. The details of such a test are given as Figs. 10, 11 and 12 in ref. 9. Successive outer yield surfaces are established by the previous loading. The stress rates in this test were chosen so that the increases in the generalized stress, T, in the totally plastic region would occur during the compression portion of the cycle, while the unloading would be confined almost entirely to the tension portion of the axial stress cycle.

Despite the large number of successive yield surfaces established, and the large differences in the location and direction of re-entry stress paths, and the cyclical rotation of principal axes from −35° to +45°, the measurements provide decisive evidence of the importance of the genesis reference configuration and of the significance of the memory properties of the outer yield surface. Throughout this complex stress

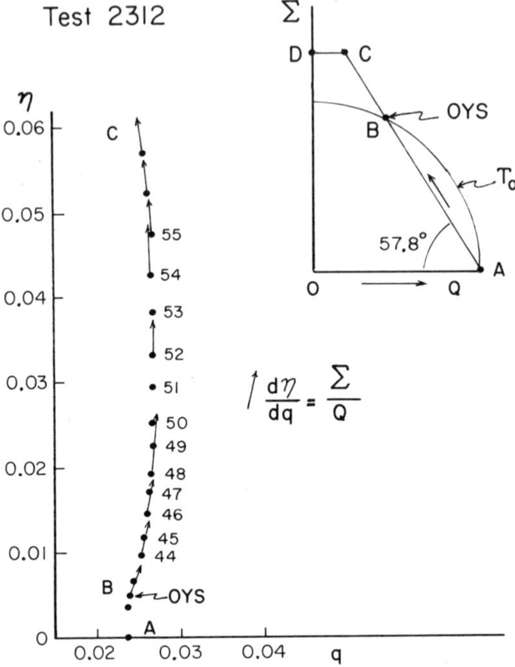

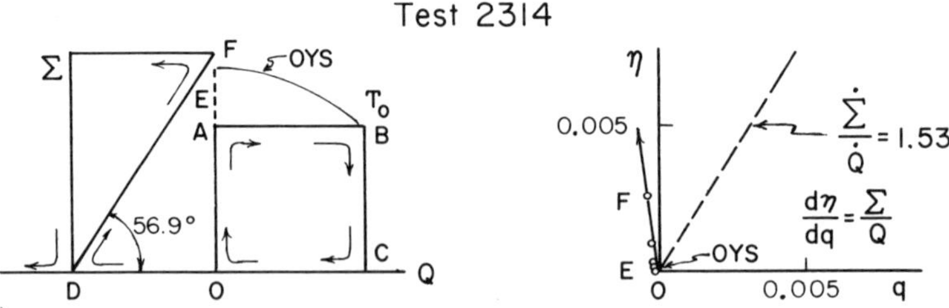

FIG. 3. Stress path and initial location for two tests for which theory (arrows) predicts that an application of torque in one direction produces a rotation in the opposite direction.

path, prediction and measurement are compatible. The cumulative absolute strain is very large. Work hardening in finite strain plasticity is indeed isotropic and von Mises in type, for successive loading surfaces.

In Fig. 12 of ref. 9, I have provided the T^2 vs Γ data for the five sufficiently large reloadings in this test. From the slopes, as predicted in eqn. (17), the parabola coefficient, $\beta = 37\ \mathrm{kg/mm^2}$ is common for all reloadings, including a final tension loading after 18 h at zero stress followed by a final torsion loading after an additional 19 h at zero stress.

All measurements described above were for dead weight loading, i.e. stress histories are assigned while strain components, displacements, and rotations are the measured unknowns. Strain rates range between 10^{-5} and $10^{-3}\ \mathrm{s}^{-1}$. At an opposite extreme, for one-dimensional wave propagation along the axis of a cylinder, at strain rates to as high as $10^4\ \mathrm{s}^{-1}$, the rotation tensor $\mathbf{R} = \mathbf{1}$. Hence from eqn. (12), $\boldsymbol{\sigma} = \mathbf{T}_\mathrm{R}^\mathrm{T}$, and from eqn. (5), $\mathrm{d}\mathbf{E} = \mathrm{d}\mathbf{F}$. Although the stresses no longer are assigned, the unknowns remain the strain and displacement. σ_1 and ε_1 are the only independent components. Dropping the subscript and integrating, the constitutive eqns. (19) reduce to

$$\sigma = \left(\sqrt{\frac{3}{2}}\right)^{3/2} \beta \varepsilon^{1/2} \tag{24}$$

Between 1951, with the discovery of the high wave speed of the incremental plastic wave [1], and the mid-1970s, the main research effort in my laboratory was the experimental study of plastic waves. There were many different types of experiments. Of particular significance was an experiment in which strain wave profiles were directly measured optically by the observation of the response of 30 720 lines per inch ruled diffraction gratings, which established that the one-dimensional nonlinear wave theory is in close correspondence with observation for each of the many metals studied. It was from that research that the parabolic response function (eqn. (24)) first evolved: when one shows from experiment that the nonlinear theory applies, the unknown governing stress–strain function follows directly, without any additional assumptions. In those experiments the measured strain rates were between 10 and $10^4\ \mathrm{s}^{-1}$.

It was from those high strain rate tests that the numerical values of the parabola coefficient β in eqns. (1)–(19) of the present theory were first identified. The difference in measured strain rates is 10^9 since the quasi-static data have strain rates as low as $10^{-5}\,s^{-1}$. For example, the annealed copper data shown above and described in ref. 6 provide a parabola coefficient of $\beta = 37{\cdot}0\,kg/mm^2$ for a strain rate of $10^{-5}\,s^{-1}$. This is within 2% of the value obtained for annealed copper by Bell and Werner in 1962 [2], from a detailed diffraction grating analysis of plastic waves for strain rates of $10^3\,s^{-1}$. The same constitutive equations for finite strains are unchanged for a range of strain rates exceeding 10 million.

Of more importance to contemporary research is the continued study of the change of volume associated with the constraint, tr $\mathbf{E} = 0$. Such changes of volume are measured. I have found that the Bridgman [11] two-dimensional compression experiment is particularly advantageous for such studies. In some instances there appears to be a hydrostatic recovery of the decreasing volume during plastic deformation, and in others, the change of volume during loading is precisely as predicted, disappearing during unloading so that from the post-deformation state one would conclude that the deformation had been isochoric. On the other hand, in alpha-brass, one measures a slow post-deformation return to the original state which requires weeks or months to be completed [12].

2. ON METAL ALLOYS

Compatibility between theory and experiment for the finite strains of 24 'as received' structural metal alloys of aluminum, steel, and copper [5] requires that one postulate a series of reference states which, as will be seen below, are related to the rotation tensor, **R**, as stable values in a material stability structure. Such states occur as five specific relative reference configurations for stress and strain tensors (eqns. (8) and (12)), which are themselves defined with respect to the properties of a rotation tensor, **R**, i.e. $\boldsymbol{\sigma} = \mathbf{R}\mathbf{T}_{R}^{T}$ and $\mathbf{1} + \mathbf{E} = \mathbf{V} = \mathbf{F}\mathbf{R}^{-1}$. To illustrate here the essential features of such a quantized material stability structure, which I have described in detail elsewhere [3], a single representative test will suffice.

In Fig. 4 is a T^2 vs Γ plot for the nonproportional loading history shown in the simultaneous tension and torsion of a thin-walled tube of annealed mild steel (see ref. 10). We note immediately that, unlike the annealed copper described above, the deformation occurs as a series of linear segments. The statistical analysis of hundreds of such measurements on 30 annealed solids [3] provided two empirical equations. The

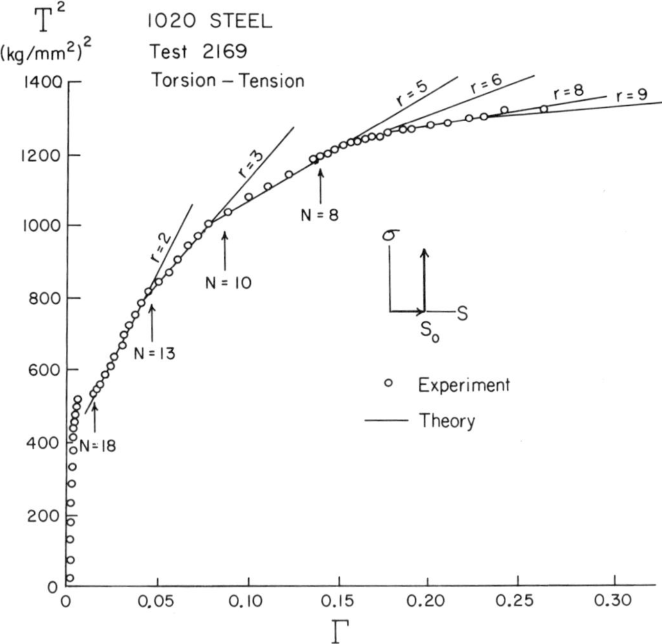

FIG. 4. An illustration of a high density transition structure for non-proportional loading.

linear segments are not arbitrary. They have specific slopes (eqn. (25)), and begin and end at specific transition strains (eqn. (26)).

$$\beta = (\tfrac{2}{3})^{r/2}\mu(0)B_0(1-\theta/\theta_m) \tag{25}$$

$$\Gamma_N = \sqrt{\frac{3}{2}}\,\varepsilon_N = \frac{1}{\sqrt{2}}\left(\frac{2}{3}\right)^{N/2} \tag{26}$$

where $N = 18$, 13, 10, 8, 6, 4, 2, and 0; β is the parabola coefficient; $r = 1, 2, 3, 4 \ldots$ represent observed deformation modes related to the quantized structure; $\mu(0)$ is the known elastic shear modulus at near zero degrees Kelvin; $B_0 = 0{\cdot}0207$ is a dimensionless universal constant common to all the metals considered; θ is the ambient temperature; and θ_m is the melting point of the solid of interest.

The integration of eqns. (22) and (23) for non-proportional loading of transition prone annealed mild steel at low strain rates provides the close correlation between measured axial strain, ε_x, and measured

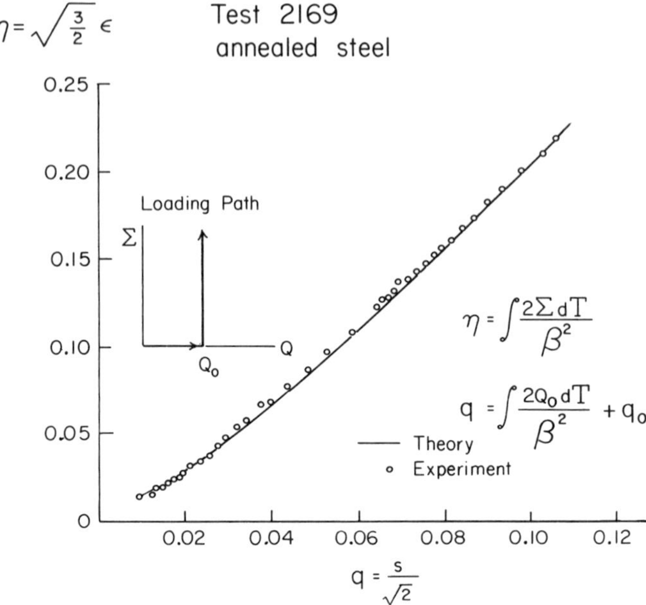

FIG. 5. A comparison of theory and experiment for the integration of eqns. (22) and (23) for the test of Fig. 4.

shear strain, s, (here η and q) in the strain space plot of Fig. 5. Clearly the theory applies.

Returning to the 'as received' structural metal alloy and its relation to this material stability structure, we note that in every instance there is a shift in the location of the origin of the T vs Γ parabola and in the magnitude of the material constant, β, which now is written β_N. Whatever the type of test, the initial location of the origin of the T vs Γ parabola is at the elastic limit of the structural metal alloy, T_Y and Γ_Y.

Introducing $\hat{T} = T - T_Y$, the relation between generalized stress and generalized strain of eqn. (17) becomes eqn. (27) and for consistency in theory, eqns. (19) must become eqns. (28) to comply with this change of origin.

$$d\Gamma = \frac{2\hat{T}\, dT}{\beta_N^2} \tag{27}$$

$$d\varepsilon_{ij} = \frac{2S_{ij}\, dT}{\beta_N^2}\frac{\hat{T}}{T} \tag{28}$$

Experiment on metal alloys reveals that this is precisely so. Theory and experiment are in accord until a value of generalized stress, T_C. This region, $T_Y \leqslant T \leqslant T_C$ is referred to as the *intermediate region.* The parabola coefficient, β_N, is a measured value for any given metal alloy. Parabolicity in the intermediate region is seen from the linearity of a $\hat{T}^2$ vs Γ plot (see refs. 5 and 8).

A continuation of this $\hat{T}^2$ vs Γ plot beyond the strain, Γ_C, corresponding to the stress, T_C, ceases to provide the straight line indicative of parabolicity. On the other hand, for $T \geqslant T_C$, a T^2 vs Γ plot provides a straight line, indicating that parabolicity again occurs and continues to failure. Hence, at T_C the origin of the T vs Γ parabola shifts to the strain abscissa, Γ. This region for $T_C \leqslant T \leqslant T_{\text{ultimate}}$ is referred to as the *totally plastic region.* For an origin of T vs Γ on the strain abscissa, eqn. (17) becomes eqn. (29) and the constitutive equations (19) become eqns (30).

$$d\Gamma = \frac{2T\,dT}{\beta_N^2} \tag{29}$$

$$d\varepsilon_{ij} = \frac{2S_{ij}\,dT}{\beta_N^2} \tag{30}$$

There remains for the totally plastic region the problem of finding the precise location of the origin of the T vs Γ parabola on the strain abscissa. For proportional loading, eqn. (29) may be integrated to provide eqn. (31):

$$T = \beta_N(\Gamma - \Gamma_C + K_N)^{1/2} \tag{31}$$

where K_N represents the strain between the unknown origin of the parabola and Γ_C. In a T^2 vs Γ plot, T_C^2 where parabolicity begins is readily measurable since, in general, a sharp change in slope is observed. The square of the parabola coefficient, β_N^2, also is readily measured as the slope found in the totally plastic domain.

Regardless of stress path, when $\Gamma = \Gamma_C$ the ratio of $T_C^2/\beta_N^2 = K_N$ is one of five constants. Each of the 24 metal alloys has as a material constant one of the following values of K_N: 0·448, 0·340, 0·250, 0·125, and 0·087. There is very little scatter among the data from one test to another and what is more important, there is very little scatter from one metal to another in the same group, as may be seen in Table I, ref. 5.

The first group of metals, $K_N = 0{\cdot}448$, contains 11 different alloys of aluminum and steel. The second, $K_N = 0{\cdot}340$, contains 7 alloys of aluminum, steel, and copper. These are the main divisions among the alloys, with 2 alloys each in the three remaining categories.

The year which followed this discovery that the finite deformation of metal alloys had a quantized structure was most perplexing; it seemed that no reason could be found for the existence of such a precise grouping. Continued additions to and study of this pattern among the metal alloys led in 1976 to the unexpected unifying discovery that the constant, $T_C^2/\beta_N^2 = K_N$, could be written as $K_N = \Gamma_N/\lambda_N$, where $\lambda_N = 1+\varepsilon_N$, or since $\Gamma_N = (\sqrt{3/2})\varepsilon_N$, the constant may be written

$$K_N = \frac{T_C^2}{\beta_N^2} = \sqrt{\frac{3}{2}}\frac{\varepsilon_N}{1+\varepsilon_N} = \frac{\sqrt{\frac{3}{2}}\varepsilon_N}{\lambda_N} \tag{32}$$

The transition strain, Γ_N, defined in terms of the genesis reference configuration is given in eqn. (26). In this context, the slopes in the T^2 vs Γ plots for 'as received' metal alloys revealed that

$$\beta_N^2 = \lambda_N^3 \beta^2 \tag{33}$$

where β is the parabola coefficient (eqn. (24)) for the annealed prime component of the metal alloy.

Thus, eqn. (31) for $\Gamma = \Gamma_C$ becomes

$$T_C = \lambda_N \beta \Gamma_{Na}^{1/2} \tag{34}$$

or, since $T_{Na} = \beta\Gamma_{Na}^{1/2}$, we may write

$$T_C = \lambda_N T_{Na} \tag{35}$$

where the subscript, a, is introduced to indicate that the stress, T_N, and the strain, Γ_N, are for the annealed parent metal. Hence, both T_{Na} and Γ_{Na} are defined in terms of the genesis reference configuration. The origin of the parabola in the totally plastic region not only is on the abscissa of the generalized strain, Γ, but also is located at a distance in the negative direction from Γ_C by a strain, K_N, in the new reference configuration given by Γ_{Na}/λ_N in the genesis reference configuration.

The common origin of the stable reference states of the as-received structural metal alloy and the quantized material stability structure of the fully annealed metal and single crystal is an empirical fact. When the fully annealed parent metal is loaded along a single parabola, β, to the appropriate transition strain, Γ_{Na}, the corresponding stress is T_{Na}. The same stable final state is achieved by the prior chemical, mechanical, and thermal metallurgical recipes which produce a stable metal alloy.

Knowing that T for the metal alloy, in a relative reference configuration, is related to T_a of the annealed prime component, as $T = \lambda_N T_a$, and with $\beta_N = \lambda_N^{3/2}\beta$, it follows that Γ for the metal alloy, in a relative reference configuration, is related to Γ_a of the annealed prime component by $\Gamma = \lambda_N^{-1}\Gamma_a$.

At this point it is reasonable to make a conjecture compatible with experiment but which obviously requires further study in continuum theory. If one introduces the series of five known relative reference states, characterized with respect to the genesis reference configuration by $\lambda_N \mathbf{R}$ and $\lambda_N^{-1}\mathbf{R}^{-1}$, then we have $\boldsymbol{\sigma} = \lambda_N \mathbf{R}\mathbf{T}_R^T$ (see eqn. (12)) and $\mathbf{V} = \lambda_N^{-1}\mathbf{F}\mathbf{R}^{-1}$ (see eqn. (5)), or $\boldsymbol{\sigma} = \lambda_N \boldsymbol{\sigma}_a$ and $d\mathbf{E} = \lambda_N^{-1}\, d\mathbf{E}_a$, as found in experiment. Eqns. (17) and (19) for the fully annealed metals are included in eqns. (29) and (30) for the structural metal alloys by introducing $\lambda_N = 1{\cdot}000$ for the genesis reference configuration, since from eqn. (33), $\beta_N^2 = \lambda_N^3\beta^2$. The finite strain response of the structural metal alloys thus is a simple extension of the general theory for the fully annealed solids.

REFERENCES

1. Bell, James F. Propagation of plastic waves in pre-stressed bars. Technical Report No. 5, U.S. Naval Contract. The Johns Hopkins University, 1951.
2. Bell, James F. and W. Meade Werner. Applicability of the Taylor theory of the polycrystalline aggregate to finite amplitude wave propagation in annealed copper, *J. appl. Phys.*, **33** (1962), 2416–2425.
3. Bell, James F. The physics of large deformation of crystalline solids, *Springer Tracts in Natural Philosophy*, Vol. 14, Springer, Berlin, 1968.
4. Bell, James F. The experimental foundations of solid mechanics, *Handbuch der Physik*, **VIa/1,** Springer, Berlin, 1973, 1–813.
5. Bell, James F. A physical basis for continuum theories of finite strain plasticity: Part I, *Arch. Rat. Mech. Anal.*, **70** (1979), 319–338.
6. Bell, James F. and Akhtar S. Khan. Finite plastic strain in annealed copper during non-proportional loading, *Int. J. Solids Struct.*, **16** (1980), 683–693.
7. Bell, James F. A physical basis for continuum theories of finite strain plasticity: Part II, *Arch. Rat. Mech. Anal.*, **75** (1981), 103–126.
8. Bell, James F. On the dynamic elastic limit, *Exp. Mech.*, **22** (1982), 270–276.
9. Bell, James F. Continuum plasticity at finite strain for stress paths of arbitrary composition and direction, *Arch. Rat. Mech. Anal.*, **84** (1983), 139–170.

10. BELL, JAMES F. Finite plastic strain in annealed mild steel during proportional and non-proportional loading, *Int. J. Solids Struct.*, **19** (1983), 857–872.
11. BRIDGMAN, P. W. Studies of plastic flow of steel, especially in two-dimensional compression, *J. appl. Phys.*, **17** (1946), 225–243.
12. HARTMAN, WILLIAM F. Propagation of large amplitude waves in annealed brass, *Int. J. Solids Struct.*, **5** (1969), 303–317.
13. MOON, HAHNGUE. An experimental study of the outer yield surface for annealed polycrystalline aluminum, *Acta Mechanica*, **24** (1976), 191–208.
14. TAYLOR, G. I. and H. QUINNEY. The plastic distortion of metals, *Phil. Trans. R. Soc., Lond.*, Ser. A, **230** (1931), 323–362.

12

On the Experimental Foundations of the Two Surface Model of Plasticity and Viscoplasticity

ARIS PHILLIPS

Department of Applied Mechanics, Yale University, New Haven, Connecticut, USA

ABSTRACT

This paper presents results obtained by the author since the time he presented a general lecture at the 1972 *Symposium on Plasticity in Warsaw. These results help put the two surface theories of Plasticity and Viscoplasticity on a firm experimental foundation. The topics to be discussed cover:* (*a*) *the yield surface and its motion due to isothermal as well as to thermal prestressing,* (*b*) *the concept of the loading surface and methods of its determination,* (*c*) *the equilibrium stress–strain curves both in their ascending and their descending forms and of the effects of temperature. The developments that occurred during the last few years will be emphasized.*

1. INTRODUCTION

In a paper [2] presented eleven years ago at the International Symposium on the Foundations of Plasticity in Warsaw, we reported on research conducted at our laboratory on Experimental Plasticity. In the present paper we shall report on some of the advances we achieved during the intervening years. The topics to be discussed will cover: the yield surface, the loading surface, the equilibrium stress–strain curves, and the effects of temperature. We shall emphasize the developments that occurred during the last few years, since the progress in our research which occurred during the period following the 1972 lecture has been reported in another of our contributions [3].

2. THE YIELD SURFACE IN THE STRESS–TEMPERATURE SPACE

Our experimental work has shown conclusively that for small strains there exists a yield surface, that is, a surface in the stress–temperature space which encloses all (and only) the region in which any stress–temperature path will produce only elastic strains. A schematic diagram of the form of the yield surface for pure aluminum before any plastic deformation has developed is shown in Fig. 1. This figure shows the yield surface in the temperature range in which our testing took place.

The yield surface can be prestressed by a stress path at constant temperature or by a temperature path at constant stress. In both cases the path must intersect the yield surface and enter deeply into the plastic region. We used both types of paths and obtained comparable results. Of course, we can also use paths in which both the stress *and* temperature vary.

A comparison of sequences of yield surfaces obtained by isothermal prestressings to sequences of yield surfaces obtained by thermal prestressings, shows that it does not make a difference to the motion and

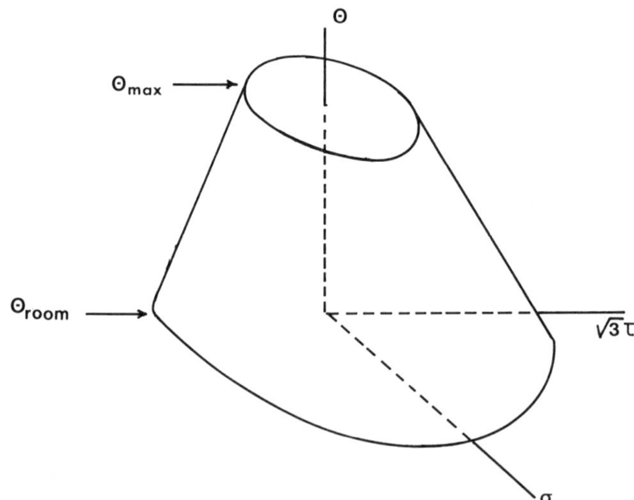

Fig. 1. Typical yield surface in the (σ, $\sqrt{3}\tau$, θ) space. The yield surface is illustrated only in the temperature region between room temperature and maximum testing temperature.

deformation of the yield surfaces whether the prestressings are thermal or isothermal. Experimental results are presented in refs. 4, 5 and 6 which indicate the above conclusion.

The question can now be raised of what happens at temperatures beyond those tested by us. An indication of what is likely to happen, in addition to the expectation that considerable creep strains will appear during testing, is seen in Fig. 2 (from ref. 1). Indeed, in the top part of this figure the portion of a yield surface which has been obtained experimentally is shown by means of a series of isothermal yield curves at increasing temperature levels. Then we draw a series of arbitrary parallel straight lines $\bar{A}A$, $\bar{B}B$, $\bar{C}C$, $\bar{D}D$, and $\bar{E}E$, which could be parallel to the prestressing direction. The intersections of these lines with the yield curves are plotted in a stress–temperature diagram as shown in the lower part of Fig. 2. We observe that these intersections produce pairs of straight lines ($\bar{A}A'$, AA'), ($\bar{B}B'$, BB'), etc., so that the intersections B', C', and D' lie on a straight line $\bar{Q}Q$ which represents a high temperature limit θ_{max} of the sequence of yield curves. In effect, the sequence of isothermal yield curves degenerates to a straight line $\bar{Q}Q$, seen in the top part of the figure. This straight line is the limiting yield curve and it occurs at the limiting temperature θ_{max}.

In Fig. 3 we see in a double logarithmic scale, for a number of experiments, the maximum temperature θ_{max}(°F), at which the limiting yield surface appears, as a function of the plastic strain (bottom line of the figure). The relationship is a linear one in a double logarithmic scale.

Another observation to be made from Fig. 3 is that the line labelled 'size of the limiting yield surface' increases nearly linearly (in double logarithmic scale) with the plastic strain. This line represents the length $\bar{Q}Q$ in Fig. 2 divided by the yield strain σ_y.

The experiments upon which the previous conclusions were based have been done with a deadload testing machine. We have also recently performed experiments on a computer controlled closed loop servohydraulic combined stress testing machine. Figure 4 shows the experimental results for isothermal tests at room temperature, when the testing machine is under load control. In this figure we see three subsequent yield curves II–IV. The loading path was OABCBD. In Figs. 5 and 6 we see the experimental results from isothermal tests at room temperature, when the testing machine is under strain control. In particular, Fig. 5 shows the straining path OBDEFGHIJK in strain space, while Fig. 6 shows the stress path. We observe that we obtained

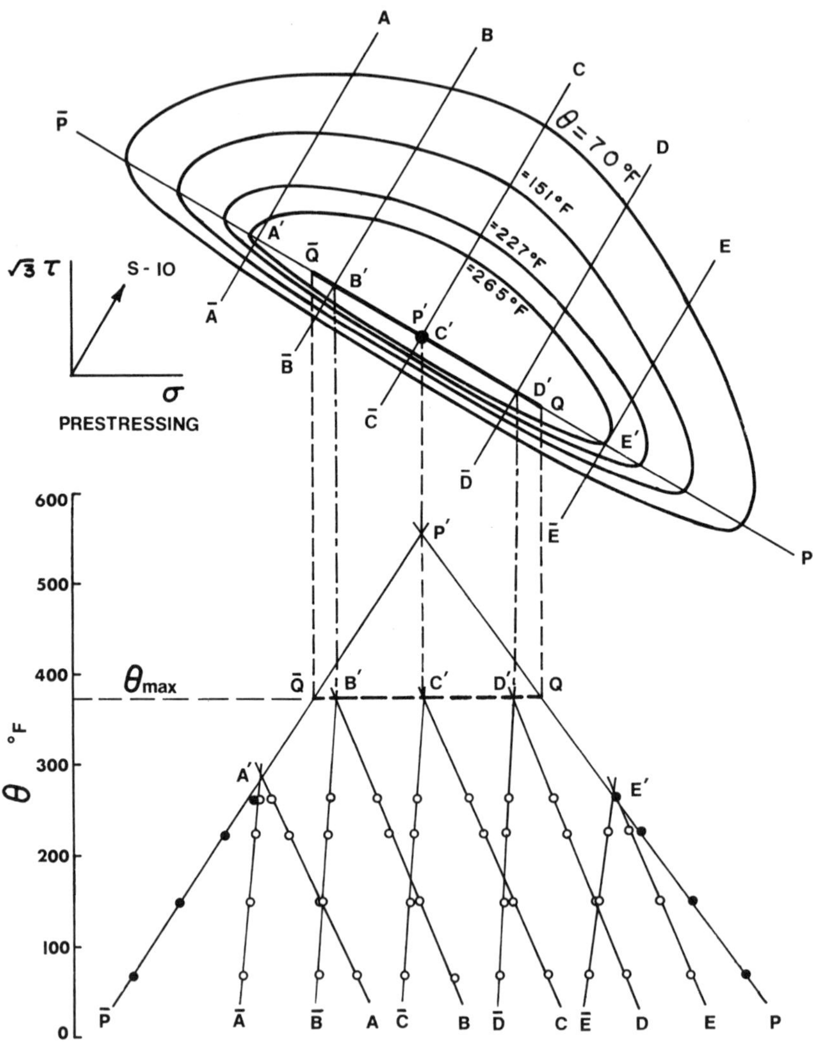

FIG. 2. The isothermal yield curves. At an appropriate temperature the yield curve degenerates into a straight line $\bar{Q}Q$.

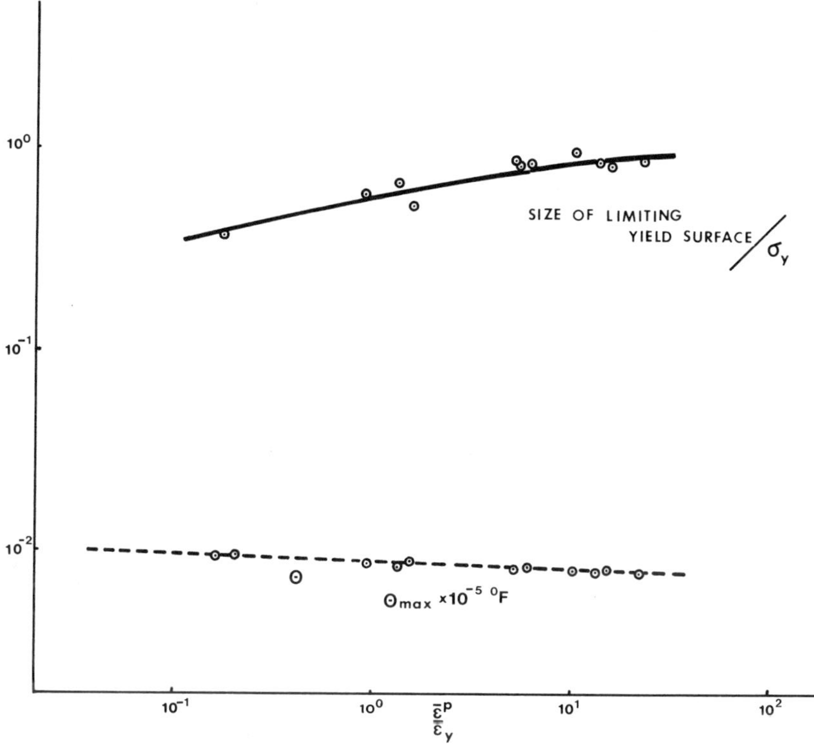

FIG. 3. The temperature θ_{max}(°F) versus the plastic strain. The size of the limiting yield surface as a function of the plastic strain.

an initial yield surface and three subsequent yield surfaces. Of particular interest are the sections BC, FG and JK which represent relaxation of stress while the strain remains constant. It will be observed that the relaxation paths are directed in the direction opposite to the plastic strain increment vectors at B, F, and J which are also shown. Additional information on these tests will be given in a forthcoming paper [7].

3. THE LOADING SURFACE

The concept of the loading surface dates from 1965 when it was presented in one of this author's papers [8]. By means of that paper

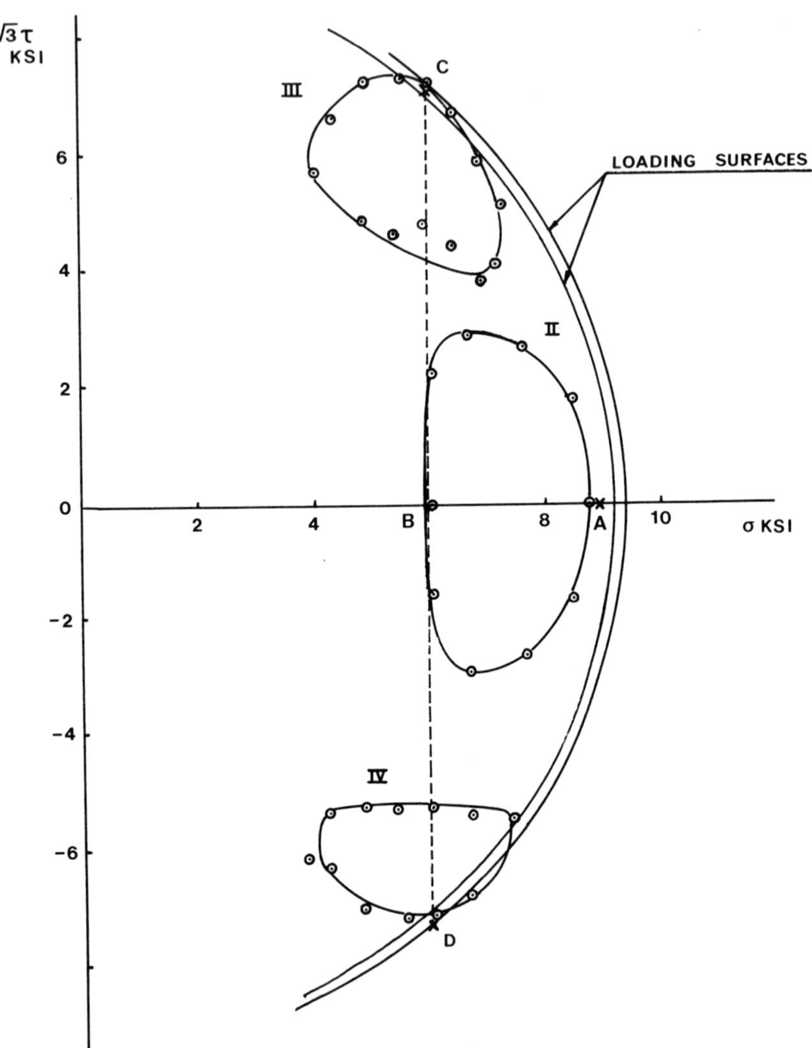

FIG. 4. An experiment with the servohydraulic testing machine under load control.

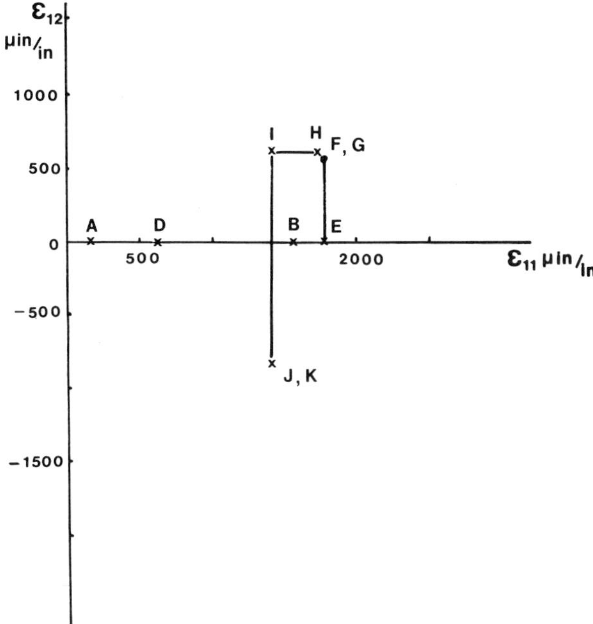

FIG. 5. An experiment with the servohydraulic testing machine under strain control. The straining path.

the conceptual foundations of the two surface theories of plasticity were introduced. In 1975 we succeeded to show experimentally the existence of the loading surface [9].

While the yield surface includes only (but completely) the elastic region, the loading surface passes through that point of the stress path reached up to that stage, which is most distant from the origin of the stress space. We have shown that the loading surface is the envelope of the yield surfaces which reach equivalent stress values equal to the stress values which are most distant from the origin. Figures 4 and 6 illustrate the meaning of the loading surface. We see in Fig. 4, for example, that the loading surface passing through C is tangential to the yield surface passing through C. It follows that the stress space can be divided into three regions, (1) an elastic region enclosed by the yield surface, (2) an intermediate plastic region where only very small strains appear and which lies outside the yield surface but is enclosed by the loading surface, and (3) a plastic region which lies outside the loading

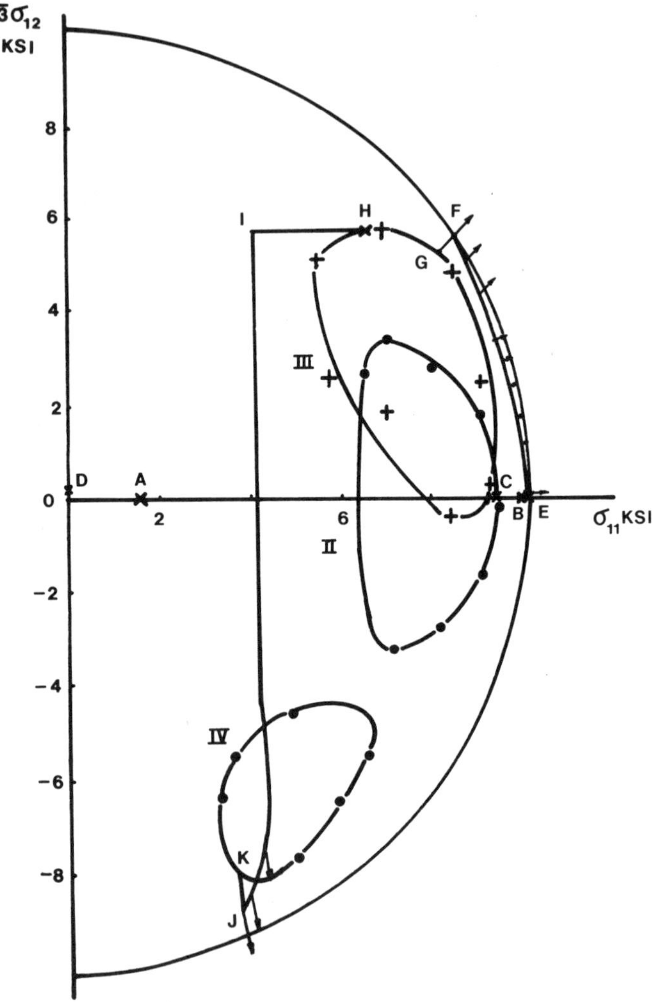

FIG. 6. An experiment under strain control. The corresponding stress plane.

surface and where larger strains appear (larger compared to those appearing in the intermediate region). This classification, by means of the magnitude of the strains which appear in a stress path, has been used to devise a new method for the determination of the loading surface [10].

We observed that the law of growth of the plastic strains when the stress path lies inside the loading surface is different from the law of growth of the plastic strains when the stress path lies outside but near the loading surface. In double logarithmic scale the two growth laws are represented by straight lines with different slopes and intercepts. The point where these two lines intersect defines a point of the loading surface. We also observed that the loading surface is relatively insensitive to modest excursions by loading paths into the region outside the loading surface. Consequently it is possible to obtain several such intersection points between pairs of straight lines for the same loading surface and thus we can locate the loading surface by means of several of its points without having to locate first several yield surfaces to which the loading surface is an envelope.

In this section we discussed loading surfaces but in effect we had to deal with loading curves since we considered only isothermal stress changes at one temperature only. We can generalize the concept of the loading curve to a genuine loading surface in the stress–temperature space for cases when the temperature varies. Since the property of the yield curve to be tangent to the loading curve is valid at all temperatures, there is no difficulty in generalizing the concept. The original definition of the loading curve as the envelope of a number of yield curves carries over to a definition of the loading surface as the envelope of a number of yield surfaces in the stress–temperature space.

4. THE EQUILIBRIUM STRESS–STRAIN CURVE

Intimately related to the concept of the yield surface is the concept of the equilibrium stress–strain curve. We introduced it in 1972 [2]. Figure 7 shows the equilibrium stress–strain curve AC corresponding to a zero stress rate. It is the sequence of equilibrium positions due to successively larger values of applied stress. In effect we assume that each increment of stress is applied only after the permanent strain due to the previous strain increments has developed fully.

In principle, we can obtain the equilibrium stress–strain curve by unloading to within the elastic range and then reloading very slowly. By successively repeating this procedure at different values of strain we can obtain several points of the equilibrium stress–strain curve. We find that we are within the elastic range when no plastic strain appears

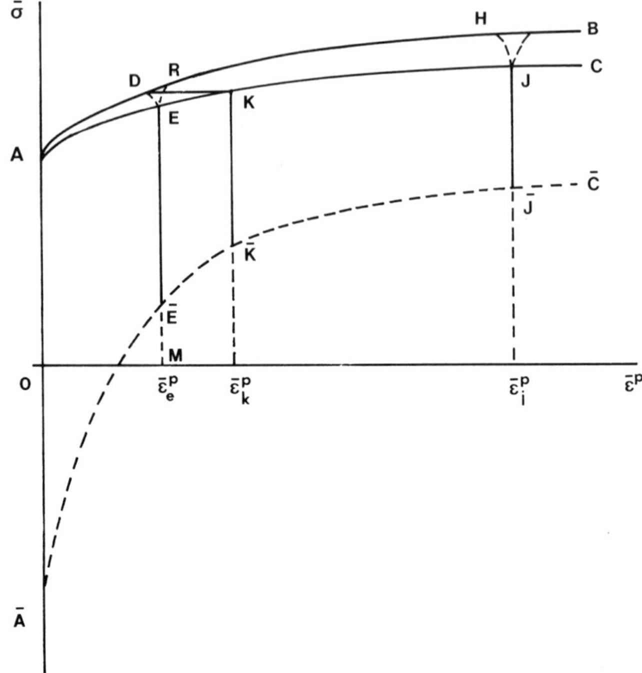

FIG. 7. The equilibrium stress–strain lines AC and $\bar{A}\bar{C}$.

while unloading. We find that we have reached the equilibrium stress–strain curve while reloading when we observe that we reached the proportional limit.

The straight lines $E\bar{E}$, $K\bar{K}$, and $J\bar{J}$ in Fig. 7 represent the yield curves at room temperature at different values of permanent strain. It follows that the equilibrium stress–strain curve AC represents the forward limits of the room temperature yield curves at different values of permanent strain. Line $\bar{A}\bar{C}$ gives the backwards limits of the room temperature yield curves. These two lines AC and $\bar{A}\bar{C}$ as a pair are obtained for an increasing permanent strain and consequently can be called *ascending equilibrium stress–strain lines* [11].

We observed previously (Fig. 2) that a higher temperature yield curve lies within the region enclosed by a lower temperature yield curve due to the same prestressing. Consequently, at a higher temperature the ascending equilibrium stress–strain curves are related to the equilibrium stress–strain curves which are generated at a lower temperature in the manner shown schematically by Fig. 8. We see that pairs of ascending equilibrium stress–strain curves of the same temperature intersect at specific values of the permanent strain. This intersection means that at each temperature there is a value of the plastic strain at which the yield curve degenerates to a straight line, such as the line $\bar{Q}Q$ in Fig. 2.

In addition to the ascending equilibrium stress–strain curves we also have *descending* equilibrium stress–strain curves. They are generated by such reloading that the stress moves downwards below the existing lower equilibrium stress–strain curve. Figure 9 gives an example from one experiment. Here, AB′C′ and A′BC are the descending upper and lower equilibrium stress–strain curves. Lines B′B and C′C represent

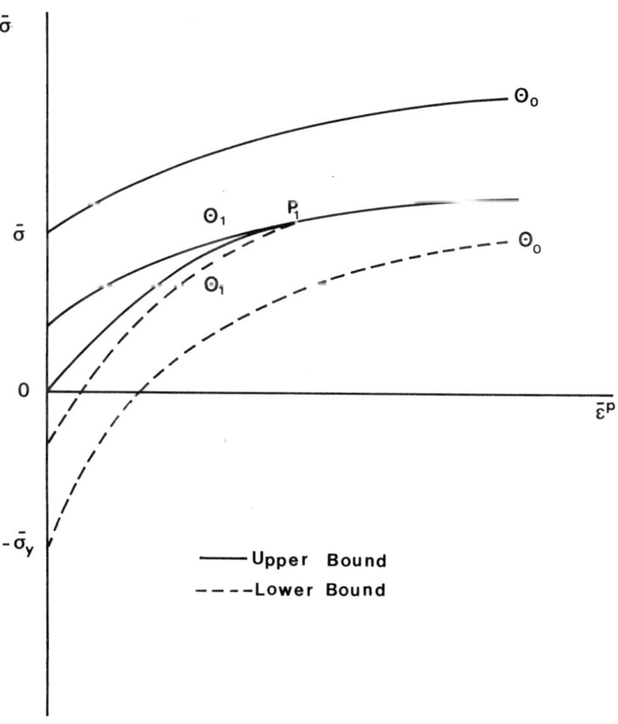

FIG. 8. The ascending equilibrium stress–strain lines at different temperatures. Observe that each two lines at the same temperature intersect at some value of the strain.

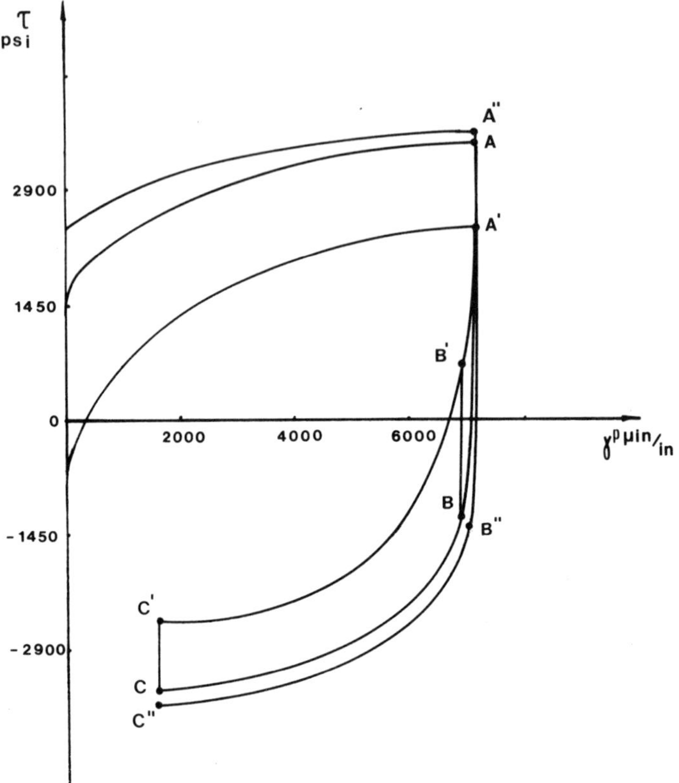

FIG. 9. The descending equilibrium stress–strain lines.

the yield surfaces during the descent in the same manner as lines $E\bar{E}$, $K\bar{K}$, etc., represented the yield surfaces during the ascent in Fig. 7. The descending equilibrium stress–strain curves at higher temperatures will be located in the region between lines AB′C′ and A′BC.

By considering pairs of descending equilibrium stress–strain lines corresponding to two different temperatures we succeeded in predicting in ref. 11 that a temperature cycle from the lower temperature to the higher temperature and back to the lower temperature will generate a plastic strain in the negative direction. Such a plastic strain was experimentally observed.

The loading surface will be determined by the most distant value of the stress from the origin which the equilibrium stress–strain curve reaches. The loading surface can be represented by a straight line parallel to the stress axis starting from the highest value of stress reached and ending at the appropriate value of stress on the negative side.

ACKNOWLEDGEMENTS

The support of this research by the National Science Foundation is gratefully acknowledged. The author wishes to thank Mr J. Macheret, graduate student at Yale University, for his help in the timely completion of this paper.

REFERENCES

1. EISENBERG, M. A., C. W. LEE, and A. PHILLIPS. Observations on the theoretical and experimental foundations of thermoplasticity, *Int. J. Solids Struct.*, **13** (1977), 1239–1255.
2. PHILLIPS, A. Experimental plasticity. Some thoughts on its present status and possible future trends. In: *Problems of Plasticity, Proc. Int. Symp. Foundations of Plasticity*, Vol. II, (Ed. A. Sawczuk), Warsaw, August 30–September 2, 1972, Noordhoff, Leyden, 1974, 193–234.
3. PHILLIPS, A. The foundations of plasticity. In: *Plasticity in Structural Engineering*, (Ed. Ch. Massonnet, W. Olszak, and A. Phillips), CISM Courses and Lectures No. 241, Springer, Wien-New York, 1980, 191–272.
4. PHILLIPS, A. and C. W. LEE, Yield surfaces and loading surfaces. Experiments and recommendations, *Int. J. Solids Struct.*, **15** (1979), 715–729.
5. PHILLIPS, A. and S. MURTHY. Yield surfaces in the stress–temperature space generated by thermal loading. In: *Mechanics of Material Behavior*, (Ed. G. J. Dvorak and R. T. Shield), Elsevier, Amsterdam, 1984, 325–344.
6. PHILLIPS, A. and W. A. KAWAHARA. The effect of thermal loading on the yield surface of aluminum, *Acta Mechanica*, **50** (1984), 249–270.
7. PHILLIPS, A. and W. Y. LU. An experimental investigation of yield surfaces and loading surfaces of pure aluminum with stress-controlled and strain-controlled paths of loading, *J. Eng. Mat. Technol.*, 1984 (in press).
8. PHILLIPS, A. and R. L. SIERAKOWSKI. On the concept of the yield surface, *Acta Mechanica*, **1** (1965), 29–32.
9. PHILLIPS, A. The inelastic theory of metal deformation. Some thoughts on its experimental foundations. In: *Workshop on Inelastic Constitutive Equations for Metals. Experimentation-Computation-Representation* (Eds. Krempl-Wells-Ludans), RPI, 1975, 211–230.
10. PHILLIPS, A. and S. MURTHY. Concepts and experimental results concerning the loading surface (to be published) 1985.

11. Phillips, A. Combined stress experiments in plasticity and viscoplasticity. The effects of temperature and time. In: *Plasticity of Metals at Finite Strains* (Eds. E. H. Lee and R. L. Mallett), Stanford University, Stanford, Calif., 1983, 230–266.

DISCUSSION

Boehler (*Institute de Mécanique de Grenoble, France*): What happens if you carry your test beyond the maximum stress point?

Phillips: We can expect that the loading surface will remain stationary and the enclosed yield surface will have degenerated into a linear element. However, we did not do any experiments at that level of stress, so that the above is only an expectation.

Boehler: Have you taken into account induced anisotropy (under large strains) in the loading and yield surfaces?

Phillips: Induced anisotropy is always present in our tests.

Boehler: Validity of the associated flow rule: Is it associated with the yield surface, the current loading surface, or the extremal loading surface (corresponding to maximum stress)? What happens when the yield surface becomes a straight line (high temperature)?

Phillips: The flow rule is associated with the yield surface and once the yield surface becomes tangential to the loading surface the flow rule becomes associated also with the loading surface. We can understand what will happen when the yield surface becomes a straight line by remembering the case of the sandhill analogy in torsion (perfect plasticity). The strain-rate vector will be normal to the straight line and a discontinuity will exist.

13

Material Rotation Effects in Tension–Torsion Testing: Experimental Results

EDWARD W. HART and YOUNG-WON CHANG

Department of Theoretical and Applied Mechanics, Cornell University, Ithaca, New York, USA

ABSTRACT

The most usual method for biaxial testing of materials is the testing of thin-walled cylinders under combined tension and torsion. Although considerable testing has been carried out in this mode, there is an important aspect of the test that has been ignored in almost all investigations. This is that the shearing deformation associated with the torsion is that of simple (engineering) shear rather than pure shear. This means that for each increment of shear strain there is also an incremental rotation of the material elements of the test specimen. An adequate analysis and representation of tension–torsion data cannot be properly carried out without taking account of the rotations. A method for such analysis is shown, and the analysis is applied to some recent test data for nickel. It is shown further that, with proper measurement and analysis, tension–torsion data can be employed to determine the values of several significant deformation parameters with good precision.

1. INTRODUCTION

Constitutive equations for inelastic deformation, that are intended for application to the analysis of engineering structures, must be fully three-dimensional. The validation of the three-dimensional formulation is commonly carried out in biaxial testing. Such testing is done by various combined loadings of thin-walled cylindrical specimens in axial tension, axial torsion, and internal pressure. The most used combinations are tension-torsion and tension–pressure. A review of the literature in biaxial testing was given recently by Hecker [4]. From that review it is evident that the largest amount of significant biaxial testing is that employing tension–torsion. It is also evident that none of the work in tension–torsion has taken account of the material rotations that accompany torsion. The rotations have a significant effect on the measured results, and such results cannot be properly related to constitutive laws unless the analysis of the results has explicitly accounted for the rotations.

An especially striking effect due to the rotations is the phenomenon of axial extension that accompanies torsion when there is no axial constraint or load. This effect, which is not to be confused with a similar effect in elastic behavior, has been known for a long time, but it has been investigated experimentally in only a very few studies. The initial report by Swift [6] described tests on solid cylindrical bars. The considerably later study by Bailey *et al.* [1] was made with hollow cylindrical specimens, but their measurements of axial extension were made only at increments of 100% shear strain by demounting and measuring the specimens. In both of these studies the authors were aware of the material rotation effects. Bailey *et al.* attempted an explanation of their results according to a theory of the effect due to Hill [5]. Hill's theory is based on the assumption of anisotropy of strain hardening coupled with concomitant rotation of the anisotropy axes. Since the anisotropy of strain hardening must be measured separately, the interpretation of results depends further on those measurements. It is not clear that the fit claimed by Bailey *et al.* is fully successful.

Another explanation for the production of axial extensions in torsion deformation was published recently by Hart [3]. This theory is based on the rotation of the internal stress that accompanies the deformation process. The magnitude of the effect depends on the way the internal stress appears in the constitutive equations for the inelastic deformation.

An additional source for the axial extension rates is any initial specimen anisotropy that might be present due to the prior deformation processing of the specimen material. This effect is related to that considered by Hill, but it can be considerably larger. Any such effect must, of course, be added to the effect of rotation of internal stress. In the present paper we shall show that for the experiments reported here there is no indication of specimen anisotropy.

We shall describe here the results of experiments in tension and torsion performed with commercial purity nickel.

2. THE PROBLEM

The test configuration of tension and torsion of thin-walled cylinders is well represented as a plane stress condition for a thin flat sheet. This is illustrated in Fig. 1. Also shown in the figure are the conventional experimental parameters $\dot{\varepsilon}$, $\dot{\gamma}$, τ, and σ. If the axial direction of the cylinder is taken as the 1-direction and the circumferential direction as the 2-axis, these parameters are, in the customary transcription, given as follows:

$$\dot{\varepsilon} \equiv \dot{\varepsilon}_{11} = \text{the axial strain rate}$$

$$\dot{\gamma} \equiv \dot{\gamma}_{12} = \text{the simple shearing rate}$$

$$\sigma \equiv \sigma_{11} = \text{the axial stress}$$

$$\tau \equiv \tau_{12} = \text{the shear stress}$$

Since the specimen shearing rate is that of simple shear, there is present a rotation rate (vorticity) $\dot{\omega}$ in addition to the pure shear rate $\dot{\varepsilon}_{12}$. Thus

$$\dot{\varepsilon}_{12} = \tfrac{1}{2}\dot{\gamma} \qquad (1)$$

$$\dot{\omega} = \tfrac{1}{2}\dot{\gamma} \qquad (2)$$

$$\dot{\omega} = \dot{\varepsilon}_{12} \qquad (3)$$

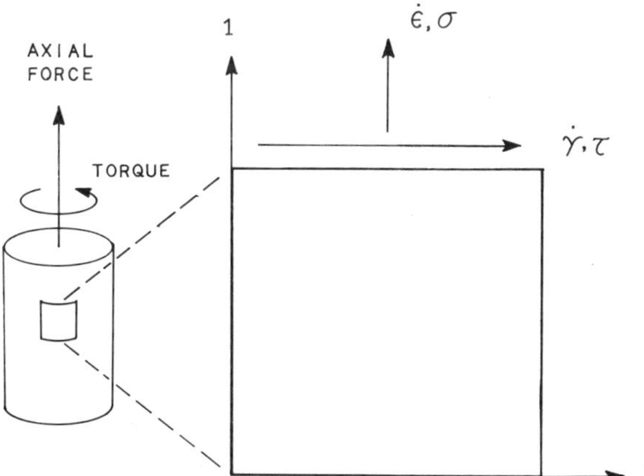

FIG. 1. Relationship between cylindrical and planar geometry.

It is the effect of $\dot{\omega}$ that is the principal subject of this investigation.

Most modern constitutive theories for inelastic deformation contain, as an internal state variable, an *internal stress* that is a tensor quantity. Since this tensor is oriented relative to the material frame of the body, any rotation of the material elements rotates the internal stress relative to the fixed external frame of reference in which the loading is specified. The effect of this rotation on the experimental observations depends, in general, on the way in which the internal stress is employed in the constitutive theory. Prediction of the effect requires knowledge of the full three-dimensional formulation of the constitutive equations for any deformation theory. We shall attempt comparison here only with the fully specified constitutive theory of Hart [2] and with the kinematic hardening theory of plasticity.

3. THEORETICAL PREDICTIONS

The theoretical derivation is given in detail in ref. 3 and is too lengthy to reproduce here. The principal idea can be stated as follows.

Given that in the constitutive equations to be applied there is a tensor internal variable **A** that is material oriented, and for which the time rate of change in the material frame $(\mathrm{d}\mathbf{A}/\mathrm{d}t)_{\mathrm{mat}}$ is given, then in the fixed laboratory frame the time rate of change $\dot{\mathbf{A}}$ is

$$\dot{\mathbf{A}} = (\mathrm{d}\mathbf{A}/\mathrm{d}t)_{\mathrm{mat}} + \dot{\omega}(\mathrm{d}\mathbf{A}/\mathrm{d}\omega) \tag{4}$$

and the quantity $\mathrm{d}\mathbf{A}/\mathrm{d}\omega$ is computed from the familiar kinematics of continuum mechanics. The application of the constitutive equations can then be carried out in the laboratory frame, in which all the loads and displacements are specified, substituting the time derivative $\dot{\mathbf{A}}$ from eqn. (4) for the material time derivative. This was done in ref. 3 for the constitutive theory of Hart [2] and for the simple kinematic hardening theory.

It was further shown in that reference that the prediction could be made less dependent on some of the details of the constitutive theory. This comes about because there is some redundancy in the experimental data.

In an experiment involving tension and torsion only two control variables can be independently specified for the test. These may be one each from the two pairs τ, $\dot{\gamma}$ and σ, $\dot{\varepsilon}$. In the experiments described here, the control parameters are $\dot{\gamma}$ and σ. Then two variables, τ and $\dot{\varepsilon}$, are measured. Now, in principle, each of these measured quantities can be predicted from the constitutive equations as functions of γ. There is, therefore, some redundant information in the two variables τ and $\dot{\varepsilon}$. If, for example, the measured values of τ during the test are taken as additional information, so that $\dot{\varepsilon}$ is now predicted from $\dot{\gamma}$, σ, *and* τ, it may be possible to eliminate some part of the constitutive equations in the computation. This was done in the computations of ref. 3 with the result that the predicted dependence of ε on γ was a formula in each case that contained only two constitutive parameters.

The predicted axial extension rate for Hart's theory is

$$\frac{\mathrm{d}\varepsilon}{\mathrm{d}\gamma}=\frac{\dot{\varepsilon}}{\gamma}=\frac{\tau-\tau_{\mathrm{f}}}{\mathcal{M}}\frac{\tau-\tau_{\mathrm{f}}}{\tau}+\frac{1}{3}\frac{\sigma}{\tau}\left[1-2\frac{\tau-\tau_{\mathrm{f}}}{\mathcal{M}}\frac{1}{\tau}\frac{\mathrm{d}\tau}{\mathrm{d}\gamma}\right] \tag{5}$$

where $\mathcal{M}$ is an anelastic modulus that must be constant throughout all the tests, and τ_{f} is a dislocation glide friction stress that depends (at a single temperature) only on the shearing rate $\dot{\gamma}$. The result for kinematic hardening is

$$\frac{\mathrm{d}\varepsilon}{\mathrm{d}\gamma}=\frac{\tau-\tau^*}{II}\left[1+2\frac{\tau^*}{H}\frac{\tau-\tau^*}{H}\right] \tag{6}$$

where τ^* is the initial yield stress, and H satisfies the equation

$$H=\frac{\mathrm{d}\tau}{\mathrm{d}\gamma}+\tau^*\frac{\tau-\tau^*}{H}-\sigma \tag{7}$$

which makes H approximately equal to $\mathrm{d}\tau/\mathrm{d}\gamma$. These predictions will be compared with experiments for which $\dot{\gamma}$ and σ are specified for each experimental run and for which $\sigma \ll \tau$.

4. THE EXPERIMENT

The experiment reported here was performed on commercial purity nickel (Ni 200).

The specimens were prepared from drawn tubing of 0·026 in (0·66 mm) wall thickness and 0·62 in (15·75 mm) outside diameter. Tube sections of 2 in (5·08 cm) length were fitted with 1040 steel end grips with a tight fitting plug in each end of each tube. The tubes were then copper brazed to the plugs leaving a free center gage length of about 0·5 in (1·25 cm). After brazing, each specimen was vacuum annealed at 1110°C for 1/2 hour. A small axial hole was drilled through one of each pair of end grips to avoid gas pressure build-up in the tube during brazing. Because of variability in brazing dimensions the actual specimen gage length was measured for each test run.

Two batches of specimen material were used. Although these had the same alloy designation, it was clear from the tests that the batches differed in solute content. We denote the specimens of each batch by the prefix A and B respectively.

The tests were performed at 25 °C in an Instron Tension–Torsion Servohydraulic testing machine. The tests were carried out under program control of an associated computer control system. Each test was done at a constant imposed torsion rate and constant axial stress. Thus the control variables were $\dot{\gamma}$ and σ. During each test a record was kept of axial extension and torque, and so the measured variables were ε and τ.

The axial extension was measured by a resistance strain gage extensometer attached to measure the same axial actuator motion as that monitored by the machine stroke control LVDT. The sensitivity of

TABLE I.
Test Parameters for All Specimens

Specimen number		σ (MPa)	$\dot{\gamma}$ ($10^{-4}\ s^{-1}$)	$\mathcal{M}$ (10^3 MPa)	τ_f (MPa)
	1	2·65	2·3	8·47	7·92
	2	2·35	2·9	8·47	8·13
	3	1·48	2·9	8·20	8·27
	4	1·10	3·0	8·34	8·41
A	5	0·21	0·3	8·41	6·61
	6	0·21	2·6	8·61	8·33
	7	0·33	6·5	8·54	9·09
	8	0·06	27·3	8·89	9·78
	9	−0·31	29·5	8·20	9·85
	1	2·27	3·3	8·82	27·77
	2	1·52	3·2	9·02	27·97
	3	0·35	2·8	8·47	28·18
B	4	0·01	3·1	8·61	28·11
	5	0·01	14·4	8·61	30·66
	6	0·01	30·2	8·75	32·59
	7	−1·52	3·1	8·41	27·35

axial extension measurement was better than 10^{-4} in ($2{\cdot}5 \times 10^{-4}$ cm). The torsion angle was measured by the same rotary LVDT as employed in machine control.

The values of $\dot{\gamma}$ used were nominally $3 \times 10^{-5}\,\mathrm{s}^{-1}$, $3 \times 10^{-4}\,\mathrm{s}^{-1}$, and $3 \times 10^{-3}\,\mathrm{s}^{-1}$. The actual values varied according to the actual gage length of each specimen. The values of σ ranged from $-1{\cdot}5$ to $+2{\cdot}65$ MPa with several runs at $\sigma = 0$. The test parameters for each run are tabulated in Table I. Each test was carried out to a maximum twist of about 30° corresponding to a simple shear of about 0·35. The corresponding cumulative axial strain was of the order of 4×10^{-3}.

5. RESULTS AND ANALYSIS

In Figs. 2 and 3 are shown the results of ε vs γ for sixteen test runs. The measured points shown are a very small selection from the data collected. The solid lines drawn through the data are smooth analytic curves fitted to the data for the purpose of calculating $d\varepsilon/d\gamma$. A selection of curves of τ vs γ is shown in Fig. 4 with smoothed curves

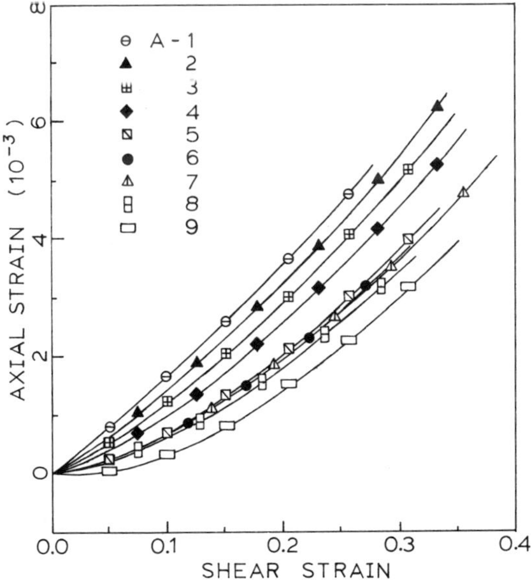

FIG. 2. Plots of ε vs γ for A specimens.

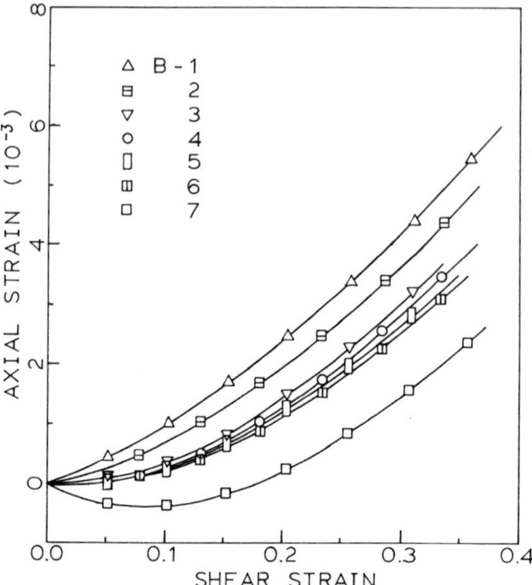

FIG. 3. Plots of ε vs γ for B specimens.

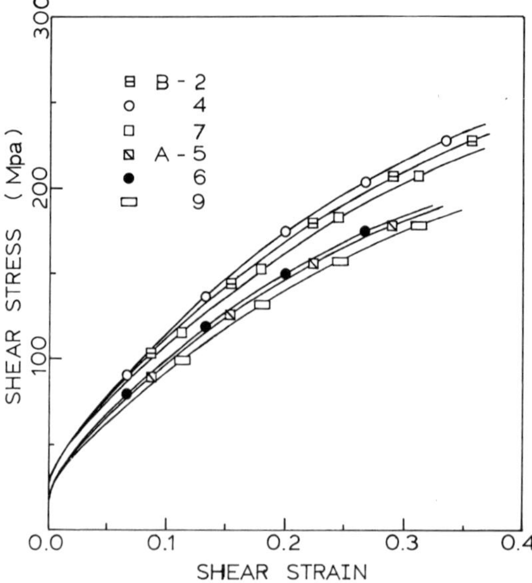

FIG. 4. A selection of experimental curves of τ vs γ.

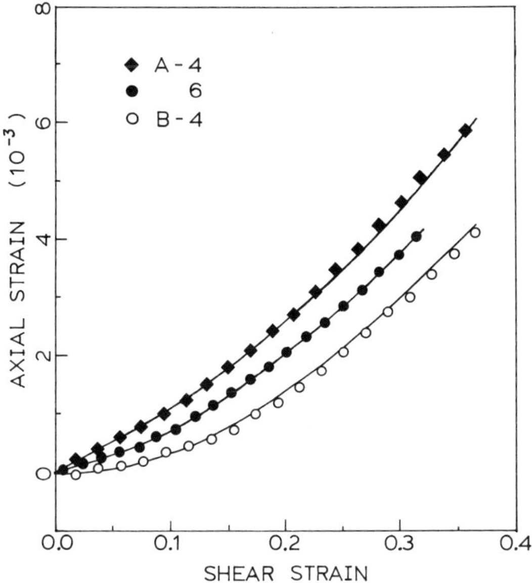

FIG. 5. Comparison between experimental data and the prediction from eqn. (5) for three experimental runs.

for determination of $d\tau/d\gamma$. It can be seen in Fig. 4 that the flow stress level for B-specimens is higher than for A-specimens.

An immediate result of the data analysis is that the kinematic hardening model predicts axial extension rates about a factor of 20 higher than those observed. That model was discarded then for further analysis.

For comparison with the predictions of Hart's model, the data was analyzed for the quantities $d\varepsilon/d\gamma$ and $d\tau/d\gamma$ as functions of γ for each specimen run. The best fit values of $\mathcal{M}$ and τ_f were determined for each run. Those values are included in Table I. The predicted curves of ε vs γ were then compared with the measured data. The fit to the data points was excellent for all experimental runs. A selection of such comparisons is shown in Fig. 5.

The best fit values of $\mathcal{M}$, as averaged for the individual runs, were $8{\cdot}46 \pm 0{\cdot}20$ GPa for group A, $8{\cdot}67 \pm 0{\cdot}20$ GPa for group B, and $8{\cdot}55 \pm 0{\cdot}22$ GPa for all specimens.

6. DISCUSSION

The predictive success of eqn. (5) is very strong confirmation of the three-dimensional formulation of Hart's constitutive equations. The strongest point is the sharp determination of the anelastic modulus $\mathcal{M}$. The success of the theory demands that $\mathcal{M}$ be a constant, independent of $\dot{\gamma}$ or stress level, and this is well satisfied by the experiment.

The friction stress τ_f is expected in the constitutive theory to be well represented by a power law dependence on $\dot{\gamma}$. Plots of τ_f vs $\dot{\gamma}$ are shown in Fig. 6. If the exponent M is given by

$$M = \mathrm{d} \ln \dot{\gamma} / \mathrm{d} \ln \tau_f$$

the result from Fig. 6 is $M = 11$. Note that the level of τ_f is different for batch B and batch A, but M is the same. This is consistent with the judgement that batch B material had a higher solute content than batch A. The parameter τ_f is quite sensitive to small differences in impurity content.

In the derivation of eqn. (5) no account was taken of possible textural anisotropy of the specimen material. A calculation of the effect of anisotropy has been carried out by the authors and will be published elsewhere. The goodness of the fit of the nickel data to eqn.

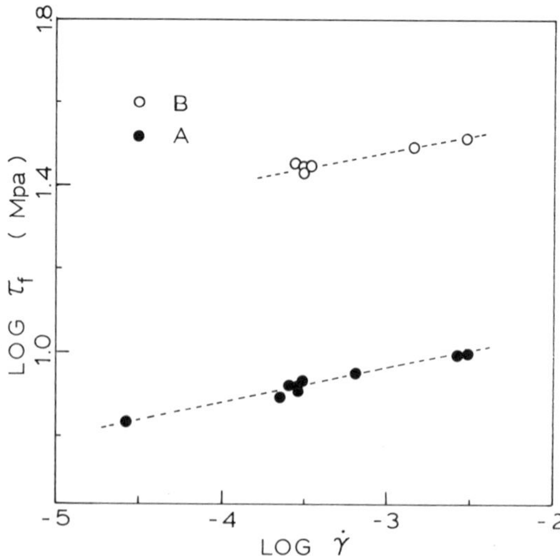

FIG. 6. Plots of τ_f vs $\dot{\gamma}$ for all test runs grouped according to material type.

(5) seems to be adequate reason to ignore possible anisotropy in the present work. The only unsatisfactory feature of the result from application of eqn. (5) is the magnitude of $\mathcal{M}$ that is reported here. That value is about a factor of $\frac{1}{2}$ or $\frac{1}{3}$ of that which would be expected from theoretical estimates. Since the theoretical estimates are themselves rather rough, and since there are not adequate uniaxial determinations available for $\mathcal{M}$, the point raised here is not a strong one. We point out, nevertheless, that, since we are dealing with real polycrystalline material and not with an ideal continuum substance, it is possible that the vorticity $\dot{\omega}$ may not be a perfect measure of the rate of rotation of the micro-elements of the crystallites. This question is best settled by a detailed micromechanical analysis of the rotation of the slip systems.

CONCLUSIONS

It has been shown that the effects of material rotations during torsion testing can be appreciable. The effects can be computed for constitutive laws that are adequately specified. The effects can be measured with good precision.

Comparison between experimental results and theoretical prediction can provide sensitive measurements of material parameters that appear in the theories.

ACKNOWLEDGEMENT

The work reported here was supported by a grant from the U.S. National Science Foundation under the Solid Mechanics Program of the Engineering Division.

REFERENCES

1. Bailey, J. A., S. L. Haas and K. C. Nawak. Anisotropy in plastic torsion, *Trans. ASME, J. Basic Engng*, **94** (1972), 231–237.
2. Hart, E. W. Constitutive relations for the non-elastic deformation of metals, *Trans. ASME, J. Engng. Mater. Technol.*, **98** (1976), 193–202.
3. Hart, E. W. The effects of material rotations in tension–torsion testing, *Int. J. Solids Struct.*, **18** (1982), 1031–1042.

4. Hecker, S. S. Experimental studies of yield phenomena in biaxially loaded metals. In: *Constitutive Equations in Plasticity*, Vol. 20, (Ed. J. A. Stricklin and K. H. Saczalski), ASME Publication AMD 1976, 1–33.
5. Hill, R. R. *The Mathematical Theory of Plasticity*, Clarendon Press, Oxford, 1950, Chap. 12, 317.
6. Swift, H. W. Length changes in metals under torsional overstrain, *Engineering*, April 4 (1947), 253–257.

14

Inelastic Work and Thermomechanical Coupling in Viscoplasticity

ERHARD KREMPL

Department of Mechanical Engineering, Rensselaer Polytechnic Institute, Troy, New York, USA

ABSTRACT

The inelastic deformation of engineering alloys was recently shown to be rate(time)-dependent at room temperature. Experiments reported herein demonstrate that inelastic strain path length is a suitable measure for the modeling of work hardening for these materials, whereas inelastic work is not appropriate. The measured deformation induced temperature change is well predicted by a previously proposed theory of thermomechanical coupling. It has been derived by neglecting the stored energy of cold work and by assuming that the internal energy expression for thermoelasticity is valid even when the mechanical deformation is inelastic.

1. INTRODUCTION

In solid mechanics it is generally assumed that the inelastic deformation of metals is rate(time)-independent at low homologous temperatures. The theory of plasticity is consequently formulated in a rate(time)-independent fashion and phenomena such as loading rate† sensitivity, creep and relaxation are excluded.

Although the existence of time-dependent inelastic deformation in metals was known before 1910 [15, 16], the concept of time-independence of inelastic deformation which was apparently for-

† Loading rate sensitivity includes stress- and strain-rate sensitivity.

malized in ref. 18 has become the predominant idealization for inelastic deformation of metals.

Recently the author and his co-workers have tested several engineering alloys at room temperature. An MTS servocontrolled axial-torsion hydraulic testing machine and strain measurement on the uniform section of solid cylindrical and tubular specimens were used consistently throughout all tests. This arrangement enables the strict enforcement of either load (stress) or displacement (strain) boundary conditions. The servocontrolled machine can therefore represent 'soft' as well as 'hard' testing machine characteristics.

In all these tests the inelastic deformation was found to be rate dependent; stress- and strain-rate sensitivity, creep and relaxation were significant in areas where the tangent moduli are much smaller than the elastic moduli [8, 9, 12–14].

The present paper deals with two aspects of the time-dependent deformation behavior of these engineering alloys, the importance of inelastic work for modeling work hardening and its dissipation into heat.

2. EXPERIMENTAL ARRANGEMENTS

Tests on the role of inelastic work in hardening were performed under displacement boundary conditions (strain control) on cylindrical specimens, see refs. 8 and 13.

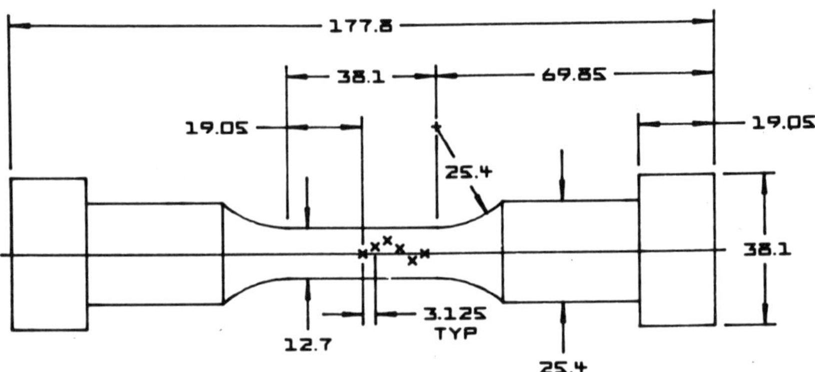

Fig. 1. Specimen for thermomechanical tests with locations of thermocouples. (All dimensions in millimeters. × denotes thermocouple.)

Deformation induced temperature changes were sensed by thermocouples spot-welded to the specimen surface. The location of the thermocouples together with the specimen dimensions used in the tests considering thermomechanical coupling are given in Fig. 1. All these tests were run on annealed Type 304 Stainless Steel at a total mechanical strain range of 1% at a frequency of 1 Hz under sinusoidal motions. Further details of testing are given in ref. 1.

In all tests a clip-on extensometer measured the strain on the uniform section of the specimens and the load cell in series provided the load signal. Both engineering stress and strain were continuously recorded in all tests. All tests were performed in air at room temperature.

3. INELASTIC WORK AS A MEASURE OF WORK HARDENING

In time(rate)-independent plasticity theory the growth of the function which models the size of the yield surface is frequently made to depend on inelastic work. In viscoplasticity the work expended to deform a material monotonically to a certain state of strain under a fixed strain path is not unique; it depends on the strain rate.

This fact, together with some of the results reported in refs. 8, 9 and 13, suggested that the role of inelastic work be investigated separately.

Figure 2 shows the results for monotonic loading. Two specimens

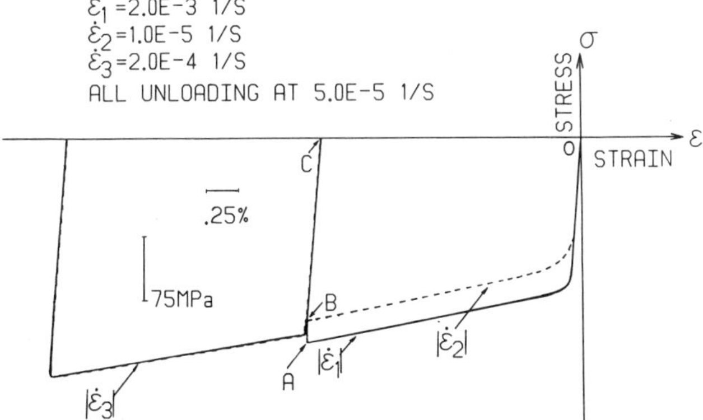

FIG. 2. Compression tests on tubular specimens using different strain rates on 0C but one strain rate starting from C, Material Type 304 Stainless Steel, fully annealed.

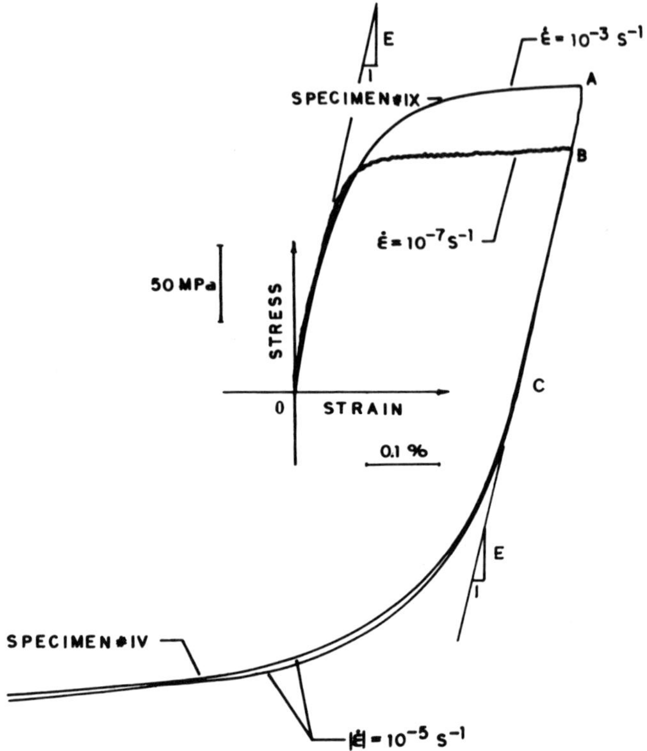

FIG. 3. Loading, unloading followed by compression of solid cylindrical specimens of Type 304 Stainless Steel.

were loaded to A and B with strain rates 10^{-5} and $10^{-3}\,s^{-1}$, respectively, and then unloaded to C. Reloading from C was accomplished with the same strain rate of $2\times10^{-4}\,s^{-1}$. Despite the different prior history the curves starting at C are identical within experimental accuracy.

In Fig. 3 strain rates up to points A and B were different. From these points on the unloading strain rate was identical for the two specimens. Again it is seen that the deformation behavior after A and B is identical in both cases.

The inelastic work† at point C in Figs. 2 and 3 is significantly

† The inelastic work at C is by the definition equal to the area OAC or OBC.

different in the two cases but the response beyond C is identical within experimental accuracy.

In contrast the inelastic strain path lengths

$$p[t] = \int_0^t (\dot{\varepsilon}^{in}_{kl}\dot{\varepsilon}^{in}_{kl})^{1/2}\, d\tau \tag{1}$$

with $\dot{\boldsymbol{\varepsilon}}^{in} = \dot{\boldsymbol{\varepsilon}} - \dot{\boldsymbol{\varepsilon}}^{el}$, where $\dot{\boldsymbol{\varepsilon}}$ and $\dot{\boldsymbol{\varepsilon}}^{el}$ are the total and elastic infinitesimal strain rates, respectively, are equal at point C.

Similar observations were made after examining some of the test results reported in refs. 8–13. As an example, reference is made to the test histories OCDEFG and $OC_1DE_1F_1G_1$ in Fig. 2 of ref. 13. From these results it is clear that the observations of Figs. 2 and 3 are not peculiar to the strain level nor to the annealed condition used in these tests. These results are rather general.

Very different materials, the austenitic stainless steel (Type 304), the Ti-7Al-2Cb-1Ta titanium alloy and the ferritic A533B pressure vessel steel, were shown to exhibit the behavior depicted in Figs. 2 and 3 [11]. At the same time these alloys did not exhibit a strain-rate history effect (SRH). (No specific tests were run in ref. 10 regarding the SRH. Nevertheless this author does not expect the ferritic steel to exhibit an SRH.) Therefore it is concluded that inelastic work is not a suitable measure of work hardening for these materials. However, the use of inelastic strain path length as a repository for work hardening is not precluded by these test results.

Theoretically, similar tests should be repeated in other states of stress and for other conditions of prestrain. Only then a complete experimental proof would be available. However, it is this author's judgement that such additional testing is not necessary. No reason exists why the behaviors shown in Figs. 2 and 3 and in ref. 11 should be peculiar to the condition or state of stress used in the tests. These conditions, with regard to the properties considered here, can be considered arbitrary.

A theory proposed by Bodner and Partom [2] uses inelastic work as a repository for work hardening. This theory will have difficulties in modeling the behavior described herein and in ref. 11. These difficulties may not be present if the materials exhibit an SRH. The behavior of some theories in simulating the present test conditions is discussed in detail in refs. 3 and 4.

4. DEFORMATION INDUCED TEMPERATURE CHANGE

4.1. Analysis

The thermomechanical analysis of inelastic deformation of metals is still very much disputed. A thermodynamic analysis is possible once the functional form of the free energy is known. However, for the case of viscoplastic behavior considered in the paper it is not known whether the free energy depends upon the temperature, strain and strain rate, or the temperature, stress and stress rate; upon internal (hidden) and as yet unidentified variables; or, in some manner, on the history of deformation. Any choice of the free energy function and of the entropy produces, through the usual derivation, restrictions on the constitutive equations of the theory. The constitutive equations so obtained may or may not predict the observed stress–strain and temperature–time behavior or may even be in conflict with them. Since entropy and internal energy are not quantities which can be directly measured in experiments, their functional form cannot be deduced directly from stress, strain and temperature measurement.

Here, as in refs. 5–7 we pursue an alternative approach which postulates the mechanical constitutive equation and the internal energy, and, therefore, makes two assumptions like the classical approaches.

In the following we assume the mechanical constitutive equation and propose an internal energy expression which satisfies the following postulates:

(i) Initial thermoelastic mechanical behavior results in initial thermoelastic temperature response.
(ii) For adiabatic conditions with no external heat supply all mechanical work should be converted into temperature change in a cycle beginning and ending at zero stress.

Because of (ii) the present theory is not capable of modeling the energy stored in the material, see ref. 1.

Following refs. 5–7 we postulate that the thermoelastic internal energy rate is given by

$$\rho\dot{e} = \rho\hat{C}\dot{\theta} + \frac{\mathrm{d}}{\mathrm{d}t}\left[\frac{1+\nu}{2E}\sigma_{ij}\sigma_{ij} + \frac{\nu}{E}\sigma_{kk}^2 + \alpha\theta\sigma_{kk}\right] \tag{2}$$

In the above, ρ = density, e = internal energy per unit mass, θ = absolute temperature, ν, E = Poisson's ratio, Young's modulus, α =

coefficient of thermal expansion, C = specific heat per unit mass, $\hat{C} = C + (3E\alpha^2\theta/\rho(1-2\nu))$, σ_{ij} = stress tensor.

The use of the first law for small deformations

$$\rho\dot{e} = \sigma_{ij}\dot{\varepsilon}_{ij} - q_{i,i} + \rho R \tag{3}$$

together with (2) results in

$$\rho\hat{C}\dot{\theta} = \sigma_{ij}\dot{\varepsilon}_{ij} - \frac{\mathrm{d}}{\mathrm{d}t}\left[\frac{1+\nu}{2E}\sigma_{ij}\sigma_{ij} - \frac{\nu}{E}\sigma_{kk}^2 + \alpha\theta\sigma_{kk}\right] + q_{i,i} + \rho R \tag{4}$$

with ε_{ij} the small strain tensor, q_i the heat flux vector, and R the heat supply per unit mass.

It is obvious that (4) satisfies (ii) above.

When constant material properties are used, (4) can be rewritten as

$$\rho\hat{C}\dot{\theta} = \sigma_{ij}\dot{\varepsilon}_{ij}^{\mathrm{in}} - \alpha\theta\dot{\sigma}_{kk} - q_{i,i} + \rho R \tag{5}$$

with

$$\dot{\varepsilon}_{ij}^{\mathrm{in}} = \dot{\varepsilon}_{ij} - \frac{1}{E}[(1+\nu)\dot{\sigma}_{ij} - \nu\,\dot{\sigma}_{kk}\delta_{ij} + E\alpha\dot{\theta}\delta_{ij}] \tag{6}$$

The adiabatic uniaxial form of (5) with $R = 0$

$$\rho\hat{C}\dot{\theta} = \sigma\dot{\varepsilon}^{\mathrm{in}} - \alpha\theta\dot{\sigma} \tag{7}$$

is used in the following analysis of experimental data.

4.2. Experimental Results

The hysteresis loops obtained during the room temperature testing of Type 304 Stainless Steel are shown in Fig. 4 for the initial cycles. The corresponding temperature changes as measured by the thermocouples, the locations of which are depicted in Fig. 1, are reproduced in Fig. 5. Note that numbering of the thermocouples starts at the midsection of the specimen. Thermocouple 1 is at the midsection and thermocouple 6 has the greatest distance from the midsection.

Figure 5 demonstrates an overall increase in temperature which reaches about 50°C after 100 s. Superposed on this overall increase is a cyclic component. It is also observed that no difference exists between the readings of the six thermocouples during the first eight or nine seconds.

A separate analysis (see Fig. 7 of ref. 1) shows that during this time the process can be considered adiabatic. Therefore an adiabatic analysis was made using the uniaxial form (7) of the heat equation. For

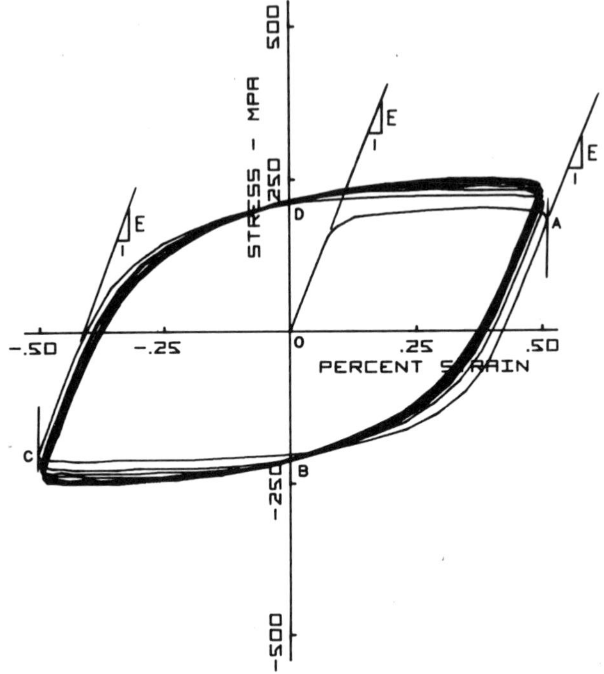

FIG. 4. Hysteresis loops obtained during strain controlled cycling of Type 304 SS using a sinusoidal forcing function of 1 Hz.

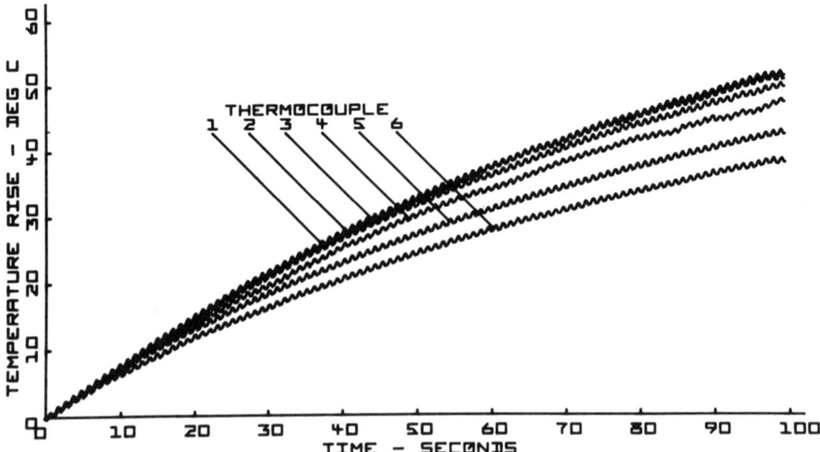

FIG. 5. Temperature rises associated with the mechanical loading shown in Fig. 4 as measured by the thermocouples.

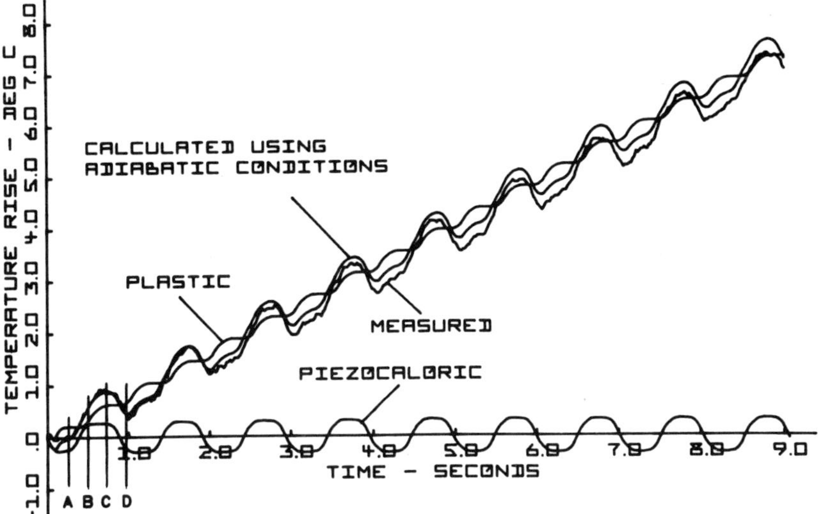

FIG. 6. Comparison of measured and predicted temperature responses under adiabatic conditions.

the properties of Type 304 SS, $\hat{C} \approx C$ and θ was considered constant during the first 9 s for the purpose of the adiabatic analysis. Note that

$$\dot{\varepsilon}^{\text{in}} = \dot{\varepsilon} - (\dot{\sigma}/E + \alpha\dot{\theta}) \tag{8}$$

is the inelastic mechanical strain rate.

The results of this analysis are depicted in Fig. 6. The curve labeled 'measured' is the reading of the center thermocouple, No. 1 in Fig. 5. The curves identified as 'piezocaloric' and 'plastic' are the contributions of the second and first terms on the right-hand side of (7), respectively. They are obtained by using the *measured stress and mechanical strain signals.* The curve identified as 'calculated using adiabatic conditions' is the sum of the two previous terms and represents the *prediction* of (7) using the actual mechanical response as input.

Comparison of the measured and of the calculated curves shows that both have the same waveshape but that there is a continuously increasing difference between the two with the calculated curve indicating a higher temperature than the measured one. This difference, which amounts to less than 1/4°C or 3% of the measured temperature after 9 s is attributed to the neglect of the stored energy of cold work in eqn. (7).

The shape of the temperature vs time curve is very well reproduced by (7); it represents the reversible elastic (piezocaloric) and the inelastic (plastic) contributions. The elastic part represents heating and cooling depending on the sign of $\dot{\sigma}$. For the majority of the cycle the inelastic contribution is positive or zero. However, small negative values are possible since the slope of the stress mechanical–strain curve can be less than the elastic one as zero stress is approached (see Fig. 4). A check of our own isothermal experiments and of hysteresis loops published by others shows that slopes smaller than the elastic one can be obtained as zero stress is approached. Even if strain rather than mechanical strain is plotted, slopes smaller than the elastic one as zero stress is approached are indicative of negative inelastic power. The experiments show that negative inelastic work is possible and $\sigma_{ij}\dot{\varepsilon}_{ij}^{in} \geqslant 0$ cannot be postulated as a physical law.†

The present experiments and others reported in ref. 1 show that the theory postulated in refs. 5–7 reproduces deformation induced temperature changes very well. Instead of the measured mechanical response, a theoretical mechanical response can be used to calculate the temperature rise; this has been done in refs. 5–7.

The heat generation terms (the first two on the right-hand side of (4) or (5)) are homogeneous of degree one in the rates. They are, therefore, rate-independent. However, when coupled with a rate-dependent constitutive law, the heat generated will depend on the rate of mechanical deformation. Further, through the usual Fourier heat conduction law, connecting $q_{i,i}$ with temperature (4) will become rate-dependent in inhomogeneous states of temperature. The temperature rise attained in the material will then strongly depend on the rate of mechanical working and the thermophysical properties. As part of ref. 1 some experiments were run at 0·01 Hz instead of 1 Hz, and no temperature rise would be observed. This fact is accounted for in (5). Therefore it is fair to say that deformation induced temperature changes are only significant in fast processes.

A modification of the theory is contemplated to account for the stored energy of cold work.

ACKNOWLEDGEMENT

This work was supported by grants from the U.S. National Science Foundation and from NASA Lewis Research Center, Cleveland, Ohio.

† Yield surfaces which do not include the origin support this conclusion. Such yield surfaces have been published among others by Phillips [17].

REFERENCES

1. ADAMS, S. L. and E. KREMPL. Thermomechanical response of 3·5 Ni-Mo-V alloy steel and Type 304 Stainless Steel under cyclic uniaxial inelastic deformation, *Res. Mech.*, **10** (1984), 295–316.
2. BODNER, S. R. and Y. PARTOM, Constitutive equations for elastic–viscoplastic strain-hardening materials, *Trans. ASME, J. appl. Mech.*, **42** (1975), 385–389.
3. CERNOCKY, E. P. Comparison of the unloading and reversed loading behavior of three viscoplastic constitutive theories, *Int. J. Non-Linear Mech.*, **17** (1982), 255–266.
4. CERNOCKY, E. P. An examination of four viscoplastic constitutive theories in uniaxial monotonic loading, *Int. J. Solids Struct.* **18** (1982), 989–1005.
5. CERNOCKY, E. P. and E. KREMPL. A theory of thermoviscoplasticity based on infinitesimal total strain, *Int. J. Solids Struct.*, **16** (1980), 723–741.
6. CERNOCKY, E. P. and E. KREMPL. A theory of thermoviscoplasticity for uniaxial mechanical and thermal loading, *J. Méc. Appl.*, **5** (1981), 293–321.
7. CERNOCKY, E. P. and E. KREMPL. A coupled theory of thermoviscoplasticity based on total strain and overstress and its predictions in monotonic torsional loading, *J. Thermal Stresses*, **4** (1981), 69–82.
8. KREMPL, E. An experimental study of room-temperature rate sensitivity, creep and relaxation of type 304 Stainless Steel, *J. Mech. Phys. Solids*, **27** (1979), 363–375.
9. KREMPL, E. The Role of Servocontrolled Testing in the Development of the Theory of Viscoplasticity Based on Total Strain and Overstress, American Society for Testing and Materials, STP 765 (March 1982), 5–28.
10. KREMPL, E. and V. V. KALLIANPUR. On the Room Temperature Rate(Time)-Dependent Behavior of A533B Class 1 Steel, RPI Report CS 81-1 (February 1981).
11. KREMPL, E. and V. V. KALLIANPUR. Some critical uniaxial experiments for viscoplasticity at room temperature, *J. Mech. Phys. Solids*, **32** (1984), 301–314.
12. KUJAWSKI, D., V. KALLIANPUR and E. KREMPL. An experimental study of uniaxial creep, cyclic creep and relaxation of AISI Type 304 Stainless Steel at room temperature, *J. Mech. Phys. Solids*, **28** (1980), 129–148.
13. KUJAWSKI, D. and E. KREMPL. The rate(time)-dependent behavior of Ti-7Al-2Cb-1Ta titanium alloy at room temperature under quasi-static monotonic and cyclic loading, *Trans. ASME, J. appl. Mech.*, **48** (1981), 55–63.
14. LIU, M. C. M. and E. KREMPL. A uniaxial viscoplastic model based on total strain and overstress, *J. Mech. Phys. Solids*, **27** (1979), 377–391.
15. LOVE, A. E. H. *A Treatise on the Mathematical Theory of Elasticity*, 4th edn, Dover Publications, New York, especially p. 117.
16. LUDWIK, P. *Elemente der Technologischen Mechanik*, Berlin, 1909.
17. PHILLIPS, A. this volume.
18. VON MISES, R. Mechanik der festen Koerper im plastisch-deformablen Zustand, *Goettinger Nachrichten, Math-Phys. Klasse* (1913), 582–592.

15

Experimental Investigations on Shakedown of Tubes

KONRAD LEERS

Institut für Mechanik, Universität Hannover, Federal Republic of Germany

WALTER KLIE

Ingenieurbüro für Maschinendynamik, Hannover, Federal Republic of Germany

JAN A. KÖNIG

Institute of Fundamental Technological Research, Warsaw, Poland

and

OSKAR MAHRENHOLTZ

Arbeitsgebiet Meerestechnik II, Technische Universität Hamburg-Harburg, Federal Republic of Germany

ABSTRACT

The shakedown experimental works were hitherto made, nearly exclusively, on small-scale models and the load variations were operated by hand. Therefore, the number of cycles applied had to be rather low. The present investigations, aimed to model the elastic–plastic response of the nuclear fuel cladding shell, on the contrary, have been made completely in an automatic way. In this manner, a large number of identical load cycles could be performed. This paper contains, in its initial sections, an outline of the state of the shakedown theory and an approximate theoretical solution of the investigated shell. The comparison of this solution with the experimental data seems to be satisfactory.

1. INTRODUCTION

Many engineering structures or structural elements, especially the elements of nuclear power plants and of nuclear and chemical reactors are exposed to simultaneous actions of loads and temperature fields. Moreover, the loads as well as the temperature fields vary, usually in a more or less cyclic way, within wide limits.

Obviously, the design of such structures or structural elements should take account of inelastic material response, in particular of plasticity effects. The most important approach is the theory of limit analysis. This theory, developed intuitively in the 1930s and theoretically and experimentally based in the 1950s, is widely used and has become recommended by many design codes. By means of methods developed by this approach, the ultimate load or the instantaneous load-carrying capacity of a structure made of a material with significant plastic properties may be determined quite accurately.

However, in the case of variable repeated loads and, especially, in the presence of temperature changes, the magnitude of the ultimate load is not the only factor characterizing the structural safety. Namely, these load or temperature changes may result in an accumulation of plastic deformations leading to excessive structural deflections (so-called incremental collapse) or in alternating plastic strain increments resulting, after a sufficient number of cycles, in so-called low-cycle fatigue and in local material failure. These phenomena may be observed even much below the ultimate load. Therefore, the structural safety requires that the plastic strain increments due to consecutive load/temperature changes should eventually cease, the structural response of further cycles being perfectly elastic. Such a stabilization of plastic deformations is called shakedown or adaptation. Figure 1 shows some cases of the above described phenomena. They show a plastic strain versus the number of load/temperature cycles.

Curve (a) indicates a perfectly elastic response. Curve (b) shows a monotonous increase of the permanent strain which stabilizes after a few cycles: shakedown. Curve (c) indicates incremental collapse. In curve (d) the plastic strain increments in subsequent cycles cancel each other but remain steady: this leads eventually to low cycle fatigue. Curve (e) is connected with an oscillating stabilization of the plastic strain. This also indicates shakedown. The vertical line (f) denotes a case in which the ultimate load has been exceeded during the first cycle. The determination of the maximum load/temperature limits allowing for adaptation is the main goal of the theory of shakedown providing an extension of the limit analysis methods to variable repeated loads and temperature fields.

Certain initial results in this field were obtained by Bleich [1] and

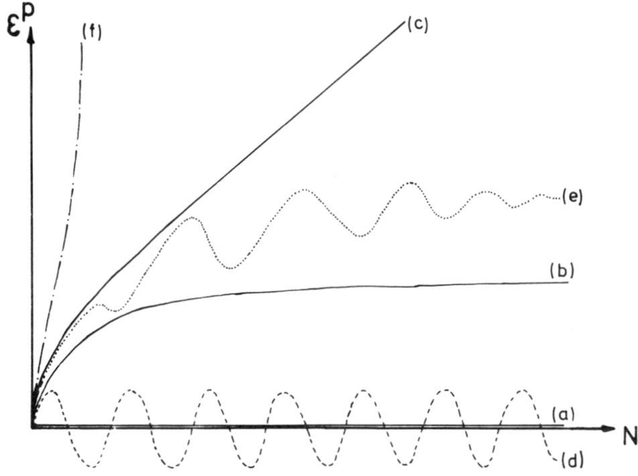

FIG. 1. Phenomena of plastic strain behavior at cyclic loading.

the fundamental lower bound theorem was given by Melan [10] already in the 1930s. Upper bounds to shakedown loads are due to Neal [11] and Koiter [6]. Čyras [3] and Maier [9] showed that the problem of determination of shakedown loads leads to a dual pair of mathematical programming problems.

All these results may be easily extended to cover the effect of thermal strains and the temperature variations of the yield stress [4, 5, 7, 12, 14].

The two fundamental shakedown theorems can be presented as follows:

(a) The statical theorem: a given structure will shake down if there exists a safety factor $\eta > 1$ and a time-independent residual stress field $\rho_{ij}(\mathbf{x})$ such that

$$\eta f[\sigma_{ij}^{\mathrm{E}}(\mathbf{x}, t) + \rho_{ij}(\mathbf{x})] \leqslant Y(\mathbf{x}, \theta) \tag{1}$$

at any instant $t > 0$ and for any point $\mathbf{x}$ of the structure. Here, θ is the temperature, Y is the yield stress, $f(\sigma_{ij}) = Y$ stands for the yield condition and σ_{ij}^{E} is the thermoelastic stress associated with the actual load and temperature.

(b) From the kinematical theorem of Koiter [6], a safety criterion against incremental collapse can be derived [5, 7]. Namely, if there are n independent agencies acting on a given structure, the instantaneous intensities of them being defined by n load/temperature factors β_k, $k = 1, \ldots, n$, then the thermoelastic stress as well as the temperature are given by linear functions of these factors:

$$\sigma_{ij}^{E}(\mathbf{x}, t) = \sum_{k=1}^{n} \beta_k(t)\sigma_{ij}^{Ek}(\mathbf{x}), \qquad \theta(\mathbf{x}, t) = \sum_{k=1}^{n} \beta_k(t)\theta^k(\mathbf{x}) \tag{2}$$

Let the range of the load/temperature variations be given by the inequalities

$$a_k \leqslant \beta_k(t) \leqslant b_k, \qquad k = 1, \ldots, n \tag{3}$$

and let the structure be loaded, additionally, by constant body forces $P_i^o(\mathbf{x})$ and surface tractions $T_i^o(\mathbf{x})$. Then, the load/temperature limits (3)—still ensuring that there will be no incremental collapse in a mechanism defined by a kinematically compatible plastic strain increment field $\Delta\bar{\varepsilon}_{ij}(\mathbf{x})$—can be determined from the following equation [5, 7]:

$$\int_V P_i^o \,\Delta\bar{u}_i \,\mathrm{d}V + \int_S T_i^o \,\Delta\bar{u}_i \,\mathrm{d}S + \int_V \sum_{k=1}^{n} \alpha_k(\mathbf{x}) J_k(\mathbf{x}) \,\mathrm{d}V = \int_V D_o(\Delta\bar{\varepsilon}_{ij}(\mathbf{x})) \,\mathrm{d}V \tag{4}$$

where

$$J_k(\mathbf{x}) = \sigma_{ij}^{Ek}(\mathbf{x}) \,\Delta\bar{\varepsilon}_{ij}(\mathbf{x}) + e(\mathbf{x})\theta^k(\mathbf{x}) D_o(\Delta\bar{\varepsilon}_{ij}(\mathbf{x}))$$
$$\alpha_k(\mathbf{x}) = \begin{cases} b_k & \text{if } J_k(\mathbf{x}) > 0 \\ a_k & \text{if } J_k(\mathbf{x}) < 0 \end{cases} \tag{5}$$
$$\Delta\bar{\varepsilon}_{ij} = \tfrac{1}{2}(\bar{u}_{i,j} + \bar{u}_{j,i})$$

and the plastic dissipation $D(\Delta\bar{\varepsilon}_{ij}, \theta)$ is

$$D(\Delta\bar{\varepsilon}_{ij}, \theta) = D_o(\Delta\bar{\varepsilon}_{ij})(1 - e\theta) \tag{6}$$

There $D_o(\Delta\bar{\varepsilon}_{ij}) = D(\Delta\bar{\varepsilon}_{ij}, 0)$ denotes the dissipation at the reference temperature. The material parameter e defines, simultaneously, the temperature dependence of the yield stress:

$$Y(\mathbf{x}, \theta) = (1 - e\theta) Y_o(\mathbf{x}, 0) \tag{7}$$

The formula (4) can be employed to solve engineering problems effectively. In the case of non-varying loads and temperatures, the formulae (1) and (4) coincide with the well-known lower and upper bounds of the ultimate load as given by the limit analysis.

As early as in the beginning of the 1950s, there were several papers published in which the shakedown theory was attempted to be verified experimentally. But even today, the situation does not seem to be satisfactory. The hitherto investigations were carried out, nearly exclusively, on small-scale models made of ductile steel, usually beams or simple plane frames, exposed to cycles of concentrated forces. There are only few reports considering corrugated plates, shell or pipe-shell connections [2, 13]. The same applies to thermal effects though in the latter case shakedown seems to be the most proper method to estimate the structural safety, at least below the excessive creep temperatures. Moreover, the investigations employed mainly standard loading equipments (presses, jacks, etc.), used in monotone loading tests controlled by hand.

The aim of the remainder of the present paper is to show an example, being a part of a broader-planned series of experiments, in which not only the cyclic loading but also the recording of the data has been fully automatized. Contrary to common practice, resulting from the existence of many finite-element computer programs, the numerical analysis of the structure considered has been simplified. Authors think that it does not make much sense to refine the computations too much if the material and structural model (perfect plasticity, geometrical linearity) is crude compared to the actual one.

2. AN APPROXIMATE SHAKEDOWN ANALYSIS OF A NUCLEAR FUEL CLADDING SHELL MODEL

Nuclear fuel cladding shells are exposed to several agencies such as internal pressure of the gas emerging from nuclear reactions, pressure due to fuel pellets which increase with time, temperature gradient across the shell wall, axial forces. When switching the reactor on and off, the magnitudes of those agencies increase or decrease. This applies, especially, to the temperature gradient intensity as well as to the internal pressure.

In the approximate shakedown analysis, performed by means of formula (4), the following assumptions have been used:

(i) The shell material is elastic, perfectly plastic obeying the generalized Tresca yield condition. The generalization consists of taking into account different magnitudes of the yield stress in longitudinal and in circumferential direction.

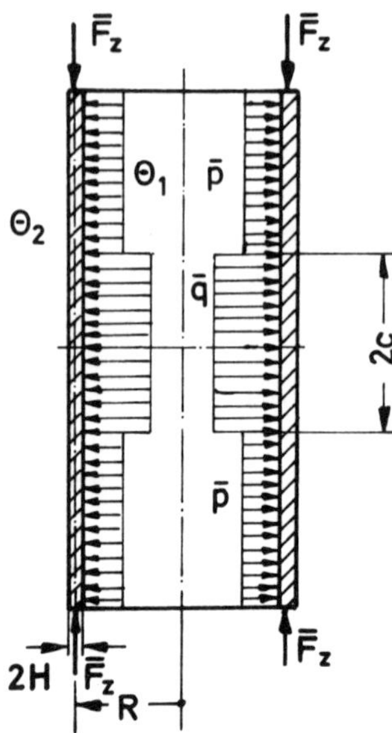

FIG. 2. Loads acting upon the tube: $\bar{F}_z$: axial force, $\bar{p}$: internal pressure, $\bar{q}$: local pressure, $2c$: length of the zone of the local pressure, θ_1: internal temperature, θ_2: external temperature, $2H$: tube wall thickness, R: tube mean diameter.

(ii) The loads acting upon the shell (Fig. 2) are:

(a) forces caused by the growth of the fuel pellets are approximated by a distributed pressure q within a length $2c$;
(b) the internal pressure p of the gas assumed to vary within zero and a certain maximum value $\bar{p}$;
(c) the axial force $\bar{F}_z$, remaining constant;
(d) the internal temperature θ_1. For the sake of safety it has been assumed to vary, independently of p, between zero and its maximum value $\bar{\theta}_1$.

(iii) No interaction has been assumed between subsequent ring forces of the pellets. Therefore, the elastic stresses have been calculated as for an infinite shell.

(iv) The Love hypothesis is assumed to hold also in the plastic range. Therefore, the axial strain ε_z and the circumferential strain ε_φ can be expressed through the longitudinal displacement u and the radial displacement w as follows:

$$\varepsilon_z = \frac{\mathrm{d}u}{\mathrm{d}z} - x\frac{\mathrm{d}^2 w}{\mathrm{d}z^2}, \qquad \varepsilon_\varphi = \frac{w}{R} \tag{8}$$

x denotes the distance from the middle shell surface, positive outwards. R is the shell middle radius.

The 'unit elastic stresses' due to internal pressure and temperature gradient are, respectively:

$$\begin{aligned} &\sigma_z^{\mathrm{E}p} = 0, \qquad \sigma_\varphi^{\mathrm{E}p} = \frac{R}{2H} \\ &\sigma_z^{\mathrm{E}\theta} = 0, \qquad \sigma_\varphi^{\mathrm{E}\theta} = \frac{E\alpha}{2(1-\nu)H}x \end{aligned} \tag{9}$$

where E is Young's modulus, ν is Poisson's ratio and α denotes the linear thermal expansion coefficient. The last formula follows from the assumption of a linear temperature distribution along the tube thickness. Two simple incremental collapse mechanisms, given in Fig. 3, have been analyzed. The corresponding displacement increments $\bar{u}$

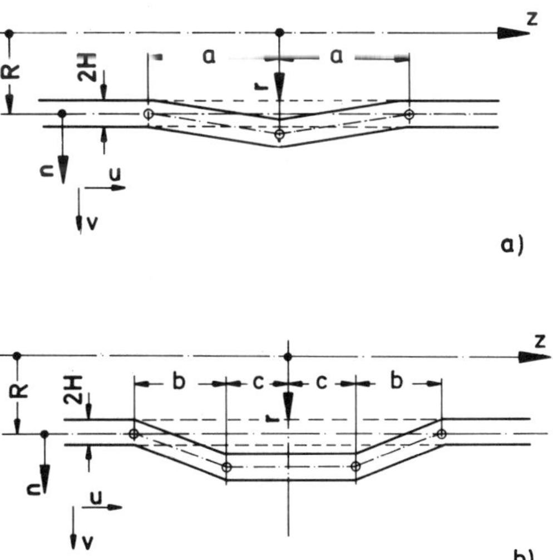

FIG. 3. Displacement fields of the incremental collapse mechanisms: (a) first mechanism, (b) second mechanism.

(longitudinal) and $\bar{v}$ (radial) are:

First mechanism (Fig. 3(a))

$$\bar{u}=\begin{cases}0 & \text{for} \quad z<-a\\ u_1-\dfrac{v_o}{a}n & \text{for} \quad -a<z<0\\ u_1+u_2+\dfrac{v_o}{a}n & \text{for} \quad 0<z<a\\ 2u_1+u_2 & \text{for} \quad a<z\end{cases} \tag{10}$$

$$\bar{v}=\begin{cases}\dfrac{v_o}{a}(z+a) & \text{for} \quad -a\leqslant z\leqslant 0\\ -\dfrac{v_o}{a}(z-a) & \text{for} \quad 0\leqslant z\leqslant a\end{cases} \tag{11}$$

Second mechanism (Fig. 3(b))

$$\bar{u}=\begin{cases}0 & \text{for} \quad z<-(b+c)\\ u_1-\dfrac{v_o}{b}n & \text{for} \quad -(b+c)\leqslant z<-c\\ u_1+u_2 & \text{for} \quad -c\leqslant z<c\\ u_1+2u_2+\dfrac{v_o}{b}n & \text{for} \quad c<z<b+c\\ 2(u_1+u_2) & \text{for} \quad b+c<z\end{cases} \tag{12}$$

$$\bar{v}=\begin{cases}\dfrac{v_o}{b}(z+b+c) & \text{for} \quad -(b+c)\leqslant z\leqslant -c\\ v_o & \text{for} \quad -c\leqslant z\leqslant c\\ -\dfrac{v_o}{b}(z-b-c) & \text{for} \quad c\leqslant z\leqslant b+c\end{cases} \tag{13}$$

By calculating the associated strain increments and the corresponding dissipated energy, after substituting into eqn. (4), one arrives after lengthy though simple calculations at the following results.

In the case of the first mechanism

$$\phi_{\max}^{(1)}=\frac{c}{a}(\bar{q}-\bar{p})+\left\{\bar{p}+\frac{E\alpha\,\Delta\bar{\theta}H}{4(1-\nu)R}-\frac{2Y_{\varphi}H}{R}\right\}=0 \tag{14}$$

where

$$a=\left[\frac{-2\bar{F}_z v_o+(\bar{q}-\bar{p})c^2-\dfrac{F_z^2}{Y_z}+4H^2Y_z}{-\bar{p}-\dfrac{E\alpha\,\Delta\bar{\theta}H}{4(1-\nu)R}+\dfrac{2Y_\varphi H}{R}}\right]^{1/2}\qquad \Delta\bar{\theta}=\bar{\theta}_1-\bar{\theta}_2 \quad (15)$$

In the case of the second mechanism

$$\phi_{\max}^{(2)}=\left\{\bar{p}+\frac{E\alpha\,\Delta\bar{\theta}H}{4(1-\nu)R}-\frac{2Y_\varphi H}{R}\right\}+\frac{c}{b}\left\{q+\frac{E\alpha\,\Delta\bar{\theta}H}{4(1-\nu)R}-\frac{2Y_\varphi H}{R}\right\}\leqslant 0 \quad (16)$$

where

$$b=\left[\frac{-2\bar{F}_z v_o-\dfrac{\bar{F}_z^2}{Y_z}+2H^2Y_z}{-\bar{p}-\dfrac{E\alpha\,\Delta\bar{\theta}H}{4(1-\nu)R}+\dfrac{2Y_\varphi H}{R}}\right]^{1/2} \quad (17)$$

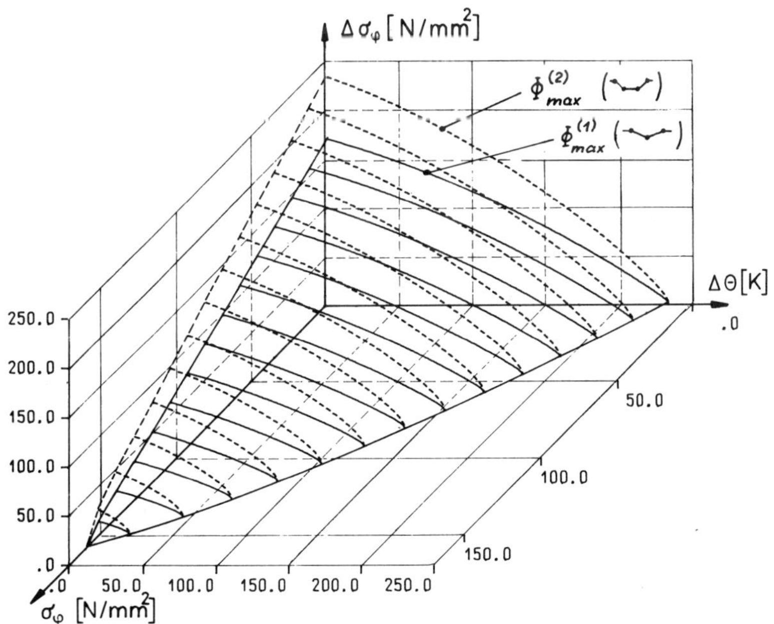

FIG. 4. The shakedown surface.

In both cases θ_m is here an average temperature of the shell wall. The magnitudes of the incremental collapse mechanism parameters a, b, u_1/v_o, u_2/v_o have been obtained from the condition that the function ϕ assumes an extremum with respect to all of them:

$$\frac{\partial \phi^{(1)}}{\partial a} = \frac{\partial \phi^{(1)}}{\partial \left(\frac{u_1}{v_o}\right)} = \frac{\partial \phi^{(1)}}{\partial \left(\frac{u_2}{v_o}\right)} = 0 \tag{18}$$

$$\frac{\partial \phi^{(2)}}{\partial b} = \frac{\partial \phi^{(2)}}{\partial \left(\frac{u_1}{v_o}\right)} = \frac{\partial \phi^{(2)}}{\partial \left(\frac{u_2}{v_o}\right)} = 0 \tag{19}$$

The safe (shakedown) domain in the $\bar{p}, \bar{q}, \Delta\bar{\theta}$ space is bounded by the $\bar{p}=0$, $\bar{q}-\bar{p}=0$, $\Delta\bar{\theta}=0$ planes and by the respective surface (14) or (16) (see Fig. 4) in which

$$\sigma_\varphi = \bar{p}\frac{R}{2H}, \qquad \Delta\sigma_\varphi = (\bar{q}-\bar{p})\frac{R}{2H} \tag{20}$$

3. THE ARRANGEMENT OF THE EXPERIMENTAL INVESTIGATIONS

The shakedown experiments of model tubes have been made on cold-drawn tube specimens made of the aluminium alloy Al Mg Si 0·5, of radius $R = 23{\cdot}65$ mm and of wall thickness $2H = 1{\cdot}0$ mm. The experimental arrangement is shown in Fig. 5. The following loads could be applied simultaneously to the specimen:

(a) an internal pressure produced by a hydraulic pump;
(b) a radial load applied at some arbitrary position along the tube by means of a ring-force membrane, connected to another pump;
(c) a longitudinal tensile or compressive force excited by a hydraulic cylinder;
(d) radial temperature gradient produced by a heating element along the tube axis.

The variations of the loads are produced by servo pressure relief valves. The sequence of the valves to be switched on and off in every cycle as well as the respective instances can be programmed and then

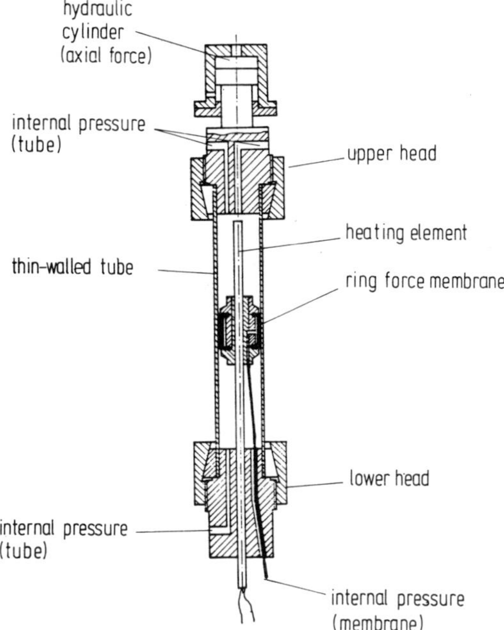

FIG. 5. The experimental arrangement.

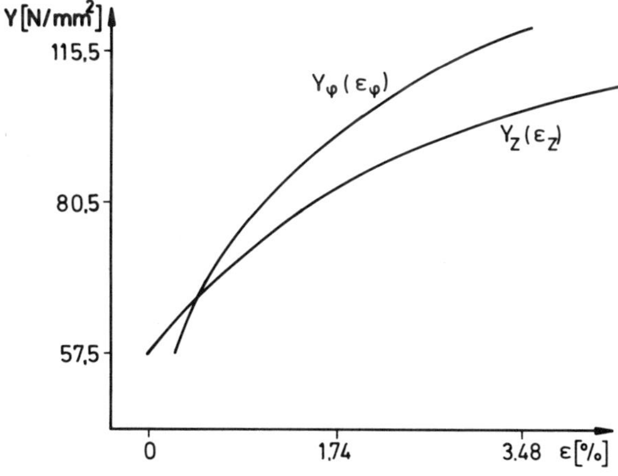

FIG. 6. The stress–strain curves in the axial and circumferential directions in the case of aluminum AlMgSi 0·5 cold-drawn tubes, annealed at the recrystallization temperature 440°C. External diameter 50 mm, wall thickness 2 mm.

executed in an automatic way. The duration of one cycle is practically arbitrary. In such a way a pretty high number (by now up to 2400) of identical consecutive load cycles can be executed. The radial displacements of the tube are measured by means of inductive displacement transducers and recorded by means of an X-Y plotter.

Additional initial tensile tests have shown some differences in the stress–strain curves in the radial and longitudinal directions. Their results are shown in Fig. 6. For comparison with the theoretical formulae derived in the Section 2, the conventional 0·2% yield stresses Y_φ and Y_z have been taken from those tests.

4. RESULTS OF THE HITHERTO COMPLETED EXPERIMENTAL INVESTIGATIONS

Though the experimental arrangement described in Section 3 allows for investigations at various internal temperatures, the investigations have been performed at a constant temperature usually slightly higher than the room temperature. One of the typical load variations within one cycle is shown in Fig. 7.

Typical data obtained from the X-Y plotter are shown in Fig. 8. Curve (a), obtained at slightly lower load amplitudes, clearly indicates shakedown. Curve (b) corresponds to the case of incremental collapse resulting at higher amplitudes. Each one of such data corresponds to a point in the $\bar{p}$, $\bar{q}$ plane (for $\bar{\theta}_1 \approx 25°C$). The former belongs to the

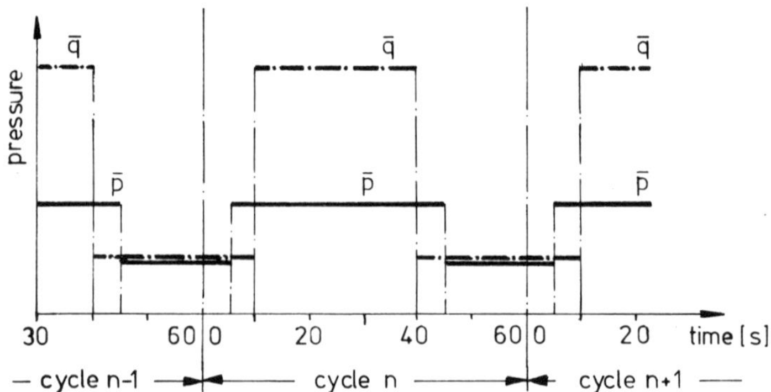

FIG. 7. The development of the internal pressure $\bar{p}$ and of the local pressure $\bar{q}$ during a loading cycle.

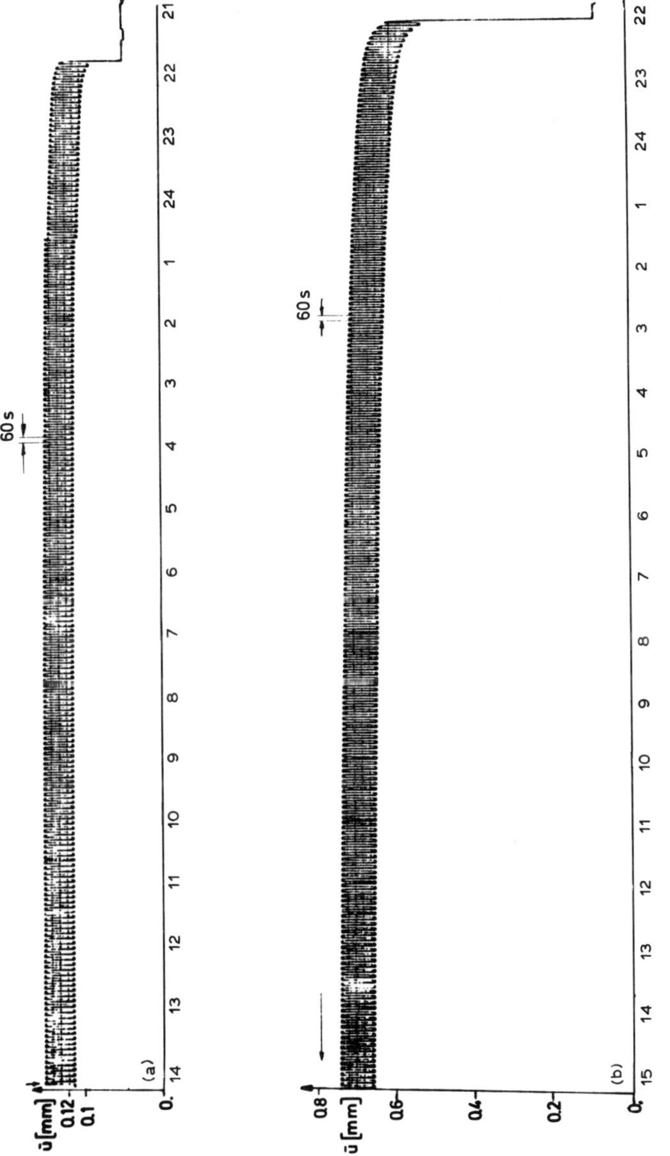

FIG. 8. The radial displacement $\bar{u}$ versus time: (a) at the pressure of 95 bars, (b) at the pressure of 102 bars.

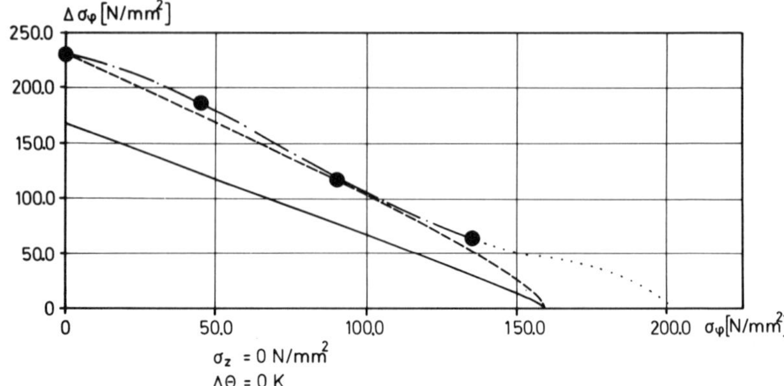

FIG. 9. Results of the shakedown analysis compared with the experimental data (temperature gradient $\Delta\bar{\theta} = 0$–2°C). —— first mechanism; ---- second mechanism; -·-·-·- experiment.

actual safe domain, the latter lies outside. Some of the extremal safe points are shown in Fig. 9. Extremal means here that no shakedown could have been obtained for the $\bar{p}$, $\bar{q}$ parameters being only slightly higher than those shown in that figure. Figure 9 shows the results of experimental investigations and Fig. 10 gives data obtained in one particular series. The corresponding (constant) parameters were:

$$\sigma_\varphi = \bar{p}\frac{R}{2H} = 45{\cdot}0\ \text{N/mm}^2, \qquad \sigma_z = 0{\cdot}5 \div 0{\cdot}7\ \text{N/mm}^2, \qquad \bar{\theta} = 0 \div 2K$$

The only variable parameter was the local pressure $\bar{q}$. The shell radial deflection $\bar{u}_{max}$ shown in Fig. 10 has been obtained under the following conditions:

(a) If the deflection increment rate per cycle $\partial\bar{u}/\partial N$ became lower than 4×10^{-6} mm/cycle, the experiment was terminated and the corresponding final deflection $\bar{u}_{max}$ measured.

(b) The deflection increment rate falls down below 10×10^{-6} mm/cycle and then becomes higher again. The experiment was stopped after a sufficiently long time or after the shell failure. The last measured deflection is taken as $\bar{u}_{max}$.

It seems worthwhile to mention that in all the series of the experimental investigations hitherto completed a characteristic point (marked in

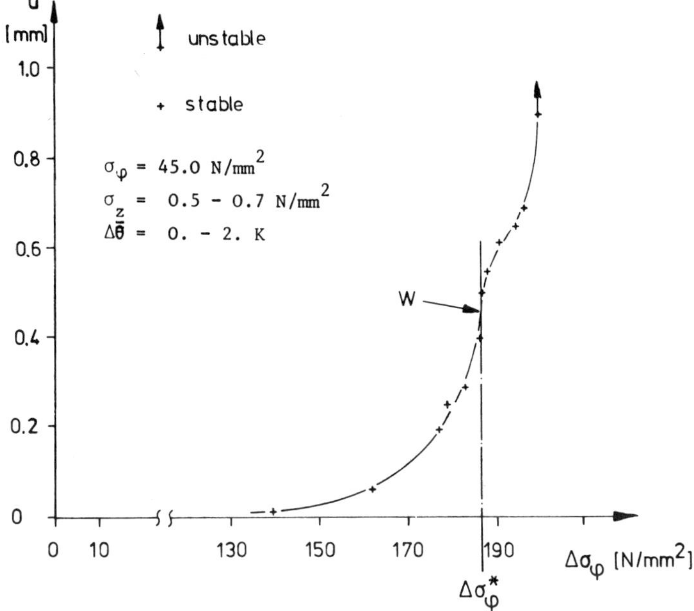

FIG. 10. Results of a series of experiments.

Fig. 10 with W) has been observed in which the curvature of the $\bar{u}_{max} - \Delta\sigma_\varphi$ curve vanishes. The corresponding magnitude of the load amplitude seems to be the experimental shakedown limit.

5. CONCLUSIONS

(i) Another example of experimental shakedown investigations in the case of a complex stress/strain state has been presented.

(ii) The existence of the phenomena of shakedown has been confirmed. The permanent deformations recorded were clearly plastic deformations and not creep deformations. The latter appeared but were negligibly small in comparison with the former ones.

(iii) Figure 11 presents the shell deformation obtained in one of the experiments compared with the first or second mechanism of incremental collapse. Similar results have been obtained in all

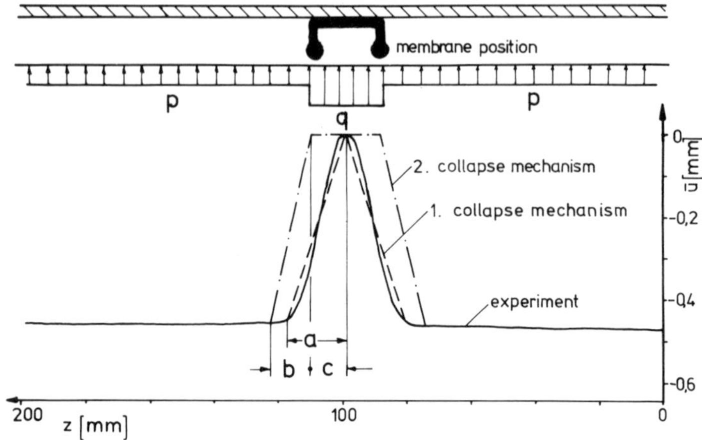

FIG. 11. The experimental radial displacement $\bar{u}$ compared with the fields of the theoretical incremental collapse mechanisms. ---- first mechanism; -·-·-·- second mechanism; —— experiment.

the hitherto made series of investigations. It should be stressed that no curve-fitting procedure took place. The theoretical results followed from the optimization according to the formulae (18) or (19). According to the kinematical shakedown theorem one could expect that the experimental points in Fig. 9 and the actual deformed shell form should follow the first mechanism rather than the second one. However, the model employed in the shakedown analysis did not take into account the geometrical effects and the material strain-hardening. These two effects are, probably, responsible for the fact that the second incremental collapse mechanism has occurred, actually. Only a proper extension of the model employed could answer this question.

(iv) The results obtained may suggest that the shakedown approach, though based on a very simple material model and (in our case) additionally simplified by assuming the Tresca yield condition and a very simple mechanism of incremental collapse, could be applied to analyze some practical engineering problems with an adequate accuracy.

REFERENCES

1. Bleich, H. Über die Bemessung statisch unbestimmter Stahlwerke unter Berücksichtigung des elastisch-plastischen Verhaltens des Baustoffes, *Bauingenieur,* **13** (1932), 261–267.
2. Ceradini, G. and C. Gavarini. Steel orthotropic plates under alternate loads, *J. Struct. Div. Proc. ASCE,* **101** (1975), 2015–2026.
3. Čyras, A. Optimization theory in limit analysis of a solid deformable body (in Russian), Mintis, Vilno, 1971.
4. De Donato, O. Second shakedown theorem allowing for cycles of both loads and temperature, *Ist. Lombardo Sci. Lett.* (*A*), **104** (1970), 265–277.
5. Gokhfeld, D. A. and D. F. Cherniavsky. *Limit Analysis of Structures at Thermal Cycling,* Sijthoff and Noordhoff, Groningen, 1980.
6. Koiter, W. T. A new general theorem on shakedown of elastic–plastic structures, *Proc. Kon. Ned. Ak. Wet.,* **BS9** (1956), 14–34.
7. König, J. A. On the incremental collapse criterion accounting for temperature dependence of yield point stress, *Arch. Mech.,* **31** (1979), 317–325.
8. Leers, K. Entwicklung eines Konzeptes zur experimentellen und theoretischen Shake-Down-Analyse von Rohren unter zyklischer mechanischer und thermischer Beanspruchung. Diplomarbeit, Universität Hannover, Institut für Mechanik, 1980 (unpublished).
9. Maier, G. Shakedown theory in perfect elastoplasticity with associated and non-associated flow-laws, *Meccanica,* **6** (1969), 250–260.
10. Melan, E. Theorie statisch unbestimmter Systeme aus ideal plastischem Baustoff, *Sitzber. Akad. Wiss. Wien, IIa,* **145** (1936), 195–218.
11. Neal, B. G. The behaviour of framed structures under repeated loading, *Q. J. appl. Math.,* **4** (1951), 78–84.
12. Prager, W. Shakedown in elastic–plastic media subjected to cycles of load and temperature. *Symp. Plasticità, Bologna* (1956), 239–244.
13. Proctor, E. and R. F. Flinders. Shakedown investigations on partial penetration welded nozzles in a spherical shell, *Nucl. Eng. Design,* **8** (1968), 171–185.
14. Rozenblum, V. I. On shakedown analysis on uneven heated elastic–plastic bodies (in Russian), *PMTF* **No. 5** (1965), 98–101.

Section 3

MATHEMATICAL METHODS OF ELASTOPLASTICITY

16

Local and Global Aspects in the Mathematical Theory of Plasticity†

PIERRE M. SUQUET

Université de Montpellier II and GRECO CNRS GDE, France

ABSTRACT

The attention is focused on three recent theories which contributed to the development of the mathematical theory of plasticity. The common point of these theories is the evidence or the use of the 'generalized standard' form of the encountered constitutive laws. Section 2 is devoted to homogenization. It demonstrates how elementary microscopic laws might give rise to standard macro-laws involving internal variables. The two next sections are devoted to a thermodynamical and mathematical discussion of these laws, as well as from a local standpoint or from a global one.

1. INTRODUCTION

The Mathematical Theory of Plasticity is the title of Hill's famous book written in the early 1950s [15]. Thirty years later, in 1983, after the remarkable and correlated development of computer sciences and of applied mathematics, after the emergence of continuum thermodynamics, this expression concerns more the field of convex analysis rather than that of Prandtl's nets. The past ten years have evidenced the benefit that mechanics could derive from collaborating with closely related fields and especially with applied analysis. The three sections of

† Part of this work was done while the author was at Mécanique Théorique, 4 place Jussieu, 75230 Paris Cédex 05, France

this paper refer to three theories elaborated in this spirit during the last ten years. Proceeding gradually from the particular case to the general one, they intend to demonstrate that theories with internal variables constitute a *rational approach* to continuum and structural mechanics.

Section 2 is devoted to *homogenization*, i.e. to theories which enable one to derive the global properties of highly heterogeneous media. The main point of the section goes as follows: consider two (or more) elastic perfectly plastic materials, aggregate them into a basic cell and repeat periodically this basic cell. Then the constitutive law of the mixture, derived by homogenization, requires the introduction of internal (micro-structural) variables.

Once the need for models with internal variables is emphasized we discuss, with the help of continuum thermodynamics, the theoretical *structure* of these models. The examples constitute the main point of Section 3 and show how this structure can be used to derive simple but efficient engineering models, endowed with the same mathematical characteristics.

In the last section the notion of standard law is generalized to structures where global variables (such as averaged strains, geometrical parameters of the structure, etc.) are under consideration. A rapid survey of the encountered variational problems obviates the mathematical common points in the discussion of the evolution of various systems or phenomena: plasticity, damage or rupture.

Since convexity is a common underlying feature of most of the paper, a small appendix is devoted to a non-polemical discussion of its 'hegemony'.

The three points under consideration here will be discussed in the context of infinitesimal strains. Another point, treated elsewhere in this Symposium, could have been an extension to finite strains. However, mathematical studies at finite strain are essentially as of now in the domain of 'Plasticity to-morrow'.

2. HOMOGENIZATION AND PLASTICITY

2.1. The Four Steps of Macro–Micromechanics

In the deterministic discussion of the overall properties of heterogeneous media the *first step* is to define a representative volume element (r.v.e) $\mathcal{V}$ small enough to distinguish the microscopic heterogeneities,

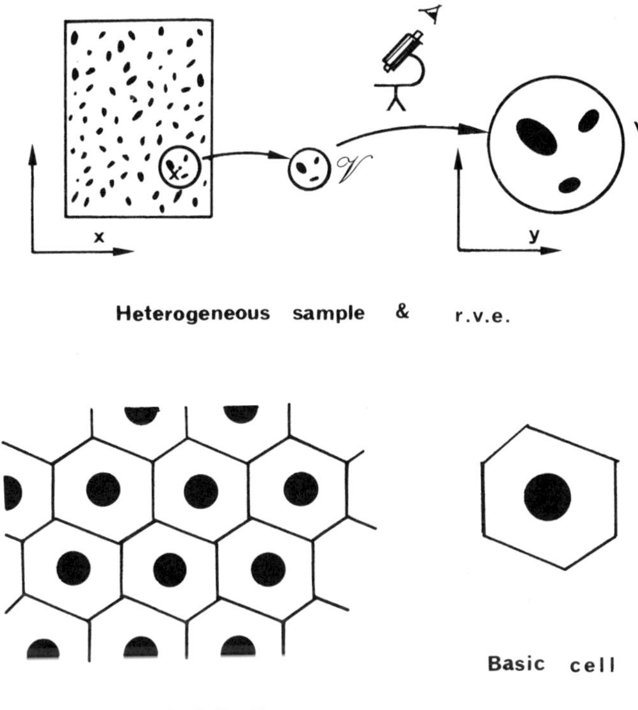

FIG. 1. Representative volume element: basic cell.

yet large enough to represent the overall behavior of the heterogeneous medium: a suitable rescaling (amplitude $1/\delta$) maps $\mathscr{V}$ onto an enlarged volume V where all the heterogeneities can easily be distinguished.

It turns out that the problem contains two scales: the 'macroscopic' scale x, on which all the macro-quantities depend and the 'microscopic'† scale y on which the micro-quantities depend. Consideration of the latter scale gives rise to highly oscillating fields in the heterogeneous body.

The *second step* of the analysis is the definition of macroscopic quantities from microscopic ones. This is achieved through an averag-

† Large enough to be in a position to apply continuum mechanics.

ing process: for a field f defined in the heterogeneous body we set

$$F(x)=\langle f\rangle(x)=\frac{1}{\mathrm{vol}\,(x+\mathcal{V})}\int_{x+\mathcal{V}} f(\zeta)\,\mathrm{d}\zeta=\frac{1}{\mathrm{vol}\,(\mathcal{V})}\int_{\mathcal{V}} f(x+\xi)\,\mathrm{d}\xi$$

$$=\frac{1}{\mathrm{vol}\,V}\int_{V} f(x+y\delta)\,\mathrm{d}y=\frac{1}{\mathrm{vol}\,(V)}\int_{V} f(x,y)\,\mathrm{d}y \qquad (1)$$

where $f(x, y)=f(x+y\delta)$.

Most of the macro-quantities are the average in the preceding sense of the micro-ones.† It is the case for the stress and strain tensors, and for all the thermodynamic functions which are usually assumed to be additive functions: internal and free energy, entropy, dissipation, mass per unit volume . . .

$$\mathbf{\Sigma}=\langle\boldsymbol{\sigma}\rangle, \qquad \mathbf{E}=\langle\boldsymbol{\varepsilon}\rangle, \qquad \bar{\rho}=\langle\rho\rangle, \qquad \bar{\rho}W=\langle\rho w\rangle‡ \qquad D=\langle d\rangle,\ \bar{\rho}S=\langle\rho s\rangle \qquad (2)$$

Remark 1: 1. If the constituents of the heterogeneous medium are not perfectly bonded (especially if voids or cracks are present) the averaging process is valid for *extended fields* defined on the defects [35].

2. We are not considering here purely statistical descriptions of the r.v.e. (Kröner [19], MacCoy [21]) or approximate ones (self-consistent models by Hill [16], Zaoui [37], Hashin's spheres model).

The *third step* in constructing a theory of homogenization is to define a *localization* procedure, i.e. a way of deriving microscopic fields from macroscopic ones. For this purpose the average relations (2) and the microscopic constitutive law must be taken into account together with other conditions arising from the equilibrium equations and from geometrical considerations. In order to derive the first condition we note that according to (1), the micro-stress field $\boldsymbol{\sigma}(x, y)$ satisfies

$$\mathrm{div}_y\,\boldsymbol{\sigma}(x,y)=\delta\,\mathrm{div}\,\boldsymbol{\sigma}(x+\delta y)=\delta(\rho\ddot{\mathbf{u}}-\rho\mathbf{f})§ \qquad (3)$$

As far as the phenomena under consideration occur in a low frequency range (when compared to the size of the r.v.e.), the second member of (3) can be neglected. We obtain a microequilibrium equation

$$\mathrm{div}_y\,\boldsymbol{\sigma}=0 \qquad (4)$$

† This assertion suffers counter examples: elastic coefficients, plastic strain, plastic work.

‡ Macro and micro fields are respectively denoted by capital or minus letters.

§ Div_y = divergence with respect to the y variable.

We are not yet in a position to determine the micro-fields $\boldsymbol{\sigma}$, $\boldsymbol{\varepsilon}$ from the macro ones $\boldsymbol{\Sigma}$, $\mathbf{E}$, since we are missing suitable boundary conditions on ∂V. These boundary conditions arise from the geometrical arrangement of the heterogeneities. They should closely reproduce the boundary conditions of the r.v.e. V in the composite. Thus they should characterize the *in situ* state of stress and strain within the heterogeneous medium. They very often reduce to assuming that the micro-strain or stress is uniform on ∂V (hence equal to the macro-strain or stress); the boundary ∂V sometimes tends to infinity. Together with (4) this set of boundary conditions constitutes the *macro–micro localization conditions* (mml). It is highly desirable for mechanical reasons (average of the micro-work = macro-work) that the following *Hill's macro-homogeneity condition* holds true for all the fields $\boldsymbol{\sigma}^*$ and $\boldsymbol{\varepsilon}^* = \boldsymbol{\varepsilon}(\mathbf{u}^*)$ satisfying the mml conditions

$$\langle \boldsymbol{\sigma}^* \boldsymbol{\varepsilon}^* \rangle = \boldsymbol{\Sigma}^* \mathbf{E}^* \tag{5}$$

The *fourth* and last step consists of the homogenization procedure itself. The micro-constitutive law is known and the relationships between micro- and macro-fields have been established. It now remains to relate the macro-fields. This is easily done in a linear context but turns out to be a difficult task in non-linear problems.

We now describe in detail a special method of homogenization valid† for periodic media.

2.2. Periodic Media

The case of periodic media is of special interest in view of the large number of 'repetitive' structures encountered in industry. The choice of the r.v.e. is readily made: V is chosen to be the basic cell of the periodic structure (cf. Fig. 1). The localization conditions are directly derived from the geometry of the composite: away from the boundary of the sample the stress and strain fields conform at the micro-level to the periodic character of the geometry:

$$\boldsymbol{\sigma}(x, y)\boldsymbol{\varepsilon}(\mathbf{u}(x, y)) \text{ are } V\text{-periodic functions of } y\ddagger \tag{6}$$

From a mathematical standpoint u belongs to the space of fields with

† We emphasize that the method described can be rendered rigorous through an asymptotic analysis [2] [29] [35]. In this sense the theory is exact.

‡ We shall omit in the sequel the dependence of the micro-fields on the x variable.

Periodic Deformation. Elements of this space can be split into a linear and a periodic part

$$\mathrm{PD}(V)=\{\mathbf{u}\in H^1(V)^3,\ u_i=E_{ij}y_j+v_i,\ \mathbf{v}\in H^1_{\mathrm{per}}(V)^3\}\dagger \tag{7}$$

The constant tensor $\mathbf{E}$ occurring in (7) is precisely the macroscopic strain tensor associated with $\mathbf{u}$ by (2).

The stress field $\boldsymbol{\sigma}$ belongs to the following space of microscopically self-equilibrated fields:

$$S^{o}_{\mathrm{per}}(V)=\{\boldsymbol{\sigma}\in L^2(V)^9_s,\ \mathrm{div}_y\,\boldsymbol{\sigma}=0,\ \boldsymbol{\sigma}\,.\,\mathbf{n}(y)\text{ opposite on opposite sides of }V\}$$

Therefore the mml conditions for periodic media reduce to (4) and (6), i.e.

$$\boldsymbol{\sigma}\in S^{o}_{\mathrm{per}}(V),\qquad \mathbf{u}\in\mathrm{PD}(V) \tag{8}$$

Let us emphasize that these mml conditions satisfy Hill's macro-homogeneity condition.

2.2.1. *An Example: Elastic Perforated Media*

Let us denote by V^* the solid part of the basic cell V and by $\mathbf{a}(y)=(a_{ijkh}(y))$ its elastic coefficients. The main point of the procedure is the localization process: for a given macro-strain state $\mathbf{E}$ what is the induced micro-strain? The equations of the problem are:

$$\left.\begin{array}{l}\boldsymbol{\sigma}(y)=\mathbf{a}(y)\boldsymbol{\varepsilon}(u)\\ \boldsymbol{\sigma}\in S^{o}_{\mathrm{per}}(V),\qquad \mathbf{u}\in\mathrm{PD}(V),\ \text{macro-strain}=\mathbf{E}\end{array}\right\} \tag{9}$$

Using the definition of PD(V), we split $\mathbf{u}$ into a linear part $\mathbf{Ey}$ and a periodic part $\mathbf{v}$. Then by Hill's macrohomogeneity condition (5) we get

$$\langle\boldsymbol{\sigma}\boldsymbol{\varepsilon}(\mathbf{v}^*)\rangle=\boldsymbol{\Sigma}\mathbf{E}^*=0\text{ for every }\mathbf{v}^*\text{ in }H^1_{\mathrm{per}}(V)^3\text{ since }\mathbf{E}^*=0$$

Therefore $\mathbf{v}$ is the solution of the following variational problem

$$\left.\begin{array}{l}\mathbf{v}\in H^1_{\mathrm{per}}(V)^3\text{ and for every }\mathbf{v}^*\text{ in }H^1_{\mathrm{per}}(V)^3\\ \langle\mathbf{a}\boldsymbol{\varepsilon}(\mathbf{v})\boldsymbol{\varepsilon}(\mathbf{v}^*)\rangle=-\langle\mathbf{a}\mathbf{E}\boldsymbol{\varepsilon}(\mathbf{v}^*)\rangle\end{array}\right\} \tag{10}$$

The second member of (10) can be identified as a concentrated loading on the boundary of the heterogeneities. Because of the condition of periodicity, (10) is not a classical boundary value problem. However it is possible to prove that it admits a solution [6]. Since (10) is a linear problem with respect to E, its solution v can be decomposed along *the*

† $H^1_{\mathrm{per}}(V)=\{\text{periodic elements of }H^1(V)\}$.

basis formed with the 6 following elementary solutions $\boldsymbol{\chi}_{ij}$

$\mathbf{E} = E_{ij}\mathbf{I}_{ij}$ where $\mathbf{I}_{ij}$ is defined as

$$(I_{ij})_{kh} = \tfrac{1}{2}(\delta_{ik}\delta_{jh} + \delta_{ih}\delta_{jk})$$

$\mathbf{v} = E_{ij}\boldsymbol{\chi}_{ij}$ where $\boldsymbol{\chi}_{ij}$ is solution of (10) with $\mathbf{E} = \mathbf{I}_{ij}$

$$\boldsymbol{\varepsilon}(\mathbf{u}) = \mathbf{E} + \boldsymbol{\varepsilon}(\mathbf{v}) = (\mathbf{I}_{ij} + \boldsymbol{\varepsilon}(\boldsymbol{\chi}_{ij}))E_{ij} = (\mathbf{I} + \boldsymbol{\varepsilon}(\boldsymbol{\chi}))\mathbf{E} \tag{11}$$

This last equality completes *the localization process*: the micro-strain $\boldsymbol{\varepsilon}(\mathbf{u})$ can be computed in terms of the macro- one $\mathbf{E}$. The forthcoming example (see Fig. 2) gives a few numerically determined elementary fields $\boldsymbol{\chi}_{ij}$.

The fourth step (i.e. homogenization) is readily done through the averaging of the micro-constitutive law

$$\begin{aligned} \boldsymbol{\sigma} &= \mathbf{a}\boldsymbol{\varepsilon}(\mathbf{u}) = \mathbf{a}(\mathbf{I} + \boldsymbol{\varepsilon}(\boldsymbol{\chi}))\mathbf{E} \\ \boldsymbol{\Sigma} &= \langle\boldsymbol{\sigma}\rangle = \langle\mathbf{a}(\mathbf{I} + \boldsymbol{\varepsilon}(\boldsymbol{\chi}))\rangle\mathbf{E} \\ \mathbf{a}^{\text{hom}} &= \langle\mathbf{a}(\mathbf{I} + \boldsymbol{\varepsilon}(\boldsymbol{\chi}))\rangle \end{aligned} \tag{12}$$

Remark 2: 1. Other mml conditions lead to the same type of elastic boundary value problem with any kind of 'classical' boundary conditions. Hill's equality (5) was the main point used there.

2. A localization with respect to the stress can be performed. For a given macro-stress $\boldsymbol{\Sigma}$ we can express the micro-stress $\boldsymbol{\sigma}(y)$ as a linear function of $\boldsymbol{\Sigma}$ [34]

$$\boldsymbol{\sigma}(y) = \mathbf{C}(y)\boldsymbol{\Sigma} \tag{13}$$

We are able to find the actual micro-stress state induced by a macro-one and to detect possible micro-stress concentrations. This is the fundamental goal of the localization process.

3. The example under consideration in Figs. 2, 3 and 4 originates in the work of Litewka and Sawczuk [20] where anisotropic damage was to be modelled. The basic cell, together with a few, numerically computed, strain and stress localizations are shown. The agreement with experimental data is good and shows how well the homogenization theory accounts for anisotropic behavior.

2.3. Rigid-plastic Constituents

In this section we assume that the constituents are rigid plastic and that they follow the normality rule

$$\boldsymbol{\sigma}(y) \in P(y), \qquad (\dot{\boldsymbol{\varepsilon}}(y), \boldsymbol{\sigma}^* - \boldsymbol{\sigma}(y)) \leq 0 \qquad \forall\, \boldsymbol{\sigma}^* \in P(y), \qquad \forall\, y \in V \tag{14}$$

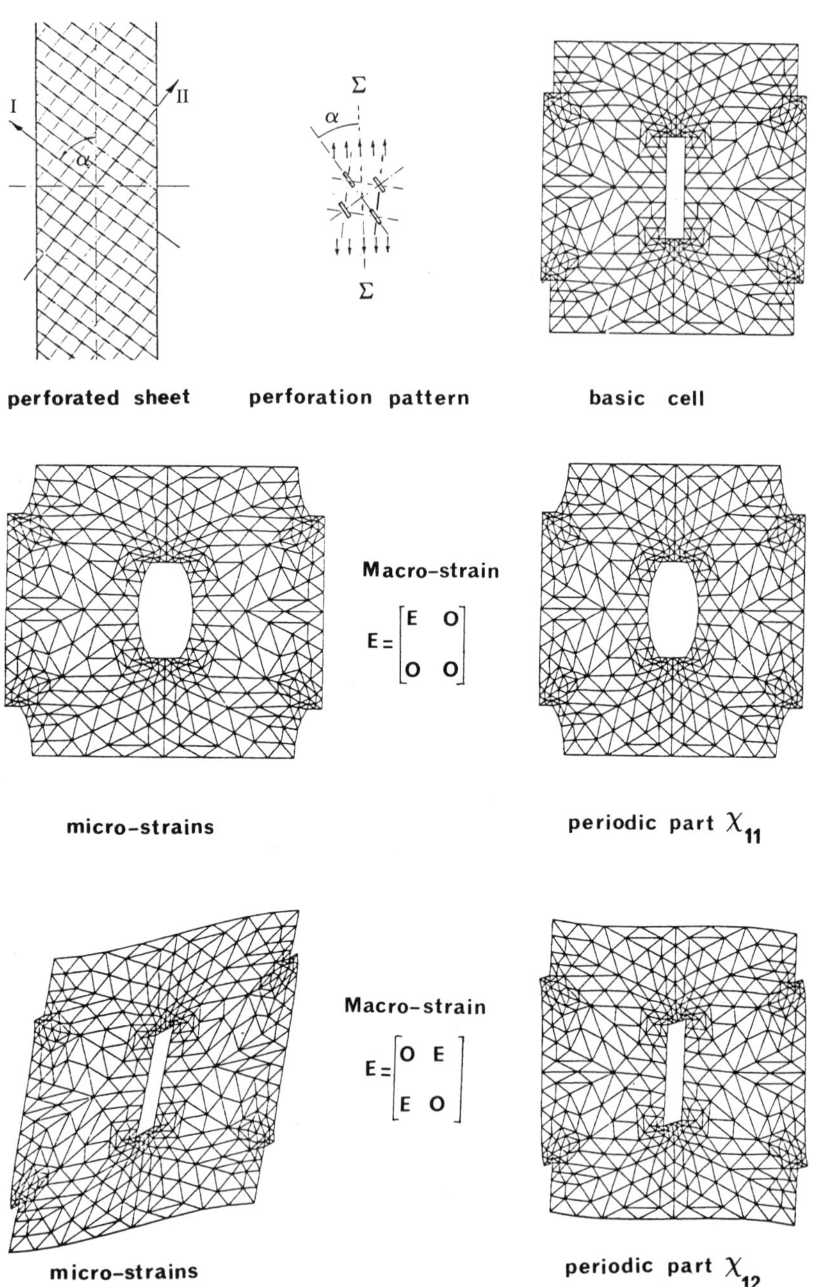

FIG. 2. Strain localization in perforated sheet.

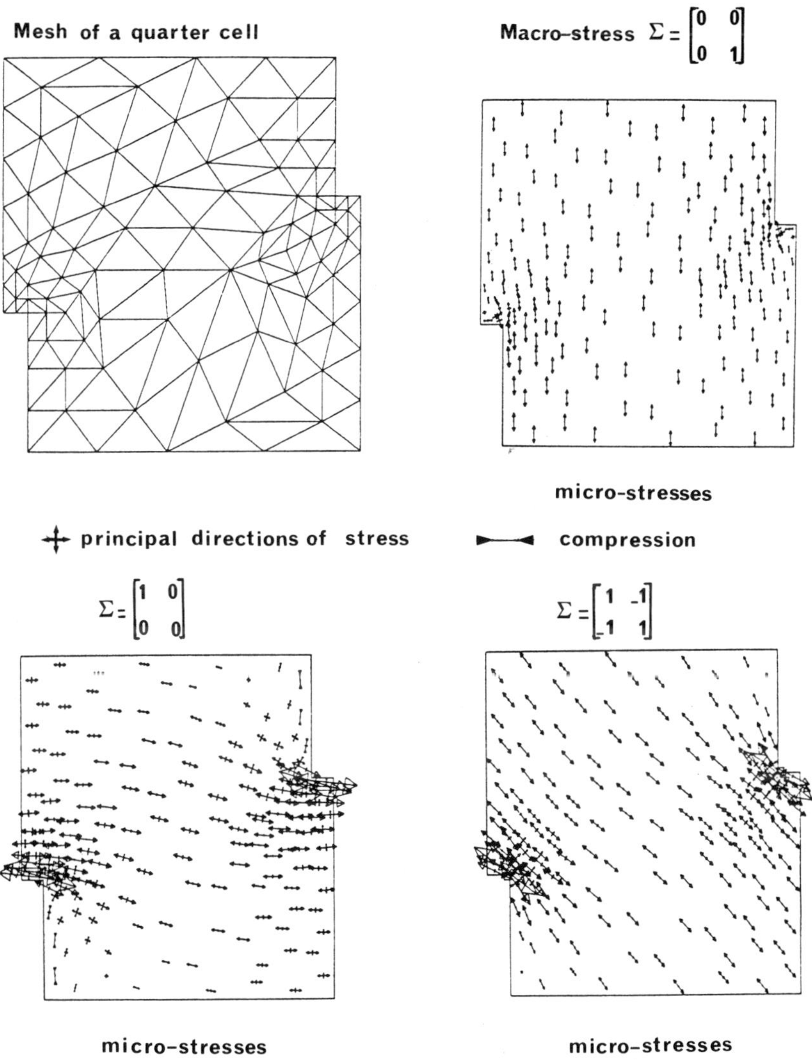

FIG. 3. Localization of stresses.

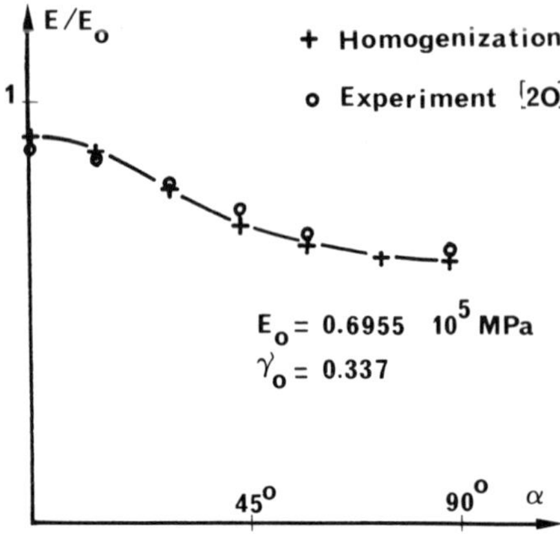

FIG. 4. Young's modulus $E(\alpha)$ in the direction of traction.

Since the micro-stress field is constrained, its average, the macro-stress $\mathbf{\Sigma}$, must be constrained too. Any admissible macro-stress $\mathbf{\Sigma}^*$ is necessarily related to a microscopic plastically admissible stress field satisfying the mml conditions. In the periodic case this leads to the following definition of the *macro-yield locus*

$$P^{\text{hom}} = \{\mathbf{\Sigma}^* |\, \exists \boldsymbol{\sigma}^*(\text{y}),\ \boldsymbol{\sigma}^* \in S^{o}_{\text{per}}(V),\ \langle \boldsymbol{\sigma}^* \rangle = \mathbf{\Sigma}^*,\ \boldsymbol{\sigma}^*(\text{y}) \in P(\text{y})\ \forall \text{y} \in V\} \tag{15}$$

We claim that the normality rule holds true at the macro-level, i.e.

$$\mathbf{\Sigma} \in P^{\text{hom}}, \qquad (\dot{\mathbf{E}}, \mathbf{\Sigma}^* - \mathbf{\Sigma}) \leqslant 0 \qquad \forall\, \mathbf{\Sigma}^* \in P^{\text{hom}} \tag{16}$$

Indeed $\mathbf{\Sigma}$ clearly belongs to P^{hom}. Moreover to any $\mathbf{\Sigma}^*$ element of P^{hom} we can associate a micro-stress state $\boldsymbol{\sigma}^*(\text{y})$ satisfying (15). Then by virtue of (14) and with the help of Hill's condition

$$0 \geqslant \langle \dot{\boldsymbol{\varepsilon}}(\text{y}),\ \boldsymbol{\sigma}^*(\text{y}) - \boldsymbol{\sigma}(\text{y}) \rangle = (\dot{\mathbf{E}}, \mathbf{\Sigma}^* - \mathbf{\Sigma}) \tag{17}$$

Remark 3: 1. Once more, Hill's condition plays a central role in the proof of the macro-normality while the periodicity is a technical point.

2. If we assume that the bonds between the constituents are plastic, i.e. that the stress vector at the interface S is constrained

$$\boldsymbol{\sigma} \,.\, \mathbf{n}(z) \in P'(z) \qquad \forall z \in S \tag{18}$$

Then the macro-yield locus is given by

$$P^{\text{hom}} = \{\boldsymbol{\Sigma} \mid \exists \boldsymbol{\sigma}^*(y) \in S^o_{\text{per}}(V),\ \langle \boldsymbol{\sigma}^* \rangle = \boldsymbol{\Sigma}^*,\ \boldsymbol{\sigma}(y) \in P(y),$$
$$\boldsymbol{\sigma} \,.\, \mathbf{n}(z) \in P'(z) \forall y,\ z \in V \times S\}$$

3. If stratified materials are under consideration, proper use of the mml conditions permits the identification† of P^{hom} as:

$$P^{\text{hom}} = \{\boldsymbol{\Sigma} \mid \boldsymbol{\Sigma} = c_\alpha \boldsymbol{\Sigma}^\alpha,\ \boldsymbol{\Sigma}^\alpha_{i3} = \boldsymbol{\Sigma}_{i3},\ i = 1, 2, 3, \boldsymbol{\Sigma}^\alpha \in P^\alpha\}$$

where P^α and c_α respectively denote the yield locus and the concentration of each constituent (the directions y_1 and y_2 are the invariant directions of the stratification).

4. For the practical determination of P^{hom}, a limit analysis problem on the r.v.e. V is to be solved: the loading parameters are the 6 independent components of the macro-stress tensor. This limit load analysis can be performed either through a static approach (as suggested by (15)) or through a kinematic one: P^{hom} is equal to the intersection ($\dot{\mathbf{E}}$ varying) of the half spaces defined by

$$\left\{ \boldsymbol{\Sigma} \text{ such that } \boldsymbol{\Sigma}\dot{\mathbf{E}} \leqslant \inf_{\substack{\mathbf{u} \in PD(V) \\ \text{macro-strain } \dot{\mathbf{E}}}} \langle \pi(y, \boldsymbol{\varepsilon}(\mathbf{u})) \rangle = \pi^{\text{hom}}(\dot{\mathbf{E}}) \right\} \tag{19}$$

where $\pi(y, .)$ and π^{hom} are respectively the support functions of the sets $P(y)$ and P^{hom}. If $\boldsymbol{\Sigma}$ and $\mathbf{E}$ denote the *actual* macro-stress and strain we have

$$\boldsymbol{\Sigma} = \frac{\partial \pi^{\text{hom}}}{\partial \dot{\mathbf{E}}}(\dot{\mathbf{E}}) \tag{20}$$

The preceding definition of P^{hom} using π^{hom} has been used in various settings. For instance, in the context of porous materials, Gurson [12] proposed to compute $\pi^{\text{hom}}(\dot{\mathbf{E}})$ by mean of a Riesz's approximation. V was chosen as a cylinder or a sphere and uniformity of the strain on ∂V was the assumed mml conditions. The velocity fields entering the Riesz approach were inspired from elastic solutions of the problem or from previous Rice and Tracey's work. Since the infimum in (19) is not necessarily obtained for these fields the value obtained for π^{hom} is overestimated and the set P^{hom} is approximated by the outside (upper

† This result was first established by De Buhan [5] in the two-dimensional case with another method of homogenization.

bound). In this sense Gurson's model overestimates the carrying capacity of porous metals.

5. We shall see in the following paragraphs a simple way of determining P^{hom} using an elasto–plastic problem.

2.4. Elasto–Plastic Constituents

Elasto–plastic constituents give rise to a more complex homogenization problem, since the residual micro-stresses due to the incompatibility of the anelastic micro-strains, have a micro-stored elastic energy, and therefore induce a hardening of the material. For a suitable description of this hardening an infinite number of internal variables (namely the whole set of the anelastic micro-strains) is required (see ref. 33). This kind of theoretical result is of little practical importance. However more specific information can be obtained on the macro-response to a specified loading and in the case of approximate models involving a finite number of internal variables.

Consider first the case of the macro-response to a specified loading. As a typical example we assume that the macro-strain rate $\dot{\mathbf{E}}$ can be kept constant (equal to $\dot{\mathbf{E}}^o$) and we investigate the macro-stress response. If we take the elastic part of the behavior to be linear, the problem of localization (determination of the micro-fields) becomes

$$\left.\begin{aligned} &\mathbf{A}(y)\dot{\boldsymbol{\sigma}}(y)+\dot{\boldsymbol{\varepsilon}}^p(y)=\boldsymbol{\varepsilon}(\dot{\mathbf{u}}(y))=\boldsymbol{\varepsilon}(\dot{\mathbf{v}})+\dot{\mathbf{E}}^o\\ &\boldsymbol{\sigma}\in S^o_{\text{per}}(V),\qquad \mathbf{v}\in H^1_{\text{per}}(V)^3\\ &\boldsymbol{\sigma}\in\mathcal{P}=\{\boldsymbol{\sigma}\mid\boldsymbol{\sigma}(y)\in P(y)\forall y\}\end{aligned}\right\}\tag{21}$$

Use of the maximal work principle yields the following variational formulation for $\boldsymbol{\sigma}$

$$\left.\begin{aligned} &\boldsymbol{\sigma}\in S^o_{\text{per}}(V)\cap\mathcal{P},\qquad \forall\boldsymbol{\sigma}^*\in S^o_{\text{per}}(V)\cap\mathcal{P}:\\ &\langle\mathbf{A}\dot{\boldsymbol{\sigma}},\boldsymbol{\sigma}^*-\boldsymbol{\sigma}\rangle\geqslant\dot{\mathbf{E}}^o\langle\boldsymbol{\sigma}^*-\boldsymbol{\sigma}\rangle=(\dot{\mathbf{E}}^o,\boldsymbol{\Sigma}^*-\boldsymbol{\Sigma})\end{aligned}\right\}\tag{22}$$

Solving (22) and averaging the micro-stress response gives the macro-response $\boldsymbol{\Sigma}(t)$.

If $\boldsymbol{\sigma}$ has a limit $\boldsymbol{\sigma}_\infty$,† as t tends to $+\infty$, we get

$$\left.\begin{aligned} &\boldsymbol{\Sigma}_\infty\in P^{\text{hom}},\qquad \forall\boldsymbol{\Sigma}^*\in P^{\text{hom}}:\\ &0\geqslant(\dot{\mathbf{E}}^o,\boldsymbol{\Sigma}^*-\boldsymbol{\Sigma}_\infty)\end{aligned}\right\}\tag{23}$$

† Such a result can be proved under geometrical assumptions on $P(y)$ using Haraux's result [14] on asymptotic behavior of solutions of evolution equations.

Then $\mathbf{\Sigma}_\infty$ is on the boundary of P^{hom} and $\dot{\mathbf{E}}^{\text{o}}$ is an outer normal vector to P^{hom} at $\mathbf{\Sigma}_\infty$: we thus obtain another way of determining P^{hom} by solving an elasto–plastic problem with periodicity conditions and a fixed loading concentrated on the interfaces of the constituents (this last part can be easily deduced from the study of (21)).

2.5. Approximate Models

Knowledge of the actual macro-constitutive law requires knowledge of an infinite number of internal variables, namely the whole set of anelastic micro-strains. However some simplified models, with piece-wise constant anelastic micro-strains can be proposed. For the sake of simplicity we shall assume in the sequel that the basic cell is made of two constituents, a matrix and a fiber, and that the anelastic micro-strains are constant on each of them:

$$\boldsymbol{\varepsilon}^{\text{P}}(y) = \mathbf{E}_{\text{m}}^{\text{P}}\theta_{\text{m}}(y) + \mathbf{E}_{\text{f}}^{\text{P}}\theta_{\text{f}}(y) \tag{24}$$

where $\theta_{\text{m}}(y) = 1$ in the matrix, 0 in the fiber (similar definition for θ_{f}). The micro-constitutive law is now

$$\boldsymbol{\varepsilon}(\mathbf{u}) = \mathbf{E} + \boldsymbol{\varepsilon}(\mathbf{v}) = \mathbf{A}\boldsymbol{\sigma} + \boldsymbol{\varepsilon}^{\text{P}} = \mathbf{A}\boldsymbol{\sigma} + \mathbf{E}_{\text{m}}^{\text{P}}\theta_{\text{m}} + \mathbf{E}_{\text{f}}^{\text{P}}\theta_{\text{f}} \tag{25}$$

Setting $\mathbf{a} = \mathbf{A}^{-1}$ we get

$$\left.\begin{array}{l} \boldsymbol{\sigma} = \mathbf{a}\boldsymbol{\varepsilon}(\mathbf{v}) + \mathbf{a}\mathbf{E} - \mathbf{a}\mathbf{E}_{\text{m}}^{\text{P}}\theta_{\text{m}} - \mathbf{a}\mathbf{E}_{\text{f}}^{\text{P}}\theta_{\text{f}} \\ \boldsymbol{\sigma} \in S_{\text{per}}^{\text{o}}(V), \qquad \mathbf{v} \in H_{\text{per}}^{1}(V)^3 \end{array}\right\} \tag{26}$$

The problem (26) bears a strong resemblance to (10), and since it is linear with respect to $\mathbf{E}$, $\mathbf{E}_{\text{m}}^{\text{P}}$, $\mathbf{E}_{\text{f}}^{\text{P}}$ its solution $\mathbf{v}$ can be split into

$$\mathbf{v} = \mathbf{E}\boldsymbol{\chi} + \mathbf{E}_{\text{m}}^{\text{P}}\boldsymbol{\chi}_{\text{m}}^{\text{P}} + \mathbf{E}_{\text{f}}^{\text{P}}\boldsymbol{\chi}_{\text{f}}^{\text{P}} \tag{27}$$

where $\boldsymbol{\chi}$ is the array of elastic localization fields defined previously by (10) and (11). $\boldsymbol{\chi}_{\text{m}}^{\text{P}}$ and $\boldsymbol{\chi}_{\text{f}}^{\text{P}}$ are solutions of

$$\left.\begin{array}{l} \boldsymbol{\chi}_{\text{m}}^{\text{P}} \in H_{\text{per}}^{1}(V)^3 \text{ and for every } \mathbf{v}^* \in H_{\text{per}}^{1}(V)^3 \\ \langle \mathbf{a}\boldsymbol{\varepsilon}(\boldsymbol{\chi}_{\text{m}}^{\text{P}})\boldsymbol{\varepsilon}(\mathbf{v}^*)\rangle = \langle \mathbf{a}\theta_{\text{m}}\boldsymbol{\varepsilon}(\mathbf{v}^*)\rangle \end{array}\right\} \tag{28}$$

(similar definition for $\boldsymbol{\chi}_{\text{f}}^{\text{P}}$).

We get the micro-strain as

$$\boldsymbol{\varepsilon}(\mathbf{u}) = \mathbf{E}(\mathbf{I} + \boldsymbol{\varepsilon}(\boldsymbol{\chi})) + \mathbf{E}_{\text{m}}^{\text{P}}\boldsymbol{\varepsilon}(\boldsymbol{\chi}_{\text{m}}^{\text{P}}) + \mathbf{E}_{\text{f}}^{\text{P}}\boldsymbol{\varepsilon}(\boldsymbol{\chi}_{\text{f}}^{\text{P}}) \tag{29}$$

We recall that the micro free energy amounts to

$$\rho w(\boldsymbol{\varepsilon}(\mathbf{u}), \boldsymbol{\varepsilon}^{\text{P}}) = \tfrac{1}{2}\mathbf{a}(\boldsymbol{\varepsilon}(\mathbf{u}) - \boldsymbol{\varepsilon}^{\text{P}})(\boldsymbol{\varepsilon}(\mathbf{u}) - \boldsymbol{\varepsilon}^{\text{P}})$$

and that the macro free energy is the average of the micro one. Thus with the help of (24) and (29), the free energy can be expressed solely in terms of E, $E^{\mathrm{P}}_{\mathrm{m}}$, $E^{\mathrm{P}}_{\mathrm{f}}$:

$$\bar{\rho}W(\mathbf{E}, \mathbf{E}^{\mathrm{P}}_{\mathrm{m}}, \mathbf{E}^{\mathrm{P}}_{\mathrm{f}}) = \langle \rho w(\boldsymbol{\varepsilon}(\mathbf{u}), \boldsymbol{\varepsilon}^{\mathrm{P}}) \rangle \tag{30}$$

In the general case of heterogeneous elastic properties it should be noted that the macro-stress can be derived from (25) and (29). We obtain

$$\begin{aligned}\boldsymbol{\Sigma} &= \langle \boldsymbol{\sigma} \rangle = \langle \boldsymbol{\sigma}(\boldsymbol{\varepsilon}(\boldsymbol{\chi}) + \mathbf{I}) \rangle \dagger \\ &= \mathbf{a}^{\mathrm{hom}}\mathbf{E} - c_{\mathrm{f}}\langle \mathbf{a}(\boldsymbol{\varepsilon}(\boldsymbol{\chi}) + \mathbf{I}) \rangle_{\mathrm{f}} \mathbf{E}^{\mathrm{P}}_{\mathrm{f}} - c_{\mathrm{m}} \langle \mathbf{a}(\boldsymbol{\varepsilon}(\boldsymbol{\chi}) + \mathbf{I}) \rangle_{\mathrm{m}} \mathbf{E}^{\mathrm{P}}_{\mathrm{m}}\end{aligned}$$

where c_{f}, c_{m} are the concentrations of each constituent and where $\langle \, \rangle_{\mathrm{f}}$ denotes the average on the fiber. Therefore the anelastic macro-strain amounts to:

$$\begin{aligned}\mathbf{E}^{\mathrm{P}} &= \mathbf{A}^{\mathrm{hom}}[c_{\mathrm{f}}\langle \mathbf{a}(\boldsymbol{\varepsilon}(\boldsymbol{\chi}) + \mathbf{I}) \rangle_{\mathrm{f}} \mathbf{E}^{\mathrm{P}}_{\mathrm{f}} + c_{\mathrm{m}} \langle \mathbf{a}(\boldsymbol{\varepsilon}(\boldsymbol{\chi}) + \mathbf{I}) \rangle_{\mathrm{m}} \mathbf{E}^{\mathrm{P}}_{\mathrm{m}}] \\ &\neq c_{\mathrm{f}}\mathbf{E}^{\mathrm{P}}_{\mathrm{f}} + c_{\mathrm{m}}\mathbf{E}^{\mathrm{P}}_{\mathrm{m}}\end{aligned}$$

The anelastic macro-strain *is not the average of the micro one.* Indeed it can be proved that

$$\mathbf{E}^{\mathrm{P}} = \langle \mathbf{C}\boldsymbol{\varepsilon}^{\mathrm{P}} \rangle = \mathbf{A}^{\mathrm{hom}} \langle \mathbf{a}(\boldsymbol{\varepsilon}(\boldsymbol{\chi}) + \mathbf{I}) \boldsymbol{\varepsilon}^{\mathrm{P}} \rangle \tag{31}$$

where $\mathbf{C}$ is the stress localization tensor defined in (13).

In the case of *homogeneous elastic coefficients* the array of elastic correctors $\boldsymbol{\chi}$ vanishes and the following simplifications occur:

$$\mathbf{E}^{\mathrm{P}} = c_{\mathrm{f}}\mathbf{E}^{\mathrm{P}}_{\mathrm{f}} + c_{\mathrm{m}}\mathbf{E}^{\mathrm{P}}_{\mathrm{m}} = \langle \boldsymbol{\varepsilon}^{\mathrm{P}} \rangle$$

$$\bar{\rho}W(\mathbf{E}, \mathbf{E}^{\mathrm{P}}_{\mathrm{m}}, \mathbf{E}^{\mathrm{P}}_{\mathrm{f}}) = \tfrac{1}{2}\mathbf{a}(\mathbf{E} - \mathbf{E}^{\mathrm{P}})(\mathbf{E} - \mathbf{E}^{\mathrm{P}}) + \left(\frac{c_{\mathrm{f}}c_{\mathrm{m}}}{2} - \alpha\right)\mathbf{a}(\mathbf{E}^{\mathrm{P}}_{\mathrm{m}} - \mathbf{E}^{\mathrm{P}}_{\mathrm{f}})(\mathbf{E}^{\mathrm{P}}_{\mathrm{m}} - \mathbf{E}^{\mathrm{P}}_{\mathrm{f}}) \tag{32}$$

where $\alpha = c_{\mathrm{f}}\langle \boldsymbol{\varepsilon}(\boldsymbol{\chi}^{\mathrm{P}}_{\mathrm{f}}) \rangle_{\mathrm{f}} = -c_{\mathrm{m}} \langle \boldsymbol{\varepsilon}(\boldsymbol{\chi}^{\mathrm{P}}_{\mathrm{f}}) \rangle_{\mathrm{m}}$

It should be noted that the *hardening* term of (32)

$$\left(\frac{c_{\mathrm{f}}c_{\mathrm{m}}}{2} - \alpha\right)\mathbf{a}(\mathbf{E}^{\mathrm{P}}_{\mathrm{m}} - \mathbf{E}^{\mathrm{P}}_{\mathrm{f}})(\mathbf{E}^{\mathrm{P}}_{\mathrm{m}} - \mathbf{E}^{\mathrm{P}}_{\mathrm{f}})$$

is always positive. This positive hardening is precisely the elastic energy of the stresses due to the difference of the anelastic strains in the fiber and in the matrix.

† We make use of Hill's condition.

Anticipating Section 3 we can observe that the macro state variables are $\mathbf{E}$, $\mathbf{E}^{\mathrm{P}}_{\mathrm{m}}$, $\mathbf{E}^{\mathrm{P}}_{\mathrm{f}}$. Computing the associated thermodynamical forces, namely the partial derivatives of $\bar{\rho}W$, yields:

$$\bar{\rho}\frac{\partial W}{\partial \mathbf{E}}=\mathbf{\Sigma},\qquad \bar{\rho}\frac{\partial W}{\partial \mathbf{E}^{\mathrm{P}}_{\mathrm{m}}}=\langle\boldsymbol{\sigma}\rangle_{\mathrm{m}}\overset{\mathrm{def}}{=}\mathbf{\Sigma}_{\mathrm{m}},\qquad \bar{\rho}\frac{\partial W}{\partial \mathbf{E}^{\mathrm{P}}_{\mathrm{f}}}=\langle\boldsymbol{\sigma}\rangle_{\mathrm{f}}=\mathbf{\Sigma}_{\mathrm{f}} \tag{33}$$

Moreover we derive the following relations from the normality law averaged on each constituent:

$$\left.\begin{aligned}(\dot{\mathbf{E}}^{\mathrm{P}}_{\mathrm{m}}, \mathbf{\Sigma}^*-\mathbf{\Sigma}_{\mathrm{m}})=\langle\dot{\boldsymbol{\varepsilon}}^{\mathrm{P}}, \mathbf{\Sigma}^*-\boldsymbol{\sigma}\rangle_{\mathrm{m}}\leqslant 0 \quad \text{for every } \mathbf{\Sigma}^*\in P_{\mathrm{m}}\\ (\dot{\mathbf{E}}^{\mathrm{P}}_{\mathrm{f}}, \mathbf{\Sigma}^{**}-\mathbf{\Sigma}_{\mathrm{f}})=\langle\dot{\boldsymbol{\varepsilon}}^{\mathrm{P}}, \mathbf{\Sigma}^{**}-\boldsymbol{\sigma}\rangle_{\mathrm{f}}\leqslant 0 \quad \text{for every } \mathbf{\Sigma}^{**}\in P_{\mathrm{f}}\end{aligned}\right\} \tag{34}$$

where P_{m}, P_{f} denote the yield loci of the constituents.

We are thus in the following position: starting from elementary (elastic perfectly plastic) constituents, we derived a macro-model including a larger number of internal variables. This model does not reduce to a classical one. However it satisfies the normality rule in a generalized sense. Indeed if we set

$$\boldsymbol{\alpha}=(\mathbf{E}^{\mathrm{P}}_{\mathrm{m}}, \mathbf{E}^{\mathrm{P}}_{\mathrm{f}}),\qquad \mathbf{A}=(\mathbf{\Sigma}_{\mathrm{m}}, \mathbf{\Sigma}_{\mathrm{f}}),\qquad P=P_{\mathrm{m}}\times P_{\mathrm{f}}$$

the macro constitutive law becomes

$$\begin{aligned}&\mathbf{\Sigma}=\bar{\rho}\frac{\partial W}{\partial \mathbf{E}}(\mathbf{E},\boldsymbol{\alpha}),\qquad \mathbf{A}=-\bar{\rho}\frac{\partial W}{\partial \boldsymbol{\alpha}}(\mathbf{E},\boldsymbol{\alpha}) \quad \text{equations of state}\\ &\left.\begin{aligned}&\mathbf{A}\in P \text{ and for every } \mathbf{A}^* \text{ in } P:\\ &(\dot{\boldsymbol{\alpha}}, \mathbf{A}^*-\mathbf{A})\leqslant 0\end{aligned}\right\} \quad \text{complementary laws}\end{aligned} \tag{35}$$

The macro-model is a *Generalized Standard Material* (GSM)

Remark 4: In this section the thermal coupling at the micro-scale has been neglected. In particular the question of deriving macro-thermodynamics (including a suitable definite of the macro temperature) from micro ones has been avoided. This important point is partly discussed in ref. 7 but remains largely open.

3. LOCAL GSM THEORY

The present section is devoted to the purely macroscopic study of Generalized Standard Materials (GSM) with the help of macroscopic thermodynamics.† The existence of such materials was established in the previous section where the occurrence in purely macro models of internal (micro-structural) variables was explained by macro–micro mechanics.

3.1. Generalized Standard Materials

The deformation of a continuous medium is a particular thermodynamical process which must be achieved in agreement with the two fundamental laws of thermodynamics. We assume here that the whole history of the medium is contained in the current value of a finite set of *state variables* $\boldsymbol{\chi}$ and that the thermostatic concepts of entropy s, temperature T, internal energy u, free energy w, extend to thermodynamical evolutions.‡ As far as the medium undergoes infinitesimal transformations,§ the state variables can be easily identified: they involve the infinitesimal strain ε and other physico-chemical variables $\boldsymbol{\alpha}$ (*internal variables*)

$$\boldsymbol{\chi} = (\boldsymbol{\varepsilon}, \boldsymbol{\alpha}), \qquad u = u(\boldsymbol{\chi}, s), \qquad w = w(\boldsymbol{\chi}, T) \ldots \tag{36}$$

Applying the first and second law of thermodynamics yields the Clausius–Duhem inequality

$$d = \boldsymbol{\sigma}^{\mathrm{IR}}\dot{\boldsymbol{\varepsilon}} + \mathbf{A}\dot{\boldsymbol{\alpha}} - \mathbf{q}\frac{\boldsymbol{\nabla} T}{T} \geqslant 0 \tag{37}$$

where d denotes the total dissipation, $\mathbf{q}$ the heat flux, $\boldsymbol{\sigma}^{\mathrm{IR}}$ is the irreversible part of the stress while $\mathbf{A}$ is the array of the thermodynamical forces associated with $\boldsymbol{\alpha}$

$$\boldsymbol{\sigma}^{\mathrm{IR}} = \boldsymbol{\sigma} - \boldsymbol{\sigma}^{\mathrm{R}}, \qquad \boldsymbol{\sigma}^{\mathrm{R}} = \rho\frac{\partial w}{\partial \boldsymbol{\varepsilon}}, \qquad \mathbf{A} = -\rho\frac{\partial w}{\partial \boldsymbol{\alpha}} \quad \text{(equations of state)} \tag{38}$$

Assuming classically that the thermal and mechanical dissipations are decoupled yields their positivity in any real evolution

$$d_1 = \boldsymbol{\sigma}^{\mathrm{IR}}\dot{\boldsymbol{\varepsilon}} + \mathbf{A}\dot{\boldsymbol{\alpha}} \geqslant 0 \qquad d_2 = -\mathbf{q}\frac{\nabla T}{T} \geqslant 0 \tag{39}$$

† Following the analysis of Halphen and Nguyen [13].
‡ This constitutes the basic assumption of the local accompanying state model [1] [9].
§ For finite strains see the accounts of refs. 13, 22 and 30.

The complementary constitutive laws relating $Y = (\dot{\boldsymbol{\varepsilon}}, \dot{\boldsymbol{\alpha}}, \mathbf{q})$ and $\mathbf{X} = (\boldsymbol{\sigma}^{\mathrm{IR}}, \mathbf{A}, -\nabla T/T)$ must satisfy (39). In classical thermodynamics of irreversible processes [10] these laws are supposed to be linear and Onsager's relations imply the symmetry of the involved matrix:

$$Y_i = L_{ij} X_j \qquad L_{ij} = L_{ji} \tag{40}$$

A natural generalization of (40) consists in assuming that a mechanical and a thermal potential of dissipation $\varphi(\dot{\boldsymbol{\varepsilon}}, \dot{\boldsymbol{\alpha}})$, $\varphi_{\mathrm{th}}(\mathbf{q})$ exist, such that†

$$\boldsymbol{\sigma}^{\mathrm{IR}} = \frac{\partial \varphi}{\partial \dot{\boldsymbol{\varepsilon}}}(\dot{\boldsymbol{\varepsilon}}, \dot{\boldsymbol{\alpha}}), \qquad \mathbf{A} = \frac{\partial \varphi}{\partial \dot{\boldsymbol{\alpha}}}(\dot{\boldsymbol{\varepsilon}}, \dot{\boldsymbol{\alpha}}), \qquad -\frac{\nabla T}{T} = \frac{\partial \varphi_{\mathrm{th}}}{\partial \mathbf{q}}(\mathbf{q}) \tag{41}$$

If moreover the thermodynamic functions $u(\boldsymbol{\varepsilon}, \boldsymbol{\alpha}, s)$ and $\varphi(\dot{\boldsymbol{\varepsilon}}, \dot{\boldsymbol{\alpha}})$, $\varphi_{\mathrm{th}}(\mathbf{q})$ are positive convex functions of their arguments the material is said to be a Generalized Standard Material (GSM) [13].

Remark 5: 1. The positivity of d_1 and d_2 (4) directly follows from the convexity of φ and φ_{th}.

2. Taking $\varphi_{\mathrm{th}} = 1/2k\ |\mathbf{q}|^2$ yields the Fourier's law:

$$\mathbf{q} = -\mathbf{k}\frac{\nabla T}{T} \simeq -\frac{\mathbf{k}}{T_0}\boldsymbol{\nabla}\theta \quad \text{where} \quad \theta = T - T_0.$$

3. Introducing the Legendre–Fenchel transforms φ^* and φ^*_{th} of φ and φ_{th} one gets an equivalent formulation of (41)

$$\dot{\boldsymbol{\varepsilon}} = \frac{\partial \varphi^*}{\partial \boldsymbol{\sigma}^{\mathrm{IR}}}(\boldsymbol{\sigma}^{\mathrm{IR}}, \mathbf{A}), \qquad \dot{\boldsymbol{\alpha}} = \frac{\partial \varphi^*}{\partial \mathbf{A}}(\boldsymbol{\sigma}^{\mathrm{IR}}, \mathbf{A}), \qquad \mathbf{q} = \frac{\partial \varphi^*_{\mathrm{th}}}{\partial\left(-\dfrac{\nabla T}{T}\right)}\left(-\frac{\nabla T}{T}\right) \tag{42}$$

4. The rate independent materials constitute an important subclass of GSM. For these materials the law (41) does not depend on the scale of time, i.e. $\partial\varphi$ is homogeneous of degree 0 with respect to $(\dot{\boldsymbol{\varepsilon}}, \dot{\boldsymbol{\alpha}})$. Thus φ is homogeneous of degree 1 with respect to $(\dot{\boldsymbol{\varepsilon}}, \dot{\boldsymbol{\alpha}})$. A simple argument of convex analysis implies that φ^* is necessarily the indicator function of a convex set P. Furthermore the complementary laws (42)

† If the potentials are not differentiable, the following relations must be understood on the sense of subdifferentials (see any standard book on Convex Analysis).

expressed in a generalized way form the principle of maximal dissipation:†

There exists a closed convex set P such that

$$\left.\begin{array}{l}\varphi^*(\boldsymbol{\sigma}^{\mathrm{IR}}, \mathbf{A}) = 0 \quad \text{if} \quad (\boldsymbol{\sigma}^{\mathrm{IR}}, \mathbf{A}) \in P, \ +\infty \quad \text{otherwise} \\ (\boldsymbol{\sigma}^{\mathrm{IR}}, \mathbf{A}) \in P \quad \text{and for every } (\boldsymbol{\sigma}^*, \mathbf{A}^*) \in P \\ (\dot{\boldsymbol{\varepsilon}}, \boldsymbol{\sigma}^* - \boldsymbol{\sigma}^{\mathrm{IR}}) + (\dot{\boldsymbol{\alpha}}, \mathbf{A}^* - \mathbf{A}) \leq 0 \end{array}\right\} \tag{43}$$

P is the generalized plasticity locus.

Most of the time φ does not depend on $\dot{\boldsymbol{\varepsilon}}$ and (43) reduces to $\mathbf{A} \in P$ and for every $\mathbf{A}^* \in P$ $(\dot{\boldsymbol{\alpha}}, \mathbf{A}^* - \mathbf{A}) \leq 0$.

In the framework of GSM, rate independence is equivalent to the principle of maximal dissipation.

3.2. Examples

The GSM theory is confirmed, not only by its mathematical structure, but essentially by the number of classical or non-classical situations abiding by it. A few of them are briefly illustrated here through three (isothermal) examples.

3.2.1. *Damage of Ductile Metals* [27]

Ductile fracture in metals involves considerable damage at crack tips, through nucleation, growth and coalescence of voids initiated by inclusions. In order to account for this effect an internal variable describing the damage is introduced into a model of plasticity with hardening. For the sake of simplicity both effects (damage and hardening) are assumed to be isotropic. We set:

$$\begin{array}{c}\boldsymbol{\alpha} = (\boldsymbol{\varepsilon}^{\mathrm{p}}, D, p) \\ \rho w(\boldsymbol{\varepsilon}, \boldsymbol{\alpha}) = \frac{1}{2}(\boldsymbol{\varepsilon} - \boldsymbol{\varepsilon}^{\mathrm{p}})\mathbf{a}(\boldsymbol{\varepsilon} - \boldsymbol{\varepsilon}^{\mathrm{p}}) + h(p) + m(D)\end{array} \tag{44}$$

The equations of state are

$$\boldsymbol{\sigma}^{\mathrm{R}} = \mathbf{a}(\boldsymbol{\varepsilon} - \boldsymbol{\varepsilon}^{\mathrm{p}}), \qquad R = -\frac{\mathrm{d}h}{\mathrm{d}p}(p), \qquad Y = -\frac{\mathrm{d}m}{\mathrm{d}D}(D) \tag{45}$$

The effect of porosity on the elastic moduli has been neglected since we are mainly interested in the ductile behavior of the material. The presence of microvoids causing damage results in a sensitivity to pressure of the plasticity criterion. More precisely we assume that the

† Sometimes referred to as 'principle of maximal work'.

plasticity locus is given in the space of generalized stresses as:

$$P=\{\boldsymbol{\sigma}^{\mathrm{IR}}=0, \mathbf{A}=(\boldsymbol{\sigma}, Y, R) \text{ satisfies: } J_2(\boldsymbol{\sigma})+\mathbf{R}+Yg(\sigma_{\mathrm{m}})\leqslant 0\}\dagger \tag{46}$$

The potential φ^* is the indicator function of P and the complementary laws are derived from the assumption of GSM

$$\dot{\boldsymbol{\varepsilon}}^{\mathrm{pD}}=\lambda\frac{\boldsymbol{\sigma}^{\mathrm{D}}}{J_2(\boldsymbol{\sigma})}, \qquad \dot{\varepsilon}^{\mathrm{p}}_{\mathrm{m}}=\lambda Yg'(\sigma_{\mathrm{m}}), \qquad \dot{p}=\lambda, \qquad \dot{D}=\lambda g(\sigma_{\mathrm{m}}), \qquad \lambda\geqslant 0 \tag{47}$$

p is identified as the cumulated deviatoric plastic strain.

The form of g can be derived through elementary considerations. Because D is related to the volumic fraction of microvoids, it is thus related to the density ρ of the ductile material

$$D=D(\rho), \quad \text{i.e.} \quad \dot{D}=\frac{\mathrm{d}D}{\mathrm{d}\rho}(\rho)\dot{\rho}=\lambda g(\sigma_{\mathrm{m}}) \tag{48}$$

If we neglect the change of volume due to the elastic part of the strain, conservation of mass yields:

$$\dot{\rho}+3\rho\dot{\varepsilon}^{\mathrm{p}}_{\mathrm{m}}=0 \tag{49}$$

and, after due account of the complementary laws, we obtain:

$$\frac{g'(\sigma_{\mathrm{m}})}{g(\sigma_{\mathrm{m}})}=\frac{1}{Y}\frac{\dot{\varepsilon}^{\mathrm{p}}_{\mathrm{m}}}{\dot{D}}=-\frac{1}{3\rho\dfrac{\mathrm{d}D}{\mathrm{d}\rho}(\rho)Y} \tag{50}$$

But $Y=-(\mathrm{d}m/\mathrm{d}D)(D(\rho))$ is a function of ρ only. Thus the two sides of equality (50) are constant and homogeneous to the inverse of a stress (i.e. of the form c/σ_0) where σ_0 is the yield stress of the undamaged material. Integrating (50) yields:

$$g(\sigma_{\mathrm{m}})=\mu \exp\left(\frac{C\sigma_{\mathrm{m}}}{\sigma_0}\right)$$

Coming back to the complementary laws we get

$$\dot{D}=\lambda\mu \exp\left(\frac{C\sigma_{\mathrm{m}}}{\sigma_0}\right) \tag{51}$$

† $J_2(\boldsymbol{\sigma})=\sqrt{\sigma^{\mathrm{D}}_{ij}\sigma^{\mathrm{D}}_{ij}}$. $\boldsymbol{\sigma}^{\mathrm{D}}$ is the deviatoric part of $\boldsymbol{\sigma}$, σ_{m} is its spheric part.

The expansion rate of the damage (i.e. of the void fraction) is proportional to the exponential of the triaxiality σ_m/σ_0. It is remarkable that this result, established here through a completely macroscopic analysis, with the simple assumption (49), has been derived through a more precise micromechanical study by Rice and Tracey [26] for high triaxial stress states and by Gurson [12].† These latter studies also provide a precise estimate of C and ν.

The reader is referred to Rousselier [27] for an extension of the theory in the domain of finite strains, together with a study of the stability of the material described by this model.

3.2.2. *Viscoplasticity*

The previous example was concerned with rate independent laws. The effect of time can however be taken into account if one cares to consider viscoplastic laws. No account is given here for damage effects although it would be easy to supply one. Kinematic hardening is described by a tensorial internal variable $\boldsymbol{\beta}$. Therefore

$$\boldsymbol{\alpha} = (\boldsymbol{\varepsilon}^{\mathrm{P}}, \boldsymbol{\beta}, p)$$

$$\rho w(\boldsymbol{\varepsilon}, \boldsymbol{\alpha}) = \tfrac{1}{2}(\boldsymbol{\varepsilon} - \boldsymbol{\varepsilon}^{\mathrm{P}})\mathbf{a}(\boldsymbol{\varepsilon} - \boldsymbol{\varepsilon}^{\mathrm{P}}) + h(p) + k(\boldsymbol{\beta})$$

The state laws read as

$$\mathbf{A}_{\boldsymbol{\varepsilon}^{\mathrm{P}}} = \boldsymbol{\sigma}^{\mathrm{R}} = \mathbf{a}(\boldsymbol{\varepsilon} - \boldsymbol{\varepsilon}^{\mathrm{P}}), \qquad R = -\frac{\mathrm{d}h}{\mathrm{d}p}(p), \qquad \mathbf{B} = -D_{\boldsymbol{\beta}} k(\boldsymbol{\beta})$$

If we denote by P the preceding yield locus (46), the viscoplastic potential φ_{n} is

$$\varphi_n(\mathbf{A}) = \frac{\mu}{n+1}(j_{\mathrm{P}}(\mathbf{A}))^{n+1} = \frac{\mu}{n+1}\left(\frac{J_2(\boldsymbol{\sigma} + \mathbf{B}) + R}{\sigma_0}\right)^{n+1}$$

where j_{P} is the gauge function of P.

The complementary laws are

$$\dot{\boldsymbol{\varepsilon}}^{\mathrm{P}} = \frac{\mu}{\sigma_0}\left(\frac{J_2(\boldsymbol{\sigma} + \mathbf{B}) + R}{\sigma_0}\right)^n \frac{\boldsymbol{\sigma}^{\mathrm{D}} + \mathbf{B}^{\mathrm{D}}}{J_2(\boldsymbol{\sigma} + \mathbf{B})}, \qquad \dot{\boldsymbol{\beta}} = \dot{\boldsymbol{\varepsilon}}^{\mathrm{P}}$$

$$\dot{p} = \frac{\mu}{\sigma_0}\left(\frac{J_2(\boldsymbol{\sigma} + \mathbf{B}) + R}{\sigma_0}\right)^n \tag{52}$$

† Gurson established a dependence on cosh (σ_m/σ_0). But for high triaxial states (underlying assumption here) cosh$\simeq$exp.

Fremond and Friaa [8] introduced a generalization of this law which turned out to be useful in limit analysis problems.

3.2.3. *Cyclic Viscoplasticity*

Under cyclic loadings the previous model fails to properly describe experimental data. Chaboche [4] obtained a good agreement with experimental data through the following modification of the law (52)

$$\dot{\boldsymbol{\varepsilon}}^{\mathrm{p}} = \frac{\partial \varphi_{\mathrm{n}}}{\partial \boldsymbol{\sigma}}(\mathbf{A}), \qquad \dot{p} = \frac{\partial \varphi_{\mathrm{n}}}{\partial R}, \qquad \dot{\boldsymbol{\beta}} = \frac{\partial \varphi_{\mathrm{n}}}{\partial \mathbf{B}} + \eta \mathbf{B}\dot{p} \tag{53}$$

The term $\eta \mathbf{B}\dot{p}$ is likely to alter the standard form of the law. However, if we admit a dependence of the potential on the state variables, it is still a standard law. Indeed† let us set

$$\tilde{\varphi}_{\mathrm{n}}(\mathbf{A}, \boldsymbol{\alpha}) = \frac{\mu}{n+1}\left(\frac{J_2(\boldsymbol{\sigma}+\mathbf{B}) + R + \frac{\eta}{2}\mathbf{B}\,.\,\mathbf{B} - \frac{\eta}{2}D_{\boldsymbol{\beta}}k\,.\,D_{\boldsymbol{\beta}}k}{\sigma_0}\right)^{n+1}$$

The additional term $(\eta/2)\mathbf{B}\,.\,\mathbf{B} - (\eta/2)D_{\boldsymbol{\beta}}k\ D_{\boldsymbol{\beta}}k$ vanishes in all real evolutions, according to the equations of state. But the new complementary laws are

$$\dot{\boldsymbol{\varepsilon}}^{\mathrm{p}} = \frac{\partial \tilde{\varphi}_{\mathrm{n}}}{\partial \boldsymbol{\sigma}}(\mathbf{A}, \boldsymbol{\alpha}) = \frac{\partial \varphi_{\mathrm{n}}}{\partial \boldsymbol{\sigma}}(\mathbf{A}), \qquad \dot{p} = \frac{\partial \tilde{\varphi}_{\mathrm{n}}}{\partial R}(\mathbf{A}, \boldsymbol{\alpha}) = \frac{\partial \varphi_{\mathrm{n}}}{\partial R}(\mathbf{A})$$

$$\dot{\boldsymbol{\beta}} = \frac{\partial \tilde{\varphi}_{\mathrm{n}}}{\partial \mathbf{B}}(\mathbf{A}, \boldsymbol{\alpha}) = \frac{\partial \varphi_{\mathrm{n}}}{\partial \mathbf{B}}(\mathbf{A}) + \frac{\mu\eta}{n+1}\left(\frac{J_2(\boldsymbol{\sigma}+\mathbf{B}) + R}{\sigma_0}\right)^{n}.\,\mathbf{B} = \frac{\partial \varphi_{\mathrm{n}}}{\partial \mathbf{B}}(\mathbf{A}) + \eta \mathbf{B}\dot{p}$$

which is exactly the desired law (53).

Comments. Chaboche's model leads us to address the following mathematical problem: a general form of complementary laws could be

$$\dot{\boldsymbol{\alpha}} = h(\mathbf{A}, \boldsymbol{\alpha})$$

Is it possible to add to h a suitable function $g(\mathbf{A}, \boldsymbol{\alpha})$, vanishing in every real evolution, i.e.

$$g\left(-\rho\frac{\partial W}{\partial \boldsymbol{\alpha}}, \boldsymbol{\alpha}\right) = 0$$

† This remark is due to J. L. Chaboche.

and giving to the law a standard form

$$h(\mathbf{A}, \boldsymbol{\alpha}) + g(\mathbf{A}, \boldsymbol{\alpha}) = \frac{\partial \varphi^*}{\partial \mathbf{A}} (\mathbf{A}, \boldsymbol{\alpha})?$$

4. GLOBAL GSM THEORY

4.1. Global Variables

When dealing with structures (i.e. with a whole body) it can be essential to consider state variables $\underset{\sim}{\boldsymbol{\alpha}}$ which are not locally defined. For instance they can be geometrical parameters of the structure (think of optimal design or of cracked bodies), averages of local variables (think of shells or of homogenization), free boundaries, etc. Considering such *global variables* requires the introduction of a global thermodynamical formalism.

We assume that the system is endowed with a potential energy $F(\mathbf{u}, \underset{\sim}{\boldsymbol{\alpha}})$ where $\mathbf{u}$ is the displacement field in the structure. We say that a set of global variables $\underset{\sim}{\boldsymbol{\alpha}}$ is complete if the specification of this set of variables together with the specification of the loading suffices to determine the displacement field in the structure. For such a set of global variables we can express the total energy in terms of $\underset{\sim}{\boldsymbol{\alpha}}$

$$\mathbf{u} = \mathbf{u}(\underset{\sim}{\boldsymbol{\alpha}}), \qquad W(\underset{\sim}{\boldsymbol{\alpha}}) = F(\mathbf{u}(\underset{\sim}{\boldsymbol{\alpha}}), \underset{\sim}{\boldsymbol{\alpha}}) \tag{54}$$

We define global thermodynamic forces as

$$\underset{\sim}{\mathbf{A}} = -\frac{\partial W}{\partial \underset{\sim}{\boldsymbol{\alpha}}} \tag{55}$$

and the global GSM assumption is the following: there exists a convex functional $\Phi^*(\underset{\sim}{\mathbf{A}})$ such that

$$\dot{\underset{\sim}{\boldsymbol{\alpha}}} = \frac{\partial \Phi^*}{\partial \underset{\sim}{\mathbf{A}}} (\underset{\sim}{\mathbf{A}})\dagger \tag{56}$$

Once more we shall emphasize the role played by materials that satisfy the principle of maximal dissipation, for which Φ^* is the indicator function of a convex set P:

$$\underset{\sim}{\mathbf{A}} \in P \quad \text{and for every} \quad \underset{\sim}{\mathbf{A}}^* \in P, \quad (\dot{\underset{\sim}{\boldsymbol{\alpha}}}, \underset{\sim}{\mathbf{A}}^* - \underset{\sim}{\mathbf{A}}) \leqslant 0 \tag{57}$$

† This equality is to be understood in the sense of subdifferentials for non-differentiable Φ^*.

4.1.1. *Examples*

(a) *Cracks* (Fig. 5(a)). Consider in a two-dimensional context a linear crack of length $\underset{\sim}{\boldsymbol{\alpha}}$ in an elastic body. $\underset{\sim}{\boldsymbol{\alpha}}$ is a global state variable of the structure since, once $\underset{\sim}{\boldsymbol{\alpha}}$ is known and the loading is specified the displacement field is derived by solving a classical elastic problem

$$W(\underset{\sim}{\boldsymbol{\alpha}}) = \underset{\substack{\mathbf{u}^*=0 \text{ on } \Gamma_0 \\ [u_2^*]\geqslant 0 \text{ on } \Gamma_{\alpha}}}{\text{Min}} \int_{\Omega_{\alpha}} \tfrac{1}{2}\mathbf{a}\boldsymbol{\varepsilon}(\mathbf{u}^*)\boldsymbol{\varepsilon}(\mathbf{u}^*)\,\mathrm{d}x - \int_{\Gamma_1} \mathbf{F}\mathbf{u}^*\,\mathrm{d}s \tag{58}$$

The force $\underset{\sim}{\mathbf{A}}$ is the energy release rate. If we define a set P as

$$P = \{\underset{\sim}{\mathbf{A}}^* \mid \underset{\sim}{\mathbf{A}}^* \leqslant \gamma\} \tag{59}$$

the corresponding rate independent law is a law of brittle fracture.

(b) *Damage* (Fig. 5(b) and (c)). Bui and Erlacher [3] introduced a

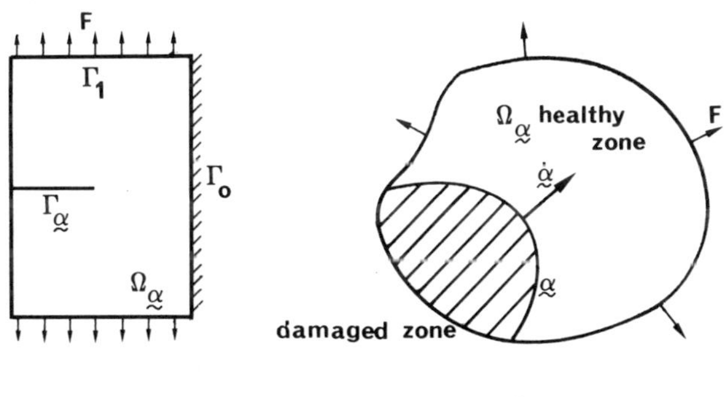

c.

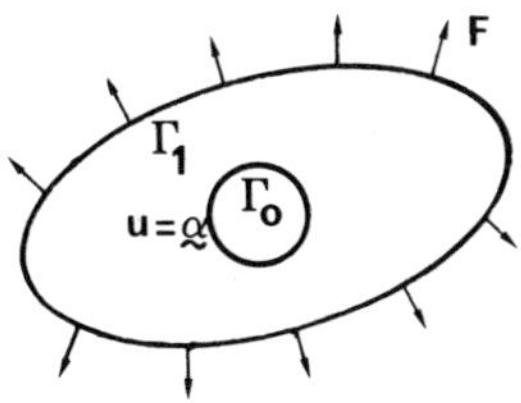

FIG. 5. Global variables.

law of total damage which goes as follows: Consider a structure which contains a 'healthy' linearly elastic zone and a totally damaged zone which is unable to support any load ($\boldsymbol{\sigma}=0$ in it). The damage front $\underset{\sim}{\boldsymbol{\alpha}}$ (free boundary) is a global variable of the structure since, once it is specified the displacement field in the healthy zone is derived as the solution of a classical elastic problem, while the front $\underset{\sim}{\boldsymbol{\alpha}}$ is stress-free

$$W(\underset{\sim}{\boldsymbol{\alpha}}) = \underset{u^*}{\text{Min}} \int_{\Omega_\alpha} \tfrac{1}{2}\mathbf{a}\boldsymbol{\varepsilon}(\mathbf{u}^*)\boldsymbol{\varepsilon}(\mathbf{u}^*)\,\mathrm{d}x - \int_{\Gamma_1} \mathbf{F}\mathbf{u}^*\,\mathrm{d}s$$

The force $\underset{\sim}{\mathbf{A}}$ is the density of elastic energy on the damage front, $\dot{\underset{\sim}{\boldsymbol{\alpha}}}$ is the normal velocity of displacement of the front. If we choose a convex set P in the form (59) the rate independent law (57) yields

$$\begin{aligned} &\dot{\underset{\sim}{\boldsymbol{\alpha}}}(x)=0 \quad \text{if} \quad \tfrac{1}{2}\mathbf{a}\boldsymbol{\varepsilon}(\mathbf{u}(x))\boldsymbol{\varepsilon}(\mathbf{u}(x))<\gamma \\ &\dot{\underset{\sim}{\boldsymbol{\alpha}}}(x)=\lambda\mathbf{n}(x)\ \lambda\geqslant 0 \quad \text{if} \quad \tfrac{1}{2}\mathbf{a}\boldsymbol{\varepsilon}(\mathbf{u}(x))\boldsymbol{\varepsilon}(\mathbf{u}(x))=\gamma \end{aligned} \tag{60}$$

In the one-dimensional case, the stress–strain relation obtained is plotted in Fig. 5(c).

(c) *Homogenization.* Considering the basic cell as a (micro)structure we claim that in the approximate model under consideration in Section 2.5, the set $\underset{\sim}{\boldsymbol{\alpha}}=(\mathbf{E}, \mathbf{E}^{\mathrm{P}}_{\mathrm{m}}, \mathbf{E}^{\mathrm{P}}_{\mathrm{f}})$ is a global state variable for the (micro)structure. Indeed, once it has been specified we are able to derive the displacement in the (micro)structure by (29), and the global energy by (30). The set of forces $\underset{\sim}{\mathbf{A}}$ is $(\boldsymbol{\Sigma}, \boldsymbol{\Sigma}_{\mathrm{m}}, \boldsymbol{\Sigma}_{\mathrm{f}})$ and the rate independent law (57) is (34). We can consider additional geometrical parameters of the microstructure: porosity of voids, surface of cracks, shape of inclusions. These micro-global variables become, through the homogenization process, macro-local variables.

(d) *Friction.* With the notation of Fig. 5(d), the displacement field $\underset{\sim}{\boldsymbol{\alpha}}$ on the part Γ_0 of $\partial\Omega$ is a global variable for an elastic body: once $\underset{\sim}{\boldsymbol{\alpha}}$ is specified, the displacement in the structure is the solution of

$$W(\underset{\sim}{\boldsymbol{\alpha}}) = \underset{\mathbf{u}^*=\underset{\sim}{\boldsymbol{\alpha}} \text{ on } \Gamma_0}{\text{Min}} \int_{\Omega} \tfrac{1}{2}\mathbf{a}\boldsymbol{\varepsilon}(\mathbf{u}^*)\boldsymbol{\varepsilon}(\mathbf{u}^*)\,\mathrm{d}x - \int_{\Gamma_1} \mathbf{F}\mathbf{u}^*\,\mathrm{d}s \tag{61}$$

The force associated with $\underset{\sim}{\boldsymbol{\alpha}}$ is $-\boldsymbol{\sigma}\mathbf{n}|_{\Gamma_0}$. Let us set

$$P=\{\underset{\sim}{\mathbf{A}}^* \mid |\underset{\sim}{\mathbf{A}}^*_{\mathrm{T}}|\leqslant k\}\dagger$$

† $\underset{\sim}{\mathbf{A}}_{\mathrm{T}}$ is the tangential part of the vector $\underset{\sim}{\mathbf{A}}$.

The rate independent law (57) yields Tresca's friction law $\dot{u}_N = 0$, $\dot{\mathbf{u}}_T = 0$ if $|(\boldsymbol{\sigma}\,.\,\mathbf{n})_T| < k$, $\dot{\mathbf{u}}_T = -\lambda(\boldsymbol{\sigma}\,.\,\mathbf{n})_T$, $\lambda \geqslant 0$ if $|(\boldsymbol{\sigma}\,.\,\mathbf{n})_T| = k$.

4.2. Evolution Problems for GSM

The set of eqns. (55) and (56) can be written:

$$\dot{\underset{\sim}{\boldsymbol{\alpha}}} - \frac{\partial \Phi^*}{\partial \underset{\sim}{\mathbf{A}}}\left(-\frac{\partial W}{\partial \underset{\sim}{\boldsymbol{\alpha}}}(\underset{\sim}{\boldsymbol{\alpha}}, t)\right) = 0, \qquad \underset{\sim}{\boldsymbol{\alpha}}(0) = \boldsymbol{\alpha}_0 \tag{62}$$

A similar evolution equation can be stated for $\underset{\sim}{\mathbf{A}}$. It is to be noticed that the equations governing the evolution of phenomena as different as plasticity, damage, brittle fracture, friction, etc., are similar. The discussion of (62) in a general context has not been done and is an open problem. Let us just mention two simple cases where existence and uniqueness of a solution of (62) can be easily established:

(a) If $\dfrac{\partial \Phi^*}{\partial \underset{\sim}{\mathbf{A}}}$ and $\dfrac{\partial W}{\partial \underset{\sim}{\boldsymbol{\alpha}}}$ are Lipschitz operators (this is the case of viscoelasticity) eqn. (62) reduces to a differential equation. Global existence and uniqueness of a solution are proved with the help of the Cauchy–Lipschitz theorem.

(b) If $\underset{\sim}{\boldsymbol{\alpha}} \to \dfrac{\partial \Phi^*}{\partial \underset{\sim}{\mathbf{A}}}\left(-\dfrac{\partial W}{\partial \underset{\sim}{\boldsymbol{\alpha}}}(\underset{\sim}{\boldsymbol{\alpha}}, t)\right)$ is a continuous coercive operator from a reflexive Banach space V into its dual space V' where $V \subsetneq H \subsetneq V'$ (H is a Hilbert space), and provided that it exhibits a smooth dependence on t, a standard theorem on evolution equations ensures the global existence and the uniqueness of a solution of (62).

Consider for instance a linearly elastic solid lying on a support Γ_0 with viscous friction (notation of Fig. 4):

$$\dot{u}_N = 0, \qquad \dot{\mathbf{u}}_T = -\lambda(\boldsymbol{\sigma}\,.\,\mathbf{n})_T \quad \text{on} \quad \Gamma_0 \tag{63}$$

Considering $\underset{\sim}{\boldsymbol{\alpha}} = \mathbf{u}_T$ as a global variable, we identify the associated force as $\underset{\sim}{\mathbf{A}} = (-\boldsymbol{\sigma}(\underset{\sim}{\boldsymbol{\alpha}})\,.\,\mathbf{n})_T\,|_{\Gamma_0}$ where $\boldsymbol{\sigma}(\underset{\sim}{\boldsymbol{\alpha}})$ is derived from $\mathbf{u}(\underset{\sim}{\boldsymbol{\alpha}})$ by the elastic constitutive law. $\mathbf{u}(\underset{\sim}{\boldsymbol{\alpha}})$ is obtained as the solution of an elastic problem similar to (61).†

Therefore the viscous friction law is a global GSM law with

$$\Phi^*(\underset{\sim}{\mathbf{A}}) = \frac{\lambda}{2}|\underset{\sim}{\mathbf{A}}|^2$$

† The constraint for the minimization is now $u_N = 0$, $\mathbf{u}_T = \underset{\sim}{\boldsymbol{\alpha}}$.

Setting $V=H^{1/2}(\Gamma_0)^2$, $H=L^2(\Gamma_0)^2$, $V'=H^{-1/2}(\Gamma_0)^2$, we see that

$$\frac{\partial \Phi^*}{\partial \underset{\sim}{\mathbf{A}}}\left(-\frac{\partial W}{\partial \underset{\sim}{\boldsymbol{\alpha}}}(\underset{\sim}{\boldsymbol{\alpha}}, t)\right)=-\lambda(\boldsymbol{\sigma}\,.\,\mathbf{n})_{\mathrm{T}}(\underset{\sim}{\boldsymbol{\alpha}})$$

It can be easily shown that

$$|(\boldsymbol{\sigma}\,.\,\mathbf{n})_{\mathrm{T}}(\underset{\sim}{\boldsymbol{\alpha}}_1)-(\boldsymbol{\sigma}\,.\,\mathbf{n})_{\mathrm{T}}(\underset{\sim}{\boldsymbol{\alpha}}_2)|_{V'}\leqslant C\,|\underset{\sim}{\boldsymbol{\alpha}}_1-\underset{\sim}{\boldsymbol{\alpha}}_2|_V$$
$$((\boldsymbol{\sigma}\,.\,\mathbf{n})_{\mathrm{T}}(\underset{\sim}{\boldsymbol{\alpha}}_1)-(\boldsymbol{\sigma}\,.\,\mathbf{n})_{\mathrm{T}}(\underset{\sim}{\boldsymbol{\alpha}}_2),\underset{\sim}{\boldsymbol{\alpha}}_1-\underset{\sim}{\boldsymbol{\alpha}}_2)_{V',V}\geqslant C'\,|\underset{\sim}{\boldsymbol{\alpha}}_1-\underset{\sim}{\boldsymbol{\alpha}}_2|_V^2$$

which establishes the continuity and coercivity of the operator under consideration. If the external loading (F) depends smoothly on t, the above mentioned theorem ensures existence and uniqueness of a solution for viscous friction.

Specific attention must be paid to materials that satisfy the principle of maximal dissipation (57):

$$\dot{\underset{\sim}{\boldsymbol{\alpha}}}\in\partial I_{\mathrm{P}}(\underset{\sim}{\mathbf{A}}) \tag{64}$$

The evolution problem, which now consists of the set of equations (55) and (64) is a highly non-linear one, and as such it remains unsolved† in the general case.

Debating of the existence of the rates‡ $\dot{\underset{\sim}{\boldsymbol{\alpha}}}$ and $\dot{\underset{\sim}{\mathbf{A}}}$ is an easier task.

4.3. Variational Principle for $\dot{\underset{\sim}{\boldsymbol{\alpha}}}$ [25]

$\dot{\underset{\sim}{\boldsymbol{\alpha}}}$ is the solution of the following variational problem

$$\left.\begin{aligned}&\dot{\underset{\sim}{\boldsymbol{\alpha}}}\in\partial I_{\mathrm{P}}(\underset{\sim}{\mathbf{A}})\quad\text{and for every}\quad \underset{\sim}{\boldsymbol{\alpha}}^*\in\partial I_{\mathrm{P}}(\underset{\sim}{\mathbf{A}})\\&\left(\frac{\partial^2 W}{\partial\underset{\sim}{\boldsymbol{\alpha}}^2}(\underset{\sim}{\boldsymbol{\alpha}},t)\dot{\underset{\sim}{\boldsymbol{\alpha}}},\underset{\sim}{\boldsymbol{\alpha}}^*-\dot{\underset{\sim}{\boldsymbol{\alpha}}}\right)\geqslant-\left(\frac{\partial^2 W}{\partial\underset{\sim}{\boldsymbol{\alpha}}\,\partial t}(\underset{\sim}{\boldsymbol{\alpha}},t),\underset{\sim}{\boldsymbol{\alpha}}^*-\dot{\underset{\sim}{\boldsymbol{\alpha}}}\right)\end{aligned}\right\} \tag{65}$$

and therefore has the following variational property:

$\dot{\underset{\sim}{\boldsymbol{\alpha}}}$ minimizes among all admissible rates $\underset{\sim}{\boldsymbol{\alpha}}^*\in\partial I_{\mathrm{P}}(\underset{\sim}{\mathbf{A}})$ the functional

$$\frac{1}{2}\frac{\partial^2 W}{\partial\underset{\sim}{\boldsymbol{\alpha}}^2}(\underset{\sim}{\boldsymbol{\alpha}},t)\underset{\sim}{\boldsymbol{\alpha}}^*\underset{\sim}{\boldsymbol{\alpha}}^*+\frac{\partial^2 W}{\partial\underset{\sim}{\boldsymbol{\alpha}}\,\partial t}(\underset{\sim}{\boldsymbol{\alpha}},t)\underset{\sim}{\boldsymbol{\alpha}}^* \tag{66}$$

Proof [24] [25]. By the principle of maximal dissipation

$$(\underset{\sim}{\mathbf{A}}(t)-\underset{\sim}{\mathbf{A}}^*,\underset{\sim}{\boldsymbol{\alpha}}^*)\geqslant 0\quad\text{for every}\quad \underset{\sim}{\boldsymbol{\alpha}}^*\in\partial I_{\mathrm{p}}(\underset{\sim}{\mathbf{A}}),\qquad \underset{\sim}{\mathbf{A}}^*\in P \tag{67}$$

† Moreau [23] and Brezis proved the existence and the uniqueness of a solution for $\underset{\sim}{\mathbf{A}}$, provided that $W(\underset{\sim}{\boldsymbol{\alpha}})$ is a quadratic function.

‡ Since $\underset{\sim}{\boldsymbol{\alpha}}(t)$ is not necessarily derivable with respect to t, the rates $\dot{\underset{\sim}{\boldsymbol{\alpha}}}$, $\dot{\underset{\sim}{\mathbf{A}}}$ are the *right* derivatives $\dfrac{\mathrm{d}^+\underset{\sim}{\boldsymbol{\alpha}}}{\mathrm{d}t}, \dfrac{\mathrm{d}^+\underset{\sim}{\mathbf{A}}}{\mathrm{d}t}$ ('future' rates).

Taking $\underset{\sim}{\mathbf{A}}^* = \underset{\sim}{\mathbf{A}}(t+\mathrm{d}t)$ in (67) and dividing by $\mathrm{d}t$ yields

$$(\dot{\underset{\sim}{\mathbf{A}}}, \underset{\sim}{\boldsymbol{\alpha}}^*) \leqslant 0 \quad \text{for every} \quad \underset{\sim}{\boldsymbol{\alpha}}^* \in \partial I_P(\underset{\sim}{\mathbf{A}})$$

On the other hand it can be shown [24] that

$$(\dot{\underset{\sim}{\mathbf{A}}}, \dot{\underset{\sim}{\boldsymbol{\alpha}}}) = 0$$

Therefore

$$0 \geqslant (\underset{\sim}{\mathbf{A}}, \underset{\sim}{\boldsymbol{\alpha}}^* - \dot{\underset{\sim}{\boldsymbol{\alpha}}}) = -\left(\frac{\partial^2 W}{\partial \underset{\sim}{\boldsymbol{\alpha}}^2}(\underset{\sim}{\boldsymbol{\alpha}}, t)\dot{\underset{\sim}{\boldsymbol{\alpha}}}, \underset{\sim}{\boldsymbol{\alpha}}^* - \dot{\underset{\sim}{\boldsymbol{\alpha}}}\right) - \left(\frac{\partial^2 W}{\partial \underset{\sim}{\boldsymbol{\alpha}}\, \partial t}(\underset{\sim}{\boldsymbol{\alpha}}, t), \underset{\sim}{\boldsymbol{\alpha}}^* - \dot{\underset{\sim}{\boldsymbol{\alpha}}}\right)$$

which completes the proof of (66).

4.4. Variational Principle for $\dot{\mathbf{A}}$

In most examples $\underset{\sim}{\mathbf{A}}$ satisfies a linear constraint of the type:

$$\underset{\sim}{\mathbf{A}} \in S(f) = \{\underset{\sim}{\mathbf{A}}^* \mid \mathrm{L}(\underset{\sim}{\mathbf{A}}^*) = \mathbf{f}\} \tag{68}$$

where $\mathbf{f}$ is a constant quantity, L is a linear operator. Therefore the use of a Lagrange multiplier accounting for (68) permits inversion of the relation (55) between $\underset{\sim}{\boldsymbol{\alpha}}$ and $\underset{\sim}{\mathbf{A}}$:

$$\underset{\sim}{\boldsymbol{\alpha}} = \frac{\partial W^*}{\partial \underset{\sim}{\mathbf{A}}}(-\underset{\sim}{\mathbf{A}}, t) + \underset{\sim}{\mathbf{e}}(t)$$

where W^* is the Legendre transform of W, e is the Lagrange multiplier orthogonal to $S(\mathbf{0})$ (take $f-0$ in (68)). Thus

$$\dot{\underset{\sim}{\boldsymbol{\alpha}}} = -\frac{\partial^2 W^*}{\partial \underset{\sim}{\mathbf{A}}^2}\dot{\underset{\sim}{\mathbf{A}}} + \frac{\partial^2 W^*}{\partial \underset{\sim}{\mathbf{A}}\, \partial t} + \dot{\underset{\sim}{\mathbf{e}}}$$

It is to be noticed that because of the linearity of L, $\dot{\underset{\sim}{\mathbf{A}}}$ belongs to $S(\dot{\mathbf{f}})$ and that $\underset{\sim}{\mathbf{e}}$ is orthogonal to $S(\mathbf{0})$. Moreover, since $\underset{\sim}{\mathbf{A}}$ must belong to P, $\dot{\underset{\sim}{\mathbf{A}}}$ must belong to the projecting cone $\dot{P}(\underset{\sim}{\mathbf{A}})$. Thus we get the following variational formulation for $\dot{\underset{\sim}{\mathbf{A}}}$

$$\dot{\underset{\sim}{\mathbf{A}}} \in \dot{P}(\underset{\sim}{\mathbf{A}}) \cap S(\dot{\mathbf{f}}) \quad \text{and for every} \quad \underset{\sim}{\mathbf{A}}^* \in \dot{P}(\mathbf{A}) \cap S(\dot{\mathbf{f}}):$$

$$\left(\frac{\partial^2 W^*}{\partial \underset{\sim}{\mathbf{A}}^2}(-\underset{\sim}{\mathbf{A}}, t)\dot{\underset{\sim}{\mathbf{A}}}, \underset{\sim}{\mathbf{A}}^* - \dot{\underset{\sim}{\mathbf{A}}}\right) \geqslant \left(\frac{\partial^2 W^*}{\partial \underset{\sim}{\mathbf{A}}\, \partial t}(-\underset{\sim}{\mathbf{A}}, t), \underset{\sim}{\mathbf{A}}^* - \dot{\underset{\sim}{\mathbf{A}}}\right) \tag{69}$$

Therefore $\dot{\underset{\sim}{\mathbf{A}}}$ has the following variational property: $\dot{\underset{\sim}{\mathbf{A}}}$ minimizes among all admissible rates $\underset{\sim}{\mathbf{A}}^* \in \dot{P}(\underset{\sim}{\mathbf{A}}) \cap S(\dot{\mathbf{f}})$ the following functional

$$\frac{1}{2}\frac{\partial^2 W^*}{\partial \underset{\sim}{\mathbf{A}}^2}(-\underset{\sim}{\mathbf{A}}, t)\underset{\sim}{\mathbf{A}}^*\underset{\sim}{\mathbf{A}}^* - \frac{\partial^2 W^*}{\partial \underset{\sim}{\mathbf{A}}\, \partial t}(-\underset{\sim}{\mathbf{A}}, t)\underset{\sim}{\mathbf{A}}^* \tag{70}$$

4.4.1. *Example*

Consider an elasto–plastic body clamped on its boundary and submitted to a volume loading f. The field of anelastic strains $\underset{\sim}{\boldsymbol{\alpha}} = (\boldsymbol{\varepsilon}^{\mathrm{p}}(x))_{x\in\Omega}$ is a global variable for the structure since, once it is known, the displacement in the whole body is known as the solution of

$$W(\underset{\sim}{\boldsymbol{\alpha}}) = \underset{\mathbf{u}^*=0 \text{ on } \partial\Omega}{\mathrm{Min}} \frac{1}{2}\int_\Omega \mathbf{a}(\boldsymbol{\varepsilon}(\mathbf{u}^*) - \underset{\sim}{\boldsymbol{\alpha}})(\boldsymbol{\varepsilon}(\mathbf{u}^*) - \underset{\sim}{\boldsymbol{\alpha}})\,\mathrm{d}x - \int_\Omega \mathbf{f}\mathbf{u}^*\,\mathrm{d}x$$

The force $\underset{\sim}{\mathbf{A}}$ associated with $\underset{\sim}{\boldsymbol{\alpha}}$ is easily identified as the field of the stress tensor. The class of materials obeying (57) reduces to elastic perfectly plastic bodies. The variational principle (70) is known as the Hodge–Prager variational principle [18], and since the operator

$$\frac{\partial^2 W^*}{\partial \underset{\sim}{\mathbf{A}}^2} = \underset{\sim}{\mathbf{a}}^{-1}$$

is positive definite, this principle ensures the existence and the uniqueness of the stress rate in a space of square integrable fields.

The variational principle (66) is known as Greenberg's principle [11]. However the involved operator

$$\frac{\partial^2 W}{\partial \underset{\sim}{\boldsymbol{\alpha}}^2} \tag{71}$$

is not definite, and neither the existence nor the uniqueness of strain rate $\dot{\underset{\sim}{\boldsymbol{\alpha}}} = (\dot{\boldsymbol{\varepsilon}}^{\mathrm{p}})$ can be proved in a classical framework. Due to the lack of coercivity of the above mentioned operator (71) the functional (66) grows linearly with $\dot{\underset{\sim}{\boldsymbol{\alpha}}}^*$ at infinity. Therefore its natural space of definition merely requires $\dot{\underset{\sim}{\boldsymbol{\alpha}}}$ to be integrable and not square integrable, which would classically be the case in this kind of variational problem. In the case of perfect plasticity this remark has led a few authors [31, 32] to introduce in 1977 the space of vector fields with Bounded Deformation:

$$\mathrm{BD}(\Omega) = \{\mathbf{u} \mid \mathbf{u} = (u_i),\ u_i \in L^1(\Omega),\ \varepsilon_{ij}(\mathbf{u}) \in M^1(\Omega) 1 \leqslant i, j \leqslant 3\}$$

where $M^1(\Omega)$ is the space of bounded measures on Ω. BD(Ω) provides the good functional framework to prove the existence of a solution of the variational problem (66). One has to notice that elements of BD(Ω) in general, and solutions of (66) in particular, can be discontinuous fields even before the limit load of the structure is reached. This mathematical anomaly has been previously noticed through a completely different approach by Zyczkowski [38].

5. CONCLUSION

This paper intends to focus the attention on the progress achieved during the last decade in the field of mathematical plasticity. This progress results in a better understanding of the structure of the constitutive laws and of the mathematical properties of the related boundary value problems. Several open mathematical problems have been addressed which will probably be solved within the next ten years with the help of one of the tools presented here: homogenization, general standard materials, global variables.

ACKNOWLEDGEMENTS

Several discussions with G. Francfort, P. Germain and Nguyen Quoc Son have helped the author to clarify his mind on many of the subjects treated here, and are gratefully acknowledged.

REFERENCES

1. Bataille, J. and J. Kestin. Irreversible processes and physical interpretation of rational thermodynamics, *J. Non-Equil. Thermodyn.* **4** (1979), 229–258.
2. Bensoussan, A., J. L. Lions and G. Papanicolaou. *Asymptotic Analysis for Periodic Structures*, North-Holland, Amsterdam, 1980.
3. Bui, H. D. and A. Erlacher. Propagation dynamique d'une zone endommagée dans un solide élastique fragile en mode III et en régime permanent, *C.r. Acad. Sci. Paris, B*, **290,** (1980), 273–276.
4. Chaboche, J. L. Sur l'utilisation des variables d'état internes. In: *Prob. non linéaires de Mécanique.* P.W.N., Varsovie, 1980, 138–159.
5. De Buhan, P. Homogénéisation en calcul à la rupture: le cas du matériau composite multicouche, *C.r. Acad. Sci., Paris, II*, **296,** (1983), 933–936.
6. Duvaut, G. Analyse fonctionnelle et mécanique des milieux continus. In: *Theoretical and Applied Mechanics*, (ed. W. Koiter), North-Holland, Amsterdam, 1976, 119–132.
7. Francfort, G., Nguyen Quoc Son and P. Suquet. Thermodynamique et lois de comportement thermomécanique homogénéisées, *C.r. Acad. Sci., Paris, II*, **296** (1983), 1007–1010.
8. Fremond, M. and A. Friaa. Les méthodes statique et cinématique en calcul à la rupture et en analyse limite, *J. Méc. Théor. appl.*, **1** (1982), 881–905.
9. Germain, P., Nguyen Quoc Son and P. Suquet. Continuum thermodynamics, *J. appl. Mech.*, **105** (1983), 1010–1019.

10. Glansdorf, P. and I. Prigogine *Structure, Stability and Fluctuations*, Wiley, New York, 1981.
11. Greenberg, G. Complementary minimum principles for an elastic plastic material, *Q. J. appl. Math.*, **7** (1948), 85.
12. Gurson, A. Continuum theory of ductile rupture by void nucleation and growth, I, *Trans. ASME J. Engng* (1977), 2–15.
13. Halphen, B. and Nguyen Quoc Son. Sur les matériaux standard généralisés, *J. Méc.*, **14** (1975), 39–63.
14. Haraux, A. *Nonlinear Evolution Equations*, Lecture Notes in Math no. 841, Springer, Berlin, 1981.
15. Hill, R. *The Mathematical Theory of Plasticity*, Clarendon Press, Oxford, 1950.
16. Hill, R. A self consistent mechanics of composite materials, *J. Mech. Phys. Solids*, **13** (1965), 213–222.
17. Hill, R. Aspects of invariance in solid mechanics, *Adv. appl. Mech.*, **18** (1978), 1–75.
18. Hodge, P. and W. Prager. A variational principle for plastic materials with strain hardening, *J. Math. Phys.*, **27** (1949), 1.
19. Kröner, E. *Continuum Statistical Mechanics*, Springer Verlag, Berlin, 1971.
20. Litewka, A. and A. Sawczuk. Experimental evaluation of the overall anisotropic material response at continuous damage (to be published in Drucker Anniversary volume), Elsevier, Amsterdam, 1983.
21. Maccoy, J. Macroscopic response of continua with random microstructures. *Mech. Today*, **5,** 1–40.
22. Mandel, J. Sur la définition de la vitesse de déformation élastique et sa relation avec la vitesse de contraintes, *Int. J. Solids Struct.*, **17** (1981), 873–878.
23. Moreau, J. J. Evolution problem associated with a moving convex set in a Hilbert space, *J. diff. Equat.*, **26** (1977), 347–374.
24. Nguyen Quoc Son. Bifurcation et stabilité des systèmes irréversibles obéissant au principe de dissipation maximale, *J. Méc. Théor. appl.*, **3** (1984), 41–61.
25. Nguyen Quoc Son. This volume, pp. 399–412.
26. Rice, J. and D. Tracey. On the ductile enlargement of voids in triaxial stress fields, *J. Mech. Phys. Solids*, **17** (1969), 201–217.
27. Rousselier, G. Finite deformations constitutive relations including ductile fracture damage. In: *Three-dimensional Constitutive Relations*, North-Holland, Amsterdam, 1981, 331–355.
28. Salencon, J. and A. Tristan Lopez. Analyse de stabilité de remblais sur sols cohérents anisotropes (to be published in *Comportement Plastique des Solides Anisotropes* (Ed. J. P. Boehler), CNRS, Paris, 1984.
29. Sanchez Palencia, E. *Non Homogeneous Media and Vibration Theory*, Lecture Notes in Physics no. 127, Springer, Berlin, 1980.
30. Stolz, C. Contribution à l'étude des grandes transformations en élastoplasticité. Thèse E.N.P.C., Paris, 1982.
31. Strang, G. A family of model problems in plasticity, *Colloque INRIA* 1977, Lecture Notes in Math. 704, 292–308.

32. SUQUET, P. Sur un nouvel espace fonctionnel pour les équations de la Plasticité, *C.r. Acad. Sci., Paris, A*, **286** (1978), 1129–1132.
33. SUQUET, P. Approach by homogenization of some linear and nonlinear problems in Solid Mechanics, to be published in *Comportement Plastique des Solides Anisotropes.* (Ed. J. P. Boehler), CNRS, Paris, 1984.
34. SUQUET, P. Une méthode duale en homogénéisation: application aux milieux élastiques, *J. Méc. Théor. Appl.*, numéro spécial (1982), 79–98.
35. SUQUET, P. *Plasticité et Homogénéisation*, Thèse de doctorat d'Etat, Université Paris VI, 1982.
36. TELEGA, J. J. Derivation of variational principles for rigid plastic materials obeying non associated flow laws, *Int. J. Engng Sci.*, **20** (1982), 913–933.
37. ZAOUI A. and M. BERVEILLER. Methodes self consistent en mécanique des Solides, *15ᵉᵐᵉ Congrès du Groupe Français de Rhéologie*, Ed. E.N.P.C., 1982.
38. ZYCZKOWSKI, M. *Combined Loadings in the Theory of Plasticity*, PWN, Varsovie, 1981.

APPENDIX: HEGEMONY OF CONVEXITY?

This paper might appear to the reader as an anthem to convexity if we were not aware of the elementary and necessary restrictions to be brought into the theory.

(a) Convexity requires the space of states to be a *linear vector space.* Consequently it is inadequate for elasticity at finite strains: for instance, in an incompressible material the principal strains λ_i must satisfy

$$(1+\lambda_1)(1+\lambda_2)(1+\lambda_3)=1$$

which is a non-convex constraint, both on the λ_i and on the displacement field. Other incompatibilities (mainly with the frame indifference principle) have been pointed out by Hill [17]. However this argument fails in plasticity, which is concerned with strain rates, hence with linear functions of the velocity fields when computed in the correct configuration.

(b) Convexity is closely related to stability, each of these notions mainly containing the other one. On the one hand Drucker's work shows that material stability implies convexity of the yield surface and normality of the flow rule. On the other hand Hill [17] and Nguyen Quoc Son [24] established for a plastic material exhibiting a convex yield surface and obeying the normality rule, a stability criterion satisfied under reasonable assumptions for the hardening. Consequently micro and macro instabilities could need a non-convex investigation.

(c) Convexity is unable to account for the behavior of frictioning systems (Coulomb's law) and of soils. For materials obeying a non-associated flow rule Telega [36] used Sewell's account to extend the problem to a convex one and to establish variational principles. When the normality law fails to hold together with the convexity of the yield locus, other tools are to be developed: this was done by Salencon and Tristan Lopez [28] who extended the notion of limit analysis.

Finally we do not claim universal validity for convexity. But addressing the question 'convex or non-convex' defines the proper nature of convexity: it is a reference property.

17

Inverse Problems in Structural Elastoplasticity: A Kalman Filter Approach

S. Bittanti[1], G. Maier[2] and A. Nappi[2]

[1] *Department of Electronics,* [2] *Department of Structural Engineering, Technical University (Politecnico), Milan, Italy*

ABSTRACT

Parameters governing local inelastic deformability in an elastic–plastic discrete structural model are determined, in the sense of Bayesian stochastic estimation, on the basis of information on meaningful displacements in the inelastic response of the structure to a given quasistatic loading history. This structural identification or 'inverse' problem is solved by an extended Kalman filter method. The formulation of the solution procedure consists of the following phases (corresponding to subsections of Section 3): state space representation of the model; linearization of the state equation; linearization of the output equation; weighted least square approach to the parameter estimation; recursive parameter estimation; extended Kalman filter equations; iterated extended Kalman filter. Numerical examples concerning frames illustrate and test the methodology adopted.

NOTATION

Bold-face symbols denote matrices (and column vectors). A tilde means transposed, $\mathbf{0}$ a matrix of all zero entries. Vector inequalities apply componentwise. The matrix of the derivatives of vector $\mathbf{y}$ with respect to $\mathbf{x}$ will be indicated by $\partial\mathbf{y}/\partial\tilde{\mathbf{x}}$, thus ordering columns according to the $\mathbf{x}$ components. The subscript t/τ means estimate of a given variable at time t based on data up to time τ.

1. INTRODUCTION

The problem studied in this paper can be described as follows. A discrete elastic–plastic model of a structure is built up for analysis purposes, but some parameters which govern the local inelastic deformability are unknown or uncertain. A prototype of the structure is subjected to a given, suitably chosen, quasi-static loading history, during which some aspects of the overall structural response (such as meaningful displacements and/or rotations) are measured. The question is to determine the unknown or uncertain parameters included in the mathematical model, on the basis of the available experimental data.

The engineering motivation of this question is two-fold: (a) the 'calibration' of models with respect to some particularly uncertain assumptions: the improved model can then be used as an analysis tool for subsequent design purposes; (b) the indirect assessment of possible local structural changes or damages which might not be detectable by direct inspections.

The above problem can be regarded as an 'inverse' problem in nonlinear structural mechanics or a particular case of 'structural identification'.

Parameter identification in the above sense of linear vibrating structures is the object of a fairly abundant recent literature, surveyed, for example, in refs. 10, 11 and 17. The relevant research activity was motivated primarily by mechanical, nuclear, earthquake and offshore technology.

In quasi-static situations, arising primarily in civil structural and geotechnical engineering, elastoplastic identification problems have been studied in some previous papers both in a deterministic and in a Bayesian-stochastic context [3, 8, 9, 16]. Generally, the experimental information has been used in a 'batch' way, i.e. by minimizing an 'error' function defined as the norm of the difference between the vector of all experimental data and the vector of the corresponding model-predicted values.

In this paper the problem of identifying or 'estimating' parameters related to the local 'plastic' deformability is investigated in a stochastic context, resorting to the well established methodology of Kalman filtering [1, 12, 13]. This methodology is intended to exploit sequentially the experimental information gathered along the evolution of a system under a history of external actions. As such, it appears to fit ideally the path-dependent, irreversible nature of plastic structural response and the sequential character of the measurements usually performed in a statical test. To the authors' knowledge, previous use of filtering techniques in solid mechanics concerns laboratory tests on creeping materials [2] and hydrodynamic coefficients in a dynamic model of offshore structures [18]. In introducing this paper the following authoritative statement (by Balakrishnan in *The Mathematical Intelligencer*, Vol. 1, No. 2, 1978) is worth quoting: 'the Kalman filter is a data-processor designed to filter signal from noise. Its "application" does not mean putting numbers into formulae; indeed any real discussion of a Kalman filter application requires no less than a case study'.

2. THE STRUCTURAL MODEL

The elastic–plastic, quasi-static, small deformation response of a truss-like structure or a frame (cf. Fig. 1(a) and Section 4) to a small load increment, say from instant t to $t+1$ of an ordering variable or 'time', is often determined on the basis of the following idealizations and mathematical description. The inelastic, irreversible deformations are assumed to be flexural only and confined to a discrete number, say c, of 'critical sections' in the form of plastic rotations p^i. With reference to a single yield mode j ($j=1$ positive, $j=2$ negative direction) of the ith critical section and to Fig. 1(b), one writes for time $t+1$, being all variables known at t

$$\varphi_{t+1}^{i,j}=(-1)^j Q_{t+1}^i-(R_j^i+H_j^i\lambda_t^{i,j})-H_j^i\Delta\lambda_{t+1}^{i,j}\leqslant 0, \qquad \Delta\lambda_{t+1}^{i,j}\geqslant 0 \quad \text{(1a)}$$

$$\text{if} \quad \varphi_{t+1}^{i,j}<0 \quad \text{then} \quad \Delta\lambda_{t+1}^{i,j}=0; \qquad \text{if} \quad \Delta\lambda_{t+1}^{i,j}>0 \quad \text{then} \quad \varphi_{t+1}^{i,j}=0 \quad \text{(1b)}$$

or, equivalently:

$$\varphi_{t+1}^{i,j}\Delta\lambda_{t+1}^{i,j}=0 \quad \text{(1b')}$$

where: φ = yield function; Q = bending moment; R = original yield moment; H = hardening modulus (positive constant for linear hardening); λ = plastic multiplier; Δ means incremental. Equations (1) can be re-written in matrix notation for all yield modes ($j=1,2$; $i=1,\ldots,c$) simultaneously, setting $\tilde{\mathbf{N}}=\text{diag}\,[1,-1]$, $\mathbf{H}=\text{diag}\,[H^{i,j}]$:

$$\boldsymbol{\varphi}_{t+1}=\tilde{\mathbf{N}}\mathbf{Q}_{t+1}-\mathbf{R}-\mathbf{H}\boldsymbol{\lambda}_t-\mathbf{H}\,\Delta\boldsymbol{\lambda}_{t+1}\leqslant\mathbf{0}; \qquad \tilde{\boldsymbol{\varphi}}_{t+1}\Delta\boldsymbol{\lambda}_{t+1}=0 \quad (2)$$

The consistency conditions (1b) are equivalent to the orthogonality requirement of $\boldsymbol{\varphi}_{t+1}$ and $\Delta\boldsymbol{\lambda}_{t+1}$ since these are vectors of sign-constrained variables.

The c-vector of the plastic rotations $\mathbf{p}$ and bending moments $\mathbf{Q}$ in all critical sections can be expressed in the forms:

$$\mathbf{p}_{t+1} = \mathbf{N}(\boldsymbol{\lambda}_t + \Delta\boldsymbol{\lambda}_{t+1}), \qquad \mathbf{Q}_{t+1} = \mathbf{Q}^{\mathrm{E}}_{t+1} + \mathbf{Z}\mathbf{p}_{t+1} \tag{3}$$

where: $\mathbf{Q}^{\mathrm{E}}$ = elastic bending moments calculated by any linear analysis procedure and taken as data at any t; $\mathbf{Z}$ = symmetric, negative-semidefinite, $c \times c$-matrix of elastic influence coefficients on moments in the critical sections due to imposed rotations there, to be primarily calculated once for all by linear analysis.

Substituting the former of eqns. (3) in the latter and this into eqn. (2), we obtain:

$$\boldsymbol{\varphi}_{t+1} = \tilde{\mathbf{N}}\mathbf{Q}^{\mathrm{E}}_{t+1} - \mathbf{A}(\boldsymbol{\lambda}_t + \Delta\boldsymbol{\lambda}_{t+1}) - \mathbf{R} \tag{4a}$$

$$\boldsymbol{\varphi}_{t+1} \leqslant \mathbf{0}, \qquad \Delta\boldsymbol{\lambda}_{t+1} \geqslant \mathbf{0}, \qquad \tilde{\boldsymbol{\varphi}}_{t+1}\Delta\boldsymbol{\lambda}_{t+1} = 0 \tag{4b}$$

where:

$$\mathbf{A} \equiv \mathbf{H} - \tilde{\mathbf{N}}\mathbf{Z}\mathbf{N} \tag{4c}$$

The relation set (4) mathematically represents a 'linear complementarity problem' (LCP) associated with matrix $\mathbf{A}$, of order $2c$, in the variables $\boldsymbol{\varphi}_{t+1}$, $\Delta\boldsymbol{\lambda}_{t+1}$; mechanically, it governs the inelastic incremental response of the structural model to given load increments (contained in $\mathbf{Q}^{\mathrm{E}}_{t+1}$). The vector $\mathbf{u}$ of the displacements (or rotations) susceptible to measurements can be expressed through the same model in the form:

$$\mathbf{u}_{t+1} = \mathbf{u}^{\mathrm{E}}_{t+1} + \mathbf{G}(\boldsymbol{\lambda}_t + \Delta\boldsymbol{\lambda}_{t+1}) \tag{5}$$

where: $\mathbf{u}^{\mathrm{E}}$ = elastic displacements and $\mathbf{G}$ = influence coefficient matrix, both calculated by linear analysis, ignoring the possibility of inelastic yielding.

The parameters (included in vector $\mathbf{P}$) to be identified in the following on the basis of experimental information on these displacements, will be bending resistances R^i_j and/or hardening moduli H^i_j. They define characteristics of local (sectional) plastic deformability in the adopted structural model. This particular model, centered on the LCP formulation, has been devised [14] and frequently used for a broad category of discretized structures and continua, covered by (4) merely by re-interpretation of symbols (see e.g. refs. 5 and 15). Despite the limitations (to frames with plastic hinges, linear nonactive hardening, etc.) here adopted for convenience and brevity, the same generality holds, clearly, for this first investigation on filtering methodologies for solving inverse problems of structural plasticity.

3. DERIVATION OF THE EXTENDED KALMAN FILTER RELATIONS

The ordinative time t will be regarded as an integer variable running over the measurement sequence $1, 2, \ldots, t, t+1, \ldots, T$. We also assume that the unit time-change corresponds to a loading step small enough to verify the holonomic (path-independence) hypothesis for each individual step. The natural state variable of the system will be the plastic multipliers included in vector $\boldsymbol{\lambda}_t$. The 'state equations' describing the system evolution relate their values at instant $t+1$ to those at t, to the parameter vector $\mathbf{P}_t$ at t and (though not explicitly indicated) to the change $\Delta\mathbf{Q}^{\mathrm{E}}_{t+1}$ of the elastic stress response to external actions. The (time-independent) parameters $\mathbf{P}_t$ will be considered as additional state variables, thus reducing their identification to a state estimation, in the spirit of the 'extended' Kalman filter methodology [1, 7]. Therefore, the 'state equations' read:

$$\boldsymbol{\lambda}_{t+1} = \boldsymbol{\lambda}_t + \Delta\boldsymbol{\lambda}_{t+1} = \boldsymbol{\lambda}_t + \mathbf{f}(\boldsymbol{\lambda}_t, \mathbf{P}_t) \tag{6}$$

$$\mathbf{P}_{t+1} = \mathbf{P}_t \tag{7}$$

where $\mathbf{f}$ represents the functional dependence implicit in the model specified in Section 2. Vector $\boldsymbol{\mu}_t$ of experimental data from measurements at t concerning meaningful nodal displacements (or rotations) $\mathbf{u}_t$ is conceived as a random variable:

$$\boldsymbol{\mu}_t = \mathbf{u}_t(\boldsymbol{\lambda}_t, \mathbf{P}_t) + \mathbf{n}_t \tag{8}$$

In this 'output equation' the 'noise' term $\mathbf{n}_t$ represents the measurement error. As usual, we assume (denoting by E [ ] the expected value):

$$\mathbf{E}[\mathbf{n}_t] = \mathbf{0}; \qquad \mathbf{E}[\mathbf{n}_t \tilde{\mathbf{n}}_\tau] = \mathbf{V} \quad \text{for} \quad t = \tau; \quad = \mathbf{0} \quad \text{for} \quad t \neq \tau \tag{9}$$

In other terms, $\mathbf{n}_t$ has zero expected value and zero cross-correlation between $\mathbf{n}_t$ and $\mathbf{n}_\tau$ for any $\tau \neq t$ ('white' noise). The covariance matrix $\mathbf{V}$ will be a diagonal matrix of (positive) variances, since the measures are conceived here as uncorrelated random variables.

The nonlinear dependence (6) is linearized around a given vector $(\bar{\boldsymbol{\lambda}}_t, \bar{\mathbf{P}}_t)$ by a truncated Taylor expansion of function $\mathbf{f}$:

$$\Delta\boldsymbol{\lambda}_{t+1} = \mathbf{f}(\bar{\boldsymbol{\lambda}}_t, \bar{\mathbf{P}}_t) + \left[\frac{\partial \mathbf{f}}{\partial \bar{\boldsymbol{\lambda}}_t} \,\vdots\, \frac{\partial \mathbf{f}}{\partial \bar{\mathbf{P}}_t}\right]_{\bar{\boldsymbol{\lambda}}_t, \bar{\mathbf{P}}_t} \begin{Bmatrix} \boldsymbol{\lambda}_t - \bar{\boldsymbol{\lambda}}_t \\ \mathbf{P}_t - \bar{\mathbf{P}}_t \end{Bmatrix} \tag{10}$$

The first term on the r.h.s. represents the distribution of plastic strains at instant $t+1$ if the system state at t is identified by $(\bar{\boldsymbol{\lambda}}_t, \bar{\mathbf{P}}_t)$ and the load change from t to $t+1$ gives rise to the known incremental elastic stress vector $\Delta\mathbf{Q}^{\mathrm{E}}_{t+1}$. With these data, the LCP (4) governing the plastic response provides the solution $\bar{\boldsymbol{\varphi}}_{t+1}$, $\Delta\bar{\boldsymbol{\lambda}}_{t+1}$. Partition these two vectors as follows, according to the way in which complementarity is fulfilled by each pair of corresponding components:

$$\bar{\boldsymbol{\varphi}}_{t+1} = \begin{Bmatrix} \bar{\boldsymbol{\varphi}}'_{t+1} = \mathbf{0} \\ \bar{\boldsymbol{\varphi}}''_{t+1} < \mathbf{0} \\ \bar{\boldsymbol{\varphi}}'''_{t+1} = \mathbf{0} \end{Bmatrix}, \qquad \Delta\bar{\boldsymbol{\lambda}}_{t+1} = \begin{Bmatrix} \Delta\bar{\boldsymbol{\lambda}}'_{t+1} > \mathbf{0} \\ \Delta\bar{\boldsymbol{\lambda}}''_{t+1} = \mathbf{0} \\ \Delta\bar{\boldsymbol{\lambda}}'''_{t+1} = \mathbf{0} \end{Bmatrix} \tag{11}$$

The determination of the derivatives in (10) calls for the solution of the LCP (4) at the perturbated state $\bar{\boldsymbol{\lambda}}_t + \delta\boldsymbol{\lambda}_t$, $\bar{\mathbf{P}}_t + \delta\mathbf{P}_t$, δ denoting infinitesimal perturbation. Neglecting higher order infinitesimals and taking into account the solution $\bar{\boldsymbol{\varphi}}_{t+1}$, $\Delta\bar{\boldsymbol{\lambda}}_{t+1}$ to the undisturbed LCP (barred symbols), the perturbed LCP reads:

$$\boldsymbol{\varphi}_{t+1} = \bar{\boldsymbol{\varphi}}_{t+1} - \bar{\mathbf{A}}\{\delta\boldsymbol{\lambda}_t + \delta\Delta\boldsymbol{\lambda}_{t+1}\} - \delta\mathbf{A}\{\bar{\boldsymbol{\lambda}}_{t+1} + \Delta\bar{\boldsymbol{\lambda}}_{t+1}\} - \delta\mathbf{R} \leqslant \mathbf{0} \tag{12a}$$

$$\Delta\bar{\boldsymbol{\lambda}}_{t+1} + \delta\Delta\boldsymbol{\lambda}_{t+1} \geqslant \mathbf{0}, \qquad \tilde{\boldsymbol{\varphi}}_{t+1}\{\Delta\bar{\boldsymbol{\lambda}}_{t+1} + \delta\Delta\boldsymbol{\lambda}_{t+1}\} = 0 \tag{12b}$$

Adopt for the solution of (12), marked by double bar, the partition shown in (11) based on the solution of the unperturbated LCP. The stability of the solution of LCP (4) with respect to data perturbation implies for the solution to (12) that:

$$\begin{aligned} \Delta\bar{\boldsymbol{\lambda}}'_{t+1} + \delta\Delta\boldsymbol{\lambda}'_{t+1} > \mathbf{0} \quad &\text{and, hence,} \quad \bar{\bar{\boldsymbol{\varphi}}}' = \mathbf{0} \\ \bar{\bar{\boldsymbol{\varphi}}}''_{t+1} < \mathbf{0} \quad &\text{and, hence,} \quad \delta\Delta\boldsymbol{\lambda}''_{t+1} = \mathbf{0} \end{aligned} \tag{13}$$

No a priori conclusion of this type is possible on subvectors $\bar{\bar{\boldsymbol{\varphi}}}'''_{t+1}$ and $\delta\Delta\boldsymbol{\lambda}'''_{t+1}$. Henceforth we will assume

$$\boldsymbol{\varphi}'''_{t+1} = \mathbf{0} \tag{14}$$

and consider these subvectors as included in $\bar{\bar{\boldsymbol{\varphi}}}'_{t+1}$ and $\delta\Delta\boldsymbol{\lambda}'_{t+1}$. Denote by n' and n'' the number of components of $\boldsymbol{\varphi}'_{t+1}$ and $\boldsymbol{\varphi}''_{t+1}$, respectively and partition matrices and vectors in (12) accordingly, e.g.

$$\bar{\mathbf{A}} = \begin{bmatrix} \bar{\mathbf{A}}' \\ \bar{\mathbf{A}}'' \end{bmatrix} = \begin{bmatrix} \bar{\mathbf{A}}_1' & \bar{\mathbf{A}}_2' \\ \bar{\mathbf{A}}_1'' & \bar{\mathbf{A}}_2'' \end{bmatrix}; \qquad \bar{\mathbf{A}}_1' = n' \times n' \text{ matrix}, \qquad \bar{\mathbf{A}}_2' = n' \times n'', \text{ etc.} \tag{15}$$

Noting that $\delta\mathbf{A} = \delta\mathbf{H}$, write:

$$\delta\mathbf{A}\bar{\boldsymbol{\lambda}}_{t+1} \equiv \delta\mathbf{H}\bar{\boldsymbol{\lambda}}_{t+1} = \text{diag}\,[\bar{\boldsymbol{\lambda}}_{t+1}]\delta\mathbf{h} \tag{16}$$

where: $\mathbf{h}$ = vector of the diagonal entries of $\mathbf{H}$; diag $[\bar{\boldsymbol{\lambda}}_{t+1}]$ = diagonal matrix of the components of $\bar{\boldsymbol{\lambda}}_{t+1}$. In eqn. (12a), taking account of (13)–(16), generates the equation:

$$\bar{\mathbf{A}}'\delta\boldsymbol{\lambda}_t + \bar{\mathbf{A}}_1'\delta\Delta\boldsymbol{\lambda}_{t+1}' + \text{diag}\,[\bar{\boldsymbol{\lambda}}_{t+1}' + \Delta\bar{\boldsymbol{\lambda}}_{t+1}']\delta\mathbf{h}' + \delta\mathbf{R}' = \mathbf{0} \tag{17}$$

Equation (17), solved with respect to $\delta\Delta\boldsymbol{\lambda}_{t+1}'$ (det $\bar{\mathbf{A}}_1' \neq 0$), provides the primed upper part of the derivative matrices in (10). The lower parts are zero submatrices. Specifically:

$$\left[\frac{\partial \mathbf{f}}{\partial \tilde{\boldsymbol{\lambda}}_t}\right]_{\bar{\boldsymbol{\lambda}}_t, \bar{\mathbf{P}}_t} = \begin{bmatrix} -(\bar{\mathbf{A}}_1')^{-1}\bar{\mathbf{A}}' \\ \hline \mathbf{0} \end{bmatrix} \tag{18}$$

$$\left[\frac{\partial \mathbf{f}}{\partial \tilde{\mathbf{R}}}\right]_{\bar{\boldsymbol{\lambda}}_t, \bar{\mathbf{P}}_t} = \begin{bmatrix} -(\bar{\mathbf{A}}_1')^{-1} \\ \hline \mathbf{0} \end{bmatrix}, \qquad \left[\frac{\partial \mathbf{f}}{\partial \tilde{\mathbf{h}}}\right]_{\bar{\boldsymbol{\lambda}}_t, \bar{\mathbf{P}}_t} = \begin{bmatrix} -(\bar{\mathbf{A}}_1')^{-1}\,\text{diag}\,[\bar{\boldsymbol{\lambda}}_t'] \\ \hline \mathbf{0} \end{bmatrix} \tag{19}$$

Turning now to the 'output equation' (8), since the measurable displacements $\mathbf{u}_{t+1}$ do not explicitly depend on the parameters $\mathbf{P}_{t+1}$, but only on $\boldsymbol{\lambda}_{t+1}$ through (4), in view of (6), we obtain:

$$\mathbf{u}_{t+1}(\boldsymbol{\lambda}_{t+1}, \mathbf{P}_{t+1}) = \mathbf{u}_{t+1}^{\mathrm{E}} + \mathbf{G}\boldsymbol{\lambda}_t + \mathbf{G}\Delta\boldsymbol{\lambda}_{t+1}(\boldsymbol{\lambda}_t, \mathbf{P}_t) \tag{20}$$

This is linearized below around vector $(\bar{\boldsymbol{\lambda}}_t, \bar{\mathbf{P}}_t)$. The required derivatives ('sensitivity matrix' $\mathbf{L}$) can be directly calculated using eqns. (19):

$$\left[\frac{\partial \mathbf{u}_{t+1}}{\partial \tilde{\mathbf{P}}_t}\right]_{\bar{\boldsymbol{\lambda}}_t, \bar{\mathbf{P}}_t} = \left[\frac{\partial \mathbf{u}_{t+1}}{\partial \Delta\tilde{\boldsymbol{\lambda}}_{t+1}}\right]_{\bar{\boldsymbol{\lambda}}_t, \bar{\mathbf{P}}_t} \left[\frac{\partial \Delta\boldsymbol{\lambda}_{t+1}}{\partial \tilde{\mathbf{P}}_t}\right]_{\bar{\boldsymbol{\lambda}}_t, \bar{\mathbf{P}}_t} = -\mathbf{G}'(\bar{\mathbf{A}}_2')^{-1}[\mathbf{I} \,\vdots\, \text{diag}\,[\bar{\boldsymbol{\lambda}}_t']] \equiv \mathbf{L} \tag{21}$$

where $\mathbf{I}$ = identity matrix; $\mathbf{G}'$ = submatrix of $\mathbf{G}$ given by its first n' columns.

Suppose first that the following information only is available: (i) a single set of observations $\boldsymbol{\mu}_1$ (under loads $\mathbf{F}_1$); (ii) the error covariance matrix $\mathbf{V}$; (iii) the a priori knowledge about the parameters defined by their expected values $\bar{\mathbf{P}}_0$ and by their (diagonal, positive definite) covariance matrix $\bar{\mathbf{W}}_0^{\mathrm{P}}$. In this case the discrepancy between the available information and the values provided by the model ('error function') is naturally defined by 'weighted least squares', i.e. attributing to each measure a 'weight' equal to the inverse of the relevant variance [6]:

$$\mathscr{E}(\mathbf{P}) = \{\tilde{\boldsymbol{\mu}}_1 - \tilde{\mathbf{u}}(\mathbf{P})\}\mathbf{V}^{-1}\{\boldsymbol{\mu}_1 - \mathbf{u}(\mathbf{P})\} + \{\tilde{\mathbf{P}} - \tilde{\bar{\mathbf{P}}}_0\}\mathbf{W}_0^{\mathrm{P}^{-1}}\{\mathbf{P} - \bar{\mathbf{P}}_0\} \qquad (22)$$

The nonlinear dependence $\mathbf{u}(\mathbf{P})$ can be linearized around some value $\bar{\mathbf{P}}$ using (21):

$$\mathbf{u}(\mathbf{P}, \bar{\mathbf{P}}) = \mathbf{u}(\bar{\mathbf{P}}) + \mathbf{L}(\bar{\mathbf{P}})\{\mathbf{P} - \bar{\mathbf{P}}\} \qquad (23)$$

Minimization of the quadratic form in $\mathbf{P}$ obtained by substituting (23) with $\bar{\mathbf{P}} = \mathbf{P}_0$ into (22) leads to the estimate:

$$\hat{\mathbf{P}}(\bar{\mathbf{P}}_0) = \bar{\mathbf{P}}_0 + \mathbf{M}(\bar{\mathbf{P}}_0)\{\boldsymbol{\mu}_1 - \mathbf{u}(\bar{\mathbf{P}}_0)\} \qquad (24)$$

having set (as 'mapping matrix'):

$$\mathbf{M}(\bar{\mathbf{P}}_0) \equiv \mathbf{W}_0^{\mathrm{P}}\tilde{\mathbf{L}}(\bar{\mathbf{P}}_0)[\mathbf{L}(\bar{\mathbf{P}}_0)\mathbf{W}_0^{\mathrm{P}}\tilde{\mathbf{L}}(\bar{\mathbf{P}}_0) + \mathbf{V}]^{-1} \qquad (25)$$

The uncertainty associated with this estimate is described by:

$$\hat{\mathbf{W}}(\bar{\mathbf{P}}_0) = \mathbf{W}_0^{\mathrm{P}} - \mathbf{M}(\bar{\mathbf{P}}_0)\mathbf{L}(\bar{\mathbf{P}}_0)\mathbf{W}_0^{\mathrm{P}} \qquad (26)$$

which is obtained, as shown in the appendix, by evaluating the covariance of $\hat{\mathbf{P}}(\bar{\mathbf{P}}_0)$, conceived as the estimator of $\mathbf{P}$, starting from the random vector $\boldsymbol{\mu}_1$.

In view of the nonlinearity of $\mathbf{u}(\mathbf{P})$ the above linear estimation procedure should be iterated:

$$\hat{\mathbf{P}}_{i+1}(\hat{\mathbf{P}}_i) = \mathbf{P}_0 + \mathbf{M}(\hat{\mathbf{P}}_i)\{\boldsymbol{\mu}_1 - \mathbf{u}(\mathbf{P}_0, \hat{\mathbf{P}}_i)\} \quad (i = 0, 1, 2, \ldots) \qquad (27)$$

For $i = 0$, eqn. (27) coincides with (24). The iterative process will be carried out until $|\hat{\mathbf{P}}_{i+1} - \hat{\mathbf{P}}_i|$ does not exceed a given tolerance. This 'iterated Bayesian estimation' exploiting the available information in a 'batch' fashion, has been applied and discussed, for example, in refs. 3 and 16.

Considering now a sequence of observations $\boldsymbol{\mu}_t$ at $t = 1, 2, \ldots, T$ along a loading path, we shall make a sequential use of this information flow in the sense of Kalman filtering. The a priori knowledge on the state variables $\boldsymbol{\lambda}$ consists of $\bar{\boldsymbol{\lambda}}_0 = \mathbf{0}$, $\mathbf{W}_0^{\lambda} \equiv E[\boldsymbol{\lambda}\tilde{\boldsymbol{\lambda}}] = \mathbf{0}$ (i.e. the loading process starts from an elastic state). Also the expected value $\bar{\mathbf{P}}_0$ of $\mathbf{P}$ and the associated covariance matrix $\mathbf{W}_0^{\mathrm{P}}$ are given. $(\hat{\mathbf{P}}_{1/0}, \hat{\boldsymbol{\lambda}}_{1/0})$ will denote the state prediction at t_1 based on the a priori information; $\mathbf{P}_{1/1}$ and $\boldsymbol{\lambda}_{1/1}$ the state estimates at t_1 obtained on the basis of the first set of measures $\boldsymbol{\mu}_1$. Consider the error function:

$$\mathscr{E}(\mathbf{P}_{1/1}, \boldsymbol{\lambda}_{1/1}) = \{\tilde{\boldsymbol{\mu}}_1 - \tilde{\mathbf{u}}(\mathbf{P}_{1/1})\}\mathbf{V}^{-1}\{\boldsymbol{\mu}_1 - \mathbf{u}(\mathbf{P}_{1/1})\} + \{\tilde{\mathbf{P}}_{1/1} - \tilde{\hat{\mathbf{P}}}_{1/0} \,\vdots\, \tilde{\boldsymbol{\lambda}}_{1/1} - \tilde{\hat{\boldsymbol{\lambda}}}_{1/0}\}\mathbf{W}_{1/0}^{-1}\begin{Bmatrix}\mathbf{P}_{1/1} - \hat{\mathbf{P}}_{1/0} \\ \boldsymbol{\lambda}_{1/1} - \hat{\boldsymbol{\lambda}}_{1/0}\end{Bmatrix} \tag{28}$$

$\mathbf{W}_{1/0}$ being the covariance matrix of the state prediction error. The optimal estimates $\hat{\mathbf{P}}_{1/1}$ and $\hat{\boldsymbol{\lambda}}_{1/1}$, obtained by minimizing (28), are given by:

$$\begin{Bmatrix}\hat{\mathbf{P}}_{1/1} \\ \hat{\boldsymbol{\lambda}}_{1/1}\end{Bmatrix} = \begin{Bmatrix}\hat{\mathbf{P}}_{1/0} \\ \hat{\boldsymbol{\lambda}}_{1/0}\end{Bmatrix} + \mathbf{M}_{1/0}(\hat{\mathbf{P}}_{1/0})\{\boldsymbol{\mu}_1 - \mathbf{u}(\hat{\mathbf{P}}_{1/0})\} \tag{29}$$

$$\mathbf{M}_{1/0}(\hat{\mathbf{P}}_{1/0}) = \mathbf{W}_{1/0}\tilde{\mathbf{S}}(\hat{\mathbf{P}}_{1/0})[\mathbf{S}(\hat{\mathbf{P}}_{1/0})\mathbf{W}_{1/0}\tilde{\mathbf{S}}(\hat{\mathbf{P}}_{1/0}) + \mathbf{V}]^{-1} \tag{30}$$

In (30), the sensitivity matrix is given by:

$$\mathbf{S}(\hat{\mathbf{P}}_{1/0}) = [\mathbf{L}(\hat{\mathbf{P}}_{1/0}) \,\vdots\, \mathbf{0}] \tag{31}$$

The a posteriori uncertainty is characterized by the covariance matrix of the state error $\{\tilde{\hat{\mathbf{P}}}_{1/1} - \tilde{\hat{\mathbf{P}}}_{1/0}, \tilde{\hat{\boldsymbol{\lambda}}}_{1/1} - \tilde{\hat{\boldsymbol{\lambda}}}_{1/0}\}$:

$$\mathbf{W}_{1/1}(\hat{\mathbf{P}}_{1/0}) = \mathbf{W}_{1/0} - \mathbf{M}_{1/0}(\hat{\mathbf{P}}_{1/0})\mathbf{S}(\hat{\mathbf{P}}_{1/0})\mathbf{W}_{1/0} \tag{32}$$

In view of the available a priori information, a natural choice of $\{\hat{\mathbf{P}}_{1/0}, \hat{\boldsymbol{\lambda}}_{1/0}\}$ is:

$$\begin{Bmatrix}\hat{\mathbf{P}}_{1/0} \\ \hat{\boldsymbol{\lambda}}_{1/0}\end{Bmatrix} = \begin{Bmatrix}\bar{\mathbf{P}}_0 \\ \mathbf{0}\end{Bmatrix} \tag{33}$$

Matrix $\mathbf{W}_{1/0}$ is partitioned as follows:

$$\mathbf{W}_{1/0} = \begin{bmatrix}\mathbf{W}_{1/0}^{PP} & \mathbf{W}_{1/0}^{P\lambda} \\ \mathbf{W}_{1/0}^{\lambda P} & \mathbf{W}_{1/0}^{\lambda\lambda}\end{bmatrix} \tag{34}$$

where: $\mathbf{W}_{1/0}^{PP}$ and $\mathbf{W}_{1/0}^{\lambda\lambda}$ (of the same dimensions of vectors $\mathbf{P}$ and $\boldsymbol{\lambda}$, respectively) are the covariance matrices of the prediction errors of the parameters and of the plastic multipliers; $\mathbf{W}_{1/0}^{P\lambda} = \tilde{\mathbf{W}}_{1/0}^{\lambda P}$ = cross-covariance matrix between the two prediction errors. $\mathbf{W}_{1/0}^{PP}$ can be identified with the a priori covariance matrix $\mathbf{W}_0^{P}$. Taking account of (6) and (10), observe that:

$$\boldsymbol{\lambda}_1 = \boldsymbol{\lambda}_0 + \mathbf{f}(\bar{\boldsymbol{\lambda}}_0, \bar{\mathbf{P}}_0) + \left[\frac{\partial \mathbf{f}}{\partial \tilde{\boldsymbol{\lambda}}_0}\right]_{\bar{\boldsymbol{\lambda}}_0, \bar{\mathbf{P}}_0}\{\boldsymbol{\lambda}_0 - \bar{\boldsymbol{\lambda}}_0\} + \left[\frac{\partial \mathbf{f}}{\partial \tilde{\mathbf{P}}_0}\right]_{\bar{\boldsymbol{\lambda}}_0, \bar{\mathbf{P}}_0}\{\mathbf{P}_0 - \bar{\mathbf{P}}_0\} \tag{35}$$

$$E[\boldsymbol{\lambda}_1] = \bar{\boldsymbol{\lambda}}_0 + \mathbf{f}(\bar{\boldsymbol{\lambda}}_0, \bar{\mathbf{P}}_0) \tag{36}$$

$$\boldsymbol{\lambda}_1 - \mathbf{E}[\boldsymbol{\lambda}_1] = \{\boldsymbol{\lambda}_0 - \bar{\boldsymbol{\lambda}}_0\} + \left[\frac{\partial \mathbf{f}}{\partial \tilde{\boldsymbol{\lambda}}_0}\right]_{\bar{\boldsymbol{\lambda}}_0, \bar{\mathbf{P}}_0}\{\boldsymbol{\lambda}_0 -- \bar{\boldsymbol{\lambda}}_0\} + \left[\frac{\partial \mathbf{f}}{\partial \tilde{\mathbf{P}}_0}\right]_{\bar{\boldsymbol{\lambda}}_0, \bar{\mathbf{P}}_0}\{\mathbf{P}_0 - \bar{\mathbf{P}}_0\} \tag{37}$$

Since $\boldsymbol{\lambda}_0 = \bar{\boldsymbol{\lambda}}_0 = \mathbf{0}$ with probability 1 (w.p.1), it follows that:

$$\boldsymbol{\lambda}_1 - E[\boldsymbol{\lambda}_1] = \left[\frac{\partial f}{\partial \tilde{\mathbf{P}}_0}\right]_{\bar{\boldsymbol{\lambda}}_0, \bar{\mathbf{P}}_0} \cdot \{\mathbf{P}_0 - \bar{\mathbf{P}}_0\}, \quad \text{(w.p.1)} \tag{38}$$

$$\mathbf{W}_{1/0}^{\lambda\lambda} = E[\{\boldsymbol{\lambda}_1 - E[\boldsymbol{\lambda}_1]\}\{\tilde{\boldsymbol{\lambda}}_1 - E[\tilde{\boldsymbol{\lambda}}_1]\}] = \left[\frac{\partial \mathbf{f}}{\partial \tilde{\mathbf{P}}_0}\right]_{\bar{\boldsymbol{\lambda}}_0, \bar{\mathbf{P}}_0} \mathbf{W}_0^{\mathrm{P}} \left[\frac{\partial \tilde{\mathbf{f}}}{\partial \mathbf{P}_0}\right]_{\bar{\boldsymbol{\lambda}}_0, \bar{\mathbf{P}}_0} \tag{39}$$

$$\mathbf{W}_{1/0}^{\mathrm{P}\lambda} = E[\{\mathbf{P}_0 - \bar{\mathbf{P}}_0\}\{\tilde{\boldsymbol{\lambda}}_1 - E[\tilde{\boldsymbol{\lambda}}_1]\}] = \mathbf{0} \tag{40}$$

In view of (24) and (29)–(33), the optimal estimate $\hat{\mathbf{P}}_{1/1}$ supplied by these recursive-type expressions does coincide with the one achieved by adopting the weighted least-squares criterion (22) as error function.

The same rationale which led to the state estimate $\{\hat{\mathbf{P}}_{1/1}, \hat{\boldsymbol{\lambda}}_{1/1}\}$ as a function of the one-step-ahead prediction $\{\hat{\mathbf{P}}_{1/0}, \hat{\boldsymbol{\lambda}}_{1/0}\}$, can be applied now to derive the following update equation at a *generic* time instant t:

$$\begin{Bmatrix} \hat{\mathbf{P}}_{t/t} \\ \hat{\boldsymbol{\lambda}}_{t/t} \end{Bmatrix} = \begin{Bmatrix} \hat{\mathbf{P}}_{t/t-1} \\ \hat{\boldsymbol{\lambda}}_{t/t-1} \end{Bmatrix} + \mathbf{M}_{t/t-1}(\hat{\boldsymbol{\lambda}}_{t-1/t-1}, \hat{\mathbf{P}}_{t-1/t-1}, \hat{\mathbf{P}}_{t/t-1})\{\boldsymbol{\mu}_t - \mathbf{u}(\hat{\mathbf{P}}_{t/t-1})\} \tag{41}$$

$$\begin{aligned} \mathbf{M}_{t/t-1}&(\hat{\boldsymbol{\lambda}}_{t-1/t-1}, \hat{\mathbf{P}}_{t-1/t-1}, \hat{\mathbf{P}}_{t/t-1}) \\ &= \mathbf{W}_{t/t-1}(\hat{\boldsymbol{\lambda}}_{t-1/t-1}, \hat{\mathbf{P}}_{t-1/t-1})\tilde{\mathbf{S}}(\hat{\mathbf{P}}_{t/t-1}) \\ &\quad \times [\mathbf{S}(\hat{\mathbf{P}}_{t/t-1})\mathbf{W}_{t/t-1}(\hat{\boldsymbol{\lambda}}_{t-1/t-1}, \hat{\mathbf{P}}_{t-1/t-1})\tilde{\mathbf{S}}(\hat{\mathbf{P}}_{t/t-1}) + \mathbf{V}]^{-1} \end{aligned} \tag{42}$$

$$\mathbf{S}(\hat{\mathbf{P}}_{t/t-1}) = [\mathbf{L}(\hat{\mathbf{P}}_{t/t-1}) \mid \mathbf{0}] \tag{43}$$

The a posteriori covariance matrix $\mathbf{W}_{t/t}(\hat{\boldsymbol{\lambda}}_{t-1/t-1}, \hat{\mathbf{P}}_{t-1/t-1}, \hat{\mathbf{P}}_{t/t-1})$ is determined from the prediction uncertainty described by the covariance matrix $\mathbf{W}_{t/t-1}(\hat{\boldsymbol{\lambda}}_{t-1/t-1}, \hat{\mathbf{P}}_{t-1/t-1})$ as follows:

$$\begin{aligned} \mathbf{W}_{t/t}(\hat{\boldsymbol{\lambda}}_{t-1/t-1}, \hat{\mathbf{P}}_{t-1/t-1}, \hat{\mathbf{P}}_{t/t-1}) &= \mathbf{W}_{t/t-1}(\hat{\boldsymbol{\lambda}}_{t-1/t-1}, \hat{\mathbf{P}}_{t-1/t-1}) \\ -\mathbf{M}_{t/t-1}(\hat{\boldsymbol{\lambda}}_{t-1/t-1}, \hat{\mathbf{P}}_{t-1/t-1}, \hat{\mathbf{P}}_{t/t-1})&\mathbf{S}(\hat{\mathbf{P}}_{t/t-1})\mathbf{W}_{t/t-1}(\hat{\boldsymbol{\lambda}}_{t-1/t-1}, \hat{\mathbf{P}}_{t-1/t-1}) \end{aligned} \tag{44}$$

Equations (41)–(44) are the generalizations of (29)–(32). However, matrix $\mathbf{W}_{t/t-1}(\hat{\boldsymbol{\lambda}}_{t-1/t-1}, \hat{\mathbf{P}}_{t-1/t-1})$ can be directly computed from the data only at $t=1$. In general, a further recursive equation is necessary in order to evaluate this matrix on the basis of $\mathbf{W}_{t-1/t-1}(\hat{\boldsymbol{\lambda}}_{t-2/t-2}, \hat{\mathbf{P}}_{t-2/t-2}, \hat{\mathbf{P}}_{t-1/t-2})$. Precisely, in view of (6) and (7), the state prediction can be expressed as:

$$\begin{Bmatrix} \mathbf{P}_{t/t-1} \\ \boldsymbol{\lambda}_{t/t-1} \end{Bmatrix} = \begin{Bmatrix} \mathbf{P}_{t-1/t-1} \\ \boldsymbol{\lambda}_{t-1/t-1} + \mathbf{f}(\boldsymbol{\lambda}_{t-1/t-1}, \mathbf{P}_{t-1/t-1}) \end{Bmatrix} \tag{45}$$

By means of the linearization (10), one obtains:

$$\mathbf{W}_{t/t-1}(\hat{\boldsymbol{\lambda}}_{t-1/t-1}, \hat{\mathbf{P}}_{t-1/t-1}) = \mathbf{D}\mathbf{W}_{t-2/t-2}(\hat{\boldsymbol{\lambda}}_{t-2/t-2}, \hat{\mathbf{P}}_{t-2/t-2}, \hat{\mathbf{P}}_{t-1/t-2})\tilde{\mathbf{D}} \tag{46}$$

where:

$$\mathbf{D} \equiv \left[\begin{array}{c|c} \mathbf{0} & \mathbf{I} \\ \hline \dfrac{\partial \mathbf{f}}{\partial \tilde{\mathbf{P}}_t} & \dfrac{\partial \mathbf{f}}{\partial \tilde{\boldsymbol{\lambda}}_t} \end{array}\right]_{\hat{\boldsymbol{\lambda}}_{t-1/t-1},\ \hat{\mathbf{P}}_{t-1/t-1}} \tag{47}$$

Recursions (41)–(44) are the extended Kalman-filter equations for the structural elastoplasticity problem in question. Since the classical Kalman papers [12, 13], the theory and applications of Kalman filtering are the subject of an extensive literature. Recently, a special issue of the *IEEE Transactions on Automatic Control* [19] has been devoted entirely to the applications of Kalman filtering. As general basic references, the interested reader is referred to refs. 1, 6 and 7.

A number of Kalman-filtering techniques have been suggested to cope with system nonlinearities. Use is made here of the so-called 'iterated' extended Kalman filter [1, 7], characterized by the fact that the iteration from t to $t+1$ is performed through a number of intermediate iterations by subsequent linearizations about the new estimate. Precisely, let $\{\mathbf{P}^{(i)}_{t/t}, \boldsymbol{\lambda}^{(i)}_{t/t}\}$, $i = 1, 2, \ldots, \nu_t$, be the state estimate at time t and subiteration (i). Then eqns. (41)–(43) are substituted by:

$$\begin{Bmatrix} \hat{\mathbf{P}}^{(i)}_{t/t} \\ \hat{\boldsymbol{\lambda}}^{(i)}_{t/t} \end{Bmatrix} = \begin{Bmatrix} \hat{\mathbf{P}}_{t/t-1} \\ \hat{\boldsymbol{\lambda}}_{t/t-1} \end{Bmatrix} + \mathbf{M}_{t/t-1}(\hat{\boldsymbol{\lambda}}_{t-1/t-1}, \hat{\mathbf{P}}_{t-1/t-1}, \hat{\mathbf{P}}^{(i)}_{t/t}) \times \{\boldsymbol{\mu}_t - \mathbf{u}(\hat{\mathbf{P}}^{(i-1)}_{t/t}) - \mathbf{L}(\hat{\mathbf{P}}^{(i-1)}_{t/t})\{\hat{\mathbf{P}}_{t/t-1} - \hat{\mathbf{P}}^{(i-1)}_{t/t}\}\} \tag{48}$$

$$\left.\begin{aligned} \mathbf{M}_{t/t-1}(\hat{\boldsymbol{\lambda}}_{t-1/t-1}, \hat{\mathbf{P}}_{t-1/t-1}, \hat{\mathbf{P}}_{t/t}) &= \mathbf{W}_{t/t-1}(\hat{\boldsymbol{\lambda}}_{t-1/t-1}, \hat{\mathbf{P}}_{t-1/t-1})\mathbf{S}(\hat{\mathbf{P}}^{(i-1)}_{t/t}) \\ &\times [\mathbf{S}(\hat{\mathbf{P}}^{(i-1)}_{t/t})\mathbf{W}_{t/t-1}(\hat{\boldsymbol{\lambda}}_{t-1/t-1}, \hat{\mathbf{P}}_{t-1/t-1})\tilde{\mathbf{S}}(\hat{\mathbf{P}}^{(i-1)}_{t/t}) + \mathbf{V}]^{-1} \end{aligned}\right\} \tag{49}$$

$$\mathbf{S}(\hat{\mathbf{P}}^{(i-1)}_{t/t}) = [\mathbf{L}(\hat{\mathbf{P}}^{(i-1)}_{t/t}) \mid \mathbf{0}] \tag{50}$$

where:

$$\hat{\mathbf{P}}^{(0)}_{t/t} = \hat{\mathbf{P}}_{t/t-1}, \qquad \hat{\mathbf{P}}^{(\nu_t)}_{t/t} = \hat{\mathbf{P}}_{t/t}, \qquad \hat{\boldsymbol{\lambda}}^{(\nu_t)}_{t/t} = \hat{\boldsymbol{\lambda}}_{t/t} \tag{51}$$

For each t, the number ν_t of subiterations can be established by comparing subsequent estimates of the parameters. In the examples of Section 4, the subiteration cycle is terminated when the variation of each parameter becomes $\leqslant 10^{-3}$ times the parameter value. The Kalman-filtering procedure ends when all data have been processed, i.e. at $t = T$ and $i = \nu_T$.

4. EXAMPLES AND COMPARISONS

Some of the numerical experiments carried out to test the proposed identification procedure concerned the frame and loading condition shown in Fig. 1(a), with the 12 marked critical sections at the column ends. These are assumed to exhibit the symmetric plastic behaviour depicted in Fig. 1(b): substantially perfectly plastic, with very small

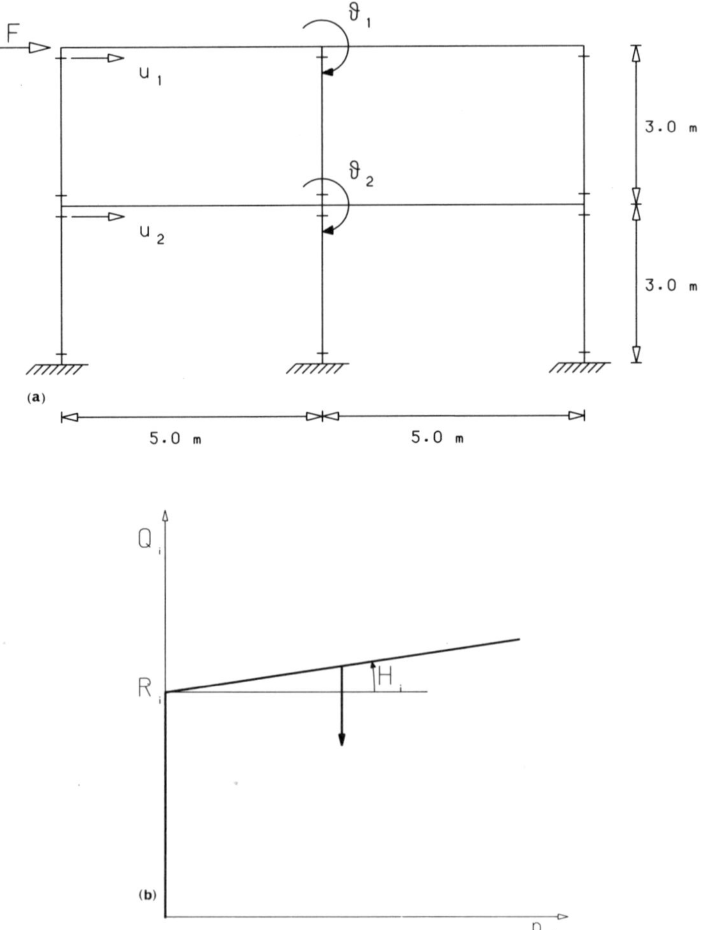

FIG. 1. (a) Frame used for the numerical tests and (b) assumed constitutive law for the critical sections.

fictitious hardening H introduced to improve the numerical performance of the procedure. Neglecting shear and extensional deformations, the only sectional characteristics which intervene in the analysis are elastic bending stiffness $E\mathcal{J}$, and yield limits R. They have been assigned the following values in kN m^2 and kN m, respectively: left column: $E\mathcal{J}_1 = 2100$; $R_1 = 40$; mid column: $E\mathcal{J}_2 = 1050$; $R_2 = 30$; right column: $E\mathcal{J}_3 = 2100$, $R_3 = 40$ horizontal girders $E\mathcal{J} = 8400$; $R = 80$. The analysis under the load increasing from zero up to $F_{max} = 72$ kN ($F_E = 61.14$ being the elastic limit load) and then decreasing to $F_{min} = -72$ kN was performed on the basis of a finite element discretization of the elastic beams between critical sections, by means of the code STRUPL 1 [4]. This analysis was taken as a simulated experiment, i.e. the displacements and rotations displacements thus calculated were regarded as measured quantities μ. Two displacements are plotted in Fig. 2, which specifies also in abscissae the eight load levels at which 'measurements' are made. Data and main results of three identification examples are presented in Table I.

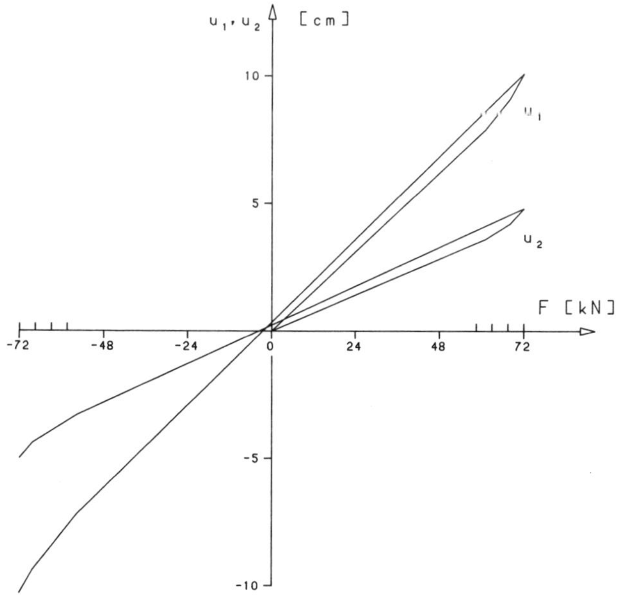

FIG. 2. Displacements u_1 and u_2 versus load in the simulated experiments.

TABLE I
Data and results of three tests of yield limit identification

Case	*Measured displacements*	*Initial estimates*			*Standard deviations*			*Variances of measure errors*	*Final estimates*			*Final standard deviations*		
		P_1	P_2	P_3	σ_1	σ_2	σ_3		P_1	P_2	P_3	σ_1	σ_2	σ_3
1	u_1u_2	32	24	32	10	10	10	10^{-7}	39·4	29·7	39·3	7·1	0·84	7·1
2	u_1u_2	32	24	32	10	10	10	10^{-5}	39·1	29·5	39·1	7·4	2·6	7·4
3	$\begin{Bmatrix} u_1u_2 \\ \theta_1\theta_2 \end{Bmatrix}$	32	24	32	10	10	10	10^{-7}	39·8	29·6	39·1	5·0	0·26	5·1

Case 1

The two measured displacements u_1, u_2 are associated with the variance $\sigma^2 = 10^{-7}\,m^2$ corresponding to a standard deviation $\sigma = 0{\cdot}316$ mm. Roughly speaking, this means considering a maximum instrument error of about ± 1 mm: in fact, if a Gaussian distribution is assumed, the probability that the error takes a value within the range $\pm 3\sigma \simeq \pm 1$ mm is over 99%. The parameters to identify are the bending moments P_1, P_2, P_3 of the left, central and right columns, respectively. The a priori information on them consists of the best estimated values (in kN) and of the standard deviations given in Table I.

A typical estimation path is illustrated in Fig. 3, where each triangle marks the end of a recursive step (i.e. the final estimate after using the available information at a given load factor). The plot of Fig. 3 concerns the estimate of the yield limit for the mid column, R_2.

Table II gives some details about the whole estimation process. The second column of the table specifies the number of iterations required in order to account for nonlinearities at each load factor. The subsequent columns indicate the parameters and the standard deviations estimated after processing each set of measurements sequentially. The figures in brackets (last three columns of the table) refer to the

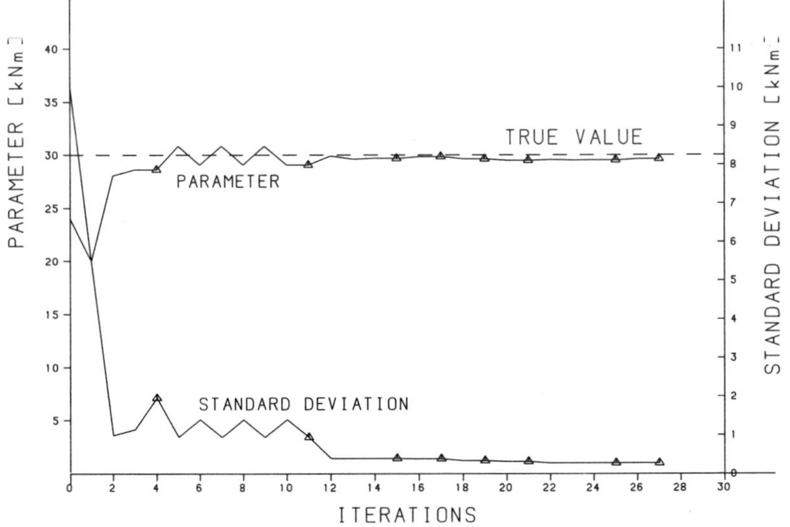

FIG. 3. Estimation path for parameter P_2 and its standard deviation (case 1).

TABLE II
Estimation process for the first case

Load factor	*Iterations*	P_1 (*kN m*)	P_2 (*kN m*)	P_3 (*kN m*)	σ_1 (*kN m*)	σ_2 (*kN m*)	σ_3 (*kN m*)
1·3	7	37·2	28·6	37·2	7·2 (0)	5·2 (0)	7·3 (0)
1·4	7	38·6	29·1	38·6	7·1 (1)	2·8 (2)	7·2 (1)
1·5	3	38·6	29·9	38·4	2·1 (1)	1·2 (2)	2·2 (1)
1·6	2	38·8	30·0	38·7	2·1 (2)	1·2 (4)	2·2 (2)
−1·3	2	39·1	29·6	39·1	2·1 (0)	1·0 (2)	2·1 (0)
−1·4	2	39·2	29·6	39·1	2·1 (0)	0·9 (2)	2·1 (1)
−1·5	2	39·3	29·6	39·1	2·1 (0)	0·8 (2)	2·1 (1)
−1·6	2	39·4	29·7	39·3	2·1 (2)	0·8 (4)	2·1 (1)

numbers of active yield modes corresponding to each parameter along the identification process.

Case 2

The assumed error variances are two orders of magnitude greater than in Case 1. As a consequence, an increased uncertainty is found in the parameter estimates (higher values of their final standard deviations).

Case 3

The error variances coincide with the variances of the first case. However, the set of 'measured' variables now includes also the rotations θ_1 and θ_2 (in radians) at the central nodes, as shown in Fig. 1(a). Such an increase of the available information results in an improvement in the estimation accuracy of the parameters.

It can be noticed that in Table I the estimates of the flexural strength P_2 of the central columns are more accurate than those for the other columns as the relevant variances are much smaller than those of P_1 and P_3. This is explained by the circumstance that the loading process 'activates' a higher number of yielding modes related to P_2 than to P_1 and P_3 (see figures in brackets in Table II) and, hence, 'more information' concerning P_2 than P_1 and P_3 is available and processed in the identification procedure.

A numerical test has been performed with the purpose of comparing the Kalman-filter algorithm with a more traditional estimation technique based on single set of measurements. It has been assumed that displacements u_1 and u_2 have been measured for $F = 72$ kN only. The parameter estimation process has been carried out on this basis starting from the same initial data of the first case (Table I). The following final estimates have been obtained: $P_1 = 38{\cdot}2$ kN, $P_2 = 32{\cdot}1$ kN, $P_3 = 38{\cdot}2$ kN. By comparing these values with the final estimates of Table I for the first case, the results of the Kalman filter approach appear to be definitely superior. Clearly, this is due to the fact that a sequence of data collected along a well chosen load history capable of activating a number of yield modes contains more information on these modes than a single set of measures. In principle, a sequence of data on the evolution of a system can be used for stochastic identification purposes in a 'batch' way rather than recursively, as in above. The batch approach of Section 3.4 can be easily extended to multiple sets of measurements and would lead to an iterated linear estimator fully analogous to eqn. (27), except that $\{\boldsymbol{\mu}_1 - \mathbf{u}(\mathbf{P}_0, \hat{\mathbf{P}}_i)\}$ should be replaced by the much larger vector containing all vectors $\{\boldsymbol{\mu}_k - \mathbf{u}_k(\mathbf{P}_0, \hat{\mathbf{P}}_i)\}$ with index k covering over all load conditions under which the measurements had been performed. The number of columns of the mapping matrix $\mathbf{M}(\bar{\mathbf{P}}_0)$ and the order of the matrix to invert in eqn. (25) would greatly increase. The present recursive Kalman filter method requires instead the inversion of much smaller matrices at equal amounts of measure sets to process.

5. CONCLUSIONS

In this paper a recursive, Kalman-filter type set of relations has been established for the stochastic estimation of yield limits in an elastoplastic structural model on the basis of a sequence of experimental data concerning the static response to given loads and of a priori 'best guesses' on the parameters to identify. This typical 'inverse' problem in structural elastoplasticity has been solved with reference to a simple frame and computer-simulated experiments and thus the feasibility of the method has been demonstrated (not yet, however, its practicality for real life large-size problems). In general, the method, though conceptually more complex, computationally compares well with respect to 'batch' ways of using available information for identification purposes in the same context.

ACKNOWLEDGEMENTS

The study expounded in this paper is part of a research project supported by MPI.

REFERENCES

1. ANDERSON, B. D. O. and J. B. MOORE, *Optimal Filtering*, Prentice-Hall, New York, 1979.
2. CARMICHAEL, D. G. The state estimation problem in experimental structural mechanics, *Third Int. Conf. on Applications of Statistics and Probability to Civil Engineering*, University of New South Wales, Jan. 1979.
3. CIVIDINI, A. M., G. MAIER, and A. NAPPI. Parameter estimation of a static geotechnical model using a Bayes' approach, *Int. J. Rock Mech. Mining Sci.*, **20,** (1983), 215–226.
4. COHN, M. Z., F. ERBATUR, and A. FRANCHI. STRUPL 1 User's Manual, Solid Mechanics Division, University of Waterloo, Waterloo, Ontario, Canada, 1982.
5. COHN, M. Z. and G. MAIER. (Eds) Engineering plasticity by mathematical programming, *Proc. NATO Advanced Study Inst.*, University of Waterloo, Pergamon Press, New York, 1979.
6. EIKHOFF, P. *System Identification*, John Wiley, Chichester, 1983.
7. GELB, A. (Ed.) *Applied Optimal Estimation*, MIT Press, Cambridge, Mass., 1974.
8. MAIER, G., F. GIANNESSI and A. NAPPI. Indirect identification of yield limits by mathematical programming, *Engineering Structures*, **4** (1982), 86–98.
9. GIODA, G. and G. MAIER. Direct search solution of an inverse problem in elastoplasticity: identification of cohesion, friction angle and in situ stress by pressure tunnel tests, *Int. J. Num. Meth.*, **15,** (1980), 1823–1848.
10. HART, G. C. and J. T. P. YAO. System identification in structural dynamics, *J. Eng. Mech. Div., Proc. ASCE,* **103** (1977), 1089–1104.
11. IBAÑEZ, P. Identification of dynamic parameters of linear and non-linear structural models from experimental data, *Nucl. Eng. Design*, **25** (1972), 30.
12. KALMAN, R. E. A new approach to linear filtering and prediction problems, *Trans. ASME, J. Basic Engng*, **82** (1960), 35–45.
13. KALMAN, R. E. and R. S. BUCY. New results in linear filtering and prediction theory, *Trans. ASME. J. Basic Engng*, **83**D (1961), 95–108.
14. MAIER, G. A matrix structural theory of piecewise-linear plasticity with interacting yield planes, *Meccanica*, **5** (1970), 55–66.
15. MAIER, G. and A. NAPPI. On the unified framework provided by mathematical programming to plasticity. In: *Mechanics of Material Behaviour*, (Ed. G. J. Dvorak and R. T. Shield), Elsevier, Amsterdam, 1983.
16. MAIER, G., A. NAPPI and A. CIVIDINI. Statistical identification of yield limits in piecewiselinear structural models. In: *Proc. Int. Conf. on Computational Methods and Experimental Measurements*, ISCME (Washington, D.C., July 1982), (Ed. G. C. Keramidas and C. A. Brebbia), Springer, Berlin, 1982, 812–829.
17. NATKE, H. G. *Einführung in Theorie und Praxis der Zeitreben-und Modalanalyse*, Vieweg, Braunschweig, 1983.
18. SHINOZUKA, M., C. YUN and R. VAIKATIS. Dynamic analysis of fixed

offshore structures subjected to wire generated waves, *J. Struct. Mech.*, **5** (1977), 135–146.
19. *IEEE Transactions on Automatic Control*, **AC-28,** No. 3, March 1983.

APPENDIX: A PROOF OF EQN. (26)

The problem is to estimate vector **P** by observing the variable vector

$$\boldsymbol{\mu}_1 = \mathbf{u}(\bar{\mathbf{P}}_0) + \mathbf{L}(\bar{\mathbf{P}}_0)\{\mathbf{P} - \bar{\mathbf{P}}_0\} + \mathbf{n}_1 \tag{52}$$

where: $\bar{\mathbf{P}}_0$ expected value of **P**; $\mathbf{n}_1$ = random variable with zero expected value and covariance matrix **V**; $\mathbf{n}_1$ and **P** are assumed uncorrelated, i.e. $E[\{\mathbf{P} - \bar{\mathbf{P}}_0\}\tilde{\mathbf{n}}_1] = \mathbf{0}$.

Denoting by $\boldsymbol{\Lambda}_{\mu\mu}$ and $\boldsymbol{\Lambda}_{P\mu}$ the covariance matrix of $\boldsymbol{\mu}_1$ and the cross-covariance matrix between **P** and $\boldsymbol{\mu}_1$, respectively, we have:

$$\mathbf{E}[\boldsymbol{\mu}_1] = \mathbf{u}(\bar{\mathbf{P}}_0) \tag{53}$$

$$\Lambda_{\mu\mu} = E[\{\boldsymbol{\mu}_1 - E[\boldsymbol{\mu}_1]\}\{\tilde{\boldsymbol{\mu}}_1 - E[\tilde{\boldsymbol{\mu}}_1]\}] = \mathbf{L}(\bar{\mathbf{P}}_0)\mathbf{W}_0^P\tilde{\mathbf{L}}(\bar{\mathbf{P}}_0) + \mathbf{V} \tag{54}$$

$$\Lambda_{P\mu} = E[\mathbf{P} - E[\mathbf{P}]\}\{\tilde{\boldsymbol{\mu}}_1 - E[\tilde{\boldsymbol{\mu}}_1]\}] = \mathbf{W}_0^P\tilde{\mathbf{L}}(\bar{\mathbf{P}}_0) \tag{55}$$

In view of (3a) and (4a), estimator reads

$$\hat{\mathbf{P}}(\bar{\mathbf{P}}_0) = \bar{\mathbf{P}}_0 + \boldsymbol{\Lambda}_{P\mu}\boldsymbol{\Lambda}_{\mu\mu}^{-1}\{\boldsymbol{\mu}_1 - \mathbf{u}(\bar{\mathbf{P}}_0)\} \tag{56}$$

The covariance matrix of the a posteriori error $\mathbf{P} - \hat{\mathbf{P}}(\bar{\mathbf{P}}_0)$ can now be shown to be given by (26). In fact:

$$\begin{aligned} &E[\{\mathbf{P} - \hat{\mathbf{P}}(\bar{\mathbf{P}}_0)\}\{\tilde{\mathbf{P}} - \tilde{\hat{\mathbf{P}}}(\bar{\mathbf{P}}_0)\} \\ &\quad = E[\{\{\mathbf{P} - \bar{\mathbf{P}}_0\} - \boldsymbol{\Lambda}_{P\mu}\boldsymbol{\Lambda}_{\mu\mu}^{-1}\{\boldsymbol{\mu}_1 - \mathbf{u}(\bar{\mathbf{P}}_0)\}\}\{\{\tilde{\mathbf{P}} - \tilde{\bar{\mathbf{P}}}_0\} - \{\tilde{\boldsymbol{\mu}}_1 - \tilde{\mathbf{u}}(\bar{\mathbf{P}}_0)\}\boldsymbol{\Lambda}_{\mu\mu}^{-1}\tilde{\boldsymbol{\Lambda}}_{P\mu}\}] \\ &\quad = \mathbf{W}_0^P + \boldsymbol{\Lambda}_{P\mu}\boldsymbol{\Lambda}_{\mu\mu}^{-1}\boldsymbol{\Lambda}_{\mu\mu}\boldsymbol{\Lambda}_{\mu\mu}^{-1}\tilde{\boldsymbol{\Lambda}}_{P\mu} - 2\boldsymbol{\Lambda}_{P\mu}\boldsymbol{\Lambda}_{\mu\mu}^{-1}\tilde{\boldsymbol{\Lambda}}_{P\mu} = \mathbf{W}_0^P - \boldsymbol{\Lambda}_{P\mu}\boldsymbol{\Lambda}_{\mu\mu}^{-1}\tilde{\boldsymbol{\Lambda}}_{P\mu} \end{aligned} \tag{57}$$

Substituting (54) and (55) in (57), eqn. (26) is arrived at.

18

Singular Perturbations in Plasticity: Application to Plane-strain Slip-line Fields†

M. SAYIR and H. FROMMER

Institut für Mechanik, ETH, Zürich, Switzerland

ABSTRACT

An application of the method of singular perturbations to problems involving plane-strain slip-line fields in ideal plasticity is presented. We start with a problem involving special geometrical and dynamical boundary conditions and leading to a particularly simple standard solution. We consider this solution as 'zeroth' order approximation of a whole family of further problems with slightly modified geometrical and dynamical conditions, using the modifications as perturbation parameters leading to asymptotic solutions of the modified problems. To illustrate the general method, its explicit application to compression of a rigid plastic block between two flat plates and to sheet drawing is shortly discussed.

1. INTRODUCTION

There is now a well-known catalogue of simple solutions for problems involving plane-strain slip-line fields in ideal plasticity with possible applications in metal forming and soil mechanics [6, 7, 8]. Several analytical and semi-numerical methods have also been proposed in order to extend this catalogue to more involved cases [1, 2, 9]. In the following, an application of the method of singular perturbations to the above-mentioned class of problems will be presented. This application

† This paper is based on the doctoral thesis [3] OF H.F. M.S. acted as his adviser.

should be of interest not only because it leads to analytical approximate solutions which can be used as a basis for further, more extensive numerical studies, but also because it illustrates some basic aspects of geometrical and multiple parameter perturbations applied to systems of quasi-linear hyperbolic differential equations.

2. METHOD OF SOLUTION

We start with a problem involving very special geometrical and dynamical boundary conditions and leading to a particularly simple and well-known standard solution. We modify slightly the boundary conditions and the geometrical shape, so that new families of problems can be defined. We then apply the method of singular perturbations with small dimensionless parameters corresponding to the slight changes introduced and obtain analytical asymptotic solutions for each family of problems.

Consider for example the compression of a rigid plastic block of height h between two rigid plates of width $w = h$ under plane-strain conditions (Fig. 1).

A very simple slip-line field gives the exact solution for the yield point and the onset of plastic deformations. The slip-lines AB, BC are discontinuity lines of the velocities. In the triangular domain ABC with uniform distribution of velocity and stress, the material moves downward (in A'BC' upwards) whereas BCC' and AA'B move sideways with

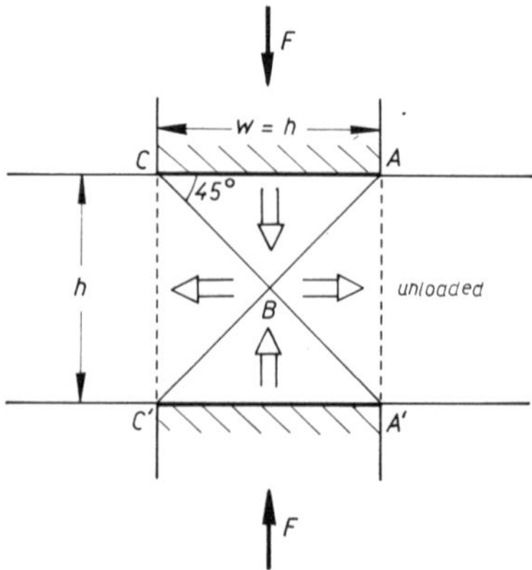

FIG. 1. Basic problem of 'zeroth' order.

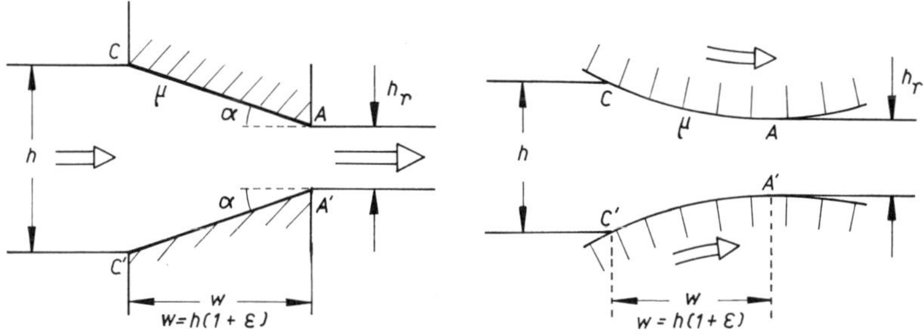

FIG. 2. Sheet drawing.

FIG. 3. Strip rolling.

the same value of the velocity. CC′ and AA′ may be considered as discontinuity lines for the stress, separating the unloaded regions of the block from the regions with uniformly distributed stress. This simple solution may be used as a starting point for several families of problems. In a first step, the width of the plates can be increased or reduced by an amount εh with $\varepsilon \ll 1$, maintaining the contact between block and plates frictionless. In a second step friction can be introduced with a coefficient μ which may be assumed to be of the same order as ε. In the latter case we obtain a two-parameter perturbation.

In a further step we can modify the shape of the block and obtain a problem of sheet-drawing with the angle of the die as a third parameter of perturbation (Fig. 2).

The wedge-shaped die can even be allowed to be curved so that the asymptotic solutions can be adapted to problems of strip rolling (Fig. 3).

Using other simple cases as 'zeroth' order approximations one can also obtain asymptotic solutions for a number of further families of problems. For example the two well-known special cases for deep drawing and extrusion of Fig. 4 can be used as basic 'zeroth' order solutions for corresponding more involved applications with modified geometry.

One may even go further and introduce with corresponding small parameters new physical effects such as strain rate dependence, strain hardening, elasticity, temperature dependence, etc. (this aspect will be reported elsewhere).

We now illustrate the general method by shortly discussing two explicit solutions based on Fig. 1 as 'zeroth' approximation (see ref. 3 for more details).

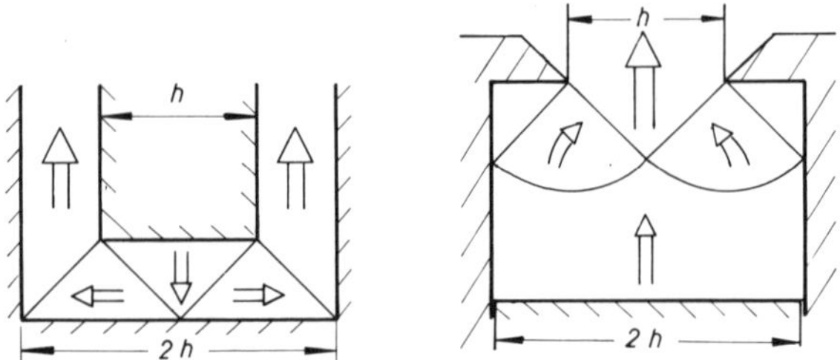

FIG. 4. Deep drawing and extrusion.

3. COMPRESSION OF A RIGID PLASTIC BLOCK OF HEIGHT h BETWEEN TWO RIGID PLATES OF WIDTH $w \gtrless h$

3.1. Basic Relations

The perturbation problem with $\varepsilon := (w-h)/h$ and the coefficient of friction $\mu =: \mu^* \varepsilon$ as small parameters, is formulated in the characteristic co-ordinates of the basic problem of Fig. 1 duly modified to allow for $w \gtrless h$. We designate the characteristic co-ordinates by (ξ, η), where $\eta =$ const. corresponds as usual to a slip-line of the first family and $\xi =$ const. to a slip-line of the second family (see for example ref. 6). The explicit shape of the orthogonal slip-lines will be given analytically in a cartesian frame of reference (x, y) by two functions $x(\xi, \eta)$, $y(\xi, \eta)$. Calling ϕ the angle between a first slip-line $\eta =$ const. and the x-axis, one determines the functions $x(\xi, \eta)$, $y(\xi, \eta)$ by integrating

$$\begin{aligned} y_{,\xi} - \tan\phi \,.\, x_{,\xi} &= 0 \\ y_{,\eta} + \cot\phi \,.\, x_{,\eta} &= 0 \end{aligned} \qquad (1)$$

(partial differentiation is denoted by ','). In Fig. 5, the x- and y-axes of the cartesian frame have been chosen along the lines of symmetry of our problem. Besides, non-dimensional co-ordinates have been introduced, so that the total (non-dimensional) thickness of the block is 2 and the total width of the plate 2 $(1+\varepsilon)$, where $\varepsilon \gtrless 0$.

In order to integrate (1), one needs the function $\phi(\xi, \eta)$. Equilibrium and yield conditions lead to

$$\begin{aligned} (p+2\phi)_{,\xi} &= 0 \\ (p-2\phi)_{,\eta} &= 0 \end{aligned} \qquad (2)$$

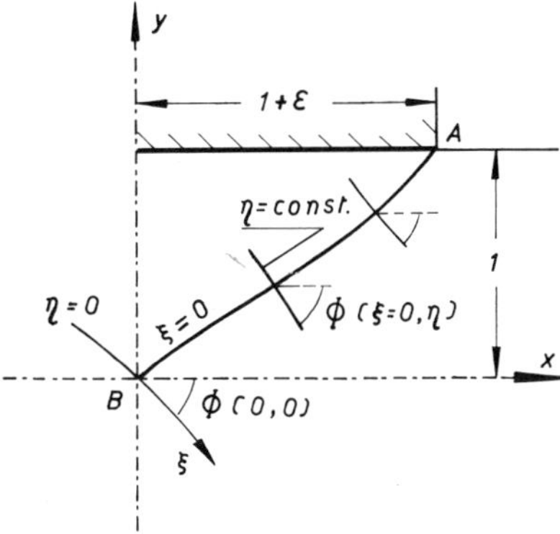

FIG. 5. Modified problem of Fig. 1.

where 'p' is the compressive stress component along the slip-lines, duly non-dimensionalized by using the yield stress in shear as reference. The velocity field in the deforming zone satisfies Geiringer's kinematic relations

$$u_{,\xi} - v \cdot \phi_{,\xi} = 0 \tag{3.1}$$

$$v_{,\eta} + u \cdot \phi_{,\eta} = 0 \tag{3.2}$$

where u and v are the velocity components tangential to the slip-lines.

The sets (1), (2), (3) constitute a system of quasilinear hyperbolic differential equations for the unknown functions x, y, p, ϕ, u, v of the characteristic co-ordinates ξ, η. Unfortunately, the boundary conditions of our problem (as in almost all problems of this type) do not correspond to one of the three classical boundary value problems of Riemann for which straightforward solutions may be found. The usual approach uses a certain amount of guesswork which has to be confirmed a posteriori. One of the advantages of the perturbation approach presented here is that it is perfectly deductive and leads by consistent logical steps to the desired solution.

3.2. Regular Perturbation

The unknown functions ϕ, p, u, v, x, y are expanded in terms of ε, for example

$$\phi(\xi, \eta, \varepsilon) = \phi_0(\xi, \eta) + \varepsilon\phi_1(\xi, \eta) + \varepsilon^2\phi_2(\xi, \eta) + \cdots$$

TABLE I

ϕ_0	p_0	u_0	v_0		x_0	y_0
$-\pi/4$	1	1	$\xi<0$	-1	$\frac{1}{2}(\xi+\eta)$	$\frac{1}{2}(-\xi+\eta)$
			$\xi>0$	1		

The terms of 'zeroth' order are known from the basic problem of Fig. 1, they are given explicitly in Table I. Clearly, the characteristic line corresponding to $\xi=0$ is a line of discontinuity for the tangential velocity v (line AB of Figs. 1 or 5).

The basic equations (1), (2) and (3) as well as the corresponding boundary conditions must be satisfied for each set of coefficients of ε, $\varepsilon^2, \ldots,$ if the limiting process $\varepsilon \to 0$ is considered. For example, the following set of relations is obtained for the coefficients of first order:

$$\begin{aligned} (p_1+2\phi_1)_{,\xi} &= 0; \qquad (p_1-2\phi_1)_{,\eta}=0; \\ u_{1,\xi}-v_0\,.\,\phi_{1,\xi}; &\qquad v_{1,\eta}+u_0\,.\,\phi_{1,\eta}=0; \\ (y_1+x_1)_{,\xi}=2\phi_1 x_{0,\xi}; &\qquad (y_1-x_1)_{,\eta}=2\phi_1 x_{0,\eta} \end{aligned} \tag{4}$$

The corresponding boundary conditions of first order can easily be formulated at the axes $x=0$, $y=0$, where, due to symmetry, shear stresses and transversal velocity components vanish. This leads to

$$\begin{aligned} \phi_1=0, \quad u_1+v_1=0, \quad x_1=0, \quad \xi=-\eta \quad \text{on} \quad By \\ \phi_1=0, \quad u_1-v_1=0, \quad y_1=0, \quad \xi=\eta \quad \text{on} \quad Bx \end{aligned} \tag{5}$$

At the contact line $y=1$ between plate and block we assume that the frictional coefficient μ is of the same order of magnitude as ε and set $\mu=|\varepsilon|\,\mu^*$. The boundary conditions of first order at this interface AC then become

$$\begin{aligned} &\phi_1=-\mu^* \quad \text{if} \quad v_{x1}=u_1+v_1+2\phi_1\neq 0 \\ &|\phi_1|\leq \mu^* \quad \text{if} \quad v_{x1}=0 \\ &u_1-v_1=0, \qquad y_1=0, \qquad \eta=2+\xi \end{aligned} \tag{6}$$

At the line of discontinuity $\xi=0$ the quantities ϕ_1, p_1, u_1 have to be continuous, and somewhere in the domain $\xi>0$ the velocity field should become uniform and correspond to the horizontal translation of the rigid part of the block. Thus (using also the incompressibility of the material)

$$u_1-v_0\phi_1=1 \tag{7}$$

along the curve separating the deforming domain from the rigid one. The total force should also vanish on this curve.

It can be shown by straightforward arguments that conditions (4)–(7) lead to an unacceptable discontinuity in u_1 along $\xi=0$. Thus, the regular perturbation expansions mentioned above become singular in the vicinity of $\xi=0$. Therefore a boundary layer develops along this line.

3.3. Singular Perturbations, Boundary Layers and Domains

We introduce a 'boundary layer variable'

$$\xi^* := \xi/|\varepsilon| \tag{8}$$

'magnifying' the domain around AB, and consider here ϕ, p, u, v, x, y as functions of (ξ^*, η). The boundary layer is sharply limited on both sides of $\xi=0$ by slip-lines given as

$$\xi=\xi_a=\xi_{a1}|\varepsilon|+\xi_{a2}\varepsilon^2+\ldots \qquad (\xi_a<0)$$
$$\xi=\xi_b=\xi_{b1}|\varepsilon|+\xi_{b2}\varepsilon^2+\ldots \qquad (\xi_b>0)$$

Thus the characteristic 'thickness' of the boundary layer around $\xi=0$ would be $\xi_b+|\xi_a|$, a value which must be determined eventually from the full asymptotic solution. The situation can be visualized in the characteristic plane $\xi\eta$ as shown in Fig. 6. Whereas the actual shape of the slip-lines in the Bxy plane of Fig. 5 will be modified with each step of approximation, the characteristic image in the $\xi\eta$-plane of Fig. 6 remains essentially the same. Only the characteristic 'thickness' of the boundary layer gets more accurate as approximations of higher order are considered. Along the limits of the boundary layer, 'matching' with the regular solution of (3.2) is obtained by requiring continuity for ϕ,

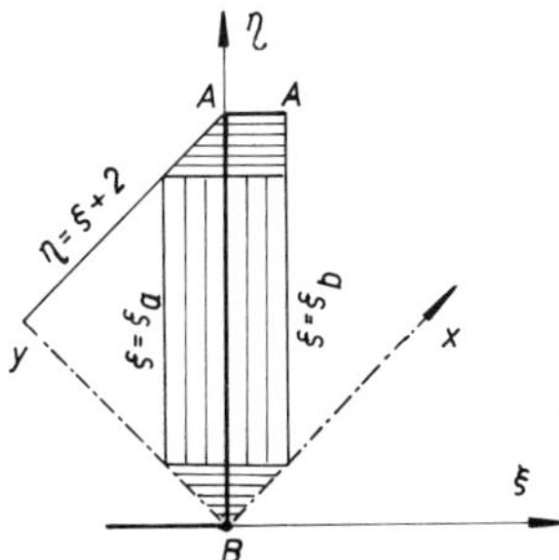

FIG. 6. Mapping into the characteristic plane.

p, u whereas v may be discontinuous. The basic relations to be integrated are identical to (4), with only ξ replaced by ξ^*. Once continuity conditions along $\xi = \xi_a$, ξ_b are satisfied, it can be shown by straightforward arguments that neither boundary conditions (5) nor (6) and (7) can be fulfilled unless new boundary domains around the corners B and A are introduced. Thus, some additive integration functions remain undetermined and can be used to satisfy matching conditions with the boundary domains around A and B. To 'magnify' the region around B, a further boundary domain co-ordinate

$$\eta^* := \eta / |\varepsilon| \tag{9}$$

is needed in addition to (8). The basic equations in ξ^* and η^* are again the same. Boundary conditions (5) can now be fulfilled as well as the continuity requirements on ϕ, p, v (u may be discontinuous) along the characteristic $\eta^* = \xi_b$ $(= -\xi_a)$. Thus, matching with the previous boundary layer solution is ensured.

A third boundary layer around A must be considered in order to satisfy (6) and (7). The corresponding boundary domain co-ordinate is

$$\eta^{**} := (\eta - 2)/|\varepsilon| \tag{10}$$

the corner A having the characteristic co-ordinate $\eta = 2$ for all $0 \leqslant \xi \leqslant \xi_b$ (see Fig. 6). The procedure of solution is similar to the one in the boundary layer around B. All boundary and continuity requirements along the lines of discontinuity $\xi = \xi_a$, 0, ξ_b; $\eta = \xi_b$, $2 - \xi_b$ can now be satisfied and the complete asymptotic solution B is obtained. There are three possibilities:

(1): $\varepsilon < 0$. The block is thicker than the width of the plate. For all μ the block adheres to the plate at the interface AC (no slippage). The solution of first order confirms results given by Prandtl and Hill [6].

(2): $\varepsilon > 0$, $\mu \geqslant \mu_m$. The solution for wider plates with 'sufficient friction' to provide for complete adhesion (no slippage) along AC given by Hill [6] is confirmed. The minimum value μ_m of the coefficient of friction can be determined thanks to the present asymptotic procedure and depends on the width/thickness ratio. Calculations including second order terms in ε lead to

$$\mu_m = \tfrac{1}{2}\varepsilon - \tfrac{3}{8}\varepsilon^2 + \ldots \tag{11}$$

(3): $\varepsilon > 0$, $0 \leqslant \mu \leqslant \mu_m$. In this case with weaker friction, slippage between block and plate occurs in a region AM of the interface (and in

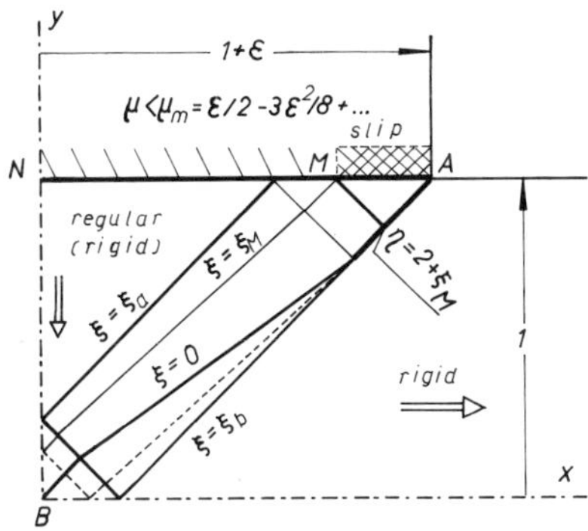

FIG. 7. 'Boundary domains' of case (3), linear approximation.

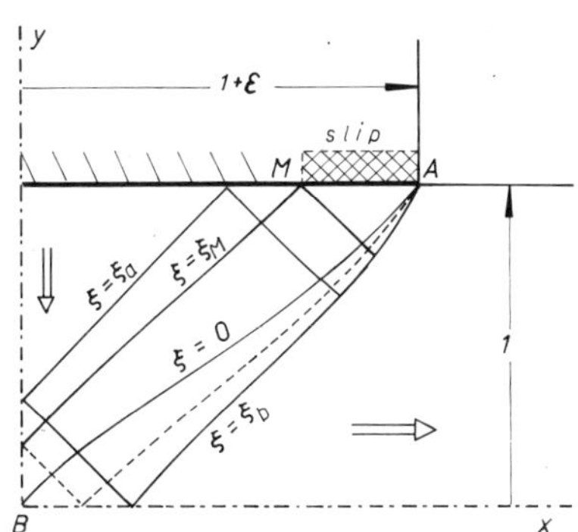

FIG. 8. Higher order correction of Fig. 7.

its symmetrical counterpart CN). The interior part MN is free of slippage (Fig. 7). The complete solution for the first order coefficients is given in Table II. The corresponding special slip-lines are indicated in Fig. 7.

At this first level of approximation the characteristic slip-lines are still straight.

Adding second order corrections, one obtains more accurate shapes (Fig. 8). The explicit expression for the characteristic co-ordinates of the boundary limits is

$$-\xi_a = +\xi_b = \varepsilon + [\tfrac{3}{4} - 2\mu^* + 2(\mu^*)^2]\varepsilon^2 + \dots \tag{12}$$

The limit ξ_M of the slippage region is given by

$$\xi_M = -\varepsilon(1-2\mu^*) - (\tfrac{3}{4} - 3\mu^*)\varepsilon^2 - \dots \tag{13}$$

(where $\mu^* := \mu/\varepsilon$). More details can be found in ref. 3. For the special frictionless case $\mu = 0$ a numerical solution given by Green [4] is confirmed.

4. SHEET DRAWING

Compression by curvilinear dies of weak curvature can also be studied with the perturbation procedure described above. Thus, the boundary AC can be slightly changed to

$$y(x, \varepsilon) = 1 + \varepsilon y_1(x) + \varepsilon^2 y_2(x) + \dots \tag{14}$$

TABLE II

		ϕ_1	p_1	x_1	y_1		u_1		v_1
Regular domain		0	0	0	0	$\xi<\xi_a$	0		0
						$\xi_b<\xi$	1		1
Boundary domain around B:	$\xi_a<\xi<0$	$-\frac{1}{2}(\eta^*+\xi^*)$	$2-\eta^*+\xi^*$	$\frac{1}{2}(\xi^*+\eta^*)$	$\frac{1}{2}(-\xi^*+\eta^*)$	$\eta<-\xi_M$	$2\mu^*$ $-1+\frac{3}{2}\eta^*+\frac{1}{2}\xi^*$	$\xi<\xi_M$	$\frac{1}{2}(\xi^*+\eta^*)$
$0\leqslant\eta<\xi_b$								$\xi_M<\xi<0$	$2\mu^*+1+\frac{1}{2}\eta^*+\frac{3}{2}\xi^*$
$\xi^*:=\xi/\varepsilon$; $\eta^*:=\eta/\varepsilon$	$0<\xi<\xi_b$	$-\frac{1}{2}(\eta^*-\xi^*)$	$2-\eta^*-\xi^*$			$\eta>-\xi_M$	$\frac{1}{2}(\xi^*+\eta^*)$	$0<\xi<-\xi_M$	$2\mu^*-1+\frac{1}{2}\eta^*+\frac{3}{2}\xi^*$
								$\xi>-\xi_M$	$\frac{1}{2}(\xi^*+\eta^*)$
Boundary layer around $\xi=0$:	$\xi_a<\xi<0$	$-\frac{1}{2}(1+\xi^*)$	$1+\xi^*$	$\frac{1}{2}\xi^*$ $+\frac{1}{4}\eta(1+\xi^*)$	$-\frac{1}{2}\xi^*$ $-\frac{1}{4}\eta(1+\xi^*)$		$\frac{1}{2}(1+\xi^*)$	$\xi<\xi_M$	$\frac{1}{2}(1+\xi^*)$
$\xi_a<\xi<\xi_b$ $\xi_b<\eta<2+\xi_M$								$\xi_M<\xi<0$	$-2\mu^*+\frac{3}{2}(1+\xi^*)$
$\xi^*:=\xi/\varepsilon$	$0<\xi<\xi_b$	$-\frac{1}{2}(1-\xi^*)$	$1-\xi^*$	$\frac{1}{2}\xi^*$ $+\frac{1}{4}\eta(1-\xi^*)$	$-\frac{1}{2}\xi^*$ $-\frac{1}{4}\eta(1-\xi^*)$			$0<\xi<-\xi_M$	$2\mu^*-\frac{1}{2}(1-3\xi^*)$
								$\xi>-\xi_M$	$\frac{1}{2}(1+\xi^*)$
Boundary domain around A:	$\xi_a<\xi<0$	$-\mu^*$ $+\frac{1}{2}(\eta^{**}-\xi^*)$	$-2\mu^*$ $+2+\eta^{**}+\xi^*$	$1+\xi^*+\eta^{**}$ $(\xi_a^*\leqslant\eta^{**}\leqslant0)$	$\eta^{**}-\xi^*$ $(\xi_a^*\leqslant\eta^{**}\leqslant0)$		$-\mu^*$ $+1+\frac{1}{2}(\xi^*+\eta^*)$	$\xi<\xi_M$	$\mu^*-\frac{1}{2}(\eta^{**}-\xi^*)$
$2+\xi_M<\eta<2$								$\xi_M<\xi<0$	$-\mu^*+1-\frac{1}{2}\eta^{**}+\frac{3}{2}\xi^*$
$\xi^*:=\xi/\varepsilon$; $\eta^{**}:=(\eta-2)/\varepsilon$	$0<\xi<\xi_b$	$-\mu^*$ $+\frac{1}{2}(\eta^{**}+\xi^*)$	$-2\mu^*$ $+2+\eta^{**}-\xi^*$	$1+\eta^{**}$ $(\xi_a^*\leqslant\eta^{**}\leqslant0)$	η^{**} $(\xi_a^*\leqslant\eta^{**}\leqslant0)$			$0<\xi<-\xi_M$	$\mu^*-\frac{1}{2}(\eta^{**}+\xi^*)$
								$\xi>-\xi_M$	$-\mu^*-1-\frac{1}{2}\eta^{**}+\frac{3}{2}\xi^*$

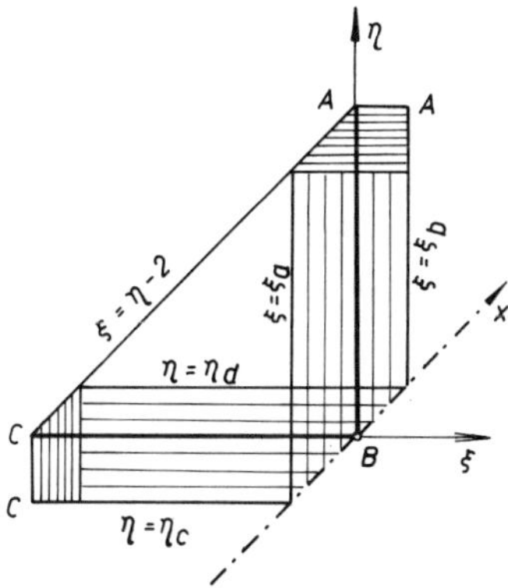

FIG. 9. 'Boundary domains' for sheet-drawing.

The problem of sheet drawing corresponds to $y_1'(x) = \text{const.} \neq 0$ along C. As mentioned in Section 2, curved-shaped dies can also be considered with the help of the second approximation ($y_2'(x) \neq \text{const.}$). The general procedure follows closely the details of the previous section.

(1) The terms of 'zeroth' order are given in Table I.

(2) The regular perturbation fails to meet continuity requirements along the lines BA ($\xi = 0$) and BC ($\eta = 0$), so that the corresponding solution is confined to the interior of the domain BAC (Fig. 9), maintaining the exterior part in stationary states of uniform translations corresponding to the reduction ratio. Due to the friction condition with permanent slippage along AC and to the change of shape of AC with respect to Fig. 1, the regular asymptotic solution is not trivial as in Section 3. It contains significant perturbation terms even in the first order approximation level.

(3) Boundary layers develop along AB and AC which have to be completed with boundary domains around A and C to allow for fulfillment of the corresponding boundary conditions around these corners (Fig. 9). The boundary domain around B is generated automatically as a combination of the layers around AB and BC.

(4) Depending on the relation between friction, length of the die and thickness reduction ratio, 4 different cases may arise. These cases have been sketched in Fig. 10 to visualize the differences in the slip-line fields of the boundary layers. Details of the

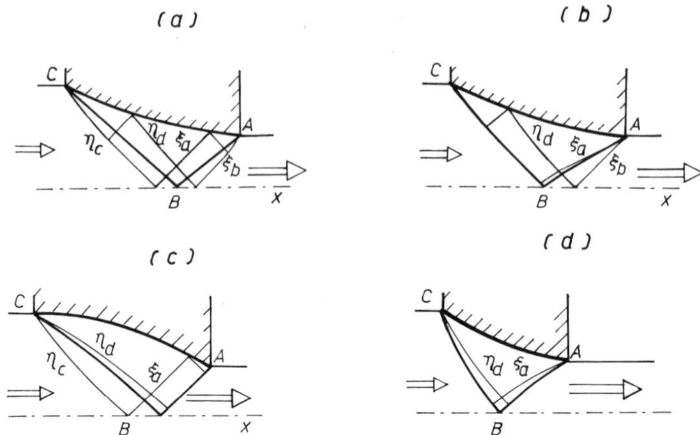

FIG. 10. The four cases of sheet-drawing.

complete analytic solution including second order terms can be found in ref. 3. Case (b) has been suggested by Hill and Tupper [5] 'for not too large thickness reductions and not too long dies'. Case (a) is mentioned as a 'possible solution for large thickness reductions and long dies' by Green [4].

REFERENCES

1. COLLINS, I. F. Geometric properties of some slip-line fields for compression and extrusion, *J. Mech. Phys. Solids*, **16** (1968), 137–152.
2. EWING, D. J. F. A series-method for constructing plastic slip-line fields, *J. Mech. Phys. Solids*, **15** (1967), 105–114.
3. FROMMER, H. Ein Lösungsverfahren für Umformprobleme mit ebenem Verformungszustand, *Diss. ETH*, Nr. 6323 (1979).
4. GREEN, A. P. A theoretical investigation of the compression of a ductile material between smooth flat dies, *Phil. Mag.*, **42** (1951), 900–918.
5. HILL, R. and S. J. TUPPER. A new theory of the plastic deformation in wire-drawing, *J. Iron Steel Inst.*, **159** (1948), 353–359.
6. HILL, R. *The Mathematical Theory of Plasticity*, Oxford University Press, 1950.
7. JOHNSON, W., R. SOWERBY, and J. B. HADDOW. *Plane Strain Slip-line Fields*, Edward Arnold, London, 1970.
8. SALENCON, J. *Théorie de la Plasticité pour les Applications à la Mécanique des Sols*, Eyrolles, Paris, 1974.
9. SAYIR, M. Zur Fortsetzungsaufgabe des Prandtlschen Stempelproblems und einiger damit verwandter Fälle, *ZAMP*, **20** (1969), 298–330.

19

New Developments in the Theory of Dynamic Shakedown

CASTRENZE POLIZZOTTO

Istituto di Scienza delle Costruzioni, Università di Palermo, Italy

ABSTRACT

This paper reports the results obtained by the author in the field of dynamic shakedown theory while extending them to damped systems. Discrete models are considered, constituted by elastic-perfectly plastic elements whose plastic behaviour is described by piecewise linear plasticity laws. The structure is supposed to be subjected to a fully specified (deterministic) loading history, the loads being represented by nodal forces and/or by element imposed strains. The displacements are treated as infinitesimal. The purely elastic response of the structure, which is crucial for any shakedown problem, will be represented here in an explicit matrix integral form in terms of the given loads. Such a representation requires knowledge of the dynamic characteristics of the structure provided through a modal analysis, is centered on the Duhamel integral, and proves to be particularly notable not only in dynamic shakedown but also in the wider field of dynamic plasticity.

1. INTRODUCTION

The shakedown theory provides criteria and analytical tools for the determination of the shakedown safety factor, i.e. the maximum loading multiplier for which shakedown occurs. This is crucial for the assessment of the structural behaviour under varying loads within the range of time-independent plasticity. For loads *below* the shakedown safety factor, after an initial elastic–plastic phase, the structure will respond to subsequent loads in a purely elastic manner, while the plastic deformation produced, though being limited, may however prove to be excessive and cause the structure to fail for, for example, unserviceability, low-cycle fatigue, local fractures, etc. For loads *above* the shakedown safety factor, the elastic–plastic phase will never cease, plastic deformation will not stop being produced and this will surely lead the structure to failure through either incremental collapse (some plastic strain components increase unlimitedly), or alternating plasticity collapse (plastic strain components, though remaining small, show many sign reversions). In both the above cases, the assessment of plastic deformation is crucial in order to make the structure escape failure and to this aim approximate analytical procedures, mostly in the form of bounding techniques, are usually thought to be sufficient, at least for preliminary design purposes. Recent survey papers [3, 5, 13, 20] give a quite complete account of the topic until the late 1970s.

The research work related to the above topic has recently addressed its efforts mainly towards the following goals:

(a) *Shakedown theory.* Within this field, the present author [22] has contributed to the improvement of the dynamic shakedown theory for elastic-perfectly plastic undamped structures through the introduction of some new concepts, such as 'minimum adaptation time' (MAT) and dynamic 'admissible plastic strain cycle' (APSC). Further, randomness of the loadings and other uncertainties have been considered in the framework of probabilistic theories [1].

(b) *Bounding techniques.* Bounding techniques which are applicable independently of whether the structure shakes down or not have been provided by the present author for quasi-static loadings [21], as well as for dynamic loadings [23].

(c) *Inadaptation collapse modes.* Criteria to characterize the two collapse modes associated with inadaptation, i.e. with lack of shakedown, have been provided in the case of quasi-static loads [9, 11], while no attempts have been made so far within the dynamic field, to the author's knowledge. Instability caused by progressively accumulating plastic strain has also been studied [12, 19].

The present paper aims to report the results in ref. 22 in the field of dynamic shakedown theory (point (a) above) while extending them to damped systems. Discrete models will be considered, constituted by elastic-perfectly plastic elements whose plastic behaviour is described by piecewise linear plasticity laws [14, 15]. The structure is supposed to be subjected to a fully specified (deterministic) loading history, the loads being represented by nodal forces and/or by element imposed strains (the latter caused, for instance, by thermal shocks). As in [22], the displacements are treated as infinitesimal. The purely elastic response of the structure, which is crucial for any shakedown problem, will be represented here in an explicit matrix integral form in terms of the given loads [22, 23]; this representation, which requires the knowledge of the dynamic characteristics of the structure provided through a modal analysis and which is centered on the Duhamel integral [6], proves to be particularly interesting only in dynamic shakedown but also in the wider field of dynamic plasticity.

2. PRELIMINARIES AND POSITION OF THE PROBLEM

Let a discrete (or discretized) elastic-perfectly plastic structure be loaded by nodal imposed forces, $\mathbf{F}(t)$, and by element imposed strains, $\boldsymbol{\theta}(t)$, both actions being specified functions of time $t \geqslant 0$. Finding the structural response to these loads in terms of nodal displacements, $\mathbf{u}(t)$, and element stresses, $\boldsymbol{\sigma}(t)$, constitutes a problem of dynamic plasticity which in general is a rather tremendous solution task. Making use of a piecewise linear description of the plastic behaviour [14, 15], the relevant governing equations are as follows:

$$\mathbf{M\ddot{u}} + \mathbf{V\dot{u}} + \mathbf{Ku} - \mathbf{C}^{\mathrm{T}}\mathbf{Dp} = \mathbf{F}(t) + \mathbf{C}^{\mathrm{T}}\mathbf{D}\boldsymbol{\theta}(t) \tag{1}$$

$$\boldsymbol{\sigma} = \mathbf{DCu} - \mathbf{Dp} - \mathbf{D}\boldsymbol{\theta}(t), \qquad \dot{\mathbf{p}} = \mathbf{N}\dot{\boldsymbol{\lambda}} \tag{2}$$

$$\mathbf{f} = \mathbf{N}^{\mathrm{T}}\boldsymbol{\sigma} - \mathbf{k} \leqslant \mathbf{0}, \qquad \dot{\boldsymbol{\lambda}} \geqslant \mathbf{0} \tag{3}$$

$$\mathbf{f}^{\mathrm{T}}\dot{\boldsymbol{\lambda}} = \dot{\mathbf{f}}^{\mathrm{T}}\dot{\boldsymbol{\lambda}} = 0 \tag{4}$$

Here $\mathbf{p}$ is the vector of plastic strains, $\mathbf{C}$ is the compatibility matrix, $\mathbf{D}$ is the matrix of elastic coefficients, $\mathbf{K} = \mathbf{C}^{\mathrm{T}}\mathbf{DC}$ is the (elastic) stiffness matrix, $\mathbf{M}$ and $\mathbf{V}$ the mass and damping matrices; dots indicate time derivatives while $(\ldots)^{\mathrm{T}}$ means 'transpose of $(\ldots)$'; the set of inequalities on the left of eqn. (3) represents the system elastic domain in the form of a hyperpolyhedron specified through the unit external normals of the yield faces (collected in the constant matrix $\mathbf{N}$) and through their distances from the stress space origin (collected in the constant vector $\mathbf{k}>\mathbf{0}$); the second of eqns. (2) and eqns. (3) and (4) account for the usual flow rules of associated plasticity theory [14, 15, 18]. To the set of eqns. (1)–(4), which hold for all $t\geqslant 0$, the following 'initial' conditions must be appended:

$$\mathbf{u}=\mathbf{u}_0, \qquad \dot{\mathbf{u}}=\dot{\mathbf{u}}_0, \qquad \mathbf{p}=\mathbf{p}_0, \quad \text{at} \quad t=0 \tag{5}$$

where $\mathbf{u}_0$, $\dot{\mathbf{u}}_0$, $\mathbf{p}_0$ are given vectors.

If the response of the structure is such that plastic deformation is produced only during a first finite time period whilst at all subsequent times no further plastic deformation is produced, the structure—by definition—shakes down (into a purely elastic state), or also adapts (to the loads), and 'shakedown', or 'adaptation', is the term used to indicate such an occurrence. Shakedown on its own implies that the overall plastic deformation produced is finite, while the exact amount of this deformation remains unknown, just as the adaptation time, i.e. the time at which plastic deformation stops being produced, remains unknown.

A criterion for dynamic shakedown was first given by Ceradini [4] and subsequently generalized in the following form [2, 16]:

Shakedown Theorem
A necessary and sufficient condition for dynamic shakedown is that there exist a finite time, $r\geqslant 0$, and some initial conditions, $(\hat{\mathbf{u}}_0, \hat{\dot{\mathbf{u}}}_0, \hat{\mathbf{p}}_0)$, such that the purely elastic stress response to the given load history associated with these initial conditions, $\hat{\boldsymbol{\sigma}}(t)$, is inside the yield surface at every time subsequent to r, i.e.

$$\mathbf{N}^{\mathrm{T}}\hat{\boldsymbol{\sigma}}(t)-\mathbf{k}\leqslant\mathbf{0}, \qquad \forall\, t\geqslant r \tag{6}$$

For the sake of convenience, let the stress $\hat{\boldsymbol{\sigma}}(t)$ be given the form

$$\hat{\boldsymbol{\sigma}}(t)=\boldsymbol{\sigma}^{\mathrm{E}}(t)+\boldsymbol{\sigma}^{\mathrm{F}}(t)+\boldsymbol{\rho} \tag{7}$$

where, by definition, $\boldsymbol{\sigma}^{\mathrm{E}}(t)$ is the purely elastic stress response of the (damped) system to the loads $\mathbf{F}(t)$, $\boldsymbol{\theta}(t)$ with *arbitrary, but fixed, initial conditions*, $\boldsymbol{\sigma}^{\mathrm{F}}(t)$ is the stress history associated with a free motion or natural vibration of the (damped) structure considered as purely elastic and $\boldsymbol{\rho}$ is a time-independent self-stress vector. Moreover, let the following transformation of the time variable be made, i.e.

$$t=r+\tau, \qquad \tau\geqslant 0 \tag{8}$$

The inequality (6) then takes on the form

$$\mathbf{N}^{\mathrm{T}}[\boldsymbol{\sigma}^{\mathrm{E}}(r+\tau)+\boldsymbol{\sigma}^{\mathrm{F}}(\tau)+\boldsymbol{\rho}]-\mathbf{k}\leqslant\mathbf{0}, \qquad \tau\geqslant 0 \tag{9}$$

in which $\boldsymbol{\sigma}^{E}(r+\tau)$, $\tau \geqslant 0$, is the backward-truncated elastic stress response, that is the elastic stress response $\boldsymbol{\sigma}^{E}(t)$ truncated backward at $t=r$, while $\boldsymbol{\sigma}^{F}(\tau)$ is a free-motion stress associated with initial conditions specified at $\tau=0$. A finite time $r \geqslant 0$ such that the inequality (9) is satisfied for some $\boldsymbol{\sigma}^{F}(\tau)$ and $\boldsymbol{\rho}$ will be called separation time in the following.

The generalized Ceradini theorem can thus be given the following equivalent form.

Shakedown Theorem—Alternative Form

A necessary and sufficient condition for dynamic shakedown is that there exist a separation time, or, in other words, there exist a finite time, $r \geqslant 0$, a free-motion stress, $\boldsymbol{\sigma}^{F}(\tau)$, and a time-independent self-stress, $\boldsymbol{\rho}$, such that superposition of these stresses to the backward-truncated elastic stress response, $\boldsymbol{\sigma}^{E}(r+\tau)$, produces total stresses inside the yield surface at all subsequent times, $\tau \geqslant 0$.

We can now state the following theorems.

Theorem 1

The capacity of a structure to shake down is independent of the initial conditions; in other words, if a structure is able to shake down under the loads $\mathbf{F}(t)$, $\boldsymbol{\theta}(t)$ with the initial conditions $(\mathbf{u}_0, \dot{\mathbf{u}}_0, \mathbf{p}_0)$, it will also shake down if the same loads are associated with any other initial conditions $(\mathbf{u}_0', \dot{\mathbf{u}}_0', \mathbf{p}_0')$.

This theorem is the extension to damped systems of that given in ref. 22 and the proof given there holds true also in the present case. It is worth noting that, while a modification of the initial conditions leaves unaltered the ability of the structure to shake down, on the contrary it will affect the adaptation time, t_a, as well as the amount of plastic deformation produced.

It was proved in ref. 22 that the existence of a separation time, r, implies that every time $t>r$ is a separation time too and that the shortest separation time, r^*, constitutes a lower bound to the adaptation time, i.e. $r^* \leqslant t_a$. Since, for an assigned structure and load history, t_a depends on the initial conditions only, r^* will constitute the 'minimum adaptation time' (MAT) of the structure subjected to the specified load history. The following theorem was given in ref. 2.

Theorem 2

For a structure subjected to a given load history, there exist some initial conditions such that the structure adapts within the minimum possible time and this minimum coincides with the shortest separation time.

3. REPRESENTATION OF THE DYNAMIC ELASTIC RESPONSE

Let the structure, considered as purely elastic, be a n-degree-of-freedom system and let ω_i, ζ_i and $\mathbf{\Phi}_i$, $(i=1,2,\ldots,n)$, be its natural frequencies, damping ratios and vibration mode shapes, respectively. The following orthogonality conditions hold [6]:

$$\mathbf{\Phi}_i\mathbf{M}\mathbf{\Phi}_j=\delta_{ij},\qquad \mathbf{\Phi}_i^{\mathrm{T}}\mathbf{V}\mathbf{\Phi}_j=2\zeta_i\omega_i\delta_{ij},\qquad \mathbf{\Phi}_i^{\mathrm{T}}\mathbf{K}\mathbf{\Phi}_j=\omega_i^2\delta_{ij} \tag{10}$$

where δ_{ij} is the Kronecker symbol.

3.1. Response to the Loads with Zero Initial Conditions

The (elastic) displacement response to the loads $\mathbf{F}(t)$ and $\boldsymbol{\theta}(t)$ is usually represented by means of the Duhamel integral [6], i.e.

$$\mathbf{u}_1^{\mathrm{E}}(t)=\sum_{i=1}^{n}\mathbf{\Phi}_i(\omega_i^{\mathrm{D}})^{-1}\int_0^t P_i(\bar{t})\exp\left[-\zeta_i\omega_i(t-\bar{t})\right]\sin\omega_i^{\mathrm{D}}(t-\bar{t})\,\mathrm{d}\bar{t} \tag{11}$$

where $P_i(\bar{t})$ is the ith ‘normal’ load, i.e.

$$P_i(\bar{t})=\mathbf{\Phi}_i^{\mathrm{T}}\mathbf{F}(\bar{t})+\mathbf{\Phi}_i^{\mathrm{T}}\mathbf{C}^{\mathrm{T}}\mathbf{D}\boldsymbol{\theta}(\bar{t}) \tag{12}$$

and $\omega_i^{\mathrm{D}}=\omega_i\sqrt{1-\zeta_i^2}$ is the ith damped frequency. Then, setting

$$\chi_i(\tau)=(\omega_i^{\mathrm{D}})^{-1}\exp\left(-\zeta_i\omega_i\tau\right)\sin\omega_i^{\mathrm{D}}\tau,\qquad \tau\geqslant 0 \tag{13}$$

$$\mathbf{H}(\tau)=\sum_{i=1}^{n}\mathbf{\Phi}_i\mathbf{\Phi}_i^{\mathrm{T}}\chi_i(\tau) \tag{14}$$

enables eqn. (11) to take on the form:

$$\mathbf{u}_1^{\mathrm{E}}(t)=\int_0^t\mathbf{H}(t-\bar{t})[\mathbf{F}(\bar{t})+\mathbf{C}^{\mathrm{T}}\mathbf{D}\boldsymbol{\theta}(\bar{t})]\,\mathrm{d}\bar{t} \tag{15}$$

$\mathbf{H}(t-\bar{t})$ is shown to be a matrix which transforms the unit impulse (or the unit momentum) applied at time $\bar{t}$ into the corresponding displacement response at a subsequent time t.

Further, introducing the new quantities:

$$\psi_i(\tau)=\bar{\omega}_i^2 \exp(-\zeta_i\omega_i\tau)\left\{\cos\omega_i^{\mathrm{D}}\tau+\frac{\zeta_i}{\sqrt{1-\zeta_i^2}}\sin\omega_i^{\mathrm{D}}\tau\right\} \tag{16}$$

$$\mathbf{G}(\tau)=\sum_{i=1}^{n}\mathbf{\Phi}_i\mathbf{\Phi}_i^{\mathrm{T}}\psi_i(\tau) \tag{17}$$

$$\mathbf{A}_{\mathrm{d}}(\tau)=\mathbf{K}^{-1}-\mathbf{G}(\tau),\qquad \mathbf{L}_{\mathrm{d}}(\tau)=\mathbf{A}_{\mathrm{d}}(\tau)\mathbf{C}^{\mathrm{T}}\mathbf{D} \tag{18}$$

and noting that $\mathrm{d}\mathbf{G}(\tau)/\mathrm{d}\tau=-\mathbf{H}(\tau)$ and that $\mathbf{G}(0)=\mathbf{K}^{-1}$ enables eqn. (15) to be rewritten in the equivalent form

$$\mathbf{u}_1^{\mathrm{E}}(t)=\int_0^t\{\dot{\mathbf{A}}_{\mathrm{d}}(t-\bar{t})\mathbf{F}(\bar{t})+\dot{\mathbf{L}}_{\mathrm{d}}(t-\bar{t})\mathbf{\theta}(\bar{t})\}\,\mathrm{d}\bar{t} \tag{19}$$

which, through an integration by parts, transforms as

$$\mathbf{u}_1^{\mathrm{E}}(t)=\int_0^t\{\mathbf{A}_{\mathrm{d}}(t-\bar{t})\dot{\mathbf{F}}(\bar{t})+\mathbf{L}_{\mathrm{d}}(t-\bar{t})\dot{\mathbf{\theta}}\}\,\mathrm{d}\bar{t} \tag{20}$$

Here we have assumed $\mathbf{F}(0)=\mathbf{0}$, $\mathbf{\theta}(0)=\mathbf{0}$, but eqn. (20) can be easily shown to hold also for $\mathbf{F}(0)\neq\mathbf{0}$ and $\mathbf{\theta}(0)\neq\mathbf{0}$ [23]. It is worth mentioning that the matrices $\mathbf{A}_{\mathrm{d}}(t-\bar{t})$ and $\mathbf{L}_{\mathrm{d}}(t-\bar{t})$ in eqn. (20) transform, respectively, the unit force and the unit imposed strain applied at time $\bar{t}$ into the corresponding displacement responses at a subsequent time t. The velocity response can be derived from eqn. (16) by differentiating with respect to t, i.e.

$$\dot{\mathbf{u}}_1^{\mathrm{E}}(t)=\int_0^t\mathbf{H}(t-\bar{t})[\dot{\mathbf{F}}(\bar{t})+\mathbf{C}^{\mathrm{T}}\mathbf{D}\dot{\mathbf{\theta}}(\bar{t})]\,\mathrm{d}\bar{t} \tag{21}$$

The stress response associated with $\mathbf{u}_1^{\mathrm{E}}(t)$, i.e. $\mathbf{\sigma}_1^{\mathrm{E}}(t)=\mathbf{D}\mathbf{C}^{\mathrm{T}}\mathbf{u}_1^{\mathrm{E}}(t)$, is

$$\mathbf{\sigma}_1^{\mathrm{E}}(t)=\int_0^t\{\mathbf{L}_{\mathrm{d}}^{\mathrm{T}}(t-\bar{t})\dot{\mathbf{F}}(\bar{t})+\mathbf{Z}_{\mathrm{d}}(t-\bar{t})\dot{\mathbf{\theta}}(\bar{t})\}\,\mathrm{d}\bar{t} \tag{22}$$

where we have set

$$\mathbf{Z}_{\mathrm{d}}(\tau)=\mathbf{D}\mathbf{C}\mathbf{A}_{\mathrm{d}}(\tau)\mathbf{C}^{\mathrm{T}}\mathbf{D}-\mathbf{D} \tag{23}$$

The matrices $\mathbf{L}_{\mathrm{d}}^{\mathrm{T}}(t-\bar{t})$ and $\mathbf{Z}_{\mathrm{d}}(t-\bar{t})$ transform, respectively, the unit force and the unit imposed strain applied at time $\bar{t}$ into the corresponding stress responses at a subsequent time t.

From eqns. (20) and (22) it follows that the matrices $\mathbf{A}_{\mathrm{d}}(t-\bar{t})$, $\mathbf{L}_{\mathrm{d}}(t-\bar{t})$ and $\mathbf{Z}_{\mathrm{d}}(t-\bar{t})$ constitute the dynamic counterparts of the matrices $\mathbf{A}=\mathbf{K}^{-1}$, $\mathbf{L}=\mathbf{AC}^{\mathrm{T}}\mathbf{D}$, $\mathbf{Z}=\mathbf{DCAC}^{\mathrm{T}}\mathbf{D}-\mathbf{D}$ which hold within the framework of statics.

3.2. Response to Imposed Initial Conditions

Imposing displacements $\mathbf{u}_0$ and velocities $\dot{\mathbf{u}}_0$ at $t=0$ produces the displacement response [6]

$$\mathbf{u}_2^{\mathrm{E}}(t)=\sum_{i=1}^{n}\mathbf{\Phi}_i \exp(-\zeta_i\omega_i t)\{a_i \cos\omega_i^{\mathrm{D}}t+b_i \sin\omega_i^{\mathrm{D}}t\} \tag{24}$$

where a_i, b_i are constants related to $\mathbf{u}_0$ and $\dot{\mathbf{u}}_0$ through

$$\sum_{j=1}^{n}\mathbf{\Phi}_j a_j=\mathbf{u}_0, \qquad \sum_{j=1}^{n}\mathbf{\Phi}_j\omega_j\{b_j\sqrt{1-\zeta_j^2}-a_j\zeta_j\}=\dot{\mathbf{u}}_0 \tag{25}$$

Through a matrix left-multiplication of the first equality of the latter equations by $\mathbf{\Phi}_i^{\mathrm{T}}\mathbf{K}$ and of the second one by $\mathbf{\Phi}_i^{\mathrm{T}}\mathbf{M}$, we obtain, respectively,

$$\omega_i^2 a_i=\mathbf{\Phi}_i^{\mathrm{T}}\mathbf{K}\mathbf{u}_0, \qquad \omega_i^{\mathrm{D}}b_i-\zeta_i\omega_i a_i=\mathbf{\Phi}_i^{\mathrm{T}}\mathbf{M}\dot{\mathbf{u}}_0 \tag{26}$$

Solving the latter equations with respect to a_i and b_i and substituting the resulting expressions into eqn. (24) then gives

$$\mathbf{u}_2^{\mathrm{E}}(t)=\mathbf{G}(t)\mathbf{K}\mathbf{u}_0+\mathbf{H}(t)\mathbf{M}\dot{\mathbf{u}}_0 \tag{27}$$

$$\boldsymbol{\sigma}_2^{\mathrm{E}}(t)=\mathbf{DC}[\mathbf{G}(t)\mathbf{K}\mathbf{u}_0+\mathbf{H}(t)\mathbf{M}\dot{\mathbf{u}}_0] \tag{28}$$

which represent the displacement and stress response in terms of the given initial conditions.

4. SAFETY FACTOR FOR LIMITED MAT

Let $\xi>0$ denote a multiplier affecting the given loads $\mathbf{F}(t)$, $\boldsymbol{\theta}(t)$ and let ξ^* be the maximum value of ξ such that, for every $\xi\leqslant\xi^*$, the structure possesses a MAT not greater than a given value, r. Since the MAT is a separation time and hence r is a separation time too, we can determine ξ^* by maximizing ξ under the constraints in eqn. (9), i.e.

$$\xi^*=\max_{(\xi,\boldsymbol{\sigma}^{\mathrm{F}},\boldsymbol{\sigma})}\{\xi \,\|\, \mathbf{N}^{\mathrm{T}}[\xi\boldsymbol{\sigma}^{\mathrm{E}}(r+\tau)+\boldsymbol{\sigma}^{\mathrm{F}}(\tau)+\boldsymbol{\rho}]-\mathbf{k}\leqslant\mathbf{0},\ \forall\tau\geqslant 0\} \tag{29}$$

where $\boldsymbol{\rho}$ must satisfy the condition $\mathbf{C}^{\mathrm{T}}\boldsymbol{\rho}=\mathbf{0}$ and $\boldsymbol{\sigma}^{\mathrm{F}}(\tau)$ is (see eqn. (28))

$$\boldsymbol{\sigma}^{\mathrm{F}}(\tau)=\mathbf{DC}[\mathbf{G}(\tau)\mathbf{K}\mathbf{c}+\mathbf{H}(\tau)\mathbf{M}\dot{\mathbf{c}}], \qquad \tau\geqslant 0 \tag{30}$$

$\mathbf{c}$ and $\dot{\mathbf{c}}$ being arbitrary initial displacement and velocity vectors.

Using the Lagrangian multiplier method [17], the following Lagrangian functional is to be considered in order to obtain the relevant optimality conditions, i.e.

$$\Psi = -\alpha\xi + \int_0^{t_1} \dot{\boldsymbol{\mu}}^{\mathrm{T}}\{\mathbf{N}^{\mathrm{T}}[\xi\boldsymbol{\sigma}^{\mathrm{E}}(r+\tau)+\boldsymbol{\sigma}^{\mathrm{F}}(\tau)+\boldsymbol{\rho}]-\mathbf{k}\}\,\mathrm{d}\tau + \int_0^{t_1} \dot{\mathbf{q}}^{\mathrm{T}}\{-\boldsymbol{\sigma}^{\mathrm{F}}(\tau)+\mathbf{DC}[\mathbf{G}(\tau)\mathbf{Kc}+\mathbf{H}(\tau)\mathbf{M}\dot{\mathbf{c}}]\}\,\mathrm{d}\tau - \mathbf{v}^{\mathrm{T}}\mathbf{C}^{\mathrm{T}}\boldsymbol{\rho} \quad (31)$$

where $\dot{\boldsymbol{\mu}} \geqslant \mathbf{0}$, $\dot{\mathbf{q}}$ and $\mathbf{v}$ are appropriate Lagrangian vector variables, α is some positive constant and t_1 is a subsequent time set sufficiently large to cover the entire loading history. Taking the first variation of Ψ gives:

$$\left.\begin{array}{l} \boldsymbol{\varphi} = \mathbf{N}^{\mathrm{T}}[\xi^*\boldsymbol{\sigma}^{\mathrm{E}}(r+\tau)+\boldsymbol{\sigma}^{\mathrm{F}}(\tau)+\boldsymbol{\rho}]-\mathbf{k} \leqslant \mathbf{0} \\ \dot{\boldsymbol{\mu}} > \mathbf{0}, \qquad \boldsymbol{\varphi}^{\mathrm{T}}\dot{\boldsymbol{\mu}} = 0, \qquad \dot{\mathbf{q}} = \mathbf{N}\dot{\boldsymbol{\mu}}, \qquad \tau \geqslant 0 \end{array}\right\} \quad (32)$$

$$\int_0^{t_1} \{\boldsymbol{\sigma}^{\mathrm{E}}(r+\tau)\}^{\mathrm{T}}\dot{\mathbf{q}}(\tau)\,\mathrm{d}\tau = \alpha \quad (33)$$

$$\int_0^{t_1} \mathbf{G}(\tau)\mathbf{C}^{\mathrm{T}}\mathbf{D}\dot{\mathbf{q}}(\tau)\,\mathrm{d}\tau = \mathbf{0}, \qquad \int_0^{t_1} \mathbf{H}(\tau)\mathbf{C}^{\mathrm{T}}\mathbf{D}\dot{\mathbf{q}}(\tau)\,\mathrm{d}\tau = \mathbf{0} \quad (34)$$

$$\int_0^{t_1} \dot{\mathbf{q}}(\tau)\,\mathrm{d}\tau = \mathbf{Cv} \quad (35)$$

which are the optimality conditions relative to the problem (29).

Equations (32)–(35) describe a limit purely elastic process constituted by a sequence of instantaneous elastic–plastic states, in each of which a plastic strain rate $\dot{\mathbf{q}}$, complying with the plastic flow-rule, is associated with the allowable stress, $\boldsymbol{\sigma}^{\mathrm{a}}(\tau) = \xi^*\boldsymbol{\sigma}^{\mathrm{E}}(r+\tau)+\boldsymbol{\sigma}^{\mathrm{F}}(\tau)+\boldsymbol{\rho}$, which is unaffected by the plastic strain previously produced. The above process can be viewed as one of simple production of plastic strain, with no stress redistribution, leading to a final collapse mechanism described by the displacement vector $\mathbf{v}$.

Setting $\tau = t_1 - \bar{t}$ in eqns. (34) and writing $\dot{\mathbf{q}}(\bar{t})$ instead of $\dot{\mathbf{q}}(t_1-\bar{t})$ enables eqns. (34) to be rewritten as

$$\int_0^{t_1} \mathbf{G}(t_1-\bar{t})\mathbf{C}^{\mathrm{T}}\mathbf{D}\dot{\mathbf{q}}(\bar{t})\,\mathrm{d}\bar{t} = \mathbf{0}, \qquad \int_0^{t_1} \mathbf{H}(t_1-\bar{t})\mathbf{C}^{\mathrm{T}}\mathbf{D}\dot{\mathbf{q}}(\bar{t}) = \mathbf{0} \quad (36)$$

where $0 \leqslant \bar{t} \leqslant t_1$; the latter relationships, through eqns. (18), (20) and (21), enable one to state that the plastic strain rate history $\dot{\mathbf{q}}(\bar{t})$, if (dynamically) imposed on the structure considered as purely elastic, will produce a motion $\mathbf{v}(\bar{t})$ which leads the structure from the initial configuration, in which $\mathbf{v}(0)=\dot{\mathbf{v}}(0)=\mathbf{0}$, to a final configuration in which

$$\mathbf{v}(t_1)=\mathbf{K}^{-1}\mathbf{CDq}(t_1), \qquad \dot{\mathbf{v}}(t_1)=\mathbf{0} \tag{37}$$

that is a configuration which coincides with the one that would be produced by the total accumulated plastic strain $\mathbf{q}(t_1)=\int_0^{t_1}\dot{\mathbf{q}}(\bar{t})\,\mathrm{d}\bar{t}$ *applied statically;* moreover, since $\mathbf{q}(t_1)$ is, in virtue of eqn. (35), congruent with the displacements v, the aforementioned final configuration identifies with the final collapse mechanism, i.e. $\mathbf{v}(t_1)=\mathbf{v}$. In the following, a plastic strain rate history, $\dot{\mathbf{q}}(\tau)$, which satisfies eqns. (34) and (35), will be called *admissible plastic strain cycle* (APSC). The latter is easily shown to satisfy the following orthogonality condition:

$$\int_0^{t_1}\{\boldsymbol{\sigma}^{\mathrm{F}}(\tau)\}^{\mathrm{T}}\dot{\mathbf{q}}(\tau)\,\mathrm{d}\tau=0 \tag{38}$$

with respect to any free-motion stress history $\boldsymbol{\sigma}^{\mathrm{F}}(\tau)$ [22].

A problem which is dual of the problem (29) is as follows [17]:

$$\alpha\xi^*=\min_{(\boldsymbol{\mu}^{\mathrm{c}})}\left\{\mathbf{k}^{\mathrm{T}}\int_0^{t_1}\dot{\boldsymbol{\mu}}^{\mathrm{c}}\,\mathrm{d}\tau\,\middle\|\,\int_0^{t_1}\{\boldsymbol{\sigma}^{\mathrm{E}}(r+\tau)\}^{\mathrm{T}}\mathbf{N}\dot{\boldsymbol{\mu}}^{\mathrm{c}}\,\mathrm{d}\tau=\alpha,\ \dot{\boldsymbol{\mu}}^{\mathrm{c}}\geqslant\mathbf{0}\right\} \tag{39}$$

in which the strain rate history $\dot{\mathbf{q}}^{\mathrm{c}}=\mathbf{N}\dot{\boldsymbol{\mu}}^{\mathrm{c}}$ belongs to the class of APSC's.

In the above discussion, the problem (29) was treated as a true linear programming problem since questions arising in relation to the infinite number of constraints (and possibly to the improper integrals when $t_1\to\infty$) were ignored. However, following this heuristic line of reasoning, some properties of the function $\xi^*=\xi^*(r)$ can be assessed.

First of all, $\xi^*(r)$ proves to be a nondecreasing function of r because increasing r amounts to decreasing the number of constraints of the problem (29). In particular, ξ^* will be constant, i.e. independent of r, in the case of a periodic loading, for the backward-truncated elastic stress response, $\boldsymbol{\sigma}^{\mathrm{E}}(r+\tau)$, assumed to be the steady-state response, remains the same when r is increased by a Δr equal to a multiple of the load period and, as a result, ξ^* remains the same at these values of r and hence at all r's.

The maximum value of $\xi^*(r)$ with respect to r, given either by

$$X=\max_{(r)}\xi^*(r) \tag{40a}$$

or, equivalently, by

$$X = \lim_{r \to t_1} \xi^*(r) \tag{40b}$$

represents what is usually known as the 'shakedown safety factor' (i.e. for any $\xi < X$ the structure shakes down). It is worth noting that X proves to be unbounded when the load history has a finite duration, $t_f < t_1 < +\infty$, for taking $r > t_f$ causes $\boldsymbol{\sigma}^E(r+\tau)$ to be a free-motion stress history and, by virtue of eqn. (38), the dual problem (39) proves to be unfeasible (i.e. $\alpha \to 0$), hence both $\xi^*(r)$ and $X \geqslant \xi^*(r)$ prove to be unbounded. As a consequence, the shakedown safety factor, X, can actually be finite only for loading histories of truly unlimited duration. Loading histories of infinite duration are rarely encountered in practice; they happen to be considered when the load is periodic or it results from the indefinite repetition of a given set of short-duration excitations [23].

It is worth mentioning that the value $\xi^*(0)$ corresponds to a zero MAT; in other words the structure, if subjected to loads amplified by $\xi^*(0)$ and to suitable initial conditions, is able to remain elastic during the entire loadings process, i.e. for $t \geqslant 0$.

5. THE MINIMUM ADAPTATION TIME (MAT)

According to Theorem 2, the MAT of a structure subjected to a specified load history amplified by a fixed factor ξ can be computed by solving the following problem:

$$r^* = \min_{(r, \boldsymbol{\sigma}^F, \boldsymbol{\rho})} \{ r \,\|\, \mathbf{N}^T[\xi \boldsymbol{\sigma}^E(r+\tau) + \boldsymbol{\sigma}^F(\tau) + \boldsymbol{\rho}] - \mathbf{k} \leqslant \mathbf{0}, \ \forall \tau \geqslant 0 \} \tag{41}$$

where $\boldsymbol{\sigma}^F$ is given by eqn (30) and $\boldsymbol{\rho}$ satisfies $\mathbf{C}^T\boldsymbol{\rho} = \mathbf{0}$. Using the same procedure as in Section 4, we consider the Lagrangian functional

$$\begin{aligned} \Psi_1 = \beta r &+ \int_0^{t_1} \dot{\boldsymbol{\mu}}^T \{ \mathbf{N}^T[\xi \boldsymbol{\sigma}^E(r+\tau) + \boldsymbol{\sigma}^F(\tau) + \boldsymbol{\rho}] - \mathbf{k} \} \, d\tau \\ &+ \int_0^{t_1} \dot{\mathbf{q}}^T \{ -\boldsymbol{\sigma}^F(\tau) + \mathbf{DC}[\mathbf{G}(\tau)\mathbf{Kc} + \mathbf{H}(\tau)\mathbf{M}\dot{\mathbf{c}}] \} \, d\tau - \mathbf{v}^T\mathbf{C}^T\boldsymbol{\rho} \end{aligned} \tag{42}$$

where we have used the same symbols as in eqn. (31) to indicate the Lagrangian variables and β is a new positive constant. The stationarity conditions are

$$\left.\begin{array}{l}\boldsymbol{\varphi} = \mathbf{N}^{\mathrm{T}}[\xi\boldsymbol{\sigma}^{\mathrm{E}}(r^* + \tau) + \boldsymbol{\sigma}^{\mathrm{F}}(\tau) + \boldsymbol{\rho}] - \mathbf{k} \leqslant \mathbf{0} \\ \dot{\boldsymbol{\mu}} \geqslant \mathbf{0}, \qquad \boldsymbol{\varphi}^{\mathrm{T}}\dot{\boldsymbol{\mu}} = 0, \qquad \dot{\mathbf{q}} = \mathbf{N}\dot{\boldsymbol{\mu}}, \qquad \tau \geqslant 0\end{array}\right\} \tag{43}$$

$$\int_0^{t_1} \{\dot{\boldsymbol{\sigma}}^{\mathrm{E}}(r^* + \tau)\}^{\mathrm{T}}\dot{\mathbf{q}}(\tau)\,\mathrm{d}\tau = -\beta \tag{44}$$

$$\int_0^{t_1} \mathbf{G}(\tau)\mathbf{C}^{\mathrm{T}}\mathbf{D}\dot{\mathbf{q}}(\tau)\,\mathrm{d}\tau = \mathbf{0}, \qquad \int_0^{t_1} \mathbf{H}(\tau)\mathbf{C}^{\mathrm{T}}\mathbf{D}\dot{\mathbf{q}}(\tau)\,\mathrm{d}\tau = \mathbf{0} \tag{45}$$

$$\int_0^{t_1} \dot{\mathbf{q}}(\tau)\,\mathrm{d}\tau = \mathbf{C}\mathbf{v} \tag{46}$$

Since $\dot{\mathbf{q}}(\tau)$ is an admissible plastic strain cycle, from the equality $\boldsymbol{\varphi}^{\mathrm{T}}\dot{\boldsymbol{\mu}} = 0$ there follows, through an integration over $(0, t_1)$,

$$\xi\int_0^{t_1} \{\boldsymbol{\sigma}^{\mathrm{E}}(r^* + \tau)\}^{\mathrm{T}}\dot{\mathbf{q}}(\tau)\,\mathrm{d}\tau = \mathbf{k}^{\mathrm{T}}\int_0^{t_1} \dot{\boldsymbol{\mu}}(\tau)\,\mathrm{d}\tau \tag{47}$$

For $\xi(0) \leqslant \xi \leqslant X$, the r.h. side of this equation is positive, eqns. (43)–(47) prove to be equivalent to eqns. (32)–(35) and thus $\xi = \xi^*(r^*)$, i.e. the fixed multiplier ξ coincides with the safety factor for MAT not greater than r^*. Since for $\xi < \xi^*(0)$ the optimal value r^* must vanish, the function $r^* = r^*(\xi)$ proves to be a nondecreasing function of ξ. For $\xi > X$, the problem (41) is unfeasible.

When the structure, through the application of suitable initial conditions, adapts within the shortest possible time, r^*, the real stress response $\boldsymbol{\sigma}^*$ at times subsequent to r^* can be set in the form

$$\boldsymbol{\sigma}^* = \boldsymbol{\sigma}^{\mathrm{E}}(r^* + \tau) + \boldsymbol{\sigma}^{\mathrm{F}*}(\tau) + \boldsymbol{\rho}^*, \qquad \tau \geqslant 0 \tag{48}$$

where $\boldsymbol{\sigma}^{\mathrm{E}}(r^* + \tau)$ is the backward-truncated elastic stress response, while the free-motion stress $\boldsymbol{\sigma}^{\mathrm{F}*}(\tau)$ and the time-independent self-stress $\boldsymbol{\rho}^*$ are the consequence of the elastic–plastic process which has taken place before r^*. Applying Drucker's postulate, we can write

$$(\boldsymbol{\sigma}^{\mathrm{a}} - \boldsymbol{\sigma}^*)^{\mathrm{T}}\dot{\mathbf{q}} \geqslant 0, \qquad \tau \geqslant 0 \tag{49}$$

where $\boldsymbol{\sigma}^{\mathrm{a}} = \boldsymbol{\sigma}^{\mathrm{E}}(r^* + \tau) + \boldsymbol{\sigma}^{\mathrm{F}}(\tau) + \boldsymbol{\rho}$, while $\boldsymbol{\sigma}^{\mathrm{F}}(\tau)$, $\boldsymbol{\rho}$, $\dot{\mathbf{q}}(\tau)$ satisfy eqns. (43)–(46) and hence also eqns. (32)–(35). An integration of (49) over the interval $(0, t_1)$ produces

$$\int_0^{t_1} (\boldsymbol{\sigma}^{\mathrm{a}} - \boldsymbol{\sigma}^*)^{\mathrm{T}}\dot{\mathbf{q}}\,\mathrm{d}\tau = \int_0^{t_1} (\boldsymbol{\sigma}^{\mathrm{F}} - \boldsymbol{\sigma}^{\mathrm{F}*})^{\mathrm{T}}\dot{\mathbf{q}}\,\mathrm{d}\tau + (\boldsymbol{\rho} - \boldsymbol{\rho}^*)^{\mathrm{T}}\int_0^{t_1} \dot{\mathbf{q}}\,\mathrm{d}\tau \tag{50}$$

and the latter, recalling eqns. (38) and (46), becomes

$$\int_0^{t_1} (\boldsymbol{\sigma}^{\mathrm{a}} - \boldsymbol{\sigma}^{*})^{\mathrm{T}} \dot{\mathbf{q}} \, \mathrm{d}\tau = 0 \tag{51}$$

As a result, eqn. (49) applies with the equality sign at all times, i.e.

$$(\boldsymbol{\sigma}^{\mathrm{a}} - \boldsymbol{\sigma}^{*})^{\mathrm{T}} \dot{\mathbf{q}} = 0, \qquad \tau \geqslant 0 \tag{52}$$

This result enables one [9, 18, 24] to assert that in the regions where yield limits are reached, the stresses $\boldsymbol{\sigma}^{\mathrm{a}}$ of the elastic–plastic stress-redistribution-free process described by eqns. (32)–(35), or by eqns. (43)–(46), are equal to the stresses $\boldsymbol{\sigma}^{*}$ of the real process which takes place in the structure when the imposed initial conditions are such that the structure adapts within the shortest possible time (except at flat regions of the yield surface where $\boldsymbol{\sigma}^{\mathrm{a}}$ may differ from one another by a stress vector normal to the yield plane).

The discussion of the mathematical nature of the problem (41) is not given here for brevity; we only mention that lower and upper bounds to r^* can be given in the form $r^- \leqslant r^* \leqslant r^+$, where r^- and r^+ are, respectively, nonseparation and separation times.

6. KINEMATICAL THEOREMS

According to a previous definition, an admissible plastic strain cycle (APSC), $\dot{q} = \mathbf{N}\dot{\boldsymbol{\mu}}(\tau)$, $\dot{\boldsymbol{\mu}}(\tau) \geqslant \mathbf{0}$, satisfies eqns. (34) and (35). Such a strain cycle possesses features which are worth stressing further, namely it is 'globally congruent' (see eqn. (35)) and 'globally quasi-static' (see eqns. (34)). These expressions refer to the circumstances that the motion resulting from the application of this strain cycle on the structure considered as purely elastic will tend to take the shape of a collapse mechanism which proves to be the quasi-static elastic response to the plastic strains accumulated in the cycle. The above definition of APSC coincides with Koiter's [10] in the case of quasi-static shakedown. Some general properties of the APSC were given in ref. 22.

The following theorems can be proved.

Theorem 3. Kinematical theorem for later adaptation time
A necessary and sufficient condition in order that a structure subjected to a specified load history has a MAT greater than an assigned time, r, is that there exist some admissible plastic strain cycle, $\dot{\boldsymbol{\mu}}^{\mathrm{c}}(\tau) \geqslant \mathbf{0}$, which

satisfies the inequality

$$\int_0^{t_1} \{\boldsymbol{\sigma}^{\mathrm{E}}(r+\tau)\}^{\mathrm{T}} \mathbf{N} \dot{\boldsymbol{\mu}}^{\mathrm{c}}(\tau)\, \mathrm{d}\tau > \mathbf{k}^{\mathrm{T}} \int_0^{t_1} \dot{\boldsymbol{\mu}}^{\mathrm{c}}(\tau)\, \mathrm{d}\tau \tag{53}$$

Proof. If the MAT is greater than r, the latter is not a separation time and the problem (29), solved with this r, will give $\xi^*(r)<1$. Further, the dual problem (39) will provide an APSC, $\dot{\boldsymbol{\mu}}^{\mathrm{c}}(\tau)$, such that

$$\mathbf{k}^{\mathrm{T}} \int_0^{t_1} \dot{\boldsymbol{\mu}}^{\mathrm{c}}(\tau)\, \mathrm{d}\tau = \alpha\xi^*(r) = \xi^*(r) \int_0^{t_1} \{\boldsymbol{\sigma}^{\mathrm{E}}(r+\tau)\}^{\mathrm{T}} \dot{\mathbf{q}}(\tau)\, \mathrm{d}\tau \tag{54}$$

which yields the inequality

$$\mathbf{k}^{\mathrm{T}} \int_0^{t_1} \dot{\boldsymbol{\mu}}^{\mathrm{c}}(\tau)\, \mathrm{d}\tau < \int_0^{t_1} \{\boldsymbol{\sigma}^{\mathrm{E}}(r+\tau)\}^{\mathrm{T}} \dot{\mathbf{q}}(\tau)\, \mathrm{d}\tau \tag{55}$$

coincident with eqn. (53). So the necessity part is proved.

Then, let us suppose that there exists an APSC, $\dot{\boldsymbol{\mu}}^{\mathrm{c}}(\tau) \geqslant \mathbf{0}$, such as to satisfy the inequality (53). In consideration that the r.h. side of this inequality is positive, the said APSC turns out to be feasible for the problem (39) and thus one can write

$$\mathbf{k}^{\mathrm{T}} \int_0^{t_1} \dot{\boldsymbol{\mu}}^{\mathrm{c}}(\tau)\, \mathrm{d}\tau \geqslant \alpha\xi^* = \xi^* \int_0^{t_1} \{\boldsymbol{\sigma}^{\mathrm{E}}(r+\tau)\}^{\mathrm{T}} \mathbf{N} \dot{\boldsymbol{\mu}}^{\mathrm{c}}(\tau)\, \mathrm{d}\tau \tag{56}$$

where $\xi^* = \xi^*(r)$ is the safety factor for a MAT not greater than r. Comparing eqn. (56) with eqn. (53) then gives $\xi^*(r)<1$, that is r is not a separation time and as a result the MAT is greater than r. So the proof has been completed.

Theorem 4. Kinematical theorem for earlier adaptation time
A necessary and sufficient condition for a structure subjected to a specified load history to have a MAT smaller than an assigned time, r, is that the inequality

$$\int_0^{t_1} \{\boldsymbol{\sigma}^{\mathrm{E}}(r+\tau)\}^{\mathrm{T}} \mathbf{N} \dot{\boldsymbol{\mu}}^{\mathrm{c}}(\tau)\, \mathrm{d}\tau < \mathbf{k}^{\mathrm{T}} \int_0^{t_1} \dot{\boldsymbol{\mu}}(\tau)\, \mathrm{d}\tau \tag{57}$$

be satisfied for every APSC, $\dot{\boldsymbol{\mu}}^{\mathrm{c}}(\tau) \geqslant \mathbf{0}$.

Proof. This theorem is easily proved by means of Theorem 3, for eqn. (57) is a negation of eqn. (53).

In order to obtain a kinematical theorem which characterizes inadaptation, we have to take $t_1 \rightarrow +\infty$.

Theorem 5. Kinematical theorem for inadaptation
A necessary and sufficient condition in order that a structure subjected to a given load history of unlimited duration does not shake down is that there exist an APSC, $\dot{\boldsymbol{\mu}}(\tau) \geqslant \mathbf{0}$, which satisfies

$$\lim_{r \rightarrow +\infty} \int_0^{+\infty} \{\boldsymbol{\sigma}^{\mathrm{E}}(r+\tau)\}^{\mathrm{T}} \mathbf{N} \dot{\boldsymbol{\mu}}^{\mathrm{c}}(\tau)\, \mathrm{d}\tau > \mathbf{k}^{\mathrm{T}} \int_0^{+\infty} \dot{\boldsymbol{\mu}}^{\mathrm{c}}(\tau)\, \mathrm{d}\tau \tag{58}$$

The proof of this theorem is as for Theorem 3 applied to a diverging value of r. The theorem holds also if the sign $\geqslant 0$ is substituted into the inequality (58).

Theorem 6. Kinematical theorem for adaptation
A necessary and sufficient condition for a structure to shake down under a given loading history of unlimited duration is that the following inequality

$$\lim_{r \rightarrow +\infty} \int_0^{t_\infty} \{\boldsymbol{\sigma}^{\mathrm{E}}(r+\tau)\}^{\mathrm{T}} \mathbf{N} \dot{\boldsymbol{\mu}}^{\mathrm{c}}(\tau)\, \mathrm{d}\tau < \mathbf{k}^{\mathrm{T}} \int_0^{+\infty} \dot{\boldsymbol{\mu}}^{\mathrm{c}}(\tau)\, \mathrm{d}\tau \tag{59}$$

be satisfied for every APSC, $\dot{\boldsymbol{\mu}}^{\mathrm{c}}(\tau) > \mathbf{0}$.

This theorem is a consequence of Theorem 5.

To close this section, let us remark that the kinematical theorems given here are expressed only in terms of the given loading history (through the relevant backward-truncated elastic stress response), as well as of the trial APSC. The initial conditions, which are involved in the previously known kinematical formulations [7, 8] are ruled out from the present formulation.

7. FINAL REMARKS AND CONCLUSIONS

We have presented the most important results previously reported in [22] as regards dynamic shakedown of discrete elastic-perfectly plastic unadapted systems; we have also extended them to include damped systems, showing that all the theorems continue to hold true.

The main points of the present new developments of dynamic

shakedown can be summarized as follows:

(i) The initial conditions (displacements, velocities, strains) do not affect the ability of a structure to adapt to a given loading history, but a change of them will have an influence on the amount of plastic deformation produced during the adaptation process, as well as on the duration of this process (adaptation time). Therefore, every structure, associated with a given load history, will possess a particular minimum adaptation time (MAT), which is the adaptation time for suitably chosen initial conditions. Further, this has suggested the concept of safety factor with respect to the ability of the structure to shake down before an assigned time (safety factor for limited MAT). The maximum value of this safety factor with respect to this assigned time coincides with the usual shakedown safety factor. The latter is shown to be unbounded when the loading history has a finite duration.

(ii) In quasi-static shakedown, Koiter's admissible plastic strain cycles (APSC) are motivated by the requirement of being globally orthogonal to any time-independent self-stress distribution; analogously, in dynamic shakedown, the APSC's are motivated by the requirement of being globally orthogonal to any time-independent self-stresses *and* to any free-motion stress distributions. From this there follows that in dynamics an APSC must possess additional features with respect to statics; in fact, in dynamics, such a strain cycle, in addition to being 'globally congruent' as in statics, is also 'globally quasi-static'. This more precise definition of APSC has permitted us to formulate new kinematical theorems in which the initial conditions are no longer involved.

In order to obtain the above results, a systematic use of the dynamic characteristics of the structure (such as displacement modal shapes and natural frequencies) has been made, giving also an explicit representation of the elastic displacement and stress responses to a given load history. Such a representation is useful not only in shakedown theory, but also in the wider field of dynamic plasticity.

For the determination of the safety factor for limited MAT, two problems dual of one another have been presented; if a suitable time discretization is adopted, they will take the forms of dual linear programming problems. In these problems, the concepts of 'separation time' (i.e. a time after which the structure is expected to satisfy Ceradini's theorem) and of 'backward-truncated' elastic stress response play a crucial role. The results herein presented show that the dynamic shakedown theory is still being improved on and that it deserves further study in the near future.

ACKNOWLEDGEMENT

This paper has been written as part of a research project sponsored by National Research Council (C.N.R.) of Italy, Structural Engineering Committee (G.I.S.).

REFERENCES

1. AUGUSTI, G. and A. BARATTA. Limit and shakedown analysis of structures with stochastic strengths. In: *2nd Int. Conf. on Structural Mechanics and Reactor Technology*, SMIRT, Berlin, 1973.
2. CAPURSO, M. Some upper bound principles for plastic strains in dynamic shakedown of elastoplastic structures, *J. Struct. Mech.*, **7** (1979), 1–20.
3. CAPURSO, M., L. CORRADI, and G. MAIER. Bounds on deformations and displacements in shakedown theory. In: *Matériaux et Structures sous Chargement Cyclique*, Ass. Amicale des Ingénieurs Anciens Elèves, E.N.P.C., Paris, 1979, 231–244.
4. CERADINI, G. Sull'adattamento dei corpi elastoplastici soggetti ad azioni dinamiche, *Giornale del Genio Civile*, **107** (1969), 239–250.
5. CERADINI, G. Dynamic shakedown in elastic–plastic bodies, *J. Engng Mech.*, Div., Proc. ASCE, EM3 (June 1980), 481–499.
6. CLOUGH, R. W. and J. PENZIEN. *Dynamics of Structures*, McGraw-Hill, New York, 1975.
7. CORRADI, L. and G. MAIER. Inadaptation theorems in the dynamics of elastic-workhardening structures, *Ingenieur-Archiv*, **43** (1973), 44–57.
8. CORRADI, L. and G. MAIER. Dynamic inadaptation theorems for elastic-perfectly plastic continua, *J. Mech. Phys. Solids*, **22** (1974), 401–413.
9. GOKHFELD, D. A. and D. E. CHERNIASKY. *Limit Analysis of Structures at Thermal Cycling*, Sijthoff & Nordhoff, The Hague, 1980.
10. KOITER, W. T. General theorems for elastic–plastic solids. In: *Progress in Solid Mechanics*, Vol. 1 (Ed. I. N. Sneddon and R. Hill), North Holland, Amsterdam, 1964, 167–221.
11. KÖNIG, J. A. On stability of the incremental collapse process, *Archiwum Inzynierii Ladowej*, **26** (1980), 219–229.
12. KÖNIG, J. A. Stability of the incremental collapse, paper contributed to the Euromech Colloquium 174 on *Inelastic Structures under Variable Loads*, University of Palermo, Oct. 10–14, 1983.
13. KÖNIG, J. and G. MAIER. Shakedown analysis of elastoplastic structures: a review of recent developments, *Nucl. Engng Design*, **66** (1981), 81–95.
14. MAIER, G. Piecewise linearization of yield criteria in structural plasticity, *SM Archives*, **1** (1976), 239–281.

15. MAIER, G. A matrix structural theory of piecewise linear elastoplastic with interacting yield planes, *Meccanica*, **V** (1970), 54–66.
16. MAIER, G. and E. VITIELLO. Bounds on plastic strains and displacements in dynamic shakedown of workhardening structures, *J. appl. Mech.*, **41** (1974), 434–440.
17. MANGASARIAN, O. L. *Nonlinear Programming*, McGraw-Hill, New York, 1969.
18. MARTIN, J. B. *Plasticity: Fundamentals and General Results*, M.I.T. Press, Cambridge, Mass., 1975.
19. NGUYEN, Q. S. Flambage par déformations plastiques cumulées sous charge cyclique additionnelle, *J. Méc. Appliquée*, **2** (1983), 351–374.
20. POLIZZOTTO, C. A unified treatment of shakedown theory and related bounding techniques, *SM Archives*, **7** (1982), 19–75.
21. POLIZZOTTO, C. Bounding principles for elastic–plastic–creeping solids loaded below and above the shakedown limit, *Meccanica*, **17** (1982), 143–148.
22. POLIZZOTTO, C. Dynamic shakedown by modal analysis, *Meccanica* **19** (1984), 133–144.
23. POLIZZOTTO, C. A bounding technique for dynamic plastic deformation of damping systems, *Nucl. Engng Design*, 1984 (in press).

20

Reliability of Rigid–Plastic Structures under Stochastic Dynamic Loading

KRZYSZTOF DOLINSKI

Institute of Fundamental Technological Research, Warsaw, Poland

ABSTRACT

Rigid–plastic structures subjected to stochastically varying loads are considered. The excursions of the stochastic load vector beyond a limit surface cause changes of the structural state, in particular they affect plastic deformations of the structure. Since the plastic displacements are assumed to be to some extent the failure indicators which control the serviceability of the structure, the probability that the vector of plastic displacements does not leave a safe domain means structural reliability.

A random process of variations in structural displacements is modelled as a Markov process. Some parameters of the load excursions are defined to describe the load–displacement relation and the first approximation technique is used to determine the transition matrix of the displacement process. The established probability of the first outcrossing of the displacement process out of the safe region is a function of time. An example is presented where the reliability of a portal frame is investigated for different load parameters.

1. INTRODUCTION

In dynamic reliability problems the event that a random time-dependent process leaves a safe region defines the structural failure. In the case of linear structural behaviour the first outcrossing of the stresses or strains beyond a given boundary is admitted as the structural failure criterion. In this case the strains or stresses play the rôle of failure indicators. They vary continuously in time for a continuously varying loading. In calculations of the failure probability some methods for the first passage probability calculation [4, 5, 16] are extensively used.

Many engineering structures have to sustain some extreme load conditions during their service time. The load intensities of, for example, strong earthquakes, hurricanes or high sea waves may appear much greater than those to which the structure is usually assumed to be subjected. It can be sometimes unreasonable and too expensive to design a structure in the elastic regime even for occasionally occurring loads of great intensity but relative short duration. Therefore, some permanent plastic deformations are allowed to occur during the structural life-time and their magnitudes define the serviceability of the structure. Hence, the structure is said to fail if the vector of the permanent deformations leaves an admissible region. This region is called the safe domain.

It is commonly recognized that the loads as quoted above are of random nature and they should be considered as random processes. In a nonelastic case the load process can be continuous in time but the failure indicators, for example, permanent displacements, are no longer continuous. They can vary within some finite time intervals only when the intensity of random loads becomes sufficiently high to produce some plastic deformations. This problem was investigated only in a few papers [2, 7, 15, 18] for the stochastic loading, where it was assumed that there was one mode of structural failure and one-dimensional failure indicator.

In this paper a multimode failure and multidimensional failure indicators are considered for rigid–plastic structures. The problem formulation is given in the next section whereas some assumptions and simplifications concerning the stochastic properties of loading are presented in the third section. The solution method for the probability of structural failure is proposed in the fourth section and in the next one an example of a rigid–plastic frame is considered.

2. FORMULATION OF THE PROBLEM

The problem formulation is essentially based on the concept of the structural reliability analysis given by Bolotin [1]. Originally three spaces are considered: of loads, of states and of quality. It is assumed that some random changes of the load vector cause some respective changes of the structural state so that a vector describing the quality of the structure varies until it reaches a given boundary. This event means the structural failure.

Applying this concept to rigid–plastic structures it is useful to modify it by some specifications:

(a) the quality space is the space $\mathscr{U}$ of plastic displacements, u_k, $k = 1, 2, \ldots, N_u$, which are chosen at some critical points of the structure;

(b) for a rigid–plastic structure there exists a limit surface, $S_L(\mathbf{x}) = 0$, in the space $\mathscr{X}$ of loads, x_i, $i = 1, 2, \ldots, N_L$. The structure starts to deform at the time instant t° only if the load vector $\mathbf{X}(t)$ leaves the passive region $\mathscr{P}_L = \{\mathbf{x} : S_L(\mathbf{x}) < 0\}$. Deformation takes place within the time interval $[t^\circ, t']$ while $\mathbf{X}(t)$ is outside the region $\mathscr{P}_L$. The deformation continues after the return of $\mathbf{X}(t)$ to $\mathscr{P}_L$ until $t = t^f$ when noninertial forces are no longer necessary to assure equilibrium of the structure. The structure remains in rest afterwards up to the next excursion of the load vector beyond the passive domain $\mathscr{P}_L$, i.e.:

$$\begin{aligned} &t^\circ \text{ denotes an instant of time when} \\ &\quad S_L[\mathbf{X}(t^\circ)] = 0 \text{ and } \dot{S}_L[\mathbf{X}(t^\circ)] > 0 \\ &t' \text{ denotes an instant of time when} \\ &\quad S_L[\mathbf{X}(t')] = 0 \text{ and } \dot{S}_L[\mathbf{X}(t')] < 0 \\ &t^f \text{ denotes an instant of time when} \\ &\quad S_L[\mathbf{X}(t^f)] < 0 \text{ and } \dot{\mathbf{U}}(t^f) = 0 \end{aligned} \tag{1}$$

(c) there is a direct relation

$$U_k(t) = \mathscr{F}^k_{[0,t]}[\mathbf{X}(\tau)] \tag{2}$$

between the load vector $\mathbf{X}(\tau)$ until the time instant t and the plastic displacements $U_k(t)$ so that it is not necessary to consider separately the space of the structural states;

(d) in the space $\mathscr{U}$ the safe region is defined

$$\mathscr{U}_S = \{\mathbf{u} : S_U(\mathbf{u}) < 0\} \tag{3}$$

where $S_U(\mathbf{u}) = 0$ denotes the failure surface imposed on the plastic displacements by the actual serviceability requirements.

The structural reliability $R(t)$ is now calculated as the probability that the vector of plastic displacements, $\mathbf{U}(t)$, remains within the time interval $[0, t]$ in the safe region $\mathcal{U}_S$

$$R(t) = \mathbb{P}[\mathbf{U}(\tau) \in \mathcal{U}_S; \tau \in [0, t]] \tag{4}$$

It is still the first passage problem for the stochastic vector $\mathbf{U}(t)$ but the stochastic properties of $\mathbf{U}(t)$ depend on the random excursions of the stochastic load vector $\mathbf{X}(t)$ beyond the passive region $\mathcal{P}_L$. The structural reliability analysis consists therefore of two subproblems:

determination of the stochastic properties of the vector $\mathbf{U}(t)$ involving random excursions of the load vector $\mathbf{X}(t)$;
calculation of the first crossing for $\mathbf{U}(t)$.

3. ASSUMPTIONS AND SIMPLIFICATIONS

Two principal assumptions concerning the stochastic properties of the load vector, $\mathbf{X}(t)$, are made in this paper:

(a) $\mathbf{X}(t)$ is a Gaussian process with stationary components;
(b) all components are twice differentiable with probability one.

Moreover, the limit surface, $S_L(\mathbf{x}) = 0$, is assumed to be piece-wise linear and defines a convex passive domain $\mathcal{P}_L$ so that

$$\mathcal{P}_L = \left\{ \mathbf{x} : \bigwedge_{j=1}^{M} Z_j(\mathbf{x}) < 0 \right\} \tag{5}$$

where

$$Z_j(\mathbf{x}) = a_{jo} + \sum_{i=1}^{N_L} a_{ji} x_i; \quad \text{for} \quad j = 1, 2, \ldots, M \tag{6}$$

Furthermore, it is assumed that within the long time interval $[0, T]$ the structure possesses a high static reliability level R_S in the sense that

$$R_S = \min_{[0,T]} \mathbb{P}[\mathbf{X}(t) \in \mathcal{P}_L] = 1 - \varepsilon; \quad \text{for} \quad 0 < \varepsilon \ll 1 \tag{7}$$

Assumption (7) assures the excursions to be very rare events of relatively short duration. Thus, the general expression (2) can be replaced by a sum of increments of the plastic displacements due to the separate excursions

$$\mathbf{U}(t) = \sum_{n=1}^{N(t)} \Delta \mathbf{U}^{(n)} \tag{8}$$

where $N(t)$ denotes a random number of excursions within a given time interval $[0, t]$ and

$$\Delta U_k^{(n)} = \mathscr{F}^{k(n)}_{[t^o,t^f]}[\mathbf{X}(\tau)] \tag{9}$$

Any direct use of eqn. (9) is not possible in the probability calculation. The stochastic character of the load path during the excursion makes expressions of this functional type impracticable. However, using the results obtained by Youngdahl [19] and Krajcinovic [12] the final plastic displacements can be related to some excursion parameters. These parameters are here defined as follows:

duration of the structural motion $\tau^f = t^f - t^o$
upcrossing point, $\mathbf{X}(t^o)$, that determines on which of the limit hyperplanes, $Z_j(\mathbf{x}) = 0$, the excursion begins;
downcrossing point, $\mathbf{X}(t')$, that determines on which of the limit hyperplanes the excursion ceases;
equivalent level of the load vector during an excursion

$$\mathbf{X}^{oE} = \frac{1}{t' - t^o} \int_{t^o}^{t'} \mathbf{X}(\tau)\, d\tau \tag{10}$$

equivalent level of the load vector during the post-excursion structural motion

$$\mathbf{X}'^F = \frac{1}{t^f - t'} \int_{t'}^{t^f} \mathbf{X}(\tau)\, d\tau \tag{11}$$

The relation (9) becomes now a function depending on the excursion parameters which are random variables

$$\Delta U_k^{(n)} = \varphi_k^{(n)}[\tau^f, \mathbf{X}(t^o), \mathbf{X}(t'), \mathbf{X}^{oE}, \mathbf{X}'^E] \tag{12}$$

4. METHOD OF SOLUTION

All possible load excursions beyond the piece-wise linear limit surface are divided into three groups:

(a) single-plane excursions, S1, which begin and cease on the same, j_1, limit hyperplane, i.e.:

$$Z_{j_1}(t^o) = 0 \wedge \bigwedge_{j \neq j_1} Z_j(t^o) < 0 \wedge Z_{j_1}(t') = 0 \wedge \bigwedge_{j \neq j_1} Z_j(t') < 0 \tag{13}$$

(b) double-plane excursions, D11′, which begin on the j_1 limit hyperplane and cease on the *neighbouring*, $j_{1'}$, hyperplane, i.e.:

$$Z_{j_1}(t^o)=0\wedge\bigwedge_{j\neq j_1} Z_j(t^o)<0\wedge Z_{j_{1'}}(t')=0\wedge\bigwedge_{j\neq j_{1'}} Z_j(t')<0 \quad (14)$$

(c) other excursions which overflow the neighbouring hyperplane to cease on the further one. These excursions are neglected in the further considerations because they are quite unlikely due to the assumption (7).

In order to determine the magnitude of the plastic displacements due to the load excursion beyond the limit surface the mode approximation technique [13, 14] is applied. Its application is not a necessary condition in this method. Any other approach can be successfully applied provided that the displacement increments given the excursion considered are able to be expressed in the following form:

$$\Delta U_k^{(n)}|_{S1}=\psi_k^{(n)}|_{S1}(\tau^f, Z_{j_1}^{oE}, Z_{j1}'^{E}) \quad (15)$$

$$\Delta U_k^{(n)}|_{D11'}=\psi_k^{(n)}|_{D11'}(\tau^f, Z_{j_1}^{oE}, Z_{j_{1'}}^{oE}, Z_{j_1}'^{E}, Z_{j_{1'}}'^{E}) \quad (16)$$

where

$$Z_{\cdot}^{oE}=\frac{1}{t'-t^o}\int_{t^o}^{t'} Z_{\cdot}[\mathbf{X}(t)]\,\mathrm{d}t \quad (17)$$

$$Z_{\cdot}'^{E}=\frac{1}{t^f-t'}\int_{t'}^{t^f} Z_{\cdot}[\mathbf{X}(t)]\,\mathrm{d}t \quad (18)$$

The arguments in eqns. (15) and (16) are random variables and the functions are moreover conditioned by the type of excursion defined in eqns. (13) and (14), respectively.

Because of the linearity of the limit surface and the stationarity of the load components all arguments in eqns. (15) and (16) are defined by excursions of stationary Gaussian processes $Z_j(t)=Z_j[\mathbf{X}(t)]$. Since the excursion level is high (eqn. (7)), it is justified to assume a parabolic approximation of the excursion path

$$Z_{\cdot}(t)=Z_{\cdot}^o+\dot{Z}_{\cdot}^o\,.\,(t-t^o)+\ddot{Z}_{\cdot}^o\,.\,(t-t^o)/2 \quad (19)$$

where $Z_{\cdot}^o, \dot{Z}_{\cdot}^o, \ddot{Z}_{\cdot}^o$ denote, respectively, the value of the function $Z_{\cdot}(t)$, and its first and second derivative at the time $t=t^o$. Extending the parabola (19) until the time $t=t^f$ it is easily seen that the arguments in eqns. (15) and (16) become functions of the appropriate vectors $[Z_{\cdot}^o, \dot{Z}_{\cdot}^o, \ddot{Z}_{\cdot}^o]$, so that one can rewrite them in the form

$$\Delta U_k^{(n)}|_{S1}=\psi_k^{(n)}|_{S1}(\dot{Z}_{j_1}^o, \ddot{Z}_{j_1}^o) \quad (20)$$

$$\Delta U_k^{(n)}|_{D11'}=\psi_k^{(n)}|_{D11'}(\dot{Z}_{j_1}^o, \ddot{Z}_{j_1}^o, Z_{j_1}^o, \dot{Z}_{j_1}^o, \ddot{Z}_{j_{1'}}^o) \quad (21)$$

By the definition of the double-plane excursion (eqn. (14)), the calculation of the time t' appears to be, in general, difficult to carry out. Therefore, the double-plane excursions are described in this paper in the following form

$$\mathrm{D11'} \to Z^{\circ}_{j_1}=0 \wedge \bigwedge_{j \neq j_1} Z^{\circ}_{j}<0 \wedge Z''_{j_1}=0 \wedge Z''_{j_{1'}}>0 \tag{22}$$

where $Z''_{\cdot}=Z_{\cdot}(t'')$, and t'' denotes an instant of time at which the excursion beyond the hyperplane $z_{j_1}=0$ ceases but the random function $Z_{j_{1'}}[X(t'')]$ still remains positive (above the hyperplane $z_{j_{1'}}=0$). It means that all excursions around the corners of the passive domain (where more than two hyperplanes meet) will be considered more than once. These excursions are however very infrequent and their participation in the probability distribution of $\mathbf{\Delta U}$ is negligible.

Conditional probability distributions of the kth component of the vector of plastic displacement increment for the nth excursion given the excursion type now take the following forms:

$$F^{(n)}_{k}(u_k \mid \mathrm{S1})=\mathbb{P}[\psi^{(n)}_{k}|_{\mathrm{S1}}(\dot{Z}^{\circ}_{j_1}, \ddot{Z}^{\circ}_{j_1}) \leqslant u_k \mid Z^{\circ}_{j_1}=0, Z'_{j_1}=0, \mathbf{Z}^{\circ}_{j(1)}<\mathbf{0}, \mathbf{Z}'_{j(1)}<\mathbf{0}] \tag{23}$$

$$F^{(n)}_{k}(u_k \mid \mathrm{D11'})=\mathbb{P}[\psi^{(n)}_{k}|_{\mathrm{D11'}}(\dot{Z}^{\circ}_{j_1}, \ddot{Z}^{\circ}_{j_1}, Z^{\circ}_{j_{1'}}, \dot{Z}^{\circ}_{j_{1'}}, \ddot{Z}^{\circ}_{j_{1'}}) \leqslant u_k \mid Z^{\circ}_{j_1}=0, Z''_{j_1}=0, \mathbf{Z}^{\circ}_{j(1)}<\mathbf{0}, \mathbf{Z}''_{j(1)}<\mathbf{0}] \tag{24}$$

whcrc

$$\mathbf{Z}^{\cdot}_{j(1)}=[Z^{\cdot}_{1}, \ldots, Z^{\cdot}_{j_{l-1}}, Z^{\cdot}_{j_{l+1}}, \ldots, Z^{\cdot}_{M}] \tag{25}$$

The conditioning events in eqns. (23) and (24) can be considered as separate if the overlapping of the corner excursion events is neglected, as mentioned above. The probability distribution of an increment of the plastic displacement vector for the nth excursion can be expressed as a sum of probabilities

$$F^{(n)}(\mathbf{u})=\Bigg\{\sum_{l=1}^{M} \mathbb{P}\Bigg[\bigwedge_{k=1}^{N_u} \psi^{(n)}_{k}|_{\mathrm{S1}} \leqslant u_k, \mathbf{Z}^{\circ}_{j(1)}<\mathbf{0}, \mathbf{Z}'_{j(1)}<\mathbf{0} \mid Z^{\circ}_{j_1}=0, Z'_{j_1}=0\Bigg] \cdot f_{Z^{\circ}_{j_1} Z'_{j_1}}(0,0)+\sum_{\substack{l=1 \\ l \neq 1'}}^{M} \sum_{l'=1}^{M} \mathbb{P}\Bigg[\bigwedge_{k=1}^{N_u} \psi^{(n)}_{k}|_{\mathrm{D11'}} \leqslant u_k, \mathbf{Z}^{\circ}_{j(1)}<\mathbf{0}, Z''_{j_{1'}}>0 \mid Z^{\circ}_{j_1}=0, Z''_{j_1}=0\Bigg] \cdot f_{Z^{\circ}_{j_1} Z''_{j_1}}(0,0)\Bigg\} \Bigg/ \sum_{l=1}^{M} \mathbb{P}[\mathbf{Z}^{\circ}_{j(1)}<\mathbf{0} \mid Z^{\circ}_{j_1}=0\Bigg] \cdot f_{Z^{\circ}_{j_1}}=(0) \tag{26}$$

The calculation of the probabilities in eqn. (26) is based on the first order approximation technique [3, 8, 9]. For every probability term the problem is transformed into the space $\mathcal{Y}$ of standard normal independent random variables $\mathbf{Y}=[Y_1, Y_2, \ldots, Y_K]$ so that the following equalities are satisfied:

for single-plane excursions, $l=1, \ldots, M$:

$$\mathbb{P}[h_k^{(n)}|_{S1}(\mathbf{Y})\leqslant 0]=\mathbb{P}[\psi_k^{(n)}|_{S1}(\dot{Z}_{j_1}^{o}, \ddot{Z}_{j_1}^{o})-u_k\leqslant 0] \quad \text{for } k=1, \ldots, N_u$$
$$\mathbb{P}[h_{N_u+1}^{(n)}|_{S1}(\mathbf{Y})\leqslant 0]=\mathbb{P}[\mathbf{Z}_{j(1)}^{o}<0, \mathbf{Z}'_{j(1)}<0 \mid Z_{j_1}^{o}=0, Z'_{j_1}=0] \tag{27}$$

for double-plane excursions, $l, l'=1, \ldots, M, l=1'$:

$$\mathbb{P}[h_k^{(n)}|_{D11'}(\mathbf{Y})\leqslant 0]=\mathbb{P}[\psi_k^{(n)}|_{D11'}(\dot{Z}_{j_1}^{o}, \ddot{Z}_{j_1}^{o}, Z_{j_{1'}}^{o}, \dot{Z}_{j_{1'}}^{o}, \ddot{Z}_{j_{1'}}^{o})-u_k\leqslant 0]$$
$$\mathbb{P}[h_{N_u+1}^{(n)}|_{D11'}(\mathbf{Y})\leqslant 0]=\mathbb{P}[\mathbf{Z}_{j(1)}^{o}\langle 0, Z''_{j_{1'}}\rangle 0 \mid Z_{j_1}^{o}=0, Z''_{j_1}=0] \tag{28}$$

The conditions (27) and (28) are assured, for example, by the Rosenblatt transformation [10, 17]. Due to this transformation the functions $h_k^{(n)}|_{S1}(\mathbf{y})=0$ and $h_k^{(n)}|_{D11'}(\mathbf{y})=0$ are defined in the space $\mathcal{Y}$ and the probability measures of the respective sets in the original space and in the space $\mathcal{Y}$ are equal to each other.

This approach leads to a very typical problem in the reliability analysis of structural systems. For every probability term in eqn. (26) and for any given vector of plastic displacement increment $\Delta\mathbf{u}$ the probability that some functions of independent standard normal random variables are less than zero, has to be found. An approximate solution of this problem can be given by the first order approximation technique where the hypersurfaces $h_k^{(n)}|_{S1}(\mathbf{y})=0$ or $h_k^{(n)}|_{D11'}(\mathbf{y})=0$ for $k=1, \ldots, N_u, N_u+1$ are replaced by the hyperplanes that are tangent to these hypersurfaces in the points $\mathbf{y}_k^{o}$ which are the closest ones to the origin in $\mathcal{Y}$. Since the random vectors $\mathbf{Y}$ are normal any linear combination of their components yields a normal random variable. Thus, after the linearization in $\mathcal{Y}$ space the problem involves some normal correlated random variables only. In the last few years methods have been developed to solve such a problem and they yield estimates or bounds of the exact solution [4, 6].

In this paper it is assumed that the increment of plastic displacements does not depend on their history. Moreover, the actual value of the displacements affects neither the limit surface nor any possible increment of the plastic deformations. Thus, the displacement vector can be modelled as a homogeneous Markov process. The safe domain $\mathcal{U}_S$ in the space of plastic displacements is discretized so that a finite number of such states, NS, is considered and the state $NS+1$ defines the structural failure where the vector **U** takes any value outside of the safe domain. For every state $\mathbf{u}_{ns}$, $ns = 1, \ldots, NS+1$, the value of the probability distribution function (eqn. (26)), is looked for. These probabilities enable us to build the transition matrix $\mathbf{P}^{(1)}$ whose elements (i, j) denote the probabilities that after a single load excursion the vector of plastic displacements changes its value from $\mathbf{u}_i$ to $\mathbf{u}_j$, i, $j = 1, 2, \ldots, NS+1$.

Due to the assumption (7) the number of the excursions of the load vector $\mathbf{X}(t)$ beyond the limit surface $S_L(\mathbf{x}) = 0$ within a time interval $[0, t]$ can be approximated by the Poisson distribution

$$\mathbb{P}[N = n \mid t] = \frac{(\nu t)^n}{n!} \exp(-\nu t) \tag{29}$$

where ν denotes the outcrossing mean rate of the random vector $\mathbf{X}(t)$ out of the passive domain $\mathcal{P}_L$.

Using the Markov properties of a random vector **U** and the deterministic initial condition

$$\mathbb{P}[\mathbf{U}(0) = \mathbf{0}] = 1 \tag{30}$$

the probability that the structure does not fail after n load excursions can be expressed by the $(1, NS+1)$st element $P^{(n)}_{1,NS+1}$ of the matrix $\mathbf{P}^{(n)}$

$$\mathbf{P}^{(n)} = \underbrace{\mathbf{P}^{(1)} \ldots \mathbf{P}^{(1)} \ldots \mathbf{P}^{(1)}}_{n \text{ times}} \tag{31}$$

in the form

$$\mathbb{P}[\mathbf{U} \in \mathcal{U}_S \mid N = n] = 1 - P^{(n)}_{1,NS+1} \tag{32}$$

Combining eqns. (29) and (32) one obtains the structural reliability $R(t)$ as a function of time

$$R(t) = \sum_{n=0}^{\infty} \mathbb{P}[N = n \mid t] \,.\, (1 - P^{(n)}_{1,NS+1}) = 1 - \exp(-\nu t) \sum_{n=1}^{\infty} \frac{(\nu t)^n}{n!} P^{(n)}_{1,NS+1} \tag{33}$$

5. NUMERICAL EXAMPLE

As an example a rigid–plastic portal frame is considered (Fig. 1). The vertical permanent displacement, U_1, in mid-span of the beam and the horizontal one, U_2, at the top of columns are accepted as failure control parameters. Due to the random vertical and horizontal loads

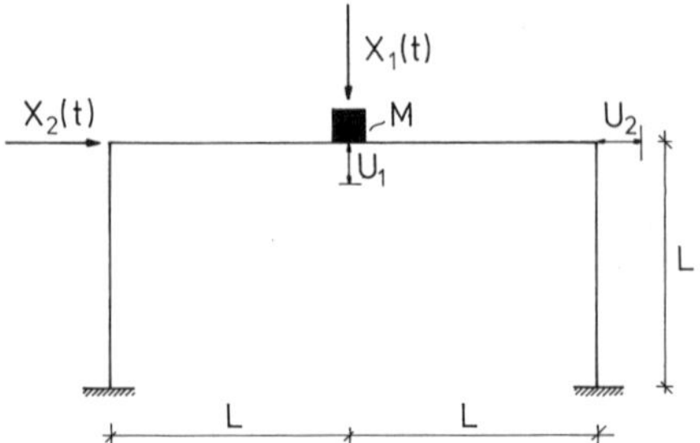

FIG. 1. Portal frame.

the permanent displacements vary randomly in time. The calculation was carried out for an ultimate displacement level $u_1^L = u_2^L = 0{\cdot}005 \,.\, L$; six combinations of the means and standard variations of the loads were considered (see Table I). The loads are assumed to be normal, stationary stochastic processes with autocorrelation functions

$$K_{X_i}(\tau) = \sigma_i^2 \exp(-\alpha_i^2 \tau^2), \qquad i = 1, 2 \tag{34}$$

The load parameters (means and standard deviations) are assumed in such a way that the static reliability levels for the respective triplet of systems are equal, i.e.

$$\mathbb{P}[\mathbf{X} \in \mathcal{P}_L] = \begin{cases} 1 - 3{\cdot}17 \times 10^{-5} & \text{for the group I} \\ 1 - 8{\cdot}04 \times 10^{-9} & \text{for the group II} \end{cases} \tag{35}$$

TABLE I

System	*Mean/N*$_o$ $\bar{X}_1$	$\bar{X}_2$	*Standard deviation/N*$_o$ 1	2	*Concentrated mass* M/N_o	*Static reliability* R_S
IA	2·0	1·5	0·250	0·200	1·00	$1-3{\cdot}17\times10^{-5}$
IB	1·0	2·0	0·400	0·300		
IC	0·5	3·0	0·200	0·250		
IIA	2·0	1·5	0·177	0·150		$1-8{\cdot}04\times10^{-9}$
IIB	1·0	2·0	0·250	0·250		
IIC	0·5	3·0	0·150	0·177		

$N_o = 30$ kN, constant.

where

$$\mathscr{P}_L = \{\mathbf{x}: z_i(\mathbf{x}) \leqslant 0, i = 1, 2, 3\}$$

and

$$z_1(\mathbf{x}) = x_1 - 4\frac{\tilde{M}_o}{L \cdot N_o}; \qquad z_2(\mathbf{x}) = x_1 + x_2 - 6\frac{M_o}{L \cdot N_o};$$

$$z_3(\mathbf{x}) = x_2 - 4\frac{M_o}{L \cdot N_o}$$

M_o denotes the yield moment in a plastic hinge.

In the actual approach the length of a life-time interval t_{R_i} of the frame is looked for. t_{R_i} denotes the time interval within which the permanent displacements remain in the safe domain $\mathscr{U}_S = \{u_1 \leqslant u_1^L, u_2 \leqslant u_2^L\}$ with a given probability R_i. The results were carried out for $R_I = 0{\cdot}999$ for the systems I and for $R_{II} = 0{\cdot}9999$ for the systems II and are shown in Figs. 2(a) and 3(a), respectively, as functions of the autocorrelation function parameter α. The times t_R are

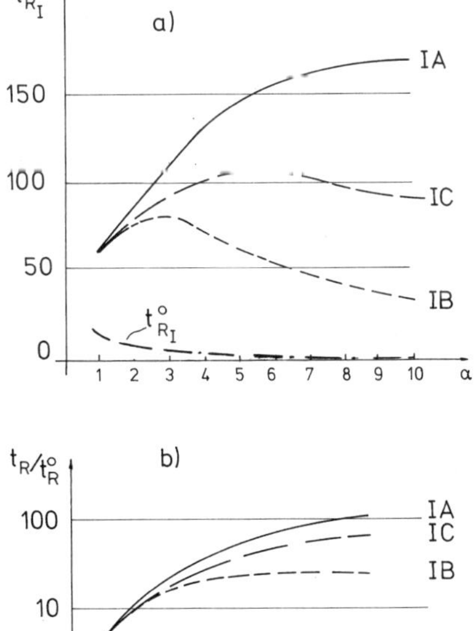

FIG. 2. Life-time of the frame for the first group of the load parameters and $R_I = 0{\cdot}999$.

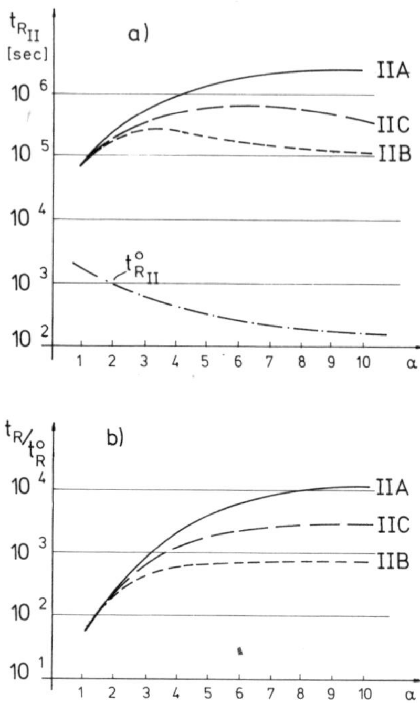

FIG. 3. Life-time of the frame for the second group of the load parameters and $R_{II}=0{\cdot}9999$.

compared with the times t^o_R to the first upcrossing of the limit surface by the load vector $X(t)$. The ratios t_R/t^o_R are shown in Figs. 2(b) and 3(b) for respective groups of systems. There is seen that the life-time of a structure, t_{R_i}, is much longer if some, even small, plastic deformations are allowed. Moreover, the reliability of structures with equal static reliability level and equal times to the first load outcrossing, $t^o_{R_i}$, can differ if a certain plastic deformation level is assumed as the failure criterion.

The autocorrelation function parameter, α, governs the frequency and duration of a load excursion so that for small α long and infrequent excursions are expected and for increasing α the excursions become more frequent and shorter. Therefore, for very small α the first load excursion can lead to structural failure and for increasing α the excursion number has to increase as well to yield the ultimate level of plastic deformations.

ACKNOWLEDGEMENT

This paper was written during the author's stay as an Alexander von Humboldt fellow with the Technical University in Munich, Germany.

The author offers his most grateful thanks to Professor G. I. Schuëller from the University of Innsbruck for some encouraging suggestions and comments relating to the preparation of the paper.

REFERENCES

1. Bolotin, V. V. Reliability theory of mechanical systems with finite number of degrees of freedom (in Russian), *Mechanika Tverdovo Tela*, No. 5, (1969), 73–81.
2. Casciati, F., L. Faravelli and A. Gobetti. Permanent deformations of rigid–plastic structures subjected to random dynamic loads, *Engng Struct.*, **1** (1979), 139–144.
3. Cornell, C. A. A first-order reliability theory for structural design, Study No. 3, Structural Reliability and Codified Design, University of Waterloo, Ontario, Canada, 1969.
4. Crandal, S. H. First-crossing probabilities of the linear oscillator, *J. Sound Vibr.*, **12** (1971), 285–299.
5. Ditlevsen, O. Extreme and first passage times with application to civil engineering, Technical University of Denmark, Copenhagen, Denmark, 1971.
6. Ditlevsen, O. Narrow reliability bounds for structural systems, *J. Struct. Mech.*, **7** (1979), 453–472.
7. Dolinski, K. Kinematic approach to reliability estimate of stochastically loaded plastic structures, *J. Struct. Mech.*, **6** (1978) 247–266.
8. Dolinski, K. First-order second moment approximation in reliability of structural systems; critical review and alternative approach, *Struct. Safety*, **1** (1983), 211–231.
9. Hasofer, A. M. and N. C. Lind. Exact and invariant second moment code format, *J. Engng Mech., ASCE,* **100** (EM1), (1974), 111–121.
10. Hohenbichler, M. and R. Rackwitz. Non-normal dependent vectors in structural safety, *J. Engng Mech., ASCE,* **107** (1981), 1227–1238.
11. Hohenbichler, M. Approximate evaluation of the multinormal distribution function, *Berichte zur Zuverleassigkeitstheorie der Bauwerke,* Technical University of Munich, Heft 58, 1981.
12. Krajcinovic, D. Dynamic analysis of clamped plastic circular plates, *Int. J. Mech. Sci.*, **14** (1972) 225–234.
13. Martin, J. B. and P. S. Symonds. Mode approximations for impulsively loaded rigid-plastic structures, *Proc. ASCE, J. Engng Mech. Div.*, **92** (1966), 46–66.
14. Lee, L. S. S. Mode response of dynamically loaded structures, *J. appl. Mech.*, **39** (1972), 904–910.

15. NIELSEN, S. R. K., J. D. SØRENSEN and P. THORF-CHRISTENSEN. Life-time reliability estimate and extreme permanent deformations of randomly excited elasto–plastic structures, Institute of Building Technology and Structural Engineering, Paper No. 15, Aalborg, Denmark, August 1981.
16. ROBERTS, J. B. The yielding behaviour of a randomly excited elasto–plastic structure, *J. Sound Vibr.*, **72** (1980), 71–85.
17. ROSENBLATT, M., Remarks on a multivariate transformation, *Ann. Math. Stat.*, **23** (1952), 470–472.
18. VENEZIANO, D. and E. H. VANMARCKE. Probabilistic seismic response of simple inelastic systems, *Proc. 5th World Conf. Earthquake Engng* (Rome, 1973).
19. YOUNGDAHL, C. K. Correlation parameters for eliminating the effect of pulse shape on dynamic plastic deformation, *J. appl. Mech.*, **37** (1970), 744–752.

Section 4

BIFURCATION AND INSTABILITY

21

On Bifurcation and Stability under Elastic–Plastic Deformation

VIGGO TVERGAARD

Department of Solid Mechanics, Technical University of Denmark, Lyngby, Denmark

ABSTRACT

Some recent developments in the analysis of uniqueness and bifurcation for time-independent elastic–plastic solids are discussed. Localization of plastic flow in shear bands as well as bifurcation into diffuse modes is considered, and special interest is taken in materials that form a vertex on subsequent yield surfaces, or materials that show a plastic dilatancy and a non-normality of the plastic flow law. For some of these materials the first critical bifurcation point is not found directly; but procedures for determining upper and lower bound estimates are discussed. Studies of post-bifurcation behaviour and imperfection-sensitivity require numerical analyses, and a few examples are presented. For materials with strain-rate sensitivity a brief discussion of stability analyses is included at the end of the paper.

1. INTRODUCTION

Much basic understanding of elastic–plastic stability problems was developed during the first half of this century in connection with studies of plastic buckling. For columns Shanley [40] explained the significance of the so-called tangent modulus load as the critical bifurcation point, at which the straight configuration loses its uniqueness, but not its stability.

A firm theoretical basis was provided by Hill's [9, 10, 13] general theory of uniqueness and bifurcation in elastic–plastic solids, which applies to bifurcation under large strain conditions as well as to the more classical area of structural buckling. This theory is based on introducing an elastic comparison solid, with the property that bifurcation in the actual elastic–plastic solid is excluded as long as uniqueness of the incremental solution is predicted for the comparison solid.

Elastic–plastic bifurcation predictions are often quite sensitive to details of the assumed constitutive law. An extreme dependence on the material description has been found for the cruciform column (Hutchinson [19]), and also for cases of plastic flow localization, where bifurcation into a shear band coincides with the loss of ellipticity of the governing incremental equations (Rice [36]).

Some elastic–plastic constitutive relations of great current interest, developed with the intention of describing the behaviour of rocks or soils, or the influence of the nucleation and growth of micro-voids in ductile metals, are characterized by non-normality of the plastic flow law. This is not accounted for by Hill's [9] bifurcation theory; but recently Raniecki and Bruhns [34] have proposed an alternative comparison solid that provides a lower bound to the first critical bifurcation point. For solids that form a vertex on subsequent yield surfaces Hill's bifurcation theory applies, with a comparison solid based on the moduli of the total loading range. In many problems of interest the correct location of the bifurcation point is predicted by this comparison solid. For other cases an upper bound estimate is suggested.

A full understanding of bifurcation behaviour requires also an analysis of the behaviour after bifurcation and of the sensitivity to small imperfections, as is provided in the elastic range by Koiter's [23] general theory of elastic stability. For the plastic range Hill's [9] bifurcation theory has been extended into the initial post-bifurcation range by an asymptotic theory due to Hutchinson [17, 19]. This theory is considerably complicated by the necessity to account for elastic unloading regions that start pointwise at bifurcation and subsequently spread into the material. In most cases loss of stability occurs subsequent to bifurcation, at some point on the post-bifurcation solution. So far, information regarding advanced post-bifurcation behaviour or imperfection-sensitivity in the plastic range relies entirely on numerical analyses.

The main focus in the present paper is on bifurcation under finite strain conditions; but a few examples of structural buckling are discussed as well. First a simple analysis of shear band formation is presented, and subsequently criteria for bifurcation into diffuse modes are discussed.

In the last section the influence of material strain-rate sensitivity on stability analyses is considered briefly. In the context of a rate-sensitive formulation the most important feature is a very strong sensitivity to small imperfections, and here bifurcation plays no central role.

2. BASIC EQUATIONS

The investigations of stability and bifurcation to be discussed here will be presented in the framework of a Lagrangian formulation of the field equations. A material point in the solid is identified by the co-ordinates x^i in the reference configuration, and the displacement components on the reference base vectors are denoted by u^i. The Lagrangian strain tensor, $\eta_{ij} = \frac{1}{2}(G_{ij} - g_{ij})$, is related to the displacements by the expression

$$\eta_{ij} = \tfrac{1}{2}(u_{i,j} + u_{j,i} + u^k_{,i} u_{k,j}) \tag{1}$$

where G_{ij} and g_{ij} are the metric tensors in the current configuration and in the reference configuration, respectively, with determinants G and g, and $(\ )_{,i}$ denotes covariant differentiation in the reference configuration. Latin indices range from 1 to 3, and the summation convention is adopted for repeated indices.

The contravariant components τ^{ij} of the Kirchhoff stress tensor on the embedded deformed co-ordinates are related to the components of the Cauchy stress tensor σ^{ij} by

$$\tau^{ij} = \sqrt{G/g}\, \sigma^{ij} \tag{2}$$

Equilibrium states satisfy the principle of virtual work

$$\int_V \tau^{ij}\, \delta\eta_{ij}\, \mathrm{d}V = \int_S T^i\, \delta u_i\, \mathrm{d}S \tag{3}$$

where V and S are the volume and surface, respectively, of the body in the reference configuration. The specified nominal tractions T^i on a surface with normal n_j in the reference state can be expressed in terms of the stresses by

$$T^i = (\tau^{ij} + \tau^{kj} u^i_{,k}) n_j \tag{4}$$

Finite strain generalizations of elastic–plastic constitutive laws are often formulated as relations between the strain increments η_{ij} and the spin-invariant Jaumann rate of the Cauchy stress tensor $\overset{\nabla}{\sigma}{}^{kl}$, which is given in terms of the convected rate $\dot{\sigma}^{kl}$ by

$$\overset{\nabla}{\sigma}{}^{kl} = \dot{\sigma}^{kl} + (G^{km}\sigma^{ln} + G^{lm}\sigma^{kn})\dot{\eta}_{mn} \tag{5}$$

The strain-rate $\dot{\eta}_{ij}$ is taken to be the sum of an elastic part $\dot{\eta}^{\mathrm{E}}_{ij}$, which is linearly related to $\overset{\nabla}{\sigma}{}^{kl}$, and a plastic part $\dot{\eta}^{\mathrm{P}}_{ij}$, which is homogeneous of degree one in $\overset{\nabla}{\sigma}{}^{kl}$. If the sum of these two expressions is inverted to yield an expression for $\overset{\nabla}{\sigma}{}^{kl}$, the incremental constitutive relations may be expressed as

$$\dot{\tau}^{ij} = L^{ijkl}\dot{\eta}_{kl} \tag{6}$$

using (5) and the incremental form of (2).

3. LOCALIZATION OF PLASTIC FLOW

In a number of different materials, ranging from ductile metals to polymers or soils, a rather sudden change from smooth deformation patterns to patterns involving localized flow in one or more narrow shear bands has been observed. Such flow localization is often a direct precursor of final material failure, since very high strains develop inside the bands without contributing much to the overall straining of the body. Theoretical investigations have shown that predictions of shear band formation are very sensitive to the constitutive law (Rice [36]).

If an initial material inhomogeneity is assumed inside a thin plane slice of material, and the stress-states inside and outside this slice of material, respectively, are assumed to remain homogeneous throughout the deformation history, a simple model problem is obtained for the study of shear band development. The normal of the thin slice of material is denoted by n_j, in a Cartesian reference co-ordinate system (Fig. 1), and quantities inside and outside the band are denoted by $(\ )^{\mathrm{b}}$ and $(\ )^{\mathrm{o}}$, respectively.

Compatibility requires no jump in tangential derivatives of the displacements u_i over the band interface, so that

$$u^{\mathrm{b}}_{i,j} = u^{\mathrm{o}}_{i,j} + c_i n_j \tag{7}$$

where c_i are parameters to be determined. Furthermore, equilibrium requires continuity of the nominal tractions T^i over the band interface

$$(T^i)^{\mathrm{b}} = (T^i)^{\mathrm{o}} \tag{8}$$

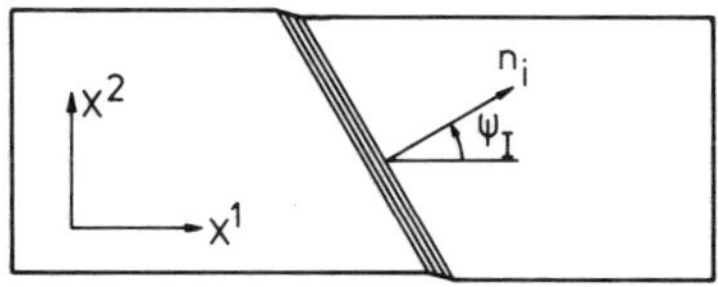

FIG. 1. Specimen in which a thin slice of material contains an initial inhomogeneity.

Using (4), the incremental form of (8) can be written as

$$[(\dot{\tau}^{ij}+\dot{\tau}^{kj}u^i_{,k}+\tau^{kj}\dot{u}^i_{,k})n_j]^{\mathrm{b}}=[(\dot{\tau}^{ij}+\dot{\tau}^{kj}u^i_{,k}+\tau^{kj}\dot{u}^i_{,k})n_j]^{\mathrm{o}} \tag{9}$$

Substitution of (1), (6) and (7) into (9) gives a set of incremental equations for $\dot{c}_i$, provided that the stress or strain history outside the band is prescribed.

If there is no imperfection, the equations for $\dot{c}_i$, resulting from (9), are homogeneous, so that non-zero solutions occur only when the determinant of the coefficient matrix vanishes. The first such bifurcation into a shear band mode, for any band inclination, coincides with the loss of ellipticity of the equations governing incremental equilibrium (see Hill [11], Rudnicki and Rice [37], and Rice [36]). Analyses of shear band bifurcation show that the classical elastic–plastic solid with a smooth yield surface is quite resistant to localization, unless the strain hardening level is very low. However, the critical strain for localization is considerably lowered by deviations from the assumptions of the classical elastic–plastic solid, such as non-normality of the plastic flow rule, plastic dilatation, or the formation of a vertex on subsequent yield surfaces.

If the thin slice of material contains an initial inhomogeneity, the gradual development of a shear band is determined by incremental solution of (9). Such studies have been made by Saje *et al.* [38] for a solid showing the plastic dilatation that is the apparent macroscopic effect of the nucleation and growth of microscopic voids in a ductile metal. Comparison with shear band bifurcation predictions based on a more detailed numerical model of a void containing solid has been carried out by Tvergaard [43]. Hutchinson and Tvergaard [22] have studied shear band growth for solids that develop a vertex on the yield surface, and have also used (9) to study the development of shear bands beyond bifurcation in cases with no inhomogeneity. It is noted that the study by Hutchinson and Tvergaard [22] for an incompressible solid covers shear band development under tensile loading as well as compressive loading.

The simple shear band analysis discussed here is based on assuming geometrical constraints such that the uniform deformations remain stable prior to localization, and thus localization is associated with a material instability. Without these constraints geometrical instabilities corresponding to diffuse bifurcation modes are possible before localization. In the following the analysis of such bifurcation into diffuse modes will be discussed, for various elastic–plastic constitutive laws.

It may be noted that a plane stress analogue of the simple shear band analysis discussed here is used to investigate the problem of necking in thin biaxially stretched metal sheets [30, 36, 41].

4. DIFFUSE BIFURCATION MODES

The analysis of bifurcation into diffuse modes will be considered here for materials where the instantaneous moduli L^{ijkl} in (6) are of the form

$$L^{ijkl} = \mathscr{L}^{ijkl} - \mu M_G^{ij} M_F^{kl} \tag{10}$$

This type of expression for the moduli follows from the assumption that the plastic part of the strain rate is given by

$$\dot{\eta}_{ij}^{\mathrm{P}} = \frac{1}{H} m_{ij}^{\mathrm{G}} m_{kl}^{\mathrm{F}} \overset{\nabla}{\sigma}{}^{kl} \tag{11}$$

with continued plastic loading requiring $1/H\, m_{kl}^{F} \overset{\nabla}{\sigma}{}^{kl} \geq 0$. Then, with the elastic stress–strain relationship given by $\overset{\nabla}{\sigma}{}^{ij} = \mathscr{R}^{ijkl} \dot{\eta}_{kl}^{\mathrm{E}}$, and the assumption $\dot{\eta}_{kl} = \dot{\eta}_{kl}^{\mathrm{E}} + \dot{\eta}_{kl}^{\mathrm{P}}$, the constitutive relationship (6) with the instantaneous moduli (10) appears, as specified by

$$M_G^{ij} = \mathscr{R}^{ijkl} m_{kl}^{G}, \qquad M_F^{kl} = m_{rs}^{F} \mathscr{R}^{rskl} \tag{12}$$

$$\mu = \begin{cases} 0, & \text{for elastic unloading} \\ \sqrt{G/g}\,[H + m_{kl}^{F} M_G^{kl}]^{-1}, & \text{for plastic loading} \end{cases} \tag{13}$$

$$\mathscr{L}^{ijkl} = \sqrt{\frac{G}{g}}\,[\mathscr{R}^{ijkl} - \tfrac{1}{2}(\sigma^{ik}G^{jl} + \sigma^{jk}G^{il} + \sigma^{il}G^{jk} + \sigma^{jl}G^{ik}) + \sigma^{ij}G^{kl}] \tag{14}$$

It is noted that the last term in the expression (14) for the elastic part of the moduli is unsymmetric. For materials with plastic incompressibility the approximation $(G/g)^{1/2} \approx 1$ and thus $\tau^{ij} \approx \sigma^{ij}$ involves little error, as the relative volume change $(G/g)^{1/2} - 1$ is entirely due to elastic strains, which remain small. In such cases the last term of (14) can be neglected, so that we have the symmetry $\mathscr{L}^{ijkl} = \mathscr{L}^{klij}$, if $\mathscr{R}^{ijkl}$ is symmetric. Then it can be advantageous to formulate the constitutive relations directly in terms of Kirchhoff stresses, as has been discussed by Hutchinson [18] and McMeeking and Rice [26].

The general theory of uniqueness and bifurcation in elastic–plastic solids developed by Hill [9, 10, 13] refers to materials with normality, $M_G^{ij} = M_F^{ij}$, and symmetric elastic moduli, $\mathscr{L}^{ijkl} = \mathscr{L}^{klij}$. The equations governing bifurcation are formulated by assuming, at any point of the loading history, that there are at least two distinct solutions $\dot{u}_i^a$ and $\dot{u}_i^b$ corresponding to a given increment of the prescribed quantity. The difference between such two solutions is denoted by $\tilde{(\ )} = (\dot{\ })^a - (\dot{\ })^b$. Then, using the incremental version of the principle of virtual work (3) the following equation is obtained

$$\int_V \{\tilde{\tau}^{ij}\,\delta\eta_{ij} + \tau^{ij}\tilde{u}^k_{,i}\,\delta u_{k,j}\}\,\mathrm{d}V - \int_S \tilde{T}^i\,\delta u_i\,\mathrm{d}S = 0 \tag{15}$$

which must be satisfied by non-zero bifurcation solutions $\tilde{(\ )}$. Here, τ^{ij} are the current stresses.

Hill [9, 10] makes use of the expression

$$I = \int_V \{\tilde{\tau}^{ij}\tilde{\eta}_{ij} + \tau^{ij}\tilde{u}^k_{,i}\tilde{u}_{k,j}\}\,\mathrm{d}V - \int_S \tilde{T}^i\tilde{u}_i\,\mathrm{d}S \tag{16}$$

to prove uniqueness. A *comparison solid*, with fixed instantaneous moduli L_c^{ijkl} in (6), is defined by choosing these moduli equal to the plastic branch of the elastic–plastic moduli (specified by (10) and (13)) for every material point currently on the yield surface, and the elastic branch elsewhere. For the comparison solid the following quadratic functional is considered

$$F = \int_V \{L_c^{ijkl}\tilde{\eta}_{ij}\tilde{\eta}_{kl} + \tau^{ij}\tilde{u}^k_{,i}\tilde{u}_{k,j}\}\,\mathrm{d}V - \int_S \tilde{T}^i\tilde{u}_i\,\mathrm{d}S \tag{17}$$

It can be proved, for $M_G^{ij} = M_F^{ij}$ and for non-negative μ, that the relation

$$\tilde{\tau}^{ij}\tilde{\eta}_{ij} \geqslant L_c^{ijkl}\tilde{\eta}_{ij}\tilde{\eta}_{kl} \tag{18}$$

is satisfied everywhere, and thus $F \leqslant I$. Clearly, a non-trivial solution of

(15) gives $I=0$, and therefore the requirement $F>0$ is a sufficient condition for uniqueness. Equality in (18), and thus $I=0$ for $F=0$, requires that both solution increments, $\dot{u}_i^a$ and $\dot{u}_i^b$, give plastic loading at all material points currently on the yield surface. When the plastic flow law admits a potential, i.e. when $M_G^{ij}=M_F^{ij}$ and $\mathscr{L}^{ijkl}=\mathscr{L}^{klij}$, the variational equation $\delta F=0$ is identical with (15) for the comparison solid.

In the case of $M_G^{ij}\neq M_F^{ij}$ the relation (18) is not satisfied for the usual comparison solid, and thus the requirement $F>0$ does not exclude bifurcation in the underlying elastic–plastic solid. Raniecki and Bruhns [34] have proposed an *alternative comparison solid*, in which the non-zero value of μ is used for every material point currently on the yield surface, both M_G^{ij} and M_F^{ij} are replaced by $\frac{1}{2}(M_G^{ij}+rM_F^{ij})/\sqrt{r}$, for $r>0$, and $\mathscr{L}^{ijkl}$ is replaced by its symmetric part $\frac{1}{2}(\mathscr{L}^{ijkl}+\mathscr{L}^{klij})$. Raniecki and Bruhns prove that (18) is satisfied by the incremental moduli of this alternative comparison solid, and thus the critical bifurcation point for this solid provides a *lower bound* to the first critical bifurcation point of the actual elastic–plastic solid. It is noted that in the special case of normality, $M_G^{ij}=M_F^{ij}$, and symmetric $\mathscr{L}^{ijkl}$ this alternative comparison solid coincides with the usual comparison solid for $r=1$, whereas smaller or larger values of r give rise to an earlier bifurcation in the alternative comparison solid.

A bifurcation found for the usual comparison solid is a possible solution for the underlying elastic–plastic solid, provided that the two solution increments, $\dot{u}_i^a$ and $\dot{u}_i^b$, can be chosen so that plastic loading takes place wherever the current stress state is on the yield surface. Thus, the usual comparison solid provides an *upper bound*. Determining the actual critical bifurcation point between the upper and lower bounds would generally be very complex. The regions, in which the non-zero value of μ is active, would have to be determined for each of the two solution increments $\dot{u}_i^a$ and $\dot{u}_i^b$, but would not necessarily coincide with the region of material currently on the yield surface.

Hill's bifurcation theory has been applied to numerous problems of structural buckling (see Hutchinson [19], Tvergaard [42] and Needleman and Tvergaard [29]), and also to a number of problems involving finite strains, such as various necking analyses. One example is the detailed study of bifurcation phenomena in the plane strain tensile test by Hill and Hutchinson [15], for an incompressible solid. An overview of these bifurcation results is shown in Fig. 2 (taken from ref. 47), where σ is the tensile stress, while μ^* and μ° are moduli. In the

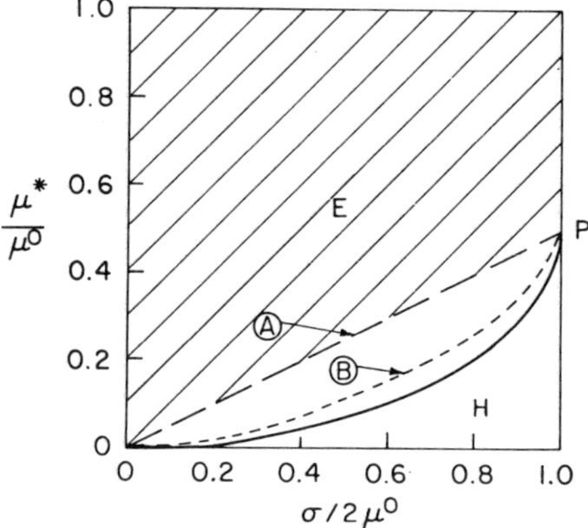

FIG. 2. Elliptic (E), parabolic (P) and hyperbolic (H) regimes for an incompressible material in plane strain tension. Bifurcation in diffuse modes is indicated (from ref. 47).

special case of J_2 flow theory $\mu^\circ = E/3$ and $\mu^* = E_t/3$; while $\mu^\circ = E_s/3$ for a hypoelastic finite strain generalization of J_2 deformation theory suggested by Stören and Rice [41]. Here, E is Young's modulus, E_t is the tangent modulus and E_s is the secant modulus. Most materials have a low value of the ratio μ^*/μ° when instabilities occur, so that diffuse bifurcation modes are first critical, ranging from long wave modes at curve A to short wave modes at curve B. For such materials localization of plastic flow, at the elliptic–hyperbolic boundary, occurs somewhat after the first critical bifurcation point.

A set of approximate constitutive relations for a ductile metal, in which microscopic voids nucleate and grow, have been suggested by Gurson [7]. The current void volume fraction is denoted by f, an equivalent tensile flow stress representing the microscopic stress-state in the matrix material is denoted by σ_M, and an approximate yield condition for the average macroscopic Cauchy stress tensor σ^{ij} is taken to be of the form

$$\Phi = \frac{\sigma_e^2}{\sigma_M^2} + 2fq_1 \cosh\left\{\frac{\sigma_k^k}{2\sigma_M}\right\} - (1 + q_1^2 f^2) = 0 \tag{19}$$

where $\sigma_e = (3s_{ij}s^{ij}/2)^{1/2}$ and $s^{ij} = \sigma^{ij} - G^{ij}\sigma^k_k/3$. Details of this constitutive description shall not be repeated here (see refs. 28, 38 and 44); but it is noted that the resulting stress–strain relationship with instantaneous moduli of the form (10) incorporates plastic dilatancy and non-normality.

Bifurcations in the plane strain tensile test have been analysed by Tvergaard [44] for the material described by the yield condition (19), in a case where non-normality of the plastic flow rule results from void nucleation controlled by the stress state. In a Cartesian co-ordinate system x^i, with the x^1-axis in the tensile direction, we search for bifurcation modes of the form

$$\tilde{u}_1 = \tilde{U}_1(x^2)\sin\frac{\pi x^1}{l}, \qquad \tilde{u}_2 = \tilde{U}_2(x^2)\cos\frac{\pi x^1}{l} \tag{20}$$

where l is the half wavelength. Substituting (20) into the Euler equations of (15) gives two ordinary differential equations for $\tilde{U}_1$ and $\tilde{U}_2$, which admit solutions of the form

$$\tilde{U}_1 = A_1 e^{\lambda x^2}, \qquad \tilde{U}_2 = A_2 e^{\lambda x^2} \tag{21}$$

For a specimen with current thickness $2h$, and for a particular set of material parameters, Fig. 3 shows the critical logarithmic strain ε_1 for bifurcation into a symmetric mode versus the thickness to wavelength ratio h/l. The figure gives both the upper bound predictions based on the usual comparison solid and the best lower bound predictions (r close to unity) obtained by the alternative comparison solid proposed by Raniecki and Bruhns [34]. The distance between these upper and lower bounds is rather large; but no better estimate of the critical bifurcation point for the actual elastic–plastic solid is known at present. Regarding shear bands it can be seen that the upper bound is the exact solution; but the same is not true for diffuse modes. However, a number of things indicate that the actual bifurcation solution will be much closer to the upper bound than to the lower bound, as has been discussed in ref. 44.

A constitutive model for the dilatant, frictional response of rocks developed by Rudnicki and Rice [37] is also of the general form (10), with non-normality of the plastic flow rule. Based on this material description Bruhns and Raniecki [3, 4] have studied the bifurcation problem governing necking or bulging in a circular cylindrical specimen, and have also found a significant difference between upper and lower bound predictions. For the same material model Needleman [27]

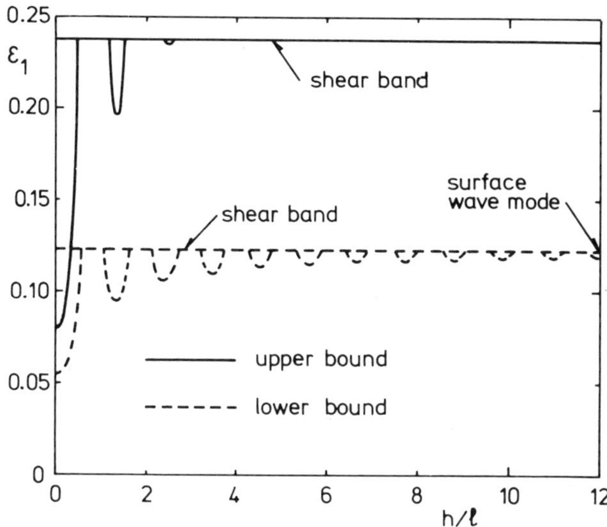

FIG. 3. Strain vs thickness to wavelength ratio at diffuse bifurcation in symmetric modes, in plane strain tension. Metal with stress controlled nucleation of micro-voids (from ref. 44).

has studied upper bound bifurcation predictions under plane strain conditions.

It may be noted that the bifurcation criteria formulated for incremental constitutive laws with moduli of the form (10) cover a rather wide class of materials. This includes materials with anisotropy resulting from crystallographic texture (Bassani [1], Hill [14]), or from an internal orientation of porosities (Litewka and Sawczuk [25]), and also the deformations of rocks or soils, for which the non-normality of the plastic flow law is frequently needed. A possible non-coaxiality of the plastic part of the strain rate and the Cauchy stress tensor, as sometimes suggested for soils [31], can also be incorporated in (10). However, the descriptions here are limited to cases for which the direction of $\dot{\eta}_{ij}^{\mathrm{P}}$, parallel with m_{ij}^{G} according to (11), is fixed at a given current stress-state, independent of the stress rate $\overset{\nabla}{\sigma}{}^{kl}$.

A full understanding of bifurcation behaviour requires also insight in the post-bifurcation solution and the sensitivity to small imperfections, as is provided in the elastic range by Koiter's general theory of elastic stability [23]. The bifurcation theory of Hill [9] has been extended into the initial post-bifurcation range by Hutchinson [17, 19], who has obtained an asymptotically exact expression for the prescribed load (or deformation) parameter λ in terms of the bifurcation mode amplitude ξ of the form

$$\lambda = \lambda_c + \lambda_1 \xi + \lambda_2 \xi^{1+\beta} + \dots \tag{22}$$

Here, λ_c is the value of λ at the first critical bifurcation point, λ_1 is positive, λ_2 is negative, and $0<\beta<1$. The asymptotic expansions of growing elastic unloading zones, etc., that lead to the expression (22) are strongly dependent on details of the constitutive law. For materials not covered by the expression (22) and for cases with small initial imperfections all current understanding of the behaviour in the vicinity of bifurcation points relies on numerical solutions.

Loss of stability of the equilibrium solution may occur prior to bifurcation, at a limit point; but often stability is retained on at least part of the post-bifurcation solution. Thus, at plastic buckling, with loading prescribed, stability is lost at a limit point on the post-bifurcation path [19, 29, 42]. Bifurcation after the load maximum is found in tensile test specimens, with the overall elongation prescribed. Here, continued necking is usually limited by fracture in the highly strained neck, rather than by an instability of the post-bifurcation solution.

5. SOLIDS THAT FORM A VERTEX ON THE YIELD SURFACE

At a vertex on a yield surface the direction of the plastic part of the strain rate is a function of the stress rate, so that the incremental moduli are not of the form (10). The occurrence of a vertex is implied by physical models of polycrystalline metal plasticity, based on the concept of single crystal slip (Hill [12], Hutchinson [16]), at least when the yield surface is defined for small offset plastic strains. The experimental evidence of vertex formation on yield surfaces is, however, conflicting (Hecker [8]).

In the context of plastic buckling it was noted more than 30 years ago that the bifurcation predictions based on J_2 deformation theory gave much better agreement with experimentally obtained buckling loads than those based on J_2 flow theory. Batdorf [2] realized that the bifurcation predictions of deformation theory can be justified by appealing to a flow theory of plasticity, which develops a vertex on the yield surface.

Since then the possibility of vertex formation has played a significant role in discussions of plastic buckling (see Hutchinson [19]), and more recently the important influence of vertex formation on the prediction of tensile instabilities has been realized [22, 37, 41].

Here, we shall focus on a phenomenological corner theory of plasticity, called J_2 corner theory, proposed by Christoffersen and Hutchinson [5]. In this theory the instantaneous moduli for nearly proportional loading are chosen equal to the J_2 deformation theory moduli, and for increasing deviation from proportional loading the moduli increase smoothly until they coincide with the elastic moduli for stress increments directed along or within the corner of the yield surface.

With M^0_{ijkl} denoting the deformation theory compliances, so that $\dot{\eta}_{ij} = M^0_{ijkl}\overset{\nabla}{\tau}{}^{kl}$, and $\mathcal{M}_{ijkl}$ denoting the linear elastic compliances, the plastic part of the compliances is $C_{ijkl} = M^0_{ijkl} - \mathcal{M}_{ijkl}$. The yield surface in the neighbourhood of the loading point is taken to be a cone in stress deviator space with the cone axis in the direction

$$\lambda^{ij} = s^{ij}(C_{mnpq}s^{mn}s^{pq})^{-1/2} \tag{23}$$

where $s^{ij} = \tau^{ij} - G^{ij}\tau^k_k/3$ is the stress deviator. A positive angular measure θ of the stress-rate direction relative to the cone axis is defined by

$$\cos\theta = C_{ijkl}\lambda^{ij}\overset{\nabla}{s}{}^{kl}(C_{mnpq}\overset{\nabla}{s}{}^{mn}\overset{\nabla}{s}{}^{pq})^{-1/2} \tag{24}$$

and a stress-rate potential at the vertex is formulated as

$$W = \tfrac{1}{2}\mathcal{M}_{ijkl}\overset{\nabla}{\tau}{}^{ij}\overset{\nabla}{\tau}{}^{kl} + \tfrac{1}{2}f(\theta)C_{ijkl}\overset{\nabla}{\tau}{}^{ij}\overset{\nabla}{\tau}{}^{kl} \tag{25}$$

From this potential the strain-rate is obtained as

$$\dot{\eta}_{ij} = \frac{\partial^2 W}{\partial\overset{\nabla}{\tau}{}^{ij}\,\partial\overset{\nabla}{\tau}{}^{kl}}\overset{\nabla}{\tau}{}^{kl} = M_{ijkl}(\theta)\overset{\nabla}{\tau}{}^{kl} \tag{26}$$

Inverting (26) and using (5) gives the θ-dependent moduli $L^{ijkl}(\theta)$ to be used in (6).

The angle of the yield surface cone is denoted θ_c, so that the transition function $f(\theta)$ in (25) is zero for $\theta_c \leqslant \theta \leqslant \pi$. In the total loading range, $0 \leqslant \theta \leqslant \theta_0$, $f(\theta)$ is unity, and in the transition region, $\theta_0 \leqslant \theta \leqslant \theta_c$, $f(\theta)$ is chosen to smoothly merge the deformation theory moduli with the elastic moduli in a way which ensures convexity of the incremental relation. The discussion in the following will refer to a function $f(\theta)$, which has been found in [5] to rather closely duplicate moduli obtained using a self-consistent model of a polycrystalline aggregate, and for which the moduli $L^{ijkl}(\theta)$ vary continuously with θ.

The cone angle θ_c is often specified in terms of a different angular measure as $\beta = (\beta_c)_{max}$, where $\cos\beta = \dot{\sigma}_e(3\overset{\nabla}{s}_{ij}\overset{\nabla}{s}^{ij}/2)^{-1/2}$.

Bifurcation is again analysed by assuming that there are at least two incremental solutions $(\dot{})^a$ and $(\dot{})^b$ corresponding to a given increment of the prescribed quantity. The equation governing a non-zero difference $(\tilde{}) = (\dot{})^a - (\dot{})^b$ between such two solutions is still (15). Sewell [39] has shown for a pyramidal vertex that a *comparison solid* defined by the total loading moduli (all slip systems active) satisfies the fundamental inequality (18). In the context of J_2 corner theory the total loading moduli are those of deformation theory, valid for $0 \leqslant \theta \leqslant \theta_o$.

For many cases of interest the fundamental solution satisfies proportional loading or nearly proportional loading everywhere, so that $\theta < \theta_0$. Then, writing the variation of the load λ with the bifurcation mode amplitude ξ initially after bifurcation on the form

$$\lambda = \lambda_c + \lambda_1 \xi + \dots \tag{27}$$

the constant λ_1 can be chosen sufficiently large so that total loading is also satisfied on the initial part of (27), as has been discussed by Needleman and Tvergaard [29]. In such cases the first critical bifurcation point for the comparison solid is identical with that of the elastic–plastic solid.

For a cruciform column the fundamental solution (uniaxial compression) satisfies $\theta = 0$. In the case of a finite total loading range, $\theta_0 > 0$, Needleman and Tvergaard [29] have found that bifurcation takes place with the smallest value of λ_1 in (27) consistent with total loading. In the limit of a thoroughly non-linear vertex description, $\theta_0 = 0$, the requirement of total loading at bifurcation gives rise to a smooth bifurcation ($\lambda_1 \to \infty$ in (27)), as has also been found for the cruciform column [29]. It is noted that at this smooth bifurcation point the two incremental solutions $(\dot{})^a$ and $(\dot{})^b$ differ for the comparison solid, as required for non-zero solutions of (15), but not for the elastic–plastic solid. Such smooth bifurcation corresponding to $\theta_0 = 0$ has also been found by Hutchinson and Tvergaard [22] in the case of shear band modes.

If the fundamental solution has $\theta_0 < \theta < \theta_c$ in some material points, the total loading moduli may result in a rather poor lower bound, since bifurcation will be governed partly by the stiffer moduli of the transition range. Here, an *alternative comparison solid* may be defined by the instantaneous moduli associated with the current values of θ on the fundamental solution (see Tvergaard [46]). A bifurcation point found for this alternative comparison solid is also a bifurcation point of the underlying elastic–plastic solid, since the moduli $L^{ijkl}(\theta)$ vary continuously with θ, and since bifurcation stress-rate directions arbitrarily close to those of the fundamental solution can be enforced by superposing a sufficiently large fundamental solution increment on the eigenmode $(\tilde{\ })$. Thus, a bifurcation point predicted by this alternative comparison solid provides an *upper bound.* Determining the actual critical bifurcation point in cases where θ exceeds θ_0 is complex, since the corresponding critical value of θ must be determined simultaneously at every material point.

A fundamental solution with values of θ exceeding θ_0 has been found in a study of the plastic buckling of axially compressed circular cylindrical shells [46]. Here, an initial geometrical imperfection has been specified in the form of the axisymmetric normal displacement

$$w = \bar{\xi} h \cos \frac{\pi x^1}{l} \tag{28}$$

where h is the shell thickness, x^1 is the axial co-ordinate, and l is the critical axial half-wave length. The growth of this axisymmetric waviness gives rise to non-proportional loading prior to the bifurcation into a non-axisymmetric mode. In Fig. 4 bifurcation points are indicated on curves of axial compressive load P versus average axial strain Δ/l, for a shell with radius to thickness ratio $R/h = 50$, and for a particular set of material parameters.

Two J_2 corner theory materials are considered in Fig. 4, one with a rather sharp vertex, $(\beta_c)_{max} = 135°$, and a finite total loading range limited by $\theta_0 = \theta_n/2$ (where $\theta_n = \theta_c - \pi/2$), and one with a less sharp vertex, $(\beta_c) = 100°$, and $\theta_0 = 0$. The bifurcations shown are upper bound predictions; but for the material with $\theta_0 = \theta_n/2$ the regions in which the fundamental solution has $\theta > \theta_0$ are relatively small, so that the bifurcations found for this material occur only slightly later than the lower bounds. For the material with $\theta_0 = 0$ the results indicate quite a significant difference between the upper and lower bound bifurcation predictions, particularly for the smaller imperfections where deviations from proportional loading are strongest. It is noted that the perfect shell has proportional loading in the fundamental solution, so that bifurcation into the axisymmetric mode is exactly governed by the total loading moduli; but subsequently, on the post-bifurcation solution, loading is strongly non-proportional prior to the secondary bifurcation into a non-axisymmetric mode.

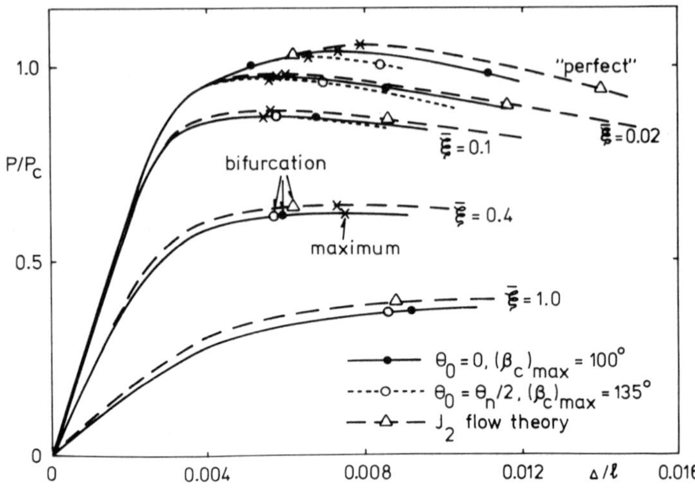

FIG. 4. Axial compressive load vs average axial strain for cylindrical shells with axisymmetric imperfections. Solid and dotted curves refer to J_2 corner theory, with upper bound bifurcation estimates indicated (from ref. 46).

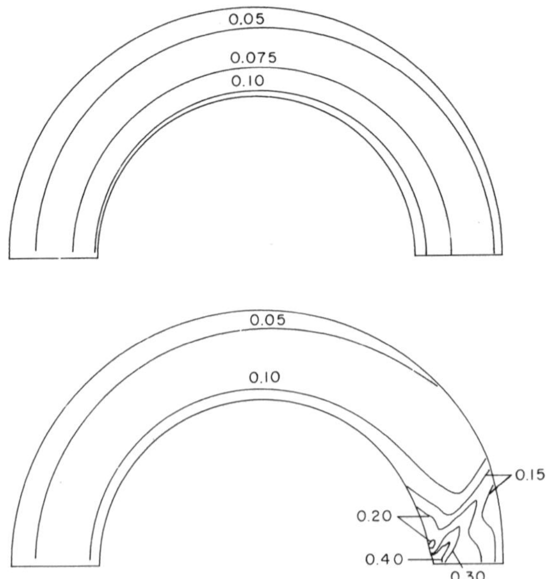

FIG. 5. Contours of constant maximum principal logarithmic strain in a thick-walled tube under external pressure, predicted by J_2 corner theory. At the last stage flow localization has developed in the neck region (from ref. 24).

Tensile instabilities at finite strains have also been investigated for materials that form a vertex on the yield surface. Thus Tvergaard *et al.* [47] have analysed the plane strain tensile test and Larsson *et al.* [24] have studied the behaviour of thickwalled tubes under internal pressure. In both cases bifurcation into a diffuse mode is found to occur first, and since there is proportional loading on the fundamental solutions, bifurcation occurs at the points predicted for the total loading moduli. The diffuse bifurcation modes develop into localized necking modes, and subsequently, on these post-bifurcation solutions, ellipticity is lost so that shear bands start to grow in the necking regions. In the numerical investigations of necking and subsequent shear band formation in refs. 24 and 47 the necking mode is triggered by a small initial imperfection. Figure 5 shows curves of constant maximum principal logarithmic strain for a pressurized tube at an early stage, just before necking, and at a later stage where a shear band has started to grow in the neck on one side of the tube.

6. EFFECT OF MATERIAL RATE-SENSITIVITY

The stability analyses considered in the previous sections apply to time-independent elastic–plastic materials. However, experimental measurements tend to show a dependence of the flow stress on the rate of strain, even at low strain-rates (Clifton [6]). It has been suggested that microstructural slip in metals and thus macroscopic inelastic deformations are inherently time dependent (Rice [35]).

When strain-rate sensitivity is taken into account, the material behaviour is represented in terms of a viscoplastic constitutive description, where the plastic part of the strain-rate $\dot{\eta}^{\mathrm{P}}_{ij}$ is a function of the current stresses and strains (but not of the stress-rate). The relationship may be of the form

$$\dot{\eta}^{\mathrm{P}}_{ij} = F(\sigma_e, \varepsilon_e)\frac{\partial \Phi}{\partial \sigma^{ij}} \tag{29}$$

where Φ is a plastic potential function, σ_e and ε_e are the effective stress and strain, respectively, and $(\dot{\ })$ denotes the time derivative. Often the function F is taken to be proportional with $[\sigma_e/g(\varepsilon_e)]^{1/m}$, where the function $g(\varepsilon_e)$ incorporates the strain hardening (at some reference strain-rate $g(\varepsilon_e)$ equals σ_e), and the rate-hardening exponent m is a measure of the degree of strain-rate sensitivity. Thus, for a rate-sensitive version of the classical J_2 flow theory an expression for the inelastic part of the strain-rate may be assumed of the form

$$\dot{\eta}_{ij}^{P} = \dot{\varepsilon}_o \left[\frac{\sigma_e}{g(\varepsilon_e)}\right]^{1/m} \frac{3}{2} \frac{s_{ij}}{\sigma_e} \tag{30}$$

where $\dot{\varepsilon}_o$ denotes the reference value of the effective strain-rate. The limiting case, $m = 0$, corresponds to time-independent plasticity; but for any positive value of m an inelastic deformation increment according to (29) requires an increment of time.

Bifurcation in a rate-sensitive elastic–plastic solid is entirely governed by the time-independent part of the deformations. Assuming the total strain-rate given by $\dot{\eta}_{ij} = \dot{\eta}_{ij}^{E} + \dot{\eta}_{ij}^{P}$, with the viscous expression (29) for $\dot{\eta}_{ij}^{P}$, the first critical bifurcation point is that predicted by linear elasticity, even for a very small degree of strain-rate sensitivity, and this elastic bifurcation point is hardly ever reached in the range of problems considered here. Thus, it turns out that the central role played by the critical bifurcation point and the post-bifurcation behaviour in the case of time-independent plasticity, is here replaced by a very strong sensitivity to small initial imperfections.

The influence of strain-rate sensitivity on tensile instabilities has been investigated by Hutchinson and Neale [20, 21] for necking in uniaxial tension and in biaxially stretched sheets, respectively. Elastic strains are neglected in these analyses, so that no bifurcation at all corresponds to the perfect case, and accordingly it is found that the critical strain for localization tends towards infinity for vanishing initial thickness inhomogeneity.

Both for uniaxial tension and for plane strain sheet necking an asymptotic expression for the critical logarithmic, axial strain, ε_c, is found of the form

$$\frac{\varepsilon_c - \varepsilon_c^o}{\sqrt{N}} \simeq \frac{m}{2\sqrt{2\bar{\xi}}} \ln\left(\frac{4\pi\bar{\xi}}{m}\right) \tag{31}$$

where N is the strain hardening exponent, $\varepsilon_c^o = N$ is the critical strain for necking according to time-independent plasticity, and $\bar{\xi}$ is the initial thickness reduction [20, 21]. This relationship, valid for $\bar{\xi} \ll 1$, $m < 2\bar{\xi}$ and small m/N, shows how $\varepsilon_c \to \varepsilon_c^o$ in the time-independent limit, with the difference $\varepsilon_c - \varepsilon_c^o$ proportional to $\sqrt{N}$ for small m. An interesting feature of these analyses is that the necking delay is independent of the rate of deformation, but depends strongly on the degree of rate-sensitivity as measured by m.

The influence of rate-sensitivity on shear band formation has been studied by Pan *et al.* [32] for a porous ductile metal, with the function (19) used as the plastic potential in (29). Also here the bifurcation discussed in connexion with (9), corresponding to no initial inhomogeneity, is irrelevant in the presence of rate-sensitivity, and the localization delay is determined by full numerical analyses of imperfection growth. Based on the same material model Needleman and Tvergaard [30] have analysed the necking delay due to rate-sensitivity in biaxially stretched sheets.

For a planar double slip model of a single crystal the influence of rate-sensitivity in the slip systems has recently been analysed by Peirce *et al.* [33]. The focus of this investigation is on necking, the formation of macroscopic shear bands and the occurrence of patchy slip. It is found that rate sensitivity of the slip mechanism does result in a significant delay of flow localization.

A recent column mode analysis for an eccentrically stiffened panel gives some insight in the influence of rate-sensitivity on plastic buckling [45]. Corresponding to a fixed speed of average axial shortening, $\dot{\varepsilon}_a = -\dot{\varepsilon}_o$, Fig. 6 gives computed load maxima versus imperfection

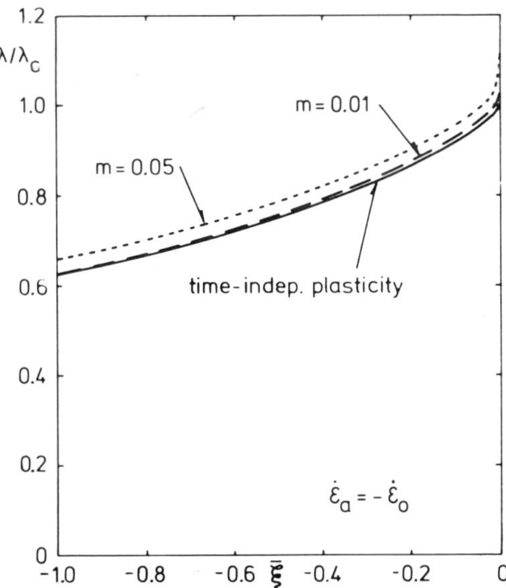

FIG. 6. Maximum load vs imperfection amplitude for a column with an asymmetric cross-section, made of a rate-sensitive material (from ref. 45).

amplitude $\bar{\xi}$. The load is normalized by the bifurcation load corresponding to time-independent plasticity and $\bar{\xi}$ denotes the imperfection amplitude relative to the flange thickness. Naturally, the maximum load attained is quite sensitive to the rate of axial compression (in contrast to the critical strain for sheet necking [21]). Therefore it should be noted that the uniaxial stress–strain curve for the time-independent plasticity model in Fig. 6 is chosen such that the three materials compared in the figure have identical response when strained uniformly at the strain-rate $\dot{\varepsilon}_o$.

For the rate-sensitive materials in Fig. 6 the first critical bifurcation point corresponds to the elastic value λ_E, which is quite high here, $\lambda_E = 2{\cdot}18\lambda_c$. Both an asymptotic analysis and numerical analyses in ref. 45 show that even for imperfections $\bar{\xi}$ as small as 10^{-7} or 10^{-12} the maximum load is at most $1{\cdot}2\lambda_c$ (for $m = 0{\cdot}05$) and thus far below the bifurcation load. The results in Fig. 6 show that although the role of bifurcation is strongly changed by accounting for strain-rate sensitivity, the classical elastic–plastic buckling predictions remain good approximations for imperfection amplitudes that are not extremely small, as long as the rate hardening exponent does not exceed values of the order of 0·01–0·05.

REFERENCES

1. Bassani, J. L. Yield characterization of metals with transversely isotropic plastic properties, *Int. J. Mech. Sci.*, **19** (1977), 651–660.
2. Batdorf, S. B. Theories of plastic buckling, *J. Aeronaut. Sci.*, **16** (1949), 405–408.
3. Bruhns, O. T. Bounds to the critical stresses in bifurcation of cylindrical specimens in the case of non-associated flow laws. In: *Stability in the Mechanics of Continua* (Ed. F. H. Schroeder), Springer, Berlin, 1982, 46–55.
4. Bruhns, O. and B. Raniecki. Ein schrankenverfahren bei verzweigungsproblemen inelastischer formänderungen, *ZAMM*, **62** (1982), T111–T113.
5. Christoffersen, J. and J. W. Hutchinson. A class of phenomenological corner theories of plasticity. *J. Mech. Phys. Solids*, **27** (1979), 465–487.
6. Clifton, R. J. Comments on microscopic mechanisms of plastic flow. In: *Plasticity of Metals at Finite Strain* (Eds. E. H. Lee & R. L. Mallett), Division of Applied Mechanics, Stanford University, 1982, 623–628.
7. Gurson, A. L. Continuum theory of ductile rupture by void nucleation and growth: Part I—Yield criteria and flow rules for porous ductile media, *J. Engng Mater. Technol.*, **99** (1977), 2–15.
8. Hecker, S. S. Experimental studies of yield phenomena in biaxially loaded metals. In: *Constitutive Equations in Viscoplasticity*, *AMD*, **20,** ASME, New York, 1976, 1–33.

9. HILL, R. A general theory of uniqueness and stability in elastic–plastic solids. *J. Mech. Phys. Solids*, **6** (1958), 236–249.
10. HILL, R. Bifurcation and uniqueness in nonlinear mechanics of continua. In: *Problems of Continuum Mechanics*, Society of Industrial and Applied Mathematics, Philadelphia, 1961, 155–164.
11. HILL, R. Acceleration waves in solids, *J. Mech. Phys. Solids*, **10** (1962), 1–16.
12. HILL, R. Generalized constitutive relations for incremental deformation of metal crystals by multislip, *J. Mech. Phys. Solids*, **14** (1966), 95–102.
13. HILL, R. Aspects of invariance in solid mechanics, *Adv. appl. Mech.*, **18** (1978), 1–75.
14. HILL, R. Theoretical plasticity of textured aggregates. *Math. Proc. Camb. Phil. Soc.*, **85** (1979), 179–191.
15. HILL, R. and J. W. HUTCHINSON. Bifurcation phenomena in the plane strain tension test. *J. Mech. Phys. Solids*, **23** (1975), 239–264.
16. HUTCHINSON, J. W. Elastic–plastic behavior of polycrystalline metals and composites, *Proc. R. Soc. Lond*, **A318** (1970), 247–272.
17. HUTCHINSON, J. W. Postbifurcation behavior in the plastic range. *J. Mech. Phys. Solids*, **21** (1973), 163–190.
18. HUTCHINSON, J. W. Finite strain analysis of elastic–plastic solids and structures. In: *Numerical Solution of Nonlinear Structural Problems* (Ed. R. F. Hartung), ASME, New York, 1973, 17.
19. HUTCHINSON, J. W. Plastic buckling, *Adv. appl. Mech.* (Ed. C. S. Yih), Vol. 14, Academic Press, New York, 1974, 67–144.
20. HUTCHINSON, J. W. and K. W. NEALE. Influence of strain-rate sensitivity on necking under uniaxial tension, *Acta Metallurgica*, **25** (1977), 839–846.
21. HUTCHINSON, J. W. and K. W. NEALE. Sheet necking–III. Strain rate effects. In: *Mechanics of Sheet Metal Forming* (Eds. D. P. Koistinen and N.-M. Wang), Plenum Publ. Corp., New York, 1978, 111–126.
22. HUTCHINSON, J. W. and V. TVERGAARD. Shear band formation in plane strain, *Int. J. Solids Struct.*, **17** (1981), 451–470.
23. KOITER, W. T. Over de stabiliteit van het elastisch evenwicht. Thesis, Delft, H. J. Paris, Amsterdam (1945). English translations (a) NASA TT-F10, 833, (1967), (b) AFFDL-TR-70-25 (1970).
24. LARSSON, M., A. NEEDLEMAN, V. TVERGAARD and B. STORÅKERS. Instability and failure of internally pressurized ductile metal cylinders, *J. Mech. Phys. Solids*, **30** (1982), 121–154.
25. LITEWKA, A. and A. SAWCZUK. A yield criterion for perforated sheets, *Ingenieur-Archiv*, **50** (1981), 393–400.
26. MCMEEKING, R. M. and J. R. RICE. Finite-element formulations for problems of large elastic–plastic deformation. *Int. J. Solids Struct.*, **11** (1975), 601–616.
27. NEEDLEMAN, A. Non-normality and bifurcation in plane strain tension and compression. *J. Mech. Phys. Solids*, **27** (1979), 231–254.
28. NEEDLEMAN, A. and J. R. RICE. Limits to ductility set by plastic flow localization. In: *Mechanics of Sheet Metal Forming* (Eds. D. P. Koistinen *et al.*), Plenum Publ. Corp., New York, 1978, 237–267.
29. NEEDLEMAN, A. and V. TVERGAARD. Aspects of plastic post-buckling

behaviour. In: *Mechanics of Solids, The Rodney Hill 60th Anniversary Volume* (Eds. H. G. Hopkins and M. J. Sewell), Pergamon Press, Oxford, 1982, 453–498.

30. Needleman, A. and V. Tvergaard. Limits to formability in rate-sensitive metal sheets. In: Mechanical Behaviour of Materials—IV (Eds. J. Carlsson and N. G. Ohlson), Pergamon Press, Oxford, 1984, 51–65.
31. Nemat-Nasser, S. Some macroscopic and microscopic bases for non-normality in finite plasticity. In: *Plasticity of Metals at Finite Strain: Theory, Computation and Experiment* (Eds. E. H. Lee & R. L. Mallett), Division of Applied Mechanics, Stanford University, 1982, 637–644.
32. Pan, J., M. Saje and A. Needleman. Localization of deformation in rate sensitive porous solids, *Int. J. Fracture*, **21** (1983), 261–278.
33. Peirce, D., R. J. Asaro and A. Needleman. Material rate dependence and localized deformation in crystalline solids. Division of Engineering, Brown University (1983).
34. Raniecki, B. and O. T. Bruhns. Bounds to bifurcation stresses in solids with non-associated plastic flow law at finite strain, *J. Mech. Phys. Solids*, **29** (1981), 153–172.
35. Rice, J. R. On the structure of stress–strain relations for time-dependent plastic deformation in metals. *J. appl. Mech.*, **37** (1970), 728–737.
36. Rice, J. R. The localization of plastic deformation. In: *Theoretical and Applied Mechanics, Proc. 14th IUTAM Congress.* (Ed. W. T. Koiter), North Holland, Amsterdam, 1977, 207–220.
37. Rudnicki, J. W. and J. R. Rice. Conditions for the localization of deformation in pressure-sensitive dilatant materials. *J. Mech. Phys. Solids*, **23** (1975), 371–394.
38. Saje, M., J. Pan and A. Needleman. Void nucleation effects on shear localization in porous plastic solids, *Int. J. Fracture*, **19** (1982), 163–182.
39. Sewell, M. J. A survey of plastic buckling. In: *Stability* (Ed. H. Leipholz), University of Waterloo Press, 1972, 85–197.
40. Shanley, F. R. Inelastic column theory, *J. Aeronaut. Sci.*, **14** (1947), 261–267.
41. Stören, S. and J. R. Rice. Localized necking in thin sheets. *J. Mech. Phys. Solids*, **23** (1975), 239–264.
42. Tvergaard, V. Buckling behaviour of plate and shell structures. In: *Theoretical and Applied Mechanics, Proc. 14th IUTAM Congress* (Ed. W. T. Koiter), North-Holland, Amsterdam, 1977, 233–247.
43. Tvergaard, V. Influence of voids on shear band instabilities under plane strain conditions, *Int. J. Fracture*, **17** (1981), 389–407.
44. Tvergaard, V. Influence of void nucleation on ductile shear fracture at a free surface, *J. Mech. Phys. Solids*, **30** (1982), 399–425.
45. Tvergaard, V. Rate-sensitivity in elastic–plastic panel buckling. Danish Center for Applied Mathematics and Mechanics, Report No. 262, (1983).
46. Tvergaard, V. Plastic buckling of axially compressed circular cylindrical shells, *Int. J. Thin-Walled Struct.*, **1** (1983), 139–163.
47. Tvergaard, V., A. Needleman and K. K. Lo. Flow localization in the plane strain tensile test, *J. Mech. Phys. Solids*, **29** (1981), 115–142.

22

Uniqueness, Stability and Bifurcation of Standard Systems

NGUYEN QUOC SON

Laboratoire de Mécanique des Solides, Ecole Polytechnique, Palaiseau, France

ABSTRACT

A general description of the evolution of a class of standard systems is given. This class includes all irreversible systems characterized by a thermodynamic potential representing the total stored energy and by a dissipative potential expression Normality and Convexity as in Plasticity. A discussion on the global behaviour of these systems under external load, such as global uniqueness, bifurcation and stability is presented. The obtained results are illustrated by simple examples in the framework of Inelastic Solids: plasticity, brittle fracture and brittle damage. In particular, it is shown that classical results of Perfect Plasticity can be directly extended into these domains.

1. INTRODUCTION

Usual modelizations of dissipative phenomena such as viscoplasticity, fracture and damage often lead to the discussion of *standard systems*. These systems are characterized by the fact that its constitutive equations can be described by two potentials; a thermodynamic potential and a dissipative potential [2], [3] which, respectively, define its reversible and irreversible behaviours by the set of associated state equations and force-flux equations. The obtained thermodynamic framework is a proper generalization of classical TIP into non-linear cases and has been intensively discussed by many authors [2].

We focus our attention here on a particular class of 'plastic-like' standard systems in the sense that the dissipative potential is homogeneous of degree 1 and thus the irreversible behaviour is described by a normality and convexity assumption in the same way as Hill's maximum work principle in plasticity. This class includes several models of friction, plasticity, brittle fracture and brittle damage. Our discussion will be limited to pure mechanical evolutions and our principal objective is to prove that a general formalism can be given to study the behaviour of these systems and to establish global results such as uniqueness, bifurcation and stability. In particular, it is shown that principal results of classical plasticity can be applied to different domains with different physical nature.

For this, firstly, evolution equations of the considered systems are given in quasi-static transformation. These abstract equations are derived simply from the fundamental law (equilibrium) and from the adopted behaviour (normality and convexity). To illustrate the description, an example is given in the context of plasticity.

In the second part, global results of the evolution are considered. A global uniqueness theorem is established, based upon an assumption of global convexity. From the study of the incremental response, Hill's method is then applied to derive non-bifurcation and stability criteria. It is shown that the second derivative of the equilibrium surface plays an important role and the obtained results may be compared to the well-known criterion of second variation in the context of elasticity.

Different examples in brittle fracture and brittle damage are given in the last part in order to illustrate the obtained results.

2. STANDARD SYSTEMS WITH NORMALITY AND CONVEXITY

(1) Let us consider a system defined by state variables (u, α) which are respectively displacement and local or global parameters describing the physical or geometrical state of the system. It is assumed that α characterizes the irreversible evolution, thus if u varies alone the transformation is purely reversible.

The system is submitted to external loading and undergoes only quasi-static transformation. For the sake of simplicity, the loading is defined by one control variable $\lambda(t)$ which is a given function of time t. Displacements must be compatible with the prescribed conditions, the set of compatible displacements depends eventually on the present value of α and λ and will be denoted by $U(\alpha, \lambda)$. The responses $u(t)$, $\alpha(t)$ under the solicitation $\lambda(t)$ obey equilibrium and constitutive equations which can be stated in the following way.

It is assumed that an energy function $E(\alpha, u, \lambda)$ can be associated with the system when state variables α, u and control variable λ are known. Energy E represents for usual structures the total potential energy resulting from global isothermal free energy (or isentropic internal energy) and from global external load potential. Equilibrium in quasi-static evolution is expressed by virtual work equation:

$$E'_u(\alpha, u, \lambda) \,.\, \delta u = 0 \tag{1}$$

These equations implicitly define equilibrium positions $u(\alpha, \lambda)$. For example, it is well known that if the elastic second variation is positive:

$$E''_{u^2}(\alpha, u, \lambda) \,.\, \delta u^2 > 0 \tag{2}$$

then by an implicit representation theorem, one (and only one) branch of equilibrium solutions $u(\alpha, \lambda)$ can be defined near an equilibrium position (α, u, λ). The implicit representation $u = u(\alpha, \lambda)$ enables us to eliminate u and to choose α as principal unknown. Let us introduce in this spirit the equilibrium energy:

$$F(\alpha, \lambda) = E[\alpha, u(\alpha, \lambda), \lambda] \tag{3}$$

which is the energy level at different equilibrium positions after elimination of displacement u.

It is assumed that the evolution of state variable α follows from the principle of maximum dissipation [7], i.e. normality and convexity as in classical plasticity.

Let A be the generalized force associated with α:

$$A = -F'_\alpha(\alpha, \lambda) \tag{4}$$

For physical reasons, the force A must belong to a domain of admissible forces C which is a convex domain in the force space $W(\alpha)$:

$$A \in C, \text{ convex of admissible forces} \tag{5}$$

Note that $0 \in \overset{\circ}{C}$ since in the natural state A must be zero and the natural state is physically admissible.

The rate-equation for state variable α is the normality law:

$$\dot{\alpha} \in N_C(A) \tag{6}$$

where $N_C(A)$ denotes the normal cone of the convex C at A. These laws of normality and convexity are conveniently condensed under the form of Hill's maximum dissipation principle as in classical plasticity:

$$(A - A^*) \,.\, \dot{\alpha} \geqslant 0 \qquad \forall\, A^* \in C \tag{7}$$

the point . denotes the force–velocity duality.

This denomination can be justified from the fact that $A \,.\, \dot{\alpha} \geqslant 0$ represents exactly the intrinsic dissipation of our system [11].

Finally, the evolution of our system is described by the set of equations (4), (6) and by an initial condition:

$$\left.\begin{aligned} A &= -F'_{\alpha}(\alpha, \lambda) \\ \dot{\alpha} &= N_C(A) \\ \alpha(0) &= \alpha_0 \end{aligned}\right\} \tag{8}$$

Remarks

(i) If F is a quadratic function and convex with respect to α, the system of eqns. (8) reduces to the equations of evolution studied by Moreau [8] and Brézis [1].

(ii) If the set of compatible displacements U does not depend on α and λ, then the variables (α, u, λ) are independent. In this case, one obtains also:

$$A = -E'_{\alpha}[\alpha, u(\alpha, \lambda), \lambda] \tag{9}$$

and it may be interesting to retain (α, u) as principal unknowns.

(2) To illustrate the preceding section, let us consider an example in the context of plasticity.

A structure of reference configuration Ω is submitted to a loading process defined by a given displacement $u^{d}(\lambda)$ and a given force $T^{d}(\lambda)$ in the complementary parts S_u and S_T of the boundary $\partial\Omega$. In the context of small deformation but finite rotation transformations, the material is assumed to be elastic–plastic and to follow Ziegler–Prager's model of kinematic hardening. To this material is associated an energy density per unit volume

$$W(\varepsilon, \varepsilon^{p}) = \tfrac{1}{2}(\varepsilon - \varepsilon^{p})L(\varepsilon - \varepsilon^{p}) + \tfrac{1}{2}\varepsilon^{p} \,.\, K \,.\, \varepsilon^{p}$$

where

$$\varepsilon_{ij} = \tfrac{1}{2}(u_{i,j} + u_{j,i} + u_{k,i}u_{k,j}) \tag{10}$$

is the strain tensor, ε^{p} is the plastic strain. The first part in the expression of $W(\varepsilon, \varepsilon^{\mathrm{p}})$ is the elastic energy and the second part the stored energy by microstructural modifications due to plastic hardening.

In this case, α is the plastic strain field defined on Ω and the set of compatible displacements depends only on λ:

$$U = \{u \mid u = u^{\mathrm{d}}(\lambda) \quad \text{on} \quad S_u\} \tag{11}$$

Energy function $E(\alpha, u, \lambda)$ is:

$$E(\alpha, u, \lambda) = \int_\Omega W(\varepsilon(u), \varepsilon^{\mathrm{p}})\,\mathrm{d}\Omega - \int_{S_T} T^{\mathrm{d}}(\lambda) \,.\, u\,\mathrm{d}S \tag{12}$$

Since $\sigma = \dfrac{\partial \omega}{\partial \varepsilon}$, eqn. (1) is effectively the usual equation of virtual work in Lagrangian description.

At least locally, the function $F(\alpha, \lambda)$ can be written as:

$$F(\alpha, \lambda) = \operatorname*{Min}_{u^* \in U} E(\alpha, u^*, \lambda) = E[\alpha, u(\alpha, \lambda), \lambda] \tag{13}$$

when the elastic second variation (2) is positive.

If plastic strain is incompressible then $\varepsilon^{\mathrm{p}}_{\mathrm{ii}} = 0$ and state variable must belong to the set of admissible fields V:

$$\alpha \in V = \{\boldsymbol{\varepsilon}^{\mathrm{p}} \mid \varepsilon^{\mathrm{p}}_{\mathrm{ii}}(x) = 0 \qquad \forall\, x \in \Omega\}. \tag{14}$$

From (12) and (13), one obtains by definition:

$$A \,.\, \delta\alpha = -F'_\alpha \,.\, \delta\alpha = \int_\Omega (\sigma' - K \,.\, \varepsilon^{\mathrm{p}}) \,.\, \delta\varepsilon^{\mathrm{p}}\,\mathrm{d}\Omega \tag{15}$$

which proves that force A is the field $\boldsymbol{\sigma}' - K.\ \boldsymbol{\varepsilon}^{\mathrm{p}}$.

The existence of a convex plastic criterion $f(\sigma' - K \,.\, \varepsilon^{\mathrm{p}}) \leqslant 0$ with normality laws:

$$\left.\begin{array}{l} f(\sigma' - K \,.\, \varepsilon^{\mathrm{p}}) \leqslant 0 \\ \dot{\varepsilon}^{\mathrm{p}} = \lambda \dfrac{\partial f}{\partial \sigma}, \qquad \lambda \geqslant 0 \quad \text{when} \quad f = 0 \\ \qquad\qquad\quad \lambda = 0 \quad \text{when} \quad f < 0 \end{array}\right\} \tag{16}$$

corresponds to the usual way to describe eqns. (5) and (6).

3. GLOBAL UNIQUENESS

The response of our system is unique if there is only one solution associated with a given solicitation $\lambda(t)$. Except in small transformations, this is a rather restricted property. The problem of global uniqueness has been intensively discussed in connection with hyperelasticity and most of the results obtained are based upon an assumption of global convexity. In a similar spirit, the following result gives a sufficient condition for global uniqueness in our context.

3.1. Proposition

If C is independent of α and if the following assumptions are satisfied:

$$\left.\begin{array}{lll}
\text{(i)} & F''_{\alpha^2}(\alpha,\lambda)\,.\,\delta\alpha^2 \text{ is positive} & \\
 & F''_{\alpha^2}(\alpha,\lambda)\,.\,\delta\alpha^2 > k\,.\,\|\delta\alpha\|^2, \quad k>0 & \\
\text{(ii)} & F''_{\alpha^2},\ F''_{\alpha\lambda},\ F'''_{\alpha^3},\ F''_{\alpha^2\lambda} \text{ are bounded:} & \\
 & |F'_{\alpha^2}\,.\,\delta\alpha\,\delta\beta| < K\,\|\delta\alpha\|\,.\,\|\delta\beta\| & K>0\\
 & |F''_{\alpha\lambda}\,.\,\delta\alpha\,\delta\lambda| < L\,\|\delta\alpha\|\,.\,|\delta\lambda| & L>0\\
 & \|F'''_{\alpha^3}\,.\,\delta\alpha\,\delta\beta\,\delta\lambda| < M\,\|\delta\alpha\|\,.\,\|\delta\beta\|\,.\,\|\delta\gamma\| & M>0\\
 & |F'''_{\alpha^2\lambda}\,.\,\delta\alpha\,\delta\beta\,\delta\lambda| < N\,\|\delta\alpha\|\,.\,\|\delta\beta\|\,.\,|\delta\lambda| & N>0\\
\text{(iii)} & \text{Load-velocity is finite:} & \\
 & |\dot\lambda\| \leqslant l &
\end{array}\right\} \qquad (17)$$

then the response is unique.

3.2. Proof

Indeed, if α_1 and α_2 denote two eventual solutions of system (8), then one obtains $\forall t$:

$$(F'_{\alpha_1} - F'_{\alpha_2})\,.\,(\dot\alpha_1 - \dot\alpha_2) \leqslant 0$$

There exists $\theta \in [0, 1]$ such that $F'_{\alpha_1} = F'_{\alpha_2} + F''_{\alpha_2^2}\,.\,\Delta\alpha + \frac{1}{2}F'''_{\alpha^3}\,.\,\Delta\alpha^2$, $\Delta\alpha = \alpha_1 - \alpha_2$ and $\alpha = \alpha_2 + \theta\,.\,\Delta\alpha$, thus the following inequality holds:

$$\Delta\alpha\,.\,F''_{\alpha_2^2}\,.\,\Delta\dot\alpha + \tfrac{1}{2}\Delta\alpha^2\,.\,F'''_{\alpha^3}\,.\,\Delta\dot\alpha \leqslant 0$$

or:

$$\left.\begin{array}{l}
\dfrac{\mathrm{d}}{\mathrm{d}t}\left(\tfrac{1}{2}\Delta\alpha\,.\,F''_{\alpha_2^2}\,.\,\Delta\alpha\right) + \tfrac{1}{2}\Delta\alpha^2\,.\,F'''_{\alpha^3}\,.\,\Delta\dot\alpha\\
-\tfrac{1}{2}\Delta\alpha^2\,.\,F'''_{\alpha_2^2\lambda}\,.\,\dot\lambda - \tfrac{1}{2}\Delta\alpha^2\,.\,F'''_{\alpha_2^3}\,.\,\dot\alpha_2 \leqslant 0
\end{array}\right\} \qquad (18)$$

On the other hand, from (6) or (7), one obtains by taking $A^* = A_i(t \pm \Delta t)$ when $\Delta t \to 0$:

$$\dot{A}_i \,.\, \dot{\alpha}_i = 0, \qquad i = 1 \text{ or } 2 \tag{19}$$

thus $F''_{\alpha_i^2} \,.\, \dot{\alpha}_i + \dot{\alpha}_i \,.\, F''_{\alpha\lambda} \,.\, \dot{\lambda} = 0$ and the following estimate holds for each solution:

$$\|\dot{\alpha}_i\| \leqslant Ll/k, \qquad i = 1 \text{ or } 2 \tag{20}$$

From (18) and (20), inequality (21) is derived:

$$k\,\|\Delta\alpha(t)\|^2 \leqslant K\,\|\Delta\alpha(o)\|^2 + X \int_0^t \|\Delta\alpha(s)\|^2\,\mathrm{d}s \tag{21}$$

where $X = 3MlL/k + Nl$ is a constant. Application of Gronwall's lemma [1] then gives the following estimate

$$\|\Delta\alpha(t)\|^2 \leqslant \frac{K}{k}\,\|\Delta\alpha(o)\|^2 \exp(Xt)$$

Since $\|\Delta\alpha(o)\| = 0$, one obtains $\|\Delta\alpha(t)\| = 0\ \forall t$, i.e. uniqueness.

In (17), (i) is the most restrictive assumption and corresponds to global convexity with respect to α.

4. RATE PROBLEM

The computation of the velocity $\dot{\alpha}$ associated with a load velocity $\dot{\lambda}$ is an important problem because of the incremental nature of eqns. (8).

This rate problem is here described in detail when the convex of admissible forces C does not depend on the present state α.

At time t, the present state of the system is assumed to be known. the normal cone $N_C(A)$ to convex C at a given force A is by definition the set:

$$N_C(A) = \{\beta \mid \beta \,.\, (A^* - A) \leqslant 0 \quad \forall\, A^* \in C\} \tag{22}$$

The rate $\dot{A}$ is a vector of the tangent cone since $A(s) \in C\ \forall\, s$, thus one obtains:

$$\dot{A} \,.\, \beta \leqslant 0 \quad \forall\, \beta \in N_C(A) \tag{23}$$

But, in the preceding paragraph we have seen that

$$\dot{A} \,.\, \dot{\alpha} = 0 \tag{24}$$

if $\alpha(t)$ is a solution of (8). Equations (23) and (24) show that the solution $\dot{\alpha}$ associated with a rate $\dot{\lambda}$ must be a solution of the following variational problem:

$$\left.\begin{array}{l}\dot{\alpha} \in N_C(A) \text{ and verifies:} \\ (\dot{\alpha}-\beta) \,.\, (F''_{\alpha^2} \,.\, \dot{\alpha} + F''_{\alpha\lambda} \,.\, \dot{\lambda}) \leqslant 0, \qquad \forall\, \beta \in N_C(A)\end{array}\right\} \tag{25}$$

The study of eqn. (25) can be done without difficulty by standard techniques of convex analysis. The following results are established.

Existence of at least one solution $\dot{\alpha}$ of (25) is obtained under the assumption of positivity of the quadratic form F'' on $N_C(A)$:

$$F''_{\alpha^2} \,.\, \delta\alpha^2 > 0 \quad \forall\, \delta\alpha \neq 0, \quad \delta\alpha \in N_C(A) \tag{26}$$

Uniqueness of the rate $\dot{\alpha}$ is obtained under more restricted conditions. The positivity of F''_{α^2} is required on a larger set which is the vectorial space $\overline{N_C(A)}$ generated by vectors of the normal cone $N_C(A)$:

$$F''_{\alpha^2} \,.\, \delta\alpha^2 > 0 \quad \forall\, \delta\alpha \neq 0, \quad \delta\alpha \in \overline{\overline{N_C(A)}} \tag{27}$$

The quadratic form F''_{α^2} plays an important role.

Remarks

(i) If (27) is satisfied, it is well known that (25) is also equivalent to a minimization problem:

$$\left.\begin{array}{l}\text{The solution } \dot{\alpha} \text{ minimizes then on } N_C(A) \text{ the functional} \\ I(\dot{\alpha}^*) = \tfrac{1}{2}\dot{\alpha} \,.\, F''_{\alpha^2} \,.\, \dot{\alpha} + \dot{\alpha} \,.\, F''_{\alpha\lambda} \,.\, \dot{\lambda}\end{array}\right\} \tag{28}$$

(ii) If U does not depend on α, λ, the variables (α, u, λ) are independent. In this case F''_{α^2}, $F''_{\alpha\lambda}$ can be obtained explicitly in terms of $E(\alpha, u, \lambda)$ as:

$$\left.\begin{array}{l}F''_{\alpha^2} = E''_{\alpha^2} - E''_{\alpha u} \,.\, E''^{-1}_{u^2} \,.\, E''_{u\alpha} \\ F''_{\alpha^2} = E''_{\alpha^2} - E''_{\alpha u} \,.\, E''^{-1}_{u^2} \,.\, E''_{u\lambda}\end{array}\right\} \tag{29}$$

and the rate problem can also be conveniently formulated with state variables $(\dot{\alpha}, \dot{u})$ [11] or, in a more classical manner, with $\dot{u}$ alone.

5. NON-BIFURCATION AND STABILITY CRITERIA

We are now mainly concerned with monotonic loading; $\lambda(t)$ is an increasing function.

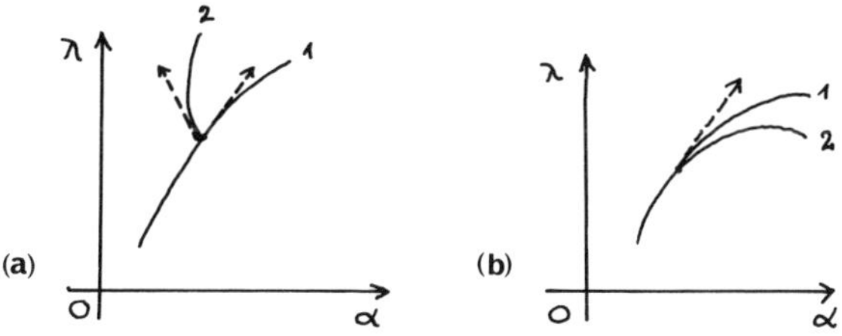

FIG. 1. (a) Non-uniqueness of $\dot{\alpha} \Rightarrow$ bifurcation. (b) Smooth bifurcation.

Let $\alpha(\lambda)$ be a continuous solution of (8). If at a load level λ the rate problem admits two solutions $\dot{\alpha}_1 \neq \dot{\alpha}_2$, bifurcation occurs by definition (Fig. 1(a)).

The situation is more embarrassing if the rate $\dot{\alpha}$ is unique. Hill's conjecture [4] is usually admitted in the sense that uniqueness of the rate $\dot{\alpha}$ is usually accepted, at least in plasticity, as a sufficient condition of non-bifurcation. In other words, smooth bifurcation as shown in Fig. 1(b) does not seem to be possible (?). This is a difficult open problem and the reader may refer to a recent paper by Triantafyllidis [15] for an interesting discussion.

Under the assumption of Hill's conjecture, inequality (27) is a sufficient condition of non-bifurcation.

The answer is much clearer for stability analysis. Indeed, the following arguments show that (26) is a sufficient condition of stability of the equilibrium (u, α, λ).

An equilibrium is stable if in any small perturbation of the system from this position, *positive* work must be done by perturbation forces. If only quasi-static perturbations are considered and $\Delta\alpha$ denotes the resulting modification of α by perturbations on time interval $[0, T]$, the energy balance is:

$$\Delta T_{\mathrm{p}} = E(\alpha + \Delta\alpha, u + \Delta u, \lambda) - E(\alpha, u, \lambda) + \Delta D$$

where ΔT_{p} is the furnished energy, D the dissipation. Since

$$\Delta D = \int_0^T \underset{\tau}{A}\, \underset{\tau}{\dot{\alpha}}\, \mathrm{d}\tau \geqslant \int_0^T A\, \underset{\tau}{\dot{\alpha}}\, \mathrm{d}\tau = A \,.\, \Delta\alpha$$

and

$$E(\alpha + \Delta\alpha, u + \Delta u, \lambda) \geqslant F(\alpha + \Delta\alpha, \lambda)$$

one obtains:

$$\Delta T_p \geqslant F(\alpha+\Delta\alpha, \lambda) - F(\alpha) - F'_\alpha \,.\, \Delta\alpha \simeq \tfrac{1}{2} F''_{\alpha^2} \,.\, \Delta\alpha^2$$

For small perturbations, the positivity of the furnished energy follows then from (26).

The criteria of stability (26) and non-bifurcation (27) present some analogy with the second variation criterion of elasticity. The second derivative of energy F''_{α^2} plays an essential role as in elasticity, the only difference is the intervention of the normal cone $N_C(A)$ and of the associated space $\overline{\overline{N_C(A)}}$.

6. EXAMPLES IN BRITTLE FRACTURE

Figure 2 presents some simple examples of propagation of cracks in brittle fracture. An elastic thin plate with two linear cracks is submitted to a displacement control test, λ is the implied displacement on symmetric parts $S_u^{\pm}$ of the boundary, α denotes the crack lengths $\alpha = (l_1, l_2)$. In small transformation the energy potential of our system is:

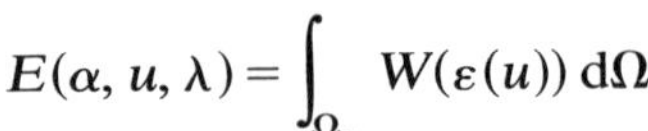

$$E(\alpha, u, \lambda) = \int_{\Omega_\alpha} W(\varepsilon(u))\, \mathrm{d}\Omega$$

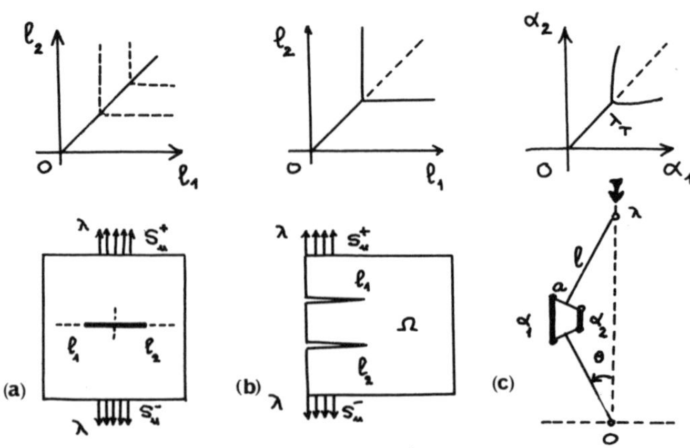

FIG. 2. (a) Stable symmetric mode (fracture). (b) Stable unsymmetric mode (fracture). (c) Shanley's model (elastoplasticity).

The set U of admissible displacements depends on the crack lengths α and on λ:

$$U(\alpha, \lambda) = \{u \mid u_i \quad \text{defined on } \Omega_\alpha, \qquad u_2 = \pm\lambda \quad \text{on} \quad S_u^\pm\}$$

The energy potential at equilibrium is:

$$F(\alpha, \lambda) = \underset{u^* \in U(\alpha, \lambda)}{\text{Min}} E(\alpha, u^*, \lambda)$$

It is well-known in brittle fracture that the generalized forces

$$G_i = -\frac{\partial F}{\partial l_i}$$

represent the energy release rates associated with crack length l_i and Griffith's law of crack propagation may be written as:

$$\left.\begin{array}{lll} \dot{l}_i = 0 & \text{if} & G_i < 2\gamma \\ \dot{l}_i \geqslant 0 & \text{if} & G_i = 2\gamma \end{array}\right\}$$

Thus the associated force $A = (G_1, G_2)$ must belong to the convex of admissible forces:

$$C = \{A^* = (G_1^*, G_2^*) \mid G_i^* \leqslant 2\gamma\}$$

and eqns. (7) are effectively verified.

The dependence of F with respect to α is rather complicated in this case. For each configuration of cracks, a stability analysis can be performed by the preceding method. For example, one can study the stability of the trivial symmetric solutions on Fig. 2(a) or 2(b). In this case, one obtains:

$$N_C(A) = \{(V_1, V_2) \mid V_1 \geqslant 0, \qquad V_2 \geqslant 0\}$$
$$\overline{\overline{N_C(A)}} = \{(V_1, V_2) \mid V_1 \in R, \qquad V_2 \in R\} = \mathbb{R}^2$$

and:

$$F'' = \begin{bmatrix} X & Y \\ Y & X \end{bmatrix}$$

matrix of eigenvalues $\mu_1 = X + Y$, $\mu_2 = X - Y$.

Non-bifurcation criterion (27) requires that $\mu_1 > 0$, $\mu_2 > 0$ thus $X > 0$, $X - Y > 0$.

Stability criterion (26) requires that $\mu_1 > 0$, $\mu_2 > -\mu_1$ thus $X > 0$, $X + Y > 0$.

One can verify that the symmetric mode of propagation is stable in Fig. 2(a) and unstable in Fig. 2(b).

We underline the fact that these examples may be compared to classical two-dimensional problems of plasticity. Figure 2(c) recalls the Shanley model of rigid bars and two plastic elements with kinematic hardening. If the rotation θ is small, the associated energy function, defined by (12), is [11]:

$$E(u, \alpha, \lambda) = \tfrac{1}{2}K\left(u_2 - \frac{a}{l}u_1 - \alpha_1\right)^2 + \tfrac{1}{2}K\left(u_2 + \frac{a}{l}u_1 - \alpha_2\right)^2$$

$$+\tfrac{1}{2}h\alpha_1^2 + \tfrac{1}{2}h\alpha_2^2 + \lambda u_2 + \lambda l\left(1 - \frac{u_1^2}{2l^2}\right)$$

where K is the elastic modulus, h the hardening modulus α_i, $i = 1, 2$ are the plastic strains.

Stability and bifurcation of the trivial symmetric solution $\alpha_1(\lambda) = \alpha_2(\lambda)$ can be obtained by the same discussion. In this case

$$X = h - \frac{1}{2}\frac{\lambda}{\lambda_E - \lambda}K, \qquad Y = \frac{1}{2}\frac{\lambda}{\lambda_E - \lambda}K, \qquad \lambda_E = 2a^2\frac{K}{l}$$

denotes the Euler critical load. Classical results are again established.

Non-bifurcation criterion (27) requires that $\lambda < \lambda_R = 2h(2h + K)\lambda_E$ the critical load of reduced modulus.

Stability criterion (26) requires that $\lambda < \lambda_T = h(h + K)\lambda_E$ the critical load of tangent modulus.

7. EXAMPLES IN BRITTLE DAMAGE

The description of brittle damage also leads to several models which can be included in our context [11]. Let us consider here a model of progressive brittle damage due to Suquet [14].

It is assumed that damage can be observed as an alteration of the elastic characteristics of a material, due to variations of one scalar variable d. In small transformations, the material admits reversible energy density $W(\varepsilon, d)$, $\sigma = \dfrac{\partial W}{\partial \varepsilon}$ is the generalized force associated with ε and $a = -\dfrac{\partial W}{\partial d}$ is associated with d. For a solid made by such material the total energy is:

$$E(\alpha, u, \lambda) = \int_\Omega W(\varepsilon, d)\, \mathrm{d}\Omega - \int_S \lambda T u \,\mathrm{d}S$$

in a force control loading. Internal parameter α denotes in this expression the scalar field **d**, defined on Ω. As in the preceding sections, equilibrium energy $F(\alpha, \lambda)$ is:

$$F(\alpha, \lambda) = \operatorname*{Inf}_{u^*} E(\alpha, u^*, \lambda) = E[\alpha, u(\alpha, \lambda), \lambda]$$

Relation $\delta F = -\int_\Omega a\, \delta d\, \mathrm{d}\Omega - \int_S \delta\lambda T\,.\,u\, \mathrm{d}S$ shows that the associated force $A = -F'_\alpha$ is the scalar field **a**, defined on Ω.

If the following constitutive equation is assumed:

$$\begin{aligned} \dot{d} &= 0 \quad \text{if} \quad a < k \\ \dot{d} &\geq 0 \quad \text{if} \quad a = k \end{aligned}$$

then (7) is verified with admissible domain:

$$C = \{A^* \mid a^*(x) \leq k \quad \forall\, x \in \Omega\}$$

in the force space.

The normal cone $N_C(A)$ associated with the present force A is

$$N_C(A) = \left\{ \boldsymbol{\delta}\mathbf{d} \;\middle|\; \begin{aligned} \delta d(x) &= 0 \quad \text{if} \quad a(x) < k \\ \delta d(x) &\geq 0 \quad \text{if} \quad a(x) = k \end{aligned} \right\}$$

and $\overline{\overline{N_C(A)}}$ is the space of function $\boldsymbol{\delta}\mathbf{d}$:

$$\overline{\overline{N_C(A)}} = \{\boldsymbol{\delta}\mathbf{d} \mid \delta d(x) = 0 \quad \text{if} \quad a(x) < k\}$$

A stress–strain curve with softening can be modelized in this way, by a proper choice of the function $W(\varepsilon, d)$, [11, 14].

REFERENCES

1. Brezis, H. *Opérateurs Monotones et Semi-groupes de Contraction dans les Espaces de Hilbert*, North-Holland, Amsterdam, 1973.
2. Germain, P. *Cours de Mécanique des Milieux Continus*, Masson, Paris, 1973.
3. Halphen, B. and Q. S. Nguyen, Sur les matériaux standards généralisés, *J. Méc.* **14** (1975), 39–63.
4. Hill, R. A general theory of uniqueness and stability in elastic plastic solids, *J. Mech. Phys. Solids*, **6** (1958), 236–49.

5. HUTCHINSON, J. Plastic buckling. In: *Advances in Applied Mechanics* (Ed. C-S. Yih), Vol. 14, Academic Press, New York, 1974, 67–144.
6. KOITER, W. T. The stability of elastic equilibrium, Thesis (1945), Traduction AFFDL-TR (1970).
7. MANDEL, J. *Cours de Mécanique*, Tome II, Gauthier-Villars, Paris, 1966.
8. MOREAU, J. J. On unilateral constraints, friction and plasticity, *Cours CIME*, Bressanone, 1973.
9. NEMAT-NASSER, S. Stability of a system of interacting cracks, *Int. J. Engng Sci.*, **16** (1978), 277–85.
10. NGUYEN, Q. S. Normal dissipativity and energy criteria in fracture, *IUTAM Symposium*, Evanston, 1978, (Ed. S. Nemat-Nasser), Pergamon Press, Oxford, 1980, 254–59.
11. NGUYEN, Q. S. Bifurcation et stabilité des systèmes irréversibles obéissant au principe de dissipation maximal, *J. Méc.*, **3** (1984), 41–61.
12. POTIER-FERRY, M. Bifurcation et stabilité pour des systèmes dérivant d'un potentiel, *J. Méc.*, **17** (1978), 579–608.
13. RICE, J. R. Mathematical analysis in the mechanics of fracture, *Fracture*, Vol. 2, (Ed. H. Liebowitz), Academic Press, New York, 1978.
14. SUQUET, P. Plasticité et homogénéisation, Thèse, Paris (1982).
15. TRIANTAFYLLIDIS, N. On the bifurcation and post bifurcation analysis of elastic–plastic solids under general prebifurcation conditions, *J. Mech. Phys. Solids*, **31** (1983), 499–510.

23

On Influence of Some Second Order Effects on the Post-yield Behaviour of Plastic Structures

M. K. DUSZEK

Institute of Fundamental Technological Research, Warsaw, Poland

and

T. ŁODYGOWSKI

Institute of Building Structures and Technology, Technical University, Poznan, Poland

ABSTRACT

In the paper various sets of conjugate and objective measures of constitutive variables are developed. The discussion of the material stability conditions both in the material (Lagrangian) and in the spatial (Eulerian) descriptions is presented. The influence of such second order effects as the material incompressibility or the change of boundary conditions due to plastic deformations on the stability at the yield-point load and on the post-yield behaviour is also investigated. The presentation is limited to the purely mechanical (isothermal) theory and quasi-static deformation processes. The considered problem is illustrated with examples of cylindrical shells and portal frames subjected to the system of proportionally increasing dead loads. Stable and unstable yield-point load regimes are discussed.

1. INTRODUCTION

The load-carrying capacity evaluated in limit analysis is simply a load intensity at which a rigid plastic structure begins to deform.

However, in the presence of structural imperfections or finite elastic displacements, the calculated load-carrying capacity may never be reached if the post-yield behaviour is predicted to be unstable. Therefore, the meaningful estimation of the load-carrying capacity must be accompanied by an analysis of the structural behaviour after the yield-point load has been attained.

The post-yield behaviour of a plastic structure depends, on the one hand, on the material properties, and on the effects of geometry changes on the other. The material is usually assumed to be strain-hardening (stable) in the range of deformations which are relevant from the engineering point of view. The influence of geometry changes, however, may be either stabilizing or destabilizing.

In this paper attention will be focused on the problems of geometric non-linearities. For clarity, the perfectly plastic material model will be assumed. Then, the geometry changes are the only reason for the structure to remain stable or become unstable.

An important question arises immediately: which description, the Lagrangian or the Eulerian, should be used when formulating the constitutive equations for the perfectly plastic material. This question should be answered in view of the consistency of the theoretical solutions with the results of suitable tests.

In the paper various sets of conjugate and objective measures of constitutive variables are discussed. The development of the material stability conditions is given both in the material (Lagrangian) and in the spatial (Eulerian) descriptions. The results are compared and discussed.

The presentation is limited to the purely mechanical (isothermal) theory and quasi-static deformation processes. The considered problem is illustrated with the examples of cylindrical shells and portal frames subjected to the system of proportionally increasing dead loads. The influence on the stability at the yield-point load and on the post-yield behaviour of such second order effects as the material incompressibility and the change of boundary conditions due to plastic deformations is also considered.

2. CONJUGATE AND OBJECTIVE MEASURES

The choice of constitutive variables is not arbitrary but guided by some invariance requirements. The commonly accepted requirements are those of conjugation and invariance under super-imposed rigid body motion.

Conjugate variables are defined by expressing the virtual work rate W per unit initial volume as a scalar product of stress and strain rate [6, 12].

In the Eulerian analysis the process of deformation can be described either in the fixed (in a physical space) or in the co-rotational co-ordinate system. If the Kirchhoff† stress tensor $\mathbf{t}$ and the deformation rate tensor $\mathbf{d}$ are used as conjugate constitutive variables of the Eulerian description, then the following representations of these variables and their time derivatives (invariant with respect to rigid body motion) can be applied.

$$t^{kl}, d_{kl}, (t^{kl})^{\nabla_{\mathrm{y}}}, (d_{kl})^{\nabla_{\mathrm{y}}} \quad \text{in fixed spatial co-ordinates} \tag{1}$$

$$\bar{t}^{KL}, \bar{d}_{KL}, (\bar{t}^{KL})^{\cdot}, (\bar{d}_{KL})^{\cdot} \quad \text{in co-rotational spatial co-ordinates} \tag{2}$$

where $(\;)^{\nabla_{\mathrm{y}}}$ and $(\;)^{\cdot}$ denote, respectively, the Jaumann and material time derivatives.

Since the sets (1) and (2) prove to be different representations of the same tensors, they lead to alternative descriptions of the same material when applied to the constitutive relations of the same forms.

The constitutive variables of the Lagrangian description, that is the second Piola–Kirchhoff stress tensor $\mathbf{S}$, the Green strain rate tensor $\dot{\mathbf{E}}$ and their time derivatives $\dot{\mathbf{S}}$, $\ddot{\mathbf{E}}$ can be written either in the initial frame of reference or in the convected co-ordinate system. Then the following sets of conjugate and objective measures may be used:

$$S^{KL}, \dot{E}_{KL}, \dot{S}^{KL}, \ddot{E}_{KL} \quad \text{in the initial frame of reference} \tag{3}$$

$$t^{kl}, d_{kl}, (t^{kl})^{\nabla_{\mathrm{O}}}, (d_{kl})^{\nabla_{\mathrm{C-R}}} \quad \text{in the convected co-ordinates} \tag{4}$$

where $(\;)^{\nabla_{\mathrm{O}}}$ and $(\;)^{\nabla_{\mathrm{C-R}}}$ denote, respectively, Oldroyd and Cotter–Rivlin time derivatives. Similarly as before, the sets (3) and (4) offer two different representations of the same tensors and therefore lead to the alternative descriptions of the same material when applied to the constitutive relations of the same forms.

† The Kirchhoff stress tensor is defined by the relation $\mathbf{t}=\dfrac{\rho_0\boldsymbol{\sigma}}{\rho}$ where $\boldsymbol{\sigma}$ is the Cauchy stress tensor and ρ_0, ρ are the initial and final mass densities.

3. MATERIAL STABILITY AND THE DEFINITIONS OF PERFECTLY PLASTIC MATERIAL IN LAGRANGIAN AND EULERIAN FORMULATIONS

In the recent literature the following two approaches to the definition of the material stability of time-independent materials are most commonly used.

The first one stems from Drucker's concept [2]. The material stability condition is then derived from the energy criterion of stability of a body or a system when the homogeneous stress and strain states are assumed. Eventually, it takes the form

$$\dot{T}^{Ki} V_{i,K} > 0 \tag{5}$$

where T^{Ki} is the first Piola–Kirchhoff stress tensor and $V_{i,K}$ stands for the velocity gradient.

The other approach consists in the derivation of the material stability condition as a consequence of assumed hardening in the incremental plasticity.

For simplicity let us restrict ourselves to the isotropic rigid–plastic material.

In the Lagrangian formulation the yield condition and the flow rule are formulated in terms of the second Piola–Kirchhoff stress tensor $\mathbf{S}$ and the Green strain rate tensor $\dot{\mathbf{E}}$:

$$F = f(\mathbf{S}) - K = 0 \tag{6}$$

$$\dot{\mathbf{E}} = \Lambda \frac{\partial f}{\partial \mathbf{S}} \tag{7}$$

where

$$\begin{aligned} \Lambda = 0 \quad &\text{if} \quad F < 0 \\ &\text{or if } F = 0 \text{ and } \dot{F} < 0; \\ \Lambda \geqslant 0 \quad &\text{if} \quad F = 0 \quad \text{and} \quad \dot{F} = 0 \end{aligned} \tag{8}$$

The material is said to be hardening or stable if $\dot{K} > 0$, perfectly plastic or neutral if $\dot{K} = 0$, softening or unstable if $\dot{K} < 0$.

The above relations result for the loading process in the following criteria of material stability:

$$\dot{\mathbf{S}}\dot{\mathbf{E}} = \dot{S}^{KL}\dot{E}_{KL} = (t^{kl})^{\nabla \circ} d_{kl} \begin{cases} > 0 & \text{for stable (hardening) material} \\ = 0 & \text{for neutral (perfectly-plastic) material} \\ < 0 & \text{for unstable (softening) material} \end{cases} \tag{9}$$

written either in the total Lagrangian or in the convected co-ordinate system.

Similar analysis made in the Eulerian description leads to the following relations:

$$F = f(\mathbf{t}) - \kappa = 0 \tag{10}$$

$$\mathbf{d} = \lambda \frac{\partial f}{\partial \mathbf{t}} \tag{11}$$

$$(\mathbf{t})^{\nabla y}\mathbf{d} = (t^{kl})^{\nabla y} d_{kl} = (\bar{t}^{KL})^{\cdot} \bar{d}_{KL} \begin{cases} >0 & \text{for stable (hardening) material} \\ =0 & \text{for perfectly plastic material} \\ <0 & \text{for unstable (softening) material} \end{cases} \tag{12}$$

Comparison of relations (5), (9) and (12) leads to the conclusion that we deal with three different kinds of plastic material. The discussion becomes even easier when the considered material stability conditions are written in terms of the same variables as it is done in the Table I.

Since the perfectly plastic material is defined as a plastic solid for which the neutral material stability condition is satisfied, the suitable constitutive relations for the materials which are perfectly plastic in

TABLE I

Description, variables / *Definitions of the material stability*	*Lagrangian* $\mathbf{S}, \dot{\mathbf{S}}, \dot{\mathbf{E}}$	*Eulerian* $\mathbf{t}, (\mathbf{t})^{\nabla y}, \mathbf{d}$
Drucker's formulation, relation (5)	$\dot{S}^{KL}\dot{E}_{KL} + S^{KL}V_{N,K}V^{N}_{,L} \geqslant 0$ or $(t^{kl})^{\nabla o}d_{kl} + t^{kl}V_{n,k}V^{n}_{,l} \geqslant 0$	$(t^{kl})^{\nabla y}d_{kl} - t^{kl}V_{k,n}V^{n}_{,l} \geqslant 0$ or $(\bar{t}^{KL})^{\cdot}\bar{d}_{KL} - \bar{t}^{KL}V_{K,N}V^{N}_{,K} \geqslant 0$
Lagrangian (material) formulation, relation (9)	$\dot{S}^{KL}\dot{E}_{KL} \geqslant 0$ or $(t^{kl})^{\nabla o}d_{kl} \geqslant 0$	$(t^{kl})^{\nabla y}d_{kl} - 2t^{kl}V_{k,n}V^{n}_{,l} \geqslant 0$ or $(\bar{t}^{KL})^{\cdot}\bar{d}_{KL} - 2\bar{t}^{KL}V_{K,N}V^{N}_{,K} \geqslant 0$
Eulerian (spatial) formulation, relation (12)	$\dot{S}^{KL}\dot{E}_{KL} + 2\dot{S}^{KL}V_{N,K}V^{N}_{,L} \geqslant 0$ or $(t^{kl})^{\nabla o}d_{kl} + 2t^{kl}V_{n,k}V^{n}_{,l} \geqslant 0$	$(t^{kl})^{\nabla y}d_{kl} \geqslant 0$ or $(\bar{t}^{KL})^{\cdot}\bar{d}_{KL} \geqslant 0$

different sense can be obtained by putting equality signs in the relations presented in Table I. Such materials shall be called perfectly plastic in Drucker's Lagrangian and Eulerian formulation respectively.

4. STRUCTURAL STABILITY

The question of structural stability depends on the material stability, boundary conditions and the geometry of the body.

Let us consider a structure referred to the total Lagrangian coordinates and subject to a system of dead loads **P** of monotonically increasing intensity

$$\mathbf{P}(\mathbf{X}, t) = \mu(t)\mathbf{P}_0(X) \tag{13}$$

where $\mu(t)$ indicates the load intensity and $\mathbf{P}_0(X)$ specifies the load distribution.

The structure is said to be stable if quasi-static motion can take place only for increasing dead loads, then $\dot{\mu} > 0$. The case of instability is associated with $\dot{\mu} < 0$, and then the structure continues to deform plastically under decreasing dead loads.

The principle of virtual work in appropriate rate form eventually yields the following expression for the rate of load intensity which constitutes the structural stability criterion [7]

$$\dot{\mu}(t) = \frac{\int_{V_0} (\dot{S}^{KL}\dot{E}_{KL} + S^{KL}V_{N,K}V^N_{,L})\,\mathrm{d}V_0}{\int_{S_0} P_0^K V_K\,\mathrm{d}S_0} > 0 \tag{14}$$

where V_0 and S_0 stand, respectively, for the initial volume and the initial surface of the structure.

According to the introduced definitions for rigid-perfectly plastic material in the Lagrangian formulation, satisfying $\dot{S}^{KL}\dot{E}_{KL} = 0$, the structural stability criterion (14) becomes

$$\dot{\mu}(t) = \frac{\int_{V_0} S^{KL}V_{N,K}V^N_{,L}\,\mathrm{d}V}{\int_{S_0} P_0^K V_K\,\mathrm{d}S_0} > 0 \tag{15}$$

in the initial frame of reference or

$$\dot{\mu}(t)=\frac{\displaystyle\int_V t^{kl}V_{m,k}V^m_{,l}\,dV}{\displaystyle\int_{S_0} P_0^K V_K\,dS_0}>0 \tag{16}$$

in the convected co-ordinate system.

For rigid-perfectly plastic material in the Eulerian formulation, the substitution of the condition $(t^{kl})^{\nabla y}d_{kl}=0$ (transformed into the variables of the Lagrangian description, Table I) into (14) yields the structural stability criterion in the form:

$$\dot{\mu}(t)=\frac{\displaystyle -\int_{V_0} S^{KL}V_{K,N}V^N_{,L}\,dV_0}{\displaystyle\int_{S_0} P_0^K V_K\,dS_0}>0 \tag{17}$$

From the comparison of conditions (15) and (17) it follows that the response of structural elements made of materials which are perfectly plastic in various senses may be quite different not only at large strains (or displacements) but even at the yield point load.

When considering thin-walled plastic structures in the range of small strains and moderately large deflections, the Kirchhoff–Love assumption of constant thickness is usually accepted whereas the incompressibility condition is ignored. Therefore, the question of the influence of material incompressibility on the post-yield behaviour of plastic structures arises as another second order geometric effect.

The influence of both effects, that is the type of definition of perfectly plastic material and the material incompressibility on the initial stability of a plastic structure will be considered on the example of a cylindrical shell.

Let us consider a thin-walled cylindrical shell subject to a rotationally symmetric deformation under uniformly distributed radial load P and the axial end load T (Fig. 1). Incompressible, rigid-perfectly plastic material in the Lagrangian formulation is assumed.

The ultimate load solution to this problem with the 'limited interaction' yield condition was obtained by P. G. Hodge [9]. The hexagon ABCDEF in Fig. 2 shows the yield-point loading curve on the plane of dimensionless loads $p=PR/2\sigma_0H$, $t=T/4\pi\sigma_0RH$ for the geometry parameter $\alpha=L^2/RH=\infty$ and the hexagon A′B′C′D′E′F′ for $\alpha=2$.

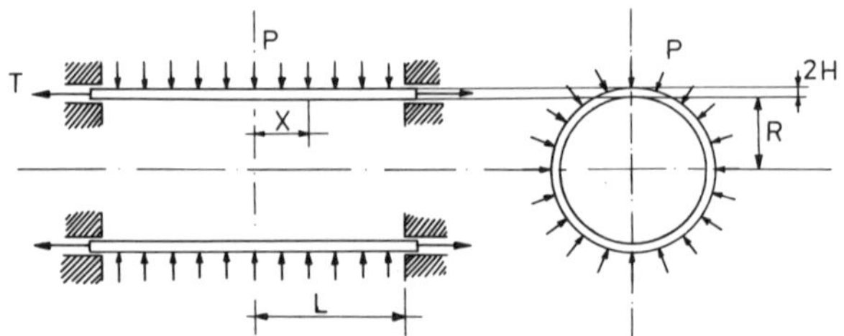

FIG. 1. Cylindrical shell under uniformly distributed radial load P and axial end load T.

The yield-point loading curves for $2<\alpha<\infty$ have similar shapes and are placed between the curves for $\alpha=2$ and $\alpha=\infty$.

Now, making use of the stress and the velocity fields furnished by the ultimate load solution, we shall determine stability and the slope of load-deflection curve at the yield-point load.

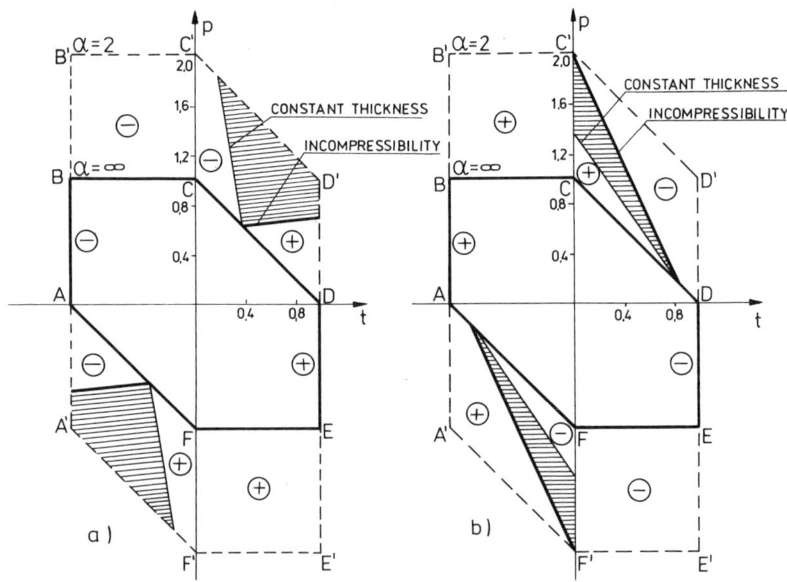

FIG. 2. The yield point loading curves and zones of stable, ⊕, and unstable, ⊖, ultimate load solution: (a) for cylindrical shells made of rigid-perfectly plastic material in the Lagrangian formulation; (b) for cylindrical shells made of rigid-perfectly plastic material in the Eulerian formulation.

For the range of loads represented by the faces CD, C′D′ in Fig. 2(a) we have

$$0 \leq t < 1, \qquad p = 1 - t + \frac{2}{\alpha} \tag{18}$$

and the load intensity, determined for the perfectly plastic material in the Lagrangian formulation by eqn. (15), becomes

$$\dot{\mu} = \frac{2\alpha \dot{W}_0 \left(3t\dfrac{R^2}{L^2} + 2t - 1 - \dfrac{5}{\alpha}\right)}{3(\alpha + 2)} \tag{19}$$

where $\dot{W}_0$ is the rate of maximum deflection. Hence

$$\text{for} \quad t > \frac{\alpha + 5}{3\dfrac{R}{H} + 2\alpha}: \qquad \dot{\mu} > 0, \quad \text{the shell remains stable} \tag{20}$$

$$\text{for} \quad t < \frac{\alpha + 5}{3\dfrac{R}{H} + 2\alpha}: \qquad \dot{\mu} < 0, \quad \text{the shell becomes unstable} \tag{21}$$

For the range of loads represented by the faces BC, B′C′

$$-1 < t < 0, \qquad p = 1 + \frac{2}{\alpha} \tag{22}$$

the load intensity (15) reduces to the form

$$\dot{\mu} = \frac{2\alpha \dot{W}_0 \left(3t\dfrac{R^2}{L^2} - 1 - \dfrac{3}{\alpha}\right)}{3(\alpha + 2)} < 0 \tag{23}$$

which means instability.

For the faces AB, A′B′ in Fig. 2, we have

$$t = -1 \tag{24}$$

and

$$\dot{\mu} = -\dot{V}_0 < 0 \tag{25}$$

where $\dot{V}_0$ is the rate of maximum displacement in the axial direction. Therefore the shell is unstable.

Similar calculations can be performed for negative values of the load p. The detailed analysis is presented in ref. 4.

The stability in the considered case depends on the ratio of the surface tractions as well as on the geometry of the shell described by the ratio R/L and the parameter α. The yield-point loading curve can be divided into two parts (Fig. 2(a, b)). One of them (denoted by the sign $\oplus$) corresponds to the stable and the other (denoted by $\ominus$) to the unstable ultimate load solutions. The results presented in Fig. 2 are calculated for $L/R = 2$.

The problem as considered above but with the simplifying assumption of constant shell thickness was considered in ref. 3 and results are also depicted in Fig. 2. The shaded zones indicate the ratios of loads p/t for which the answer whether the shell, at the yield-point load, is stable or unstable, changes if the incompressibility condition is taken into account. The differences are greater for thicker shells (small α), whereas both solutions coincided for infinitely thin shells ($\alpha = \infty$).

The solution of the stability problem for the same cylindrical shell as considered above but made of rigid-perfectly plastic material in the Eulerian formulation is presented in Fig. 2(b).

The comparison of results shown in Figs. 2(a) and 2(b) indicates that the influence of the definition of perfectly plastic material on the stability at the yield-point load is essential over the whole range of considered loads. There is no evident proof which definition of plastic material is more suitable for the description of the behaviour of shells made of mild steel, however available experimental data [1] are in agreement with the solutions obtained for the shells made of perfectly plastic material in the Lagrangian formulation.

5. THE POST-YIELD ANALYSIS OF RIGID-PLASTIC FRAMES

The steel portal frames, used primarily in tall buildings, constitute one of the most important problems in the plastic analysis of structures.

For complex frame structures, if the yield-point load is associated with the local collapse mechanism, the plastic deformations may result in such changes of the boundary conditions that the stabilizing effect of tensile forces is reduced.

By the proper choice of ratios of rigidities of beam and column rigidity ratio the unstable mechanisms can be avoided.

With the assumption that the plastic deformations are concentrated in generalized plastic hinges and the displacements do not exceed the depths of cross-sections, the structural stability criterion (for the perfectly plastic material in the Lagrangian sense) takes the form [5, 10]

$$\dot{\mu} = \frac{\sum_{i=1}^{k} N_i \dot{\Phi}_i^2 L_i}{\sum_{j=1}^{m} P_j \dot{W}_j} > 0 \tag{26}$$

where N_i, Φ_i, L_i are, respectively, the axial force, the rate of rotation and the length of the ith element; k is the number of elements, $\dot{W}_j$ denotes the displacement rate of the force in its direction.

Since the denominator in (26) is non-negative, the stability of the frame structure is ensured if

$$\sum_{i=1}^{k} N_i \Phi_i^2 L_i > 0 \tag{27}$$

The stabilizing effect of tensile forces and destabilizing effect of compressive forces is, therefore, evident.

In the post-yield analysis of plastic frames at the moderately large deflections, the mechanism of motion is usually assumed to be the same as at the yield-point load. This assumption may however result in inaccurate estimation of the post-yield behaviour.

To illustrate the problem let us consider an s-storey portal frame subject to concentrated vertical loads as shown in Fig. 3(a). The analysis will be carried for $(s-n)$th floor (Fig. 3(b)).

It follows from the limit analysis that the frame begins to deform according to the beam mechanism with generalized plastic hinges at the cross-sections 2, 2′, 3. However, at a certain deflection $\bar{W}$, the increase of axial force in the beam results in yielding of columns at the cross-sections 1, 1′.

Taking into account the second order geometrical effects as well as the influence of axial forces on the yielding of the beam and columns, the equilibrium equation eventually leads to the condition:

$$\alpha^2\gamma - 2\alpha\omega(n+1)n - n\gamma(1-\omega^2) - \beta^2\gamma n^2(1+2\omega^2) = 0 \tag{28}$$

The following dimensionless coefficients are introduced:

$$\alpha = \frac{H_c}{H_b}, \qquad \beta = \frac{H_b}{L_b}, \qquad \gamma = \frac{H_c}{L_c}, \qquad \omega = \frac{W}{H_b} \tag{29}$$

where L_c, L_b, H_c, H_b denote, respectively, the lengths of columns and

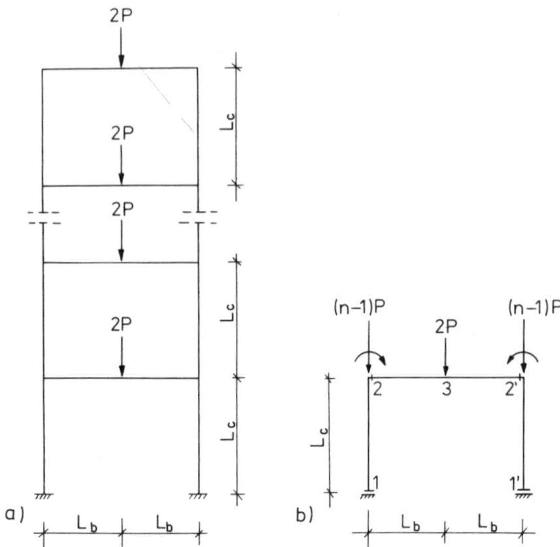

FIG. 3. (a) s-storey portal frame subjected to concentrated vertical loads; (b) $(s-n)$th floor.

beam, the depths of columns and beam. Moreover the last term in eqn. (29) being a dimensionless deflection at the midspan of the beam.

All beams and columns are assumed to have the same widths and be made of the same material.

Putting $\omega=0$ into (28) we obtain the relation between n (the number of floors) and the geometry parameters for which the hinges in the cross-sections 1, 1′, 2, 2′ and 3 develop simultaneously, namely

$$\alpha^2 = n(1+\beta^2 n) \tag{30}$$

The above relation for $\beta = 0, 1;\ 0, 15$ is illustrated in Fig. 4.

For $\alpha^2 > n(1+\beta^2 n)$ the generalized plastic hinges appear at the cross-sections 2, 2′, 3 and the beam mechanism (a) presented in Fig. 5 is realized. The stability criterion (26) applied to the geometrically non-linear solution yields

$$\dot{\mu} = \frac{4\omega\dot{\omega}}{(1+\omega^2)} \geqslant 0 \tag{31}$$

Thus, the frame is stable for the deflections not exceeding the value given by eqn. (28). When W reaches the value satisfying eqn. (28), the

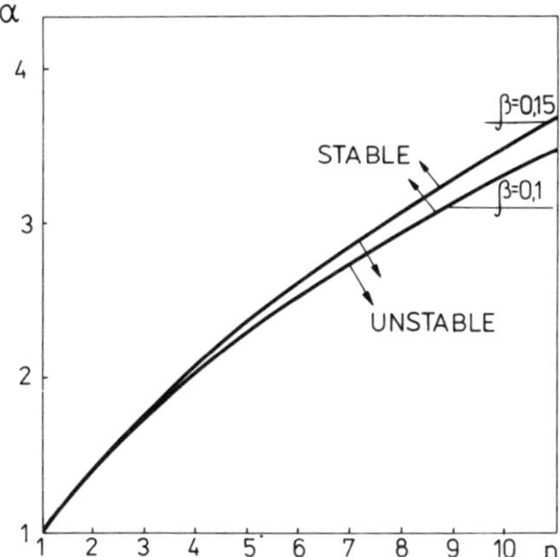

FIG. 4. The relations between the number of floors n and the geometry parameters α, β for neutrally stable limit load solution.

plastic hinges at the cross-sections 1, 1′ appear and further deformation continues according to the mechanism (b) shown in Fig. 5, at the constant axial force in the beam. The beam behaves then as a beam on moving supports [10]. The increase of destabilizing effect of compressive axial forces in columns and reduced effect of stabilizing tensile force in the beam can result in unstable behaviour of the whole frame.

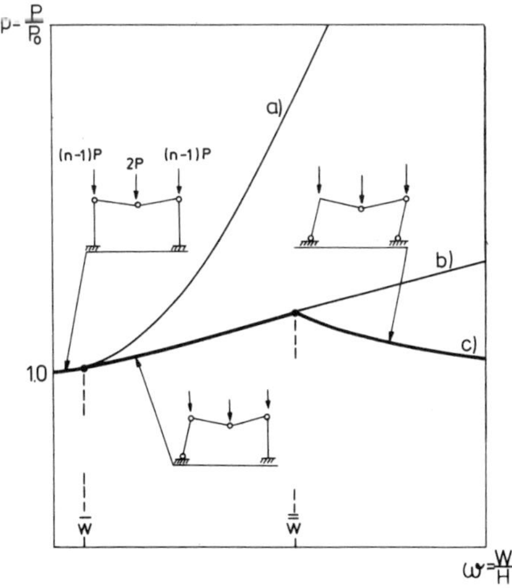

FIG. 5. Load-deflection curves for various mechanisms of motion.

If the rigidity of columns is less than, or equal to, the critical value described by eqn. (30) then, from the beginning, the mechanism of motion (b) shown in Fig. 5 takes place. The stability criterion (26) yields

$$\dot{\mu} = -2n\beta\gamma\omega\dot{\omega}^2 \leqslant 0 \tag{32}$$

and, therefore, the frame becomes unstable.

For slender frames the mechanism (b) may be next changed into combined mechanism (c) shown in the Fig. 5 which proves to be unstable.

6. THE INFLUENCE OF ELASTIC DEFORMATIONS ON THE BEHAVIOUR OF ELASTIC–PLASTIC STRUCTURES

The geometrically non-linear numerical analysis of the elastic–plastic beams and frames is worked out [11] with the use of the finite element method. The cross-sections of a structure are assumed to be divided into layers subject to either uniaxial tension or compression. The finite element is assumed to be a segment of a layer, so the distribution of plastic zones and local unloading in the process of deformation can be followed at each step. The obtained load–deflection curves for beams, columns and frames are of the shape presented in Fig. 6. From comparison of the results obtained for structures made of a rigid–plastic material and for those made of an elastic–plastic material, the following conclusions can be drawn:

(a) In the case of the stable post-yield behaviour of a rigid–plastic

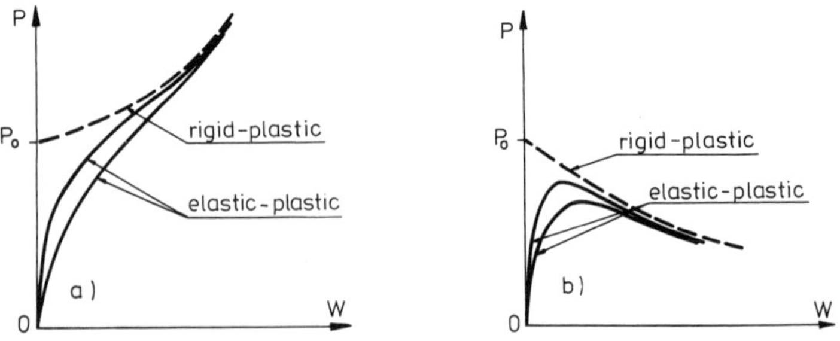

FIG. 6. The load-deflection curves for rigid–plastic and elastic–plastic structures.

structure, the elastic–plastic structure is capable of sustaining the load which exceeds the load-carrying capacity assessed by the methods of limit analysis.

(b) If the post-yield behaviour of a rigid–plastic structure is predicted to be unstable, then the maximum load which can be supported by the same structure but made of elastic–plastic material may never reach the load-carrying capacity since the structure may collapse much earlier.

Therefore the full advantage of the load-carrying capacity (from the point of view of direct applications in structural design) can be taken only with an additional analysis of the behaviour of the structure after the yield-point load has been attained.

It may happen that a structure with smaller ultimate load but stable, is in reality capable of supporting greater maximum load prior to collapse than an apparently stronger structure but proved unstable.

The rigidity ratios of the constituent elements in an optimal structure should be chosen not only with respect to the ultimate load but also with respect to the stable post yield behaviour.

REFERENCES

1. Augusti, G. and S. d'Agostino. Test of cylindrical shells in the plastic range, *J. Engng Mech. Division*, **90** *EM1* (1964) 69–81.
2. Drucker, D. C. A more fundamental approach to plastic stress, strain relations, *Proc. First U.S. Nat. Congress app. Mech., ASME* (1951), 487–491.
3. Duszek, M. K. Stability analysis of rigid plastic structures at the yield point load. In: *Buckling of Structures* (Ed. B. Budiansky), Springer-Verlag, Berlin, 1976, 106–116.
4. Duszek, M. K. and E. Alessandrini. The influence of some second order effects on the behaviour of rigid–plastic shells at the yield point load, *IPPT, Rep.*, **1,** Warsaw, 1983.
5. Duszek, M. K. and A. Sawczuk. Stable and unstable states of rigid–plastic frames at the yield-point load, *J. Struct. Mech.*, **4** (1976), 33–47.
6. Hill, R. On constitutive inequalities for simple materials I, *J. Mech. Phys. Solids*, **16** (1968), 229–242.
7. Hill, R. Stability of rigid–plastic solids, *J. Mech. Phys. Solids*, **6** (1957), 1–8.
8. Hill, R. Aspects of invariance in solid mechanics. In: *Adv. appl. Mech.*, **18** (1978), 1–75.
9. Hodge, P. G. *Limit Analysis of Rotationally Symmetric Plates and Shells*, Prentice-Hall, New Jersey, 1963.

10. Łodygowski, T. Geometrically non-linear analysis of rigid–plastic and elastic–plastic beams and frames (in Polish), *IPPT, Rep.*, **9,** Warsaw, 1982.
11. Łodygowski, T. Geometrically non-linear elastic–plastic analysis of beams and frames (in Polish), *Arch. Inż. Lądowej.*, **27** (1982), 79–97.
12. Truesdell, T. The mechanical foundations of elasticity and fluid dynamics, *J. Rational Mech. Anal.*, **1** (1952), 125–300.

24

On the Onset of Instability in Elastic–Plastic Solids

H. PETRYK

Institute of Fundamental Technological Research, Warsaw, Poland

ABSTRACT

Instability phenomena in elastic–plastic solids are investigated by using the energy-type definition of stability of a quasi-static deformation process, proposed recently by the author. A number of instability criteria are derived which provide upper bounds to the onset of various forms of instability. The conditions are specified under which these criteria lead to the same results as the criteria for bifurcation in velocities or accelerations.

1. INTRODUCTION

In many cases it is observed experimentally that the real behaviour of a loaded plastic body confirms the simplest theoretical predictions only up to some stage of the deformation and deviates remarkably from them beyond this stage. Usually the term 'instability' is employed in those circumstances, and the critical stage of the deformation is regarded as the onset of instability. There are several different forms of what is known as stability loss. Dynamic snap-through phenomena, buckling, necking and the localization of plastic flow are the types of instability commonly observed in elastic–plastic solids.

Various theoretical approaches have been used in the literature to study such phenomena (see refs. 7, 9 and 13). The most rigorous and general of them is the bifurcation theory, developed by Hill [1, 3, 6].

Recently, a postulate of stability of a quasi-static deformation process has been formulated [10, 11] which is sufficiently universal to deal with all the instability phenomena mentioned above. In the postulate the usually loosely defined term 'instability' is given a precise energy sense. A part of the present paper concerns the relations between the postulate and the bifurcation theory. However, the primary aim is to demonstrate that by assuming this single stability postulate a unified approach to instability phenomena in elastic–plastic solids becomes available. A method is presented for deriving a number of instability criteria from which the onset of various types of instability can be estimated.

2. CONSTITUTIVE EQUATIONS AND DEFORMATION PROCESSES

We are concerned with isothermal, slow deformations of a continuous body occupying in a reference configuration a space region V bounded by a piecewise regular surface S. Denote by $\mathbf{x}$ and $\boldsymbol{\xi}$ the place of a material point in space in the current and reference configuration, respectively. On a common, fixed rectangular basis used throughout this paper, the vectors $\mathbf{x}$ and $\boldsymbol{\xi}$ have the components x_i and ξ_i, $i=1, 2, 3$, referred to as the space and material co-ordinates, respectively.

In order to retain possible generality, we assume the constitutive rate equations in the general form proposed by Hill [1, 2]:

$$\dot{s}_{ij} = \frac{\partial U}{\partial(v_{j,i})}, \qquad 2U = \dot{s}_{ij} v_{j,i} \tag{1}$$

where s_{ij} are the components of the (unsymmetric) nominal stress tensor $\mathbf{s}$, related to the components of Cauchy stress tensor $\boldsymbol{\sigma}$ by the formula $x_{j,i} s_{ik} = \det(\mathbf{F})\ \sigma_{jk}$. Here and in the following the comma denotes partial derivative with respect to a material co-ordinate ξ_i and the dot denotes the material time derivative. $\mathbf{F}$ and $\mathbf{L} = \dot{\mathbf{F}}$ are the deformation gradient and the velocity gradient, in components $(\mathbf{F})_{ij} = x_{i,j}$ and $(\mathbf{L})_{ij} = v_{i,j}$, respectively. U is a continuous, continuously differentiable and piecewise continuously twice differentiable function of $\mathbf{L}$, and also a functional of the deformation history. To be definite, we assume that

$$U = U(\mathbf{L}; \mathbf{F}, \mathbf{H}) \tag{2}$$

where **H** is a finite set of arbitrary internal parameters H_γ, $\gamma = 1, 2, \ldots, N$, varying with the deformation. The evolution equations for H_γ are in the form

$$\dot{H}_\gamma = \kappa_\gamma(\mathbf{L}; \mathbf{F}, \mathbf{H}) \tag{3}$$

The material is assumed to be time-independent. Hence, for **F** and **H** defined by a prior deformation history, the functions $\dot{\mathbf{s}}(\mathbf{L})$ and $\dot{\mathbf{H}}(\mathbf{L})$ obtained from (1) and (3) must be homogeneous of degree one. $U(\mathbf{L})$ is homogeneous of degree two and can be written in the form (1). By homogeneity of the function $\dot{\mathbf{s}}(\mathbf{L})$, eqns. (1) can be rewritten as

$$\dot{s}_{ij} = C_{ijkl}(\mathbf{L})v_{l,k}, \qquad C_{ijkl} = C_{klij} \tag{4}$$

provided the second derivatives of $U(\mathbf{L})$,

$$\frac{\partial^2 U(\mathbf{L})}{\partial(v_{j,i})\,\partial(v_{l,k})} = C_{ijkl}(\mathbf{L}) \tag{5}$$

are continuous at **L**. For other **L** the moduli C_{ijkl} are not well defined by **L** itself though $\dot{\mathbf{s}}$ is always derivable from (1).

As shown by Hill [6], existence of a velocity-gradient potential U is equivalent to existence of a rate-of-strain potential W such that

$$\dot{t}_{ij} = \frac{\partial W}{\partial \dot{e}_{ij}}, \qquad 2W = \dot{t}_{ij}\dot{e}_{ij} \tag{6}$$

where $(\mathbf{t}, \mathbf{e})$ is an arbitrary conjugate pair of objective stress and strain measures. The formulae relating U and W, as well as the moduli C_{ijkl} and any objective moduli, have been explicitly given by Hill [6]. No specific pair $(\mathbf{t}, \mathbf{e})$ will be assumed here; by using the formulae mentioned, it is a simple task to rewrite all the following expressions in terms of any objective measures if needed.

The constitutive equations for a classical elastic–plastic solid obeying the normality flow rule can be considered as a particular case of eqns. (4). The moduli C_{ijkl} can assume then only two different values according to whether further plastic loading or elastic unloading takes place in an element being currently in the plastic state. More specifically, if the *current* configuration is taken as the reference one, the moduli have the form [6]

$$C_{ijkl} = \begin{cases} C^{\mathrm{p}}_{ijkl} = \mathscr{L}_{ijkl} + \sigma_{ik}\delta_{jl} - \dfrac{1}{g}\lambda_{ij}\lambda_{kl} & \text{if} \quad \lambda_{ij}v_{j,i} > 0, \\ C^{\mathrm{e}}_{ijkl} = \mathscr{L}_{ijkl} + \sigma_{ik}\delta_{jl} & \text{if} \quad \lambda_{ij}v_{j,i} < 0 \end{cases} \tag{7}$$

where $\mathscr{L}_{ijkl} = \mathscr{L}_{klij}$ are the elastic moduli referred to the Green measure of strain, δ_{ij} is the Kronecker symbol, $\boldsymbol{\lambda}$ is the (unique) outward normal to the yield surface in strain space, and g is a positive scalar parameter. By writing elastoplasticity equations in the form (4) with the moduli (7), we need not discuss the question of decomposition of deformation into plastic and elastic parts, nor make any difference between 'hardening' and 'softening' of the material.

A process of deformation of the body is described by (and will be identified with) a function $\boldsymbol{\chi}$ such that $\mathbf{x} = \boldsymbol{\chi}(\boldsymbol{\xi}, t)$, where t is a timelike parameter, for simplicity referred to as time. We will consider only continuous functions $\boldsymbol{\chi}(\boldsymbol{\xi}, t)$ but the velocities $\mathbf{v} = \partial\boldsymbol{\chi}/\partial t$ are assumed to be continuous in space only and not necessarily in time. Therefore all material or time derivatives shall be regarded as right-hand ones, and these are assumed always to exist. Moreover, the function $\boldsymbol{\chi}(\boldsymbol{\xi}, t)$ is assumed to have at least piecewise continuous third derivatives with respect to its arguments.

We assume that an initial body configuration at time t_1 and complete specification of the material properties are given. The body may be inhomogeneous, provided the material properties vary within the body in a piecewise smooth manner. The following time-dependent boundary conditions are considered. On the part S_u of the body surface S the displacement history is prescribed. On the remaining part S_T of S the nominal tractions $\mathbf{T}$ are assumed to be derivable from a given, sufficiently regular, time-dependent potential ω,

$$T_i = -\frac{\partial\omega(\mathbf{x}, \boldsymbol{\xi}, t)}{\partial x_i} \tag{8}$$

Without loss of generality, we can put $\omega \equiv 0$ at $t = t_1$. In the particular case when the tractions $\mathbf{T}$ are given directly as a function of $\boldsymbol{\xi}$ and t but not of $\mathbf{x}$, we put $\omega(\mathbf{x}, \boldsymbol{\xi}, t) = -T_i(\boldsymbol{\xi}, t) \,.\, (x_i - \chi_i(\boldsymbol{\xi}, t_1))$. Body forces are omitted for brevity.

In the following we will consider only such processes $\boldsymbol{\chi}$ which start from the given body configuration at time t_1, are compatible with the kinematical constraints on S_u and satisfy the regularity conditions specified above. Such processes will be called kinematically admissible.

3. STABILITY AND INSTABILITY CRITERIA

Let us consider a kinematically admissible theoretical process $\boldsymbol{\chi}^{\circ}$ which is intended to describe sufficiently slow deformations of the real body in some finite time interval. The process $\boldsymbol{\chi}^{\text{o}}$ will be called the fundamental process and all corresponding quantities will be distinguished by the superscript 'o'. In reality, there are always some influences not taken into account in the theoretical analysis; they are interpreted here as disturbances. Generally speaking, a fundamental process may be regarded as stable or not, depending on the magnitude of deviations from it caused by disturbances, however, many different definitions of stability can be formulated. We use the stability definition proposed previously by the author [10, 11] in which persistent disturbances are considered from a class which is sufficiently wide to produce any kinematically admissible process.

The work done by internal forces in a kinematically admissible process $\boldsymbol{\chi}$ in a time interval $[t_1, t]$ can be written in the form

$$W(\boldsymbol{\chi}, t) = \int_{t_1}^{t} \int_{V} s_{ij} v_{j,i} \, \mathrm{d}V \, \mathrm{d}\tau \tag{9}$$

where the stresses are determined by integration of the rate equations (1) and evolution equations (3) along the deformation path. The corresponding potential energy of the loading device at time t is represented by

$$W^{\text{e}}(\boldsymbol{\chi}, t) = \int_{S_{\text{T}}} \omega(\boldsymbol{\chi}(\boldsymbol{\xi}, t), \boldsymbol{\xi}, t) \, \mathrm{d}S \tag{10}$$

We introduce the energy functional

$$E(\boldsymbol{\chi}, t) = W(\boldsymbol{\chi}, t) + W^{\text{e}}(\boldsymbol{\chi}, t) \tag{11}$$

the value of which is the amount of energy which has to be supplied to the system :{body + loading device} in the time interval $[t_1, t]$ in order to realize the process $\boldsymbol{\chi}$ with vanishingly small speed, i.e. in a quasi-static manner. The corresponding disturbance measure ρ is taken in the form

$$\rho(\boldsymbol{\chi}, \boldsymbol{\chi}^{\text{o}}, t) = \sup_{\tau \in [t_1, t]} \{E(\boldsymbol{\chi}, \tau) - E(\boldsymbol{\chi}^{\text{o}}, \tau)\} \tag{12}$$

which is equivalent to that one proposed in refs. 10 and 11 since inertia forces may be included with perturbing forces. We need not specify here the measure of *distance* d $(d \geqslant 0)$ between two processes. We assume only a natural property that two processes which require a different amount of energy must differ in the sense of the measure d at some instant of time, i.e. that $E(\boldsymbol{\chi}_1, t) \neq E(\boldsymbol{\chi}_2, t)$ implies $d(\boldsymbol{\chi}_1, \boldsymbol{\chi}_2, \tau) > 0$ at some $\tau \in (t_1, t)$.

Definition [10, 11]: *A fundamental process* $\boldsymbol{\chi}^{\circ}$ *is stable in a time interval* $[t_1, t_2]$ *if and only if for every number* $\varepsilon > 0$ *there is another number* $\delta > 0$ *such that for every kinematically admissible process* $\boldsymbol{\chi}$ *and all* $t \in [t_1, t_2]$

$$\rho(\boldsymbol{\chi}, \boldsymbol{\chi}^{\circ}, t) < \delta \quad \textit{implies} \quad d(\boldsymbol{\chi}, \chi^{\circ}, t) < \varepsilon$$

A process is called unstable if it is not stable.

The definition jointly with the definitions of a process $\boldsymbol{\chi}$ and measures ρ and d form the stability postulate. It is seen from the definition that a process is regarded as stable only if every deviation from it requires some additional amount of energy to be supplied.

It can happen that a process $\boldsymbol{\chi}^{\circ}$ is stable in some time interval $[t_1, \tau_{cr}]$ but unstable in any wider interval $[t_1, t_2]$, $t_2 > \tau_{cr}$. If the stability postulate has a physical meaning then beyond the critical stage of the deformation corresponding to the time τ_{cr} the theoretical process $\boldsymbol{\chi}^{\circ}$ can no longer describe properly the real behaviour of the body. It is thus of primary interest to determine such instant τ_{cr} at which the onset of instability occurs. It is not easy to do this directly, by using the definition of stability, therefore we will derive from it a number of criteria which are more convenient in applications. In this paper we are concerned with instability criteria which provide upper bounds to the onset of instability.

We will consider a possibility of the kinematically admissible branching of the deformation path at some time τ through a difference of displacement-rate fields of some order at time τ, the difference being called branching mode.† In other words, a kinematically admissible process $\boldsymbol{\chi}$ is considered which coincides with the fundamental one up to time τ and then deviates from it. We can calculate the corresponding differences $\Delta \overset{(k)}{E}(\tau) = \overset{(k)}{E}(\tau) - \overset{(k)_{\circ}}{E}(\tau)$, $k = 1, 2, \ldots,$ of the values of kth (right-hand) time derivatives of the energy functional (11) at time τ; of course, $\Delta E(\tau) = 0$. Here and in the following the prefix Δ denotes the difference of any corresponding quantities in the processes $\boldsymbol{\chi}$ and $\boldsymbol{\chi}^{\circ}$ at the same instant of time. From the definition of stability the following criterion results.

If at time $\tau \in [t_1, t_2)$ *there is a kinematically admissible branching*

† No statical conditions are imposed on the branching mode. If the branching mode satisfies additionally the statical requirements, the usual term 'bifurcation' will be used.

mode such that for some n

$$\overset{(n)}{\Delta E}(\tau)<0 \quad and \quad \overset{(k)}{\Delta E}(\tau)=0, \qquad k=1,\ldots,n-1 \tag{13}$$

then the process χ° *is unstable in the interval* $[t_1, t_2]$ *and* $\tau_{cr} \leqslant \tau$.

In proof we need only observe that if the assumptions of the criterion are satisfied then at some $t \in (\tau, t_2)$ we have $d>0$ and $\rho=0$, that implies instability. Note that only one particular mode of branching is needed to prove instability and to obtain an upper bound to the onset of instability. The mode must correspond to a kinematically admissible branching path but need not satisfy any conditions of equilibrium.

While for practical applications the criterion (13) may be useful, we will be more interested here in necessary conditions for stability. By negation of (13) we obtain the following condition (which was in fact used in [11]). *Let for some class of kinematically admissible branching modes* $\overset{(k)}{\Delta E}$ *vanish identically at the instant of branching for all* $k \leqslant n-1$. *Then for stability of the process* χ° *in a time interval* $[t_1, t_2]$ *it is necessary that*

$$\overset{(n)}{\Delta E}(\tau) \geqslant 0 \tag{14}$$

for every mode from that class and all $\tau \in [t_1, t_2)$.

The condition (14) will usually have the form of minimum principles for functionals, or of variational inequalities, with the kinematically admissible field of displacement-rate of some order as a varied quantity. The methods of the calculus of variations can be used to derive from such minimum principles more simple criteria. Some illustrations of this are given below. The criterion (14) may be regarded as a generalization and a precise formulation of the intuitive 'minimum energy-rate hypothesis'.

Let us consider a first-order branching of the deformation path, i.e. a process χ which deviates from the process χ° through a non-zero difference $\mathbf{v}-\mathbf{v}^{\circ}$ of the velocity fields at some time τ. By differentiating (11) we obtain (cf. ref. 11)

$$\Delta \dot{E}(\tau) = J_1(\mathbf{v}) - J_1(\mathbf{v}^{\circ}) \tag{15}$$

where

$$J_1(\mathbf{v}) = \int_V s_{ij} v_{j,i}\, \mathrm{d}V - \int_{S_T} T_i v_i\, \mathrm{d}S \tag{16}$$

is a linear functional of velocity field $\mathbf{v}=\mathbf{v}(\boldsymbol{\xi})$ since the stresses $\mathbf{s}$ and tractions $\mathbf{T}$ at time τ are independent of $\mathbf{v}$. The criterion (14) with $n=1$ leads to the variational principle which holds for every stable process $\boldsymbol{\chi}^{\text{o}}$:

$$\delta J_1(\mathbf{v}^{\text{o}};\mathbf{w})=J_1(\mathbf{w})=\int_V s_{ij}^{\text{o}}w_{j,i}\,\mathrm{d}V-\int_{S_{\text{T}}} T_i^{\text{o}}w_i\,\mathrm{d}S=0 \tag{17}$$

where $\mathbf{w}=\mathbf{w}(\boldsymbol{\xi})$ is an arbitrary continuous and continuously differentiable vector field vanishing over S_{u}. This is of course the virtual work principle. This means that in a stable process the equilibrium conditions are satisfied at any instant. Hence we need not assume in advance that the fundamental process $\boldsymbol{\chi}^{\text{o}}$ corresponds to a quasi-static solution to the problem: this results from the stability postulate as a necessary condition for stability. To obtain further necessary conditions, we will from now on assume that (17) is always satisfied.

By using the Green theorem and the equilibrium conditions for $\mathbf{s}^{\text{o}}$ and taking the time derivative of (11) we obtain the useful formula

$$\Delta\dot{E}=\int_V(\Delta s_{ij})v_{j,i}\,\mathrm{d}V-\int_{S_{\text{T}}}\left\{(\Delta T_i)v_i-\Delta\frac{\partial\omega}{\partial t}\right\}\mathrm{d}S \tag{18}$$

which is valid universally, not only at the instant of branching. By differentiating (18) with respect to time and substituting in (14), we can obtain stability criteria in the form of variational inequalities. We will make use of (18) in Section 6.

4. BIFURCATION AND THE ONSET OF INSTABILITY

The incipient buckling and necking in the plastic range are usually not connected with dynamic loss of stability of equilibrium. Such phenomena are regarded here as a symptom of instability of a deformation process† in the energy sense. This is an alternative approach to the widely used Hill's bifurcation theory, therefore it is of interest to clarify the relations between them. Our aim in this section is to specify the circumstances under which both approaches lead to the same results.

† A similar point of view has been adopted by Klushnikov [8] who, however, in the general case applies in fact the bifurcation approach. In ref. 8 the connection between bifurcation and instability was only postulated.

In this section we will allow only such processes in which at any instant of time the velocities are continuously differentiable functions of place, including boundary. Consider a first-order branching of the deformation path at time τ; such branching may correspond to the initial stage of buckling or necking. From (17) or (18) it is evident that $\Delta\dot{E}(\tau)$ vanishes. By differentiating twice the expression (11) with respect to time we find [11] that

$$\tfrac{1}{2}\Delta\ddot{E}(\tau) = J_2(\mathbf{v}) - J_2(\mathbf{v}^{\circ}) \tag{19}$$

where

$$J_2(\mathbf{v}) = \int_V U(\mathbf{L})\,\mathrm{d}V - \int_{S_T} \{\dot{t}_i + \tfrac{1}{2}k_{ij}v_j\}v_i\,\mathrm{d}S \tag{20}$$

is a functional defined on the class of all admissible velocity fields taking the values prescribed over S_u at time τ. For brevity we have used in (20) a short notation

$$\dot{t}_i \equiv -\frac{\partial^2\omega}{\partial t\,\partial x_i} = \frac{\partial T_i}{\partial t}, \qquad k_{ij} = k_{ji} = -\frac{\partial^2\omega}{\partial x_i\,\partial x_j} = \frac{\partial T_i}{\partial x_j} \tag{21}$$

It is seen that the traction-rate on S_T has the form

$$\dot{T}_i \equiv -\frac{\mathrm{d}}{\mathrm{d}t}\left(\frac{\partial\omega}{\partial x_i}\right) = \dot{t}_i + k_{ij}v_j$$

this leads to a special case of the self-adjoint rate-boundary value problem considered by Hill [4]. We note that $\mathbf{v}^{\circ}$ represents a solution to this problem since by the assumption introduced at the end of Section 3, in the process $\boldsymbol{\chi}^{\circ}$ the body is at any stage in equilibrium.

By substituting (19) in the criterion (14) with $n = 2$, we arrive at the following result [11]. *For stability of the process* $\boldsymbol{\chi}^{\circ}$ *in a time interval* $[t_1, t_2]$ *it is necessary that the extremum principle*

$$J_2(\mathbf{v}) \geqslant J_2(\mathbf{v}^{\circ}) \tag{22}$$

holds at all $\tau \in [t_1, t_2)$ *and for every admissible velocity field* $\mathbf{v}$ *taking the values prescribed over* S_u *at time* τ.

In the case of the boundary conditions considered here, the functional (20) coincides with the functional examined by Hill [4, 6]. Hence (22) with strict inequality coincides with Hill's extremum principle [4] derived originally in a quite different way from the sufficiency condition for uniqueness of solution to the first-order rate-problem. Here, (22) must hold for any stable solution, unique or not.

In order to find more direct relations between the present energy approach and the bifurcation theory, we will examine now a particular but in practice the frequent case when the moduli of an elastic–plastic solid are of the form (7) and the body is deformed in the process $\boldsymbol{\chi}^{\text{o}}$ in such a way that no rapid unloading occurs. That is, we assume that $\lambda_{ij}v^{\text{o}}_{j,i}>0$ in the zone where the material is currently in the plastic state, except possibly a region of zero volume only. For brevity, as long as the (continuously differentiable) velocity field $\mathbf{v}^{\text{o}}$ has this property, the process $\boldsymbol{\chi}^{\text{o}}$ will be called regular.

At any stage of a regular process $\boldsymbol{\chi}^{\text{o}}$ the moduli C^{o}_{ijkl} are well defined by the velocity gradient $\mathbf{L}^{\text{o}}$ (except possibly a region of zero volume) and coincide, by definition, with the moduli of the 'comparison linear solid' introduced by Hill [1, 3]. Consider the functional

$$I_2(\mathbf{v}^{\text{o}};\mathbf{w})=\frac{1}{2}\int_V C^{\text{o}}_{ijkl}w_{j,i}w_{l,k}\,\mathrm{d}V-\frac{1}{2}\int_{S_{\text{T}}} k_{ij}w_iw_j\,\mathrm{d}S \tag{23}$$

defined on the class of continuous and continuously differentiable vector fields $\mathbf{w}$ vanishing over S_{u}. This is the 'exclusion functional' [6] for the comparison linear solid, whose positive definiteness excludes, by the comparison theorem, the first-order bifurcation in the actual elastic–plastic solid and implies (22) with strict inequality for $\mathbf{v}\neq\mathbf{v}^{\text{o}}$.

We prove now that for regular processes the functional (23) coincides also with the second (weak) variation of the functional $J_2(\mathbf{v})$ taken at $\mathbf{v}=\mathbf{v}^{\text{o}}$, i.e.

$$I_2(\mathbf{v}^{\text{o}};\mathbf{w})=\delta^2J_2(\mathbf{v}^{\text{o}};\mathbf{w})\overset{\text{df}}{=}\frac{1}{2}\frac{\mathrm{d}^2}{\mathrm{d}\alpha^2}J_2(\mathbf{v}^{\text{o}}+\alpha\mathbf{w})\bigg|_{\alpha=0}, \tag{24}$$

within the class $\mathbf{w}$ as above.

First of all, by using the constitutive equations in the form (4) and the equilibrium conditions for $\dot{\mathbf{s}}^{\text{o}}$, we obtain after some transformations

$$J_2(\mathbf{v})-J_2(\mathbf{v}^{\text{o}})=I_2(\mathbf{v}^{\text{o}};\mathbf{v}-\mathbf{v}^{\text{o}})+\frac{1}{2}\int_V(\Delta C_{ijkl})v_{j,i}v_{l,k}\,\mathrm{d}V \tag{25}$$

$J_2(\mathbf{v}^{\text{o}})$ does not undergo variations, and the second variation of the quadratic functional $I_2(\mathbf{v}^{\text{o}};\mathbf{v}-\mathbf{v}^{\text{o}})$ is of course $I_2(\mathbf{v}^{\text{o}};\mathbf{w})$ itself. It remains to be shown that the second variation of the last integral in (25) vanishes at $\mathbf{v}=\mathbf{v}^{\text{o}}$. To do this, we observe that ΔC_{ijkl} can be different from zero only in a part V^{u} of the plastic region where $\lambda_{ij}v_{j,i}\leqslant 0$ since $\lambda_{ij}v^{\text{o}}_{j,i}>0$ there by assumption. If this is the case then in V^{u} we have

$\Delta C_{ijkl} = (1/g)\lambda_{ij}\lambda_{kl}$ and $|\lambda_{ij}v_{j,i}| < |\alpha|\,|\lambda_{ij}w_{j,i}|$ if we put $\mathbf{v} = \mathbf{v}^{\circ} + \alpha\mathbf{w}$. Hence,

$$0 \leqslant \int_V (\Delta C_{ijkl})v_{j,i}v_{l,k}\,\mathrm{d}V = \int_{V^{\mathrm{u}}} \frac{1}{g}(\lambda_{ij}v_{j,i})^2\,\mathrm{d}V \leqslant \alpha^2 \int_{V^{\mathrm{u}}} \frac{1}{g}(\lambda_{ij}w_{j,i})^2\,\mathrm{d}V \tag{26}$$

With vanishing α the volume V^{u} must tend to zero. Thus, the second derivative with respect to α of the last term, and therefore of all terms in (26) vanishes at $\alpha = 0$. This is what we set out to prove.

As shown by Hill [4], the first variation of $J_2(\mathbf{v})$ vanishes at $\mathbf{v} = \mathbf{v}^{\circ}$; this is seen explicitly from (25). Hence, it follows that non-negativeness of the second variation:

$$I_2(\mathbf{v}^{\circ}; \mathbf{w}) \geqslant 0 \tag{27}$$

for every admissible vector field $\mathbf{w}$ vanishing over S_{u}, is necessary for (22) and therefore also necessary for stability of the process $\boldsymbol{\chi}^{\circ}$. Conversely, (27) implies (22) as can be seen from (25) with (26). Thus, for regular processes the necessary condition for stability (22) is satisfied if and only if the exclusion functional (23) is non-negative for all admissible fields $\mathbf{w}$ vanishing over S_{u}.

A connection between the instability of a regular process in the energy sense and a possibility of bifurcation is now apparent. As long as the exclusion functional (23) is positive definite, first-order bifurcation is excluded and the process may be stable. If the exclusion functional (23) can take negative values then the process is unstable and bifurcation may take place. Between these two ranges there is frequently a critical stage when the functional (23) is positive semi-definite and vanishes for some non-zero field $\mathbf{w}^*$. It is a standard result that the field $\mathbf{w}^*$ is then an eigenmode for the comparison linear solid. Hence, both the approaches discussed lead to the same linear eigenvalue problem from which the critical stage can be found.

Suppose that such field $\mathbf{w}^*$ exists at the critical stage. If there is a constant c such that $\lambda_{ij}v^{\circ}_{j,i} \geqslant c > 0$ everywhere in the current plastic zone then the velocity field $\mathbf{v}^* = \mathbf{v}^{\circ} + \alpha\mathbf{w}^*$ corresponds to the same field of moduli C°_{ijkl} as $\mathbf{v}^{\circ}$, provided $|\alpha|$ is sufficiently small [7]. $\mathbf{v}^*$ is then another solution to the actual first-order rate problem. In that case, at the critical stage when (22) still holds the primary first-order bifurcation occurs, and for the field $\mathbf{v}^*$ we have $\Delta\ddot{E} = 0$ as is seen from (19) and (25) with $\Delta C_{ijkl} = 0$. Conversely, the earliest branching mode in velocities such that $\Delta\ddot{E} = 0$ for that mode and $\Delta\ddot{E} \geqslant 0$ for others has the properties of the field $\alpha\mathbf{w}^*$ discussed above and therefore defines the instant and the mode of the primary first-order bifurcation.

However, if $\lambda_{ij}v^{\circ}_{j,i}$ tends to zero somewhere in the plastic zone, for instance when the elastic–plastic boundary is approached, then there is no such constant c as mentioned above. It can then happen that any mode of the type $\alpha\mathbf{w}^*$ introduces a change of moduli due to unloading in the region where $\lambda_{ij}v^{\circ}_{j,i}$ tends to zero, no matter how small $|\alpha|$ is; in that case no first-order bifurcation is possible at the critical stage. However, it can be shown that the second-order bifurcation can then take place instead, namely, the eigenmode $\mathbf{w}^*$ multiplied by an arbitrary coefficient can be added to the acceleration field in the fundamental process to obtain another solution to the second-order rate problem. Some remarks concerning this question can be found in ref. 8; a more detailed and precise analysis of the second-order rate problem is to be given in ref. 12.

No matter what is the order of bifurcation, from the stability condition (14) the criterion of choice of the post-bifurcation path follows immediately: the stable path must minimize at the instant of branching the lowest-order time derivative of the energy functional (11) which is not constant within the class of possible bifurcation modes.

5. INSTABILITY OF EQUILIBRIUM

In snap-through phenomena, or at the maximum load point in the case of a 'soft' loading device, a rapid transition occurs from slow deformations to a dynamic process. This is regarded here, as commonly agreed, as a symptom of instability of equilibrium. The stability postulate discussed above can be used to study also this kind of instability. To this purpose we consider the stability of equilibrium as the stability of a particular fundamental process in which the velocities vanish identically and the boundary conditions are time-independent. In that case $E(\boldsymbol{\chi}^{\circ}, t)\equiv 0$, and in the instability criteria (13) and (14) the prefix Δ can simply be omitted.

If the stability of equilibrium is considered then the necessary condition for stability (22) reduces to (cf. ref. 11)

$$\int_V U(w_{j,i})\,\mathrm{d}V-\frac{1}{2}\int_{S_T} k_{ij}w_iw_j\,\mathrm{d}S\geqslant 0 \tag{28}$$

for every continuous and continuously differentiable velocity field $\mathbf{w}$ vanishing over S_u. In the case of dead loading the second integral in (28) vanishes, and (28) with strict inequality coincides then with the criterion proposed by Hill [1] as a sufficient condition for stability of equilibrium.

The connections between the stage when $\int U(w_{j,i})\,dV \geqslant 0$ with equality for some non-zero $\mathbf{w}$ and existence of eigenstates under dead loading have been discussed in detail by Hill [5]. Analogously, if at some stage of the deformation the condition (28) is still satisfied but with equality for some non-zero $\mathbf{w}^*$, then such $\mathbf{w}^*$ is an eigenmode satisfying the rate-equilibrium equations and compatible with the homogeneous boundary conditions $\mathbf{w}=0$ on S_u and $\dot{s}_{ij}n_i = k_{ij}w_i$ on S_T. Beyond this stage (28) will in general cease to hold, implying instability of equilibrium. A connection between the loss of stability and non-uniqueness is again apparent, now in the case of time-independent boundary conditions.

In the case of a linear solid when the moduli C_{ijkl} do not depend on the actual velocity gradient, the instability criteria (22), (27) and (28) coincide. In particular, if the solid is hyperelastic, then each of these criteria is equivalent to the familiar condition of non-negativeness of the second variation of the total potential energy of the system. The present interpretation of this last condition is that it is necessary for stability.

6. LOCALIZATION OF DEFORMATION

Applying the same general energy approach as in the preceding sections, we will consider now a class of branching modes which can describe the onset of localization of deformation. Contrary to the usually examined bifurcation within a planar band of finite thickness, no additional assumptions nor approximations concerning the boundary conditions or material homogeneity are required here. In particular, the modes are kinematically exact even if the body is finite and the displacements are prescribed over its whole boundary.

In the class of branching modes we allow now the class of discontinuous acceleration fields $\Delta\dot{\mathbf{v}}$ vanishing over S_u. At the instant τ of branching the difference $\Delta\mathbf{v}$ of velocity fields is identically zero. In a typical mode from this class the discontinuity $[\![\Delta\dot{\mathbf{v}}]\!]$ of $\Delta\dot{\mathbf{v}}$ appears across some regular surface $S_D(\tau)$ lying within the body. We define the

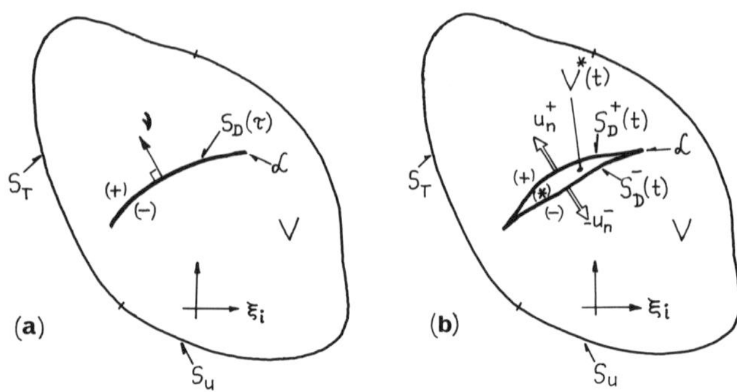

FIG. 1. Kinematically admissible mode of localization of deformation: (a) at the instant of branching, $t=\tau$, (b) at $t>\tau$.

discontinuity $[\![\cdot]\!]$ across $S_D(\tau)$ by writing $[\![\cdot]\!]=(\cdot)^+-(\cdot)^-$ where the sides $(+)$ and $(-)$ of $S_D(\tau)$ are related to the direction of the unit normal $\boldsymbol{\nu}$ to $S_D(\tau)$ as shown in Fig. 1(a). The position of $S_D(\tau)$ is chosen such that possible discontinuities: $[\![\mathbf{s}^o]\!]$, $[\![\dot{\mathbf{s}}^o]\!]$, $[\![\mathbf{L}^o]\!]$ and $[\![\dot{\mathbf{v}}^o]\!]$ vanish, so that $[\![\dot{\mathbf{v}}]\!]=[\![\Delta\dot{\mathbf{v}}]\!]$.

The branching mode must correspond to a kinematically admissible branching path. This will be possible provided the conditions discussed below are satisfied. The surface $S_D(\tau)$ splits instantaneously into two surfaces $S_D^+(t)$ and $S_D^-(t)$, $\tau\leq t$ (Fig. 1(b)), which propagate with respect to the material in opposite directions with normal speeds $u_n^+\geq 0$ and $u_n^-\leq 0$ calculated in the reference configuration. The surfaces $S_D^+(t)$ and $S_D^-(t)$ have always a common perimeter $\mathscr{L}$, taken for simplicity to be a material line, so that $u_n^+=u_n^-=0$ on $\mathscr{L}$. Within the disc-like region $V^*(t)$ bounded by the closed surface $S_D^+(t)\cup S_D^-(t)$ the velocity gradient $v_{j,i}$ on the branching path is taken to differ essentially from that in the fundamental process. From the well-known geometrical compatibility conditions [14] applied to a continuous velocity field $\mathbf{v}(t)$ we obtain that in the limit when $t\to\tau$ the difference must take the form

$$v^*_{j,i}-v^o_{j,i}=b_j\nu_i \tag{29}$$

where $v^*_{j,i}$ is the limit when $t\to\tau$ of the velocity gradient $v_{j,i}$ in the region $V^*(t)$ at a material point lying on $S_D(\tau)$, $v^o_{k,i}$ is calculated also at time τ at this point, and $\mathbf{b}$ is an arbitrary vector. In order for the branching path to be kinematically admissible, the kinematical compatibility conditions [14] on each of the surfaces $S_D^+(t)$ and $S_D^-(t)$ are yet to be satisfied. This will be possible provided

$$[\![\Delta\dot{v}_j]\!]=[\![\dot{v}_j]\!]=b_j(u_n^+-u_n^-) \tag{30}$$

For, if (30) is satisfied then there exists $\dot{\mathbf{v}}^*$, regarded as the limit when $t\to\tau$ of the acceleration $\dot{\mathbf{v}}$ in the region $V^*(t)$ at a point on $S_D(\tau)$, such that at time τ

$$[\![\dot{v}_j]\!]^+=\dot{v}_j^+-\dot{v}_j^*=b_ju_n^+, \qquad [\![\dot{v}_j]\!]^-=\dot{v}_j^*-\dot{v}_j^-=-b_ju_n^- \tag{31}$$

where $[\![\cdot]\!]^+$ and $[\![\cdot]\!]^-$ denote the jump across the surfaces $S_D^+(t)$ and $S_D^-(t)$, respectively, in the limit when $t\to\tau$ and in the order as in (31). By comparison with (29) it is seen that (31) are the required kinematical compatibility conditions, written in the limit when $t\to\tau$.

Now, we will calculate the differences $\Delta\overset{(k)}{E}(\tau)$ corresponding to the examined branching mode, assuming that the potential U as a *function* of velocity gradient varies smoothly in time. To do this, we will differentiate eqn. (18) with respect to time, taking into account the effect of propagation of the discontinuity surfaces $S_D^+(t)$ and $S_D^-(t)$, in a way used by Thomas [14]. It is easy to see that not only $\Delta\dot{E}(\tau)$ but also $\Delta\ddot{E}(\tau)$ vanishes due to $\Delta\mathbf{v}\equiv 0$. By differentiating twice eqn. (18), substituting $\Delta\mathbf{s}\equiv 0$, $\Delta\mathbf{v}=0$ and $\Delta\mathbf{L}=0$, $\Delta\dot{\mathbf{s}}=0$ except on the surface $S_D(\tau)$, and using (21), we obtain

$$\begin{aligned}\Delta\dddot{E}(\tau)=&\int_V(\Delta\ddot{s}_{ij})v_{j,i}^{\circ}\,\mathrm{d}V-\int_{S_D^+}[\![(\Delta\dot{s}_{ij})v_{j,i}]\!]^+u_n^+\,\mathrm{d}S-\int_{S_D^-}[\![(\Delta\dot{s}_{ij})v_{j,i}]\!]^-u_n^-\,\mathrm{d}S\\&-\frac{\mathrm{d}}{\mathrm{d}t}\int_{S_D^+}[\![(\Delta s_{ij})v_{j,i}]\!]^+u_n^+\,\mathrm{d}S-\frac{\mathrm{d}}{\mathrm{d}t}\int_{S_D^-}[\![(\Delta s_{ij})v_{j,i}]\!]^-u_n^-\,\mathrm{d}S\\&-\int_{S_T}\dot{t}_i(\Delta\dot{v}_i)\,\mathrm{d}S-\int_{S_T}k_{ij}(\Delta\dot{v}_i)v_j^{\circ}\,\mathrm{d}S\end{aligned} \tag{32}$$

Observe that at time τ we have

$$\begin{aligned}\frac{\mathrm{d}}{\mathrm{d}t}\int_{S_D^+}[\![(\Delta s_{ij})v_{j,i}]\!]^+u_n^+\,\mathrm{d}S&=\int_{S_D^+}\frac{\delta}{\delta t}[\![(\Delta s_{ij})v_{j,i}]\!]^+u_n^+\,\mathrm{d}S\\&=\int_{S_D^+}\frac{\delta}{\delta t}(\Delta s_{ij}^+)[\![v_{j,i}]\!]^+u_n^+\,\mathrm{d}S\\&=\int_{S_D^+}\{\Delta\dot{s}_{ij}^++(\Delta s_{ij,k}^+)\nu_ku_n^+\}[\![v_{j,i}]\!]^+u_n^+\,\mathrm{d}S=0\end{aligned} \tag{33}$$

where $\frac{\delta}{\delta t}$ denotes Thomas' derivative [14] which 'follows' the motion of the surface $S_D^+(t)$. The first equality in (33) results from the differentiation rule for integrals defined on moving surfaces since $\Delta\mathbf{s}=0$ at time τ, the second from continuity of the stresses in both processes across $S_D^+(t)$ when $u_n^+ \neq 0$, the third from the definition of the derivative $\frac{\delta}{\delta t}$. Since we have assumed $u_n^+ \geqslant 0$, the surface S_D^+ propagates into the region where the differences of stress-gradient and stress-rate in both processes are zero at time τ; this implies the last equality in (33). Analogously, the term in (32) which is the time derivative of the surface integral over S_D^- is also equal to zero.

Denote by $\dot{\mathbf{s}}^*$ the limit when $t \to \tau$ of the stress-rate in the region $V^*(t)$, related to the corresponding value $v^*_{j,i}$ of velocity gradient by the constitutive equation (1), viz. $\dot{s}^*_{ij} = \partial U(v^*_{k,l})/\partial(v_{j,i})$. At time τ we have $\Delta\dot{\mathbf{s}}^+ = \Delta\dot{\mathbf{s}}^- = 0$, so that

$$[\![(\Delta\dot{s}_{ij})v_{j,i}]\!]^+ = -[\![(\Delta\dot{s}_{ij})v_{j,i}]\!]^- = (\dot{s}^{\circ}_{ij} - \dot{s}^*_{ij})v^*_{j,i} \tag{34}$$

Consider now the volume integral in (32). By comparing the time derivatives of U written first in the form (1) and next in the form (2), we obtain

$$\tfrac{1}{2}\ddot{s}_{ij}v_{j,i} + \tfrac{1}{2}\dot{s}_{ij}\dot{v}_{j,i} = \frac{\partial U}{\partial(v_{j,i})}\dot{v}_{j,i} + \frac{\partial U}{\partial(x_{j,i})}v_{j,i} + \frac{\partial U}{\partial H_\gamma}\dot{H}_\gamma \tag{35}$$

The last two terms in (35) do not depend on accelerations. Hence, at time τ,

$$(\Delta\ddot{s}_{ij})v^{\circ}_{j,i} = \dot{s}^{\circ}_{ij}(\Delta\dot{v}_{j,i}) \tag{36}$$

except on the surface $S_D(\tau)$ of zero volume. Thus, by the Green theorem,

$$\int_V (\Delta\ddot{s}_{ij})v^{\circ}_{j,i}\,\mathrm{d}V = \int_V \dot{s}^{\circ}_{ij}(\Delta\dot{v}_{j,i})\,\mathrm{d}V = \int_{S_T} \dot{T}^{\circ}_i(\Delta\dot{v}_i)\,\mathrm{d}S - \int_{S_D} \dot{s}^{\circ}_{ij}\nu_i[\![\Delta\dot{v}_j]\!] \tag{37}$$

The first surface integral on the right-hand side of (37) and the last two surface integrals in (32) will cancel when substituting (37) into (32).

By substituting (33), (34), (37), (29) and (30) into (32), we finally obtain

$$\Delta\dddot{E}(\tau) = \int_{S_D(\tau)} 2E_U(u_n^+ - u_n^-)\,\mathrm{d}S \tag{38}$$

where

$$E_U = E_U(\boldsymbol{\xi}, \mathbf{L}^0; \mathbf{b}, \boldsymbol{\nu}) = \tfrac{1}{2}\dot{s}_{ij}^* v_{j,i}^* + \tfrac{1}{2}\dot{s}_{ij}^{\circ} v_{j,i}^{\circ} - \dot{s}_{ij}^{\circ} v_{j,i}^*$$
$$= U(v_{j,i}^{\circ} + b_j \nu_i) - U(v_{j,i}^{\circ}) - \frac{\partial U(v_{j,i}^{\circ})}{\partial(v_{j,i})} b_j \nu_i \qquad (39)$$

From the criterion (14) with $n=3$ it follows that for stability of the fundamental process $\boldsymbol{\chi}^0$ the integral (38) must be non-negative for all arbitrarily oriented S_D-surfaces of all arbitrarily small area. This can be the case only if the integrand in (38) is non-negative for all orientations of $\boldsymbol{\nu}$ and at every point of the body, by assumed piecewise regularity of the process $\boldsymbol{\chi}^{\circ}$ and of the material properties. Further $(u_n^+ - u_n^-)$ is non-negative and the vector $\mathbf{b}$ is still arbitrary. Thus, we arrive at the following result.

For the process $\boldsymbol{\chi}^{\circ}$ *to be stable in a time interval* $[t_1, t_2]$ *it is necessary that*

$$E_U(\boldsymbol{\xi}, \mathbf{L}^{\circ}; \mathbf{b}, \boldsymbol{\nu}) \geqslant 0 \qquad (40)$$

at every point $\boldsymbol{\xi}$ *of the body for all* $\tau \in [t_1, t_2)$ *and for all vectors* $\mathbf{b}$ *and* $\boldsymbol{\nu}$.

Compared to the criteria derived in the preceding sections, this criterion is local in character. If the material properties at some point of the body are such that (40) is not satisfied, then the fundamental process is unstable, no matter what the boundary conditions and material properties are elsewhere. The form of the branching mode used to derive (40) strongly suggests that the localization of deformation could be a physical symptom of that kind of instability.

If the moduli $C_{ijkl}(\mathbf{L})$ are continuous at $\mathbf{L} = \mathbf{L}^{\circ}$ then (40) written for infinitesimal $\mathbf{b}$ readily implies

$$C_{ijkl}(\mathbf{L}^{\circ}) \nu_i \nu_k b_j b_l \geqslant 0 \qquad (41)$$

for all $\boldsymbol{\nu}$ and $\mathbf{b}$. (A formal proof can be easily obtained by using the standard conditions for a minimum of the function E_U). This means that *for stability of a process it is necessary that the actual moduli satisfy the* (*semi-strong*) *ellipticity condition* (41)† (cf. refs. 13 and 9).

Now, let the body be homogeneous and homogeneously deformed in a process $\boldsymbol{\chi}^{\circ}$. Consider the matrix $\mathbf{A}$, defined by $A_{jl} = C_{ijkl}(\mathbf{L}^{\circ}) \nu_i \nu_k$. The condition (41) means that $\mathbf{A}$ is positive semi-definite for all $\boldsymbol{\nu}$. Let $\mathbf{A}$ be

† The criteria (40) and (41) could be derived directly from the minimum principle (22) by using theorems of the calculus of variations, however, without the present connection with the localization of deformation.

positive definite at the beginning of the process $\boldsymbol{\chi}^{\circ}$ and the moduli $C_{ijkl}(\mathbf{L}^{\circ})$ vary smoothly with the deformation. Then the critical stage beyond which (41) ceases to hold coincides with the first instant of time when $\det(\mathbf{A})=0$ for some $\boldsymbol{\nu}$. If the moduli $C_{ijkl}(\mathbf{L})$ are piecewise constant, this reduces to the familiar criterion for the onset of localization of plastic deformation [13], obtained from the bifurcation analysis. The present interpretation of this last criterion is that it provides an upper bound to the onset of instability of the deformation process in the energy sense.

It can be shown [12] that at the critical stage when $\det(\mathbf{A})=0$, a second-order bifurcation may take place (rather than a first-order one), the eigenmode being of the form of a discontinuous acceleration field as discussed above. This is another example of the relations between instability in the energy sense and bifurcation.

ACKNOWLEDGEMENTS

I am indebted to Professor K. Thermann for his hospitality at Dortmund University, Germany, where this paper has been written. The support of the Alexander von Humboldt Foundation is gratefully acknowledged.

REFERENCES

1. Hill, R. A general theory of uniqueness and stability in elastic–plastic solids, *J. Mech. Phys. Solids*, **6** (1958), 236–249.
2. Hill, R. Some basic principles in the mechanics of solids without a natural time, *J. Mech. Phys. Solids*, **7** (1959), 209–225.
3. Hill, R. Bifurcation and uniqueness in nonlinear mechanics of continua. In: *Problems of Continuum Mechanics* (Ed. M. A. Lavrentev), Society of Industrial and Applied Mathematics, Philadelphia, 1961, 155–164.
4. Hill, R. Uniqueness and extremum principles in self-adjoint boundary-value problems in continuum mechanics, *J. Mech. Phys. Solids*, **10** (1962), 185–194.
5. Hill, R. Eigenmodal deformations in elastic–plastic continua, *J. Mech. Phys. Solids*, **15** (1967), 371–386.
6. Hill, R. Aspects of invariance in solid mechanics. In: *Advances in Applied Mechanics*, Vol. 18 (Ed. C.-S. Yih), Academic Press, New York, 1978, 1–75.

7. HUTCHINSON, J. W. Plastic buckling. In: *Advances in Applied Mechanics*, Vol. 14 (Ed. C.-S. Yih), Academic Press, New York, 1974, 67–144.
8. KLUSHNIKOV, V. D. *Stability of Elastic-Plastic systems* (in Russian), Nauka, Moscow, 1980.
9. MILES, J. P. On necking phenomena and bifurcation solutions, *Arch. Mech.*, **32** (1980), 909–931.
10. PETRYK, H. A stability postulate for quasi-static processes of plastic deformation, *GAMM Conference*, Würzburg 1981, *Arch. Mech.*, **35** (1983), 753–756.
11. PETRYK, H. A consistent energy approach to defining stability of plastic deformation processes. In: *Stability in the Mechanics of Continua* (Ed. F. H. Schroeder), Springer, Berlin, 1982, 262–272.
12. PETRYK, H. and K. THERMANN. Second-order bifurcation in elastic–plastic solids.
13. RICE, J. R. The localization of plastic deformation. In: *Proc. 14th IUTAM Congr. Theor. Applied Mech.* (Ed. W. T. Koiter), North-Holland, Delft, 1976, 207–220.
14. THOMAS, T. Y. *Plastic Flow and Fracture in Solids*, Academic Press, New York, 1961.

Section 5

HARDENING AND MEMORY EFFECTS

25

Macroscopic Plastic Behaviour of Microinhomogeneous Materials

André Zaoui

Laboratoire PMTM, CNRS, Université Paris-Nord, Villetaneuse, France

ABSTRACT

This paper deals with the specific difficulties which arise in the field of plasticity concerning the deduction of the macroscopic behaviour from microstructural information and with the way to overcome some of them by use of a self-consistent homogenization technique. After a discussion of the general problems involved by an only partial statistical description of the microstructure, a particular formulation of the self-consistent scheme is reported. Applications are given in the case of polycrystalline materials for which special attention is paid to texture development and to the associated induced plastic anisotropy. Finally the reported formulation is discussed in order to suggest extensions towards generalized, more systematic self-consistent procedures.

1. INTRODUCTION

Plastic behaviour is known to be a complex one. One particular expression of this complexity, strictly reflecting that of nature, consists in the large number of variables which are usually needed in order to yield an explicit constitutive law which could suffer any extensive comparison with experimental data for a given material.

Such a difficulty may be overcome according to two kinds of treatment:

(a) The first one consists in a phenomenological approach using internal 'hidden' variables whose physical meaning has not to be elucidated; these parameters are only expected to allow, after adequate identification from standard experiments, a fairly good predicting ability, i.e. fairly good agreement with further experimental results.

(b) The second one proceeds from a deductive approach; it starts from some knowledge of the physical phenomena involved in the macroscopic response and works out a mechanical description of the related variables and laws of evolution. Then, dealing with available information (or adequate assumptions) on the initial microstructure of the concerned material, it has to deduce the corresponding macroscopic behaviour through some homogenization procedure.

When the second approach is used, as will be the case in the following, two quite distinct methods still remain open: either the microstructure is completely described and the solution may be an exact one, or it is only statistically defined and, generally, one can only expect to get bounds within which any solution must lie. The first method may be used for very special cases, e.g. for well oriented composite structures such as periodic media [25], or when typical particular cases are investigated. The second method is more adapted to common materials, such as usual building metals, for which only partial microstructural information can be obtained from observation. This latter point of view will be adopted in the following.

Note that, in any case, what is aimed at is the derivation of a local constitutive law for microheterogeneous media, a law which could be used for further structural analysis. This means that the matter of our investigation would appear on a larger scale as a 'macroscopic point' whereas it will be considered in what follows as an infinite medium. Furthermore, in our case, this medium is not completely, but only partly, statistically determined.

Such a deductive construction from microscopic information towards macroscopic behaviour may be, schematically, divided into three stages:

(1) First, the heterogeneous microstructure of the material must be analysed in order to identify elementary homogeneous phases, the mechanical behaviour of which must be specified in the complete, tensorial form which is needed for further analysis. According to the degree of accuracy which is expected or required and, of course, to the available physical information, such a phase identification may be more

or less coarse. For instance, in the case of a (chemically) single phase metallic polycrystal, the microheterogeneity may be considered as determined only by the crystallographic lattice misorientations between grains, so that each grain family with the same orientation can define a specific phase; now, if grain shape and size effects are considered to be significant, the elementary homogeneous phase must be restricted to a grain family with the same orientation, shape and size at the same time. Even in that case, lower scale inhomogeneities within each restricted phase are neglected, such as slight variations of residual stresses, impurity content, initial point defects or dislocations density etc. On the other hand, in the case of a two-phase metal whose phases exhibit strongly different mechanical characteristics, the microheterogeneities which still exist within both phases between grain families of the former type may be neglected and one can deal with only two homogenized phases. Obviously, the same material may be analysed according to various scales of microheterogeneity, depending on a preliminary mechanical analysis of the physical structure (e.g. see refs 4 and 6).

(2) Secondly, the statistical description of the phases in the initial state must be specified. This may be done, too, in a more or less detailed manner. According to the case, phase volume fractions only or correlation functions of higher degree may be given. In any case, one has also to take note of the evolution of the phase distribution following the changes of the mechanical state of the phases themselves. For instance, if the initial crystallographic texture function is given for a polycrystal, one has also generally to deal with the texture evolution during the plastic flow as a result of the combined crystal lattice rotations of all the grain families with increasing plastic deformation. The same is true for the grain shape changes too. Generally speaking, the statistical distribution of the phases cannot be considered as given once and for all: especially in the field of plasticity, where large irreversible deformation occurs, the inhomogeneous microstructure must be analysed as a variable state in the course of the plastic flow.

(3) Finally, one has to solve the homogenization problem itself, namely the determination of the 'homogeneous equivalent medium' behaviour. Several operations are needed for this solution, the most important of which are the definition of macroscopic mechanical variables from microscopic ones, the calculation of the eigenstress field resulting from the mechanical interactions between the phases and the determination of the microstructural evolution. It is apparent from the foregoing that the validity of the result depends not only on the quality

of the homogenization procedure, but also on the accuracy of the definition of phases and of the description of their statistical distribution.

As a matter of fact, in the common case of an only partly statistically defined microinhomogeneous structure, there is not a unique homogenization procedure which would be better than all the others, since there is not a unique solution, but, on the contrary, an infinite number of solutions to such an only partly defined statistical problem. In other words if, as usual, the statistical description of the microstructure has been stopped at some finite degree, there remains an infinite diversity of space phase distributions which are the same, from a statistical point of view, up to this finite degree, but which can differ beyond, and, consequently, an infinite number of possible overall behaviour characteristics. For instance, if one really does not know anything but the phase volume fractions of some given material, the actual properties of this material may be anything within a more or less extended range, so that there is no reason at all for anybody to claim that his solution, which has been derived according to his own homogenization procedure, is better (or worse) than his colleague's one, which results from another method, provided that both of these methods use what is really known, namely, in this case, the phase volume fractions, and these fractions only.

In what follows, the classical self-consistent homogenization method is related, discussed and applied to low temperature plasticity in Section 2. An original numerical formulation is reported, which allows a convenient treatment of arbitrary applied stress or strain paths. Applications are given for metallic polycrystals, when an isotropic approximation of elastic–plastic interactions between grains is assumed, with special attention paid to strain-hardening, texture development and plastic anisotropy (Section 3). Finally in Section 4 the reported formulation is shown to clearly point out the space phase distribution insensitivity of the classical self-consistent scheme and to suggest generalized self-consistent procedures which could rectify such a deficiency.

2. FORMULATIONS FOR THE SELF-CONSISTENT SCHEME

The self-consistent scheme is known to be a homogenization procedure which is based upon an approximation of the mechanical interactions between the phases. Instead of calculating the whole eigenstress and strain field in the microinhomogeneous medium when given stress or strain is applied at infinity, one has to solve several simpler problems each of which is concerned with a two-phase situation: on the one hand, any particular phase of the composite, gathered inside an 'inclusion' and, on the other hand, a fictitious homogeneous phase uniformly distributed in an infinite 'matrix', the behaviour of which is the (unknown) one of the homogeneous equivalent medium. This unknown behaviour, the derivation of which is the very purpose of the procedure, is finally identified from the solution of each matrix/inclusion problem (the matrix being the same and the inclusion each phase by turns) and from the average equation which correlates overall macroscopic mechanical variables with local microscopic ones.

It is soon obvious that such a procedure cannot take the spatial distribution of the various phases into account but only the shape and volume fraction of the phase domains. As a matter of fact, the self-consistent scheme can use only very poor statistical information on the microstructure. Nevertheless, it leads to only one estimate of the composite behaviour instead of a whole range of possible values lying between upper and lower bounds, as it should logically do with such partial statistical information. This apparent inconsistency has been clarified by the systematic theory of statistical continuum mechanics [22] which has proved, in the case of linear elasticity, that the self-consistent scheme was based upon an implicit statistical assumption, namely an assumption of perfect disorder. In fact, such an assumption implies a specific complete set of correlation functions which leads the associated upper and lower bounds to join at the self-consistent value. Unfortunately, the systematic theory is, for the time being, restricted to linear behaviour and the statistical foundations of the self-consistent scheme have not been really elucidated yet in the case of plasticity.

It must be admitted that the obstacles which a systematic statistical theory would have to overcome in the field of plasticity look somewhat depressing: the energy criteria which would be used in order to derive bounds for the overall behaviour still remain to be specified in a rigorous and general manner; moreover, the statistical description of the microstructure is continuously changing during the plastic flow and the correlation functions would have to be modified at every step. Consequently, plastic statistical approaches are still far from being in a full-grown state; for the time being, they are only concerned with first-order correlation functions since they deal with the phase volume fractions only. Within this restricted field, it was a long time before the hegemony of the Taylor model [26] was disputed: its basic assumption (i.e. the uniformity of the plastic strain $\boldsymbol{\varepsilon}^{\mathrm{P}}$ when the elastic strain $\boldsymbol{\varepsilon}^{\mathrm{e}}$ is neglected, or, in the extended sense proposed by Lin [23], the uniformity of the total strain $\boldsymbol{\varepsilon}$ when this is not the case) is just the plastic equivalent of the classical Voigt elastic assumption. Of course, a plastic equivalent of the Reuss model (i.e. the uniformity of stress $\boldsymbol{\sigma}$) could also be defined, but this exists only in the modified form of the Sachs model [24] in literature, for the specific case of crystalline aggregates. Except for various attempts to rectify, in a more or less empirical way, part of the extreme character of such assumptions in order to obtain a better agreement with experimental data, an alternative approach may only be found in the self-consistent way opened by Kröner [21] for the plastic case, and then systematized by Hill [16].

Hill's general treatment will first be briefly described. Let us consider an ellipsoidal inclusion (I) with instantaneous moduli $\mathbf{L}_{\mathrm{I}}$, embedded in an infinite homogeneous matrix with moduli $\mathbf{L}$, and submitted to uniform stress or strain rate ($\dot{\boldsymbol{\Sigma}}$ or $\dot{\mathbf{E}}$ resp.) at infinity. $\mathbf{L}$ and $\mathbf{L}_{\mathrm{I}}$, which are elastic–plastic moduli, may have several branches. The classical Eshelby analysis [12] leads to a uniform solution in the inclusion ($\dot{\boldsymbol{\sigma}}_{\mathrm{I}}$, $\dot{\boldsymbol{\varepsilon}}_{\mathrm{I}}$) which depends on the applied quantities through the 'overall constraint tensor' $\mathbf{L}^*$, namely:

$$\dot{\boldsymbol{\sigma}}_{\mathrm{I}} = \dot{\boldsymbol{\Sigma}} + \mathbf{L}^*(\dot{\mathbf{E}} - \dot{\boldsymbol{\varepsilon}}_{\mathrm{I}}) \tag{1}$$

with the additional relations:

$$\dot{\boldsymbol{\sigma}}_{\mathrm{I}} = \mathbf{L}_{\mathrm{I}}\dot{\boldsymbol{\varepsilon}}_{\mathrm{I}} \qquad \dot{\boldsymbol{\Sigma}} = \mathbf{L}\dot{\mathbf{E}} \tag{2}$$

After this operation has been performed for all the phases (I), the average relations between local and overall variables:

$$\dot{\boldsymbol{\Sigma}} = \langle \dot{\boldsymbol{\sigma}}_{\mathrm{I}} \rangle_{\mathrm{I}} \qquad \dot{\mathbf{E}} = \langle \dot{\boldsymbol{\varepsilon}}_{\mathrm{I}} \rangle_{\mathrm{I}} \tag{3}$$

where $\langle\ \rangle_{\mathrm{I}}$ denotes the average over the phases (I) with the volume fractions c_{I}, lead to the self-consistent equation:

$$\mathbf{L} = \langle \mathbf{L}_{\mathrm{I}}(\mathbf{L}_{\mathrm{I}} + \mathbf{L}^*)^{-1}(\mathbf{L} + \mathbf{L}^*) \rangle_{\mathrm{I}} \tag{4}$$

or, in a more explicit form:

$$\mathbf{L} = \langle \mathbf{L}^*(\mathbf{L}_{\mathrm{I}} + \mathbf{L}^*)^{-1} \rangle_{\mathrm{I}}^{-1} \langle \mathbf{L}_{\mathrm{I}}(\mathbf{L}_{\mathrm{I}} + \mathbf{L}^*)^{-1}\mathbf{L}^* \rangle_{\mathrm{I}} \tag{5}$$

The solution of this integral equation, which must be performed step by step, is made difficult by the fact that the unknown tensor $\mathbf{L}^*$

depends on **L** (and also on the aspect ratios and the orientation of the ellipsoids) in a very complex manner in the general case. A direct numerical treatment of this equation has been performed by Hutchinson [18], whereas a more analytical one has been recently proposed by Iwakuma and Nemat-Nasser [19] for plane problems at finite strains. In what follows, attention will be focused on the foundations of the self-consistent procedure itself, in order to stress, to use and, to a certain extent, to overcome its main deficiency. Even if this could be done within a more general formal framework, it will be specified in the restricted form which derives from an isotropic elastic–plastic approximation of the interactions between the phases [8]. Let us first recall the more classical Kröner approach which considers the constraint tensor $\mathbf{L}^*$ as depending only on the elastic part of **L**. If all the phases are assumed to have the same isotropic elastic moduli (shear modulus μ and Poisson's ratio ν), and the inclusions to be spherical, (1) then becomes simply:

$$\dot{\boldsymbol{\sigma}}_{\mathrm{I}} = \dot{\boldsymbol{\Sigma}} + 2\mu(1-\beta)(\dot{\mathbf{E}}^{\mathrm{P}} - \dot{\boldsymbol{\varepsilon}}_{\mathrm{I}}^{\mathrm{P}}) \tag{6}$$

where $\dot{\mathbf{E}}^{\mathrm{P}}$ and $\dot{\boldsymbol{\varepsilon}}_{\mathrm{I}}^{\mathrm{P}}$ denote the average and local plastic strain rate traceless tensors respectively and $\beta = 2(4-5\nu)/(15(1-\nu))$, so that:

$$\mathbf{L}^* = (2\mu(1-\beta)/\beta)\mathbf{I} = k\mathbf{I} \tag{7}$$

where **I** is a unit tensor. These assumptions make the solution of the integral equation (5) quite straightforward since it reduces to:

$$\mathbf{L} = -k\mathbf{I} + \langle(\mathbf{L}_{\mathrm{I}} + k\mathbf{I})^{-1}\rangle_{\mathrm{I}}^{-1} \tag{8}$$

But, due to the fact that k in (7) is of the order of 2μ, while the elastic plastic moduli in $\mathbf{L}_{\mathrm{I}}$ are lower by one or two orders of magnitude as soon as plastic flow occurs in the composite, (8) is, in practice, quite close to the Lin–Taylor equation:

$$\mathbf{L} = \langle\mathbf{L}_{\mathrm{I}}\rangle_{\mathrm{I}} \tag{9}$$

which derives from (8) when $k \to \infty$. Nevertheless, it may be noticed that, while such an elastic treatment of the matrix/inclusion interaction is an over-simplification, what makes Kröner's approximation so convenient is the isotropy assumption more than the elastic one. So, an elastic-plastic extension of Kröner's approximation may be defined [3] which leads to replace (7) by:

$$L^*_{ijkl} = \frac{\mu^*(3-5\nu^*)}{4-5\nu^*}\delta_{ij}\delta_{kl} + \frac{\mu^*(7-5\nu^*)}{2(4-5\nu^*)}(\delta_{ik}\delta_{jl} + \delta_{il}\delta_{jk}) \tag{10}$$

and (6) by:

$$\dot{\boldsymbol{\sigma}}_{\mathrm{I}} = \dot{\boldsymbol{\Sigma}} + \frac{2\mu\mu^*(7-5\nu^*)}{\mu^*(7-5\nu^*)+2\mu(4-5\nu^*)}(\dot{\mathbf{E}}^{\mathrm{P}} - \dot{\boldsymbol{\varepsilon}}_{\mathrm{I}}^{\mathrm{P}}) = \dot{\boldsymbol{\Sigma}} + \alpha\mu(\dot{\mathbf{E}}^{\mathrm{P}} - \dot{\boldsymbol{\varepsilon}}_{\mathrm{I}}^{\mathrm{P}}) \quad (11)$$

where μ^* and ν^* are the instantaneous elastic–plastic shear modulus and Poisson's ratio, respectively and δ_{ij} the Kronecker symbol.

Note that the term $2(1-\beta)$ in (6), which is of the order of 1, is now replaced by a scalar factor (α say) which, except in the elastic range or at the very commencement of the plastic flow, currently lies between 10^{-1} and 10^{-2}, so that $\alpha\mu$ is now of the same order as the moduli in $\mathbf{L}_{\mathrm{I}}$, which leads to significant differences to Taylor's predictions. Note also from (5) that such an isotropic approximation of $\mathbf{L}^*$ does not prevent the corresponding estimate of $\mathbf{L}$ from being anisotropic. Thus, the full self-consistency of the procedure has been somewhat diminished (nevertheless in a less pronounced manner than according to (7)), but the plastic relaxation effects have been taken into account and the numerical treatment is widely simplified.

What a numerical formulation still has to solve is the practical way to derive the overall response of the aggregate to any strain or stress path. A classical solution of this problem has already been given by Budiansky and Wu [7] for the case of an isotropic polycrystal when Kröner's equation (6) is used. It can be extended easily to the use of (11) in the following manner (here α will be considered as a given constant, i.e. the corresponding procedure for its calculation will be assumed as already performed).

Let (11) be written in the form:

$$\dot{\boldsymbol{\sigma}}_{\mathrm{I}} + \alpha\mu\dot{\boldsymbol{\varepsilon}}_{\mathrm{I}}^{\mathrm{P}} = \dot{\boldsymbol{\Sigma}} + \alpha\mu\dot{\mathbf{E}}^{\mathrm{P}} = \dot{\mathbf{Q}} \quad (12)$$

Here $\dot{\mathbf{Q}}$ has the same value for all the phases (I).

If the $\mathbf{Q}$-history was known, it would be easy to deal with each phase by turns, to combine (12) with the constitutive law of each phase correlating $\dot{\boldsymbol{\sigma}}_{\mathrm{I}}$ with $\dot{\boldsymbol{\varepsilon}}_{\mathrm{I}}^{\mathrm{P}}$, to calculate $\dot{\boldsymbol{\sigma}}_{\mathrm{I}}$ and $\dot{\boldsymbol{\varepsilon}}_{\mathrm{I}}^{\mathrm{P}}$ and then the averages $\dot{\boldsymbol{\Sigma}}$ and $\dot{\mathbf{E}}^{\mathrm{P}}$ so as to divide $\dot{\mathbf{Q}}$ into its stress rate and plastic strain rate components. But the $\mathbf{Q}$-history is never given a priori. If, for instance, the $\boldsymbol{\Sigma}$-history is given, it is necessary to find, at each step, the only form of $\dot{\mathbf{Q}}$ which is compatible with this history—which generally needs lengthy iterative procedures, except in the specific case of plastic isotropy, where $\dot{\boldsymbol{\Sigma}}$ and $\dot{\mathbf{E}}^{\mathrm{P}}$ (and then $\dot{\mathbf{Q}}$) have the same principal directions.

That is the reason why another formulation is needed in the general case of plastic anisotropy. This can be performed as follows [15].

Let $\dot{\mathbf{E}}^{\mathrm{P}}$ be written in the explicit form in (11):

$$\dot{\boldsymbol{\sigma}}_{\mathrm{I}}+\alpha\mu\dot{\boldsymbol{\varepsilon}}_{\mathrm{I}}^{\mathrm{P}}-\alpha\mu\sum_{\mathrm{J}}c_{\mathrm{J}}\dot{\boldsymbol{\varepsilon}}_{\mathrm{J}}^{\mathrm{P}}=\dot{\boldsymbol{\Sigma}} \tag{13}$$

where the summation is performed on all the phases (J), with the volume fraction c_{J}. Here $\dot{\boldsymbol{\Sigma}}$ is the only macroscopic quantity—here supposed to be imposed— and it lies alone on one side of the equation.

Now, the whole set of N equations (13), with $I=1$ to N, must be drawn up, with $\dot{\boldsymbol{\sigma}}_{\mathrm{I}}$ expressed in terms of $\dot{\boldsymbol{\varepsilon}}_{\mathrm{I}}^{\mathrm{P}}$, thanks to the constitutive equation of phase (I), so that the unknown $\dot{\boldsymbol{\varepsilon}}_{\mathrm{J}}^{\mathrm{P}}$ may be calculated, as well as the average $\dot{\mathbf{E}}^{\mathrm{P}}$.

This formulation, which allows us to deal with the plastic anisotropy without any iterative procedure, may be easily extended to the case of a given total strain $\mathbf{E}$-history: in (13), $\dot{\boldsymbol{\Sigma}}$ must be expressed in terms of $(\dot{\mathbf{E}}-\dot{\mathbf{E}}^{\mathrm{P}})$ in view of Hooke's law so that $\dot{\mathbf{E}}^{\mathrm{P}}$ could be transferred into the first side and $\dot{\mathbf{E}}$ could stay alone in the second side as the given macroscopic data, namely:

$$\dot{\boldsymbol{\sigma}}_{\mathrm{I}}+\alpha\mu\dot{\boldsymbol{\varepsilon}}_{\mathrm{I}}^{\mathrm{P}}+(2-\alpha)\mu\sum_{\mathrm{J}}c_{\mathrm{J}}\dot{\boldsymbol{\varepsilon}}_{\mathrm{J}}^{\mathrm{P}}=2\mu\left(\dot{\mathbf{E}}+\frac{\nu}{1-2\nu}(\mathrm{Tr}\,\dot{\mathbf{E}})\mathbf{I}\right) \tag{14}$$

where Tr $(\dot{\mathbf{E}})$ indicates the trace of $\dot{\mathbf{E}}$. The case of mixed data (i.e. some components of $\dot{\boldsymbol{\Sigma}}$ and other ones of $\dot{\mathbf{E}}$) may be treated in a similar way.

The practical merits of this formulation are, of course, balanced by the drawbacks of a simultaneous treatment of all the phases instead of a separate one, according to the classical formulation (12), so that the latter may be preferred in the case of plastic isotropy. Both of these methods have been used in the following applications which are concerned with the plasticity of polycrystals.

3. THE PLASTICITY OF POLYCRYSTALLINE MEDIA

In the case of metallic, single phase, polycrystals, if the grain shape is roughly equiaxed so that it would not be considered as a phase characteristic, the crystallographic orientation of the grains appears as the main factor to be used in order to define the 'micro-phases'. In the case of a two-phase polycrystal, this partition must be performed within each phase separately [6]. Then, the statistical description of each 'macrophase' in the initial state reduces to the texture function $f(\Omega)$, where Ω is the orientation parameter (usually, the three Euler angles of each crystal lattice with respect to exterior fixed axes): $f(\Omega) \,.\, \mathrm{d}\Omega$ denotes the volume fraction of the grains which have their lattice orientation Ω in the range $\mathrm{d}\Omega$ around Ω. This texture function, of course, must be discretized in order to deal with a finite number N of phases (I), with the volume fraction c_{I} and to be able to use the formulation which derives from eqn. (13) or (14).

Two more points must still be specified: first the plastic behaviour of the single crystals and then the calculation of the crystal lattice rotations which are responsible for the texture formation.

The single crystal plastic behaviour may be approximated by an extended Schmid law according to which an easy glide system (g), defined by its unit normal $\mathbf{n}^{\mathrm{g}}$ and slip direction $\mathbf{m}^{\mathrm{g}}$, may become active as soon as the resolved shear stress on it, τ^{g}, reaches some critical value $\tau_{\mathrm{c}}^{\mathrm{g}}$. This critical resolved shear stress depends on the resolved shear strain γ^{h} on any active (h) system of the same crystal. An assumption of linear strain-hardening leads to the relation:

$$\dot{\tau}_{\mathrm{c}}^{\mathrm{g}} = \sum_{\mathrm{h}} H^{\mathrm{gh}} \dot{\gamma}^{\mathrm{h}} \tag{15}$$

where any component (g, h) of the strain-hardening matrix H^{gh} expresses the hardening modulus on system (g) associated with the plastic activation of system (h). For FCC crystals, for example, easy glide systems are $\{111\}\langle 110\rangle$ ones. Latent hardening experimental data [13] indicate that the H^{gh} components may be divided into two groups: lower terms, H_1 say, for systems pairs (g) and (h) whose dislocations cannot form junctions between each other and higher terms, H_2 say, for stronger interacting systems pairs (junction and lock reactions). The ratio $A = H_2/H_1$ may be considered as a characteristic parameter of hardening anisotropy: it is the higher, the lower the stacking fault energy of the material.

For moderate plastic strains, the following classical relations may be used:

$$R_{ij}^{\mathrm{g}} = m_i^{\mathrm{g}} n_j^{\mathrm{g}}, \qquad R_{(ij)}^{\mathrm{g}} = \tfrac{1}{2}(R_{ij}^{\mathrm{g}} + R_{ji}^{\mathrm{g}}), \qquad \dot{\varepsilon}_{ij}^{\mathrm{p}} = \sum_{\mathrm{g}} R_{(ij)}^{\mathrm{g}} \dot{\gamma}^{\mathrm{g}} \tag{16}$$

For an applied $\dot{\mathbf{\Sigma}}$ stress path, (13), (15) and (16) lead to the following symmetrized set of linear equations for active slip systems (g) in grain

(I):

$$c_{\mathrm{I}} \sum_{\mathrm{h}} (H_{\mathrm{I}}^{\mathrm{gh}} + \alpha\mu R_{(ij)\mathrm{I}}^{\mathrm{g}} R_{(ij)\mathrm{I}}^{\mathrm{h}}) \dot{\gamma}_{\mathrm{I}}^{\mathrm{h}} - \alpha\mu \sum_{J} \sum_{\mathrm{h}} c_{\mathrm{I}} c_{\mathrm{J}} R_{(ij)\mathrm{I}}^{\mathrm{g}} R_{(ij)\mathrm{J}}^{\mathrm{h}} = c_{\mathrm{I}} R_{(ij)\mathrm{I}}^{\mathrm{g}} \dot{\Sigma}_{ij} \tag{17}$$

or, in a symbolic form:

$$\sum_{\mathrm{hJ}} A^{\mathrm{g_I h_J}} \dot{\gamma}^{\mathrm{h_J}} = c_1 R_{(ij)}^{\mathrm{g_I}} \dot{\Sigma}_{ij} \tag{18}$$

where $\mathrm{g_I}$ indicates the slip system (g) in grain (I), and similarly for $\mathrm{h_J}$.

The linear approximations which have been introduced from the beginning (namely: constant values for α and H^{gh} as well as for R_{ij}^{g}, which means that the 'geometrical' hardening effects are neglected [3]) make the set of equations (18) easy to solve by classical methods, such as Cholesky's method. Before reporting typical applications, two remarks may be made:

(1) The A^{gh} matrix in (18) is symmetrical, thanks to the multiplication of each initial equation for a system in grain (I) by the corresponding volume fraction c_{I}. The order of this matrix is modified at each step of calculation, since the step length is defined by the activation (or the passivation) of some slip system. Note that for a monotonic stress path, this order is generally increasing from step to step, with the theoretical maximum value $12N \times 12N$ for FCC crystals.

(2) The prediction, according to this model, of the crystallographic texture development, depends on the calculation of crystal lattice rotation in each grain family. As far as plastic rotations $\boldsymbol{\omega}^{\mathrm{P}}$ are concerned, we simply get:

$$\omega_{ij}^{\mathrm{P}} = \tfrac{1}{2} \sum_{\mathrm{g}} (R_{ij}^{\mathrm{g}} - R_{ji}^{\mathrm{g}}) \gamma^{\mathrm{g}} \tag{19}$$

but the derivation of the lattice rotations $\boldsymbol{\omega}^{\mathrm{L}}$ is not so easy. A convenient rule, which might only be justified from a self-consistent point of view, in the case of complete isotropy (i.e. behaviour as well as grain shape isotropy) [5], states that the total rotation is uniform in the polycrystal. In usual cases (tension, rolling . . .) where the macroscopic rotation is assigned to be zero, this means that the lattice rotation just opposes the plastic one ($\boldsymbol{\omega}^{\mathrm{L}} = -\boldsymbol{\omega}^{\mathrm{P}}$): this relation will be assumed in the following applications, in order to predict the microstructure evolution, i.e. the crystallographic texture development.

We now simply report some typical results which have been obtained by use of this model, in order to illustrate part of its potentialities. Note that the linear approximations which have been already

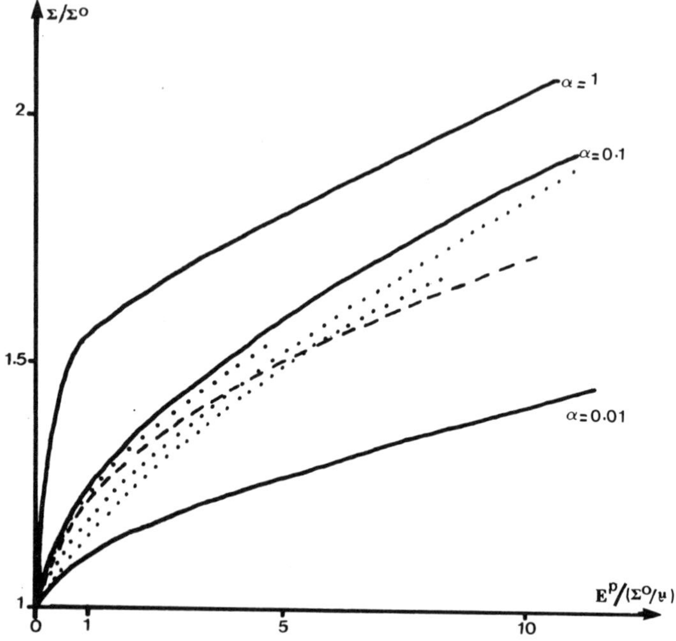

FIG. 1. Normalized theoretical (solid curve) and experimental (dotted [1] and dashed [20]) stress–strain curves.

mentioned exclude any comparison with experimental data at too high plastic strain. Nevertheless it can be seen on the following figures that, awaiting a non-linear treatment which is now in progress, the comparison may already be judged fairly good.

Figure 1 shows normalized theoretical tensile curves (Σ° is the yield stress) for a FCC polycrystal without initial texture for $A = 1{\cdot}5$ (as estimated from experimental results on copper single crystals [13]), $H_1 = \mu/250$ and $\alpha = 1$, $\alpha = 0{\cdot}1$ and $\alpha = 0{\cdot}01$. The comparison with experimental data from copper polycrystal tests [1] and with an empirical relation of Jaoul [20] deduced from various literature data indicates that the best agreement is obtained for α of the order of $0{\cdot}1$, as expected for this low plastic strain range. Figures 2(a), (b) and (c) refer to the simulation of a real case of a cold-rolled copper sheet [17], exhibiting a pronounced initial texture ({123}⟨634⟩, {112}⟨111⟩ and {011}⟨211⟩): Fig. 2(a) shows theoretical stress–strain curves for tensile tests performed at various angles θ from the rolling direction, in

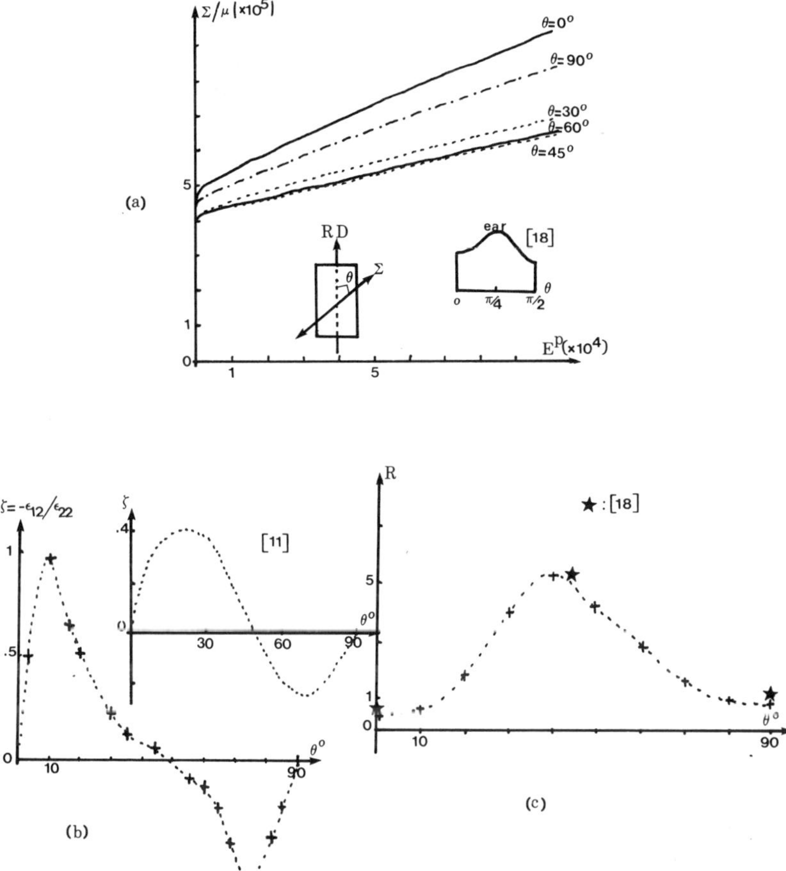

FIG. 2. Simulation of tensile tests on samples cut in a cold-rolled copper sheet.

accordance with the observed earing behaviour in deep drawing tests; Fig. 2(b) reports the angular variation with θ of a factor which indicates the deviation between stress and strain eigen-directions in such tensile tests, compared with classical experimental results [11]; Fig. 2(c) shows calculated and measured values of the Lankford coefficient as a function of θ: here the three main texture components have been considered together with an 'isotropic' component, with the volume fraction $c = 0{\cdot}2$. Figure 3(a) illustrates the rolling texture

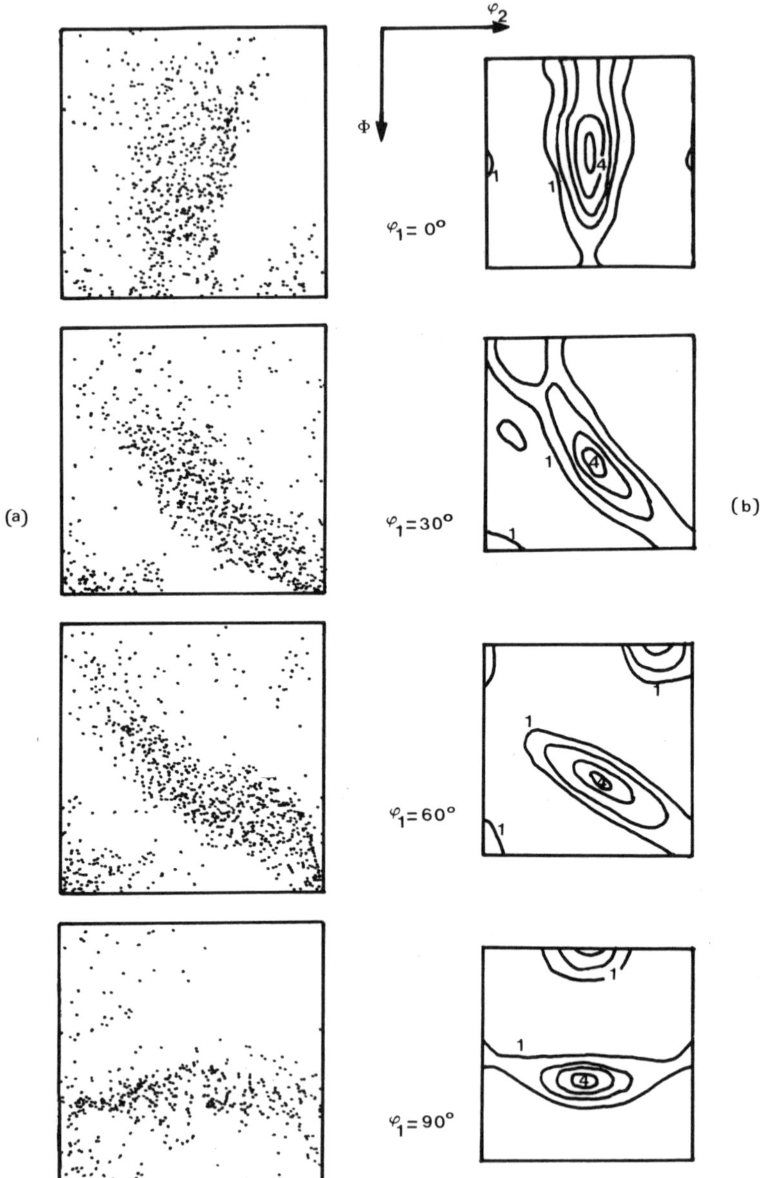

FIG. 3. Theoretical (a) and experimental (b) rolling-texture development of a BCC polycrystal.

TABLE I
Calculated ⟨100⟩/⟨111⟩ texture components ratio for different values of the intracrystalline anisotropy parameter A

A	$\% \frac{\langle 100\rangle}{\langle 100\rangle+\langle 111\rangle}$	$\% \frac{\langle 100\rangle}{\langle 111\rangle}$
1	15	18
1·5	36	56
3	38	61

development, for an initially isotropic BCC polycrystal, in the classical Bunge's (Φ, φ_2) representation: the main trends of the measured texture function of a cold rolled mild steel sheet in Fig. 3(b) [9] are already present at a lower strain (20% instead of 70%). Finally, Table I shows that the calculated ⟨100⟩/⟨111⟩ texture final components ratio of a pulled FCC polycrystal is an increasing function of the intracrystalline anisotropy parameter A, which conforms with classical experimental results [10] indicating that this ratio is a decreasing function of the stacking fault energy γ_{SF} and with the observation [13] that A and γ_{SF} have opposite variations.

4. GENERALIZED SELF-CONSISTENT SCHEMES

The numerical formulation which has been used in the foregoing applications derives from the decomposition (13) or (14), which reduces formally to:

$$\dot{\boldsymbol{\sigma}}_{\mathrm{I}} = \dot{\boldsymbol{\Sigma}} + \mathbf{L}^* \left(\sum_{\mathrm{J}} c_{\mathrm{J}} \dot{\boldsymbol{\varepsilon}}_{\mathrm{J}} - \dot{\boldsymbol{\varepsilon}}_{\mathrm{I}} \right) \tag{20}$$

This interaction law can be put in the more significant form:

$$\dot{\boldsymbol{\sigma}}_{\mathrm{I}} = \sum_{\mathrm{J}} c_{\mathrm{J}} \dot{\boldsymbol{\sigma}}_{\mathrm{I}}^{\mathrm{J}} = \sum c_{\mathrm{J}} [\dot{\boldsymbol{\Sigma}} + \mathbf{L}^* (\dot{\boldsymbol{\varepsilon}}_{\mathrm{J}} - \dot{\boldsymbol{\varepsilon}}_{\mathrm{I}})] \tag{21}$$

which indicates that $\dot{\boldsymbol{\sigma}}_{\mathrm{I}}$ may appear as the weighted average of the N different stress rates $\dot{\boldsymbol{\sigma}}_{\mathrm{I}}^{\mathrm{J}}$ which would exist in the same (I) inclusion if successively embedded in N different matrices (J), with the successive plastic strain rate $\dot{\boldsymbol{\varepsilon}}_{\mathrm{J}}^{\mathrm{P}}$ and the mechanical behaviour of each phase (J) by turns. This clearly stresses the fact that the classical self-consistent scheme is insensitive to the spatial phase distribution since, according to (21), any given phase (I) is submitted not to the actual action of another phase (J) located here or there, but to the action of a fictitious phase (J) which would be uniformly diluted all over the matrix. For instance, the 'self-consistent polycrystal' defined by (18) reduces to some homogeneous 'supercrystal', extending up to infinity, uniformly stressed and strained, with a very large number of easy glide systems (12N in the FCC case) exhibiting both intracrystalline and intergranular hardening interactions according to some generalized 'Schmid law for the polycrystal'. Two interesting consequences result from this comment:

(1) First, since a single crystal obeying the usual Schmid law is at once a standard material, this must be true, too, for this 'supercrystal', i.e. for the polycrystal as analysed by the self-consistent scheme.

(2) Accordingly, the classical extremal theorems for strain rates and stress rates are valid in that case, so that the solution of eqns. (18) and the associated inequalities for non-active systems is equivalent to the optimization of some energy expression. This may be easily derived from (18) in the case of an assigned stress rate path, since the quadratic form associated with A^{gh}, namely:

$$\sum_{g_I} \sum_{h_J} A^{g_I h_J} \dot{\gamma}^{h_J} \dot{\gamma}^{g_I} = \left(\sum_{g_I} c_I R^{g_I}_{(ij)} \dot{\gamma}^{g_I} \right) \dot{\Sigma}_{ij} = \dot{E}^{p}_{ij} \dot{\Sigma}_{ij} \tag{22}$$

is just the expression which has to be a maximum for the solution, according to Greenberg's theorem. When the A^{gh} matrix is positive definite, this solution is unique. Then, according to a procedure which has been corroborated by experimental data in the similar case of multislip in single crystals [14], the effective solution is determined as the admissible slip system combination whose associated sub-matrix is positive definite and which, at once, makes the corresponding values of (22) maximum.

So, the reported numerical formulation makes the practical self-consistent procedure easier and clearly states its validity limits. But, at the same time, it suggests one way for an extension of these limits so that some space phase distribution sensitivity could be included into the classical self-consistent scheme. This may be done, as a first step, in a very simple form by introducing directly inclusion pair effects into (20) for a given well-described microstructure: the terms $(c_J \mathbf{L}^* \dot{\boldsymbol{\varepsilon}}_J)$ have then to be replaced by more complicated ones which derive from the solution of the 'two plastic inclusions/matrix' problem in order to express pair interactions between phase domains. In the case of an isotropic elastic matrix and spherical inclusions and if the average stress state in the inclusions only is considered, the classical Eshelby's factor $2\mu(1-\beta)$ must be replaced by a fourth order tensor depending on the sphere radii and on the distance between their centres [2]. If more than two inclusions are considered, the solution is obtained by a direct superposition.

A further step would consist in the construction of true 'n-sites self-consistent schemes' which needs the solution of the more complicated problem of n inhomogeneous inclusions, for which the superposition procedure no longer holds. An approximate solution may be found when average values are considered in the inclusions only, thanks to the use of fictitious equivalent plastic inclusions. Referring to ref. 2 for further details, one gets the final solution as an extension of Hill's formula (1) in the form:

$$\dot{\boldsymbol{\sigma}}_{\mathrm{I}} = \dot{\boldsymbol{\Sigma}} + \mathbf{L}_{\mathrm{I}}^{*}(\dot{\mathbf{E}} - \dot{\boldsymbol{\varepsilon}}_{\mathrm{I}}) + \sum_{\mathrm{J}\neq\mathrm{I}} \mathbf{L}_{\mathrm{IJ}}^{*}(\dot{\mathbf{E}} - \dot{\boldsymbol{\varepsilon}}_{\mathrm{J}}) \tag{23}$$

where new pair constraint tensors $\mathbf{L}_{\mathrm{IJ}}^{*}$ have to be calculated. So, for a given nth order of graded statistical description, averaging over all the possible configurations of n inclusions finally yields an extended self-consistent equation for an n-sites approximation.

5. CONCLUSIONS

Two main conclusions may be drawn from the foregoing:

(1) Current homogenization methods for the plasticity of microinhomogeneous materials still stand at the very first step of any systematic statistical theory, using the one-point correlation functions only, i.e. the phase volume fractions. At this step, if no more statistical information is available concerning the actual space phase distribution, there is no reason to prefer this or that method, since each of them leads to one of the possible estimates of the overall moduli within the continuous range of solutions compatible with the known statistical information.

(2) Among these methods, the self-consistent one is likely to be more efficient in the case of highly disordered materials. In this case, the self-consistent scheme may be formulated in such a way that the numerical treatment is easier and can deal with a wide variety of applied stress or strain paths. Such a formulation presents the additional advantage of suggesting a systematic extension of the classical

self-consistent scheme; in contrast with the classical theory for linear elasticity which leads to more and more narrow upper and lower bounds for the expectation values of the overall moduli, such an n-sites scheme yields only one estimate at any step n, which takes all the statistical information at this step into account and assumes perfect disorder beyond. Since this assumption is introduced at a more and more delayed stage with increasing n, both these approaches are expected to meet at infinity. But these conjectures remain to be stated and proved more firmly in the near future.

REFERENCES

1. Berveiller, M. Doctoral Thesis, Université Paris-Nord, Villetaneuse (1978).
2. Berveiller, M., A. Hihi and A. Zaoui. Self-consistent schemes for the plasticity of polycrystalline and multiphase materials, *Proc. 2nd Risø Int. Symp. on Metallurgy and Materials Science*, Roskilde 1981, 145–156.
3. Berveiller, M. and A. Zaoui. An extension of the self-consistent scheme to plastically flowing polycrystals. *J. Mech. Phys. Solids*, **26** (1979), 325–344.
4. Berveiller, M. and A. Zaoui. A simplified self-consistent scheme for the plasticity of two-phase metals, *Res Mech. Letters*, **1** (1981), 119–124.
5. Berveiller, M. and A. Zaoui. Etude de l'anisotropie élastique, plastique et géométrique dans les polycristaux métalliques. In: Comportement mécanique des solides anisotropes (Ed. J. P. Boehler), Martinus Nijhoff, The Hague and Editions du CNRS, Paris 1982, 335–347.
6. Berveiller, M. and A. Zaoui. Modelling of the plasticity and the texture development of two-phase metals, *Proc. 4th Risø Int. Symp. on Deformation of Multi-phase and Particle Containing Materials*, Roskilde, 1983, (to be published).
7. Budiansky, B. and T. T. Wu. Theoretical prediction of plastic strains of polycrystals, *Proc. 4th U.S. Nat. Cong. Appl. Mech.*, 1962, 1175–1185.
8. Bui, H. D., A. Zaoui and J. Zarka. Sur le comportement élasto-plastique et viscoplastique des monocristaux et polycristaux métalliques de structure CFC. In: *Foundations of Plasticity* (Ed. A. Sawczuk), Noordhoff, Leyden, 1974, 51–75.
9. Bunge, H. J. Texturbeschreibung durch dreidimensionale Polfiguren. In: *Textures in Research and Practice* (Ed. J. Grewen and G. Wassermann), Springer, Berlin, 1969, 24–35.
10. Chin, G. Y. Tension and compression textures. In: *Textures in Research and Practice* (Ed. J. Grewen and G. Wassermann), Springer, Berlin, 1969, 51–80.
11. Crans, W. Doctoral Thesis, Delft (1967).
12. Eshelby, J. D. Elastic inclusions and inhomogeneities. In: *Progress in*

Solid Mechanics (Ed. I. N. Sneddon and R. Hill), North-Holland, Amsterdam, 1961, 87–140.

13. FRANCIOSI, P., M. BERVEILLER and A. ZAOUI. Latent hardening in copper and aluminium single crystals, *Acta Met.*, **28** (1980), 273–283.
14. FRANCIOSI, P. and A. ZAOUI. Multislip in FCC crystals: a theoretical approach compared with experimental data, *Acta Met.* **30** (1982), 1627–1637.
15. HIHI, A. Doctoral Thesis, Université Paris-Nord, Villetaneuse (1982).
16. HILL, R. Continuum micro-mechanics of elastoplastic polycrystals, *J. Mech. Phys. Solids*, **13** (1965), 89–101.
17. HIRSCH, J., R. MUSICK and K. LÜCKE. Comparison between earing, R-values and 3-dimensional orientation distribution functions of copper and brass sheets, *Proc. 5th Int. Conf. on Textures of Materials*, Aachen, 1978, II 437–446.
18. HUTCHINSON, J. W. Elastic–plastic behaviour of polycrystalline metals and composites, *Proc. R. Soc.*, **A319** (1970), 247–272.
19. IWAKUMA, T. and S. NEMAT-NASSER. Finite elastic–plastic deformation of polycrystalline metals and composites. Tech. Report N° 83-3-51, (1983), of the Earthquake Research and Engineering Laboratory, Evanston.
20. JAOUL, B. *Etude de la Plasticité et Application aux Métaux*, Dunod, Paris, 1965.
21. KRÖNER, E. Zur plastischen Verformung des Vielkristalls, *Acta Met.*, **9** (1961), 155–161.
22. KRÖNER, E. *Statistical Continuum Mechanics*, CISM Lecture Notes, no. 92, Springer, Wien, 1972.
23. LIN, T. H. Analysis of elastic and plastic strains of a FCC crystal, *J. Mech. Phys. Solids*, **5** (1957), 143.
24. SACHS, G. Zur Ableitung einer Fliessbedingung, *Z. VDI*, **72** (1928), 734–736.
25. SUQUET, P. M. Méthodes d'homogénéisation en mécanique des solides. In: *Comportements Rhéologiques et Structure des Matériaux* (Ed. C. Huet and A. Zaoui), Editions Anciens ENPC, Paris, 1981, 87–128.
26. TAYLOR, G. I. Plastic strain in metals, *J. Inst. Met.*, **62** (1938), 307–324.

26

Evolution Equations for Anisotropic Hardening and Damage of Elastic–Viscoplastic Materials

S. R. Bodner†

Department of Mechanical Engineering, Technion, Israel Institute of Technology, Haifa, Israel

ABSTRACT

Evolution equations are developed for anisotropic hardening and damage parameters which are used as internal variables in a set of elastic–viscoplastic constitutive equations. Those equations do not require a yield criterion or loading and unloading conditions and appear to be applicable over a wide range of strain rates and temperatures. Both anisotropic hardening and damage are represented by symmetric second order tensors and the current stress is taken as the directional index in the evolution equations. As an approximation for initially isotropic materials, scalar effective values of those tensors are used in the classical equation for plastic strain rate. This leads to a relatively simple 'incrementally isotropic' flow law which is considered to have a useful range of validity.

1. INTRODUCTION

A set of constitutive equations for elastic–viscoplastic materials was developed a few years ago which do not require a specific yield criterion or loading and unloading conditions [1]. At the present stage of development, the equations include isotropic hardening [2], thermal recovery of isotropic hardening [5, 13], temperature dependence [3, 12] and isotropic damage development [6]. Results of investigations

† Visiting Professor, Institut für Mechanik, ETH-Zürich, April–Sept. 1983.

on methods of including anisotropic hardening into the equations have been reported [4, 7, 15].

Proper treatment of anisotropic (directional) hardening has proven to be a difficult problem. The use of a 'back stress' or a 'kinematic' hardening variable for this purpose was not considered appropriate in a 'unified' plasticity-creep theory without a yield criterion. That parameter is, in effect, a manifestation of the influence of anisotropic hardening on stress for particular loading conditions such as those associated with determination of a yield surface. As a consequence, there are considerable difficulties in the physical interpretation of the 'back stress' and its experimental measurement. For these reasons, the previous treatments of anisotropic hardening by the author and his associates [4, 7, 15] have concentrated on considering it as part of a tensor or scalar hardening parameter which appears in the coefficient of the equation for plastic straining. For the case of initially isotropic materials and for special classes of anisotropic hardening laws, it can be shown that the general anisotropic form of the flow law with a fourth order tensor as the coefficient can be made equivalent to the 'back stress' form on an incremental basis.

In the current presentation, the evolution equations for isotropic and anisotropic hardening are independent functions. This seems necessary in order to obtain realistic asymptotic stress–strain curves under cyclic loading conditions. Anisotropic hardening is represented by a symmetric second order tensor in three-dimensional space or as a simple vector in six-dimensional stress space. As an approximation for initially isotropic materials, in the present study, an 'effective' scalar value of the anisotropic hardening tensor is obtained as the scalar product of that variable and the normalized current stress. This is added to the isotropic hardening value to make up the total hardening parameter. The equation for plastic strain rate would therefore be 'incrementally isotropic' since different values of the scalar coefficient would be realized for changes in the stress direction or sign. Such an equation should have a reasonable range of validity for initially isotropic materials. In the uniaxial stress case, the 'incrementally isotropic', the 'back stress', and the general anisotropic forms of the flow law are essentially equivalent.

A method for including isotropic damage as an additional internal variable in the constitutive equations, in the manner proposed by Kachanov, was presented in ref. 6. A new form for the damage evolution equation was suggested in ref. 6 and is described in more detail in the present paper. A numerical example based on those equations is compared with experimental results. Guided by the procedure developed for anisotropic hardening, a similar treatment of anisotropic damage is suggested. Also, a simple method for coupling the evolution equations of hardening and damage is indicated.

2. CONSTITUTIVE EQUATIONS

2.1. General Form

A large strain formulation of the elastic–viscoplastic constitutive equations was presented in ref. 1 for the case of no hardening. For simplicity, a 'small strain' formulation is given in this paper but generalization to large strains seems to offer no special difficulties. For 'small strains', the strain rates are considered decomposable into elastic (reversible) and inelastic (non-reversible) components

$$\dot{\varepsilon}_{ij} = \dot{\varepsilon}^{\mathrm{e}}_{ij} + \dot{\varepsilon}^{\mathrm{p}}_{ij} \tag{1}$$

where $\dot{\varepsilon}^{\mathrm{e}}_{ij}$ is obtained from the time derivative of Hooke's Law. It is noted that for large strains the decomposition indicated by (1) would apply to the deformation rates and not the strain rates [1]. Both strain rate components in (1) are generally non-zero since a yield condition is not prescribed. The law for the plastic strain rate is assumed to be the classical one,

$$\dot{\varepsilon}^{\mathrm{p}}_{ij} = \dot{e}^{\mathrm{p}}_{ij} = \lambda s_{ij}, \qquad \dot{\varepsilon}^{\mathrm{p}}_{kk} = 0 \tag{2}$$

where $\dot{e}^{\mathrm{p}}_{ij}$ and s_{ij} are the deviatoric plastic strain rate and the stress. For large strains, the plastic strain rate in (2) would be replaced by the plastic deformation rate.

Squaring (2) leads to

$$D^{\mathrm{p}}_2 = \lambda^2 J_2 \tag{3}$$

where D^{p}_2 and J_2 are the second invariants of the strain rate and the deviatoric stress. All aspects of inelastic deformation are assumed to be governed by a functional relationship between D^{p}_2 and J_2 which includes temperature and load history dependent variables representing properties of the inelastic state of the material. In particular, hardening, i.e. overall resistance to plastic flow, is indicated by a scalar parameter $Z(>0)$, and damage, i.e. deterioration in the ability of the material to support stress, by the term $\omega(\geqslant 0)$. A form for D^{p}_2, motivated by equations for dislocation velocity, which appears to offer good representation capability over a wide range of conditions, is

$$D_2^p = D_o^2 \exp\{-[Z^2(1-\omega)^2/3J_2]^n\} \tag{4}$$

For the case of uniaxial stress σ_{11}, eqns. (2), (3) and (4) give,

$$\dot{\varepsilon}_{11}^p = \frac{2}{\sqrt{3}}\left(\frac{\sigma_{11}}{|\sigma_{11}|}\right) D_o \exp\left\{-\frac{1}{2}\left[\frac{Z(1-\omega)}{\sigma_{11}}\right]^{2n}\right\} \tag{5}$$

Evolution equations for the internal variables Z and ω are discussed in the next sections. The coefficient D_o is the limiting strain rate in shear and experimental evidence tends to indicate that such a limit does exist. For metals, it appears to be in the range from 10^6 to $10^7\,s^{-1}$. The material constant n controls strain rate sensitivity and influences the overall level of the flow stress.

Equation (5) can be rewritten in the form,

$$\frac{\sigma_{11}}{Z(1-\omega)} = f\left(\frac{\dot{\varepsilon}_{11}^p}{D_o}, n\right) \tag{6}$$

and empirical results [3, 12] indicate that n is a strong function of temperature, e.g. $n = (a/T) + b$. A plot of (σ_{11}/Z) against $\log_{10}(\dot{\varepsilon}_{11}^p/D_o)$, taking $\omega = 0$ in this example, shows a pronounced increase in (σ_{11}/Z) at about $(\dot{\varepsilon}_{11}^p/D_o) = 10^{-2}$ for all values of n. For $D_o = 10^6\,s^{-1}$, the sharp rise in flow stress would commence at $\dot{\varepsilon}_{11}^p = 10^4\,s^{-1}$. The equation also shows a linear dependence of (σ_{11}/Z) on $\dot{\varepsilon}_{11}^p$ in the neighbourhood of the inflection point which, for $D_o = 10^6\,s^{-1}$, would be slightly above $\dot{\varepsilon}_{11}^p = 10^5\,s^{-1}$. These consequences of (5) appear to be reasonably consistent with recent experimental results at very high strain rates [8]. Pressure dependence of incompressible inelastic flow could also be represented by the equations by taking n to be an increasing function of pressure. An increase of pressure would then lead to a higher level of flow stress and decreased strain rate sensitivity.

As the temperature approaches absolute zero, n becomes large and the function f in (6) approaches unity. At very low temperature, therefore, $\sigma_{11} = Z$ and is independent of strain rate. The hardening variable Z can then be interpreted as the rate independent flow stress at very low temperature. It is expected that the variation of Z with measures of hardening, e.g. plastic work, should not be strongly dependent on temperature over the entire range.

The preceding basic equations are intended to apply for all loading conditions and the details of the material response would depend on the particular circumstances. It is noted that all aspects of inelastic response, e.g. strain rate dependent plasticity, creep, and stress relaxation, are consequences of the same set of equations.

2.2. Evolution Equations for Hardening

In the previous studies with the constitutive equations, the isotropic hardening rate was assumed to depend on the plastic work rate, $\dot{W}_p$, and thermal recovery of hardening was taken to be a standard rate process. The particular functional relationships that were used are not an intrinsic requirement of the theory but generally did lead to good agreement with experimental results, e.g. refs. 2 and 13. The evolution equation for isotropic hardening proposed in refs. 2 and 5, which served as the basis of a number of exercises, is

$$\dot{Z}^{\mathrm{I}}(t) = m_1[Z_1 - Z^{\mathrm{I}}(t)]\dot{W}_p(t) - A_1 Z_1 \left[\frac{Z^{\mathrm{I}}(t) - Z_2}{Z_1}\right]^{r_1} \tag{7}$$

$$W_p(0) = 0, \qquad Z^{\mathrm{I}}(0) = Z_o \tag{7a}$$

where m_1, Z_1, Z_2, A_1, r_1 are material constants. The quantity Z_1 is the limiting (asymptotic) value of Z^{I}, and Z_2 is the stable, minimum, value at a given temperature. Equation (7) could be integrated incrementally to obtain $Z^{\mathrm{I}}(t)$. For simple uniaxial creep, (7) would apply completely for $\dot{Z}$ and secondary creep is the condition under constant stress when $\dot{Z}(t) = 0$.

Anisotropic hardening, in this formulation, enters the equation for plastic strain rate, (4), as part of the overall scalar hardening variable Z, i.e.

$$Z = Z^{\mathrm{I}} + Z^{\mathrm{A}} \tag{8}$$

so that the basic form of the plastic flow law (2) remains unchanged. This is considered to be a reasonable approximation for initially isotropic materials undergoing induced anisotropic hardening [7]. For obtaining detailed response characteristics at a sudden change in stress path and in the case of initially anisotropic materials, the fully anisotropic form of the flow law would be required [7, 15]. In the current treatment of anisotropic hardening, the isotropic and anisotropic hardening parameters in (8) are given by independent functions which enhances the predictive capability of the equations.

As part of the calculation procedure, a reference anisotropic hardening tensor β_{ij} is introduced which is a symmetric second order tensor with the required transformation properties. The evolution equation for β_{ij} is taken to have a form similar to that for isotropic hardening but with a directional index for hardening controlled by the current stress and associated recovery terms in the direction of $-\beta_{ij}$,

$$\dot{\beta}_{ij}(t) = m_2[Z_3 u_{ij}(t) - \beta_{ij}(t)]\dot{W}_p(t) - A_2 Z_1 \left[\frac{[\beta_{kl}(t)\beta_{kl}(t)]^{1/2}}{Z_1}\right]^{r_2} \left[\frac{\beta_{ij}(t)}{[\beta_{kl}(t)\beta_{kl}(t)]^{1/2}}\right] \tag{9}$$

In (9), the initial and stable (minimum) values of β_{ij} are zero corresponding to the isotropic state, m_2, Z_3, A_2, r_2 are material constants, and u_{ij} are direction cosines based on the current stress tensor,

$$u_{ij} = \sigma_{ij}/(\sigma_{kl}\sigma_{kl})^{1/2} \tag{10}$$

Equation (9) could also be integrated over a given history of loading $\dot{W}_p(\tau)$, $u_{ij}(\tau)$, to obtain $\beta_{ij}(t)$, with τ varying from 0 to t. The scalar 'effective' anisotropic hardening parameter Z^A in (8) and (9) is obtained as the component of the current value of β_{ij} in the direction of the current stress σ_{ij}, i.e.

$$Z^A(t) = \beta_{ij}(t)u_{ij}(t) \tag{11}$$

Equations (9) and (10) state that the anisotropic hardening rate is maximum in the direction of the current stress and would vary according to a cosine law in other directions in stress space. If a 'yield surface' were obtained from these equations on the basis of appropriate loading paths, then the proposed anisotropic hardening rule would indicate a tendency towards a vertex in the stress direction and a softening and flattening of the 'surface' in the diametrically opposite direction. No cross effects will result from the proposed anisotropic hardening equations. The equations can be restated in terms of a six-dimensional stress space S_α, $\alpha = 1, \ldots, 6$, where

$$S_1 = \sigma_{11}, \qquad S_2 = \sigma_{22}, \qquad S_3 = \sigma_{33}, \qquad S_4 = \sqrt{2}\,\sigma_{12}, \quad \text{etc.} \tag{12}$$

so that

$$u_\alpha = S_\alpha \bigg/ \left(\sum_{\alpha=1}^{6} S_\alpha^2\right)^{1/2} = S_\alpha/(\tfrac{1}{3}I_1^2 + 2J_2)^{1/2} \tag{13}$$

Anisotropic hardening is then represented as a vector β_α in S_α space and the rate of hardening is in the direction of S_α. In this notation, Z^A, (11), is given by

$$Z^A(t) = \beta_\alpha(t)u_\alpha(t) \tag{14}$$

Stress is taken as the directional reference for anisotropic hardening on the basis of available experimental information and to avoid the strong cross softening effect consequent to an index based on deviatoric stress or plastic strain rate. This procedure more closely parallels Ziegler's modification of Prager's original anisotropic (kinematic) hardening rule. It is noted that a uniform hydrostatic pressure would not lead to hardening since $\dot{W}_p = 0$ due to plastic incompressibility. Although β_{ij} is symmetric, it is not traceless and therefore reducible to a scalar, i.e. an isotropic component, and a traceless symmetric tensor [14]. As a consequence, the proposed anisotropic hardening law could result in an isotropic hardening contribution under certain circumstances.

Under proportional loading conditions, $u_{ij}(t)$ is constant and $u_{ij}u_{ij} = 1$ so that $\dot{Z}^A$ from (9) and (11) would reduce to a form similar to that of $\dot{Z}^I$, (7). These functions could be combined by defining new material constants so that an isotropic hardening formulation is applicable under proportional loading conditions. Determination of all the material constants in (7) and (9) would generally require both monotonic and reversed loading tests.

For uniaxial cyclic loading, the above procedure is similar to that described in ref. 4, which showed reasonable predictive capability, with the difference that isotropic and anisotropic hardening are given here by independent functions. Greater flexibility is therefore obtained in the matching process and the asymptotic cyclic stress–strain curves are more realistic. A stable cyclic stress–strain curve would develop upon saturation of isotropic hardening. The initial condition $Z_0 < Z_1$ would lead to cyclic hardening while cyclic softening will result for $Z_0 > Z_1$. Cyclic creep and cyclic stress relaxation would be obtained from the equations under the appropriate loading conditions.

2.3. Evolution Equations for Damage

An equation for isotropic damage development was proposed in ref. 6 in integrated form. It can also be expressed in incremental form appropriate for an evolution equation, namely,

$$\dot{\omega} = \frac{P}{H}\left\{\left[\ln\left(\frac{1}{\omega}\right)\right]^{(P+1)/P}\right\}\omega\dot{Q} \tag{15}$$

Integration of (15) for constant applied stress leads to

$$\omega = \exp\left[-(H/Q)^P\right] \tag{16}$$

where

$$\dot{Q} = [\alpha\sigma^{+}_{\max} + \beta\sqrt{3J_2} + \gamma I^{+}_{1}]^{\nu} \tag{17}$$

$$\alpha + \beta + \gamma = 1 \tag{17a}$$

and, $\sigma^{+}_{\max}$ is the maximum tensile principal stress, I^{+}_{1} is the first stress invariant (positive), and α, β, γ, ν, P, H are material constants. Evolution equation (15) has the general form

$$\dot{\omega} = f_1(\omega)f_2(\sigma) \tag{18}$$

which has been used in many studies on this subject. The function $\dot{Q}$, (17), for $f_2(\sigma)$ was proposed by Leckie and Hayhurst for damage development under multi-dimensional stress states [10]. Power laws for f_1 and f_2 have generally been used in conjunction with (18) and the proposed form, (15), has not yet been critically examined. It was motivated by the characteristics of the response function (4) which is almost zero until a threshold level of the independent variable is reached and then increases until it approaches a saturation value. In this case, $\omega \to 1$ as $Q \to \infty$ but damage, as defined and used as an internal variable, loses its meaning once ω approaches unity. From the practical side, the system response would exhibit high strain rates once ω became a significant fraction of unity.

An example of creep response based on eqns. (5), (15), (16) and (17) is shown in Fig. 1 for copper at 550°C where the experimental data were obtained from ref. 9. Since the stress is constant, the equation for $\dot{Q}$, (17), becomes $Q = Kt$ and (16) can be written as

$$\omega = \exp\{-[(H/K)/t]^{P}\} \tag{19}$$

The parameter (H/K) is very close in value to the time corresponding to the point of inflection of the ω, t relation and could be regarded as a practical failure time. In obtaining the results shown in Fig. 1, H/K was set equal to 255,000 s and P to 5·8. The material constants in the equations for plastic strain rate and isotropic hardening were obtained from ref. 13 and were based on similar creep test data at other stress levels. Primary creep is also predicted in this formulation but cannot be seen on the scale of Fig. 1. The calculated curve shown in Fig. 1 was a better fit to the test data than was obtainable by the power law forms used by Kachanov and Rabatnov for $\dot{\omega}$, (18). There was insufficient test data in ref. 9 in the tertiary creep regime to provide a proper basis for exercises on the full predictive capability of the equations. Such exercises are planned in the future.

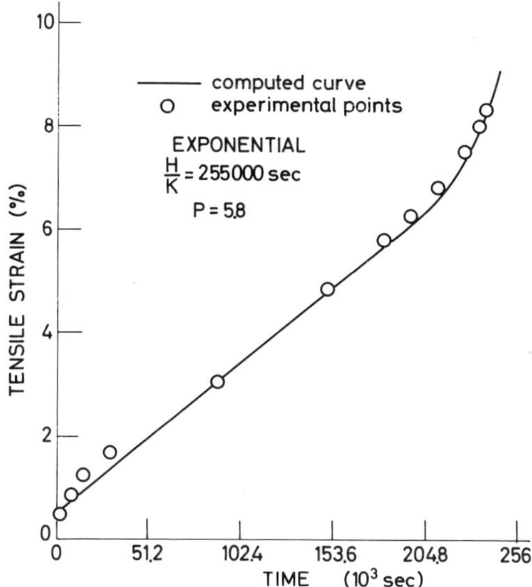

FIG. 1. Creep response of copper at 550°C for an applied tensile stress of 0·112 kbar (from ref. 9).

Plots of damage, ω, as a function of time for the same example are shown in Fig. 2 for the exponential law (19) and for two values of the power law form of (18) (designated by K–R). When the power law expression for $\dot{\omega}$ is integrated, then ω can be expressed as

$$(1-\omega)^b = 1-(t/T) \tag{20}$$

where T is the failure time corresponding to $\omega = 1$. Both b and T can be related to material constants in the equation for $\dot{\omega}$. In the numerical exercises, T was set to be 240,000 s and values of b of 10 and 33·3 were examined. The differences in the damage growth behaviour resulting from the different functional relations are evident. It is noted in Fig. 2 that the damage value computed by the exponential law at the presumed failure condition, $(H/K) = t_{cr}$, is $\omega = 0{\cdot}37$.

Since the response equation contains both hardening and damage variables, it should be applicable for general loading histories. A full reversal of torsional stress during a creep test would show an immediate jump in the reversed strain response because of the reduced directional hardening and then continuation of the creep behaviour. The influence of the stress reversal on subsequent damage development is related to the anisotropic characteristics of the damage. To

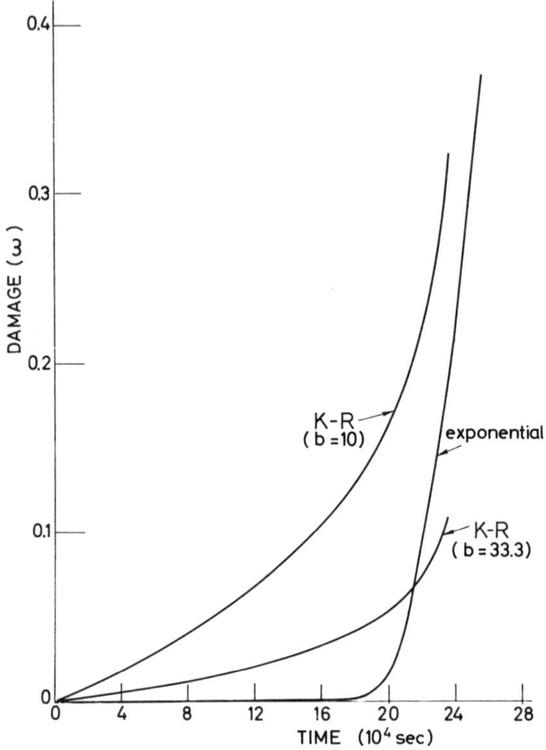

Fig. 2. Computed damage development during tensile creep of copper at 550°C, $\sigma = 0{\cdot}112$ kbar, based on two evolution equations.

some extent, damage anisotropy is already present in the above equations through (17) since a material with significant α and γ values for the stress coefficients would exhibit a reduced damage rate if the direct stress were compressive over part of the loading history.

A reversal of an applied torsional stress would not result in any change in damage development according to the above equations. Such effects are expected physically, however, due to damage occurring primarily on planes normal to the direction of the principal tensile stress. If damage development were isotropic, then a torsional stress reversal during a creep test would lead to a response curve paralleling the time extension of the original one. A damage reduction upon stress reversal due to anisotropy would result in subsequent lower strain rates and intersection of the new response curve with the time extension of the original one.

An anisotropic damage formulation similar to that proposed for anisotropic hardening should be capable of representing the main directional features of damage. That is, a symmetric second-order anisotropic damage tensor is required with an evolution law similar to (15) but with a directional character. As in the case of hardening, the stress tensor seems to be the appropriate reference for the directional index in the damage rate equation. Provision must be made to ensure that direct compressive stress does not contribute to the damage process. In essence, the evolution equations for anisotropic damage would parallel those for anisotropic hardening with an overall scalar damage parameter appearing in the equation for plastic strain rate. The suggested procedure has some similarities, as well as essential differences, to that proposed by Leckie and Onat [11]. Due to the lack of sufficient experimental data on this subject, it is difficult at this time to perform meaningful exercises to confirm or disprove a particular formulation of anisotropic damage.

In the preceding formulation, the evolution equations for Z and ω are uncoupled although some coupling enters through the equation for plastic straining. It would be more realistic to include Z in the equation for $\dot{\omega}$ since hardening tends to reduce damage growth. A method for doing this that is consistent with the other equations would be to replace the general expression for $\dot{\omega}$, (18), by

$$\dot{\omega} = f_1(\omega) f_2(\sigma/Z) \tag{21}$$

The stress parameter $\dot{Q}$, (17), would then be redefined by dividing the stress terms by the hardening variable Z. A fully coupled theory for anisotropic hardening and damage could thereby be developed.

ACKNOWLEDGEMENTS

This paper was written during the author's stay at the Institut für Mechanik, ETH-Zürich. The subject is part of a research program sponsored by the U.S. Air Force Office of Scientific Research at the Technion, Haifa, Israel (Grant AFOSR-80-0214).

REFERENCES

1. Bodner, S. R. and Y. Partom. A large deformation elastic–viscoplastic analysis of a thick-walled spherical shell, *ASME J. appl. Mech.*, **39** (1972), 741–757.
2. Bodner, S. R. and Y. Partom. Constitutive equations for elastic–viscoplastic strain-hardening materials, *ASME J. appl. Mech.*, **42** (1975), 385–389.
3. Bodner, S. R. and A. Merzer. Viscoplastic constitutive equations for copper with strain rate history and temperature effects, *ASME J. Engng Mater. Technol.*, **100** (1978), 388–394.
4. Bodner, S. R., I. Partom and Y. Partom. Uniaxial cyclic loading of elastic–viscoplastic materials, *ASME J. appl. Mech.*, **46** (1979), 805–810.
5. Bodner, S. R. Representation of time dependent mechanical behavior of Rene 95 by constitutive equations, Air Force Materials Laboratory, Wright-Patterson AFB, Report AFML-TR-79-4116 (1979).
6. Bodner, S. R. A procedure for including damage in constitutive equations for elastic–viscoplastic work-hardening materials, *Proc. IUTAM Symposium on Physical Non-linearities in Structural Analysis* (Ed. J. Hult and J. Lemaitre), Springer, Berlin, 1981, 21–28.
7. Bodner, S. R. and D. C. Stouffer. Comments on anisotropic plastic flow and incompressibility, *Int. J. Engng Sci.*, **21** (1983), 211–215.
8. Clifton, R. J. and C. H. Li. Dynamic stress–strain curves at plastic shear strain rates of $10^5\,s^{-1}$, *Proc. Conf. Am. Phys. Soc. on Shock Waves in Condensed Matter*, 1981, 360–366.
9. Feltham, P. and J. D. Meakin. Creep in face centered cubic metals with special reference to copper, *Acta Met.*, **7** (1959), 614–627.
10. Leckie, F. A. and D. R. Hayhurst. Constitutive equations for creep rupture, *Acta Met.*, **25** (1977), 1059–1070.
11. Leckie, F. A. and E. T. Onat. Tensorial nature of damage measuring internal variables. In: *Proc. IUTAM Symposium on Physical Non-linearities in Structural Analysis* (Ed. J. Hult and J. Lemaitre) Springer, Berlin, 1981, 140–155.
12. Merzer, A. and S. R. Bodner. Analytical and computational representation of high rate of strain behaviour. In: *Proc. Conf. Mechanical Properties of Materials at High Rates of Strain* (Ed. J. Harding), Institute of Physics Conference Series No. 47, London, 1979, 142–151.
13. Merzer, A. M. Steady and transient creep behavior based on unified constitutive equations, *ASME J. Engng Mater. Technol.*, **104** (1982), 18–15.
14. Onat, E. T. Representation of inelastic behavior in the presence of anisotropy and of finite deformations. In: *Recent Advances in Creep and Fracture of Engineering Materials and Structures* (Ed. B. Wilshire and D. R. J. Owen), Pineridge Press, Swansea, U.K., 1982, 231–264.
15. Stouffer, D. C. and S. R. Bodner. A constitutive model for the deformation induced anisotropic plastic flow of metals, *Int. J. Engng Sci.*, **17** (1979), 757–764.

27

On a Rational Formulation of Isotropic and Anisotropic Hardening

J. P. Boehler

Institut de Mécanique de Grenoble, St Martin d'Hères, France

ABSTRACT

The classical concepts of isotropic, kinematic and anisotropic hardening are analysed within the framework of the theory of tensor function representations. It is shown that they lead to restrictive forms for the subsequent criteria and do not account for the full range of hardening phenomena. A general concept and a unified rational formulation of isotropic and anisotropic hardening, based on general invariant transformations of the stress tensor, are proposed. Examples are developed in considering first order polynomial stress transformations and are applied to materials obeying in their initial state the von Mises and linear criteria. The classical hardening rules are involved as special cases.

1. INTRODUCTION

The aim of this work is to develop a unified and rational formulation of isotropic and anisotropic hardening within the framework of the tensor functions representation theory. We consider the evolution of the yield criteria during plastic deformations and we assume that this evolution depends on the strain history only through the present value of the plastic strain. For more complicated situations, the proposed concepts can be developed in a straightforward manner.

The classical concepts of isotropic, kinematic and anisotropic hardening are analysed in detail. It is first shown that kinematic hardening can be split into specific isotropic and anisotropic hardening, the contribution of the anisotropic part being of a very restrictive orthotropic type. Secondly, it is shown that certain materials, when subjected to isotropic plastic strain, undergo an isotropic evolution of their internal structure, which results in a modification of the shape of the initial yield condition, without appearance of anisotropic properties. Examples are given for this phenomenon, which, obviously, cannot be taken into account by the classical formulation of isotropic or kinematic hardening. Finally, the classical concepts of isotropic, kinematic and anisotropic hardening are not disjointed and do not cover the full range of hardening phenomena. Consequently, there appears a need for a more general concept and a rational formulation of isotropic and anisotropic hardening.

In our analysis of the classical formulations, we show that the proposed rules are equivalent to very restrictive transformations of the stress tensor. We define the generalized concepts of isotropic and anisotropic hardening by, respectively, general isotropic and anisotropic transformations of the stress tensor. Physically, such a formulation is based on the assumption that the modifications of the materials' internal structure during plastic deformations result in an evolution of the mechanical properties, which can be taken into account by specific transformations of the stress tensor. By application of the theorems of representations for tensor functions, it is shown that the proposed procedure enables the transformation of an arbitrary initial yield criterion into an arbitrary isotropic or anisotropic subsequent criterion.

Within the proposed concept, general isotropic and anisotropic hardening rules are developed, in considering first-order polynomial isotropic and anisotropic transformations of the stress tensor and applying the procedure to materials obeying the von Mises and linear criterion. Although the considered stress transformation and yield criteria are very simple, the proposed general rules constitute a framework, which is already wide enough to be able to account for the observed mechanical phenomena and to include the classical concepts of isotropic, kinematic and anisotropic hardening as special cases.

The essential results of the proposed rational formulation are presented in this paper. Further details and developments on the subject will be included in forthcoming papers [6], [7].

2. CLASSICAL FORMULATION OF ISOTROPIC, KINEMATIC AND ANISOTROPIC HARDENING

2.1. Isotropic Hardening

Consider first an isotropic material. The general invariant form of the initial yield criterion is given by:

$$f(\mathrm{tr}\,\mathbf{T}, \sqrt{\mathrm{tr}\,\mathbf{S}^2}, \sqrt[3]{\mathrm{tr}\,\mathbf{S}^3}) = 0; \qquad \mathbf{S} = \mathbf{T} - (\tfrac{1}{3}\,\mathrm{tr}\,\mathbf{T})\mathbf{I} \tag{1}$$

In the classical concept of isotropic hardening [3], it is assumed that the yield surface (1) undergoes a homothetic transformation of the ratio $\lambda(\mathbf{P})$ in the stress space, where $\mathbf{P}$ is the plastic strain tensor. This transformation is equivalent to a homothetic transformation of the stress tensor $\mathbf{T}$, with the ratio $1/\lambda(\mathbf{P})$. Thus, the classical isotropic hardening rule can be written in the form:

$$\left.\begin{array}{l} f(\mathrm{tr}\,\mathbf{T}, \sqrt{\mathrm{tr}\,\mathbf{S}^2}, \sqrt[3]{\mathrm{tr}\,\mathbf{S}^3}) = 0 \Rightarrow f\left(\dfrac{1}{\lambda}\,\mathrm{tr}\,\mathbf{T}, \dfrac{1}{\lambda}\sqrt{\mathrm{tr}\,\mathbf{S}^2}, \dfrac{1}{\lambda}\sqrt[3]{\mathrm{tr}\,\mathbf{S}^3}\right) = 0 \\ \text{where} \\ \lambda(\mathbf{P}) > 0; \qquad \begin{cases} \lambda(\mathbf{P}) > 1: & \text{hardening} \\ \lambda(\mathbf{P}) < 1: & \text{softening} \end{cases} \end{array}\right\} \tag{2}$$

If the material is anisotropic, the general form of the initial criterion is given by:

$$f(\mathbf{T}, \boldsymbol{\xi}_1, \boldsymbol{\xi}_2, \ldots, \boldsymbol{\xi}_p) = 0 \tag{3}$$

where the $\boldsymbol{\xi}_i$ are structural tensors, taking into account the anisotropic material behaviour [4], [5]. In order to derive the invariant form of (3), the theory of representation of anisotropic tensor functions [4], [21] is very useful, for it indicates the type and the minimal number of basic invariants involved. By application of this theory, we obtain the general invariant form of the initial criterion:

$$f(\mathrm{tr}\,\mathbf{T}, \sqrt{\mathrm{tr}\,\mathbf{S}^2}, \sqrt[3]{\mathrm{tr}\,\mathbf{S}^3}, J_1, J_2, \ldots, J_n) = 0 \tag{4}$$

where the J_k are mixed invariants of the stress deviator $\mathbf{S}$ and the structural tensors $\boldsymbol{\xi}_i$ and are homogeneous of order 1 with respect to $\mathbf{S}$. The type of anisotropy of the material is specified by the invariance group of the J_k [4].

For example, the initial yield criterion of an orthotropic material is given by:

$$f(\mathrm{tr}\,\mathbf{T}, \sqrt{\mathrm{tr}\,\mathbf{S}^2}, \sqrt[3]{\mathrm{tr}\,\mathbf{S}^3}, \mathrm{tr}\,\mathbf{M}_1\mathbf{S}, \mathrm{tr}\,\mathbf{M}_2\mathbf{S}, \sqrt{\mathrm{tr}\,\mathbf{M}_1\mathbf{S}^2}, \sqrt{\mathrm{tr}\,\mathbf{M}_2\mathbf{S}^2}) = 0$$
$$\mathbf{M}_1 = \mathbf{v}_1 \otimes \mathbf{v}_1; \qquad \mathbf{M}_2 = \mathbf{v}_2 \otimes \mathbf{v}_2 \tag{5}$$

where $(\mathbf{v}_1, \mathbf{v}_2)$ are two of the three privileged orthogonal directions of the medium.

Applying the classical concept of isotropic hardening, the subsequent criterion of an initially anisotropic material is given by:

$$f(\mathrm{tr}\,\mathbf{T}, \sqrt{\mathrm{tr}\,\mathbf{S}^2}, \sqrt[3]{\mathrm{tr}\,\mathbf{S}^3}, J_k) = 0 \;\Rightarrow\; f\left(\frac{1}{\lambda}\,\mathrm{tr}\,\mathbf{T}, \frac{1}{\lambda}\sqrt{\mathrm{tr}\,\mathbf{S}^2}, \frac{1}{\lambda}\sqrt[3]{\mathrm{tr}\,\mathbf{S}^3}, \frac{1}{\lambda}J_k\right) = 0 \tag{6}$$

Such a transformation does not modify the invariance group of the J_k. Thus, the initial type of anisotropy is preserved in the subsequent criterion.

Finally, for initially isotropic or anisotropic materials, the classical concept of isotropic hardening is equivalent to a homothetic transformation of the stress tensor:

$$\mathbf{T} \Rightarrow \bar{\mathbf{T}} = \frac{1}{\lambda(\mathbf{P})}\mathbf{T} \tag{7}$$

For this type of isotropic hardening rule, we suggest the name 'homothetic isotropic hardening' (see also Section 2.5 below).

2.2. Kinematic Hardening

In order to account for the Bauschinger effect, different kinematic hardening rules were proposed [18], [25]. The classical concept of kinematic hardening assumes that the yield surface undergoes a rigid translation $\mathbf{C}(\mathbf{P})$ in the stress space, where $\mathbf{C}(\mathbf{P})$ is a symmetric second order tensor. This transformation is equivalent to the translation $-\mathbf{C}(\mathbf{P})$ of the stress tensor:

$$\mathbf{T} \Rightarrow \bar{\mathbf{T}} = \mathbf{T} - \mathbf{C}(\mathbf{P}) \tag{8}$$

We consider first an isotropic material, with initial criterion (1). The translation tensor $\mathbf{C}(\mathbf{P})$ can be split into isotropic and deviatoric parts:

$$\mathbf{C}(\mathbf{P}) = \tfrac{1}{2}\bar{c}(\mathbf{P})\mathbf{I} + \mathbf{C}'(\mathbf{P}); \qquad \mathrm{tr}\,\mathbf{C}'(\mathbf{P}) = 0 \tag{9}$$

By the isotropic part $\frac{1}{3}\bar{c}(\mathbf{P})\mathbf{I}$, which induces a translation of $\mathbf{T}$ parallel to the isotropic axis of the stress space, the initial criterion (1) is transformed into:

$$f(\operatorname{tr}\mathbf{T}-\bar{c}(\mathbf{P}), \sqrt{\operatorname{tr}\mathbf{S}^2}, \sqrt[3]{\operatorname{tr}\mathbf{S}^3})=0 \tag{10}$$

Thus, the subsequent criterion remains isotropic.

By the deviatoric part $\mathbf{C}'(\mathbf{P})$, which induces a translation orthogonal to the isotropic subspace of the full six-dimensional stress space, the initial isotropic criterion (1) is transformed into:

$$f(\operatorname{tr}\bar{\mathbf{T}}, \sqrt{\operatorname{tr}\bar{\mathbf{S}}^2}, \sqrt[3]{\operatorname{tr}\bar{\mathbf{S}}^3})=0; \qquad \bar{\mathbf{S}}=\bar{\mathbf{T}}-(\tfrac{1}{3}\operatorname{tr}\bar{\mathbf{T}})\mathbf{I} \tag{11}$$

where the isotropic invariants of the transformed stress tensor $\bar{\mathbf{T}}$ are related to that of $\mathbf{T}$ by:

$$\left.\begin{aligned}
&\operatorname{tr}\bar{\mathbf{T}}=\operatorname{tr}\mathbf{T}\\
&\operatorname{tr}\bar{\mathbf{S}}^2=\operatorname{tr}\mathbf{S}^2-2(2c_1'+c_2')\operatorname{tr}\mathbf{M}_1\mathbf{S}-2(c_1'+2c_2')\operatorname{tr}\mathbf{M}_2\mathbf{S}+2(c_1'^2+c_2'^2+c_1'c_2')\\
&\operatorname{tr}\bar{\mathbf{S}}^3=\operatorname{tr}\mathbf{S}^3+3(c_1'+c_2')\operatorname{tr}\mathbf{S}^2-3(2c_1'+c_2')\operatorname{tr}\mathbf{M}_1\mathbf{S}^2-3(c_1'+2c_2')\operatorname{tr}\mathbf{M}_2\mathbf{S}^2\\
&\qquad-3c_2'(2c_1'+c_2')\operatorname{tr}\mathbf{M}_1\mathbf{S}-3c_1'(c_1'+2c_2')\operatorname{tr}\mathbf{M}_2\mathbf{S}+3c_1'c_2'(c_1'+c_2')
\end{aligned}\right\} \tag{12}$$

where c_1', c_2', c_3' are the principal values of $\mathbf{C}'(\mathbf{P})$ and:

$$\mathbf{M}_1=\mathbf{e}_1\otimes\mathbf{e}_1; \qquad \mathbf{M}_2=\mathbf{e}_2\otimes\mathbf{e}_2 \tag{13}$$

$(\mathbf{e}_1, \mathbf{e}_2)$ being the two first principal directions of $\mathbf{C}'(\mathbf{P})$.

Comparing (11) and (12) with (5), we can conclude that the subsequent criterion is that of an orthotropic medium, with privileged directions $\mathbf{e}_1, \mathbf{e}_2, \mathbf{e}_3$. It is worthwhile to point out that the induced orthotropy is of a very restrictive type: for a general orthotropic medium, seven independent invariants are involved in the yield criterion (5), whereas in (11) only the three combinations (12) are involved.

Finally, kinematic hardening can be split into isotropic and orthotropic parts. The isotropic part transforms an initial isotropic criterion into a subsequent isotropic criterion by a translation of the stress tensor parallel to the isotropic axis of the stress space. For this type of hardening rule, we suggest the name 'translational isotropic hardening'. The orthotropic part transforms an initial isotropic criterion into a subsequent orthotropic criterion of a very restrictive form. For this type of anisotropic hardening rule, we suggest the name 'translational orthotropic hardening'.

If the material is initially anisotropic, the translational isotropic hardening rule leaves the type of anisotropy unchanged, whereas the translational orthotropic hardening rule modifies the type of the initial anisotropy. Further details will be given in ref. 6.

2.3. Anisotropic Hardening

The analysis of Section 2.2 shows that the anisotropic part of kinematic hardening is a very restrictive type of orthotropic hardening. In fact, kinematic hardening cannot take into account most observed phenomena of anisotropic hardening [9]. Different, more elaborate anisotropic hardening rules have been proposed [1, 2, 22, 24], in order to account for change in shape and size and for rotation and translation of the initial yield criterion. The aforementioned rules are based on anisotropic linear transformations of the stress tensor. A more general approach, based on the theory of representation for tensor functions, has been developed in ref. 17 for initially isotropic and in refs. 8 and 9 for initially anisotropic material. More information will be given in ref. 6.

2.4. Hardening Phenomena which Cannot be Described by the Classical Formulations

Several important hardening phenomena cannot be taken into account by the classical approach of isotropic, kinematic and anisotropic hardening. We shall present here two examples for two different initially isotropic materials subjected to isotropic plastic strains.

The first example concerns granular materials without cohesion, obeying the isotropic linear criterion:

$$\operatorname{tr}\mathbf{T} - a\sqrt{\operatorname{tr}\mathbf{S}^2} = 0 \tag{14}$$

If such materials are subjected to isotropic compressions $\mathbf{P} = \varepsilon\mathbf{I}$, they remain isotropic, but the density increases with ε. The increase of density induces an increase in the internal friction, without appearance of cohesion. Such phenomena have been observed experimentally on sands, for example [23]. Thus, the subsequent criterion is given by:

$$\operatorname{tr}\mathbf{T} - \bar{a}(\varepsilon)\sqrt{\operatorname{tr}\mathbf{S}^2} = 0, \quad \text{with} \quad \bar{a}(0) = a; \qquad \frac{d\bar{a}}{d\varepsilon} < 0 \tag{15}$$

By the homothetic isotropic hardening rule, the initial criterion is unchanged; by translational isotropic hardening, the initial friction is unchanged and there appears non-zero cohesion (Fig. 1). Thus, the subsequent criterion (15) cannot be obtained by the classical concepts of isotropic hardening.

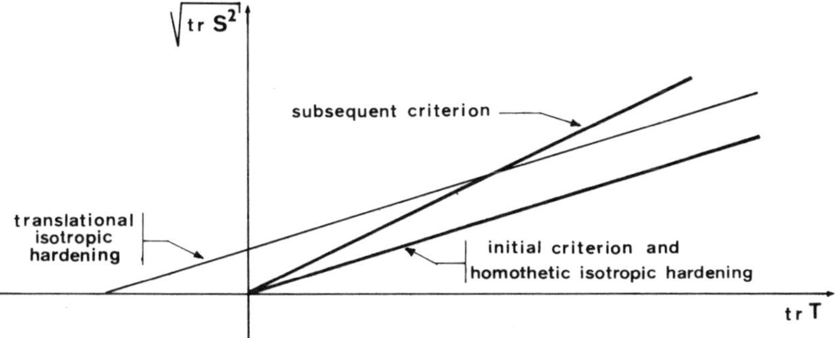

FIG. 1. Granular materials subjected to isotropic plastic strain.

The second example concerns rolled sheet-steel obeying the von Mises criterion:

$$\sqrt{\text{tr}\,\mathbf{S}^2} = \sqrt{2}\,k \tag{16}$$

For orthotropic rolled sheet-steel, a generalization of the von Mises criterion has been proposed, taking into account the initial anisotropy and the anisotropy induced by the plastic strain [8]; the theoretical predictions are well in agreement with experimental results [9]. For an initially isotropic steel subjected to a plane isotropic plastic strain $\mathbf{P} - \varepsilon\mathbf{I}$, the subsequent criterion, developed with respect to the principal plane stresses, is given by:

$$(\sigma_1^2 + \sigma_2^2)(\tfrac{2}{3} + a\varepsilon) + 2\sigma_1\sigma_2(-\tfrac{1}{3} + b\varepsilon) = 2k^2(1 + c\varepsilon) \tag{17}$$

where a, b and c are material constants. In the general case ($a + 2b \neq 0$), equation (17) represents in the (σ_1, σ_2) plane an ellipse with the same principal axes as the von Mises ellipse, but with a different size and shape.

By the homothetic isotropic hardening rule, the initial von Mises ellipse is transformed into a homothetic ellipse without change of shape; by the translational isotropic hardening rule, the initial criterion (16) is unchanged (Fig. 2). Thus, the subsequent criterion (17) cannot be obtained by the classical concepts of isotropic hardening.

In these two examples, an initial isotropic criterion is transformed into a subsequent isotropic criterion with change in size or (and) change in shape. For such hardening phenomena, we suggest the name 'isotropic hardening' in a more general sense than the classical one.

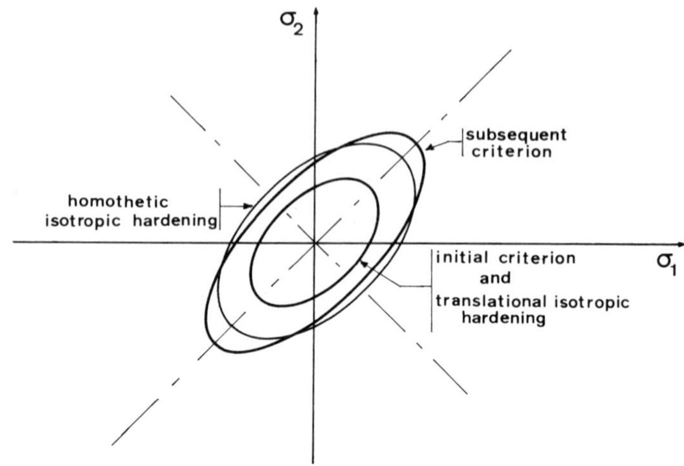

FIG. 2. Von Mises materials subjected to isotropic plastic plane strain.

That is why we suggest the name 'homothetic isotropic hardening' for the classical concept of isotropic hardening and 'translational isotropic hardening' for the isotropic part of the classical concept of kinematic hardening, in order to underline that the classical concepts are special cases of a more general formulation of isotropic hardening.

2.5. Conclusions

The classical concepts of isotropic, kinematic and anisotropic hardening:

(1) are not disjointed. Indeed, kinematic hardening is a combination of a special isotropic hardening (translational isotropic hardening) and a special anisotropic hardening (translational orthotropic hardening).
(2) give restrictive forms for the subsequent criteria.
(3) cannot take into account several important hardening phenomena.

Thus, there appears a need for a more general concept and a rational formulation of isotropic and anisotropic hardening.

3. GENERAL FORMULATION OF ISOTROPIC AND ANISOTROPIC HARDENING

3.1. Proposed General Concept

In our analysis of the classical formulation of isotropic, kinematic and anisotropic hardening, we have noticed that the proposed rules can be interpreted by very restrictive transformations of the stress tensor.

A general concept of isotropic and anisotropic hardening can be based on the assumption that the evolution of mechanical properties, due to the modifications of the material internal structure, can be taken into account by specific transformations of the stress tensor involved in the constitutive law. A similar process, in a more restrictive manner, has been proposed for the modelling of damage growth [11, 14, 15, 16].

Within the proposed scheme, the generalized concepts of isotropic and anisotropic hardening are defined by, respectively, general isotropic and anisotropic transformations of the stress tensor. The invariant forms of these general transformations will be specified by application of the theory of representation of tensor functions. The main object of Section 3 is to set up the mathematical tool involved in the proposed general concept of isotropic and anisotropic hardening. How to derive specific transformation rules of the stress tensor from microscopic observations or phenomenological considerations will not be treated here. Proposed general isotropic and anisotropic hardening rules will be presented in Section 4.

3.2. Influence of the Plastic Strain on the Hardening Rule

The influence of the plastic strain **P** has been mathematically analysed for initially isotropic [17] and anisotropic materials [8], by the application of the theorems of representation for tensor functions. The results of these analyses are: if the initial isotropic or anisotropic criterion undergoes an isotropic transformation, the involved plastic strain is necessarily isotropic; a non-zero plastic strain deviator induces an anisotropic transformation of the yield criterion.

Physically, these theoretical results can be interpreted in the following manner:

(a) A non-isotropic plastic strain induces the appearance of a preferred orientation of the material internal structure, which results in anisotropic hardening on the macroscale. For example, an initially isotropic criterion is transformed into a subsequent anisotropic criterion.

(b) Possibly there may exist materials for which a non-isotropic plastic strain induces only isotropic modifications of the internal structure. But, in this case, only the isotropic part of **P** is involved in the law of transformation.

Finally, if the hardening is isotropic, either the plastic strain is isotropic, or only its isotropic part is involved in the hardening rule. If the hardening is anisotropic, the deviatoric part of the plastic strain is necessarily different from zero. Further details will be developed in ref. 6.

3.3. Isotropic Hardening

3.3.1. *Initially Isotropic Materials*

Consider first an initially isotropic material. The invariant form of the initial criterion is given by (1). The general concept of isotropic hardening is introduced by the general isotropic transformation of the stress tensor **T**, involving the plastic strain $\varepsilon\mathbf{I}$:

$$\mathbf{T} \Rightarrow \bar{\mathbf{T}} = \bar{\mathbf{T}}(\mathbf{T}, \varepsilon\mathbf{I}) \tag{18}$$

The invariant form of (18) is given by the theory of representation for tensor functions:

$$\left.\begin{aligned} \bar{\mathbf{T}} &= \alpha_0\mathbf{I} + \alpha_1\mathbf{S} + \alpha_2\mathbf{S}^2 \\ \alpha_i &= \alpha_i(\operatorname{tr}\mathbf{T}, \sqrt{\operatorname{tr}\mathbf{S}^2}, \sqrt[3]{\operatorname{tr}\mathbf{S}^3}, \varepsilon) \end{aligned}\right\} \tag{19}$$

Employing the Cayley–Hamilton theorem, the tensorial relation (19) can be written in the form of relations between the basic isotropic invariants of **T** and those of the transformed tensor $\bar{\mathbf{T}}$:

$$\left.\begin{aligned} \operatorname{tr}\bar{\mathbf{T}} &= k_1^\varepsilon(\operatorname{tr}\mathbf{T}, \sqrt{\operatorname{tr}\mathbf{S}^2}, \sqrt[3]{\operatorname{tr}\mathbf{S}^3}) \\ \sqrt{\operatorname{tr}\bar{\mathbf{S}}^2} &= k_2^\varepsilon(\operatorname{tr}\mathbf{T}, \sqrt{\operatorname{tr}\mathbf{S}^2}, \sqrt[3]{\operatorname{tr}\mathbf{S}^3}) \\ \sqrt[3]{\operatorname{tr}\bar{\mathbf{S}}^3} &= k_3^\varepsilon(\operatorname{tr}\mathbf{T}, \sqrt{\operatorname{tr}\mathbf{S}^2}, \sqrt[3]{\operatorname{tr}\mathbf{S}^3}) \end{aligned}\right\} \tag{20}$$

where the superscript ε denotes the dependence of the functions k_i on the plastic strain.

Finally, the general rule of isotropic hardening for an initially isotropic material is given by:

$$f(\operatorname{tr}\mathbf{T}, \sqrt{\operatorname{tr}\mathbf{S}^2}, \sqrt[3]{\operatorname{tr}\mathbf{S}^3}) = 0 \Rightarrow \begin{cases} f(k_1^\varepsilon, k_2^\varepsilon, k_3^\varepsilon) = 0 \\ k_i^\varepsilon = k_i^\varepsilon(\operatorname{tr}\mathbf{T}, \sqrt{\operatorname{tr}\mathbf{S}^2}, \sqrt[3]{\operatorname{tr}\mathbf{S}^3}) \end{cases} \tag{21}$$

The versatility of the proposed concept is manifest, when considering the following properties of rule (21):

(1) Homothetic and translational isotropic hardening rules are obviously involved as very special cases.
(2) The hardening phenomena which cannot be described by the classical formulations are taken into account (see Section 4.2 below).
(3) By rule (21), a given initial isotropic criterion can be transformed into an arbitrary subsequent isotropic criterion.

The last property shows that the proposed concept of isotropic hardening is the most general. The demonstration of this property will be given in ref. 6.

3.3.2. *Initially Anisotropic Materials*

Consider now an initially anisotropic material. The invariant form of the initial criterion is given by (4). The general concept of isotropic hardening is introduced by a similar procedure: the stress tensor $\mathbf{T}$ is subjected to a general isotropic transformation by the invariant form (19).

Using Rivlin's generalization of the Cayley–Hamilton theorem [20], the three isotropic invariants of $\bar{\mathbf{T}}$ and the mixed (anisotropy) invariants $\bar{J}_k$ of $\bar{\mathbf{T}}$ and the structural tensors $\boldsymbol{\xi}_i$, can be expressed in terms of the corresponding invariants of the stress tensor $\mathbf{T}$. Such a procedure has been developed in ref. 3.

Finally, the general rule of isotropic hardening for an initially anisotropic material is given by:

$$f(\operatorname{tr}\mathbf{T}, \sqrt{\operatorname{tr}\mathbf{S}^2}, \sqrt[3]{\operatorname{tr}\mathbf{S}^3}, J_k) = 0 \Rightarrow f(\operatorname{tr}\bar{\mathbf{T}}, \sqrt{\operatorname{tr}\bar{\mathbf{S}}^2}, \sqrt[3]{\operatorname{tr}\bar{\mathbf{S}}^3}, \bar{J}_k) = 0 \tag{22}$$

Further details will be given in ref. 6. We simply mention here the main properties of the general rule (22):

(1) The transformation (19) of the stress tensor being isotropic, the transformed mixed invariants $\bar{J}_k$ admit the same invariance group as the initial mixed invariants J_k; thus, the type of anisotropy of the initial criterion is preserved for the subsequent criterion.

(2) The ‘degree of anisotropy’ varies in general. In fact, there exists a problem for a realistic definition of the ‘degree of anisotropy’. We propose the following definition: this degree is specified by the shape, the position with respect to the deviatoric subspace and the orientation of the yield surface in the stress space. Thus, the degree is preserved if and only if the transformation of the stress tensor is that of homothetic or translational isotropic hardening rules.

(3) By rule (22), a given initial anisotropic criterion can be transformed into an arbitrary subsequent anisotropic criterion of the same type of anisotropy.

3.4. Anisotropic Hardening

Within the proposed concept, the anisotropic modifications of the materials’ internal structure, which are induced by non-isotropic plastic strains, can be taken into account by the following transformation of the stress tensor:

$$\mathbf{T} \Rightarrow \bar{\mathbf{T}} = \bar{\mathbf{T}}(\mathbf{T}, \boldsymbol{\xi}_i, \mathbf{P}); \qquad \mathbf{P} - (\tfrac{1}{3}\,\mathrm{tr}\,\mathbf{P})\mathbf{I} = \mathbf{E} \neq \mathbf{0} \tag{23}$$

In (23), the tensor function $\bar{\mathbf{T}}$ is isotropic with respect to its arguments $(\mathbf{T}, \boldsymbol{\xi}_i, \mathbf{P})$, where $\mathbf{T}$ is the stress tensor, $\mathbf{P}$ the plastic strain and $\boldsymbol{\xi}_i$ the initial structural tensors of the material. It can be shown [4] that $\bar{\mathbf{T}}$ is an anisotropic function with respect to $\mathbf{T}$, the type of anisotropy being specified by the invariance group of $(\boldsymbol{\xi}_i, \mathbf{P})$, which is obviously different from the initial invariance group of the material.

Following a similar procedure to that for isotropic hardening, we use the theorems of representations for tensor functions and the generalized Cayley–Hamilton theorem, in order to obtain the general rule of anisotropic hardening:

$$f(\mathrm{tr}\,\mathbf{T}, \sqrt{\mathrm{tr}\,\mathbf{S}^2}, \sqrt[3]{\mathrm{tr}\,\mathbf{S}^3}, J_k) = 0 \Rightarrow f(\mathrm{tr}\,\bar{\mathbf{T}}, \sqrt{\mathrm{tr}\,\bar{\mathbf{S}}^2}, \sqrt[3]{\mathrm{tr}\,\bar{\mathbf{S}}^3}, \mathscr{K}_l) = 0 \tag{24}$$

where the $\mathscr{K}_l$ are mixed invariants of $\mathbf{T}$, $\boldsymbol{\xi}_i$ and $\mathbf{P}$, which take into account the initial anisotropy, the induced anisotropy and the coupling between them [8].

Further details and developments will be presented in ref. 6. The main properties of the general rule (24) are the following:

(1) A given initial anisotropic criterion can be transformed into an arbitrary subsequent anisotropic criterion, of arbitrary type and degree.

(2) The hypothesis that the hardening rule depends only on the present plastic strain results in the fact that for an initially isotropic material, the subsequent criterion is necessarily orthotropic [8]. But, if strain history is taken into account, considering for example a sequence of small plastic strains, it can be shown that the initial isotropic condition can be transformed into an arbitrary subsequent anisotropic criterion.

4. EXAMPLES

4.1. Introduction

In developing examples of hardening rules, different specific transformations of the stress tensor may be considered. We propose a general isotropic (respectively anisotropic) hardening rule, in considering a very simple transformation of the stress tensor: the isotropic (respectively anisotropic) first order polynomial transformation.

It is worthwhile to point out that this procedure furnishes a framework which is already large enough to include the classical isotropic, kinematic and proposed anisotropic hardening rules as special cases and to account for phenomena which cannot be described by the classical formulations.

4.2. Proposed General Isotropic Hardening Rule

When restricted to a first order polynomial transformation, the general invariant form (19) can be written in the following form:

$$\lambda\bar{\mathbf{T}}=\left(C+\frac{A}{3}\operatorname{tr}\mathbf{T}\right)\mathbf{I}+D\mathbf{S} \tag{25}$$

where the material coefficients λ, A, C, D are scalar-valued functions of the plastic strain ε.

Consider first a material obeying in its initial state the von Mises criterion. In this case, the coefficients involved in (25) are subjected to the initial conditions: $C(0)=A(0)=0$, $\lambda(0)=1$, so that for $\mathbf{P}=\mathbf{0}$, we have $\bar{\mathbf{T}}(\mathbf{0})=\mathbf{S}$. Employing (25) and rule (21) of isotropic hardening, the subsequent criterion is given by:

$$\operatorname{tr}(\lambda\bar{\mathbf{T}})^2=3C^2+2AC\operatorname{tr}\mathbf{T}+\tfrac{1}{3}A^2\operatorname{tr}^2\mathbf{T}+D^2\operatorname{tr}\mathbf{S}^2=2k^2\lambda^2 \tag{26}$$

In (26), the coefficients $D(\varepsilon)$ and $\lambda(\varepsilon)$ account for the change of size of the criterion, $C(\varepsilon)$ describes the translation parallel to the isotropic axis and $\dot{A}(\varepsilon)$ stands for the change in shape. Thus, the very simple stress transformation rule (25) and the application of the von Mises criterion provide a hardening rule involving the essential features of our general concept of isotropic hardening: the initial criterion can change in size, shape and position along the isotropic axis.

The functions $A(\varepsilon)$, $C(\varepsilon)$, $D(\varepsilon)$ and $\lambda(\varepsilon)$ must be further specified from physical observations or (and) phenomenological considerations. The purpose of this paper is to show the possibilities of the proposed rule. Thus, we consider the following simple case:

$$\begin{aligned} \varepsilon \geqslant 0 \qquad & \sqrt{3}\,\lambda = \sqrt{1+l\varepsilon} \qquad && \sqrt{3}\,C = \sqrt{c\varepsilon} \\ & \sqrt{3}\,A = \sqrt{(a+2d)\varepsilon} && \sqrt{3}\,D = \sqrt{1+(a-d)\varepsilon} \end{aligned} \tag{27}$$

where a, c, d, l are material constants. For $\varepsilon = 0$, we obtain the von Mises criterion:

$$\varepsilon = 0 \;\Rightarrow\; \operatorname{tr}\bar{\mathbf{T}}^2 = \operatorname{tr}\mathbf{S}^2 = 2k^2 \tag{28}$$

Employing (27) in (26) and developing with respect to the principal stresses, the subsequent criterion takes finally the form:

$$\begin{aligned} 3c\varepsilon + 2\sqrt{c(a+2d)}\,\varepsilon(\sigma_1+\sigma_2+\sigma_3) + (\tfrac{2}{3}+a\varepsilon)(\sigma_1^2+\sigma_2^2+\sigma_3^2) \\ + 2(-\tfrac{1}{3}+d\varepsilon)(\sigma_1\sigma_2+\sigma_2\sigma_3+\sigma_3\sigma_1) = 2k^2(1+l\varepsilon) \end{aligned} \tag{29}$$

In plane stress, it is easy to see that in the special case $c = 0$, the initial von Mises ellipse is transformed into a subsequent ellipse with different size and shape. Thus, the second example of isotropic hardening phenomena described in Section 2.4 is taken into account. If $c \neq 0$, the von Mises ellipse undergoes in addition a translation parallel to the isotropic axis.

Consider now a material obeying in its initial state the linear criterion:

$$\operatorname{tr}\mathbf{T} - a\sqrt{\operatorname{tr}\mathbf{S}^2} = c \tag{30}$$

Employing (25) and rule (21) of isotropic hardening, the subsequent criterion, when developed with respect to the principal stresses, is given by:

$$\begin{aligned} (3A+C)(\sigma_1+\sigma_2+\sigma_3) + a\sqrt{\{3C^2 + 2AC(\sigma_1+\sigma_2+\sigma_3)} \\ + \tfrac{1}{3}(A^2+2D^2)(\sigma_1^2+\sigma_2^2+\sigma_3^2) + \tfrac{2}{3}(A^2-D^2)(\sigma_1\sigma_2+\sigma_2\sigma_3+\sigma_3\sigma_1)\} = \lambda c \end{aligned} \tag{31}$$

Unfortunately, there is a lack of experimental results for testing the proposed rule (31) in its general form. The increase of internal friction of granular materials, when subjected to isotropic strain ε, can be described by introducing the following forms of the coefficients involved in the stress transformation (25):

$$\lambda \equiv 1; \qquad C \equiv 0; \qquad A \equiv 1; \qquad D = k(\varepsilon), \quad \text{with} \quad 0 < k(\varepsilon) \leqslant 1, \quad k(0) = 1 \tag{32}$$

Introducing the coefficients (32) in (25) and employing rule (21) of isotropic hardening for the criterion (30), we obtain for cohesionless granular materials the following subsequent criterion:

$$\operatorname{tr} \mathbf{T} - \bar{a}\sqrt{\operatorname{tr} \mathbf{S}^2} = 0, \quad \text{with} \quad \bar{a} = k(\varepsilon)a \tag{33}$$

Thus, the first example of isotropic hardening phenomena described in Section 2.4 is taken into account.

4.3. Proposed General Anisotropic Hardening Rule

4.3.1. *Initially Isotropic Material*
Consider an initially isotropic material subjected to a plastic strain $\mathbf{P}$ with a non-zero deviatoric part $\mathbf{E}$. In this case, the subsequent criterion is necessarily orthotropic [8].

The proposed general hardening rule in the case of a three-dimensional stress-tensor will be developed in ref. 7. We simply mention here the procedure employed and the main results obtained. Using the general invariant form of an orthotropic first-order polynomial tensor transformation, we derive the stress transformation rule, involving 15 material coefficients, which are scalar-valued functions of $\mathbf{P}$. Introducing the transformed stress into the isotropic von Mises criterion, we obtain a general three-dimensional hardening rule, of which the rules proposed by Yoshimura [24] and Svensson [22] are special cases.

We develop here a similar procedure in the case of plane stress. In this case, it can be shown that the representation of the transformation $\bar{\mathbf{T}} = \bar{\mathbf{T}}(\mathbf{T}, \mathbf{P})$ takes the form:

$$\begin{aligned} \bar{\mathbf{T}} &= \psi_1 \mathbf{M}_1 + \psi_2 \mathbf{M}_2 + \psi_3 \mathbf{T} \\ \psi_i &= \psi_i(\operatorname{tr} \mathbf{M}_1\mathbf{T}, \operatorname{tr} \mathbf{M}_2\mathbf{T}, \operatorname{tr} \mathbf{T}^2, \operatorname{tr} \mathbf{M}_1\mathbf{P}, \operatorname{tr} \mathbf{M}_2\mathbf{P}) \end{aligned} \tag{34}$$

where $\mathbf{M}_1 = \mathbf{v}_1 \otimes \mathbf{v}_1$, $\mathbf{M}_2 = \mathbf{v}_2 \otimes \mathbf{v}_2$; $(\mathbf{v}_1, \mathbf{v}_2)$ being the principal directions of $\mathbf{P}$ in the plane of stress.

When restricted to a first-order polynomial transformation, the invariant form (34) can be written in the form:

$$\lambda\bar{\mathbf{T}} = (C_1 + B_1 \operatorname{tr} \mathbf{M}_1\mathbf{T} + D_1 \operatorname{tr} \mathbf{M}_2\mathbf{T})\mathbf{M}_1 + (C_2 + B_2 \operatorname{tr} \mathbf{M}_1\mathbf{T} + D_2 \operatorname{tr} \mathbf{M}_2\mathbf{T})\mathbf{M}_2 + F\mathbf{T} \quad (35)$$

where the material coefficients λ, C_1, C_2, B_1, B_2, D_1, D_2, F are scalar-valued functions of $\operatorname{tr} \mathbf{M}_1\mathbf{P} = \varepsilon_1$ and $\operatorname{tr} \mathbf{M}_2\mathbf{P} = \varepsilon_2$; ε_1 and ε_2 are the principal values of $\mathbf{P}$.

Employing the stress transformation (35), different initial yield conditions may be considered. For a material obeying in its initial state the von Mises criterion, the subsequent criterion is given by:

$$\begin{aligned}(C_1^2 + C_2^2) &+ 2[C_1(B_1 + F) + C_2B_2] \operatorname{tr} \mathbf{M}_1\mathbf{T}\\ &+ 2[C_1D_1 + C_2(D_2 + F)] \operatorname{tr} \mathbf{M}_2\mathbf{T}\\ &+ (B_1^2 + B_2^2 + 2FB_1) \operatorname{tr}^2 \mathbf{M}_1\mathbf{T}\\ &+ (D_1^2 + D_2^2 + 2FD_2) \operatorname{tr}^2 \mathbf{M}_2\mathbf{T}\\ &+ 2[D_1(B_1 + F) + B_2(D_2 + F)] \operatorname{tr} \mathbf{M}_1\mathbf{T} \operatorname{tr} \mathbf{M}_2\mathbf{T}\\ &+ F^2 \operatorname{tr} \mathbf{T}^2 = 2k^2\lambda^2 \end{aligned} \quad (36)$$

where the initial values of the material coefficients are:

$$\mathbf{P} = \mathbf{0} \Rightarrow C_1 = C_2 = B_1 = D_1 = 0; \qquad B_2 = D_2 = -\tfrac{1}{3}; \qquad \lambda = F = 1 \quad (37)$$

so that for $\mathbf{P} = \mathbf{0}$, we obtain the initial isotropic von Mises criterion:

$$\mathbf{P} = \mathbf{0} \Rightarrow \operatorname{tr} \bar{\mathbf{T}}^2 = \operatorname{tr} \mathbf{S}^2 = 2k^2 \quad (38)$$

4.3.2. *Prager's Kinematic Hardening Rule*

We consider the following very simple case for the stress transformation (35):

$$\begin{aligned}&\lambda(\mathbf{P}) \equiv \mathbf{F}(\mathbf{P}) \equiv 1; \qquad B_1(\mathbf{P}) \equiv B_2(\mathbf{P}) \equiv D_1(\mathbf{P}) \equiv D_2(\mathbf{P}) \equiv 0;\\ &C_1(\mathbf{P}) = c\varepsilon_1; \qquad C_2(\mathbf{P}) = c\varepsilon_2 \Rightarrow \bar{\bar{\mathbf{T}}} = c\varepsilon_1\mathbf{M}_1 + c\varepsilon_2\mathbf{M}_2 + \mathbf{T}\end{aligned} \quad (39)$$

where c is a material constant.

Employing (39) in (36) and developing with respect to the principal stresses, the subsequent criterion takes the particular form:

$$2c(\varepsilon_1 \cos^2\theta + \varepsilon_2 \sin^2\theta)\sigma_1 + 2c(\varepsilon_1 \sin^2\theta + \varepsilon_2 \cos^2\theta)\sigma_2 + \sigma_1^2 + \sigma_2^2 = 2k^2 - c^2(\varepsilon_1^2 + \varepsilon_2^2) \quad (40)$$

where θ is the angle between the principal directions of **T** and **P**. Thus, we obtain Prager's kinematic hardening rule [18], which appears as a very simple particular case of the proposed general rule. We recall that by the kinematic hardening rule, the initially isotropic criterion is transformed into a particular orthotropic subsequent criterion (see Section 2.2).

Similar results are obtained for Ziegler's modification of Prager's hardening rule [25]. The Prager–Hodge hardening rule [13] is simply obtained by introducing $\lambda = \lambda(\varepsilon_1 + \varepsilon_2)$ in the restrictions (39), in order to add homothetic isotropic hardening to the kinematic hardening.

4.3.3. *Baltov–Sawczuk Hardening Rule*

Several proposed anisotropic hardening rules appear as special cases of the general rule (36) in plane stress. Consider for example the Baltov–Sawczuk hardening rule [2], which can be written in the following invariant three-dimensional form:

$$\operatorname{tr}(\mathbf{S} - c\mathbf{E})^2 + A \operatorname{tr}^2[\mathbf{E}(\mathbf{S} - c\mathbf{E})] = 2k^2 \tag{41}$$

where **S** and **E** are respectively the stress and plastic strain deviators; c and A are material constants.

The rule (41) involves a kinematic part, which is taken into account by the following stress transformation:

$$\mathbf{S} \Rightarrow \bar{\mathbf{S}} = \mathbf{S} - c\mathbf{E} \tag{42}$$

With respect to $\bar{\mathbf{S}}$, the rule (41) takes the invariant form:

$$\operatorname{tr} \bar{\mathbf{S}}^2 + A \operatorname{tr}^2 \bar{\mathbf{S}}\mathbf{E} = 2k^2 \tag{43}$$

Introducing the plastic strain principal values:

$$\begin{gathered} \mathbf{E} = \varepsilon_1 \mathbf{M}_1 + \varepsilon_2 \mathbf{M}_2 - (\varepsilon_1 + \varepsilon_2)\mathbf{M}_3 \\ \mathbf{M}_1 = \mathbf{v}_1 \otimes \mathbf{v}_1; \qquad \mathbf{M}_2 = \mathbf{v}_2 \otimes \mathbf{v}_2; \qquad \mathbf{M}_3 = \mathbf{v}_3 \otimes \mathbf{v}_3 \end{gathered} \tag{44}$$

where $(\mathbf{v}_1, \mathbf{v}_2, \mathbf{v}_3)$ are the principal directions of **E**, we obtain finally:

$$\begin{aligned} &2A(2\varepsilon_1 + \varepsilon_2)(\varepsilon_1 + 2\varepsilon_2) \operatorname{tr} \mathbf{M}_1\bar{\mathbf{S}} \operatorname{tr} \mathbf{M}_2\bar{\mathbf{S}} \\ &\quad + A(2\varepsilon_1 + \varepsilon_2)^2 \operatorname{tr}^2 \mathbf{M}_1\bar{\mathbf{S}} + A(\varepsilon_1 + 2\varepsilon_2)^2 \operatorname{tr} \mathbf{M}_2\bar{\mathbf{S}} + \operatorname{tr} \bar{\mathbf{S}}^2 = 2k^2 \end{aligned} \tag{45}$$

When applied to $\bar{\mathbf{S}}$, the proposed anisotropic hardening rule (36), with $C_1 \equiv C_2 \equiv 0$, can be written in the form:

$$2K_1 \operatorname{tr} \mathbf{M}_1\bar{\mathbf{S}} \operatorname{tr} \mathbf{M}_2\bar{\mathbf{S}} + K_2 \operatorname{tr}^2 \mathbf{M}_1\bar{\mathbf{S}} + K_3 \operatorname{tr}^2 \mathbf{M}_2\bar{\mathbf{S}} + K_4 \operatorname{tr} \bar{\mathbf{S}}^2 = 2k^2\lambda^2 \tag{46}$$

Identifying (45) with (46), we obtain:

$$\lambda \equiv K_4 \equiv 1; \qquad K_2 = (2\varepsilon_1 + \varepsilon_2)^2; \qquad K_3 = (\varepsilon_1 + 2\varepsilon_2)^2; \qquad K_1^2 \equiv K_2 K_3 \tag{47}$$

Thus, the Baltov–Sawczuk hardening rule appears as a special case of our proposed general rule in plane stress. The Baltov–Sawczuk rule is three-dimensional, but the involved transformation of stress has the tensorial form of a particular two-dimensional first-order polynomial transformation. A similar result is obtained for the anisotropic hardening rule proposed by Backhaus [1]. In the sense of the tensorial form of the involved stress transformation, the anisotropic hardening rules proposed by Yoshimura [24] and by Svensson [22] are more general.

4.3.4. *Initially Anisotropic Material*

Considering the procedure presented in Section 3.4, a general anisotropic hardening rule for initially anisotropic materials can be derived in a straightforward manner. In the case of an initially orthotropic material, such a procedure has already been developed and the hardening rule obtained has been tested by experimental results [10, 9]. Further details and other examples will be developed in ref. 7.

5. CONCLUSIONS

The proposed general concept results in a rational formulation of isotropic and anisotropic hardening. In the developed scheme, the generalized sense of isotropic hardening is clearly defined: a given initial isotropic criterion can be transformed into an arbitrary subsequent isotropic criterion, change of size, shape and position along the isotropic stress axis being possible for the yield surface; likewise, a given initial anisotropic criterion can be transformed into an arbitrary subsequent anisotropic criterion, but of the same type of anisotropy. All other hardening processes are involved in our general formulation of anisotropic hardening. Thus, the proposed concepts are definitely disjointed.

In the presented examples, the proposed general isotropic and anisotropic hardening rules are based on first-order polynomial isotropic and anisotropic transformations of the stress tensor and on the application of the developed procedure to the von Mises and linear criteria. Although these stress transformations and the applied criteria are very simple, the rules obtained are already general enough to be able to include the classical hardening rules and to generalize them for the description of hardening phenomena which they cannot take into account.

More general stress transformation rules and more elaborate criteria can be introduced in the proposed concept. But even for the simple scheme developed in this work, there exists a great lack of experimental results for testing fully the proposed isotropic and anisotropic hardening rules. Well organized multiaxial experiments, guided by a sound theoretical basis are needed.

The work presented shows the usefulness and the capacity of the theory of representation for tensor functions in the development of a rational and unified description of complex mechanical phenomena.

REFERENCES

1. Backhaus, G. Zur Fliessgrenze bei allgemeiner Verfestigung, *ZAMM*, **48** (1968), 99–108.
2. Baltov, A. and A. Sawczuk. A rule of anisotropic hardening, *Acta Mech.*, **1** (1965), 81–92.
3. Boehler, J. P. Contributions théoriques et expérimentales à l'étude des milieux plastiques anisotropes, Thèse de Doctorat ès-Sciences, Grenoble, 1975.
4. Boehler, J. P. Lois de comportement anisotrope des milieux continus, *J. de Mécanique*, **17** (2), (1978), 153–190.
5. Boehler, J. P. A simple derivation of representations for non-polynomial constitutive equations in some cases of anisotropy, *ZAMM*, **59** (1979), 157–167.
6. Boehler, J. P. On a general concept of isotropic and anisotropic hardening. Part I: theoretical (in preparation).
7. Boehler, J. P. On a general concept of isotropic and anisotropic hardening. Part II: examples (in preparation).
8. Boehler, J. P. and J. Raclin. Ecrouissage anisotrope des matériaux orthotropes prédéformés, *J. Mécanique Théorique et Appliquée*, Numéro Spécial, 1982, 23–44.
9. Boehler, J. P. and J. Raclin. Anisotropic hardening of prestrained rolled sheet-steel. In: *Current Advances in Mechanical Design and Production*, Proc. 2nd Cairo Un. MDP Conf., Cairo University, 1982, 483–492.
10. Boehler, J. P. and J. Raclin. A unified theoretical and experimental study of anisotropic hardening, *Trans. 6th Int. Conf. on SMIRT*, North-Holland, Amsterdam, 1982, Invited lecture L3/2.
11. Chaboche, J. L. Le concept de contrainte effective appliqué à l'élasticité

et à la viscosité en présence d'un endommagement anisotrope. In: *Mechanical Behavior of Anisotropic Solids, Proc. Euromech Coll. 115—CNRS Int. Coll.* 295 (Villard-de-Lans, June 1979), (Ed. J. P. Boehler), Editions du CNRS (Paris) and Martinus Nijhoff, The Hague, 1982, 737–760.

12. Hill, R. *The Mathematical Theory of Plasticity*, Oxford University Press, Oxford, 1950.
13. Hodge, P. G. Discussion on W. Prager paper 'A new method of analyzing stresses and strains in work-hardening plastic solids', *J. appl. Mech.*, **24** (1957), 482–483.
14. Kachanov, L. M. Time of the rupture process under creep conditions, *Izv. Akad. Nauk SSR Otd. Tekh. Nauk*, **8** (1958), 26–31 (in Russian).
15. Lemaitre, J. and J. L. Chaboche. Aspect phénoménologique de la rupture par endommagement, *J. Mécanique Appliquée*, **2** (3), 1978, 317–365.
16. Murakami, S. and N. Ohno. A continuum theory of creep and creep damage. In: *Creep in Structures, Proc. IUTAM Symp.* (Leicester, Sept. 1980), (Ed. A. R. S. Ponter and D. R. Hayhurst), Springer, Berlin, 1981, 422–444.
17. Murakami, S. and A. Sawczuk. On description of rate-independent behaviour for prestrained solids, *Arch. Mech.*, **32** (1979), 251–264.
18. Prager, W. A new method of analysing stresses and strains in work-hardening plastic solids, *J. appl. Mech.*, **23** (4), (1956), 493–496.
19. Rees, D. W. A. A survey of hardening in metallic materials. In: *Failure Criteria of Structured Media, Proc. CNRS Int. Coll. 351* (Villard-de-Lans, June 1983), in press.
20. Rivlin, R. S. Further remarks on the stress-deformation relations for isotropic materials, *J. Rat. Mech. Anal.*, **4** (1955), 681–702.
21. Spencer, A. J. M. The formulation of constitutive equations for anisotropic solids. In: *Mechanical Behavior of Anisotropic Solids, Proc. Euromech. Coll. 115-CNRS Int. Coll. 295* (Villard-de-Lans, June 1979), (Ed. J. P. Boehler), Editions du CNRS (Paris) and Martinus Nijhoff, The Hague, 1982, 3–26.
22. Svensson, N. L. Anisotropy and the Bauschinger effect in cold rolled aluminium, *J. Mech. Engng Sci.*, **8** (1966), 162–172.
23. Touati, A. Comportement mécanique des sols pulvérulents sous fortes contraintes, Thèse de Docteur-Ingénieur, Ecole Nationale des Ponts et Chaussées, Paris, 1982.
24. Yoshimura, Y. Hypothetical theory of anisotropy and the Bauschinger effect due to plastic strain history, Aero. Res. Inst., Tokyo University, **349** (1959), 224–246.
25. Ziegler, H. A modification to Prager's hardening rule, *Q. J. appl. Math.*, **17** (55), (1959), 55–65.

28

New Aspects of the Anisotropic Hardening Rule

A. BALTOV

Institute of Mechanics and Biomechanics, Sofia, Bulgaria

ABSTRACT

Quasidynamic processes of advanced plastic deformation in metals are considered. The strain rates are in the interval $10^{-6} - 1\ s^{-1}$. *The deformational anisotropy and the sensitivity of the conditions of the deformation process are taken into account. A theoretical model of anisotropic hardening for the case of large elasto–plastic strains is proposed. The mixed system of description is adopted. Material parameters of the model are functions of the factors of the conditions of plastic deformation, strain rates and temperature. As a special case a nonsymmetric yield surface is considered. An example is given to illustrate the model. In the case of fixed (during the process) principal axes of the Cauchy stress tensor an equivalent stress is introduced. A condition of micronecking initiation is proposed and the transition to fracture is discussed.*

1. INTRODUCTION

Anisotropic hardening of metals has been a subject of interest for a long time. This effect is experimentally very well investigated [3, 12, 19, 22]. Several suggestions exist concerning its theoretical description [2, 5, 8, 14, 15]. The approach considering changes of a yield surface in the stress space is very often used. In spite of the results obtained in such a way however, there are some problems of interest being studied nowadays by many authors:

(1) influence of the factors specifying the process on the deformational anisotropy [3, 5];
(2) nonsymmetry of the yield surface after partial unloading [7, 19, 21];
(3) large plastic strains [8, 9, 11, 15, 20];
(4) characteristic features of the transition from plastic deformation with anisotropic hardening to fracture [1, 6, 16, 17, 18].

The interest in the above problems is connected with the optimal design of metal forming processes.

The present paper investigates processes of advanced plastic deformation, anisotropic hardening of the material sensitivity on the metal forming procedures. A model is proposed in which the Baltov–Sawczuk anisotropic hardening rule [2] is generalized in such a way as to be able to describe the effects mentioned above. Of considerable concern are processes in which the parameters specifying conditions of the process in question are as follows: plastic deformation, strain rate and temperature. The strain rate is in the interval 10^{-6}–$1\,s^{-1}$. This enables us to consider the process as a quasidynamic one, i.e. the inertia effects are neglected and the strain rate sensitivity is taken into account by the dependence of the mechanical characteristics of the material on the parameters of the process. The material is considered to be plastically incompressible. The plastic strains are large and the elastic ones are small. The temperatures are moderate.

2. ANISOTROPIC STRAIN HARDENING

The basic problem in modelling of processes associated with large strains is the choice of a system of reference—material, spacial, mixed, etc. [23]. The material description is suitable when large deformation of structural elements is investigated, because the well known undeformed configuration of the element is used. The separation of the geometrical effects from the physical ones is easier in this description. The disadvantage is that the material measures of strain are not directly determined from the experimental data. The spacial description is suitable for investigation on plastic flow in some technological processes (rolling, extrusion, etc.). Spacial measures could be directly obtained experimentally. It is, however, more difficult to separate the geometrical from the physical effects. The mixed description has the advantages of the material one in referring the measures to the initial undeformed configuration and of the spacial one in the direct recording of the mixed measures during the experiment. Utilisation of the double tensor measures leads to some disadvantages. The mixed description is convenient for modelling since the mixed measures are easily expressed through all kinds of usual experiments (one-dimensional tension, thin-walled tubes under internal pressure, tension or torsion, etc.). This enables better specification of the basic assumptions of the model [9, 10]. This is the reason for choosing the mixed description in the paper.

The model is based on the following assumptions.

2.1. Description of the Process

A material Cartesian co-ordinate system $O_M X_1 X_2 X_3$ and a spacial Cartesian co-ordinate system $O_S x_1 x_2 x_3$ are introduced. The measures of the process are:

(a) the first Piola–Kirchhoff stress tensor T_{iK} $(i = 1, 2, 3;\ K = 1, 2, 3)$, as the measure of a stress state;

(b) the displacement gradient $H_{iK} = U_{i,K}$, as the measure of a strain state. U_i is the displacement vector;

(c) the absolute temperature θ and the specific entropy η, as measures of the thermal state;

(d) the microstress tensor T^{μ}_{iK} and the plastic part of the displacement gradient $H^{(p)}_{iK}$, as the measure of the internal structural state.

The double tensor T^{μ}_{iK} represents the microstresses occurring in the material due to plastic deformations on the microlevel [5]. The tensor $H^{(p)}_{iK} = H_{iK} - H^{(e)}_{iK}$ is the residual part of the displacement gradient, and can be determined experimentally. $H^{(e)}_{iK}$ is the corresponding elastic part which is small and coincides with the elastic strain tensor. All the measures are functions of the material co-ordinates (X_K) and of the time t. At the initial moment $t = t_0$ the body occupies the region $\Omega_0 \subset E_3$ (the three-dimensional Euclidean space). The initial material density is ρ_0.

2.2. Description of the Process Conditions

The following scalar measures of the process conditions are involved:

(a) measure of the plastic deformation:

$$\gamma = \sqrt{\tfrac{1}{2} H^{(p)}_{iK} H^{(p)}_{iK}}$$

(b) measure of the strain rate:

$$\beta = \sqrt{\tfrac{3}{2} V_{iK} V_{iK}}, \qquad V_{iK} = \dot{H}_{iK}$$

(c) temperature measure:

$$\Delta\theta = \theta - \theta_0$$

where θ_0 is the initial temperature in point (X_K).

We introduce the arranged sequence $\pi = \{\gamma, \beta, \Delta\theta\}$. The behaviour of the material depends on the combination of the values of the scalar parameters involved.

2.3. Behaviour of the Material During the Process

(a) A material is assumed to be plastically incompressible. It is approximately expressed by the relation $\delta_{iK} H_{iK}^{(p)} = 0$, where δ_{iK} is the shifter if one refers both x and X to the same Cartesian system of reference.

(b) The displacement gradient is assumed to consist of an elastic and a plastic part [9]. The plastic strain in a material description is:

$$E_{KL}^{(p)} = \tfrac{1}{2}(\delta_{jK} H_{jL}^{(p)} + \delta_{jL} H_{jK}^{(p)} + H_{jK}^{(p)} H_{jL}^{(p)}) \tag{1}$$

(c) A yield condition $F = 0$ exists. It depends on the stress T_{iK}, the microstress T_{iK}^{μ} and on the parameters of the process conditions π, i.e. $F = F(T_{iK}, T_{iK}^{\mu}, \pi)$.

(d) Plastic properties of the material are independent of the hydrostatic pressure $P = \delta_{iK} T_{iK}$. It is assumed that they also do not depend on $P^{\mu} = \delta_{iK} T_{iK}^{\mu}$.

(e) The rate of $H_{iK}^{(p)}$ is given by the plastic flow law:

$$V_{iK}^{(p)} = \mathring{\Lambda} F_{iK}, \qquad F_{iK} = \frac{\partial F}{\partial T_{iK}} \tag{2}$$

$$\mathring{\Lambda} = \begin{cases} 0 & \text{if} \quad F < 0 \quad \text{or} \quad F = 0 \quad \text{but} \quad L \leqslant 0, \\ > 0 & \text{if} \quad F = 0 \quad \text{but} \quad L > 0. \end{cases}$$

$\mathring{\Lambda}$ is obtained by the consistency condition $\mathring{F} = 0$. The loading function L in (2) is:

$$L = F_{iK} \mathring{T}_{iK} + \frac{\partial F}{\partial \beta} \mathring{\beta} + \frac{\partial F}{\partial \theta} \mathring{\theta} \tag{3}$$

(f) The rate of $H_{iK}^{(e)}$ is given by Hooke's law.

(g) The deviatoric part $S^{\mu}_{iK} = T^{\mu}_{iK} - \frac{1}{3}\delta_{iK}P^{\mu}$ is significant when describing the plastic properties. Microstresses develop during the plastic deformation [5]. Their rate depends on the rate of the parameter π, specifying the process conditions:

$$\overset{\circ}{S}{}^{\mu}_{iK} = H^{(p)}_{iK}(R \,.\, \overset{\circ}{\gamma} + P \,.\, \overset{\circ}{\theta} + D \,.\, \overset{\circ}{\beta}) + Q \,.\, V^{(p)}_{iK} \tag{4}$$

where R, P, D and Q are material functions, depending on π.

Let us consider the case of an initially isotropic material, which due to plastic deformations obtains deformational anisotropy of the plastic properties. Taking into account the assumption (d) F will depend on $S_{iK} = T_{iK} - \frac{1}{3}\delta_{iK}P$, S^{μ}_{iK} and on π. F should be an isotropic tensor scalar-valued function of the mixed tensors S_{iK} and S^{μ}_{iK}, hence it depends on their invariants: $\gamma_{ss} = S_{iK}S_{iK}$, $\gamma_{s\mu} = S_{iK}S^{\mu}_{iK}$, $\gamma_{\mu\mu} = S^{\mu}_{iK}S^{\mu}_{iK}$, i.e. $F = F(\gamma_{ss}, \gamma_{s\mu}, \gamma_{\mu\mu}, \pi)$.

Experimental data [3, 12, 19, 22] show that a quadratic form may be used for F:

$$F \equiv \tfrac{1}{2}\gamma_{\bar{s}\bar{s}} + \tfrac{1}{2}A \,.\, \gamma^2_{\bar{s}\mu} - \varphi^2 = 0 \tag{5}$$

where

$$\gamma_{\bar{s}\bar{s}} = \gamma_{ss} - 2\gamma_{s\mu} + \gamma_{\mu\mu} = \bar{S}_{iK}\bar{S}_{iK}$$
$$\gamma_{\bar{s}\mu} = \gamma_{ss} - \gamma_{s\mu} = \bar{S}_{iK}S^{\mu}_{iK}, \qquad \bar{S}_{iK} = S_{iK} - S^{\mu}_{iK}$$
$$\bar{S}_{iK} = \bar{T}_{iK} - \tfrac{1}{3}\delta_{iK}\bar{P}, \qquad \bar{P} = \delta_{iL}\bar{T}_{iL}$$

$\bar{T}_{iK}$ is the active stress tensor [5]. A and φ are the material functions, depending on π.

In the case of partial unloading, a nonsymmetry of the yield surface $F=0$ appears in the space $\{S_{iK}\}$. Moreover φ must depend also on the scalar χ [7], i.e. $\varphi = \varphi^*(\chi, \pi)$, where χ is defined as

$$\chi = \frac{\gamma_{\bar{s}\mu}}{\sqrt{\gamma_{\bar{s}\bar{s}} \cdot \gamma_{\mu\mu}}} \tag{6}$$

Taking into account (5), the expression for F_{iK} becomes

$$F_{iK} = \bar{S}_{iK} + A \,.\, \gamma_{\bar{s}\mu} \,.\, S^{\mu}_{iK} - \frac{\varphi \cdot \dfrac{\partial\varphi}{\partial\chi}}{\sqrt{\gamma_{\bar{s}\bar{s}}}\sqrt{\gamma_{\mu\mu}}} \cdot \left(S^{\mu}_{iK} - \frac{\gamma_{\bar{s}\mu}}{\gamma_{\bar{s}\bar{s}}}\bar{S}_{iK}\right) \tag{7}$$

When total unloading occurs, in the case of symmetry we have $(\partial\varphi/\partial\chi) \equiv 0$ and the last term in (7) vanishes.

In the case when $\theta = \text{const.}$ and $\beta = \text{const.}$ during the process, the loading function L is given by the following expression:

$$L = F_{iK}\dot{\bar{S}}_{iK} \tag{8}$$

while

$$\mathring{\Lambda} = h \,.\, L \tag{9}$$

where

$$h = \left| R\sqrt{\tfrac{1}{2}F_{iK}F_{iK}} \,.\, M_{iK}H^{(p)}_{iK} + QM_{jL}F_{jL} + \frac{\partial F}{\partial \gamma}\sqrt{\tfrac{1}{2}F_{iK}F_{iK}} \right|^{-1}$$

$$M_{iK} = \frac{\partial F}{\partial S^{\mu}_{iK}} \tag{10}$$

We involve an eight-dimensional space, by analogy with the five-dimensional space proposed by Ilyushin [13] in order to give a better geometrical interpretation to the condition (5) and to obtain a good approximation of the experimental data:

$$\begin{aligned} &S_1 = \sqrt{\tfrac{3}{2}}S_{x_1X_1}, \quad S_2 = \sqrt{2}(S_{x_2X_2} + \tfrac{1}{2}S_{x_1X_1}), \quad S_3 = S_{x_1X_2} \\ &S_4 = S_{x_2X_1}, \quad S_5 = S_{x_1X_3}, \quad S_6 = S_{x_3X_1} \\ &S_7 = S_{x_2X_3}, \quad S_8 = S_{x_3X_2} \end{aligned} \tag{11}$$

According to (11) the invariants involved take the form:

$$\gamma_{ss} = S_\alpha S_\alpha, \qquad \gamma_{s\mu} = S_\alpha S^{\mu}_{\alpha}, \qquad \gamma_{\mu\mu} = S^{\mu}_{\alpha}S^{\mu}_{\alpha}, \qquad (\alpha = 1, \ldots, 8) \tag{12}$$

The yield condition (5) transforms therefore into the equation of a hypersurface in the space $\{S_\alpha\}$:

$$\mathscr{A}_{\alpha\beta}\bar{S}_\alpha\bar{S}_\beta + C = 0 \tag{13}$$

where

$$\begin{aligned} &\mathscr{A}_{\alpha\beta} = \tfrac{1}{2}(\delta_{\alpha\beta} + AS^{\mu}_{\alpha}S^{\mu}_{\beta}), \qquad (\alpha, \beta = 1, 2, \ldots, 8) \\ &C = -\varphi^2, \qquad \bar{S}_\alpha = S_\alpha - S^{\mu}_{\alpha} \end{aligned} \tag{14}$$

and $\delta_{\alpha\beta}$ is the Kronecker delta. If $\varphi = \varphi(\pi)$, eqn. (13) represents a central second-order hypersurface in the space $\{\bar{S}_\alpha\}$. The material functions have to ensure therefore that this hypersurface is a hyperellipsoid, in order to represent a yield condition. This leads to the requirement that the matrix $\|\mathscr{A}_{\alpha\beta}\|$ does not degenerate and should be positive definite, i.e. $\|\mathscr{A}_{\alpha\beta}\|$ should have eight positive eigenvalues

$\lambda_\alpha > 0$, $(\alpha = 1, 2, \ldots, 8)$. If $\varphi = \varphi^*(\pi, \chi)$ we assume the following relations:

$$\varphi \equiv \begin{cases} \varphi_1(\pi) \quad \text{at} \quad \chi_1 \leqslant \chi \leqslant 2 \qquad \text{(I region } \Omega_1) \\ \varphi_2 \equiv (1-\bar{\chi})^2(2\bar{\chi}-1)\varphi_1(\pi) + \bar{\chi}^2(3-2\chi)\varphi_3(\pi) \\ \qquad\qquad\qquad \text{at} \quad \chi_2 \leqslant \chi \leqslant \chi_1, \qquad \text{(II region } \Omega_2), \\ \varphi_3(\pi) \quad \text{at} \quad -2 \leqslant \chi \leqslant \chi_2, \qquad \text{(III region } \Omega_3) \end{cases} \tag{15}$$

$$\bar{\chi} = (\chi - \chi_1)/(\chi_2 - \chi_1)$$

In the first and third regions we obtain hyperellipsoids, and in the second region a transition hypersurface. χ_1 and χ_2 define the boundaries of the regions. Thus, by means of (15), a closed complex and smooth hypersurface is constructed in $\{\bar{S}_\alpha\}$. The smoothness is ensured by the conditions:

$$\left.\frac{\partial \varphi}{\partial \chi}\right|_{\chi=\chi_1} = \left.\frac{\partial \varphi}{\partial \chi}\right|_{\chi=\chi_2} = 0 \tag{16}$$

These conditions are fulfilled if φ has the form as given in (15) [7]. The following expression:

$$\chi = \frac{\bar{S}_\alpha S^\mu_\alpha}{\sqrt{\bar{S}_\beta \bar{S}_\beta}\sqrt{S^\mu_\gamma S^\mu_\gamma}}, \qquad (\alpha, \beta, \gamma = 1, 2, \ldots, 8) \tag{17}$$

specifies χ. Obviously, χ represents the angle between the vectors S^μ_α and $\bar{S}_\alpha$. S^μ_α represents the translation of the centre of the yield surface in $\{S_\alpha\}$, due to anisotropic strain hardening. $\bar{S}_\alpha$ represents the effect of the active part of the stresses; it is obtained when subtracting the microstress. The experimental data according to ref. 19 are compared with the yield surface according to the theoretical model and this comparison is illustrated in Fig. 1. Thin-walled aluminium tubes under torsion and tension were used. The figure shows the projection of the yield surface on the plane (ξ_1, ξ_2), where:

$$\begin{aligned} \xi_1 &= 0{\cdot}0086 S_1 + 0{\cdot}0149 S_2 - 64 \\ \xi_2 &= -0{\cdot}0073 S_1 + 0{\cdot}0042 S_2 \end{aligned} \tag{18}$$

S_1 and S_2 are given in the units of N/cm^2. $H^{(p)}_{x_1x_1} = 0{\cdot}1703\%$, $H^{(p)}_{x_1x_2} = 0{\cdot}4884\%$, $S^\mu_1 = 1910$ N/cm^2, $S^\mu_2 = 3300$ N/cm^2, $\chi_1 = 0{\cdot}201$, $\chi_2 = -0{\cdot}815$, $A = 0{\cdot}351 \times 10^{-6}$ (N/cm^2)$^{-2}$, $\varphi_1 = 763$ N/cm^2, $\varphi_3 = 1310$ N/cm^2.

The comparison shows that the proposed model approximates successfully the experimental data. The material functions of the model can be determined if experiments at different π are performed.

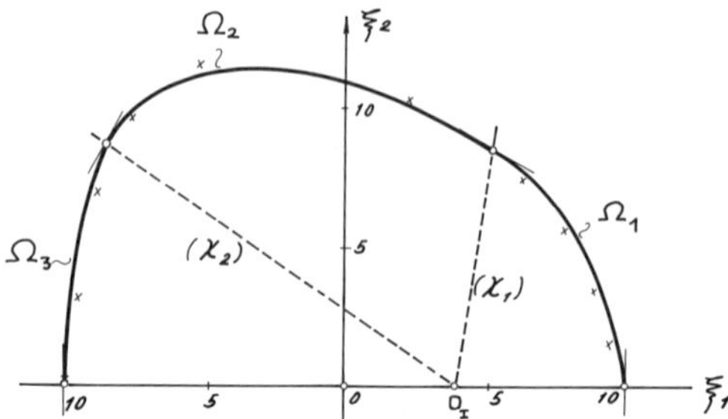

FIG. 1. Subsequent yield surface. Comparison between theoretical approximation and experimental data according to ref. 19.

It is interesting to note that the mixed description enables us to arrive at a direct approximation of the experimental data. If the transformation to material or spacial reference systems is carried out, rather complicated expressions for the yield condition are obtained.

3. EQUIVALENT STRESS

Let us consider now a quasidynamic process in a body, the principal axes of the Cauchy stress tensor of which do not change during the process. For the regions of the body in which at least one of the principal stresses is tensile, an equivalent stress $T_{(eq)}$ can be introduced. In a three-dimensional case the $T_{(eq)}$ is related to the plastic strain intensity γ similarly, as $T_{x_1X_1}$ and $H^{(p)}_{x_1X_1}$ in the case of one-dimensional tension. The equivalent stress is defined so that the yield condition (5) is in the following form:

$$T_{(eq)} = \Omega_{(p)}, \qquad \Omega_{(p)} = \sqrt{3}\varphi \tag{19}$$

and then

$$T_{(eq)} = \sqrt{\tfrac{3}{2} N_{iKjL} \bar{S}_{iK} \bar{S}_{jL}} \tag{20}$$

where

$$N_{iKjL} = \tfrac{1}{2}(\delta_{iK}\delta_{jL} + \delta_{iL}\delta_{jK} + S^{\mu}_{iK}S^{\mu}_{jL} + S^{\mu}_{iL}S^{\mu}_{jK}) \tag{21}$$

The tensile stress $\Omega_{(B)}$ at which necking begins in the specimen is of special interest. This is a critical state at which the material loses its

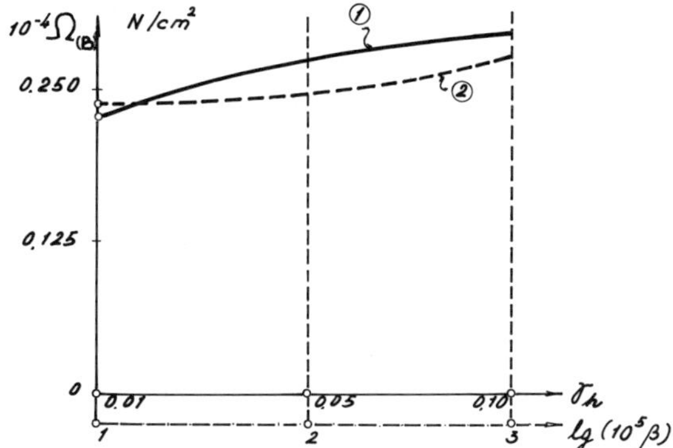

FIG. 2. (1) Relation between $\Omega_{(B)}$ and γ_h, at $\beta_h = 5\times10^{-3}\,s^{-1}$ and $\beta = 5\times10^{-3}\,s^{-1}$. (2) Relation between $\Omega_{(B)}$ and β, at $\beta_h = 5\times10^{-3}\,s^{-1}$ and $\gamma_h = 0{\cdot}04$.

internal plastic stability [6]. In the three-dimensional case, to this state corresponds a micronecking initiation in the material point neighbourhood. The condition for reaching this critical state is:

$$T_{(eq)} = \Omega_{(B)} \tag{22}$$

The corresponding material functions $A = A_{(n)}$ and $\varphi = \varphi_{(n)} = \Omega_{(B)}/\sqrt{3}$ depend on the process conditions $\pi_{(n)} \equiv \{\gamma_{(n)}, \beta, \Delta\theta\}$, where $\gamma_{(n)}$ is the value of γ from relation $T_{(eq)} - \gamma$ at $T_{(eq)} = \Omega_{(B)}$. $\gamma_{(n)}$ depends on β and $\Delta\theta$. Then $A_{(n)} = A_{(n)}(\beta, \Delta\theta)$ and $\varphi_{(n)} = \varphi_{(n)}(\beta, \Delta\theta)$. The relation $T_{(eq)} - \gamma$ is valid up to value $\gamma = \gamma_{(n)}$, as after necking the specimen cannot be used any more as a model of the infinitesimal neighbourhood of the material point.

Figure 2 illustrates the relations $\Omega_{(B)} - \gamma_h$ (at β_h and β_{fixed}) and $\Omega_{(B)} - \beta$ (at β_h and γ_{hfixed}) in the case of two successive processes, according to experimental data for steel St-3 [6]. γ_h and β_h are process condition parameters for the preliminary process, and β for the actual process.

4. CONCLUDING REMARKS

The mixed description of advanced plastic deformations has certain advantages in modelling of the material behaviour as it allows for a direct use of experimental results. For example, the direct postulation of a kinematic modelling in a material or in spacial description is not confirmed experimentally. It is more convenient to use a mixed description, when $F \equiv \frac{1}{2}\bar{S}_{iK}\bar{S}_{iK} - \varphi^2 = 0$. By an appropriate transformation its form can be obtained in any desired reference system.

With yield condition (5) there is a good possibility for approximate experimental data in the case of large plastic deformations of anisotropic hardening, and of sensitivity to the conditions of the forming process.

When transition to fracture is studied, it is appropriate to consider the micronecking initiation and nucleation together with microfissures and void initiation and growth [1]. The choice of measures for these micromechanisms is a difficult problem. The concepts applied to microfracture [16, 17, 18] may be used to describe the micronecking. This results in introducing a scalar measure for micronecking density or, by analogy with the microdamage tensors, a tensor measure.

REFERENCES

1. Ashby, H. F., C. Gandhi and D. M. R. Taplin. Fracture mechanism maps and their construction for F.C.C. metals and alloys, *Acta Met.*, **27** (1979), 699–729.
2. Baltov, A. and A. Sawczuk. A rule of anisotropic hardening, *Acta Mech.*, **I/2** (1965), 81–92.
3. Baltov, A., N. Boncheva, R. Kazandjiev, I. Radovanov and S. Vodenicharov. On the theoretical and experimental description of the mechanical behaviour of metals during metal forming processes. In: *Metal Forming Plasticity* (Ed. H. Lippmann), Springer, Berlin, 1979, 14–26.
4. Baltov, A. Modelling of inelastic deformation and microfracture processes in ductile materials. In: *Mechanics* (Ed. G. Brankov), Izd. BAN, Sofia, 1981, 531–536.
5. Baltov, A. and N. Boncheva. A rule of anisotropic hardening with taking into account preliminary and actual plastic deformations and strain rates. In: *Comportement Mecanique des Solides Anisotropes* (Ed. J. P. Boehler), Nijhoff, The Hague, 1982, 243–256.
6. Baltov, A. The effect of anisotropic hardening and strain rate on the plastic fracture of metals. In: *Le Comportement Plastique des Solides Anisotropes* (Ed. J. P. Boehler), Editions du CNRS (in press).
7. Boncheva, N., A. Baltov and S. Todorov. A rule of nonsymmetric anisotropic hardening for composite materials, *ZAMM*, **64** (1983), (in press).
8. Dafalias, Y. F. Lagrangian and Eulerian description of plastic anisotropy at large strains. In: *Le Comportement Plastique des Solides Anisotropes* (Ed. J. P. Boehler), Editions du CNRS (in press).

9. Duszek, M. K. Problems of geometrical nonlinear theory of plasticity, *Mitt. Inst. Mech.*, Ruhr University, Bochum, **21** (1980), 1–103.
10. Green, A. E. and J. E. Adkins. *Large Elastic Deformations and Non-linear Continuum Mechanics*, Clarendon Press, Oxford, 1960.
11. Green, A. E. and P. M. Naghdi. A general theory of an elastic–plastic continuum, *Arch. Rat. Mech. Anal.*, **18** (1965), 19–25.
12. Ikegami, K. A historical perspective of the experimental study of subsequent yield surfaces for metals, BISI 14420 (1976), 1–33.
13. Ilyushin, A. A. *Plasticity*, Nauka, Moscow, 1963 (in Russian).
14. Kolarov, D., A. Baltov and N. Boncheva. *Mechanics of Deformable Media*, Mir, Moscow, 1975 (in Russian).
15. Lehmann, Th. On concept of stress–strain relation in plasticity, *Acta Mech.*, **42** (1982), 263–275.
16. Lemaitre, J. and J.-L. Chaboche. Aspect phénomenologique de la rupture par endommagement, *J. Mec. appl.*, **2** (1978), 317–365.
17. Murakami, S. and N. Ohno. A continuum theory of creep and creep damage. In: *Creep in Structures* (Ed. A. R. S. Ponter and D. R. Hayhurst), Springer, Berlin, 1981, 422–444.
18. Nemat-Nasser, S., M. M. Memrabadi and T. Iwakuma. On certain macroscopic and microscopic aspects of plastic flow of ductile materials. In: *Three-dimensional Constitutive Relations and Ductile Fracture* (Ed. S. Nemat-Nasser), North-Holland, Amsterdam, 1981, 157–172.
19. Phillips, A. *The Foundation of Plasticity*, CISM, Lecture Notes, Udine, 1974.
20. Sidoroff, F. Incremental constitutive equation for large strain elasto-plasticity, *Int. J. Engng Sci.*, **20** (1982), 19–26.
21. Shiratori, E., K. Ikegami and F. Yochida. Analysis of stress–strain relations by use of an anisotropic hardening plastic potential, *J. Mech. Phys. Solids*, **27** (1979), 213–229.
22. Szczepinski, W. *Introduction to the Mechanics of Metals*, Nijhoff, The Hague, 1979.
23. Truesdell, C. and R. Toupin. The non-linear field theories of mechanics. In: *Handbuch der Physik*, **III/1**. Springer, Berlin, 1960, 226–793.

29

Shape Memory as a Thermally Activated Process

MANFRED ACHENBACH and INGO MÜLLER

Hermann-Föttinger-Institut, Technische Universität, Berlin, Federal Republic of Germany

ABSTRACT

Materials with shape memory and pseudoelasticity at low temperatures exhibit many features that are common also in plasticity, viz. yield, residual deformation and temperature dependence of yield limit. At the same time, at intermediate and high temperatures these materials are elastic, but exhibit a hysteresis. There is strong evidence that all these phenomena are macroscopic indications of an austenitic–martensitic phase transition and of twin formation within the martensite. In this paper a model is described that is capable of simulating the behaviour of memory alloys under uniaxial dynamic loading and variable external temperature. The model is exploited by the use of methods common in statistical mechanics and it explores both yielding and recovery as thermally activated processes.

1. PHENOMENA

The most striking phenomenon in a memory alloy is, of course, the memory: if a wire made of Nitinol (say) is bent at room temperature, it stays bent after the removal of the bending force; this is not different from any other wire which can be deformed plastically. However when the temperature is raised to about 60°C, the wire 'remembers' its original configuration and returns to it.

A fairly comprehensive account of the observed properties of memory alloys can be found in ref. 3. Generally speaking, a memory alloy is

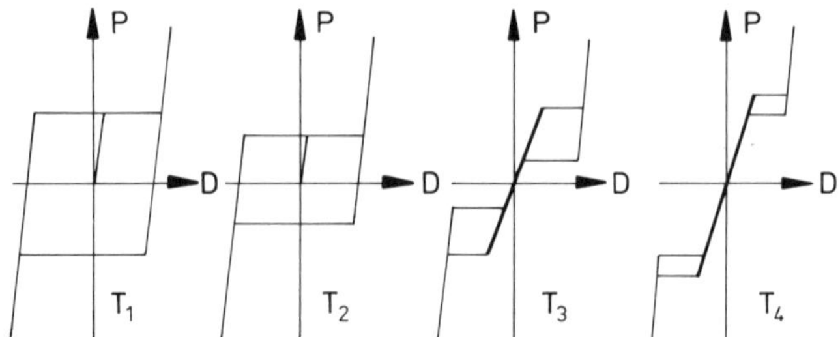

FIG. 1. Observed load–deformation curves for constant temperatures.

characterized by a strong dependence of the shape of the load–deformation curves upon temperature (see the schematic curves of Fig. 1).

At low temperatures the alloy behaves much like a plastic body with original elastic deformation at small loads, yield and creep at higher loads and residual deformation after unloading. But, unlike a plastic body, after yielding about 8% the alloy can be loaded elastically far beyond the yield load. All this is shown in the first two diagrams of Fig. 1 which also indicate that the yield load drops at an increase of temperature.

At higher temperatures the alloy exhibits pseudoelasticity. It allows only one configuration in the unloaded state and it returns to this deformation after loading and unloading. In this respect the alloy behaves elastically; it is called *pseudo*elastic, however, because the load–deformation diagrams contain a hysteresis between a yield load and a recovery load. The last two diagrams of Fig. 1 illustrate this pseudoelastic behaviour for two temperatures. The slope of the initial elastic line changes with temperature and so do yield load and recovery load. The hysteresis becomes smaller as the temperature increases.

It is clear how the diagrams of Fig. 1 show the memory effect. Indeed, if the body is loaded beyond the yield load at low temperature and then unloaded, there remains a residual deformation. After heating, however, the only configuration which the unloaded body can attain is the undeformed one and therefore the body returns to it.

While the diagrams of Fig. 1 allow us to form fairly detailed conclusions about the deformation resulting from isothermal loading and unloading of memory alloys at various temperatures, they provide incomplete information about processes in which both load and temp-

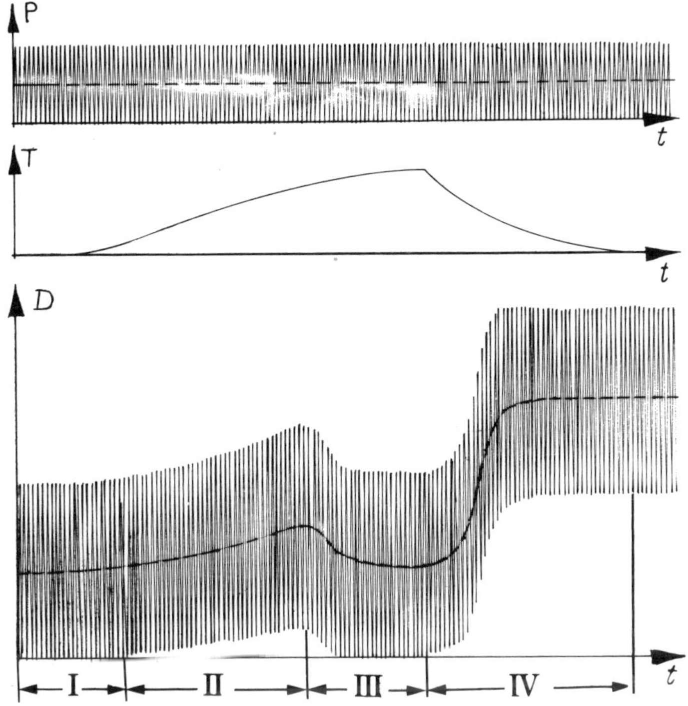

FIG. 2. Observed process under oscillating load and variable temperature [4].

erature change. As a prototype of such a process we consider the one shown in Fig. 2, where an oscillating tensile load—less than the low-temperature yield load—is applied while the temperature first increases in time and then decreases. The resulting measured deformation is shown in the lower part of Fig. 2. First the deformation oscillates along with the load, but as the temperature increases there is some residual deformation which slowly increases. At a higher temperature, however, the residual deformation vanishes and again the deformation oscillates along with the load albeit with a different amplitude than before. A sizeable increase in deformation and a subsequent oscillation about a large mean value occur when the temperature decreases from its maximum value. The dotted lines in Fig. 2 represent mean values of load and deformation.

The purpose of this paper is the qualitative understanding of all these phenomena by the description and exploitation of a model that is capable of simulating the behaviour of memory alloys. This model has grown out of the consideration of simpler models that have been proposed previously in refs. 5 and 1, where the ideas about the significance of thermal fluctuations and thermal activation were first applied to memory alloys. The key to the formulation of the model is the observation that at high temperature the unloaded memory material has a highly symmetric, austenitic lattice structure, while at low temperature it is martensitic and generally twinned. More specifically, along the thick lines in Fig. 1 the body is austenitic while it is martensitic along the thin lines. Within the hysteresis loops we have martensitic twins in different proportions at low temperatures and a mixture of austenite and one martensitic twin at high temperature.

2. DESCRIPTION OF THE MODEL

2.1. Basic Element: Lattice Particle

The basic element of the model is a lattice particle, i.e. a small piece of the metallic lattice, of which the Figs. 3 and 4 in their lower parts show three equilibrium configurations that belong to the austenitic phase A and to the martensitic twins M_+ and M_-. The particles can only deform in shear and can be considered as sheared versions of A. In this case the shear lengths are denoted by $\pm J$. Intermediate shear lengths Δ are also possible and the upper part of Fig. 3 gives the postulated form of the potential energy corresponding to such values.

According to this potential the martensitic particles may assume a stable equilibrium, while the equilibrium of the austenitic particles is

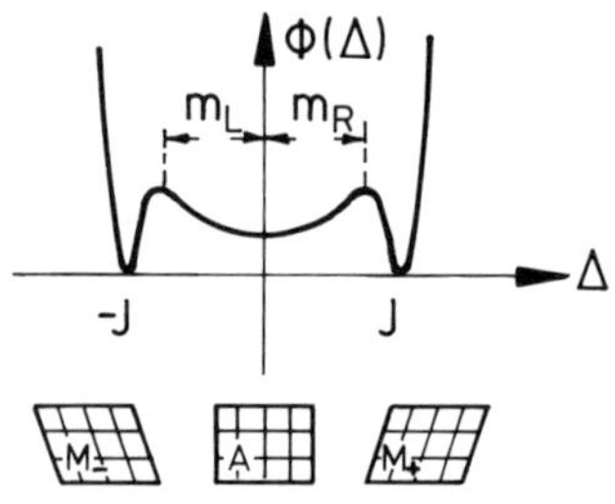

FIG. 3. Unloaded lattice particles and their potential energy.

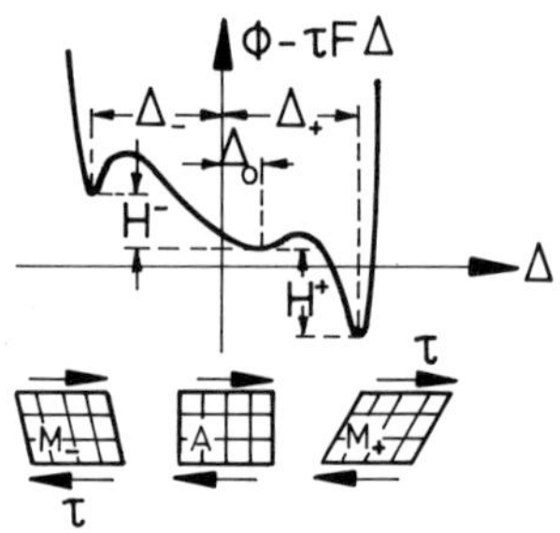

FIG. 4. Loaded lattice particles and their potential energy.

only metastable. The various equilibria are separated by potential barriers at $\Delta = m_L$ and $\Delta = m_R$.

If a lattice particle is subjected to a shear load τF, the potential energy $-\tau F \Delta$ of that load must be added to the potential energy $\Phi(\Delta)$ of the unloaded particle. τ is the shear stress and F is the area of the upper or lower surface of the particle. As a result the potential energy

$$\Phi(\Delta; \tau) = \Phi(\Delta) - \tau F \Delta \tag{1}$$

has the shape drawn in Fig. 4. One of the martensitic minima becomes much deeper than the other one and the barrier between the different potential wells may change considerably depending on the load.

2.2. The Body: A Stack of Layers of Lattice Particles

In a body the lattice particles form layers and those layers are stacked on top of each other in the manner shown in Fig. 5(a). The layers can only be sheared as a whole and their mass will be denoted by m. All loads that will be considered are in the vertical direction and therefore the indicated arrangement of the layers at 45° coincides with the direction of maximum shear stress. The shear stress τ on a lattice particle resulting from a vertical load P is given by

$$\tau = \frac{1}{\sqrt{2}} \cdot \frac{P}{NF} \tag{2}$$

where N is the number of particles in a layer and F is the area of the upper and lower surface of the particle.

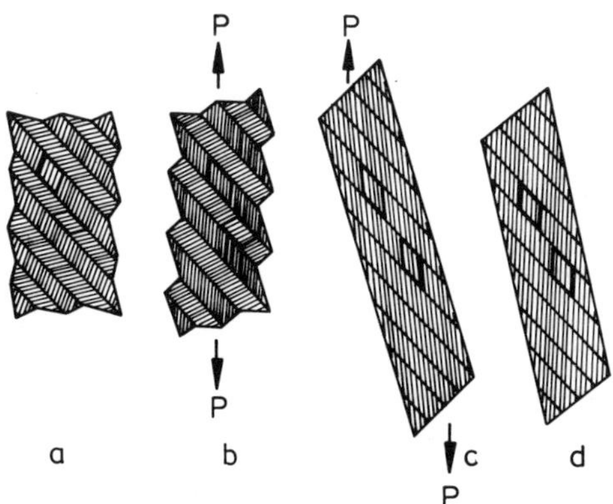

FIG. 5. The body as a stack of lattice layers and its deformation.

When the layers are sheared we assume that they retain their orientation. In this manner the vertical component of their shear length Δ is equal to $\frac{1}{\sqrt{2}}\Delta$. This component contributes to the overall vertical length of the body which determines the deformation D. We have

$$D = \frac{1}{\sqrt{2}} \sum_{i=1}^{n} \Delta_i \tag{3}$$

where the index i indicates a layer of which there are n. D is normalized here so that, due to the symmetry of $\Phi(\Delta)$, $D = 0$ holds in the unloaded body, if all particles are in the central minimum, or half of them lie in the right and left minimum.

2.3. Thermal Motion

The lattice particles do not lie still in the minima of their potential wells. Rather they are subject to thermal motion so that they fluctuate about these minima. We call a particle austenitic or martensitic of phase M_- or M_+, if its shear length lies in the range (m_L, m_R) or $(-\infty, m_L)$, $(m_R, +\infty)$ respectively.

We can expect equilibrium to prevail within the different potential wells, but the barriers between the wells are considered to be so high that we must expect a nonequilibrium distribution between the phases during the process. In that case statistical mechanics dictates the following probabilities for the occurrence of a particular shear length in the phases A and $M_\pm$ respectively:

$$\mathfrak{P}_\Delta^- = x^- \frac{e^{-\Phi(\Delta;\tau)/kT}}{\int_{-\infty}^{m_L} e^{-\Phi(\Delta;\tau)/kT}\, d\Delta}; \qquad \mathfrak{P}_\Delta^0 = x^0 \frac{e^{-\Phi(\Delta;\tau)/kT}}{\int_{m_L}^{m_R} e^{-\Phi(\Delta,\tau)/kT}\, d\Delta};$$

$$\mathfrak{P}_\Delta^+ = x^+ \frac{e^{-\Phi(\Delta;\tau)/kT}}{\int_{m_R}^{\infty} e^{-\Phi(\Delta;\tau)/kT}\, d\Delta} \tag{4}$$

The index 0 refers to austenite, while $\pm$ refer to the martensitic twins $M_\pm$. x^-, x^0 and x^+ are the fractions of particles in the phases M_-, A and M_+ respectively and of course they are constrained by the requirement

$$x^- + x^0 + x^+ = 1 \tag{5}$$

Under these circumstances there is no answer to the question which shear length a particular lattice layer may have. Rather we must calculate the predicted value of the shear length which is

$$\boldsymbol{\Delta}^{+}(\tau)=\frac{\displaystyle\int_{m_R}^{\infty}\Delta e^{-\Phi(\Delta;\tau)/kT}\,\mathrm{d}\boldsymbol{\Delta}}{\displaystyle\int_{m_R}^{\infty}e^{-\Phi(\Delta;\tau)/kT}\,\mathrm{d}\boldsymbol{\Delta}} \tag{6}$$

in the case of an M_+ particle. The contribution of all M_+ particles to the deformation D is then $Nx^+\bar{\Delta}^+$ and we may write instead of (3)

$$\frac{\sqrt{2}D}{n}=x^{-}\frac{\displaystyle\int_{-\infty}^{m_L}\Delta e^{-\Phi(\Delta;\,\tau)/kT}\,\mathrm{d}\Delta}{\displaystyle\int_{-\infty}^{m_L}e^{-\Phi(\Delta;\,\tau)/kT}\,\mathrm{d}\Delta}+x^{0}\frac{\displaystyle\int_{m_L}^{m_R}\Delta e^{-\Phi(\Delta;\,\tau)/kT}\,\mathrm{d}\Delta}{\displaystyle\int_{m_L}^{m_R}e^{-\Phi(\Delta;\,\tau)/kT}\,\mathrm{d}\Delta}$$

$$+x^{+}\frac{\displaystyle\int_{m_R}^{\infty}\Delta e^{-\Phi(\Delta;\tau)/kT}\,\mathrm{d}\Delta}{\displaystyle\int_{m_R}^{\infty}e^{-\Phi(\Delta;\tau)/kT}\,\mathrm{d}\Delta} \tag{7}$$

2.4. Transition Probabilities

Although we assume non-equilibrium between the different potential wells of the lattice particles in the layer, those wells are not entirely without communication. Indeed, occasionally the thermal motion will lift a particle across the barrier and bring it from one well into the neighbouring one. The probability for that to happen is called a transition probability and there are four of those appropriate to the four possible transitions which we denote by −0, 0−, 0+, and +0 in a self-explanatory manner.

The transition probability $\overset{-0}{\mathfrak{B}}$ (say) is assumed to be proportional to

$$\frac{e^{-\Phi(m_L;\tau)/kT}}{\displaystyle\int_{-\infty}^{m_L}e^{-\Phi(\Delta;\tau)/kT}\,\mathrm{d}\Delta} \tag{8}$$

which is the probability for an M_- particle to be on the top of the barrier at m_L. The transition probability is also proportional to the frequency with which the particles run against the barrier. This in turn is proportional to $\sqrt{kT/2\pi m}$, and the factor of proportionality, which will be denoted by A, depends on the shape of the potential wells: A is large for narrow and deep wells while it is small for wide and shallow ones. Thus we write

$$\overset{-0}{\mathfrak{P}} = A^{-}\sqrt{\frac{kT}{2\pi m}}\,\frac{\mathrm{e}^{-\Phi(m_{\mathrm{L}},\tau)/kT}}{\displaystyle\int_{-\infty}^{m_{\mathrm{L}}} \mathrm{e}^{-\Phi(\Delta;\tau)/kT}\,\mathrm{d}\Delta};$$

$$\overset{0-}{\mathfrak{P}} = A^{0}\sqrt{\frac{kT}{2\pi m}}\,\frac{\mathrm{e}^{-\Phi(m_{\mathrm{L}};\tau)/kT}}{\displaystyle\int_{m_{\mathrm{L}}}^{m_{\mathrm{R}}} \mathrm{e}^{-\Phi(\Delta;\tau)/kT}\,\mathrm{d}\Delta}$$

$$\overset{0+}{\mathfrak{P}} = A^{0}\sqrt{\frac{kT}{2\pi m}}\,\frac{\mathrm{e}^{-\Phi(m_{\mathrm{R}};\tau)/kT}}{\displaystyle\int_{m_{\mathrm{L}}}^{m_{\mathrm{R}}} \mathrm{e}^{-\Phi(\Delta;\tau)/kT}\,\mathrm{d}\Delta};$$

$$\overset{+0}{\mathfrak{P}} = A^{+}\sqrt{\frac{kT}{2\pi m}}\,\frac{\mathrm{e}^{-\Phi(m_{\mathrm{R}};\tau)/kT}}{\displaystyle\int_{m_{\mathrm{R}}}^{\infty} \mathrm{e}^{-\Phi(\Delta;\tau)/kT}\,\mathrm{d}\Delta} \tag{9}$$

We shall assume that $A^{+}=A^{-}$ holds, but other than that the As remain in the theory as adjustable parameters.

Processes that evolve only after a transition across a potential barrier are quite common, most notably in chemistry, and we talk about activated processes. Such processes may either be activated by high temperature, which enables the particles to overcome barriers, or by loading, which lowers the barriers.

Here we consider the deformation of a memory alloy as an activated process and, since the transition probabilities (9) depend on T and τ we may have both: activation by temperature and by loading.

3. SUGGESTIVE INTERPRETATION OF THE OBSERVED PROPERTIES IN TERMS OF THE MODEL

3.1. The Prevailing Phases at High and Low Temperatures in the Unloaded Body

In the unloaded body the potential energy of the lattice particles has the symmetric shape shown in Fig. 3. At a high temperature the thermal motion of the lattice particles, which is of the order of magnitude of kT, is so big that the particles cross the barriers of the potential easily. Nearly all particles will then lie in the wide and shallow central well, i.e. the body is austenitic.

As the temperature decreases, the kinetic energy of the thermal motion decreases and the particles find it difficult to surmount the energy barriers. They will still be able to jump from the central minimum to the right or left one, but once arrived it will be impossible for them to get back, because the barrier for the backward jumps is higher. Thus when a cooling body arrives at low temperatures, we expect it to be all martensitic and, because of the symmetry of $\Phi(\Delta)$, we expect M_+ and M_- to be present in equal amounts.

This is indeed what is observed and, in fact, the shape of the potential with a wide and shallow metastable minimum and narrow and deep stable lateral minimum has been postulated in order to enable us to use the above arguments.

3.2. Isothermal Load–Deformation Curves

At low temperature Fig. 5 can help us to understand the observed load–deformation curves reported in the first two diagrams of Fig. 1. The argument runs as follows.

Let the unloaded body be composed of M_+ and M_- particles in equal proportion as indicated in Fig. 5(a). When the load is applied, the M_- particles become a little steeper, the M_+ particles become flatter and all particles contribute their share $\bar{\Delta}^-(\tau)$ and $\bar{\Delta}^+(\tau)$ respectively to the deformation (Fig. 5(b)). Upon unloading the particles return to their initial shear lengths, i.e. the deformation is elastic. When the load is increased, however, there comes the point when the M_- particles flip over to become M_+ particles. This is accompanied by a large increase in shear length and the corresponding large yield shown in Fig. 5(c). Further loading just shears the M_+ particles and results in an elastic deformation. When the body is unloaded, all particles assume the mean shear length $\bar{\Delta}^+(0)$ and there results a residual deformation.

We may also describe these phenomena by use of energy arguments. In the unloaded body the potential energy is represented by the curve in Fig. 3 and at low temperature all particles are lying in either one of the lateral minima. The thermal motion is too weak to make the particles surmount the barriers. But a load τ decreases the left barrier, as shown in Fig. 4, and eventually with an increasing load it becomes so small that even at low temperature the particles can pass it. This is when the yielding occurs. It is also clear from this argument that yield occurs sooner when the temperature is higher. Indeed, at more lively thermal motion, even higher barriers can be overcome.

Thus we understand the load–deformation diagrams for low temperature qualitatively. For high temperatures the argument is similar except that austenitic particles must also be taken into consideration.

In the unloaded body at high temperature nearly all particles are austenitic as we have understood in Section 3.1. The barriers do not matter and the mean shear length is zero. The application of a load τ increases the right potential well, however, and there comes the point when that well is so deep that a particle which falls into it cannot at that temperature climb out of it again. This enables the martensite to form in the loaded body even at high temperature and, rather obviously, the higher the temperature, the bigger the load must be for martensite formation. When the load is decreased, the particles will be able to leave the M_+-well and jump back to become austenitic.

The high temperature hysteresis cannot be easily understood in this suggested manner, because the delay in reforming austenite is a question of the time scale on which the activated process evolves. We come back to this in the formal theory of Section 4.

3.3. Interpretation of a Non-isothermal Deformation

As we have seen the model provides a good intuitive understanding of the inner workings of a memory alloy and it will also enable us to explain the behaviour of such a body under an oscillating load and with variable temperature that was recorded in Fig. 2.

We have essentially the four ranges I–IV to consider in the deformation process of Fig. 2.

On I: (deformation oscillates with the load at low temperature). This is most easy to understand, because the body just moves up and down on the original elastic curve. Half of the particles are M_+, the other half are M_- and that proportion never changes. The amplitude of the oscillation is indicative of the elastic modulus: a small amplitude corresponds to a large modulus.

On II: (increasing residual deformation at higher temperature). As the load oscillates the height of the barrier cutting off the M_- particles also oscillates. This does not enable the particles to jump across the barrier as long as the thermal motion is weak. But as the temperature rises some particles will flip from M_- to M_+ during the intervals of high load when the barrier is small. Thus during those intervals creep occurs and we observe a residual deformation.

On III: (deformation oscillates with the load at high temperature). Once the temperature is so high that the body has become austenitic it moves up and down the elastic curve that passes the origin in the last two diagrams of Fig. 1.

On IV: (large increase of deformation at decreasing temperature). Since the body is loaded all the time, except for the brief moments where the $P(t)$ curve touches the time axis, its particles have a mean potential energy of the form shown in Fig. 4. As long as the temperature is high, the particles are nearly all austenitic, because their barriers do not matter. But as the temperature decreases, the highest barrier, which is one between the M_+-phase and the A-phase, becomes insurmountable. Consequently, whenever a particle falls into the M_+ potential well, it is trapped there and eventually all particles will be trapped. This accounts for the large deformation, because each particle will now contribute the large shear length $\Delta^+(\tau)$. As P and τ oscillate, so does $\bar{\Delta}^+(\tau)$ and the deformation D.

4. RATE LAWS AND RESULTS

4.1. Rate Laws for Phase Fractions and Temperature

Having seen that the model offers an intuitive understanding of the observed phenomena in memory alloys, we proceed to write down the governing equations which—after an adjustment of parameters—should give a quantitative description of the phenomena.

In order to use (7) for determination of $D(t)$, when the external load $P(t)$ and the external temperature $T_E(t)$ are given, we need to have equations for the phase factors $x^\pm$ and x^0 as well as for temperature T.

For the phase factors the equations are simple rate laws of the form

$$\begin{aligned} \dot{\mathbf{x}}^- &= -\overset{-0}{\mathfrak{P}}x^- + \overset{0-}{\mathfrak{P}}x^0 \\ \dot{x}^0 &= \overset{-0}{\mathfrak{P}}x^- - \overset{0-}{\mathfrak{P}}x^0 - \overset{0+}{\mathfrak{P}}x^0 + \overset{+0}{\mathfrak{P}}x^+ \\ \dot{x}^+ &= \overset{0+}{\mathfrak{P}}x^0 - \overset{+0}{\mathfrak{P}}x^+ \end{aligned} \tag{10}$$

where $\overset{-0}{\mathfrak{P}}$, etc., are the transition probabilities (9). The rationale behind these equations is obvious. For instance, $\dot{X}^-$ is due to particles that jump out of the potential well M_- and to particles that arrive there coming from the neighbouring potential well A; and the number of the particles leaving a well is proportional to the number in that well with the corresponding transition probability as a factor of proportionality. Of course, because of (5) only two of the three equations (10) are independent.

The differential equation for temperature is based upon the balance of internal energy whose leading terms are easily appreciated by the following simple argument:† as the particles jump across a barrier into a deeper potential well, they convert potential energy into kinetic energy and thus the temperature rises. When the temperature exceeds the external temperature T_E there is a heat flux out of the body which we take to be proportional to $T-T_E$. Thus we obtain the energy equation in the form

$$C\dot{T}=-a(T-T_E)-(\dot{x}^-H^-(\tau)+\dot{x}^+H^+(\tau)) \tag{11}$$

where C is the heat capacity, a is the coefficient of heat transfer and $H^\pm$ are the latent heats for the phase transition from $M_\pm$ to A. If $\nu=Nh$ is the number of lattice particles, we may write

$$H^\pm=\nu(\Phi(\Delta^\pm(\tau);\tau)-\Phi(\Delta^0(\tau);\tau)) \tag{12}$$

when $\Delta^{+0-}(\tau)$ denotes the shear lengths of the minima of the potential $\Phi(\Delta;\tau)$ (see Fig. 4).

Equation (11) and two of the equations (10) must now be solved for given functions $P(t)$ and $T_E(t)$. There result the functions $T(t)$ and $x^\pm(t)$, $x^0(t)$ which must be inserted into (7) to give the deformation $D(t)$. The calculation must of course be done numerically, because of the intricacy in the set of equations (10) and (11).

4.2. Specific Choice of Potential

The specific form of the potential energy $\Phi(\Delta; 0)$ which is used in the subsequent calculations is shown in Fig. 6. For numerical simplicity we

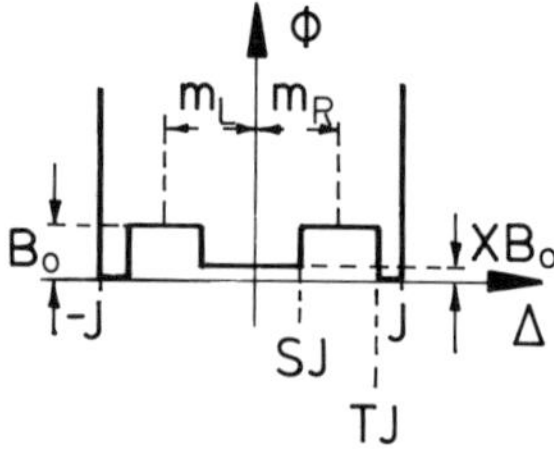

FIG. 6. Potential composed of rectangular potential wells.

have chosen a potential with rectangular potential wells. This choice of potential causes a difficulty in the selection of values for m_L and m_R

† A more careful argument is given in ref. 2, where the power of the external load is taken into account as well.

under tensile and compressive loading. This is a problem of numerical detail which we do not discuss here.

4.3. Non-dimensional Form of the Dynamic Problem

For the numerical calculations we must introduce dimensionless variables and we define

$$\hat{t}=\frac{t}{\sqrt{\dfrac{B_0}{2\pi mJ^2}}};\quad \delta=\frac{\Delta}{J};\quad \vartheta=\frac{T}{T_{\mathrm{I}}};\quad d=\frac{\sqrt{2}D}{nJ};$$
$$p=\frac{P}{B_0/(J-m_{\mathrm{L}})};\quad \varphi=\frac{\Phi}{kT_1} \tag{13}$$
$$\varepsilon=\frac{B_0}{kT_{\mathrm{I}}};\quad \mu_{\mathrm{I,R}}=\frac{m_{\mathrm{L,R}}}{J};\quad \alpha=\frac{a}{G}\sqrt{\frac{B_0}{2\pi mJ^2}},$$
$$h^{\pm}(p)=\varphi(\delta^{\pm},p)-\varphi(\delta^{0},p)$$

where T_{I} is the initial temperature and B_0 is equal to $\Phi(m_{\mathrm{L}};0)-\Phi(-J;0)$, i.e. B_0 is the height of the left barrier as seen by the M_-particles.

Insertion of these dimensionless variables into (10) and (11) gives the corresponding dimensionless equations

$$\frac{\mathrm{d}x^-}{\mathrm{d}\hat{t}}=\sqrt{\frac{\vartheta}{\varepsilon}}\left(-A^{\pm}\frac{\mathrm{e}^{-\varphi(\mu_{\mathrm{L}};\,p)/\vartheta}}{\displaystyle\int_{-\infty}^{\mu_{\mathrm{L}}}\mathrm{e}^{-\varphi(\delta;\,p)/\vartheta}\,\mathrm{d}\delta}x^-\right.$$
$$\left.+A^{0}\frac{\mathrm{e}^{-\varphi(\mu_{\mathrm{L}};\,p)/\vartheta}}{\displaystyle\int_{\mu_{\mathrm{L}}}^{\mu_{\mathrm{R}}}\mathrm{e}^{-\varphi(\delta;\,p)/\vartheta}\,\mathrm{d}\delta}(1-x^--x^+)\right)$$
$$\frac{\mathrm{d}x^+}{\mathrm{d}\hat{t}}=\sqrt{\frac{\vartheta}{\varepsilon}}\left(+A^{0}\frac{\mathrm{e}^{-\varphi(\mu_{\mathrm{R}};p)/\vartheta}}{\displaystyle\int_{\mu_{\mathrm{L}}}^{\mu_{\mathrm{R}}}\mathrm{e}^{-\varphi(\delta;p)/\vartheta}\,\mathrm{d}\delta}(1-x^--x^+)\right. \tag{14}$$
$$\left.-A^{\pm}\frac{\mathrm{e}^{-\varphi(\mu_{\mathrm{R}};p)/\vartheta}}{\displaystyle\int_{\mu_{\mathrm{R}}}^{\infty}\mathrm{e}^{-\varphi(\delta;p)/\vartheta}\,\mathrm{d}\delta}x^+\right)$$
$$\frac{\mathrm{d}\vartheta}{\mathrm{d}\hat{t}}=-\alpha(\vartheta-\vartheta_{\mathrm{E}})-\frac{\nu}{C/k}\left(\frac{\mathrm{d}x^-}{\mathrm{d}\hat{t}}h^-(p)+\frac{\mathrm{d}x^+}{\mathrm{d}\hat{t}}h^+(p)\right)$$

For given functions $p(\hat{t})$ and $\vartheta_E(\hat{t})$ the equations (14) may be solved to give $\vartheta(\hat{t})$, $x^{\pm}(\hat{t})$. Subsequent introduction of these functions into (7) or into its non-dimensional form

$$d = \frac{\int_{-\infty}^{\mu_L} \delta e^{-\varphi(\delta;\,p)/\vartheta}\,d\delta}{\int_{-\infty}^{\mu_L} e^{-\varphi(\delta;\,p)/\vartheta}\,d\delta} x^- + \frac{\int_{\mu_L}^{\mu_R} \delta e^{-\varphi(\delta;\,p)/\vartheta}\,d\delta}{\int_{\mu_L}^{\mu_R} e^{-\varphi(\delta;\,p)/\vartheta}\,d\delta}(1-x^- - x^+)$$
$$+ \frac{\int_{\mu_R}^{\infty} \delta e^{-\varphi(\delta;p)/\vartheta}\,d\delta}{\int_{\mu_R}^{\infty} e^{-\varphi(\delta;p)/\vartheta}\,d\delta} x^+ \tag{15}$$

furnishes the dimensionless deformation $d(t)$.

In the next section numerical results are given that correspond to the parameter values

$$A^0 = 10, \qquad A^{\pm} = 100, \qquad \alpha = 0{\cdot}5, \qquad \frac{\nu}{G/k} = \frac{1}{3}\frac{1}{20} \tag{16}$$

The heat capacity C is equal to $3\mathcal{N}k$ according to the law of Dulong–Petit, where $\mathcal{N}$ is the total number of atoms. Therefore $(16)_4$ implies that a lattice particle has 20 atoms.

4.4. Results

The system of eqns. (14) and (15) has been evaluated for different conditions of loading, corresponding to those which in the experimental investigation have furnished the curves of Figs. 1 and 2.

Thus Fig. 7 shows the calculated load–deformation curves of the model under an alternating tensile and compressive load which, as

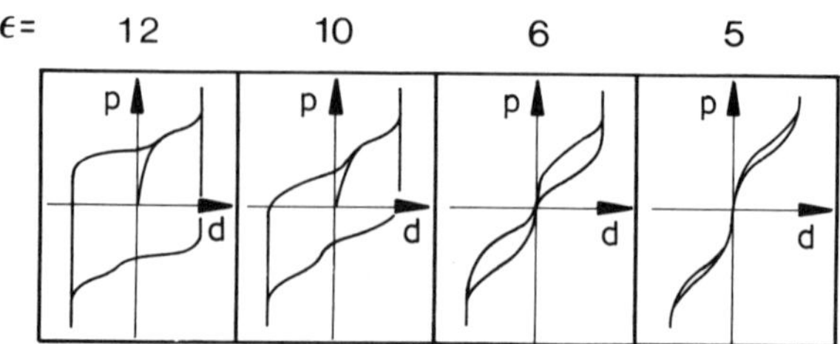

FIG. 7. Calculated load–deformation curves at constant temperatures.

function of time is represented by a zig-zag line. The external temperature is held constant so that $T_E(t) = T_I$. Thus the curves marked by a large value of ε correspond to low temperatures. These curves must be compared to those of Fig. 1 and one recognizes the qualitative agreement: At low temperature the model simulates yield and creep and residual deformation and at high temperature it exhibits the typical pseudoelastic behaviour. In both ranges of temperature the width of the hysteresis loops decreases with growing temperature. It is true, however, that the curves of Fig. 7, and in particular the width of the hysteresis loops, depend on the frequency of the applied load. The present theory must be amended in order to allow for a hysteresis in a quasi-static process. Work in that direction is in progress.

The curves of Fig. 8 show the calculated reaction of the model to the application of a constant load under an external temperature that first increases and then decreases again. The resulting deformation shows all features of the mean observed deformation recorded in Fig. 2 by the dotted line. In particular there is the creep of range II, the recovery of range III and the drastic increase of deformation in range IV, where temperature is decreased. Figure 8, apart from giving the deformation, allows some insight into the causes of the reaction of the model. Thus in its upper part it shows the phase fractions x^+ and x^0 as they change in time in response to the changes of temperature: In particular, the

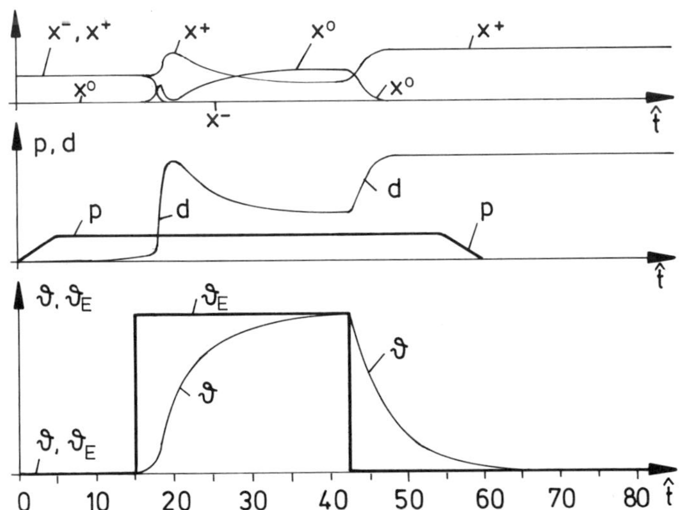

FIG. 8. Calculated process under a constant load and variable temperature.

model is primarily austenitic at high temperature. In the curve $\vartheta(\hat{t})$ Fig. 8 also exhibits the evolution of the temperature of the model as it adjusts to the external temperature and to the supply of latent heat during the phase transitions.

Comparison of the Figs. 7 and 8 with the observed behaviour of a memory alloy, as recorded in Figs. 1 and 2 shows good qualitative agreement over a wide range of temperature. A proper adjustment of parameters should enable us in the future to describe memory alloys quantitatively.

5. A REMARK ON THE THERMODYNAMICS OF MEMORY ALLOYS

The suggested interpretation of the observed properties of memory alloys in Section 3 has made use of the two competing tendencies that are present in all systems of identical elements that are subject to thermal fluctuations: on the one hand such elements tend to be evenly distributed over the range of states that is available to them and on the other hand they tend to assemble in the state of lowest energy. The temperature will determine which one of these tendencies will prevail.

Thermodynamically this conflict can be described as a competition between energy U and entropy H. Quantitatively this is expressed by the free energy

$$\Psi = U - TH \tag{17}$$

which is a function of temperature T and deformation D and which tends to a minimum at fixed temperature and deformation. By (17) this means minimal energy at low temperature, while at high temperature it means maximal entropy.

If the free energy is known, the load–deformation curves for different temperatures can be calculated by differentiation of Ψ with respect to D

$$P = \frac{\partial \Psi}{\partial D} \tag{18}$$

In refs. 6 and 7 Wilmanski and Müller have determined the free energy of the model described in Section 2. This was done by the approximate calculation of the partition function $Z = \exp\{-\Psi/kT\}$ of statistical thermodynamics. Load–deformation curves could then be

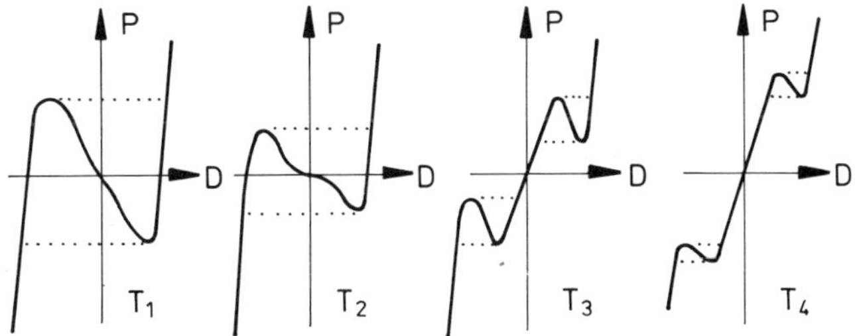

FIG. 9. Load–deformation curves calculated from statistical mechanics.

calculated from (18) and they prove to be non-monotonic, i.e. they have parts with a negative slope (see Fig. 9). Such parts characterize unstable states and in a loading–unloading experiment they are bridged by the horizontal lines of Fig. 9. In this manner statistical thermodynamics is capable of describing the observed hysteresis loops shown in Fig. 1. In fact, the shapes of the load–deformation curves of Fig. 1 are well described qualitatively as functions of temperature by statistical thermodynamics of the model.

REFERENCES

1. ACHENBACH, M. and I. MÜLLER. Creep and yield in martensitic transformation, *Ingenieur-Archiv*, **53** (1983), 73–83.
2. ACHENBACH, M., T. ATANACKOVIĆ and I. MÜLLER. A Model for Shape Memory in Biaxial Loading (in preparation).
3. DELAEY, L. and M. CHANDRASEKARAN, (Eds) *Int. Conf. on Martensitic Transformations* 82, *J. Physique*, **43**, Colloque C-4, 1982.
4. EHRENSTEIN, H. and R. LOHMANN. TU Berlin (private communication).
5. MÜLLER, I. A model for a body with shape memory, *Arch. Rat. Mech. Anal.*, **70** (1979), 61–77.
6. MÜLLER, I. and K. WILMANSKI. A model for phase transition in pseudoelastic bodies, *Il Nuovo Cimento*, **57** (1980), 283–318.
7. MÜLLER, I. and K. WILMANSKI. Memory alloys—phenomenology and ersatzmodel, *Cont. Models of Discrete Systems*, **4** (Eds. O. Brulin, R. K. T. Hsieh), North Holland, Amsterdam, 1981, 495–509.

Section 6

DAMAGE AND FAILURE

30

Creep and Creep Damage of Copper under Multiaxial States of Stress

S. MURAKAMI*

Department of Energy Engineering, Toyohashi University of Technology, Toyohashi, Japan

Y. SANOMURA

Department of Mechanical and Electrical Engineering, Tokuyama Technical College, Tokuyama, Japan

ABSTRACT

After discussing the framework of an anisotropic creep damage theory based on a symmetric second rank damage tensor describing the damaged states, specific forms of the constitutive and the evolution equation for creep and creep damage were determined by performing the creep damage tests of thin-walled copper tubes under constant uniaxial and multiaxial states of stress at 250°C. Then, the resulting equations were applied to analyse the creep damage processes under various non-steady multiaxial states of stress, and the coupling between creep and creep damage, especially the effects of damage anisotropy on creep strains and creep rupture times were elucidated. Finally, the validity, limitations and possible elaboration of the theory were discussed by comparing these results with those of the corresponding experiments.

1. INTRODUCTION

Coupling between deformation and damage of materials constitutes one of the most important domains of the current problems of inelasticity. Above all, time dependent plastic deformation, i.e. creep, of metals at elevated temperature usually induces deterioration of the

* Present address: Department of Mechanical Engineering, Nagoya University, Nagoya, Japan.

materials, which in turn not only accelerates the deformation but also brings about sudden rupture after a certain period of loading. In the case of polycrystalline metals, in particular, this phenomenon of creep damage proceeds mainly due to the nucleation and growth of microcracks or microvoids at grain boundaries [1, 2, 6, 12]. These cavities develop most significantly on grain boundaries perpendicular to the direction of maximum tensile stress, and hence the material damage of this sort usually has marked anisotropy.

The present paper is concerned with the continuum mechanics modelling of the anisotropic and coupled features of creep and creep damage in terms of tensorial damage variables, as well as with its experimental verification by means of creep damage tests of thin-walled copper tubes under non-steady multiaxial states of stress at 250°C.

2. CONSTITUTIVE AND EVOLUTION ASSUMPTIONS

2.1. Evolution Equation of Creep Damage

By postulating that the principal mechanical effect of creep damage consists in the net area reduction caused by cavity formation in materials, Murakami and Ohno [15, 16] described the damage state by means of a second rank symmetric damage tensor $\mathbf{\Omega}$ specified by the three-dimensional cavity-area density, and developed a continuum theory of creep and creep damage of polycrystalline metals. Then, the tensor $(\mathbf{I}-\mathbf{\Omega})$ may be interpreted as a linear transformation which maps the area vector of an arbitrary surface element in undamaged materials, into the corresponding net area vector of the element diminished by the material damage [16], where $\mathbf{I}$ is an identity tensor of rank two. As a result of this net area reduction, Cauchy stress $\boldsymbol{\sigma}$ is magnified to the following net stress tensor [16]

$$\mathbf{S}=\tfrac{1}{2}(\boldsymbol{\sigma}\mathbf{\Phi}+\mathbf{\Phi}\boldsymbol{\sigma}), \qquad \mathbf{\Phi}=(\mathbf{I}-\mathbf{\Omega})^{-1} \tag{1}$$

where the tensor $\mathbf{\Phi}$ is called a damage effect tensor.

Development of cavity area densities in the process of creep damage will be governed by the local stress acting on individual cavities. Thus, if we assume that the effect of the material damage on the damage growth may be represented in terms of $\mathbf{S}$ and $\mathbf{\Phi}$, the evolution equation of creep damage may have the form

$$\dot{\mathbf{\Omega}}=\mathbf{H}(\mathbf{S}, \mathbf{\Phi}, \kappa) \tag{2}$$

where $(\dot{})$ denotes the differentiation with respect to time, and κ is a strain-hardening parameter of the material matrix.

2.2. Creep Constitutive Equation of Damaged Materials

Unlike the damage growth, the deformation of damaged materials depends not only on the net area reduction caused by cavity formation, but also on its three-dimensional arrangement. Moreover, the deformation characteristics of materials are generally specified by a fourth rank tensor which transforms a stress tensor into the corresponding strain tensor. Thus, Murakami and Imaizumi defined the net stress tensor $\tilde{\mathbf{S}}$ for the constitutive equation of the damage materials in terms of a fourth rank tensor $\mathbf{\Gamma}(\mathbf{\Phi})$ specified by the damage state as follows [14]:

$$\tilde{\mathbf{S}} = \tfrac{1}{2}[\mathbf{\Gamma}\boldsymbol{\sigma} + (\mathbf{\Gamma}\boldsymbol{\sigma})^{\mathrm{T}}] \tag{3}$$

Then, they further performed model tests to discuss the validity of this notion, and elucidated that, while the damage effect on the deformation should be described by this fourth rank tensor for significant cavity volume fraction, it may be represented by a second rank tensor (Fig. 7 of ref. 14) or a scalar (Figs. 7 and 8 of ref. 14) for small values of cavity fraction.

Now, in order to avoid the complexity of the fourth rank tensor $\mathbf{\Gamma}(\mathbf{\Phi})$, and according to the experimental results of Dyson and McLean [4], we will assume that the cavity density in the material is sufficiently small, and employ a new second rank tensor $\mathbf{\Lambda}(\mathbf{\Phi})$ constructed from $\mathbf{\Phi}$ instead of $\mathbf{\Gamma}(\mathbf{\Phi})$ in eqn. (3). Then, the corresponding net stress tensor $\bar{\mathbf{S}}$ may be written in the form

$$\bar{\mathbf{S}} = \tfrac{1}{2}(\mathbf{\Lambda}\boldsymbol{\sigma} + \boldsymbol{\sigma}\mathbf{\Lambda}), \qquad \mathbf{\Lambda} = \lambda\mathbf{I} + \mu\mathbf{\Phi} + \rho\mathbf{\Phi}^2 \tag{4}$$

where $\mathbf{I}$ is an identity tensor of rank two, and λ, μ and ρ are polynomials of the scalar invariants of $\mathbf{\Phi}$. By assuming further that λ, μ and ρ are material constants, and noting the condition that $\bar{\mathbf{S}}$ should coincide with $\boldsymbol{\sigma}$ in the undamaged state, eqn. (4) leads to

$$\bar{\mathbf{S}} = \lambda\boldsymbol{\sigma} + \tfrac{1}{2}\mu(\boldsymbol{\sigma}\mathbf{\Phi} + \mathbf{\Phi}\boldsymbol{\sigma}) + \tfrac{1}{2}(1 - \lambda - \mu)(\boldsymbol{\sigma}\mathbf{\Phi}^2 + \mathbf{\Phi}^2\boldsymbol{\sigma}) \tag{5}$$

The first and the second term in the right hand side of this relation represent a scalar multiple of the Cauchy stress $\boldsymbol{\sigma}$ and the net stress tensor of eqn. (1), respectively.

If we assume that the effect of the material damage on the creep deformation may be expressed in terms of $\bar{\mathbf{S}}$ and $\mathbf{\Phi}$, the constitutive equation for creep may have the form

$$\dot{\boldsymbol{\varepsilon}}^{C} = \mathbf{G}(\bar{\mathbf{S}}, \boldsymbol{\Phi}, \kappa) \tag{6}$$

3. CREEP DAMAGE TESTS AND SPECIFICATION OF EVOLUTION AND CONSTITUTIVE EQUATIONS

3.1. Specimen and Test Condition

Creep damage tests of thin-walled copper tubes under combined tension and torsion were performed to specify the explicit forms of the evolution and constitutive equations as well as to estimate their validity. The specimens of outer diameter 21 mm, thickness 1 mm and gauge length 25 mm were machined from a tough pitch copper bar of 30 mm diameter, and were annealed in vacuum at 600°C for one hour to remove the initial anisotropy and the residual stress caused by the machining.

The tests were carried out at 250 ± 1°C by using three combined tension and torsion creep test machines under the condition of specified true stress. The temperature of the specimens was measured by two platinum–platinum 13% rhodium thermocouples attached to two locations along the gauge length of the specimens. Axial and torsional strains were measured by a shoulder type extensometer connected to circumferential collars machined on the gauge lines of the specimens, with combined use of dial indicators and differential transformers. The accuracy of the extensometer was 10×10^{-6} and 5×10^{-6} for axial and torsional strains, respectively.

3.2. Results of Uniaxial and Constant Stress Tests and their Formulation

Specification of evolution equation (2) and constitutive equation (6) requires first to determine their functional forms to describe the creep curves for uniaxial state of stress. If we employ the conventional creep laws of McVetty type [5] and Bailey–Norton type [11] based on the strain-hardening hypothesis, together with a power law damage equation proposed by Kachanov [10] and Rabotnov [18], eqns. (6) and (2) may be expressed in the following forms

$$\left.\begin{aligned} \dot{\varepsilon}^{c} &= A_1\sigma^{n_1} r \exp(-rt^*) + A_2\bar{S}^{n_2} \\ \varepsilon^{c}(t) &= A_1[\sigma(t)]^{n_1}[1-\exp(-rt^*)] + A_2[\sigma(t)]^{n_2}t^* \\ \bar{S} &= [\lambda + \mu\Phi + (1-\lambda-\mu)\Phi^2]\sigma \end{aligned}\right\} \text{(McVetty type)} \tag{7}$$

$$\left.\begin{aligned}&\dot{\varepsilon}^{c}=mA^{1/m}\kappa^{(m-1)/m}\bar{S}^{n/m},\ \dot{\kappa}=mA^{1/m}\kappa^{(m-1)/m}\sigma^{n/m}\\&\bar{S}=[\lambda+\mu\Phi+(1-\lambda-\mu)\Phi^{2}]\sigma\end{aligned}\right\}$$

(Bailey–Norton type) (8)

$$\dot{\Omega}=BS^{k}\Phi^{l},\qquad \Phi=\frac{1}{1-\Omega},\qquad S=\frac{\sigma}{1-\Omega} \tag{9a}$$

where A, A_1, A_2, B, k, l, m, n, n_1, n_2 and r are material constants. The symbol t^* in eqn. (7), on the other hand, is a fictitious time which would give the same magnitude of the strain as that of the current strain $\varepsilon^c(t)$ if the current stress $\sigma(t)$ is supposed to have acted from the beginning. For the problems of variable stress, t^* should be eliminated from eqns. $(7)_1$ and $(7)_2$. The first term of the right hand side of eqn. $(7)_1$ represents the primary creep, and is assumed to be unaffected by the creep damage.

For constant stress in particular, eqn. (9a) may be integrated under the condition $\Omega=0$ at $t=0$. Then, the resulting relation gives the following rupture time t_R in view of the fact that the rupture occurs for $\Omega=1$:

$$t_R=1/[B(k+l+1)\sigma^k] \tag{9b}$$

The material constants involved in eqns. (7) (9) may be identified by fitting them to the experimental results shown in Fig. 1. While circles in the figure stand for the experimental results, the solid and the dashed lines represent the numerical results calculated from eqns. (7), (9) and (8), (9), respectively, employing the following material constants:

$$\left.\begin{aligned}&A_1=4{\cdot}61\times10^{-6},\quad n_1=1{\cdot}98,\quad r=0{\cdot}11,\\&A_2=1{\cdot}20\times10^{-10},\quad n_2=3{\cdot}43\\&\lambda=0{\cdot}85,\quad \mu=0{\cdot}0,\quad B=5{\cdot}52\times10^{-10},\\&k=3{\cdot}46,\quad l=0{\cdot}0\end{aligned}\right\}\ \text{(McVetty type)} \tag{10}$$

$$\left.\begin{aligned}&A=4{\cdot}80\times10^{-7},\quad m=0{\cdot}424,\\&n=2{\cdot}20,\quad \lambda=0{\cdot}50\\&\mu=0{\cdot}50,\quad B=7{\cdot}11\times10^{-10},\\&k=3{\cdot}46,\quad l=-1{\cdot}0\end{aligned}\right\}\ \text{(Bailey–Norton type)} \tag{11}$$

As observed from Fig. 1, eqns. (7)–(11) describe the experimental

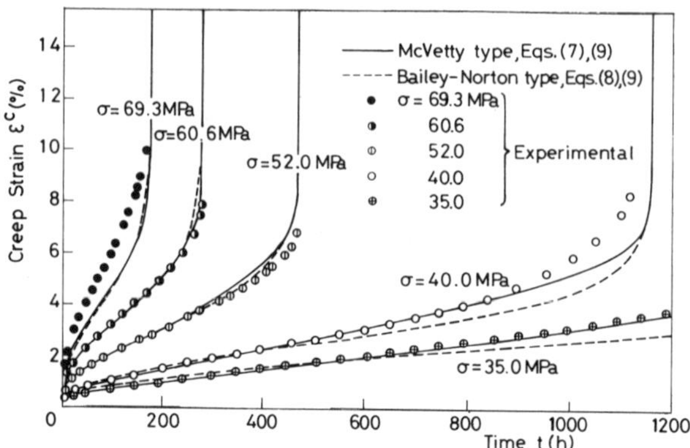

FIG. 1. Uniaxial creep curves of copper at 250°C.

results sufficiently well. Particularly, in the cases of $\sigma = 60{\cdot}6$ and $52{\cdot}0$ MPa, the experimental results have been described by both types of creep equation with the accuracy better than 10 per cent. The Bailey–Norton type equation (8) represents the primary and the tertiary creep by the balance of the continuous deceleration of creep rate due to strain-hardening and its acceleration caused by the creep damage, and hence it is apt to overestimate the effects of creep damage from the early stage of the creep process.

3.3. Stress Measure of Damage Growth under Constant Multiaxial States of Stress

In order to extend eqns. (7)–(9) to multiaxial states of stress, it is necessary to elucidate stress state dependence of the damage growth. Hence, we then performed a series of creep damage tests under constant combined tension and torsion for stress ratios $\tau/\sigma = 0$, 1/3, 1 and 3, and by specifying a constant maximum tensile stress $\sigma_1 =$ 40 MPa.

Figure 2 shows the resulting isochronous creep rupture curve, where σ_0 denotes the corresponding tensile stresses which would give the same rupture times as those of each multiaxial rupture test, and has been obtained from eqn. (9b). The circles in the figure show the experimental results, whereas the dashed and the chain lines are isochronous rupture curves calculated on the basis of constant equivalent stress criterion (von Mises type) $\sigma_{EQ} = \text{const.}$ and constant maximum tensile stress criterion $\sigma_1 = \text{const.}$ Since the experimental results

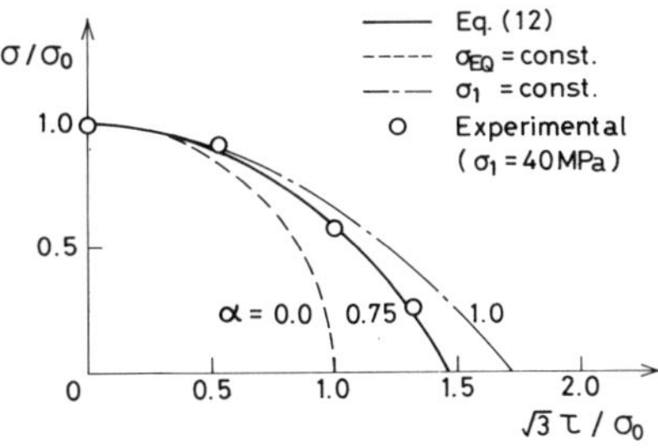

FIG. 2. Isochronous curve for creep rupture times.

are located between two curves, the stress measure of rupture may be expressed by a linear combination of σ_1 and σ_{EQ} [8, 13]:

$$\sigma^* = \alpha\sigma_1 + (1-\alpha)\sigma_{EQ} \qquad (0 \leqslant \alpha \leqslant 1) \tag{12}$$

The solid line in Fig. 2 represents the isochronous rupture curve calculated by equation (15) for $\alpha = 0.75$.

3.4. Specification of Constitutive and Evolution Equations

In the process of these creep damage tests, the axial and torsional strains in specimens were also measured. According to these results, the magnitude of creep rates was ascertained to be governed by the equivalent stress σ_{EQ} of von Mises type. Hence, if we assume that the creep theory of von Mises type combined with strain-hardening hypothesis apply for vanishing creep damage, the creep constitutive equation (6) may be specified as follows by extending eqns. (7) and (8):

$$\left.\begin{aligned} \dot{\boldsymbol{\varepsilon}}^c &= \tfrac{3}{2}[A_1 {\sigma_{EQ}}^{n_1-1} r \exp(-rt^*) \boldsymbol{\sigma}_D + A_2 {\bar{S}_{EQ}}^{n_2-1} \bar{\mathbf{S}}_D] \\ \varepsilon^c_{EQ}(t) &= A_1[\sigma_{EQ}(t)]^{n_1}[1-\exp(-rt^*)] + A_2[\sigma_{EQ}(t)]^{n_2} t^* \end{aligned}\right\} \text{(McVetty type)} \tag{13a}$$

$$\varepsilon^c_{EQ}(t) = \int_0^t [\tfrac{2}{3}\,\mathrm{tr}\,(\dot{\boldsymbol{\varepsilon}}^c)^2]^{1/2}\,dt, \qquad \sigma_{EQ} = [\tfrac{3}{2}\,\mathrm{tr}\,\boldsymbol{\sigma}_D^2]^{1/2}, \qquad \bar{S}_{EQ} = [\tfrac{3}{2}\,\mathrm{tr}\,\mathbf{S}_D^2]^{1/2} \tag{13b}$$

$$\left.\begin{aligned} \dot{\boldsymbol{\varepsilon}}^c &= \tfrac{3}{2} m A^{1/m} \kappa^{(m-1)/m} {\bar{S}_{EQ}}^{(n-m)/m} \bar{\mathbf{S}}_D \\ \dot{\kappa} &= m A^{1/m} \kappa^{(m-1)/m} {\sigma_{EQ}}^{n/m} \end{aligned}\right\} \text{(Bailey–Norton type)} \tag{14}$$

where $\boldsymbol{\sigma}_D$ and $\bar{\mathbf{S}}_D$ denote the deviatoric tensors of the Cauchy stress $\boldsymbol{\sigma}$ and the net stress tensor $\bar{\mathbf{S}}$ of eqn. (5).

According to the results of the metallurgical observations reported so far, cavities caused by creep damage develop mainly on grain boundaries perpendicular to the maximum tensile principal stress [1, 2, 6, 12]. The discussion of Section 3.3, on the other hand, showed that creep rupture times can be described by the stress measure of eqn. (12). Therefore, if we assume that the net area reduction due to cavity formation proceeds mainly on the planes perpendicular to the direction of the maximum tensile principal stress, and the rate of this cavity formation is governed by a stress measure analogous to eqn. (12), then eqn. (9a) may be extended to multiaxial states of stress to give a specialized form of the evolution equation (2) as follows:

$$\dot{\mathbf{\Omega}} = B[\beta S^{(1)} + (1-\beta)S_{\mathrm{EQ}}]^k[\mathrm{tr}\,\mathbf{\Phi}\boldsymbol{\nu}^{(1)} \otimes \boldsymbol{\nu}^{(1)}]^l \boldsymbol{\nu}^{(1)} \otimes \boldsymbol{\nu}^{(1)} \tag{15a}$$

$$S_{\mathrm{EQ}} = [\tfrac{3}{2}\,\mathrm{tr}\,\mathbf{S}_{\mathrm{D}}^2]^{1/2} \tag{15b}$$

where $\mathbf{S}_{\mathrm{D}}$ denotes the deviatoric tensor of the net stress tensor of eqn. (1), and $S^{(1)}$ and $\boldsymbol{\nu}^{(1)}$ are the maximum positive principal value of $\mathbf{S}$ and the corresponding principal direction. The symbols B, k, l and β, furthermore, stand for material constants. Equation (15) reduces to eqn. (9a) in the case of uniaxial stress.

A three-dimensional theory of creep damage has been formulated by Leckie and Hayhurst [8, 13] in terms of a scalar damage variable Ω by postulating the isotropy of the creep damage state. For the sake of comparison, if we further assume the damage isotropy, eqn. (7) may be readily extended to the multiaxial case as follows:

$$\dot{\boldsymbol{\varepsilon}}^{\mathrm{c}} = \tfrac{3}{2}[A_1 \sigma_{\mathrm{EQ}}{}^{n_1-1} r \exp(-rt^*)\boldsymbol{\sigma}_{\mathrm{D}} + A_2(\eta\sigma_{\mathrm{EQ}})^{n_2-1}\eta\boldsymbol{\sigma}_{\mathrm{D}}] \tag{16a}$$

$$\varepsilon_{\mathrm{EQ}}^{\mathrm{c}}(t) = A_1[\sigma_{\mathrm{EQ}}(t)]^{n_1}[1-\exp(-rt^*)] + A_2[\sigma_{\mathrm{EQ}}(t)]^{n_2} t^* \tag{16b}$$

$$\eta = \lambda + \mu\left(\frac{1}{1-\Omega}\right) + (1-\lambda-\mu)\left(\frac{1}{1-\Omega}\right)^2 \tag{16c}$$

$$\dot{\Omega} = B\frac{(\sigma^*)^k}{(1-\Omega)^{k+l}} \tag{16d}$$

4. ESTIMATION OF VALIDITY OF CONSTITUTIVE AND EVOLUTION EQUATIONS

4.1. Combined Tension–Torsion Tests at Constant Stress

4.1.1. *Discussion on Creep Equations of McVetty Type and Bailey–Norton Type*

The discussion of the validity of eqns. (13)–(15) requires examination of the damage behaviour under combined states of stress. Figure 3(a), (b)

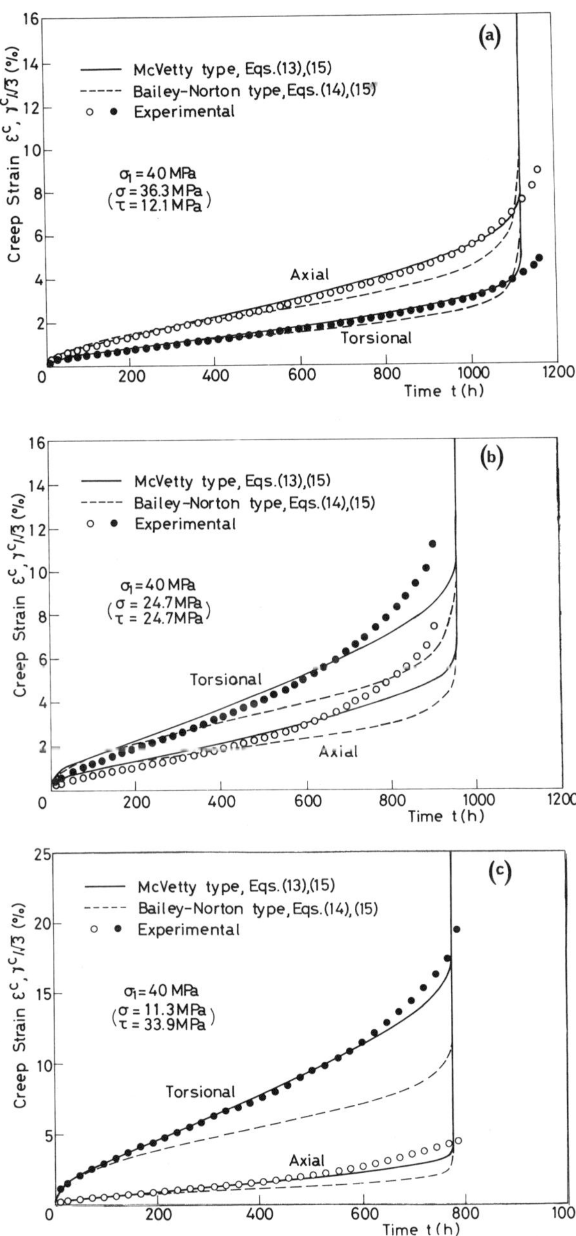

Fig. 3. Axial and torsional creep strains of constant combined tension and torsion tests ($\sigma_1 = 40$ MPa). (a) $\tau/\sigma = \frac{1}{3}$; (b) $\tau/\sigma = 1$; (c) $\tau/\sigma = 3$.

and (c) show the experimental results for creep strain components observed in the preceding tension–torsion damage tests at constant stress. The solid and the dashed lines in the figure represent the corresponding predictions of eqns. (13)–(15). As regards eqns. (13) and (14), material constants of eqns. (10) and (11), respectively, were employed again. The values of β in eqn. (15) were determined as $\beta = 0{\cdot}68$ and $0{\cdot}66$, respectively for eqns. (13) and (14), so that the predictions of eqn. (15) together with B, k, l in eqns. (13) and (14) may coincide well with the corresponding experiment.

It was confirmed in Fig. 1 that the Bailey–Norton equation (14), as well as the McVetty equation (13), combined with the evolution equation (15) described accurately the creep behaviour of the damaged materials under uniaxial constant stress. As regards the combined stress creep of Fig. 3, the McVetty equation (13) describes the creep curves of each case generally well. For larger values of the stress ratio τ/σ, in particular, though the Bailey–Norton equation (14) describes well the first part of the creep curves, it predicts considerably smaller values of strains in the secondary and tertiary stages of creep. These discrepancies originate in the differences between the stress measure for creep rate and that of damage growth, and impose an essential difficulty on the Bailey–Norton equation. This feature may be explained as follows.

The Bailey–Norton equation describes the primary and the tertiary stage of creep by the balance of strain-hardening and material deterioration, as mentioned already, and this balance is usually determined by characterizing the results of tensile creep damage tests. However, in multiaxial states of stress, while creep hardening is governed by σ_{EQ}, the damage growth is specified by σ^*. For a prescribed value of σ_{EQ}, the value of σ^* for pure torsion is smaller than that for simple tension. This means that the effect of stress increase due to material damage and hence the creep rate is unduly underestimated in the Bailey–Norton equation. In contrast, the McVetty equation (13) does not have such an inconsistency, and thus we will employ this equation exclusively in the succeeding discussions.

4.1.2. *Effects of Material Damage on Creep Deformation*

Figure 4 illustrates the creep strain trajectories observed in the preceding combined tension–torsion tests. The solid and the chain lines in this figure stand for the corresponding predictions calculated from eqns. (13) and (15), and that of isotropic damage theory (16), respectively.

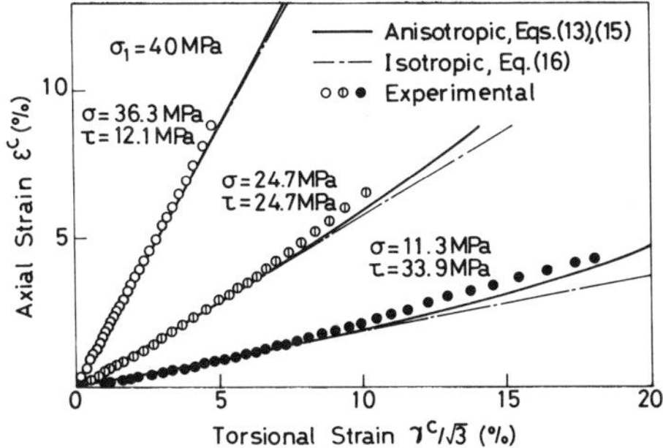

FIG. 4. Strain trajectories of constant combined tension and torsion tests.

As damage proceeds, the trajectories observed in the experiments deviate gradually from straight lines. This deviation is a consequence of the anisotropic feature of material damage, and can be adequately represented by the present theory (eqns. (13) and (15)).

Unlike eqns. (13) and (15), the isotropic damage theory (16) predicts the trajectories of straight lines. However, as observed in Fig. 4, the difference between the axial strain predicted by eqn. (16) and that of experiment is less than 20 per cent before 90 per cent of rupture time. It will be noticed from this result that as far as the damage effect on creep deformation is concerned, we may assume it isotropic except immediately before the creep rupture. This fact has been confirmed also by Leckie and Hayhurst [13] on the basis of the experimental results of Johnson *et al.* [9].

4.2. Tests under Combined Constant Tension and Reversed Torsion

4.2.1. *Discussion on Validity of Constitutive and Evolution Equations*
Anisotropic feature of creep damage is especially important when we describe the material responses under non-proportional loading. Therefore, in order to discuss the validity of the anisotropic creep damage theory (eqns. (13) and (15)), we further performed a series of creep damage tests under combined constant tension and reversed torsion. Then, for a constant stress magnitude $\sigma^* = 45$ MPa ($\alpha = 0{\cdot}75$), the changes of the principal stress direction were selected as $\Delta\theta = 30$, 60, 80°, and these stress changes were performed at $t = 480$ h in all cases.

Figure 5(a), (b) and (c) shows the variation of creep strain compo-

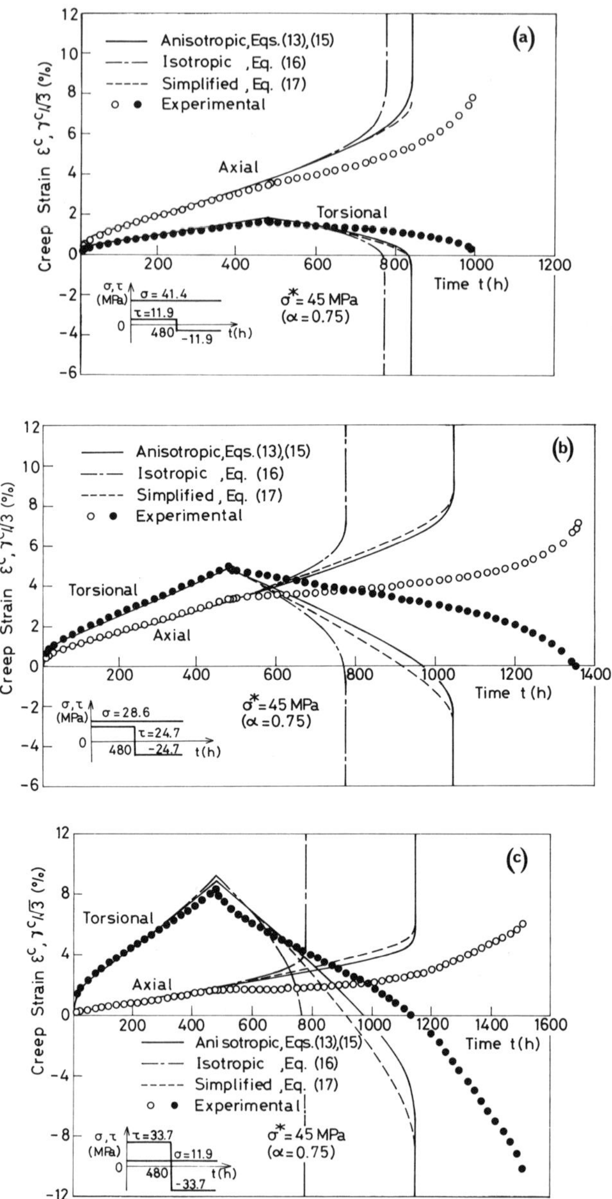

FIG. 5. Axial and torsional creep strains of combined tension and reversed torsion tests ($\sigma^* = 45$ MPa). (a) $\Delta\theta = 30°$; (b) $\Delta\theta = 60°$; (c) $\Delta\theta = 80°$.

nents observed in these experiments. The solid and the chain lines in the figure represent the predictions of the anisotropic damage theory (eqns. (13) and (15)) and those of isotropic damage theory (eqn. (16)). The dashed lines, furthermore, are the results of the simplified theory mentioned later.

Let us discuss first the rupture times of each case. Since the tests of Fig. 5(a), (b) and (c) are concerned with an identical value of $\sigma^* = 45$ MPa, the isotropic damage theory provides an identical rupture time $t_R = 774$ h for these cases. According to the results of experiment, on the other hand, the corresponding rupture times for $\Delta\theta = 30$, 60 and 80° are $t_R = 992$, 1360 and 1504 h, respectively. The reason for the prolongation of the rupture time with the increase of $\Delta\theta$ may be accounted for by a renewal of surface elements of significant damage growth induced by the stress rotation [16], and has been confirmed by micrographs of damaged copper tubes obtained by Trampczyński *et al.* [19]. Thus, it should be noted that, when the stress rotation is significant as in the case of Fig. 5(c), the isotropic damage theory (eqn. (16)) may predict the rupture times to be less than half of the experimental values, and may not be applicable to the accurate prediction of the rupture times.

The rupture times predicted by the anisotropic theory (eqns. (13) and (15)) are $t_R = 840$, 1047 and 1147 h for $\Delta\theta = 30$, 60, 80°, and hence there is a similar trend to those of experiment. Nevertheless, these times are 15–35 per cent shorter than the corresponding times of experiment. These discrepancies may be accounted for partly by the stress state dependence of the cavity growth modes in the damage process [3, 4]. However, noting that the nucleation and growth of the grain boundary cavities in creep damage process are governed mainly by the grain boundary sliding [2, 12], the discrepancies may be attributable more directly to those of creep rates as will be discussed below. This implies the necessity of elaborating the evolution equation by incorporating the contribution of the creep rates to the damage growth.

Now, let us discuss the creep response of the damaged materials. One of the marked features observed in the experimental results of Fig. 5 is that, except for the renewed primary regions, the axial and the torsional creep rates after the stress change are reduced always by 30–50 per cent in comparison with the corresponding minimum creep rates before the stress change. These phenomena were already observed by Trampczynski *et al.* [19] for the single reverse torsion as well as the repeated torsion tests in copper tubes. On the other hand, the second characteristic feature observed in the experiment is a transient increase of creep rates just after the stress reversals. This is usually referred to as the activation of dislocation movement due to stress changes [7], and is observed in many metals.

Though the present theory (eqns. (13) and (15)) describes the creep behaviour up to the stress change very well, it can neither represent the transient increase of creep rates nor the succeeding decrease of the constant creep rates. While the former phenomenon has already been incorporated into creep constitutive equations [11, 17], the latter effect, in spite of its importance, cannot be described properly by the currently available theories.

4.2.2. *Effect of Damage Anisotropy on Creep Rupture Times*

Finally, Fig. 6 shows the relation between the rupture times t_R and the angle of the principal stress rotation $\Delta\theta$, calculated for the conditions identical to those of Fig. 5. The symbol t_{σ^*} denotes the rupture time under constant combined stress of the corresponding stress ratio τ/σ.

It will be observed from the figure that, unlike the isotropic damage theory, the present theory may predict the increase of the rupture time to be about 45 per cent at its maximum as a result of the rotation of

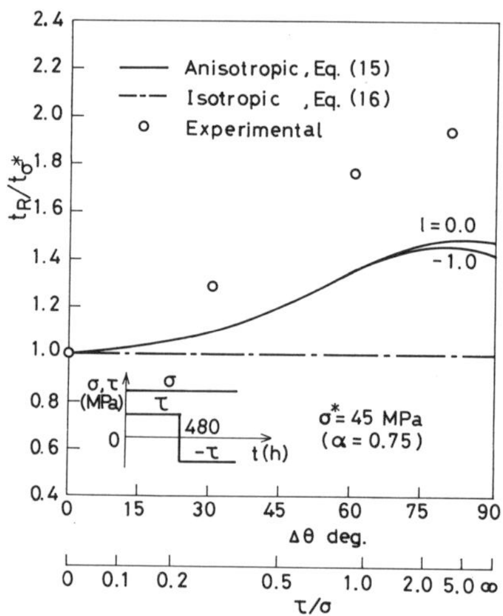

FIG. 6. Relation between creep rupture times of combined constant tension and reversed torsion tests and the angle of principal stress rotation (σ^* = 45 MPa).

the principal stress direction. However, though the experimental results showed still larger increase of t_R, this should be accounted for not only by the effect of the damage anisotropy but also by the effect of material strengthening caused by the stress reversal mentioned previously. Hence, in view of this fact and taking account of the similar tendency to that of the experimental results observed in Fig. 6, the anisotropic effect of the creep damage on the rupture process may be reasonably described by a symmetric damage tensor of rank two $\boldsymbol{\Omega}$.

4.3. A Simplified Theory for Anisotropic Creep Damage

As observed from Figs. 4 and 6, while the anisotropy of creep damage has an essential effect on creep rupture times and hence on damage evolution, its effect on creep rates may be assumed to be isotropic. This suggests that we can define a net stress tensor for a creep constitutive equation by means of an adequate scalar variable constructed with the damage tensor $\boldsymbol{\Omega}$, rather than the net stress tensor $\bar{\mathbf{S}}$ of eqn. (5). In this case, the constitutive and the evolution equation developed above may be simplified as follows:

$$\dot{\boldsymbol{\Omega}} = B[\beta S^{(1)} + (1-\beta)S_{EQ}]^k \boldsymbol{\nu}^{(1)} \otimes \boldsymbol{\nu}^{(1)} \tag{17a}$$

$$\dot{\boldsymbol{\varepsilon}}^c = \tfrac{3}{2}[A_1 \sigma_{EQ}{}^{n_1-1} r \exp(-rt^*)\boldsymbol{\sigma}_D + A_2 \bar{\bar{S}}_{EQ}{}^{n_2-1} \bar{\bar{\mathbf{S}}}_D] \tag{17b}$$

$$\varepsilon^c_{EQ}(t) = A_1[\sigma_{EQ}(t)]^{n_1}[1-\exp(-rt^*)] + A_2[\sigma_{EQ}(t)]^{n_2} t^* \tag{17c}$$

$$\bar{\bar{\mathbf{S}}} = [\xi + (1-\xi)(\tfrac{1}{3}\,\mathrm{tr}\,\boldsymbol{\Phi})]\boldsymbol{\sigma} \tag{17d}$$

where ξ is a material constant.

The predictions obtained by these equations were shown on Fig. 5 by dashed lines. The value of ξ in eqn. (17d) was determined as $\xi = 0{\cdot}1$. As seen in Fig. 4, the stress history of Fig. 5(c) corresponds to the case where the effect of the anisotropic damage on creep is the largest. In this case, the differences between the equivalent strain predicted by eqns. (13) and (15) and that of eqn. (17) are about 3 and 12 per cent, at the time of stress reversal and after 95 per cent of the rupture time, respectively. Hence, the simplified equation (17) gives almost identical results to those of eqns. (13) and (15), and is much more convenient to apply.

5. CONCLUSIONS

After discussing the continuum mechanics modelling of the anisotropic and coupled features of creep and creep damage, a series of creep damage tests of thin-walled tubes under non-proportional loading at 250°C were performed to estimate the validity of the proposed theory. The results of the present work may be summarized as follows:

(1) Selection of a proper creep equation for uniaxial tension is essential for accurate description of creep response under multiaxial states of stress. The McVetty type creep equation is more adequate than that of Bailey–Norton.
(2) The anisotropic feature of creep damage state can be described by a second rank symmetric damage tensor $\mathbf{\Omega}$ or the corresponding damage effect tensor $\mathbf{\Phi}$.
(3) Creep rupture times under non-proportional loading cannot be predicted adequately by the classical isotropic damage theory based on scalar damage variables, but may be predicted reasonably well by an evolution equation formulated in terms of a net stress tensor $\mathbf{S}$ constructed by $\mathbf{\Phi}$.
(4) In order to elaborate further the evolution equation of creep damage, it is necessary to take account of the stress state dependence of the damage growth mode as well as the contribution of creep rate to damage growth.
(5) Anisotropic effects of creep damage on creep deformation can be expressed by another net stress tensor $\bar{\mathbf{S}}$ formed by a second rank tensor function of $\mathbf{\Phi}$.
(6) As regards the damage extent concerned with the present experiment, the effect of creep damage on creep behaviour is almost isotropic, and can be represented well by a still simplified net stress $\bar{\bar{\mathbf{S}}}$ formed by a scalar invariant of $\mathbf{\Phi}$.

REFERENCES

1. Argon, A. S., I.-W. Chen and C. W. Lau. Mechanics and mechanisms of intergranular cavitation in creeping alloys. In: *Three-Dimensional Constitutive Relations and Ductile Fracture* (Ed. S. Nemat-Nasser), North-Holland, Amsterdam, 1981, 23–49.
2. Ashby, M. F. and R. Raj. Creep fracture. In: *The Mechanics and Physics of Fracture*, The Metals Society, London, 1975, 148–158.
3. Chaboche, J. L. Continuous damage mechanics—a tool to describe phenomena before crack initiation, *Nucl. Engng Design*, **64** (1981), 233–247.
4. Dyson, B. F. and D. McLean. Creep of Nimonic 80A in torsion and tension, *Metal Sci.*, **11** (1977), 37–45.

5. FINNIE, I. and W. R. HELLER. *Creep of Engineering Materials*, McGraw-Hill, New York, 1959.
6. GAROFALO, F. *Fundamentals of Creep and Creep-ruptures in Metals*, Macmillan, New York, 1965.
7. GITTUS, J. *Creep, Viscoelasticity and Creep Fracture in Solids*, Applied Science Publishers, London, 1975.
8. HAYHURST, D. R. Creep rupture under multi-axial states of stress, *J. Mech. Phys. Solids*, **20** (1972), 381–390.
9. JOHNSON, A. E., J. HENDERSON and V. D. MATHUR. Combined stress creep fracture of a commercial copper at 250 Deg. Cent, *The Engineer*, **202** (1956), 261–265, 299–301.
10. KACHANOV, L. M. On rupture time under condition of creep, *Izv. Akad. Nauk SSSR, Otd. Tekh. Nauk*, No. 8 (1958), 26–31 (in Russian).
11. KRAUS, H. *Creep Analysis*, John Wiley, New York, 1980.
12. LAGNEBORG, R. Creep: mechanisms and theories. In: *Creep and Fatigue in High Temperature Alloys* (Ed. J. Bressers), Applied Science Publishers, London, 1981, 41–71.
13. LECKIE, F. A. and D. R. HAYHURST. Creep ruptures of structures, *Proc. R. Soc., Lond.*, **A340** (1974), 323–347.
14. MURAKAMI, S. and T. IMAIZUMI. Mechanical description of creep damage state and its experimental verification, *J. Méc. Théor. Appl.*, **1** (1982), 743–761.
15. MURAKAMI, S. and N. OHNO. A constitutive equation of creep damage in polycrystalline metals, *Euromech Symposium* 111; *Constitutive Modelling in Inelasticity*, Mariánské Lázně, Sept. 1978.
16. MURAKAMI, S. and N. OHNO. A continuum theory of creep and creep damage. In: *Creep in Structures* (Ed. A. R. S. Ponter and D. R. Hayhurst), Springer, Berlin, 1981, 422–444.
17. MURAKAMI, S. and N. OHNO. A constitutive equation of creep based on the concept of a creep-hardening surface, *Int. J. Solids Struct.*, **18** (1982), 597–609.
18. RABOTNOV, YU. N. *Creep Problems in Structural Members*, North-Holland, Amsterdam, 1969.
19. TRAMPCZYŃSKI, W. A., D. R. HAYHURST and F. A. LECKIE. Creep rupture of copper and aluminium under non-proportional loading, *J. Mech. Phys. Solids*, **29** (1981), 353–374.

31

Damage Evolution and its Influence on Metal Forming

M. SUÉRY, J. M. JALINIER and J. RAPHANEL

Institut National Polytechnique de Grenoble, Saint Martin d'Hères, France

ABSTRACT

Industrial metals used in forming processes are alloys made up of a single phase or of multiphases and generally containing precipitates and inclusions. During plastic deformation both in hot or cold forming, the heterogeneous nature of the material leads to stress concentrations which induce void nucleation. This nucleation may occur by several mechanisms (decohesion, fragmentation . . .). Experimental observations of damage and its evolution are presented. They are analysed through a single model for plastic deformation controlled growth of cavities which is based on the Rice and Tracey model. The influence of damage on the macroscopical material behavior is shown for some examples: limiting strain and effect of the sheet thickness in cold forming, fracture strain and stress softening in superplastic deformation.

1. INTRODUCTION

Several metal forming processes involve large deformation of thin sheets either at low temperature (deep-drawing, stamping, etc.) or at high temperature (superplastic thermoforming). These sheets are industrial polycrystalline materials. They may sometimes be made up of several phases with similar volume fractions, but they more often consist of a single phase matrix with either small inclusions or precipitates. The latter are usually introduced in order to increase the strength of the material and in some cases also to inhibit grain coarsening during hot-forming. The plasticity of such an aggregate is heterogeneous owing to the different crystallographic orientations of the grains, rheology of each phase and presence of particles. Stress concentrations consequently occur, leading to void nucleation. During deformation, these voids are growing and create internal damage which influences service properties and hastens the fracture of the material. The study of damage is thus of particular importance for the understanding of the material capability to deform.

Our research in this field has been focused on the characterization of damage and its influence on the forming limit diagrams during deep-drawing of steels and aluminium alloys, together with the analysis of cavitation during superplastic deformation of two-phase brasses.

The actual stress state and strain history of the sheets during such forming operations are complex so that fundamental studies have to be carried out under a variety of conditions (such as tension, equibiaxial stretching, plane strain) which partially simulate actual situations, in order to modelize damage and analyse its influence on material behavior.

The aim of this paper is to present some experimental results concerning cavity formation and growth occurring during cold and hot plastic deformation, to propose an analysis of these results which is based on a model introduced by Rice and Tracey [21] and finally to discuss some macroscopic effects due to internal damage.

2. DESCRIPTION OF DAMAGE

2.1. Experimental Procedures

Damage may be characterized by direct methods such as microscopy and density variation measurements or by indirect methods which measure a change of some material parameter, such as Young's modulus. Indirect methods require additional assumptions since damage may not be the only phenomenon involved in the variation which is measured. Direct methods provide a better insight. They may be local methods of observation such as optical or scanning electron microscopy or global methods such as the measure of relative density variations using a high precision microscale.

A traditional experimental technique is microscopy which is carried out on representative metallographic cuts of the sample. They can provide a direct image of the size, shape distribution and location of the cavities. The chemical nature of the particles may be determined by X-ray analysis. Special care however has to be taken when preparing the sample so that cavities are not filled or artifically widened during mechanical or chemical polishing [4, 16]. It also remains necessary to relate the data obtained on a cross-section of the material to a three-dimensional image. This relation cannot in general be established without making strong assumptions on the shape and distribution of the voids.

Relative density variations allow a global measure of damage. These measures however do not assess initial damage of the material since they give a difference between the density of a deformed sample and that of an undeformed one. The relation with the volume fraction of cavity is expressed by:

$$-\frac{\Delta d}{d_o}=\frac{C_v-C_{v_o}}{1-C_{v_o}}$$

where C_{v_o} is the initial volume fraction of cavities.

2.2. Observations

Damage has been observed under several testing conditions of temperature and applied stresses. Several damage mechanisms may occur:

(1) failure of particles (Fig. 1(a));
(2) decohesion around inclusions (Fig. 1(b)).

These two mechanisms depend on the relative values of the failure stress of the particle and the matrix. Furthermore, elongated particles tend to break while spherical ones would lead to decohesion [1, 3, 8, 19, 28].

Another mechanism involves decohesion between grains by failure of the grain boundary or phase boundary (Fig. 1(c)) [6, 9, 10, 20, 25]. In some instances, initial damage exists before the start of deformation. This is due to the previous history of the material (e.g. rolling) and may be inhomogeneously distributed in the material (Fig. 1(d)).

Qualitative observations may show the growth and coalescence of cavities (Fig. 1(e), (f)). Quantitative metallography may also give some information about the evolution of the number of cavities, the growth rate and shape factor with strain and strain path [24]. Figure 2a shows a typical result concerning the evolution of damage measured by relative density changes with strain during cold deformation of copper

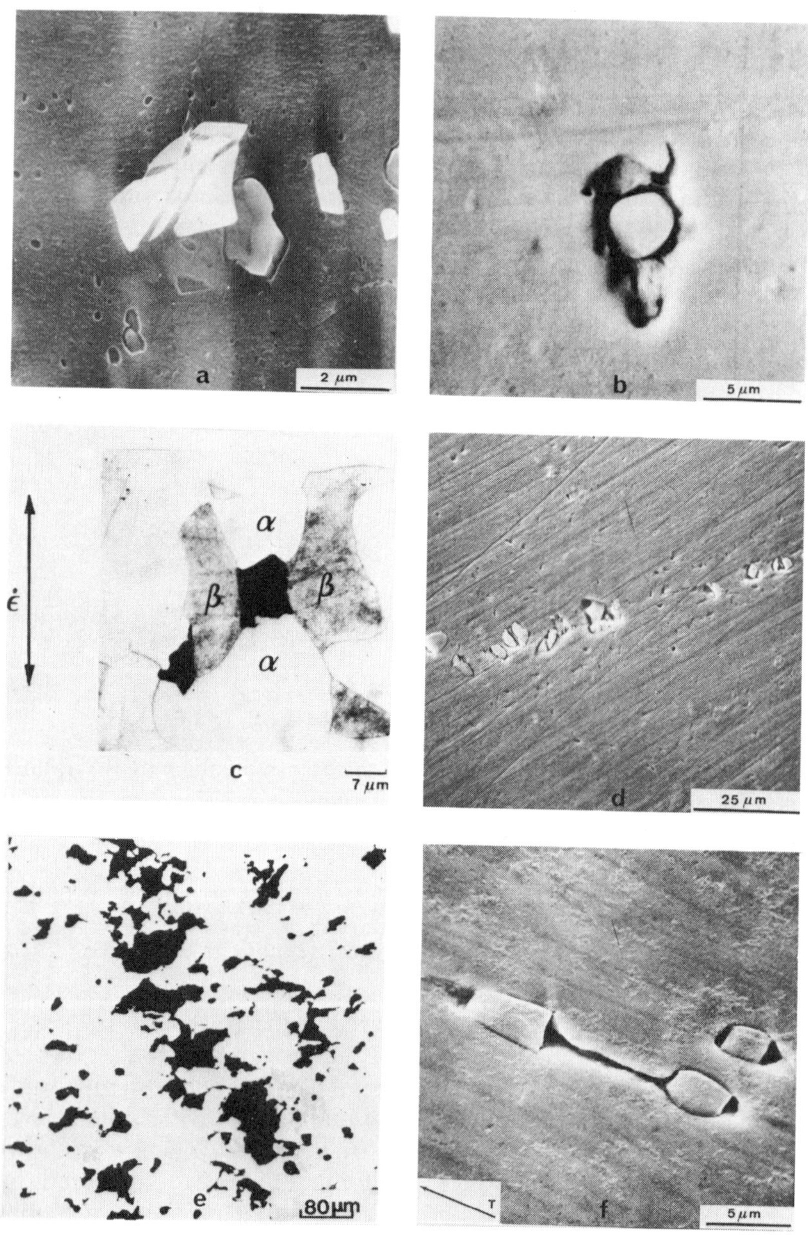
a
2 μm
b
5 μm
α
β
β
α
$\dot{\epsilon}$
c
7 μm
d
25 μm
e
80μm
T
f
5 μm

for several strain paths and different distributions of particles. The volume fraction of cavities increases with equivalent strain and is more important for equibiaxial strain paths than for simple tension. The distribution of particles also affects the cavitation level [22]. It is to be noted that the relative density change is of the order of 10^{-3} for an equivalent strain of 0·5.

Figure 2b presents plots of cavity volume fractions versus strain for tensile superplastic deformation of two-phase brasses with different compositions. For these materials which are very prone to cavitation, the cavity volume fraction may reach large values (up to a few per cent) and is particularly sensitive to the volume fraction of the harder α-phase. Such large values are however associated with large elongations of several hundred per cent which are made possible only by the high plastic stability due to large values of the strain rate sensitivity parameter. The cavitation level also increases with strain rate and grain size of the material.

3. ANALYSIS OF CAVITY GROWTH

After nucleation, cavity growth may be controlled either by vacancy diffusion toward the cavity or plastic deformation of the surrounding region. The relative importance of each mechanism varies greatly according to the thermomechanical conditions of forming. Plastic deformation is always the main mechanism for cold forming whereas diffusion processes are predominant at high temperature, low strain rates and for relatively small cavities. However, since large strains are involved in superplastic metal forming, plastic deformation controlled growth is the only relevant mechanism.

Consequently, the analysis of cavity growth requires a model for the growth of a void in a plastic matrix submitted to biaxial stress states. A model for the growth of a spherical void in a non-hardening plastic

---

FIG. 1. Observations of damage. (a) Failure of the particle in 3003 Al alloy after uniaxial tension at room temperature (J. H. Schmitt Thèse D.I). (b) Decohesion in copper after equibiaxial stretching at room temperature. (c) Damage at the phase boundary in an α/β brass after uniaxial tension in hot forming. (d) Alignment of particles in a tough pit copper. (e) Coalescence of voids at grain boundary after uniaxial tension in hot forming of an α/β brass. (f) Coalescence of voids around particles after uniaxial tension at room temperature of a tough pit copper.

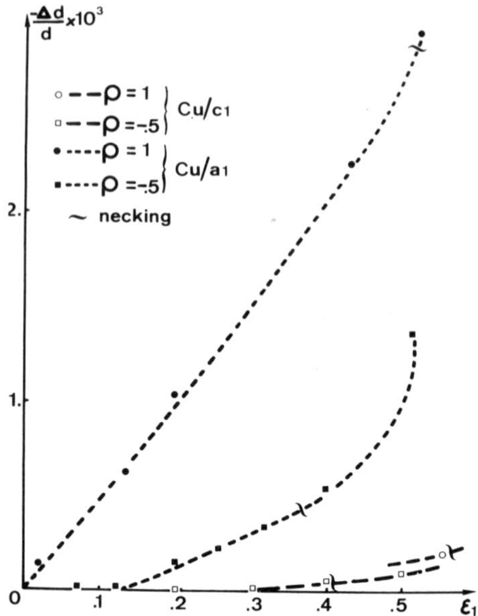

FIG. 2a. Influence of the strain path ρ and of the inclusion content for two types of copper: tough pit copper Cu/a1 and oxygen free copper Cu/c1.

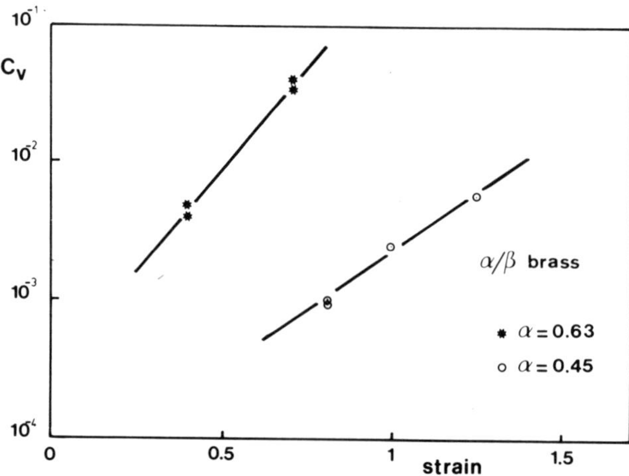

FIG. 2b. Influence of the second phase volume fraction on the volume fraction of nucleated cavities for α/β brass.

matrix has been developed by Rice and Tracey [21]; it has been chosen here in order to predict the evolution of cavity volume fraction with strain during cold metal forming and high temperature superplastic forming. The restrictive assumptions of the model will be discussed later and some simple extensions will be proposed based on physical considerations.

According to the model, the change in volume and shape of a spherical cavity of current radius R is given by

$$\frac{\dot{R}_i}{R} = C\dot{\varepsilon}_{\mathrm{i}}^{\infty} + D\dot{\varepsilon}_{\mathrm{e}}^{\infty} \tag{1}$$

where $\dot{R}_i$ is the radial velocity of the void surface in the ith direction, $\dot{\varepsilon}_{\mathrm{i}}^{\infty}$ the applied strain rate far from the void, C a strain concentration factor, $\dot{\varepsilon}_{\mathrm{e}}^{\infty}$ the applied equivalent strain-rate and D the volume growth rate. The first term $C\dot{\varepsilon}_{\mathrm{i}}^{\infty}$ measures the shape changes of the void without volumic growth, the second term $D\dot{\varepsilon}_{\mathrm{e}}^{\infty}$ is a spherical term associated with isotropic volume expansion.

For a von Mises material, D is given by the following expression.

$$D = 0{\cdot}558 \sinh \mu + \nu 0{\cdot}008 \cosh \mu$$

with

$$\mu = \frac{1}{2}\frac{1+\alpha}{\sqrt{1-\alpha+\alpha^2}} \tag{2}$$

and

$$\nu = -\frac{3\rho}{2+\rho} \quad \text{for} \quad -0{\cdot}5 \leqslant \rho \leqslant 1$$

where for a linear stress path

$$\alpha = \frac{\sigma_2}{\sigma_1} = \text{const.}$$

and

$$\rho = \frac{\varepsilon_2}{\varepsilon_1} = \text{const., at } \varepsilon_1 \geqslant \varepsilon_2$$

D is here a constant for any given linear stress path.

If C is assumed to remain constant, integration of (1) leads to:

$$\mathrm{Log}\,\frac{R_{\mathrm{i}}}{R_{\mathrm{o}}} = A_{\mathrm{i}}\varepsilon_{\mathrm{i}}^{\infty} \tag{3}$$

where the A_i are defined by:

$$A_i = C + \frac{2}{\sqrt{3}} D(1+\rho+\rho^2)^{1/2} \frac{\varepsilon_1^\infty}{\varepsilon_i^\infty} \tag{4}$$

so that

$$A_1 = C + \frac{2}{\sqrt{3}} D(1+\rho+\rho^2)^{1/2}$$

$$A_2 = C + \frac{2}{\sqrt{3}} D \frac{(1+\rho+\rho^2)^{1/2}}{\rho}$$

$$A_3 = C - \frac{2}{\sqrt{3}} D \frac{(1+\rho+\rho^2)^{1/2}}{1+\rho}$$

Equations (4) express the strain concentration factors around the cavity for each direction.

In plane strain $\rho = 0$, and A_2 is not defined so that another formulation of (3) is preferred

$$\text{Log} \frac{R_i}{R_o} = B_i \varepsilon_e^\infty \tag{5}$$

where the B_i are defined by:

$$B_i = C \frac{\varepsilon_i^\infty}{\varepsilon_e^\infty} + D \tag{6}$$

In (6), D may be simplified for a certain domain of applications. It is then easy to express the volume V of the cavities (if V_o is the initial volume):

$$\text{Log} \frac{V}{V_o} = 3D\varepsilon_e^\infty \tag{7}$$

If the cavities in the material are all of the same size and if their number remains constant:

$$\text{Log} \frac{C_v}{C_{v_o}} = 3D\varepsilon_e^\infty \tag{8}$$

where C_v is the volume fraction of the cavities.

Equations (5)–(8) describe the evolution of cavities which are assumed to pre-exist in the material (with volume fraction C_{v_o}) and to be, at least initially, spherical. When one combines eqns. (2) and (8) one can predict that the cavity volume fraction is larger in equibiaxial expansion than in tension for the same equivalent strain; this prediction is in agreement with experimental results.

We shall now propose a few extensions to the basic model so that it may be applicable, even with some necessary approximations, to a wider range of situations. The model has been developed for non-hardening materials, which is fully justified for superplastically deforming materials (provided that no excessive grain growth occurs). For strain hardening materials, the model may be applied in a step by step procedure, where at each step the material is taken as perfectly plastic with an updated yield stress.

Continuous nucleation is not considered in the derivation of eqn. (8). The increase of the number of voids can be taken into account through a rate of nucleation which is defined as:

$$N = \frac{\mathrm{d}n}{\mathrm{d}\varepsilon_e} \tag{9}$$

where n is the number of cavities as a function of the current equivalent strain ε_e.

It is assumed that a void nucleates with a given volume V_o, for example, the size of the particle in a decohesion process. We then write the current value V of the volume of the cavity:

$$V = V_o f(\varepsilon_e - \varepsilon_e^o) \tag{10}$$

with a growth function f where ε_e is the current strain and ε_e^o the strain at which nucleation starts. For a material with an initial number of voids n_o and nucleation rate N, the current volume fraction of voids at any strain is given by

$$C_v = C_v^o + \int_o^{\varepsilon_e^1} V_o N(\varepsilon) f(\varepsilon_e^1 - \varepsilon)\,\mathrm{d}\varepsilon \tag{11}$$

where C_v^o is the volume fraction due to the initial n_o voids at the current strain ε_e^1:

$$C_v^o = n_o V_o f(\varepsilon_e^1) \tag{12}$$

With hot deformation, cavities usually do not pre-exist but are growing at first by vacancy diffusion up to a critical radius R_c, after which plasticity controlled growth is predominant. This critical radius is attained after a strain $\varepsilon_{ec}^{\infty}$ so that eqn. (5) has to be modified as follows:

$$\text{Log}\,\frac{R_i}{R_c} = B_i(\varepsilon_e^{\infty} - \varepsilon_{ec}^{\infty}) \tag{13}$$

This formulation has been used by Shang and Suéry [23]. Other physical phenomena have been considered which were not initially in the basic model.

When the volume fraction of cavities becomes relatively large as during superplastic deformation (Fig. 2b), the material will deform faster than in the absence of cavities. This situation was at first considered [7, 26] in the case of uniaxial tension of a single phase material containing uniformly distributed cavities. It was later adapted for the special case of two-phase α/β brasses [23].

It was shown indeed that owing to the quasi-undeformability of the α-phase, the cavity growth is controlled by the plastic strain of the softer β-phase [5] so that in eqn. (8) the strain has to be replaced by the strain in the softer phase. Taking into account all these considerations and introducing an initial strain at the onset of the plasticity controlled growth mechanism (as in eqn. (13)), the equation for proportional growth of cavitation level with total strain in the softer phase for uniaxial tension was found to be:

$$\text{Log}\,\frac{C_v}{C_{v_c}} = 3D(\varepsilon_S^{\infty} - \varepsilon_{S_c}^{\infty}) - (\varepsilon_{ST}^{\infty} - \varepsilon_S^{\infty}) \tag{14}$$

The subscript S refers to the softer β-phase, $\varepsilon_{ST}^{\infty}$ is the total strain which takes into account the presence of cavities and is given by:

$$\varepsilon_{ST}^{\infty} = \varepsilon_S^{\infty} + \text{Log}\,[C_{v_c}(\exp\,(3D(\varepsilon_S^{\infty} - \varepsilon_{S_c}^{\infty})) - 1) + 1] \tag{15}$$

These equations describe the cavitation in an ideal material consisting only of the softer phase. With the presence of the non-deformable harder phase, it is further assumed that, when the specimen is tested with a prescribed strain rate $\dot{\varepsilon}_T^{\infty}$ the relation

$$\dot{\varepsilon}_{ST}^{\infty} = \frac{1}{G}\,\dot{\varepsilon}_T^{\infty} \tag{16}$$

holds, so that:

$$\varepsilon_{ST}^{\infty} = \int \frac{1}{G}\,\dot{\varepsilon}_T^{\infty}\,dt$$

where G is a parameter which may be a function of the testing conditions and of the microstructure of the material, namely the phase proportions and the presence of the cavities. If G is a function only of the volume fraction of the non-deformable phase as can be assumed for low deformation levels, it identifies with $g(\alpha)$ which represents the strain rate amplifying effect of the harder phase on the flow behavior of the softer phase in the superplastic range. This function g was introduced [27] for a material without cavities. It is a rapidly decreasing function of the α-phase volume fraction. For small volume fraction of cavities, $\varepsilon_{ST}^{\infty} \simeq \varepsilon_{S}^{\infty}$ and

$$\varepsilon_{S}^{\infty} = \frac{1}{g(\alpha)} \varepsilon_{T}^{\infty}$$

so that eqn. (14) becomes:

$$\text{Log} \frac{C_{v}}{C_{v_c}} = \frac{3D}{g(\alpha)} (\varepsilon_{T}^{\infty} - \varepsilon_{Tc}^{\infty}) \tag{17}$$

This equation is in very good agreement with experimental results (Fig. 2b), up to a cavity volume fraction of about 0·5% after which it overestimates the cavitation level because G can no longer be identified with $g(\alpha)$ and also eqn. (15) has to be used to compute $\varepsilon_{ST}^{\infty}$. This is linked to macroscopical effects which will be considered in the following section. The basic model for cavity growth was developed in order to describe the evolution of a spherical void in a plastic matrix for a small strain increment. After this step, the cavity is no longer spherical and the model should not be used. It has however been assumed implicitly through this analysis that either the cavity shapes do not change drastically so that a sphere remains a good approximation or that the change may be taken into account by a change in the strain concentration parameter C (eqn. (1)) which can become directional.

4. MACROSCOPICAL EFFECTS OF DAMAGE

The generation and growth of cavities in a plastic material change mechanical properties and increase the tendency towards instability by reducing the values of the strain hardening coefficient and the strain rate sensitivity. Such effects have been analysed by several authors [13, 15] and were found to be more important for equibiaxial stretching than for plane strain or uniaxial tension.

Softening may also occur for sufficiently large volume fractions of cavities and it is consequently observed mainly during superplastic deformation. It may be expressed for two-phase materials such as α/β brasses through the previously introduced G function which is then written as $G(\alpha, C_v)$ and which must increase with strain. This corresponds to an apparent decrease of the strain concentration factor of the α-phase. As the volume fraction or size of cavities becomes very large however, coalescence takes place and leads to apparent hardening and fracture without external necking (Fig. 3). Such a mode of failure occurs only for alloys with a high volume fraction of the harder phase whereas localized necking is observed for alloys with a low volume of harder phase.

The effect of the internal damage on the occurrence of plastic instability and on the level of the forming limit can be taken into account through a statistical model [14]. In this model, the probability of geometrical defect due to alignments of voids in the thickness direction is given (in the case of damage by decohesion) by:

$$P(\xi) = \begin{bmatrix} \nu_o \\ \nu_o(1-\xi) \end{bmatrix} C_v^{\nu_o(1-\xi)}(1-C_v)^{\nu_o\xi} \tag{18}$$

where ξ characterizes the defect as the ratio of the effective thickness in the vicinity of the defect to the thickness in the undamaged region; $D = 1-\xi$ measures the damage and increases with strain. ν_o is the ratio

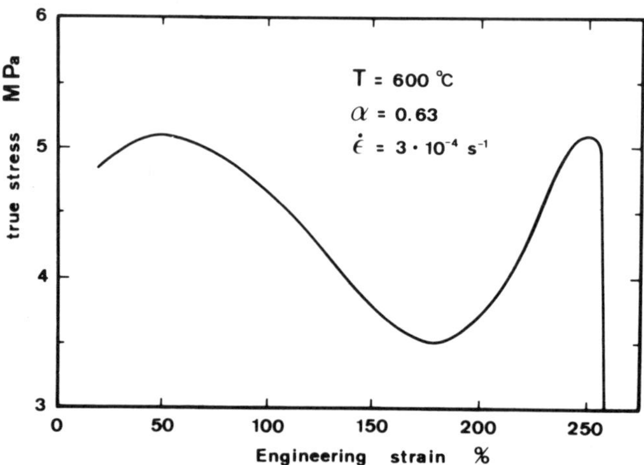

FIG. 3. Softening effect due to the growth of damage in an α/β brass.

of the initial thickness of the sheet to the initial radius of the spherical void and C_v is the current volume fraction of voids.

The critical defect however is not the most probable defect but the one with the greatest amplitude of damage having a sufficiently high probability of existence so that the defects can interact. Introducing a threshold probability derived from experimental observations of defect interactions [2], the critical defect for a given volume fraction of voids can be calculated and the step by step evolution of this defect is obtained. Such an evolution due to internal damage can be introduced in a plastic instability calculation using a two zone material [11, 12, 17, 18].

As the volume fraction of voids depends on the strain and strain path, the evolution of the critical defect and the forming limit is found to depend upon these parameters. The shape and position of the forming limit diagram is therefore affected by internal damage.

The probability function has been found to depend also on ν_o as defined in eqn. (18). It has been observed that damage increases with

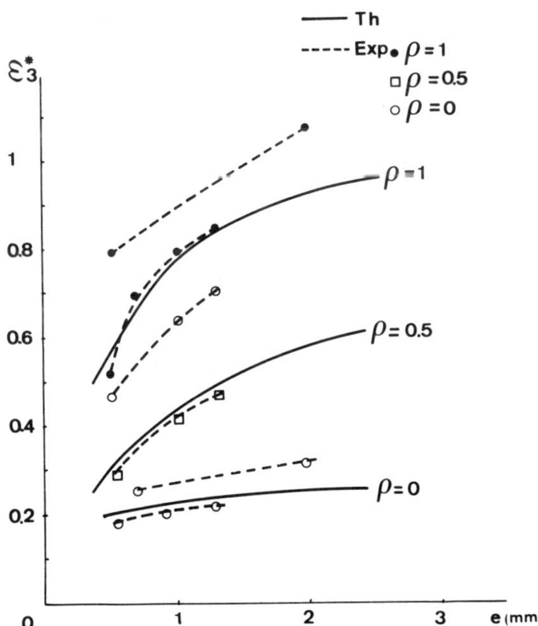

FIG. 4. Prediction of the influence of the thickness of a sheet on the limiting thickness strain.

decreasing values of ν_o for a given volume fraction of voids C_v. This means that for a given thickness of sheet metal and a given C_v, a large number of small voids is less critical than a small number of large cavities. Another characteristic of the dependence of damage on ν_o is that a given size of voids is less dangerous in thicker sheets, at constant C_v.

On Fig. 4 the thickness effect is calculated and compared with experiments. It can be seen that the effect is larger in equibiaxial stretching than in plane strain and does not exist in uniaxial tension (for damage decohesion). These theoretical results are in complete agreement with experimental observations and the internal damage appears as one important parameter controlling the thickness effect, among other factors such as strain distribution changes.

5. CONCLUSION

In cases where the growth of damage is controlled by a strain concentration phenomenon, in hot forming as well as cold forming, a single model can be used in order to describe local changes in volume and shape of the cavities. This model is an extension of a model for growth of a spherical cavity in a non-hardening plastic matrix for small strains. It would therefore be necessary to improve this model by taking account of the large changes in shape of the voids at large plastic strains and the rheology of the matrix, in particular strain and strain rate hardening. This study has been carried out using a random distribution of voids; the effect of an oriented distribution on the macroscopical properties has to be considered. An influence on the anisotropy of ductile fracture and on the development of plastic instability is expected.

ACKNOWLEDGEMENTS

The authors would like to thank Professor H. M. Shang, National University of Singapore, for his contribution to the analysis of cavitation in superplastic alloys. The collaboration of J. H. Schmitt, on damage by failure and its influence on the forming limit, and of F. Barlat, on observations of void distribution and growth in dual-phase steels, is gratefully acknowledged.

REFERENCES

1. Argon, A. S., J. Im and R. Safoglu. Cavity formation from inclusions in ductile fracture, *Met. Trans.*, **6A** (1975), 825–837.
2. Asceirad, O. and D. V. Wilson. Effects of microstructure on forming limits in biaxial stretching, *X-th I.D.D.R.G. Congress*, Porculiss Press, London, 1978, 155–166.
3. Ashby, M. F. Work hardening of dispersion-hardened crystals, *Phil. Mag.*, **14** (1966), 1157–1178.
4. Barlat, F., J. M. Jalinier and J. H. Schmitt. Observation de l'endommagement—polissage par bombardement ionique. *Matériaux et Tech.*, to be published 1984.
5. Belzunce, J. and M. Suéry. Normalization of cavitation in superplastic α/β brasses with different phase proportions, *Scrip. Met.*, **15** (1981), 895–898.
6. Cane, J. and G. W. Greenwood. The nucleation and growth of cavities in iron during deformation at elevated temperatures, *Met. Sci.*, **9** (1975), 55–60.
7. Edward, G. H. and M. F. Ashby. Intergranular fracture during power-law creep, *Acta Met.*, **27** (1979), 1505–1518.
8. Eshelby, J. D. The determination of the elastic field of an elliptical inclusion and related problems, *Proc. R. Soc.*, **A241** (1957), 376–396.
9. Gandhi, C. and M. F. Ashby. Fracture mechanism maps for materials which cleave, *Acta Met.*, **27** (1979), 1565–1602.
10. Hull, D. and D. E. Rimmer. *Phil. Mag.*, **4** (1959), 673–681.
11. Hutchinson, J. W. and K. W. Neale. Sheet necking II: time independent behavior. In: *Mechanics of Sheet Metal Forming*, Plenum Press, New York, London, 1978, 127–153.
12. Hutchinson, J. W. and K. W. Neale. Sheet necking III: strain-rate effects. In: *ibid.*, 269–285.
13. Jalinier, J. M., B. Christodoulou, B. Baudelet and J. Jonas. The four stages of flow localization in tensile samples containing geometrical defects, *Formability, ASM, AIME* (1977), 29–45.
14. Jalinier, J. M. and J. H. Schmitt. Damage in sheet metal forming: II. plastic instability, *Acta Met.*, **30** (1982), 1799–1809.
15. Jonas, J. J. and B. Baudelet. Effect of crack and cavity generation in tensile stability, *Acta Met.*, **25** (1977), 43–50.
16. Lehtinen, B. and A. Melander. An ion polishing technique to reveal voids at particles, *Metallography*, **13** (1980), 283–287.
17. Marciniak, Z. Limit strains in the processes of stretch-forming sheet metals, *Int. J. Mech. Sci.*, **9** (1967), 609–620.
18. Marciniak, Z. Sheet metal forming limits. In: *Mechanics of Sheet Metal Forming*, Plenum Press, New York, London, 1978, 215–235.
19. Melander, A. Void initiation in a random distribution of spherical particles, *Mater. Sci. Engng*, **37** (1979), 137–142.
20. Raj, R. and M. F. Ashby. Intragranular fracture at elevated temperature, *Acta Met.*, **23** (1975), 653–666.

21. RICE, J. R. and D. M. TRACEY. On the ductile enlargement of voids in triaxial stress field, *J. Mech. Phys. Solids*, **17** (1969), 201–217.
22. SCHMITT, J. H., J. M. JALINIER and B. BAUDELET. Analysis of damage and its influence on the plastic properties of copper, *J. Mater. Sci.*, **16** (1981), 95–101.
23. SHANG, H. M. and M. SUÉRY. Modelisation of cavitation in two-phase superplastic alloys under uniaxial stress systems, *Met. Sci.*, **18** (1984), 143–152.
24. SOUZANOBREGA, C., B. FIDELIS DA SILVA, G. FERRAN, J. M. JALINIER and B. BAUDELET. Effect of the inclusion content on damage generation during sheet metal forming of low carbon steel, *Mém. Sci. Rev. Met.*, **3** (1980), 293–302.
25. SPEIGHT, M. V. and W. BEERE. Vacancy potential and void growth on grain boundaries, *Met. Sci.*, **9** (1975), 190–191.
26. STOWELL, M. J. Cavity growth in superplastic alloys, *Met. Sci.*, **14** (1980), 267–272.
27. SUÉRY, M. and B. BAUDELET. Hydrodynamical behavior of a two-phase superplastic alloy: α/β brass, *Phil. Mag.*, **41A** (1980), 41–64.
28. TANAKA, K., T. MORI and T. NAKAMURA. Cavity formation at the interface of a spherical inclusion in a plastically deformed matrix, *Phil. Mag.*, **18** (1970), 267–279.

32

Plasticity Approach to the Mechanics of Softening and Ductile Fracture of Metals

W. SZCZEPIŃSKI

Institute of Fundamental Technological Research, Warsaw, Poland

ABSTRACT

Several theoretical and experimental models of the possible mechanisms of ductile fracture of metals are discussed. In each of these models a system of cracks of sufficiently weak grain boundaries is assumed to exist in the material. In the theoretical analysis of these mechanisms a rigid–plastic model is assumed. It is shown that in experimental models the interaction of hardening and softening factors leads to the distinctly marked instabilities in the stress–strain diagrams.

1. INTRODUCTION

The mechanics of softening and ductile fracture of metals is very complex and still not fully examined. Depending on the temperature, the rate of deformation and the structure of the metal various mechanisms may be responsible for the softening and fracture. Coalescence of voids, interaction between variously oriented cracks leading to local internal microdecohesion contribute to the progressing process of ductile fracture which may be considered as the factor responsible for the softening effect. In polycrystalline metals intercrystalline void formation and intergranular sliding may lead to the softening and finally to the ductile fracture of the aggregate. These phenomena cannot be analysed in terms of the classical fracture mechanics based on the assumption of the elastic behaviour of the material. On the other hand the possibilities of application of the methods of the theory of plasticity to the analysis of the phenomena of softening and ductile fracture have been much less examined.

The aim of the paper is to present an attempt to obtain a deeper insight into the mechanics of softening and ductile fracture of metals. Several possible theoretical and experimental models of the mechanisms of ductile fracture are presented and discussed. In each of these models a system of cracks or sufficiently weak grain boundaries is assumed to exist in the body. In the theoretical analysis of these mechanisms a rigid–plastic model of the material is assumed.

Plastic behaviour of metals is described by various strain hardening rules, for example by isotropic, kinematic or various mixed hardening rules (for references see for example ref. 7). Such hardening models may be expressed by the yield condition

$$f(\sigma_{ij}-\alpha_{ij})-\sigma_p(\alpha)=0 \tag{1}$$

accounting for the translation and expansion of the initial yield surface in the stress space. In this expression α_{ij} and α are tensorial and scalar hardening parameters respectively.

It is evident, however, that in the advanced stages of plastic deformation the softening effects mentioned above may play an important role in the global behaviour of metals. Thus the yield criterion should include not only the hardening parameters α but also softening parameters β connected with the reduction of the effective cross-sectional area. Generally such yield condition may be written in the form

$$f(\sigma_{ij}, \alpha, \beta)=0 \tag{2}$$

In the following sections we will discuss certain models of the possible mechanisms of the evolution of the softening parameter β.

2. INTERNAL MICRONECKING

The mechanism of internal necking has been discussed in some works. Rogers [8] compared the tensile ductility of two theoretical models of a perfectly plastic material. One is a solid cylindrical rod and the other is a cable consisting of a bundle of cylindrical strands. Each of the strands necks down to a point in the same manner as the rod. However, total elongation of the bundle is much smaller than that of the rod. This idea

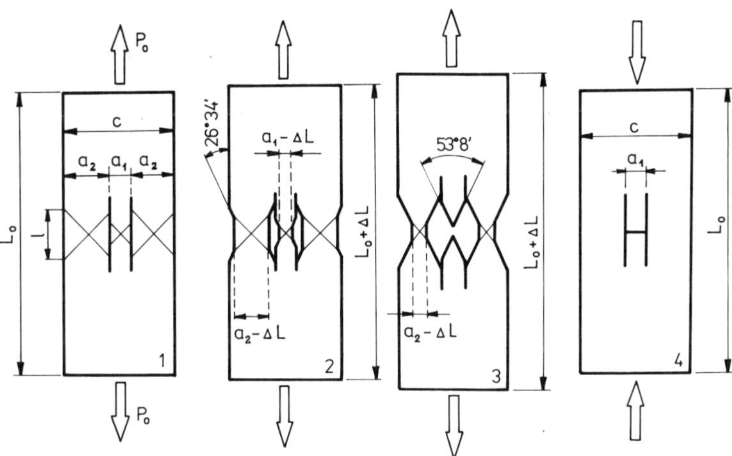

FIG. 1. Necking and ductile internal fracture in the plane strain rigid–plastic model with two slits.

will be used below in a study of the evolution of the damage inside the material.

Let us consider an idealized plane strain model of the material with defects in the form of two slit cracks parallel to the direction of the uniaxial tensile stress. The initial configuration is shown as the stage 1 in Fig. 1. The central strip of width a_1 is narrower than the two outer strips of width a_2.

It is evident that if the elastic–plastic material is assumed there is no stress concentration in the elastic stage of the loading history. However, during the plastic stage of deformation there appears strong strain concentration caused by the necking of the strips between the cracks. In order to estimate the amount of this strain concentration we will assume that the model is made of a rigid–perfectly plastic material. Each of the strips separated by the cracks necks down according to the slip-line solution [4]. Initial configuration of slip-lines is shown in the left-hand figure (stage 1). Plastic yielding begins when the pulling force per unit thickness reaches the value $P_o = c\sigma_{p1}$, where σ_{p1} is the yield locus of the material. Stage 2 in Fig. 1 shows how the strips separated by the slits begin to neck. Stage 3 illustrates how the central strip has been separated into two parts due to the necking down to a point, while the outer strips of larger width necked down to a certain degree only without separation.

Let us consider stage 2 of deformation. If the total deformation of the model is equal to ΔL, the lateral contraction of all three strips will be equal to $-\Delta L$. Thus the conventional lateral strain in the central strip may be written as

$$\varepsilon_{a_1} = -\frac{\Delta L}{a_1} \tag{3}$$

and in the two outer strips is equal to

$$\varepsilon_{a_2} = -\frac{\Delta L}{a_2} \tag{4}$$

Comparing both expressions we obtain the internal strain concentration factor

$$f = \frac{\varepsilon_{a_1}}{\varepsilon_{a_2}} = \frac{a_2}{a_1} \tag{5}$$

which is valid for intermediate stages of deformation when the central strip has not been separated. Comparing the lateral strain in the central strip with that of the block without slits we obtain for the intermediate stage of deformation another definition of the strain concentration factor

$$f^* = \frac{\varepsilon_{a_1}}{\varepsilon_c} = \frac{c}{a_1} \tag{6}$$

related to the width of the body. Such a strain concentration factor may reach very high values.

Consider now the compression process of the model from the configuration shown as stage 3 to stage 4. Idealizing the compression process we obtain the configuration shown as stage 4 in Fig. 1. The two separated parts of the central strip have been flattened leading to the formation of an additional horizontal crack between the two initial parallel slits. Comparing the initial structure of the model (stage 1) with the final one (stage 4) we may define the factor of relative reduction of cross-sectional area as

$$\beta = \frac{a_1}{c} \tag{7}$$

which may be interpreted as one of the possible factors of the softening of the material.

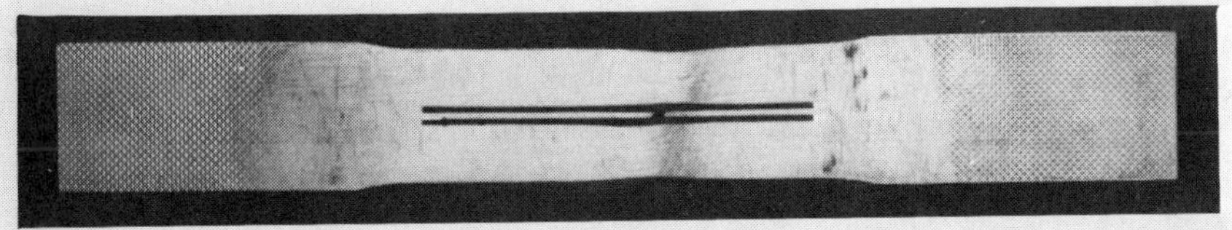

FIG. 2. Internal fracture in the experimental model with two prepared slits under tension.

Since the analysis presented above is based upon the assumption of the idealized rigid–plastic model of the material simple experiments were performed in order to check whether the softening due to the local micronecking may occur in real commercial metals. Figure 2 shows the specimen of an AlMg2 aluminium alloy deformed plastically in uniaxial tension. Two parallel slits were prepared according to the scheme shown in the left-hand figure in Fig. 1. It is clearly visible that the necking process is only slightly advanced in the two wide strips while the narrow central strip suffered decohesion. This simple experiment shows that the theoretical analysis based on the slip-line technique has a real physical significance. Further experimental results concerning the progressing evolution of the factor β of relative reduction of cross-sectional area due to internal necking have been previously presented in ref. 9.

An example of the force–elongation diagram for the specimen with four prepared slits is shown in Fig. 3. These slits form five strips of unequal widths which are shown in the inset in the figure. The thickness of the specimen was 15 mm. Thus conditions of plastic deformation were close to those of plane strain. In the final sector of the diagram there appear sharp steps corresponding to the consecutive fracture of necking strips. The process of necking begins at the same moment in all strips but according to the slip-line analysis shown in Fig. 1 the fracture takes place consecutively depending on the width of the strip.

Completely different is the course of the internal micronecking mechanism for another configuration of defects which in idealized form is simulated in the specimens prepared according to Fig. 4. Such configuration of the prepared slits simulates two parallel cracks accompanied in the central part by narrow voids. In the elastic state of deformation there appears local concentration of stresses, which is

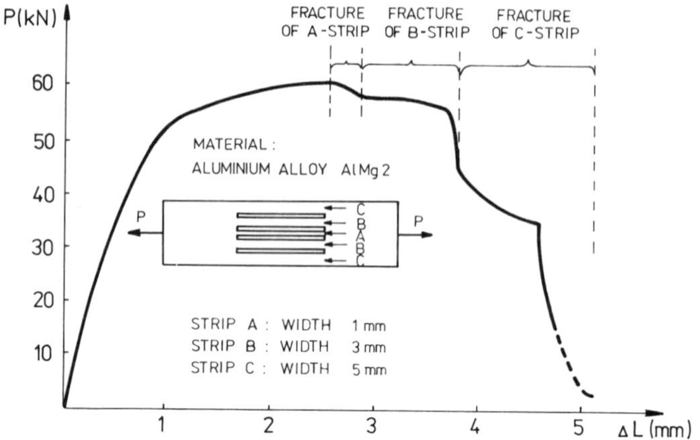

FIG. 3. Tensile force–elongation diagram for the experimental model with four prepared slits.

defined by the stress concentration factor

$$f=\frac{\sigma_2}{\sigma_1}=\frac{1}{\dfrac{l_1}{l}+\left(1-\dfrac{l_1}{l}\right)\dfrac{b_1}{b}} \tag{8}$$

Stresses in the narrow central part are

$$\sigma_2=\frac{\sigma}{\dfrac{b_1}{c}+\dfrac{2a}{c}\left[\dfrac{l_1}{l}+\left(1-\dfrac{l_1}{l}\right)\dfrac{b_1}{b}\right]} \tag{9}$$

When stresses $\sigma=\dfrac{P}{c}$ (P is the pulling force per unit thickness) reach a certain value this central part begins to yield plastically. The incipient

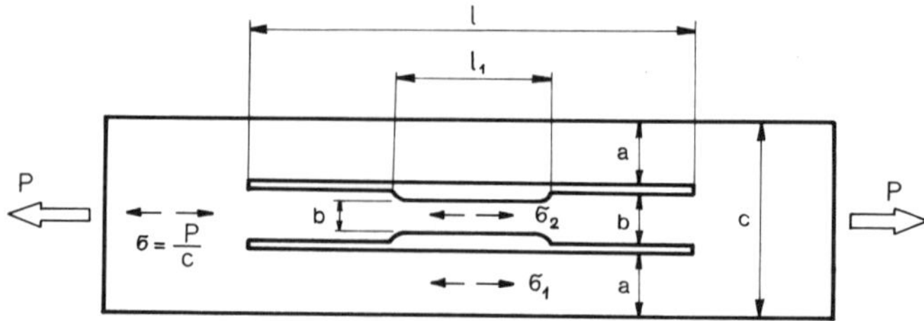

FIG. 4. Experimental model simulating two parallel cracks accompanied by narrow voids.

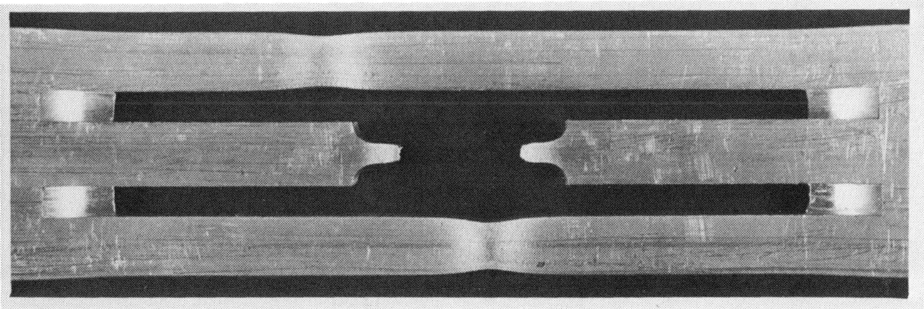

FIG. 5. Internal fracture in the experimental model prepared according to Fig. 4 and pulled in tension.

plastic flow is accompanied by small plastic deformations of the central part, because outer parts of the model still remain in the elastic state. The large plastic flow begins when stresses in the outer strips also reach the yield point. Since the specimen is made of an AlMg2 aluminium alloy displaying a strain hardening effect, further increase of the pulling force causes uniform deformation in the strips, which for a total elongation Δl is in the narrow strip equal to $\varepsilon_1 = \Delta l/l_1$ and in the outer strips is equal to $\varepsilon_2 = \Delta l/l$. Thus the uniform deformations are not equal and the central strip begins to neck much earlier than two outer strips. This is clearly seen in Fig. 5, which shows one of the tested specimens in a very advanced stage of deformation. The two outer

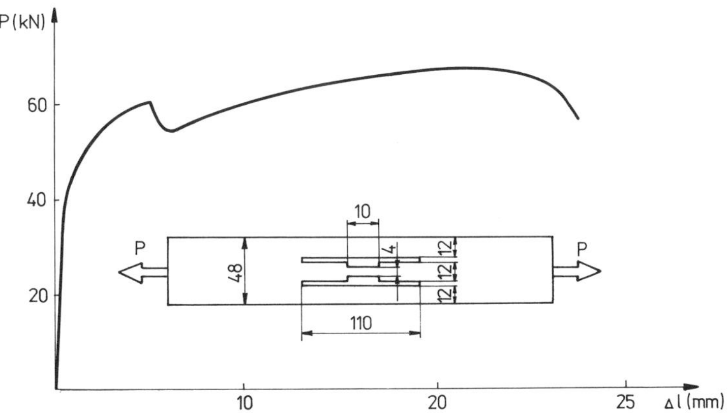

FIG. 6. Tensile force–elongation diagram for the experimental model prepared according to Fig. 5.

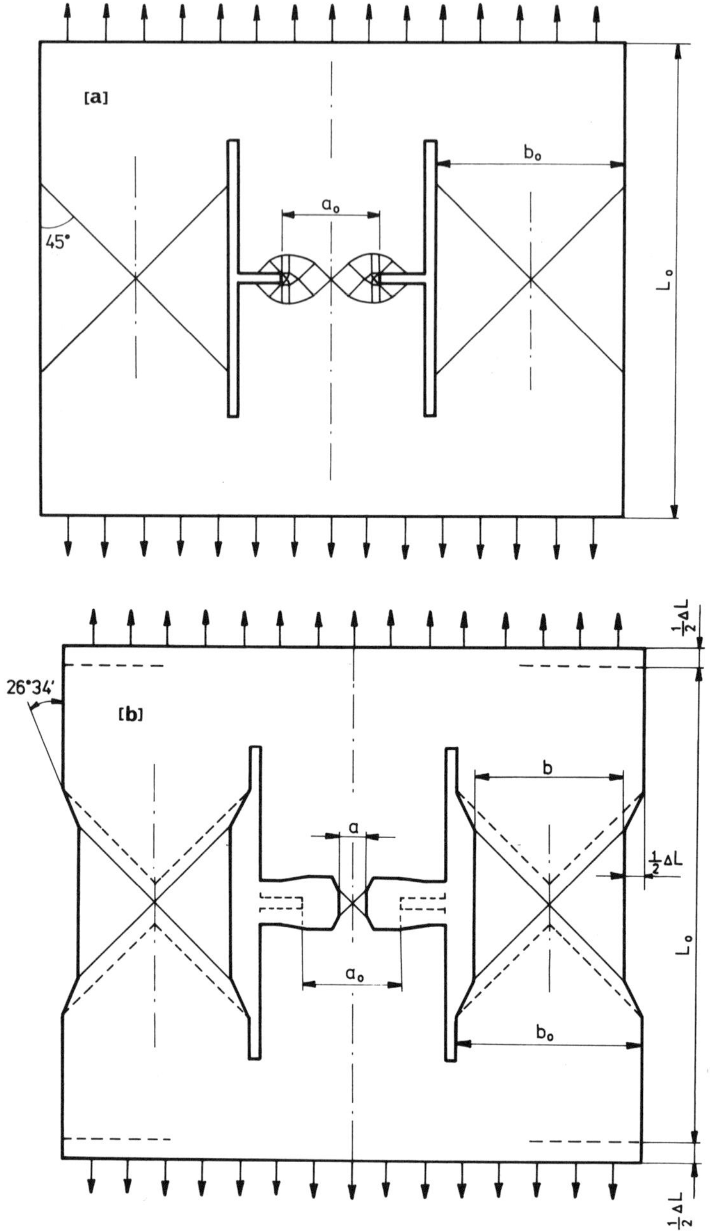

FIG. 7. Slip-line analysis of the process of deformation in a theoretical rigid–plastic plane strain model simulating two parallel cracks with short branches. (a) Initial configuration; (b) advanced stage of deformation.

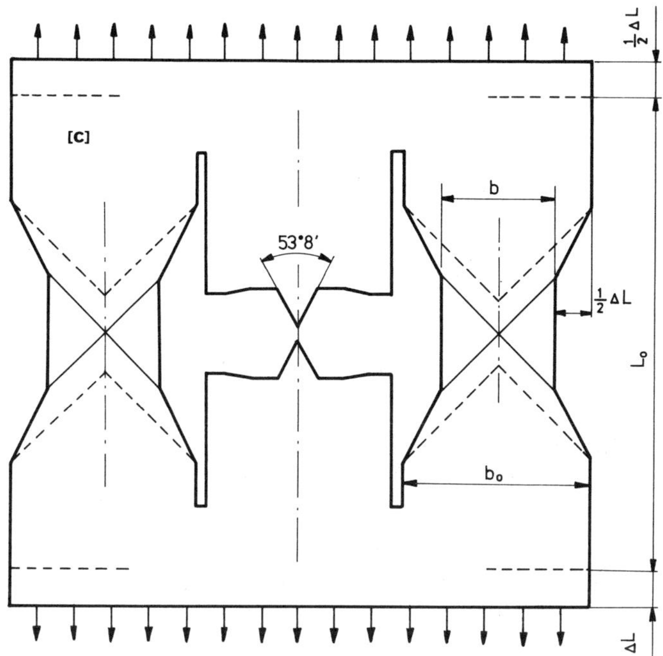

FIG. 7.— *contd.* (c) Fracture of the central part.

strips begin to neck, while the central narrow part has been separated into two parts much earlier.

The force–elongation diagram for one of the specimens of this kind is shown in Fig. 6. The sudden decrease in the pulling force in the initial portion of the diagram is connected with the necking and fracture of the central part of the specimen. This simple experiment indicates that special configurations of defects simulated in our specimen may lead to periodically unstable behaviour of plastically deformed metals connected with the reduction of the cross-sectional area.

3. INTERNAL DUCTILE MICROFRACTURE

Consider now still another configuration of defects in the form of two parallel cracks with short branches forming the internal notch. Figure 7(a) shows a model with such a configuration in the idealized form.

Assuming the rigid–perfectly plastic material of the model we may analyse the deformation process of such model with the use of slip-line technique. Initial configuration of slip-lines is shown in Fig. 7(a). The two outer strips neck down according to the scheme presented in Fig. 1. The theoretical solution to the process of deformation in the central strip is identical with that for notched bars (see, for example, refs. 5 and 6).

FIG. 8. Two stages of deformation of the experimental model simulating the configuration shown in Fig. 7(a).

Figure 7(b) shows the intermediate stage of the process of plastic deformation. Cavities of complex shape have been formed on both sides of the central strip. Further advancement of the process leads to separation of the central strip (Fig. 7(c)). In this slip-line solution, in which the material is assumed to have no hardening, the necking in the outer strips begins simultaneously with the plastic yielding of the notched part.

In the prepared specimen made of an AlMg2 aluminium alloy both outer strips suffer considerable uniform plastic deformation before necks begin to form. This is clearly seen in Fig. 8, in which two stages of plastic deformation are shown. The right-hand photograph shows the initial stage of the formation of necks.

One of the force–elongation diagrams for this type of specimen is shown in Fig. 9. Sharp instability in the initial part of the diagram is connected with the fracture of the notched strip. After a sudden drop the carrying capacity of the specimen has recovered due to the strain hardening effect in outer strips. Note that the effect observed here is similar to that shown in Fig. 6 for another idealized configuration of defects. The two idealized models of defects simulate the possible interaction between the hardening and softening parameters α and β

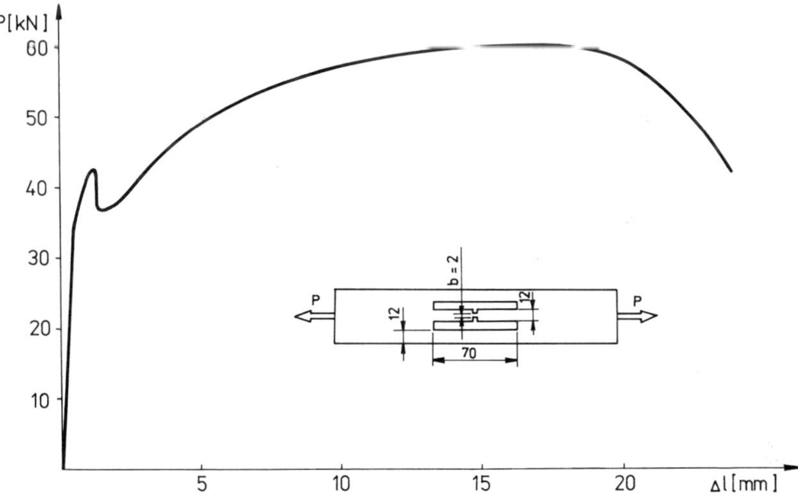

FIG. 9. Tensile force–elongation diagram for the experimental model shown in Fig. 8.

appearing in the yield condition (2), which may occur during plastic deformation of metals.

4. MODELS OF INTERCRYSTALLINE SLIDING AND FRACTURE

The possible role of intercrystalline sliding and fracture as a factor responsible for both the softening and hardening of metals will be discussed here with the use of an experimental model prepared according to the scheme presented in Fig. 10. The grain boundaries are simulated by rows of holes prepared in a block of a ductile AlMg2 aluminium alloy. The block is compressed between two steel platens. In a much wider study which will be published in a separate paper specimens with various angles $\varphi = 0°$, $10°$ and $20°$ and various ratios $\xi = (s-d)/s$ were tested. For the sake of brevity experimental results for the particular values $\varphi = 10°$ and $\xi = 0{\cdot}233$ only will be presented here. Figure 11 presents the initial portion of the force–shortening diagram for the compressed specimen. The unstable behaviour of the specimen during compression test is clearly visible. Sudden drops in the force have been caused by fracturing along the rows of holes inclined at the angle $\varphi = 10°$ to the vertical axis.

In Fig. 12 is presented the specimen compressed up to the end-point

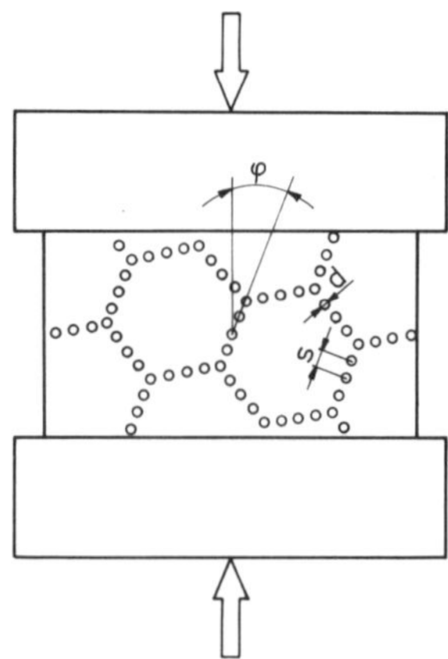

FIG. 10. Experimental model with the rows of holes simulating weak grain boundaries.

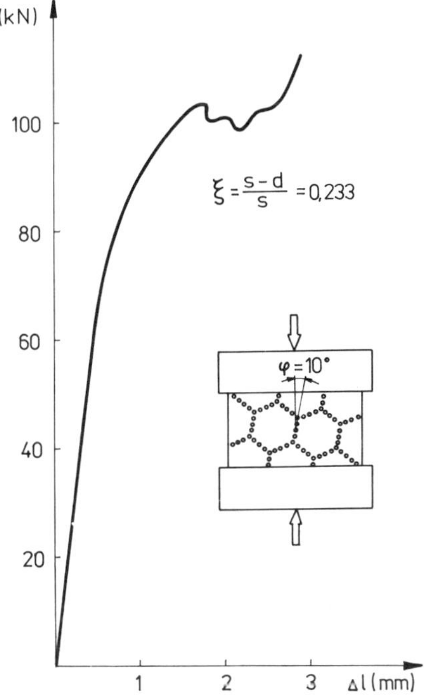

FIG. 11. Compressive force–shortening diagram for the experimental model shown in Fig. 10.

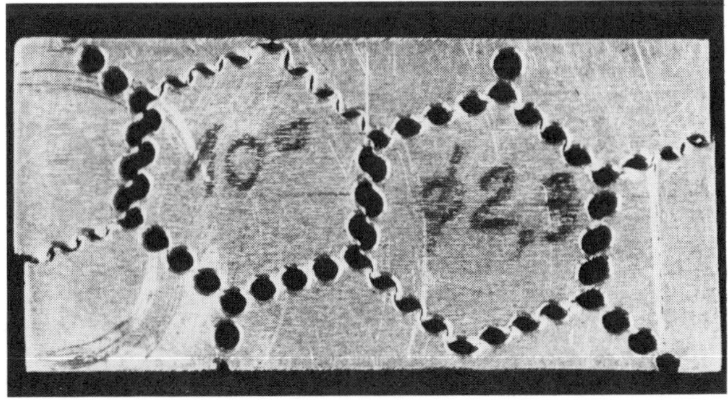

FIG. 12. The experimental model from Fig. 11 after compression test.

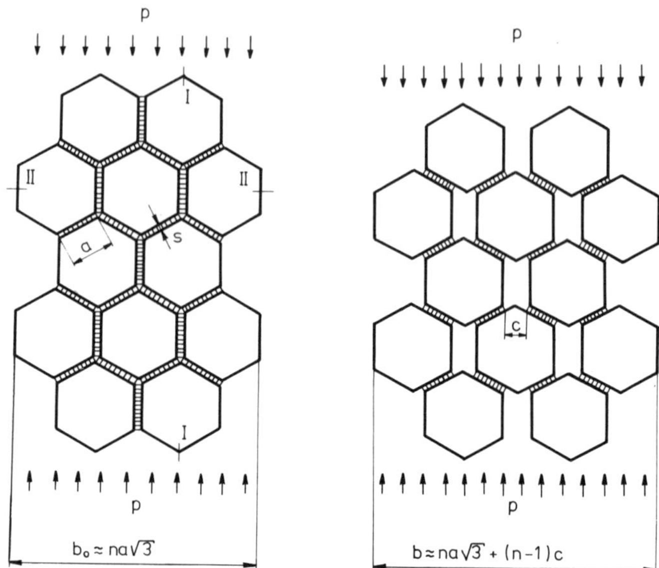

FIG. 13. Theoretical idealized model of a polycrystalline material.

of the diagram shown in Fig. 11. Separation of the ligaments between the holes along the most exposed rows of holes is clearly visible. Along the rows of holes slightly inclined to the horizontal direction the shear deformation is observed.

In order to explain such a behaviour of our experimental model let us consider a theoretical model composed of n hexagons in horizontal rows and m hexagons in vertical rows as shown in Fig. 13. Such a configuration corresponds to $\varphi = 0$. The hexagons are assumed to be joined together by thin layers of the thickness s simulating the grain boundaries. Let the model be loaded by uniaxial compression similarly to the previous experimental model.

Note that similar theoretical models were considered in several works, for example by Kelly [3] and Dyson [2], in the analysis of the growth of grain boundary cavities during creep. A similar model was used by Drucker [1] for the analysis of the strength of sintered carbides.

The hexagons in our model are taken as rigid and the ratio s/a is assumed to be small. The material of the layers between hexagons is taken as rigid–plastic. Figure 14 presents on an enlarged scale the

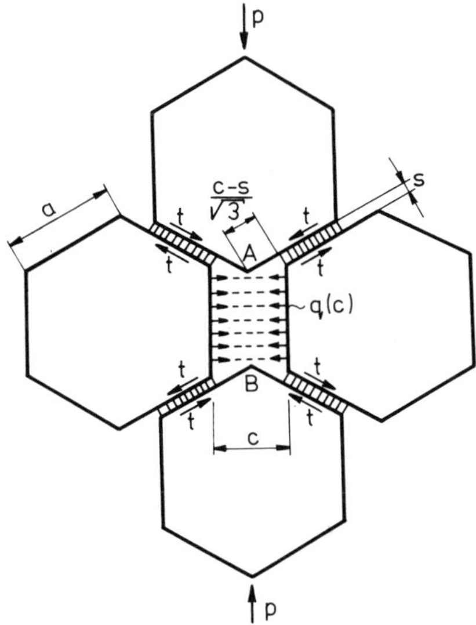

FIG. 14. Deformation mode of the model from Fig. 13.

configuration of four hexagons after deformation defined by the distance c between vertical edges $(c > s)$. Let us assume that in each layer between the hexagons there exists a system of voids or cracks perpendicular to the edge of the respective hexagon. Thus the layer may be treated as a system of individual elements of limited strength in tension. Therefore the stress $q(c)$ along the boundary AB may be taken as uniformly distributed up to the limit value c_o of the distance c between the edges of the hexagons by which the cohesion forces decrease to zero. It means that $q(c_o) = 0$ and moreover $q(c = s) = q_o$.

Let us assume a linear strain hardening relation between the shear stress t on the inclined surfaces of hexagons and the average strain ε in the lateral direction of the form

$$t = t_o(1 + r\varepsilon) \tag{10}$$

where r is the strain hardening factor and

$$\varepsilon = \frac{c - s}{s + a\sqrt{3}}$$

Writing the equation of virtual work for small deformations $(s < c < c_o)$

we obtain the following equation for the incipient plastic flow

$$pn(m-1)=[m(n-2)+(m-1)(n-1)]q(c) + \frac{4}{\sqrt{3}}(m-1)(n-1)(1+r\varepsilon)(1-\varepsilon)t_o \qquad (11)$$

The cohesion force along AB changes its value from q_o at the beginning of the deformation process $c=s$ to zero for $c=c_o$. Thus the first member of the right hand side of the yield condition (11) changes its value from

$$[m(n-2)+(m-1)(n-1)]q_o$$

for $c=s$ to zero for $c=c_o$.

Our model displays, therefore, the phenomenon of the upper and lower yield point at the beginning of the deformation process. This is shown in Fig. 15 in which the stress–strain diagram is presented.

For the advanced stage of deformation (see ref. 10) we obtain the

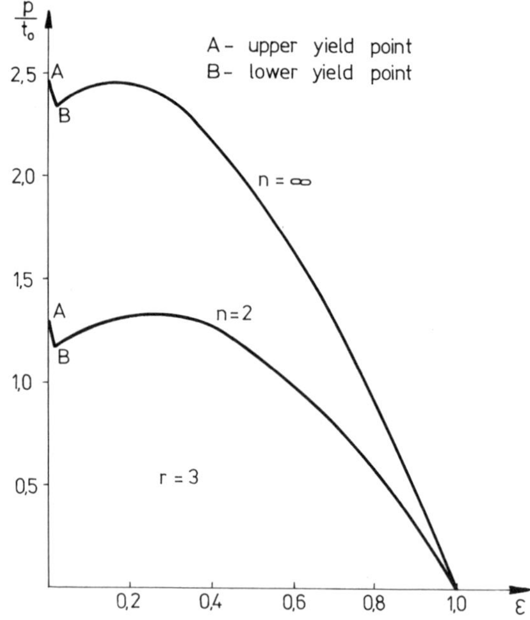

FIG. 15. Stress–strain diagram for the theoretical model from Fig. 13.

following expression for the yield stress

$$p = \frac{4}{\sqrt{3}} \frac{(n-1)}{n+(n-1)\varepsilon}(1+r\varepsilon)(1-\varepsilon)t_o \tag{12}$$

which is graphically presented in Fig. 15, for $r=3$ and $a=2$ or $n=\infty$ respectively. These stress–strain diagrams display two kinds of instability. At the beginning of the diagrams there appears a remarkable sudden decrease in stresses connected with the decohesion process along the vertical boundaries between hexagons. The second type of instability appearing in the advanced stage of deformation results from the reduction of the contact area on the inclined boundaries.

Comparing these theoretical diagrams with the diagram obtained for the experimental model (Fig. 11), we see that the two models, the theoretical and the experimental one, predict unstable behaviour at the beginning of the deformation process and then the recovery of the carrying capacity due to the strain hardening on the sliding surfaces.

5. CONCLUSIONS

Basic continuum concepts concerning plastic deformation and fracture have been analysed in numerous works in terms of microstructure of the material. Simple models discussed in the present work allow us to demonstrate and study the possible contribution of internal microfrac ture to the complex phenomena connected with the plastic deformation of metals.

REFERENCES

1. DRUCKER, D. C. Engineering and continuum aspects of high-strength materials. In: *High Strength Materials*, Chapter 21 (Ed. V. F. Zackay), Proc. 2nd Berkeley International Material Conference on High Strength Materials, Berkeley, 1964, John Wiley, New York.
2. DYSON, B. F. Constraints of diffusional cavity growth rates, *Met. Sci.* (Oct. 1976), 349–353.
3. KELLY, D. A. A two-dimensional model of creep cavitation failure in copper subject to biaxial stress systems at 250°C, *Met. Sci.* (Feb. 1976), 57–62.
4. LEE, E. H. Plastic flow in a V-notched bar pulled in tension, *J. appl. Mech.*, **19** (1952), 331–336.
5. LEE, E. H. Plastic flow in a rectangularly notched bar subjected to tension, *J. appl. Mech.*, **21** (1954), 140–146.

6. Lee, E. H. and A. J. Wang. Plastic flow in deeply notched bars with sharp internal angles, *Proc. 2nd U.S. Nat. Congr. Appl. Mech.* (1954), 489–497.
7. Mróz, Z. On generalized kinematic hardening rule with memory of maximal prestress, *J. Méc. Appl.*, **5** (1981), 241–260.
8. Rogers, C. The effect of material variables on ductility. In: *Ductility*, Papers presented at the Seminar of the ASM, October 14–15, 1967, ASM, Metal Park, Ohio, 1968.
9. Szczepiński, W. On the mechanism of local internal necking as a factor of the process of ductile fracture of metals, *J. Méc. Théor. et Appl.*, Numero spécial (1982), 161–174.
10. Szczepiński, W. On experimental two-dimensional models of intercrystalline sliding and fracture in polycrystalline metals, *Arch. Mech.*, **34** (1982), 503–514.

33

A Five-parameter Mixed Hardening Model for Concrete Materials

W. F. Chen and D. J. Han

School of Civil Engineering, Purdue University, West Lafayette, Indiana, USA

ABSTRACT

Recent research in structural concrete has been moving toward the development of refined constitutive equations for concrete materials for finite element analysis of reinforced concrete structures. The objective of this study is to develop a comprehensive elastic–plastic-fracture stress–strain relation to simulate the concrete behavior during loading and failure under three dimensional stress conditions. In this development, the Willam–Warnke five-parameter model is used as the yielding and failure surface. Based on the theories of elasticity and plasticity, the elastic–plastic incremental stress–strain relationship for concrete materials is developed, by using a mixed hardening rule, and by using the associated as well as a nonassociated flow rule.

1. INTRODUCTION

The developments of modern computers and applications of the finite element method to structural engineering in general have led to a new research field in concrete technology in particular, i.e. the numerical analysis of reinforced concrete structures. The material modeling under the multiaxial stress conditions plays a key role in this development in simulating the complex phenomena that take place in a real structure during loading and collapse. Hence, the objective of this research is to develop a comprehensive three-dimensional elastic–plastic-fracture stress–strain relation for finite element analysis of reinforced concrete structures. Although significant progress on this subject has been made in recent years [3, 4, 5, 8, 14], much more remains to be done. This includes for example, the improvement of the constitutive relation for concrete and the numerical simulation of concrete fracturing, among others. Herein, based on the classical theory of plasticity, a constitutive law using a five-parameter failure model for concrete materials is presented.

The classical theory of plasticity is well founded on a physical as well as on a mathematical basis with a long history of successful practical applications involving metals. This wealth of experience and familiarity had led to attempts to apply the theory to frictional materials, such as concrete, rock and soils. Within the framework of plasticity theory, one of the most important assumptions is that of the existence of a yield surface. As more experimental data became available for concrete, the limitations on the assumed shape for the yield and failure surfaces became apparent. It has been generally agreed that on the π-plane the circular cross-sectional shape of von Mises criterion must be replaced by a bulged triangular shape. Thus, the third invariant of the stress deviatoric tensor J_3 must be included in the yield function in addition to the first invariant of stress tensor I_1 and the second invariant of stress deviatoric tensor J_2. So, the general yield function can be expressed in the form

$$f(I_1, J_2, J_3) = 0 \tag{1}$$

The use of three invariants I_1, J_2, J_3 instead of six stress components in eqn. (1) implies of course the assumption of isotropy which is generally considered to be a reasonable one for concrete materials in the pre-failure regions of the material.

Various yield and failure criteria for concrete materials are discussed in the recent book by Chen [3]. Models involving three stress invariants in different forms have been proposed by Willam and Warnke [15], Ottosen [11], and Hsieh, Ting and Chen [8], among others. Among them, the Willam-Warnke five-parameter model appears to be the most appropriate one for defining the shape of a triaxial failure surface; it is therefore adopted here for the present development of a triaxial constitutive relation for concrete materials.

2. WILLAM–WARNKE FIVE-PARAMETER FAILURE SURFACE [15]

The five-parameter model is illustrated graphically in Fig. 1(a) for its cross-sections in the meridian plane and in Fig. 1(b) for its cross-sections in the deviatoric plane. In the meridian plane, it has curved tensile and compressive meridians as can be expressed by the quadratic parabola in the form normalized by the uniaxial compressive strength of concrete f_c

$$\begin{aligned}\sigma_m &= a_0 + a_1 r_t + a_2 r_t^2 \\ \sigma_m &= b_0 + b_1 r_c + b_2 r_c^2\end{aligned} \tag{2}$$

where, for convenience, σ_m, r_c and r_t are also used to denote the nondimensional stress quantities σ_m/f_c, r_c/f_c and r_t/f_c respectively, $\sigma_m = \frac{1}{3}I_1$, mean stress; r_t, r_c = stress components perpendicular to the hydrostatic axis at $\theta = 0°$ and $\theta = 60°$, respectively. Here, θ is the angle of similarity defined in eqn. (4). a_0, a_1, a_2, b_0, b_1, b_2 are material constants.

These two meridians intersect the hydrostatic axis at the same point. It follows that

$$a_0 = b_0 \tag{3}$$

The remaining five parameters can be determined by five tests given in Table I.

In Fig. 1(a), the five test points required for the determination of the five material constants are indicated by a cross within a circle. Once these two meridians have been determined from a set of experimental data, the failure surface is then constructed by connecting them, using an ellipsoidal surface.

In the deviatoric plane, the failure curves having a threefold symmetry, as shown in Fig. 1(b), are expressed by a portion of an elliptic curve in the following form.

$$r(\sigma_m, \theta) = \frac{2r_c(r_c^2 - r_t^2)\cos\theta + r_c(2r_t - r_c)[4(r_c^2 - r_t^2)\cos^2\theta + 5r_t^2 - 4r_c r_t]^{1/2}}{4(r_c^2 - r_t^2)\cos^2\theta + (r_c - 2r_t)^2} \tag{4}$$

where

$r = \sqrt{(2J_2)}$, stress component on the deviatoric plane

$\theta = \frac{1}{3}\arccos\left(\frac{3\sqrt{(3)}}{2}\frac{J_3}{J_2^{3/2}}\right)$, angle of similarity

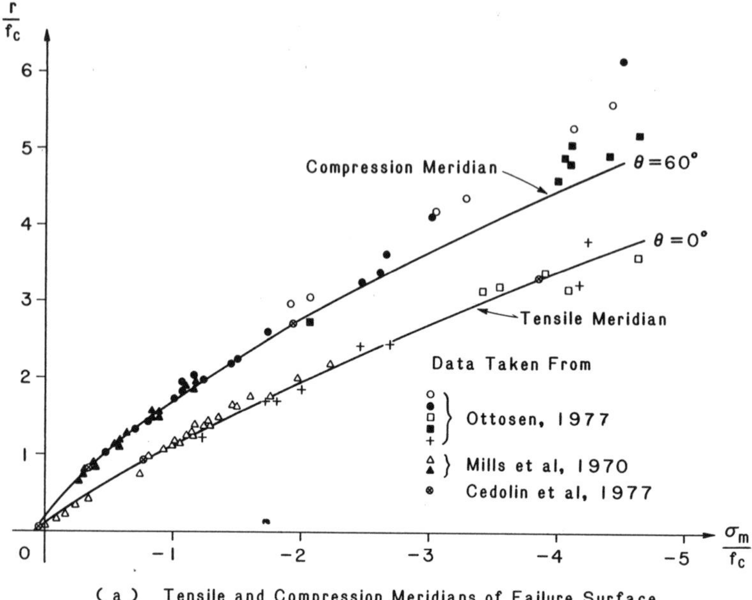

(a) Tensile and Compression Meridians of Failure Surface

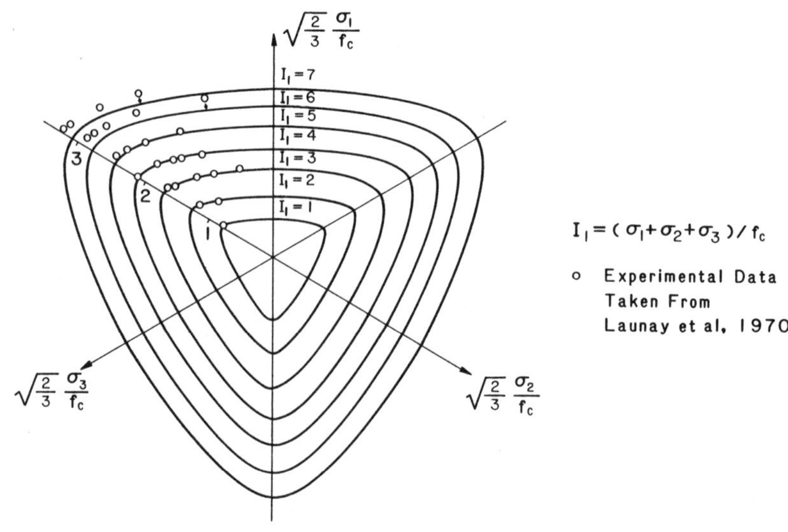

(b) Deviatoric Sections of Failure Surface

FIG. 1. Failure criterion by Willam–Warnke's five-parameter model ($a_0 = b_0 = 0{\cdot}1025$, $a_1 = -0{\cdot}8403$, $a_2 = -0{\cdot}0910$, $b_1 = -0{\cdot}4507$, $b_2 = -0{\cdot}1018$).

TABLE I

Tests	*Failure stress points* σ_m	r	$\theta°$
1. Uniaxial tension	$\frac{1}{3}f_t$	$\sqrt{\frac{2}{3}}f_t$	0
2. Biaxial compression	$-\frac{2}{3}f_{bc}$	$\sqrt{\frac{2}{3}}f_{bc}$	0
3. High-compressive-stress point 1	σ_{mt}	r_t	0
4. Uniaxial compression	$-\frac{1}{3}f_c$	$\sqrt{\frac{2}{3}}f_c$	60
5. High-compressive-stress point 2	σ_{mc}	r_c	60

To simplify the expression below, we rewrite eqn. (4) as

$$r=\frac{s+t}{v} \tag{5}$$

where

$$s=s(\sigma_m,\theta)=2r_c(r_c^2-r_t^2)\cos\theta \tag{6}$$
$$t=t(\sigma_m,\theta)=r_c(2r_t-r_c)u^{1/2} \tag{7}$$
$$u=u(\sigma_m,\theta)=4(r_c^2-r_t^2)\cos^2\theta+5r_t^2-4r_tr_c \tag{8}$$
$$v=v(\sigma_m,\theta)=4(r_c^2-r_t^2)\cos^2\theta+(r_c-2r_t)^2 \tag{9}$$

From eqns. (2), r_t and r_c can be expressed as

$$r_t=\frac{-a_1-\sqrt{(a_1^2-4a_2(a_0-\sigma_m))}}{2a_2} \tag{10}$$

$$r_c=\frac{-b_1-\sqrt{(b_1^2-4b_2(b_0-\sigma_m))}}{2b_2} \tag{11}$$

Equations (2), (3), (4) or eqns. (5)–(11) completely define the failure criterion of Willam–Warnke's five-parameter model.

In Fig. 1, this model is compared to the triaxial test data, and a close agreement can be observed for both hydrostatic and deviatoric sections. It can be seen that this model reproduces the principal features of the triaxial failure surface of concrete. It is smooth, convex everywhere. It has the desired characteristics that the failure curves are nearly triangular for tensile and small compressive stresses and become increasingly bulged (more circular) for higher compressive stresses. In view of the fluctuations of experimental results, there is little need for further refinements of this model. Thus, we choose this surface as the basic form for yielding, loading and failure surfaces.

3. HARDENING RULE

The initial yield criterion and the failure criterion define the limits of the elastic region and plastic region respectively. The criterion for initial yielding may be assumed to have a similar shape to that of the failure criterion but with a reduced size (see Fig. 2(a)). The mixed hardening rule is used here to define the growth and motion of the subsequent yield surfaces during plastic loading.

The mixed hardening rule proposed by Hodge [7] is a combination of isotropic and kinematic hardening. The isotropic hardening rule proposed by Hill [6] assumes a uniform expansion of the initial yield surface. It applies mainly to monotonic proportional loadings. In the present model, the yield function and subsequent loading function for isotropic hardening can be formulated as

$$f(\sigma_{ij}, \varepsilon^{p}_{ij}, k) = r - k\left(\frac{s+t}{v}\right) = 0 \tag{12}$$

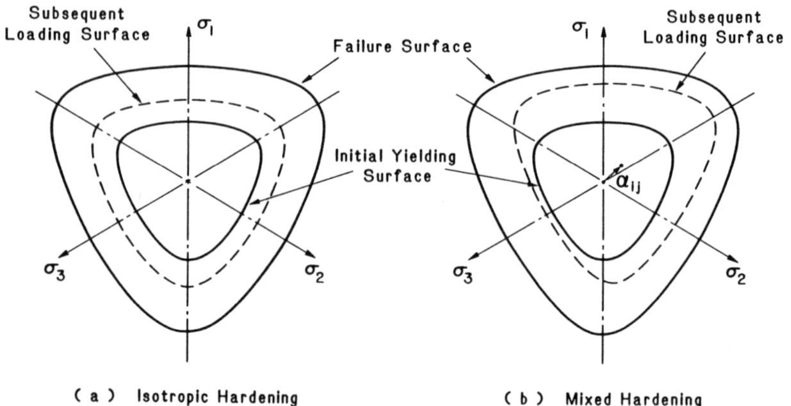

FIG. 2. Failure, initial yielding and subsequent loading curves in deviatoric plane (σ_m = const.).

where

$$k = \frac{\tau}{f_c} \tag{13}$$

in which τ is related to the effective stress, f_c, the ultimate strength of concrete under uniaxial compression. For $\tau = \tau_y$, eqn. (12) represents the initial yielding condition, while $\tau = f_c$, then $k = 1$, eqn. (12) becomes the failure function. An illustration of isotropic hardening is provided in Fig. 2(a).

The hardening behavior postulated in the kinematic hardening rule as proposed by Prager [13], assumes that during plastic deformation the loading surface translates as a rigid body in the stress space, maintaining the size, shape, and orientation of the yield surface. The primary aim of this rule is to provide a means to account for the Bauschinger effect. Since the loading surface translates in the stress space, its equation has the form

$$f(\sigma_{ij} - \alpha_{ij}, \varepsilon^{p}_{ij}, k) = \bar{r} - k\left(\frac{\bar{s} + \bar{t}}{\bar{v}}\right) = 0 \tag{14}$$

where α_{ij} are the co-ordinates of the center of the loading surface, which changes with plastic deformation, and $\bar{r}$, $\bar{s}$, $\bar{t}$, $\bar{v}$ are calculated by eqns. (5)–(11) and by using $\bar{\sigma}_{ij} = \sigma_{ij} - \alpha_{ij}$ instead of σ_{ij}.

If k is taken as a constant, eqn. (14) represents the loading function of pure kinematic hardening. But if k is defined by eqn. (13), which describes the expansion of the yield surface, then eqn. (14) represents the loading function of mixed hardening (see Fig. 2(b)).

In the present model, we consider the effective plastic strain increment as the sum of isotropic and kinematic parts,

$$d\varepsilon_p = d\varepsilon_p^i + d\varepsilon_p^k \tag{15}$$

where $d\varepsilon_p$ is the effective plastic strain increment defined as

$$d\varepsilon_p = \sqrt{d\varepsilon^p_{ij} d\varepsilon^p_{ij}} \tag{16}$$

and if we introduce a constant parameter M to define the isotropic hardening effect

$$d\varepsilon_p^i = M d\varepsilon_p \tag{17}$$

then, the remaining effective plastic strain increment

$$d\varepsilon_p^k = (1 - M) d\varepsilon_p \tag{18}$$

is due to the kinematic hardening.

The effective plastic strain increment $d\varepsilon_p^i$ is related to the isotropic hardening function through the concept of effective stress-effective plastic strain which makes it possible to extrapolate from a simple uniaxial compression test into the multi-dimensional situation. The effective plastic strain increment $d\varepsilon_p^k$ is related to the kinematic hardening rule of Ziegler [16] as

$$d\alpha_{ij} = c(1-M)d\varepsilon_p(\sigma_{ij} - \alpha_{ij}) \tag{19}$$

where c is the work hardening constant, characteristic for a given material.

In this mixed hardening model for concrete, the weighting coefficient M $(0<M<1)$ is introduced to allow the freedom of selecting different proportions of isotropic and kinematic effects in the mixed model. M can also have a negative value, so that isotropic softening can also be considered. The advantages of using the concept of mixed hardening have been demonstrated by Axelsson and Samuelsson [1] in describing the loading cycles of metals and by Hsieh *et al.* [8] in describing the reversed loading behavior of concrete materials.

4. CONSTITUTIVE EQUATIONS

4.1. Normality and Flow Rule

So far, the initial yield surface, failure surface, and loading surface given by the choice of a specific hardening rule have been discussed. The stress–strain relation for concrete in finite element analysis can now be formulated based on the classical theory of plasticity.

The necessary connection between the loading function f and the stress–strain relation is made by the flow rule. Introducing a plastic potential function $g(\sigma_{ij}, \varepsilon_{ij}^p, k)$ in analogy with ideal-fluid-flow problems, we express the flow rule as

$$d\varepsilon_{ij}^p = d\lambda \frac{\partial g}{\partial \sigma_{ij}} \tag{20}$$

where $d\lambda \geqslant 0$ is a scalar hardening parameter and the gradient of the potential surface $\partial g/\partial\sigma_{ij}$ defines the direction of the plastic-strain-increment vector $d\varepsilon_{ij}^p$. If the plastic potential surface g has the same shape as the current loading surface, i.e.

$$g(\sigma_{ij}, \varepsilon_{ij}^p, k) = f(\sigma_{ij}, \varepsilon_{ij}^p, k) \tag{21}$$

the flow rule is termed associated and eqn. (20) takes the form

$$d\varepsilon_{ij}^{p} = d\lambda \frac{\partial f}{\partial \sigma_{ij}} \tag{22}$$

i.e. the plastic flow develops along the normal to the loading surface. This normality condition assures a unique solution for a given boundary value problem using the stress–strain relation developed on the basis of eqn. (22).

4.2. Stress–Strain Relations

The total strain increment is taken to be the sum of the elastic and plastic increments

$$d\varepsilon_{ij} = d\varepsilon_{ij}^{e} + d\varepsilon_{ij}^{p} \tag{23}$$

According to Hooke's law, the stress increment is determined by

$$d\sigma_{ij} = D_{ijkl}^{e}(d\varepsilon_{kl} - d\varepsilon_{kl}^{p}) \tag{24}$$

where

$$D_{ijkl}^{e} = 2G\left(\delta_{ik}\delta_{jl} + \frac{\nu}{1-2\nu}\delta_{ij}\delta_{kl}\right) \tag{25}$$

While plastic flow takes place, the consistency condition

$$df = 0 \tag{26}$$

must hold. In mixed hardening, the loading function eqn. (14) is relevant. The total differential of this function is

$$df = \frac{\partial f}{\partial \sigma_{ij}} d\sigma_{ij} + \frac{\partial f}{\partial \alpha_{ij}} d\alpha_{ij} + \frac{\partial f}{\partial k}\frac{\partial k}{\partial \tau}\frac{d\tau}{d\varepsilon_{p}^{i}} d\varepsilon_{p}^{i} = 0 \tag{27}$$

Substituting eqn. (24) for $d\sigma_{ij}$, eqn. (19) for $d\alpha_{ij}$, eqn. (17) for $d\varepsilon_{p}^{i}$, into eqn. (27) and noting that

$$\frac{\partial f}{\partial \alpha_{ij}} = -\frac{\partial f}{\partial \sigma_{ij}}, \qquad \frac{d\tau}{d\varepsilon_{p}^{i}} = \bar{H} \tag{28}$$

where τ is the effective stress and $\bar{H}$ is a plasticity modulus, eqn. (27) becomes

$$df = \frac{\partial f}{\partial \sigma_{ij}} D_{ijkl}^{e}(d\varepsilon_{kl} - d\varepsilon_{kl}^{p}) - \frac{\partial f}{\partial \sigma_{ij}} c(1-M)\bar{\sigma}_{ij}\, d\varepsilon_{p} + \frac{\partial f}{\partial \tau}\bar{H}M\, d\varepsilon_{p} = 0 \tag{29}$$

Using the flow rule (20) and the effective plastic strain definition (16) in eqn. (29), and solving for $d\lambda$, we have

$$d\lambda = \frac{1}{h}\frac{\partial f}{\partial \sigma_{rs}} D^{e}_{rskl}\, d\varepsilon_{kl} \tag{30}$$

where

$$h = \frac{\partial f}{\partial \sigma_{mn}} D^{e}_{mnrs}\frac{\partial g}{\partial \sigma_{rs}} + \left[c(1-M)\frac{\partial f}{\partial \sigma_{ij}}\bar{\sigma}_{ij} - \bar{H}M\frac{\partial f}{\partial \tau}\right] \times \left(\frac{\partial g}{\partial \sigma_{mn}}\frac{\partial g}{\partial \sigma_{mn}}\right)^{1/2} \tag{31}$$

Substitution of eqn. (30) into eqn. (20) yields the expression for plastic strain. Further, using eqn. (24) leads to the constitutive equation

$$d\sigma_{ij} = (D^{e}_{ijkl} + D^{p}_{ijkl})\, d\varepsilon_{kl} \tag{32}$$

where the plastic stiffness tensor has the form

$$D^{p}_{ijkl} = -\frac{1}{h} H^{*}_{ij} H_{kl} \tag{33}$$

and

$$H^{*}_{ij} = D^{e}_{ijmn}\frac{\partial g}{\partial \sigma_{mn}} \tag{34}$$

$$H_{kl} = \frac{\partial f}{\partial \sigma_{rs}} D^{e}_{rskl} \tag{35}$$

Once the loading function f and plastic potential function g are defined for a mixed-hardening elasto–plastic material, the constitutive relation can be directly derived from eqns. (31)–(35) in a rather straightforward manner.

5. CONSTITUTIVE LAW BASED ON THE ASSOCIATED FLOW RULE

Now, both the loading function f and plastic potential function g are taken as in eqn. (14). Noting that

$$\frac{\partial f}{\partial \bar{\sigma}_{ij}} = \frac{\partial f}{\partial \sigma_{ij}}$$

the derivative $\partial f/\partial \sigma_{ij}$ can be expressed as

$$\frac{\partial f}{\partial \sigma_{ij}} = B_0 \delta_{ij} + B_1 s_{ij} + B_2 t_{ij} \tag{36}$$

where $B_0 = \partial f/\partial I_1$, $B_1 = \partial f/\partial J_2$, $B_2 = \partial f/\partial J_3$; δ_{ij} is the Kronecker delta; s_{ij} the deviatoric stress tensor; J_3 is the third invariant of stress deviatoric tensor; and t_{ij} is defined as follows

$$t_{ij} = s_{ik}s_{kj} - \tfrac{2}{3}J_2\delta_{ij} \tag{37}$$

Carrying out the mathematical derivation from eqns. (5)–(11) gives

$$\begin{aligned} B_0 = -\frac{r'_t}{3v}\{&4r(r_c - 2r_t) + 8rr_t \cos^2\theta - k[4r_c r_t \cos\theta - 2r_c u^{1/2} \\ &+ r_c(2r_t - r_c)u^{-1/2}(4r_t \cos^2\theta - 5r_t + 2r_c)]\} \\ &+ \frac{r'_c}{3v}\{2r(r_c - 2r_t) + 8rr_c \cos^2\theta - k[2(3r_c^2 - r_t^2)\cos\theta \\ &+ 2(r_t - r_c)u^{1/2} + 2r_c(2r_t - r_c)u^{-1/2}(2r_c \cos^2\theta - r_t)]\} \end{aligned} \tag{38}$$

where

$$r'_t = \frac{dr_t}{d\sigma_m} = \frac{1}{a_1 + 2a_2 r_t} \tag{39}$$

$$r'_c = \frac{dr_c}{d\sigma_m} = \frac{1}{b_1 + 2b_2 r_c} \tag{40}$$

and

$$\begin{aligned} B_1 = \frac{1}{r} - \frac{3\sqrt{3}}{2}\frac{(r_c^2 - r_t^2)}{v(4\cos^2\theta - 1)}\frac{J_3}{J_2^{5/2}} \\ \times\{4r\cos\theta - kr_c[1 + 2u^{-1/2}(2r_t - r_c)\cos\theta]\} \end{aligned} \tag{41}$$

$$B_2 = \frac{\sqrt{3}(r_c^2 - r_t^2)}{v(4\cos^2\theta - 1)J_2^{3/2}}\{4r\cos\theta - kr_c[1 + 2u^{-1/2}(2r_t - r_c)\cos\theta]\} \tag{42}$$

Substitution of eqns. (36)–(42) into eqns. (31) and (35) leads to

$$\begin{aligned} h = 2G&\left[3B_0^2\frac{1+\nu}{1-2\nu} + 2B_1^2 J_2 + 6B_1 B_2 J_3 + B_2^2(J_4 - \tfrac{4}{3}J_2^2)\right] \\ &+ \left[c(1-M)(\bar{B}_0\bar{I}_1 + 2\bar{B}_1\bar{J}_2 + 3\bar{B}_2\bar{J}_3) - \frac{\partial f}{\partial \tau}\bar{H}M\right][3B_0^2 \\ &+ 2B_1^2 J_2 + 6B_1 B_2 J_3 + B_2^2(J_4 - \tfrac{4}{3}J_2^2)]^{1/2} \end{aligned} \tag{43}$$

in which

$$J_4 = s_{ik}s_{kj}s_{il}s_{lj}$$

and

$$H_{ij} = H^*_{ij} = 2G\left(B_0\frac{1+\nu}{1-2\nu}\delta_{ij} + B_1 s_{ij} + B_2 t_{ij}\right) \tag{44}$$

Then, the plastic stiffness tensor has a symmetric form as follows

$$D^{\mathrm{p}}_{ijkl} = -\frac{1}{h}H_{ij}H_{kl} \tag{45}$$

6. CONSTITUTIVE LAW BASED ON A NONASSOCIATED FLOW RULE

Herein, we use eqn. (14) as the loading function f but use the Drucker–Prager surface as the plastic potential function g, i.e.

$$g = \alpha I_1 + \sqrt{J_2} - k^* = 0 \tag{46}$$

where α and k^* are constants in the stress space, but they could be functions of plastic deformation. With the expression (46), the derivative of g becomes

$$\frac{\partial g}{\partial \sigma_{ij}} = \alpha\delta_{ij} + \frac{1}{\sqrt{2}\,r}s_{ij} \tag{47}$$

Substitution of eqn. (47) and eqns. (36)–(42) into eqns. (31) and (34) gives

$$h = 2G\left(3B_0\alpha\frac{1+\nu}{1-2\nu} + \frac{r}{\sqrt{2}}B_1 + \frac{3B_2}{\sqrt{2}\,r}J_3\right) + \left[c(1-M)(\bar{B}_0\bar{I}_1 + 2\bar{B}_1\bar{J}_2 + 3\bar{B}_2\bar{J}_3) - \bar{H}M\frac{\partial f}{\partial \tau}\right]\sqrt{3\alpha^2 + \tfrac{1}{2}} \tag{48}$$

$$H^*_{ij} = 2G\left(\alpha\frac{1+\nu}{1-2\nu}\delta_{ij} + \frac{1}{\sqrt{2}\,r}s_{ij}\right) \tag{49}$$

Now, the plastic stiffness tensor D^{p}_{ijkl} is expressed by eqn. (33), where the tensors H^*_{ij} and H_{kl} are given by eqn. (49) and eqn. (44), respectively. It can be seen that the stiffness tensor D^{p}_{ijkl} is no longer symmetric.

To implement the incremental stress–strain relationship for finite element analysis, it is convenient to write eqns. (44) and (49) explicitly as

$$H_{xx} = 2G\left[B_0 \frac{1+\nu}{1-2\nu} + B_1 s_{xx} + B_2(s_{xx}^2 + s_{xy}^2 + s_{xz}^2 - \tfrac{2}{3}J_2)\right], \text{ etc.}$$
$$H_{yz} = 2G[B_1 s_{yz} + B_2(s_{xy}s_{xz} + s_{yy}s_{yz} + s_{yz}s_{zz})], \text{ etc.} \tag{50}$$

and

$$H_{xx}^* = 2G\left(\alpha \frac{1+\nu}{1-2\nu} + \frac{1}{2\sqrt{J_2}} s_{xx}\right), \text{ etc.}$$
$$H_{yz}^* = \frac{G}{\sqrt{J_2}} s_{yz}, \text{ etc.} \tag{51}$$

Then, the stress–strain relation is written as

$$\{d\sigma\} = ([\mathscr{D}]^e + [\mathscr{D}]^p)\{d\varepsilon\}, \tag{52}$$

where

$$\{d\sigma\} = (d\sigma_x, d\sigma_y, d\sigma_z, d\tau_{yz}, d\tau_{xz}, d\tau_{xy})^T$$
$$\{d\varepsilon\} = (d\varepsilon_x, d\varepsilon_y, d\varepsilon_z, d\gamma_{yz}, d\gamma_{xz}, d\gamma_{xy})^T \tag{53}$$

and

$$[\mathscr{D}]^p = -\frac{1}{h}[H^*]^T[H] \tag{54}$$

in which

$$[H] = [H_{xx}, H_{yy}, H_{zz}, H_{yz}, H_{xz}, H_{xy}]$$
$$[H^*] = [H_{xx}^*, H_{yy}^*, H_{zz}^*, H_{yz}^*, H_{xz}^*, H_{xy}^*] \tag{55}$$

7. EFFECTIVE STRESS AND EFFECTIVE STRAIN

Based on the five-parameter loading function (eqn. (12)), the effective stress σ_e and effective strain increment $d\varepsilon_p$ are defined so that the single σ_e–ε_p curve would be calibrated on the uniaxial compressive stress–strain curve.

Obviously, the hardening parameter k in eqn. (12) is related to the uniaxial compressive stress σ by the equation

$$k = \frac{\tau}{f_c} = \frac{\sigma}{\sqrt{\frac{3}{2}}r_c} \tag{56}$$

Then, we can define σ_e as

$$\sigma_e = \sqrt{\frac{3}{2}} \frac{r_c}{f_c} \tau = \sqrt{\frac{3}{2}} r_c k \tag{57}$$

The corresponding effective strain increment $d\varepsilon_p$ is defined in terms of the plastic work per unit volume in the form

$$dW^p = \sigma_e \, d\varepsilon_p \tag{58}$$

On the other hand, we have

$$dW^p = \sigma_{ij} \, d\varepsilon_{ij}^p = \sigma_{ij} \, d\lambda \frac{\partial g}{\partial \sigma_{ij}} \tag{59}$$

Noting that

$$d\lambda = \frac{\sqrt{d\varepsilon_{ij}^p \, d\varepsilon_{ij}^p}}{\sqrt{\dfrac{\partial g}{\partial \sigma_{kl}} \dfrac{\partial g}{\partial \sigma_{kl}}}} \tag{60}$$

So, $d\varepsilon_p$ can be expressed as

$$d\varepsilon_p = D\sqrt{d\varepsilon_{ij}^p \, d\varepsilon_{ij}^p} \tag{61}$$

where

$$D = \frac{\dfrac{\partial g}{\partial \sigma_{ij}} \sigma_{ij}}{\sqrt{\frac{3}{2}} r_c k \sqrt{\dfrac{\partial g}{\partial \sigma_{kl}} \dfrac{\partial g}{\partial \sigma_{kl}}}} \tag{62}$$

For the associated flow rule, using eqn. (36) for $\partial g/\partial \sigma_{ij}$, we have

$$D = \frac{B_0 I_1 + 2B_1 J_2 + 3B_2 J_3}{\sqrt{\frac{3}{2}} r_c k [3B_0^2 + 2B_1^2 J_2 + 6B_1 B_2 J_3 + B_2^2 (J_4 - \frac{4}{3} J_2^2)]^{1/2}} \tag{63}$$

For the nonassociated flow rule, using eqn. (47) for $\partial g/\partial \sigma_{ij}$, we have

$$D = \frac{\alpha I_1 + \sqrt{J_2}}{\sqrt{\frac{3}{2}} r_c k \sqrt{3\alpha^2 + 1/2}} \tag{64}$$

Comparing eqn. (57) and eqn. (61) to variable τ and eqn. (16) respectively, and noting eqn. (28), the plastic modulus $\bar{H}$ should be expressed as

$$\bar{H} = H\left[\frac{D}{\sqrt{\frac{3}{2}\frac{r_c}{f_c}}}\right] \tag{65}$$

where H is the slope of the experimental uniaxial compressive σ–ε_p curve.

REFERENCES

1. AXELSSON, K. and A. SAMUELSSON. Finite element analysis of elastic–plastic materials displaying mixed hardening, *Int. J. Num. Mech. Engng*, **14** (1979), 211–225.
2. CEDOLIN, L., Y. R. J. CRUTZEN and S. DEI POLI. Triaxial stress–strain relationship for concrete, *J. Engng Mech. Div., ASCE*, **103** (EM3), Proc. Paper 12969 (June 1977), 423–439.
3. CHEN, W. F. *Plasticity in Reinforced Concrete*, McGraw-Hill, New York, 1982.
4. CHEN, W. F. (Chairman), Z. P. BAZANT, O. BUYUKOZTURK, T. Y. CHANG, D. DARWIN, T. C. Y. LIU and K. J. WILLAM. Constitutive relations and failure theories. In: *The State-of-the-art Committee Report on Finite Element Analysis of Reinforced Concrete Structures* (Chapter 2), ASCE Special Publication 1982, 34–148.
5. CHEN, W. F. and H. SUZUKI. Constitutive models for concrete, *Computers & Structures*, **12** (1980), 23–32.
6. HILL, R. *The Mathematical Theory of Plasticity*, Clarendon Press, Oxford, 1950.
7. HODGE, P. G., Jr. Discussion [of Prager (1956)], *J. appl. Mech.*, **23** (1957), 482–484.
8. HSIEH, S. S., E. C. TING and W. F. CHEN. A plastic-fracture model for concrete, *Int. J. Solids Struct.*, **18(3)** (1982), 181–197.
9. LAUNAY, P. and H. GACHON. Strain and Ultimate Strength of Concrete Under Triaxial Stresses, Special Publication, SP-34, ACI, 1, 1970, 269–282.
10. MILLS, L. L. and R. M. ZIMMERMAN. Compressive strength of plain concrete under multiaxial loading conditions, *ACI J.*, **69** (10), (Oct. 1970), 802–807.
11. OTTOSEN, N. S. A failure criterion for concrete, *J. Engng Mech. Div., ASCE*, **103** (EM4), Proc. Paper 13111 (August 1977), 527–536.
12. OTTOSEN, N. S. Nonlinear analysis of pull-out test, *J. Engng Mech. Div., ASCE*, **107** (ST4), Proc. Paper 16197 (April 1981), 127–141.
13. PRAGER, W. The theory of plasticity: a survey of recent achievements, *Proc. Inst. mech. Engs*, **169** (1955), 3–19.
14. SUZUKI, H. and W. F. CHEN, Elastic-plastic-fracture analysis of concrete structures, *Computers & Structures*, **16** (1983), 697–705.

15. Willam, K. J. and E. P. Warnke. Constitutive models for the triaxial behavior of concrete, *IABSE Seminar for Concrete Structures Subjected to Triaxial Stresses*, Bergamo, Italy (May 1974), *IABSE Proceedings*, **19** (1975), 1–30.
16. Ziegler, H. A modification of Prager's hardening rule, *Q. J. appl. Math.*, **17** (1959), 55–65.

34

Crack Band Theory and Microplane Model for Concrete

ZDENĚK P. BAŽANT

Center for Concrete and Geomaterials, The Technological Institute, Northwestern University, Evanston, Illinois, USA

ABSTRACT

This paper describes recent developments at Northwestern University in the modeling of inelastic deformations of concrete due to progressive microcracking, as well as their use in fracture analysis. A review of the crack band theory is presented and the stress–strain relations for progressive microcracking are analysed. For the purpose of general loading, the microplane model which is an analog of Taylor–Batdorf–Budianski's slip theory of plasticity, is described and its use in numerical structural analysis is discussed. It is also shown that the crack band theory along with a microplane model generalized for crack unloading is capable of describing the resistance of cracks in concrete to shear. Some comparisons with test data are demonstrated.

1. INTRODUCTION

Concrete is a material in which the inelastic behavior is dominated, except in conditions of very high hydrostatic pressure, by progressive microcracking and frictional crack shear, rather than plastic slip. Nevertheless, the basic ideas of plasticity theory are very useful and can be adapted to the formulation of stress–strain relations for concrete.

First, the paper will focus on the crack band theory for fracture of concrete, in which stress–strain relations with strain-softening are used to describe the behavior of concrete in the fracture process zone, i.e. a zone of microcracking which precedes the formation of complete fracture and is quite large for concrete. Subsequently, development of a more general constitutive relation which is applicable in general situations and is analogous to Taylor–Batdorf–Budianski's slip theory of plasticity, is described. The method of numerical integration, which is particularly important for the microplane model, is briefly reviewed and the capability to represent strain-softening is demonstrated by some comparisons with test data. Finally, adaptation of the model for the modeling of crack shear, including friction, crack dilatancy and development of transverse pressure under restraint, is outlined. Since the formulation of the crack band theory is just about to appear in one of the journals in the field [11], only a brief summary is included in the present text, and emphasis is given to the microplane model and its use.

The paper concentrates on describing recent developments at Northwestern University, which the author achieved jointly with B. Oh, P. Gambarova, J. K. Kim, and P. Pfeiffer. A comprehensive review of the field cannot be included in the present lecture, and the reader is referred to other works (ref. 11 or 5).

2. CRACK BAND THEORY

In brittle heterogeneous materials such as concrete or rocks, the fracture forms through a process of progressive microcracking, during which the macroscopic stress decreases at increasing overall strain, i.e. the material strain-softens. This type of fracture may be efficiently modeled with the crack band approach, in which the material behavior in the fracture process zone is described by a suitable strain-softening triaxial stress–strain relation. At the same time, the width of the crack front must be considered to be a material property related to the fracture energy of the material. The value of the latter equals the area under the tensile stress–strain curve (work dissipated per unit volume) times the width of the crack-band front.

This approach is particularly suitable for finite element analysis, in which the smeared cracking is modeled as a change of material stiffness in the direction normal to the cracks. However, in contrast to the traditional smeared cracking approach, now widely used in large finite element codes, the finite element size cannot be chosen arbitrarily but must equal the characteristic width of the crack band front, since otherwise the correct energy dissipation due to fracture would not occur. However, as a convenient artifice, it is possible to use larger finite elements provided that the strain-softening slope and possibly also the strength limit are adjusted so as to ensure the correct energy dissipation due to fracture [3, 5, 11].

A suitable triaxial stress–strain relation of the total strain type (deformation theory type) has been recently formulated and has been shown to lead to satisfactory agreement with essentially all existing fracture test data available in the literature [3, 11]. This stress–strain relation is, however, limited to situations in which the direction of the maximum principal stress does not significantly rotate during the fracture formation.

This is not so in certain important situations, especially various dynamic problems. Here, a longitudinal wave may produce only a partial tensile fracture (i.e. distributed microcracking) and the fracture may be completed subsequently when a shear wave arrives, causing a principal tensile stress in a different direction. For such situations of progressive fracturing, it is necessary to develop a triaxial strain-softening stress–strain relation which is path-dependent and is formulated incrementally. A model called the microplane model, first briefly outlined in ref. 10, is developed to fill this need.

3. MICROPLANE MODEL

In the microplane model, the constitutive properties are characterized by a relation between the stresses and strains acting within the microstructure on planes of various orientation, called the microplanes. This formulation involves no tensorial invariance restrictions on the microplane level. These restrictions are automatically satisfied by a combination of planes of all possible orientations within the material. For example, in the case of isotropy, each orientation must be equally frequent. Thus, one circumvents the difficulty of setting up a general nonlinear constitutive equation in terms of invariants.

The idea of defining the inelastic behavior independently on planes of different orientation within the material, and then in some way superimposing the inelastic effects from all planes, appeared in Taylor's work [35] on plasticity of polycrystalline metals. Batdorf and Budianski [2] formulated the slip theory of plasticity, in which the stresses acting on various planes of slip are obtained by resolving the macroscopic applied stress, and the plastic strains (slips) from all planes are then superimposed. The same superposition of inelastic strains was used in the so-called multilaminate models of Zienkiewicz *et al.* [37] and Pande *et al.* [27, 28] and in many works on the plasticity of polycrystals. While in the previous works dealing with plasticity of polycrystals [2, 13, 18, 19, 21, 23, 24, 32, 35] or soils [12, 14] the stresses on various microplanes were assumed to be equal to the resolved macroscopic stress, the present model uses a similar assumption for part of the total strains.

The resultants of the stresses acting on the microplanes over unit areas of the macroscopic continuum will be called the microstresses s_{ij}, and the strains of the macroscopic continuum accumulated from the deformations on the microplanes will be called the microstrains, e_{ij}. With regard to the interaction between the micro- and macro-levels, one may introduce the following basic hypotheses.

Hypothesis I. The normal microstrain e_n which governs the progressive development of cracking on a microplane of any orientation is equal to the resolved macroscopic strain tensor e_{ij} for the same plane, i.e.

$$e_n = n_i n_j e_{ij} \tag{1}$$

in which latin lower case subscripts refer to cartesian co-ordinates x_i ($i = 1, 2, 3$); n_i = direction cosines of the unit normal $\mathbf{n}$ of the microplane; and the repeated latin lower case subscripts indicate a summation over 1, 2, 3.

Hypothesis II. The stress relaxation due to all microcracks normal to $\mathbf{n}$ is characterized by assuming that the microstress s_n on the microplane of any orientation is a function of the normal microstrain e_n on the same plane, i.e.

$$s_n = (2\pi/3)F(e_n) \tag{2}$$

The factor $(2\pi/3)$ is introduced just for convenience, as it will later cancel out.

The second hypothesis is similar to that made for shear microstresses and microstrains in the slip theory of plasticity. Hypothesis I is, however, opposite. There are three reasons for Hypothesis I.

(1) Using resolved stresses rather than resolved strains on the mic-

roplanes would hardly allow describing strain-softening, since in this case there are two strains corresponding to a given stress but only one stress corresponding to a given strain.

(2) The microstrains must be stable when the macrostrains are fixed. It has been found numerically that, in the case of strain-softening, the model becomes unstable if resolved stresses rather than strains are used.

(3) The use of resolved strains rather than resolved stresses seems to reflect the microstructure of a brittle aggregate material more realistically. The use of resolved stresses is reasonable for polycrystalline metals in which local slips scatter widely while the stress is roughly uniformly distributed throughout the microstructure. By contrast, in a brittle aggregate material consisting of hard inclusions embedded in a weak matrix, the stresses are far from uniform, having sharp extremes at the locations where the surfaces of aggregate pieces are nearest. The deformation of the thin layer of matrix between two aggregate pieces, which yields the major contribution to inelastic strain, is determined chiefly by the relative displacements of the centroids of the two aggregate pieces, which roughly correspond to the macroscopic strain. The microplanes may be imagined to represent the thin layers of matrix and the bond planes between two adjacent aggregate pieces, since microcracking is chiefly concentrated there.

In Hypothesis II, the shear stiffness on individual microplanes is neglected. This assumption is probably acceptable for tensile loading. One must admit, however, that eqn. (2) is also justified by its simplicity. A more general formulation which considers both the normal and shear stiffnesses on the individual microplanes has been also developed [4], but it is more complicated.

In the slip theory of plasticity, the material property on a slip of any direction is described as a functional dependence of the shear stress on the plastic shear strain. By analogy, it was tried first to use in eqn. (2) inelastic strain e''_n instead of total strain e_n, or inelastic stress s''_n instead of total stress s_n, and then superimpose the elastic strains or stresses upon those resulting from the microplane model. This did not, however, produce satisfactory response curves. The fact that the stress–strain relation for the microplanes (eqn. (2)) includes elastic response is a basic difference from the slip theory of plasticity [2].

The aforementioned difference, however, causes a certain limitation which must be overcome. If the shear stiffness on individual microplanes is neglected, then the initial elastic Poisson ratio is always obtained as 0·25. This value is not realistic for concrete. An adjustment of the Poisson ratio to a correct value, such as 0·16–0·20, may be achieved by superimposing on the total strains from the microplane model an additional elastic strain e^{a}_{ij} produced by the same stresses σ_{ij}, i.e.

$$\varepsilon_{ij} = \varepsilon^{a}_{ij} + e_{ij} \tag{3}$$

where ε_{ij} is the tensor of total macroscopic strain and e_{ij} is the tensor of strains corresponding to the microplane model. Since the Poisson ratio of concrete is relatively close to 0·25, the required correction ε^{a}_{ij} is relatively small in magnitude compared to e_{ij}, and so one does not have to worry too much about the nonphysical justification of eqn. (3). Alternatively, one would have to include elastic shear stiffnesses on the microplanes [4].

4. TANGENTIAL STIFFNESS MATRIX

Equilibrium conditions may be expressed by means of the principle of virtual work:

$$\delta W^{c} = \tfrac{4}{3}\pi\sigma_{ij}\,\delta e_{ij} = 2\int_S s_n\,\delta e_n f(\mathbf{n})\,dS \tag{4}$$

in which S represents the surface of a unit hemisphere, the factor $(4\pi/3)$ is due to integrating over the volume of a sphere of radius 1, and $dS = \sin\phi\, d\theta\, d\phi$ (Fig. 2(a)). Note that we do not need to integrate over the entire surface of the sphere, since the values of σ_n or e_n are equal at any two diametrically opposite points. Function $f(\mathbf{n})$ defines the relative frequency of the planes of various orientations $\mathbf{n}$, contributing to inelastic stress relaxation, and may be used to characterize anisotropy. For $f(\mathbf{n}) = 1$ we have an isotropic solid.

Substituting eqns. (1) and (2) into eqn. (3), we get

$$\sigma_{ij}\,\delta e_{ij} = \int_S F(e_n) n_i n_j\,\delta e_{ij} f(\mathbf{n})\,dS,$$

and because this must hold for any δe_{ij}, we must have

$$\sigma_{ij} = \int_0^{2\pi}\int_0^{\pi/2} F(e_n) n_i n_j f(\mathbf{n}) \sin\phi\, d\phi\, d\theta \tag{5}$$

Therefore,

$$d\sigma_{ij} = \int_0^{2\pi} \int_0^{\pi/2} F'(e_n)\, de_n n_i n_j f(\mathbf{n}) \sin\phi \, d\phi \, d\theta \tag{6}$$

Furthermore, according to eqn. (1), $de_n = n_k n_m de_{km}$, and so

$$d\sigma_{ij} = D^c_{ijkm}\, de_{km} \tag{7}$$

in which

$$D^c_{ijkm} = \int_0^{2\pi} \int_0^{\pi/2} a_{ijkm} F'(e_n) f(\mathbf{n}) \sin\phi \, d\phi \, d\theta \tag{8}$$

with

$$a_{ijkm} = n_i n_j n_k n_m \tag{9}$$

D^c_{ijkm} may be called the tangent stiffnesses of the microplane system. Note that the sequence of subscripts of D^c_{ijkm} is immaterial; therefore, there are only six independent values of incremental stiffnesses.

The compliance corresponding to the additional elastic strain σ^a_{ij} must satisfy isotropy conditions, and so

$$C^a_{ijkm} = \frac{1}{9K^a}\, \delta_{ij}\, \delta_{km} + \frac{1}{2G^a}\, (\delta_{ik}\, \delta_{jm} - \tfrac{1}{3}\, \delta_{ij}\, \delta_{km}) \tag{10}$$

in which K^a and G^a are specific bulk and shear moduli which cannot be less than the actual initial bulk and shear moduli K and G.

In view of eqn. (3), we may now write the incremental stress–strain relation as

$$d\sigma_{ij} = D_{ijkm}\, d\varepsilon_{ij} \tag{11}$$

with

$$[D_{ijkm}] = [(D^{c^{-1}})_{ijkm} + (C^a_{ijkm})]^{-1} \tag{12}$$

Poisson ratios greater than 0·25 can be achieved with $1/K^a = 0$ and $G^a > 0$, while those less than 0·25 can be achieved with $1/G^a = 0$ and $K^a > 0$. The latter is the case for concrete. Let us now determine the value of K^a needed to achieve the desired Poisson's ratio ν. Let superscripts c and a distinguish between the values corresponding to D^c_{ijkm} and C^a_{ijkm}. For uniaxial stress we have $\varepsilon_{11} = \sigma_{11}/9K^a + \sigma_{11}/E^c$ and $\varepsilon_{22} = \sigma_{11}/9K^a - \nu^c \sigma_{11}/E^c$ in which $\nu^c = 1/4$ and $E^c = 2\pi E_n/5$, $E_n = F'(0)$ = initial normal stiffness for the microplane, Since $\varepsilon_{22} = -\nu\varepsilon_{11}$, we

(a)

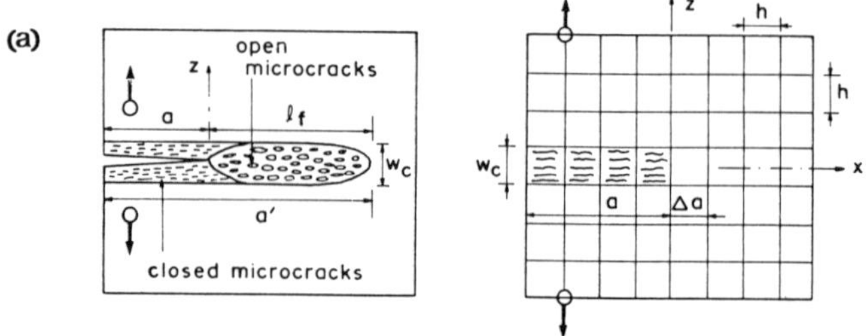

(b)

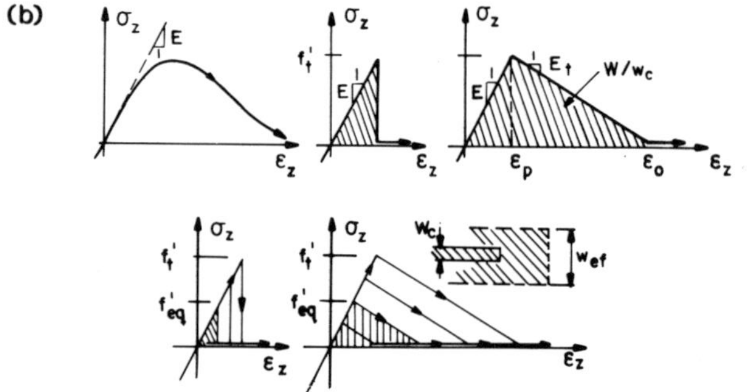

(c)

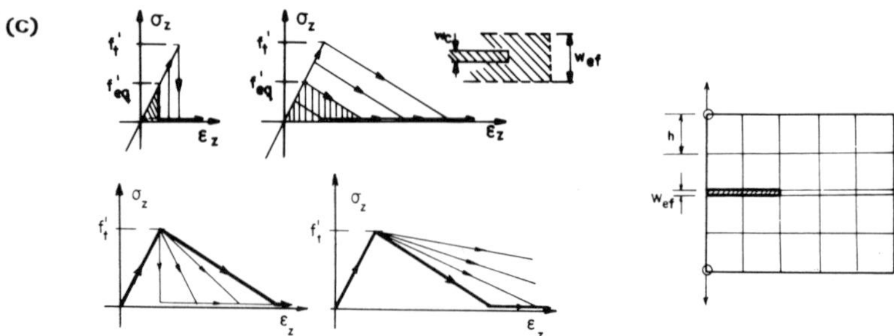

FIG. 1. Illustration of crack band model for fracture (after Bažant and Oh [10]). (a) Finite element simulation; (b) strain-softening diagrams used; (c) modifications to preserve same fracture energy for a different element size.

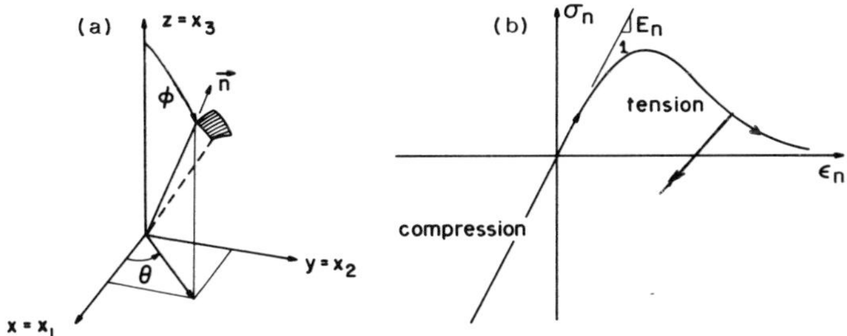

FIG. 2. Spherical co-ordinate system for the microplane model and the stress–strain relation for the microplane.

must have

$$K^{a} = \frac{1+\nu}{9(\nu^{c}-\nu)} E^{c} \text{ (for } \nu \leqslant \nu^{c}) \tag{13}$$

This is, of course, under the assumption that $1/G^{a} = 0$.

The stress–strain relation for the microplanes, relating σ_n to ε_n, must describe cracking all the way to complete fracture, at which σ_n reduces to zero. In view of the kinematics visualized in Fig. 1(b), it is clear that σ_n as a function of ε must first rise, then reach a maximum, and then gradually decline to zero. We choose the final zero value to be attained asymptotically, since no precise information exists on the final strain at which $\sigma_n = 0$, and since a smooth curve is convenient computationally. The following expressions were used in computations [9] (Fig. 2(b):

for $\varepsilon_n \geqslant 0$:

$$\sigma_n = E_n e_n \mathrm{e}^{-(k e_n{}^{p})} \tag{14}$$

for $\varepsilon_n \leqslant 0$:

$$\sigma_n = E_n e_n \tag{15}$$

in which E_n, k, and p are positive constants; $k = 1{\cdot}8 \times 10^{7}$, $p = 2$.

5. NUMERICAL INTEGRATION

The integral in eqn. (8) has to be evaluated numerically, approximating it by a finite sum:

$$D^{c}_{ijkm} = 4\pi \sum_{\alpha=1}^{N} w_{\alpha} [a_{ijkm} F'(e)]_{\alpha} \tag{16}$$

in which α refers to the values at certain numerical integration points on the surface of a unit sphere (i.e. certain directions), and w_α are the coefficients (or weights) associated with the integration points, such that $\sum w_\alpha = 0{\cdot}5$, because the integration is carried out over a half of the sphere surface.

Since, in finite element programs for incremental loading the numerical integration needs to be carried out a great number of times, a very efficient numerical integration formula is needed. For the slip theory of plasticity, the integration was performed using a rectangular grid in the θ–ϕ plane. This formula is, however, computationally inefficient because the integration points are crowded near the poles, and also because in the θ–ϕ plane the singularity arising from the pole takes away the benefit from the use of a higher-order integration formula.

Optimally, the integration points should be distributed over the spherical surface as uniformly as possible. A perfectly uniform subdivision is obtained when the microplanes normal to the α-directions are the faces of a regular polyhedron. A regular polyhedron with the most faces is the icosahedron, for which $N = 10$ (half the number of faces). Such a numerical integration formula was proposed by Albrecht and Collatz [1].

Numerical experience revealed, however, that a ten-point integration formula does not give sufficient accuracy, and so formulas with nonuniformly distributed points and unequal weights have to be used. A review of available Gaussian-type integration formulas for the surface of a sphere may be found in Stroud [34].

Numerical calculations indicated that for the nonlinear softening range the use of rather accurate formulas is necessary. Stroud [34] lists formulas with 13 and 16 points for a half sphere, but they are found to be too crude in the strain-softening range although they perform very accurately in the initial linear elastic range. The accuracy may be judged by considering a uniaxial tensile test and calculating the response for various directions of the tensile stress with regard to the set of integration points. If the formula were exact, the same response would be obtained for any orientation, and so the spread of the response curves for various orientations is a measure of integration error. If an error of about 5% is deemed acceptable, then the most efficient integration formula is a 21-point formula derived in ref. 16, which is of 9th degree, i.e. it integrates exactly all polynomials up to the 9th degree. The points of this formula lie at the vertices and mid-edges of an icosahedron. Slightly more accurate for the aforementioned calculations appears to be a 9th degree formula with 25 points per half-sphere, derived by McLaren [34, 25]. In contrast to the 21-point formula, this formula is symmetric with regard to cartesian co-ordinate planes. This symmetry is advantageous in the case of plane stress or axisymmetric stress because the integrated function does not have to be evaluated at all 25 points but only at 16 points, while no such reduction is possible for the aforementioned 21-point formula. A still more accurate formula, derived in ref. 16, also possesses the same symmetry, is of 13th degree, and involves 2×37 points. For plane or axisymmetric stress, this can be reduced to 19 points. A highly accurate 13th degree formula with 61 points per half-sphere was also given in ref. 16.

The directions of integration points are illustrated in Fig. 3 for the formulas mentioned above. Also shown are the response curves for uniaxial stress acting at various directions (a, b, c, d, . . .) with regard to the set of integration points. The spread of these curves characterizes the range of error.

6. NUMERICAL ALGORITHM

The following numerical algorithm may be used for the microplane model in each loading step.

(1) Determine $e_n^{(\alpha)}$ from eqns. (2) and (3) for all directions $\alpha = 1, \ldots, N$. In the first iteration of the loading step, use ε_{ij} for the end of the previous step, and in subsequent iterations use the value of ε_{ij} determined for the mid-step in the previous iteration. In structural analysis, repeat this for all finite elements and for all integration points within each finite element.

(2) For all directions $\mathbf{n}^{(\alpha)}$, evaluate $F'(e_n)$ for use in eqn. (2). Also check for each direction whether unloading occurs, as indicated by violation of the condition $s_n \Delta e_n \geqslant 0$. If violated, replace $F'(e_n)$ with the unloading stiffness (which may be approximately taken as E_n; however, a better expression exists).

(3) Evaluate D_{ijkm}^c from eqn. (8) and D_{ijkm} from eqn. (12). In structural analysis, repeat this for all elements and all integration points in each element.

(4) When solving stress–strain curves, calculate then the increments of unknown stresses and unknown strains from eqn. (11). In finite

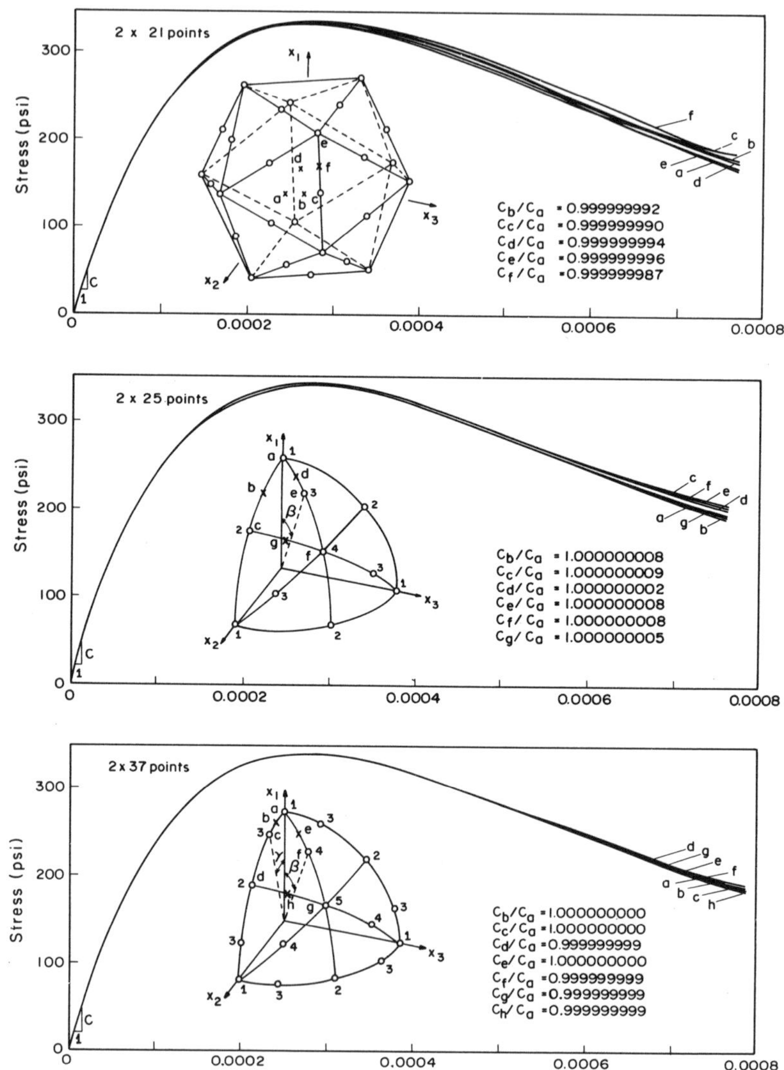

FIG. 3. Arrangements of nodes for some numerical integration formulas and the response curves for uniaxial stress of various orientations with regard to the set of nodes (after Bažant and Oh [10]).

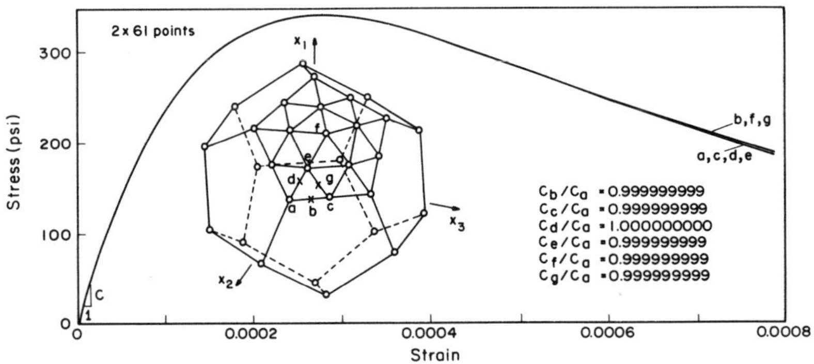

FIG. 3—*contd.*

element structural analysis, solve the increments of nodal displacements from the given load increments.

(5) Calculate the increments of ε_{ij} and σ_{ij} for all elements and all integration points in each element.

(6) Advance to the next iteration of the same loading step, or advance to the next loading step.

In simulating uniaxial tensile loading of fixed direction, the unloading criterion is not important since the only unloading occurs at moderate compressive stresses, for which a perfectly elastic unloading may be assumed.

The microplane model can be calibrated by comparison with direct tensile tests which cover the strain-softening response. Such tests, which can be carried out in a very stiff testing machine and on sufficiently small test specimens, have been performed by Evans and Marathe [15] as well as others [17, 20, 33]. Optimal values of the three parameters of the model, E_n, k, and p, have been found [9] so as to achieve the best fits of the data of Evans and Marathe. Some of these fits are shown as the solid lines in Fig. 4, and the data are shown as the dashed lines. A better test of the model would, of course, be a tensile test under rotating principal stress directions, but such tests have not yet been performed.

Note that with this theory one has only two material parameters, E_n and k, to determine by fitting test data. Trial and error approach is sufficient for that.

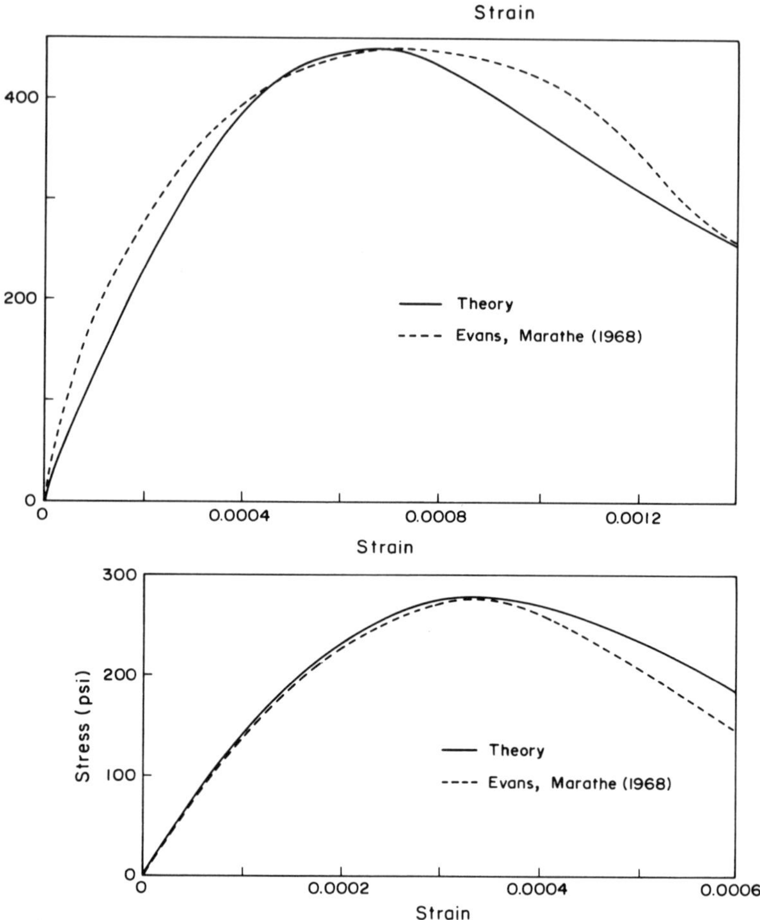

FIG. 4. Fits of Evans and Marathe's test data for direct tension tests by the microplane model (after Bažant and Oh [10]).

7. SHEAR IN CRACKED CONCRETE

The microplane model just described appears capable of modeling also the resistance of cracks in concrete for shear, characterized by crack friction (aggregate interlock effect) and dilatancy. For this purpose, the model needs to be enhanced by more realistic σ_n–ε_n curves for unload-

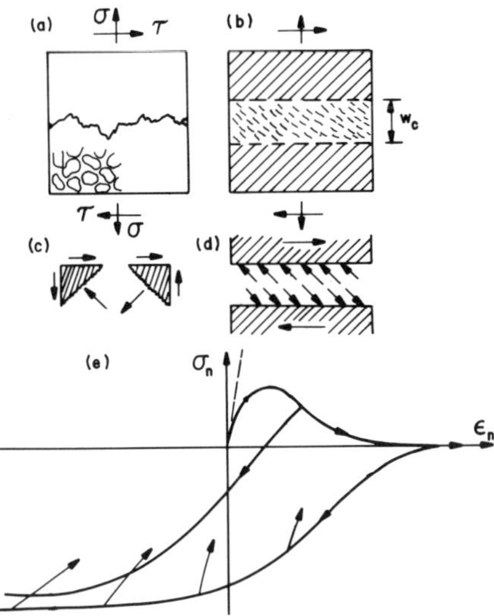

FIG. 5. Illustrations for the microplane model applied to crack shear (after Bažant and Gambarova [7]).

ing, and if cyclic shearing is considered, then also for reloading. This was done in ref. 6.

Test data are available only for shear loading of blocks (Fig. 5) that have been previously fully cracked in tension [22, 26, 29, 30, 31, 36]. Even though a finite separation is evident from the relative displacement of the blocks, the resistance to shear is not zero, not even at the beginning of shear. Obviously there must be some contacts between the opposite surfaces even after the tensile stress normal to the crack has already been reduced to zero.

Using the microplane model, we treat the distinct crack in a rectangular test specimen as a band of a certain finite width, w_c. This is probably not too unrealistic in view of the roughness of the crack surface, as well as the fact that concrete near the crack must have been microcracked during its previous tensile loading that produced the crack. In numerical simulation, one starts with intact concrete and implements first uniaxial tensile loading in direction z until the stress σ_z is reduced to zero (in practice, to $0{\cdot}001\ f_t'$). Subsequently, either a shear stress τ_{xz} or a shear strain γ_{xz}, depending on the conditions of simulated test, is gradually applied in small increments (see Fig. 5(b)).

Doing this, the normal strain ε_n on the microplanes inclined at +45° (Fig. 5(c)) is increased, and so σ_n remains zero on these microplanes. However, ε_n on the microplanes inclined at −45° (Fig. 5(c)) is decreased, and so contraction (unloading) occurs on those microplanes. For contraction, the normal stiffness is non-zero. Therefore, shear produces in the crack band a set of inclined compression forces illustrated in Fig. 5(d). These forces have a component along the crack, representing crack friction, and a component normal to the crack, representing the pressure opposing dilatancy. If such a pressure is not generated by the support conditions, then a simultaneous expansion (dilatancy) occurs so as to reduce the normal force component to zero.

Figure 5(e) shows the unloading σ_n–ε_n curves that have been used in ref. 6, in which analytical expressions for these unloading curves may be found. Typical response curves which have been simulated with the microplane model are shown in Fig. 6, where they are compared with the data points obtained by Walraven and Reinhardt [31, 36], and by Paulay and Loeber [29]. In ref. 29 it was shown that this model can fit essentially all the existing data on aggregate interlock or crack shear.

Thus we have a model which correctly describes both the tensile strain-softening up to full fracture and the crack shear in fully fractured

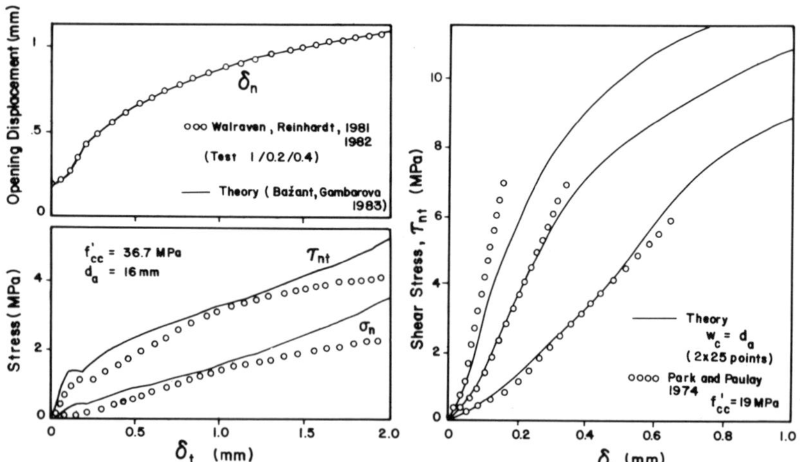

FIG. 6. Examples of fits of crack shear data of Walraven and Reinhardt [31, 36] and Park and Paulay [30] by the microplane model (after Bažant and Gambarova [7]).

concrete. It may be expected that the model would also represent the shear resistance of partially cracked concrete, and perhaps also shear fracture. Another possible use is biaxial (spatial) nonproportional shear loading, i.e. loading of the crack plane by shear stresses τ_{xz} and τ_{yz} (or shear strains γ_{xz} and γ_{yz}) which do not increase in proportion.

8. CONCLUSION

The microplane model in which the strains on the individual microplanes are assumed to be the resolved components of the macroscopic strain appears capable of simulating realistic tensile stress–strain curves with strain-softening and with a reduction of stresses all the way to zero. Combined with the blunt crack band concept, in which the strain-softening is restricted to occur over a region of a certain characteristic front width in the material, the microplane model should give a realistic representation of fracture under general loading histories, including those in which the principal stress directions rotate. As an example of such application, the microplane model may be applied to describe the crack shear phenomenon, including crack friction, dilatancy, and development of transverse normal stress.

In this manner, we are approaching a formulation that can describe strain-softening in direct tension tests, fracture test data, and crack shear data with one and the same model. Such a unified formulation is particularly desirable for finite element analysis.

ACKNOWLEDGEMENT

Partial support under Air Force Office of Scientific Research Grant No. AFOSR83-0009 is gratefully acknowledged. Thanks are also due to Mary Hill for her meticulous typing.

REFERENCES

1. Albrecht, J. and L. Collatz. Zur Numerischen Auswertung mehrdimensionaler Integrale, *Z. Angewandte Mathematik und Mechanik*, **38**(1/2), 1–15.

2. Batdorf, S. B. and B. Budianski. *A Mathematical Theory of Plasticity Based on the Concept of Slip*, NACA TN1871, April 1949.
3. Bažant, Z. P. Crack band model for fracture of geomaterials, *4th Int. Conf. on Numerical Methods in Geomechanics* (Ed. Z. Eisenstein) University of Alberta, Edmonton, Vol. **3,** 1982, 1137–1152.
4. Bažant, Z. P. Microplane Model for Strain Controlled Inelastic Behavior, Report No. 83–3/448m, Center for Concrete and Geomaterials, Northwestern University, Evanston, IL 60201, March 1983; to appear in: *Constitutive Equations for Engineering Materials* (Ed. C. Desai), John Wiley, London.
5. Bažant, Z. P. Mechanics of Fracture and Progressive Cracking in Concrete Structures, Report No. 83–2/428m, Center for Concrete and Geomaterials, Northwestern University, Evanston, IL, Feb. 1983; to appear in: *Fracture Mechanics Applied to Concrete Structures* (Ed. J. C. Sih) Martinus Nijhoff, The Hague, 1984.
6. Bažant, Z. P. and P. Gambarova, Rough cracks in reinforced concrete, *J. Structural Div., Proc. ASCE,* **106,** No. ST4 (1980), 819–842, Paper 15330; Discussion pp. 2579–2581, Closure 1981, pp. 1377–1388.
7. Bažant, Z. P. and P. Gambarova. Crack Shear in Concrete Crack Band Microplane Model, Report No. 83–9/679c, Center for Concrete and Geomaterials, Northwestern University, Evanston, IL., Sept. 1983.
8. Bažant, Z. P. and B. H. Oh. Efficient Numerical Integration on the Surface of a Sphere, Report No. 83–2/4283, Center for Concrete and Geomaterials, Northwestern University, Evanston, IL 60201.
9. Bažant, Z. P. and B. H. Oh. Model of Weak Planes for Progressive Fracture of Concrete and Rock, Report No. 83–2/428m, Center for Concrete and Geomaterials, Northwestern University, Evanston, IL, Feb. 1983.
10. Bažant, Z. P. and B. Oh. Microplane model for fracture analysis of concrete structures, *Proc. Symp. Interaction of Non-nuclear Munitions with Structures,* U.S. Air Force Academy, Colorado Springs, Colorado, May 1983, 49–55.
11. Bažant, Z. P. and B. H. Oh. Crack band theory for fracture of concrete, *Materials and Structures* **16** (1983), 155–179 (based on Concrete Fracture via Stress–Strain Relations, Center for Concrete and Geomaterials, Report No. 81–10/665, October 1981, Northwestern University).
12. Bažant, Z. P., K. Ozaydin and R. J. Krizek. Micromechanics model for creep of anisotropic clay, *J. Engng Mech. Div., ASCE,* **101** (1975), 57–78.
13. Budianski, B. and T. T. Wu. Theoretical prediction of plastic strains of polycrystals, *Proc. 4th U.S. Nat. Congress of Appl. Mechanics,* ASME, New York, 1962, 175–1185.
14. Calladine, C. R. A microstructural view of the mechanical properties of saturated clay, *Geotechnique,* **21** (1971), 391–415.
15. Evans, R. H. and M. S. Marathe. Microcracking and stress–strain curves for concrete in tension, *Materials and Structures,* No. 1, 1968, 61–64.
16. Finden, C. Spherical Integration, Dissertation, Cambridge University, 1961.
17. Heilmann, H. G., H. H. Hilsdorf and K. Finsterwalder. Festigkeit

und Verformung von Beton unter Zugspanungen, *Deutscher Ausschuss für Stahlbeton*, Heft 293, W. Ernst & Sohn, West Berlin, 1969.

18. HILL, R. Continuum micromechanics of elastoplastic polycrystals, *J. Mech. Phys. Solids*, **13** (1965), 89–101.
19. HILL, R. Generalized constitutive relations for incremental deformations of metal crystals by multislip, *J. Mech. Phys. Solids*, **14** (1966), 95–102.
20. HUGHES, B. P. and G. P. CHAPMAN. The complete stress–strain curve for concrete in direct tension, *Bulletin RILEM*, No. 30 (1966), 95–97.
21. KRÖNER, E. Zur Plastischen Verformung des Vielkristalls, *Acta Met.*, **9** (1961), 155–161.
22. LAIBLE, J. P., R. N. WHITE and P. GERGELY. Experimental Investigation of Shear Transfer Across Cracks in Concrete Nuclear Containment Vessels, American Concrete Institute, Special Publ. SP53, Detroit 1977, 203–226.
23. LIN, T. H. and M. ITO. Theoretical plastic stress–strain relationship of a polycrystal, *Int. J. Engng Sci.*, **4** (1966), 543–561.
24. LIN, T. H. and M. ITO. Theoretical plastic distortion of a polycrystalline aggregate under combined and reversed stresses, *J. Mech. Phys. Solids*, **13** (1965), 103–115.
25. MCLAREN, A. D. Optimal numerical integration on a sphere, *Math. Comput.*, **17** (1963), 361–383.
26. MATTOCK, A. H. The Shear Transfer Behavior of Cracked Monolithic Concrete Subject to Cyclical Reversing Shear, Report SM7404, Dept. of Civil Engng, University of Washington, Seattle, Nov. 1974.
27. PANDE, G. N. and K. G. SHARMA. Multi-laminate model of clays—a numerical evaluation of the influence of rotation of the principal stress axes, Report, Dept. of Civil Engng, University College of Swansea, U.K., 1982; see also *Proc. Symp. on Implementation of Computer Procedures and Stress–Strain Laws in Geotechnical Engineering* (Ed. C. S. Desai and S. K. Saxena), Acorn Press, Durham, N. C., 1981, 575–590.
28. PANDE, G. N. and W. XIONG. An improved multi-laminate model of jointed rock masses, *Proc. Int. Symposium on Numerical Models in Geomechanics* (Ed. R. Dungar, G. N. Pande, and G. A. Studer), Balkema, Rotterdam, 1982, 218–226.
29. PAULAY, T. and P. J. LOEBER. Shear Transfer by Aggregate Interlock, American Concrete Institute Special Publ. SP42, Detroit 1974, 1–15.
30. PAULAY, T., R. PARK, and M. H. PHILLIPS. Horizontal construction joints in cast-in-place reinforced concrete. In: *Shear in Reinforced Concrete*, Vol. **2,** American Concrete Institute Special Publ. SP42, Detroit 1974.
31. REINHARDT, H. W. and J. E. WALRAVEN. Crack in concrete subject to shear, *J. Structural Division ASCE*, **108** (1982), 207–224.
32. RICE, J. R. On the structure of stress–strain relations for time-dependent plastic deformation of metals, *J. appl. Mech.* **37** (1970), 728–737.
33. RÜSCH, H. and H. HILSDORF. Deformation characteristics of concrete under axial tension, *Vorunterschunge*, Bericht Nr. 44, Munich, May, 1963.
34. STROUD, A. H. *Approximate Calculation of Multiple Integrals*, Prentice Hall, Englewood Cliffs, N.J., 1971, 296–302.
35. TAYLOR, G. I. Plastic strain in metals, *J. Inst. Metals*, **62** (1938), 307–324.

36. Walraven, J. C. and H. W. Reinhardt. Theory and experiments on the mechanical behavior of cracks in plain and reinforced concrete subjected to shear loading, *HERON Journal*, **26,** No. 1A, Dept. of Civil Engng, Delft University of Technology, Delft, 1981.
37. Zienkiewicz, O. C. and G. N. Pande. Time-dependent multilaminate model of rocks—a numerical study of deformation and failure of rock masses, *Int. J. Numer. Analyt. Methods Geomechanics*, **1** (1977), 219–247.

35

Periodic Hexagonal Array Models for Plasticity Analysis of Composite Materials

GEORGE J. DVORAK* and JAN L. TEPLY

Department of Civil Engineering, University of Utah, Salt Lake City, Utah, USA

ABSTRACT

A class of micromechanical models for inelastic analysis of fibrous and particulate composites is described. Periodic hexagonal arrays of fibers or particles, separated by matrix interlayers, form the basis of the models. A representative volume element is selected from the periodic array and subdivided into 10 finite elements. Piecewise uniform fields are introduced through appropriate shape functions. Displacement and equilibrium finite element models are used to obtain estimates of local instantaneous stress and strain fields, as well as upper and lower bounds on instantaneous moduli of elastic–plastic composites.

1. INTRODUCTION

Elastic behavior of both fibrous and particulate composites is reasonably well understood. Evaluation of overall thermoelastic properties and of averaged local stress and strain fields can be routinely made, either in terms of rigorous bounds, or through self-consistent estimates of elastic moduli [6–10, 16]. The situation is quite different in the case of elastic–plastic aggregates. A bounding technique has not been developed, and the self-consistent method often gives erroneous results [3]. Approximate estimates of instantaneous compliances and local fields can be derived only from very simple models of the microstructure, such as the vanishing fiber diameter model for fibrous systems [4].

* Present address: Rensselaer Polytechnic Institute, Troy, New York, USA.

More accurate modelling techniques are required, especially in applications to metal matrix composites with low yield stress and high strength, where plastic straining is the predominant deformation mode.

In the present paper certain initial steps are taken toward development of a class of micromechanical models for elastic–plastic fibrous, particulate, and hybrid composite systems, which are particularly suitable for use in numerical analysis of composite structures. The models are based on variants of a periodic hexagonal array geometry. Appropriate representative volume elements are chosen, and subdivided into a small number of finite elements. Then upper and lower bounds on instantaneous overall stiffnesses are obtained from displacement and equilibrium models of the finite element method. The numerical procedure can be incorporated as a subroutine into a larger general purpose program.

2. GEOMETRY OF MICROSTRUCTURE

2.1. Geometry of Periodic Hexagonal Array Model

In metal matrix composite systems reinforced with large-diameter continuous filaments, such as boron or silicon carbide, the fibers are often carefully aligned and laid in regular intervals during fabrication. The transverse cross-section of a unidirectional composite then contains a nearly periodic array of circular fibers. Such periodic array geometry, square or hexagonal, facilitates the choice of small representative volume elements which are repeated throughout the volume. Further simplification of transverse geometry can be achieved by altering the cross-section of the fiber, from circular to hexagonal. When arranged in a periodic hexagonal array, the fibers are separated by layers of matrix (Fig. 1). Both phases can be divided into subregions of piecewise uniform deformation. This procedure, which will be described in detail later, is particularly useful in nonlinear analysis of the fibrous medium.

The transverse geometry shown in Fig. 1 forms a basis for development of the Periodic Hexagonal Array (PHA) model. A similar geometry was used by Drucker [2] and others [1] in evaluation of limit loads of particulate and fibrous composites.

Consider that the area of the composite hexagon (Fig. 1) is equal to

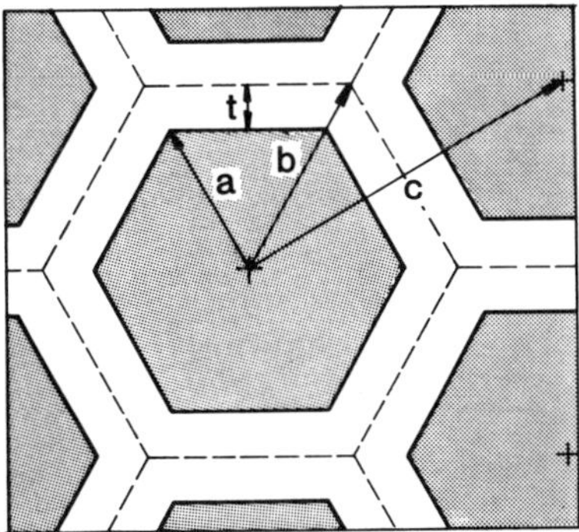

FIG. 1. Characteristic dimensions of periodic hexagonal array.

unity. Then, the dimensions become, in terms of fiber volume fraction c_f:

$$a = \frac{2\sqrt{c_f}}{\sqrt[4]{3^3}} \qquad t = \frac{1-\sqrt{c_f}}{\sqrt[4]{3}} \tag{1}$$

$$b = \frac{2}{\sqrt[4]{3^3}} = \frac{c}{\sqrt{3}} \qquad c = \frac{2}{\sqrt[4]{3}} = \sqrt{3}\, b \tag{2}$$

Our ultimate goal is to determine the instantaneous overall stiffness and compliance tensors of the composite medium. To make the overall properties meaningful, it is necessary to find them when a uniform, overall stress or strain field is applied to the aggregate. This can be accomplished by prescribing either tractions or boundary displacements

$$\bar{\mathbf{t}} = \bar{\boldsymbol{\sigma}}\bar{\mathbf{n}}, \qquad \bar{\mathbf{u}} = \bar{\boldsymbol{\varepsilon}}\bar{\mathbf{x}} \tag{3}$$

where $\bar{\mathbf{u}}$ and $\bar{\mathbf{t}}$ are the boundary displacements and traction vectors, respectively, $\bar{\mathbf{x}}$ is the position vector of boundary points, $\bar{\mathbf{n}}$ is the unit outward normal; $\bar{\boldsymbol{\sigma}}$ and $\bar{\boldsymbol{\varepsilon}}$ are tensors of uniform stress and strain.

Evaluation of the overall properties is most conveniently made by analyzing the response of a representative volume element (RVE) of the microstructure. In periodic microstructures, the RVE can be defined as a subregion which has the following properties:

(a) When repeated in a certain pattern, it continuously covers the entire macroscopic volume of the composite.

(b) When the composite is loaded by uniform stresses or strains (3), the local stresses and strains are identical within each RVE in the macroscopic volume.

It is clear from the outset that the composite hexagon shown in Fig. 1 may be selected as RVE. Let the co-ordinates of an arbitrary point Y in a selected basic composite hexagon be given as $y=[y_1, y_2, s]^T$. The hexagonal periodicity of the array suggests that all points X with co-ordinates $\mathbf{x}=[x_1, x_2, s]^T$, such that

$$\mathbf{x}=\mathbf{y}+\mathbf{c} \tag{4}$$

where

$$\mathbf{c}=\frac{c}{2}\left(i\sqrt{3}, j, \frac{2s}{c}\right)^T,$$

are equivalent to Y, or, more specifically, have identical local stress and strain states when the composite is loaded according to (3). The symbols i and j in (4) represent integers, both must be odd or even at the same time; the distance c is given by (2). Since the expressions (3) are invariant under transformation (4) the composite hexagon satisfies requirement (b) above. It is seen from Fig. 1 that requirement (a) is satisfied as well. However, the composite hexagon is not a particularly convenient choice, because appropriate boundary conditions for the RVE, compatible with (3) are not readily available.

A more suitable selection of representative volume elements is indicated in Fig. 2. The periodic hexagonal array is now covered by a

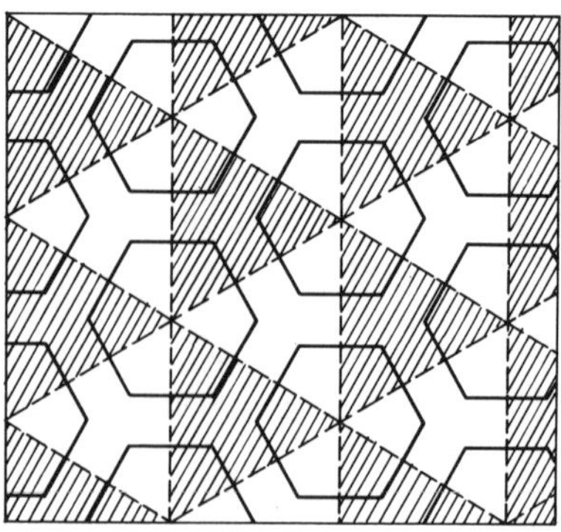

FIG. 2. Mesh of triangular representative volume elements.

triangular mesh consisting of shaded and unshaded triangles. It is probably obvious that (4) is valid for each of the two sets of triangles, either shaded or unshaded.

To admit the composite triangle as a RVE, it is necessary to show that the local stress and strain fields in the two sets of triangles are identical, as required by (b) above. There is proof on the grounds that the local stress and strain fields in one set of triangles are invariant with respect to the transformation

$$\mathbf{x}' = -\boldsymbol{\delta}(\mathbf{y}+\mathbf{c}) = \mathbf{y}' + \mathbf{c}' \tag{5}$$

where $\boldsymbol{\delta}$ is the Kronecker symbol δ_{ij}. This transformation converts the shaded set into unshaded, and vice versa. When (5) is applied to (3), it follows that

$$\bar{\mathbf{t}}' = (-\boldsymbol{\delta})\bar{\boldsymbol{\sigma}}(-\boldsymbol{\delta}^{\mathrm{T}})\bar{\mathbf{n}}' = \bar{\boldsymbol{\sigma}}\bar{\mathbf{n}}', \qquad \bar{\mathbf{u}}' = (-\boldsymbol{\delta})\bar{\boldsymbol{\varepsilon}}(-\boldsymbol{\delta}^{\mathrm{T}})\bar{\mathbf{x}}' = \bar{\boldsymbol{\varepsilon}}\bar{\mathbf{x}}' \tag{6}$$

Therefore, local fields are invariant with respect to transformation (4), as well as (5). That satisfies requirement (b) above; the composite triangular prism, represented by the shaded or unshaded area of unit thickness ($0<x_3<1$) can be admitted as a representative volume element.

The choice of RVE offers certain advantages, which can be illustrated with the help of Fig. 3. Let the vector $\mathbf{c}$ be selected so that

$$\mathbf{c} = \bar{\mathbf{c}} = \frac{c}{2}\left(\sqrt{3}, 1, \frac{2s}{c}\right)$$

and let (4) relate identical points X, Y, or X′, Y′, in two different RVEs. Then displacements $\mathbf{u}$ of these points are:

$$\mathbf{u}_{\mathrm{Y}} - \mathbf{u}_{\mathrm{X}} = \bar{\boldsymbol{\varepsilon}}\bar{\mathbf{c}}, \qquad \mathbf{u}'_{\mathrm{Y}'} - \mathbf{u}'_{\mathrm{X}'} = \bar{\boldsymbol{\varepsilon}}\bar{\mathbf{c}}' \tag{7}$$

It follows that displacements of the vertices V and V′ in Fig. 3 are:

$$\mathbf{u}_{\mathrm{V}'} - \mathbf{u}_{\mathrm{V}} = \bar{\boldsymbol{\varepsilon}}\bar{\mathbf{c}}', \qquad \mathbf{u}'_{\mathrm{V}} - \mathbf{u}'_{\mathrm{V}'} = \bar{\boldsymbol{\varepsilon}}\bar{\mathbf{c}} \tag{8}$$

Also, it can be shown [14] that the displacement of the midpoint M≡M′ is

$$\mathbf{u}_{\mathrm{M}} = \tfrac{1}{2}(\mathbf{u}_{\mathrm{P}} + \mathbf{u}_{\mathrm{P}'}) = \tfrac{1}{2}(\mathbf{u}_{\mathrm{V}} + \mathbf{u}_{\mathrm{V}'}) \tag{9}$$

where the points P and P′ are equidistant from M.

Similarly,

$$\mathbf{u}_{\mathrm{M}} - \mathbf{u}_{\mathrm{V}} = \tfrac{1}{2}\bar{\boldsymbol{\varepsilon}}\bar{\mathbf{c}} \qquad \mathbf{u}'_{\mathrm{M}} - \mathbf{u}'_{\mathrm{V}} = \tfrac{1}{2}\bar{\boldsymbol{\varepsilon}}\bar{\mathbf{c}}' \tag{10}$$

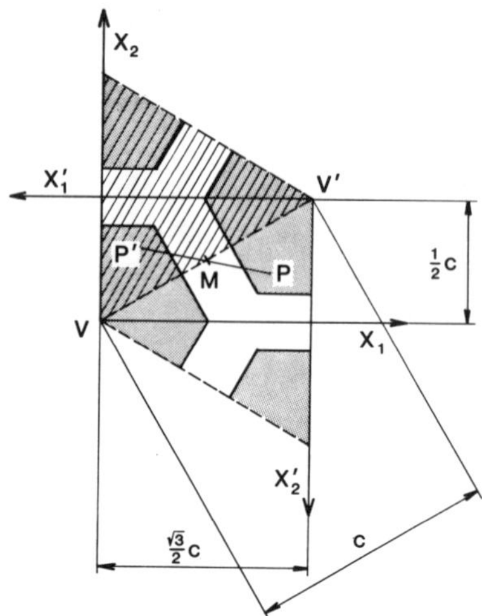

FIG. 3. Two adjacent representative volume elements and their local co-ordinate systems.

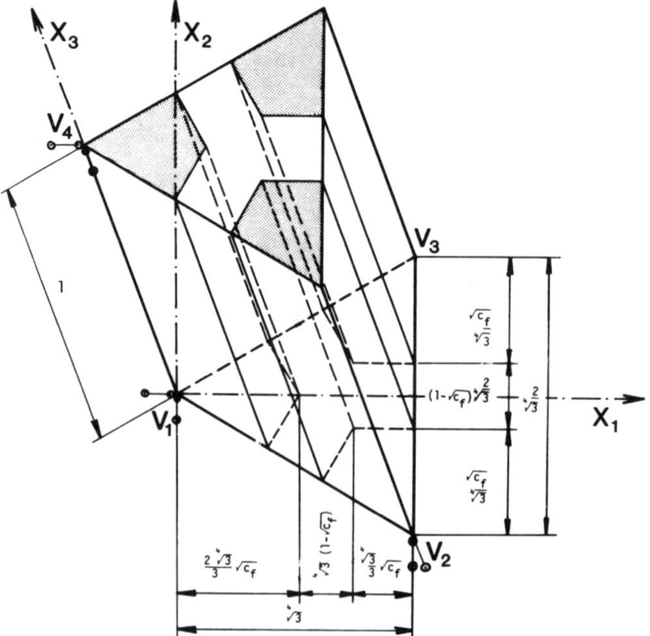

FIG. 4. Dimensions of representative volume element.

Equations (9) and (10) indicate a particularly simple way in which the overall strain $\bar{\boldsymbol{\varepsilon}}$ can be related to displacements of the vertices V, V′, and of the midpoints M of the sides of the triangular RVE. In each side of the triangular RVE, these three points are translated during deformation in the same way as they would be in a uniformly strained solid. Displacements of other points on the side V–V′ can be determined from (9) if points P and P′ are chosen on the RVE boundary. The RVE is redrawn in Fig. 4, with dimensions derived from (1) and (2).

2.2. Modified Geometries of the Periodic Hexagonal Array Model

Applications of the PHA model are not limited to fibrous composites. With simple modifications, the model can be utilized in analysis of elastic–plastic behavior of composites reinforced by ribbons, whiskers, needles, and flat or nearly spherical particles. To accomplish this, it is only necessary to modify the geometry of Figs. 1 and 2 by a simple affine transformation, as shown in Figs. 5–7. It is probably obvious that the transformation does not change the results (9) to (10). Hence, the vertices of the triangular RVE, Fig. 5, and midpoints M of the sides are translated during deformation in the same way as they would be in a uniformly strained solid.

The modified geometry may be immediately applied to composites reinforced by aligned ribbons (Fig. 6). Modelling of discontinuously reinforced composites requires another modification of RVE

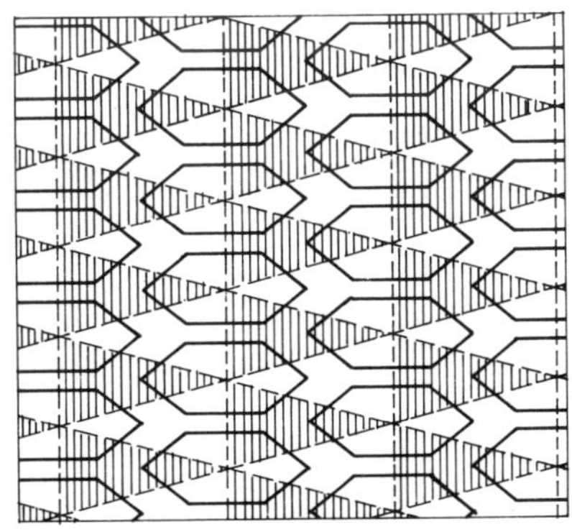

FIG. 5. Transformed geometry of periodic hexagonal array.

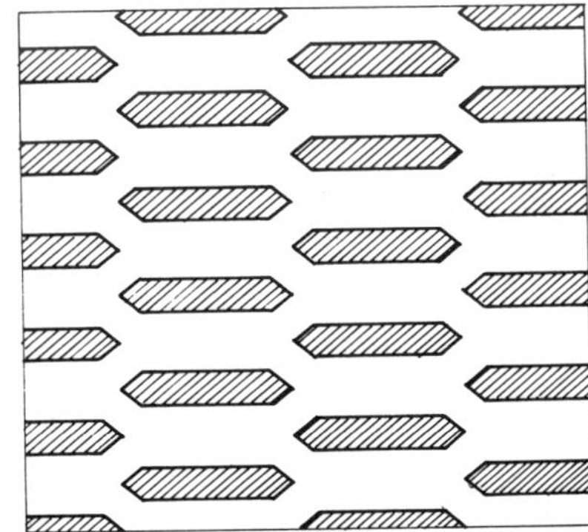

FIG. 6. Whiskers represented by transformed periodic hexagonal array.

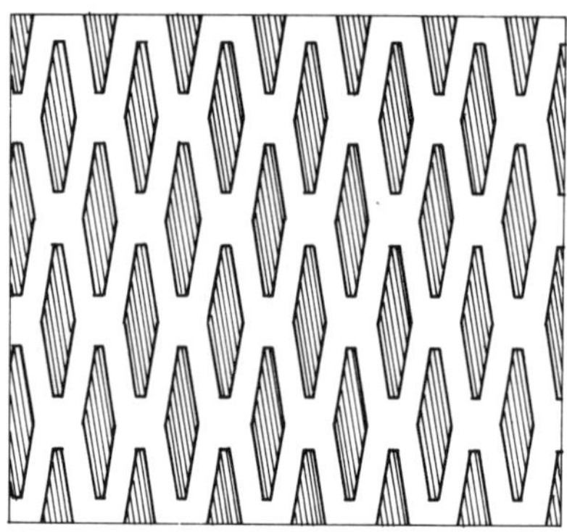

FIG. 7. Needles represented by transformed periodic hexagonal array.

geometry. A finite thickness layer is chosen in Figs. 5–7, such that $0 \leqslant x_3 \leqslant h$. In addition, if a more realistic model is desired, this layer of discontinuously reinforced composite may be overlayed by a layer of matrix which is subdivided in the same way as the composite RVE (Fig. 8). With suitable boundary conditions, the overlay RVE may represent a macroscopic volume of a composite medium. In many instances, analysis of RVEs taken from layers of discontinuous reinforcement shown in Figs. 5–7, without overlay, should provide adequate estimates of overall properties.

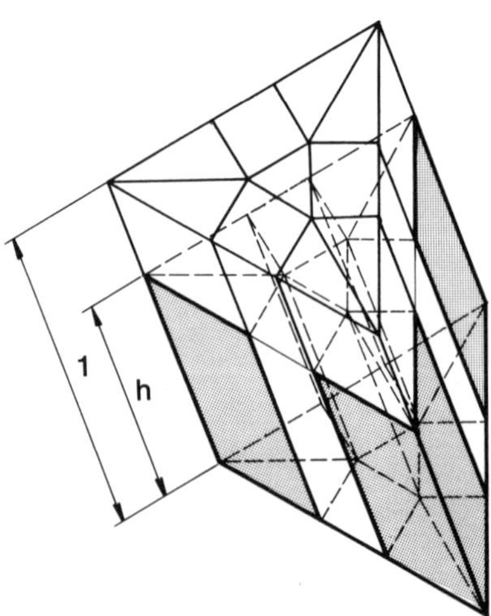

FIG. 8. Representative volume element with matrix overlay.

It has not escaped our attention that many other geometries can be represented by further modifications of the periodic hexagonal array model. The overlay technique can be extended to include not only matrix layers, but also composite layers. Thus, composites with any desired combination of reinforcements, or with selected distributions of reinforcement spacing and orientation can be modelled. Of course, the representative volume elements of such material combinations may not possess the simplicity of the original RVEs, shown above.

The model can also be applied to analysis of porous and cracked media when stiffness of the reinforcement is made to vanish.

3. BOUNDS ON INSTANTANEOUS OVERALL MODULI

Estimates of upper and lower bounds on instantaneous overall moduli of the Periodic Hexagonal Array model will be now determined by finite element analysis of the RVE. Upper bounds are obtained from the displacement approach, and the minimum principle for plastic strain rates, or minimum potential energy in the elastic case. Lower bounds on moduli follow from the equilibrium model, static minimum principle in plasticity, or the elastic minimum complementary energy theorem.

Each bound is obtained in three separate steps.

(a) The composite RVE is replaced by an equivalent homogeneous volume (EHV) of material with unknown instantaneous overall moduli $\mathbf{L}$ of the composite medium. Appropriate boundary conditions are applied, and corresponding energy changes are evaluated in the EHV in terms of $\mathbf{L}$.

(b) The composite RVE is divided into subelements (k) which have known local instantaneous moduli $\mathbf{L}_k$. Boundary conditions of the type used in step (a) are applied, and corresponding extremum energy changes in the RVE are evaluated using the finite element method. These energy changes are made equal to those obtained under similar boundary conditions for the EHV. This yields an expression for evaluation of the overall instantaneous stiffness $\mathbf{L}$ in terms of RVE geometry and local instantaneous moduli $\mathbf{L}_k$.

(c) Relevant energy principles are introduced to show that the results obtained in step (b) provide either upper or lower bounds on $\mathbf{L}$.

3.1. Instantaneous Material Properties

In what follows we shall restrict our attention to elastic and elastic–plastic materials. Let $\bar{\boldsymbol{\sigma}}_n$ and $\bar{\boldsymbol{\varepsilon}}_n$ denote uniform overall strains and stresses applied to the composite, at the end of loading step n, as in (3). Let $\boldsymbol{\sigma}_n^{\mathrm{k}}$ and $\boldsymbol{\varepsilon}_n^{\mathrm{k}}$ be the corresponding local quantities in element k. Strain increments during the loading step $(n+1)$ are

$$\Delta\bar{\boldsymbol{\varepsilon}} = \bar{\boldsymbol{\varepsilon}} - \bar{\boldsymbol{\varepsilon}}_n, \qquad \Delta\boldsymbol{\varepsilon}^{\mathrm{k}} = \boldsymbol{\varepsilon}^{\mathrm{k}} - \boldsymbol{\varepsilon}_n^{\mathrm{k}} \tag{11}$$

where $\Delta\bar{\boldsymbol{\varepsilon}}$, $\Delta\boldsymbol{\varepsilon}^{\mathrm{k}}$ are the instantaneous strain magnitudes. Then, the current stresses are equal to

$$\bar{\boldsymbol{\sigma}}(\boldsymbol{\varepsilon}) = \bar{\boldsymbol{\sigma}}_n + \mathbf{L}\,\Delta\bar{\boldsymbol{\varepsilon}}, \qquad \boldsymbol{\sigma}^{\mathrm{k}}(\boldsymbol{\varepsilon}^{\mathrm{k}}) = \boldsymbol{\sigma}_n^{\mathrm{k}} + \mathbf{L}_{\mathrm{k}}\,\Delta\boldsymbol{\varepsilon}^{\mathrm{k}} \tag{12}$$

where $\mathbf{L}$, and $\mathbf{L}_{\mathrm{k}}$, are the overall, and local, instantaneous stiffness tensors, respectively. On occasion, we shall write

$$\bar{\boldsymbol{\varepsilon}}(\bar{\boldsymbol{\sigma}}) = \bar{\boldsymbol{\varepsilon}}_n + \mathbf{M}\,\Delta\bar{\boldsymbol{\sigma}}, \qquad \boldsymbol{\varepsilon}^{\mathrm{k}}(\boldsymbol{\sigma}^{\mathrm{k}}) = \boldsymbol{\varepsilon}_n^{\mathrm{k}} + \mathbf{M}_{\mathrm{k}}\,\Delta\boldsymbol{\sigma}^{\mathrm{k}} \tag{13}$$

where $\bar{\mathbf{M}} = \bar{\mathbf{L}}^{-1}$, and $\mathbf{M}_{\mathrm{k}} = \mathbf{L}_{\mathrm{k}}^{-1}$, if the inverses exist.

3.2. Upper Bounds on L

3.2.1. *Energy Change in EHV*

Consider a macroscopic volume of a fibrous composite subjected to boundary conditions (3). Assume that the composite has been replaced by an effective homogeneous medium with instantaneous stiffness $\mathbf{L}$ of the composite, and that a volume element which is geometrically similar to the RVE (Fig. 4) has been separated from the effective medium. This element will be referred to as the Equivalent Homogeneous Volume (EHV), its volume $V=1$.

Suppose that strain or displacement increments $\Delta\bar{\boldsymbol{\varepsilon}}$, or $\Delta\bar{\mathbf{u}}$, have been applied to the composite EHV, according to (3). The corresponding energy change is equal to

$$\Delta\bar{\pi} = \int_V \int_{\bar{\boldsymbol{\varepsilon}}_n}^{\bar{\boldsymbol{\varepsilon}}_n + \Delta\bar{\boldsymbol{\varepsilon}}} \bar{\boldsymbol{\sigma}}^{\mathrm{T}}\,\mathrm{d}\bar{\boldsymbol{\varepsilon}}\,\mathrm{d}V - \int_{S_{\mathrm{F}}} \int_{\bar{\mathbf{u}}_n}^{\bar{\mathbf{u}}_n + \Delta\bar{\mathbf{u}}} \bar{\mathbf{t}}\,\mathrm{d}\bar{\mathbf{u}}\,\mathrm{d}S$$

The stress $\bar{\boldsymbol{\sigma}}$ can be taken from (12), and $\Delta\bar{\pi}$ can be integrated to yield

$$\Delta\bar{\pi} = \bar{\boldsymbol{\sigma}}_n^{\mathrm{T}}\Delta\bar{\boldsymbol{\varepsilon}} + \tfrac{1}{2}\Delta\bar{\boldsymbol{\varepsilon}}^{\mathrm{T}} L \Delta\bar{\boldsymbol{\varepsilon}} - \int_{S_{\mathrm{F}}} (\bar{\mathbf{t}}_n + \Delta\bar{\mathbf{t}})^{\mathrm{T}}\,\Delta\bar{\mathbf{u}}\,\mathrm{d}S \tag{14}$$

Since the overall and local stresses and strains are equal in the homogeneous EHV, the top bars in (13) can be removed. The expression then represents $\Delta\bar{\pi}$ in terms of local quantities.

To evaluate (14) in EHV, it is advantageous to regard the EHV as a single finite element with vertices V_1, V_2, V_3, V_4 (Fig. 4). Let $\bar{\mathbf{a}}$ be the vector of nodal displacements of vertices V_i, and $\bar{\mathbf{N}}$ the matrix of linear shape functions. Then,

$$\bar{\mathbf{u}} = \bar{\mathbf{N}}\bar{\mathbf{a}} \tag{15}$$

and

$$\bar{\boldsymbol{\varepsilon}} = \mathbf{G}\bar{\mathbf{u}} = \mathbf{G}\bar{\mathbf{N}}\bar{\mathbf{a}} = \bar{\mathbf{B}}\bar{\mathbf{a}} \tag{16}$$

where $\mathbf{G}$ is a linear operator.

The strains are unique only if rigid body motion components are excluded from $\bar{\mathbf{a}}$, and $\bar{\mathbf{B}}$ is nonsingular.

$$\Delta\bar{\pi} = \bar{\boldsymbol{\sigma}}_n^{\mathrm{T}}\bar{\mathbf{B}}\,\Delta\bar{\mathbf{a}} + \tfrac{1}{2}\Delta\bar{\mathbf{a}}^{\mathrm{T}}\bar{\mathbf{B}}^{\mathrm{T}}\mathbf{L}\bar{\mathbf{B}}\,\Delta\bar{\mathbf{a}} - (\bar{\mathbf{F}}_n + \Delta\bar{\mathbf{F}})^{\mathrm{T}}\,\Delta\bar{\mathbf{a}} \tag{17}$$

The last term in (17) is equal to the last term in (14), the $\bar{\mathbf{F}}_n$ and $\Delta\bar{\mathbf{F}}$ are vectors of nodal forces which are equivalent to surface tractions $\bar{\mathbf{t}}_n$ and $\Delta\bar{\mathbf{t}}$, respectively. These nodal forces and tractions are in equilibrium with $\bar{\boldsymbol{\sigma}}_n$ and $\Delta\bar{\boldsymbol{\sigma}}$. Since $\boldsymbol{\sigma} = \bar{\boldsymbol{\sigma}}$ in the EHV, one can write the virtual work equation:

$$(\bar{\mathbf{F}}_n + \Delta\bar{\mathbf{F}})^{\mathrm{T}}\,\delta\bar{\mathbf{a}} = \int_V (\bar{\boldsymbol{\sigma}}_n^{\mathrm{T}} + \Delta\bar{\boldsymbol{\sigma}}^{\mathrm{T}})\bar{\mathbf{B}}\,\delta\bar{\mathbf{a}}\,\mathrm{d}V \tag{18}$$

where $\delta\bar{\mathbf{a}}$ is arbitrary, and $V = 1$. Hence,

$$\bar{\mathbf{F}}_n = \bar{\boldsymbol{\sigma}}_n^{\mathrm{T}}\bar{B}, \qquad \Delta\bar{\mathbf{F}}^{\mathrm{T}} = \Delta\bar{\boldsymbol{\sigma}}^{\mathrm{T}}\bar{\mathbf{B}} \tag{19}$$

Therefore, the final expression for energy change $\Delta\bar{\pi}$ in the EHV is:

$$\Delta\bar{\pi} = \tfrac{1}{2}\Delta\bar{\mathbf{a}}^{\mathrm{T}}\bar{\mathbf{B}}^{\mathrm{T}}\mathbf{L}\bar{\mathbf{B}}\,\Delta\bar{\mathbf{a}} - \Delta\bar{\boldsymbol{\sigma}}^{\mathrm{T}}\bar{\mathbf{B}}\,\Delta\bar{\mathbf{a}} \tag{20}$$

3.2.2. *Energy Change in RVE*

Evaluation of $\Delta\bar{\pi}$ in the RVE is quite similar to that outlined in eqns. (15)–(20) above. However, the RVE is not represented by a single element as the EHV was. Instead, its volume is subdivided into subelements. Figure 9 shows a very simple choice. In each subelement, an interpolation of the displacement $\mathbf{u}$ is designed in such a way that $\mathbf{u}$ is a continuous function in RVE. Appropriate choice of subelements and shape functions is important if the model is to be kept simple. Still, some heavy algebra is involved, hence discussion of the subject is presented elsewhere [14].

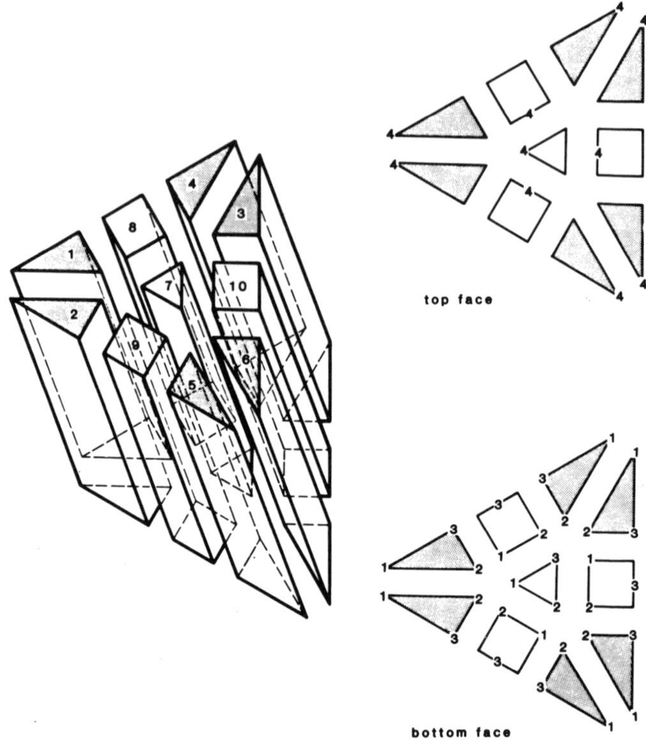

FIG. 9. Subelements of representative volume element and their nodal points for upper bounds.

Evaluation of the stiffness matrix $\mathbf{K}$ of the entire RVE is done by assembling all element contributions to the energy change $\Delta\pi$. In order to find $\Delta\pi$ in terms $\Delta\bar{\mathbf{a}}$ and $\Delta\mathbf{F}$, it is necessary to evaluate the variation of $\Delta\pi$ with respect to internal nodal displacements $\Delta\mathbf{a}^{\mathrm{r}}$. The minimum condition eliminates $\Delta\mathbf{a}^{\mathrm{r}}$, and the result then is:

$$\min(\Delta\pi)\big|_{\Delta\mathbf{a}^{\mathrm{r}}} = \tfrac{1}{2}\Delta\bar{\mathbf{a}}^{\mathrm{T}}\mathbf{K}\,\Delta\bar{\mathbf{a}} - \Delta\mathbf{F}^{\mathrm{T}}\,\Delta\bar{\mathbf{a}} \tag{21}$$

It is now possible to evaluate overall instantaneous moduli of the RVE through a comparison of (20) and (21). By definition,

$$\Delta\bar{\pi} = \min(\Delta\pi)\big|_{\Delta\mathbf{a}^{\mathrm{r}}} \tag{22}$$

and also

$$\Delta\bar{\boldsymbol{\sigma}}^{\mathrm{T}}\bar{\mathbf{B}}\,\Delta\bar{\mathbf{a}} = \Delta\mathbf{F}^{\mathrm{T}}\,\Delta\bar{\mathbf{a}} \tag{23}$$

i.e. the EHV and the RVE are subjected to identical boundary conditions. Therefore, it is only necessary to make equal the first terms on the right hand sides of (20) and (21). This immediately yields the matrix of instantaneous overall moduli of the EHV, and of the composite aggregate

$$\mathbf{L} = (\bar{\mathbf{B}}^{\mathrm{T}})^{-1}\mathbf{K}\bar{\mathbf{B}}^{-1} \tag{24}$$

Also, since $\bar{\boldsymbol{\sigma}} = L\bar{\boldsymbol{\varepsilon}}$, one obtains from (21) and (23)

$$\Delta\mathbf{F} = \mathbf{K}\,\Delta\bar{\mathbf{a}} \tag{25}$$

which is, of course, the minimum condition for (21).

This concludes evaluation of L by the displacement approach.

3.2.3. *Evaluation of Upper Bounds on* L

In order to establish that **L** found in (24) is an upper bound on the actual stiffness $\mathbf{L}^*$ of the composite medium, we utilize the minimum principle for strain rates. Martin [11] gives essentially the following form:

$$\int_V \tfrac{1}{2}\Delta\bar{\mathbf{a}}^{\mathrm{T}}\mathbf{K}\,\Delta\bar{\mathbf{a}}\,\mathrm{d}V - \int_{S_F} \Delta\mathbf{F}^{\mathrm{T}}\,\Delta\bar{\mathbf{a}}\,\mathrm{d}S \geqslant \int_V \tfrac{1}{2}\Delta(\bar{\mathbf{a}}^*)^{\mathrm{T}}\mathbf{K}^*\,\Delta\bar{\mathbf{a}}^*\,\mathrm{d}V - \int_{S_F} \Delta(\mathbf{F}^*)^{\mathrm{T}}\,\Delta\bar{\mathbf{a}}^*\,\mathrm{d}S \tag{26}$$

where $\Delta\mathbf{a}^*$ are surface displacements rates on S_F in the actual state, $\Delta\bar{\mathbf{a}}$ are derived from a kinematically admissible field.

Suppose that the RVE is subjected to stress boundary conditions (3), i.e. nodal force increments are specified such that $\Delta\bar{\mathbf{F}} = \Delta\mathbf{F}^*$. Hence, from (25):

$$\mathbf{K}\,\Delta\bar{\mathbf{a}} = \mathbf{K}^*\,\Delta\bar{\mathbf{a}}^* \tag{27}$$

on S_F, where external forces are presented. The left hand side of (26) can be rewritten with (25), (27) as a surface integral on S_F. Comparison of the integrands leads to

$$-\tfrac{1}{2}(\Delta\mathbf{F}^*)^{\mathrm{T}}\,\Delta\bar{\mathbf{a}} \geqslant -\tfrac{1}{2}(\Delta\mathbf{F}^*)^{\mathrm{T}}\,\Delta\bar{\mathbf{a}}^* \tag{28}$$

and

$$\Delta\bar{\mathbf{a}}_i \leqslant \Delta\bar{\mathbf{a}}_i^* \tag{29}$$

When **K** in (25) is transformed into its diagonal form, $\mathbf{K} = \mathrm{diag}\,k_i$, then it follows from (27) and (29) that

$$\operatorname{diag} k_i \geqslant \operatorname{diag} k_i^* \tag{30}$$

Since **K** and **L** are similar,

$$\operatorname{diag} l_i \geqslant \operatorname{diag} l_i^* \tag{31}$$

In the elastic case, (26) is replaced by the minimum potential energy theorem. The proof is similar (cf. Veubeke [15]), and the results (30) and (31) are again recovered.

3.3. Lower Bounds on L

3.3.1. *Energy Change in EHV*

As in Section 3.2, we assume that the composite has been replaced by an effective homogeneous medium, and that a unit volume element which is geometrically similar to the RVE (Fig. 4) has been separated.

Suppose that a stress or traction increment $\Delta\bar{\boldsymbol{\sigma}}$ or $\Delta\bar{\mathbf{t}}$ has been added to the current values of $\bar{\boldsymbol{\sigma}}_n$, $\bar{\mathbf{t}}_n$ applied to the composite EHV, in agreement with (3). The corresponding energy change is equal to

$$\Delta\pi_{\mathrm{c}}^{\mathrm{e}} = \int_{V^{\mathrm{e}}} \int_{\bar{\boldsymbol{\sigma}}_n}^{\boldsymbol{\sigma}_n+\Delta\boldsymbol{\sigma}} \bar{\boldsymbol{\varepsilon}}^{\mathrm{T}}\, \mathrm{d}\bar{\boldsymbol{\sigma}}\, \mathrm{d}V - \int_{S_{\mathrm{u}}^{\mathrm{e}}} \int_{\bar{\mathbf{t}}}^{\bar{\mathbf{t}}_n+\Delta\bar{\mathbf{t}}} \bar{\mathbf{u}}^{\mathrm{T}}\, \mathrm{d}\bar{\mathbf{t}}\, \mathrm{d}S \tag{32}$$

As in evaluation of $\Delta\bar{\pi}$ of EHV in (14), the $\Delta\pi_{\mathrm{c}}$ in (32) can be made by the finite element method. However, a different element is used, as discussed next.

3.3.2. *Energy Change in RVE*

Evaluation of lower bounds on instantaneous moduli of the RVE is best done with the equilibrium model which was originally described by Veubeke [15]. A constant stress field $\boldsymbol{\sigma}^{\mathrm{k}}$ is assumed inside each subelement. Equilibrium between subelements is satisfied in terms of inter-element tractions. Energy change in each subelement is found from (32), taken only over the subelement. The complete energy change $\Delta\pi_{\mathrm{c}}$ in the RVE is then obtained by summation of $\Delta\pi_{\mathrm{c}}^{\mathrm{k}}$. Note that $\bar{\boldsymbol{\sigma}}$ and $\bar{\mathbf{u}}$ in (32) are assumed to be independent fields. The latter may not be compatible, but it is continuous at nodes placed at middle points of the sides of EHV, or of RVE subelements. Figure 10 shows a simple collection of subelements. Then (32) is a form of a mixed energy principle (Zienkiewicz [17]).

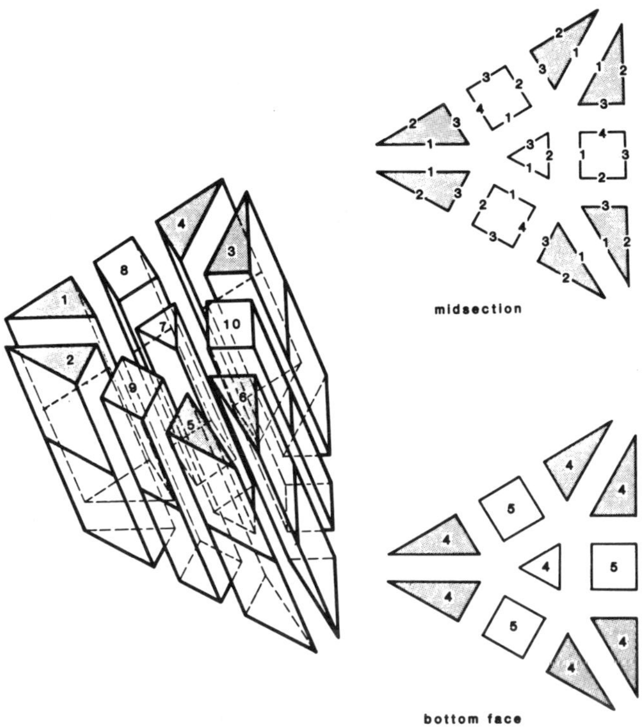

FIG. 10. Subelements of representative volume element and their nodal points for lower bounds.

Substitution of constitutive equation (13) into (32) yields

$$\Delta\pi_{\mathrm{c}}^{\mathrm{e}} = \int_{V^{\mathrm{e}}} (\boldsymbol{\sigma}_m^{\mathrm{k}} + \tfrac{1}{2}\,\Delta\boldsymbol{\sigma}^{\mathrm{k}})^{\mathrm{T}} \mathbf{M}\,\Delta\boldsymbol{\sigma}^{\mathrm{k}}\,\mathrm{d}V - \int_{S_{\mathrm{u}}^{\mathrm{e}}} (\mathbf{t}_n^{\mathrm{k}} + \Delta\mathbf{t}_n^{\mathrm{k}})^{\mathrm{T}}\,\Delta\mathbf{u}\,\mathrm{d}S \tag{33}$$

When an appropriate discretization with continuous stresses is applied to independent fields $\boldsymbol{\sigma}$ and $\mathbf{n}$ in the RVE, the compliance $\mathbf{M}$, or stiffness $\mathbf{L}$ of EHV, can be calculated from the equation

$$\Delta\bar{\pi}_{\mathrm{c}} = \min(\Delta\pi_{\mathrm{c}})|_{\Delta\mathbf{a}^{\mathrm{r}}} \tag{34}$$

which is analogous to (22).

The procedure which leads to evaluation of (34) through a finite element equilibrium model, with piecewise constant stresses in subelements, is formally identical to a modified displacement approach. This analogy, which was discussed by Veubeke [15], Strand and Fix [13], and Zienkiewicz [17] suggests that the overall RVE stiffness in (34) can be obtained in exactly the same way as **K** in (21), providing that the assumed displacement field is linear, incompatible, except at nodes placed at the element mid-sides. This new displacement field is introduced into the analysis of Section 3.2 through appropriate element shape functions **N**. Selection of these shape functions is described in ref. 14.

3.3.3. *Evaluation of Lower Bounds on* **L**

The minimum principle for stress rates (Martin [11]) suggests that the energy change (32) attains an absolute minimum when the stresses and internal reactions on S_u are equal to those of the actual state. As in (26), the principle can be written to read:

$$\int_V \tfrac{1}{2}\Delta\bar{\mathbf{a}}^T\mathbf{K}\,\Delta\bar{\mathbf{a}}\,\mathrm{d}V - \int_{S_u}\Delta\mathbf{F}^T\,\Delta\bar{\mathbf{a}}\,\mathrm{d}S \geqslant \int_V \tfrac{1}{2}\Delta(\bar{\mathbf{a}}^*)^T\mathbf{K}\,\Delta\bar{\mathbf{a}}^*\,\mathrm{d}V - \int_{S_u}\Delta(\mathbf{F}^*)^T\,\Delta\bar{\mathbf{a}}^*\,\mathrm{d}S \quad (35)$$

where $\Delta\mathbf{F}$ are now a part of an assumed equilibrium field, and $\Delta\mathbf{F}^*$ are the actual internal reaction rates on S_u.

If displacements (3) are prescribed on the RVE, then $\Delta\bar{\mathbf{a}} = \Delta\bar{\mathbf{a}}^*$ on S_u. Now, from (25):

$$\Delta\mathbf{F}^* = \mathbf{K}^*\,\Delta\bar{\mathbf{a}}^*, \qquad \Delta\mathbf{F} = \mathbf{K}\,\Delta\bar{\mathbf{a}} \quad (36)$$

In analogy with the steps leading to (28) one obtains from (35) the inequality of integrands on S_u:

$$-\Delta\mathbf{F}^T\,\Delta\bar{\mathbf{a}} \geqslant -\Delta(\mathbf{F}^*)^T\,\Delta\bar{\mathbf{a}}^* \quad (37)$$

and

$$\Delta\mathbf{F} \leqslant \Delta\mathbf{F}^* \quad (38)$$

Hence, from (35), after diagonalization of **K**

$$\mathrm{diag}\;k_i \leqslant \mathrm{diag}\;k_i^* \quad (39)$$

and also after diagonalization of **L**:

$$\mathrm{diag}\;l_i \leqslant \mathrm{diag}\;l_i^* \quad (40)$$

The bounding procedure has been developed with regard to computational efficiency in ref. 14. One or both bounds can be used in calculation of element stiffness matrices in larger structural analysis programs. Numerical evaluation of upper bounds on instantaneous stiffnesses of elastic–plastic composites with the simple mesh of Fig. 9 may lead to erroneous results which are caused by overconstraint of the RVE solution domain. Under such circumstances, which were encountered in small transverse tension and shear loading modes, the stress increments in elements are predominantly hydrostatic in the plastic range. The RVE does not deform plastically in shear, and the overall stiffnesses are only slightly smaller than the respective elastic values. To remedy this difficulty, which is similar to that observed by

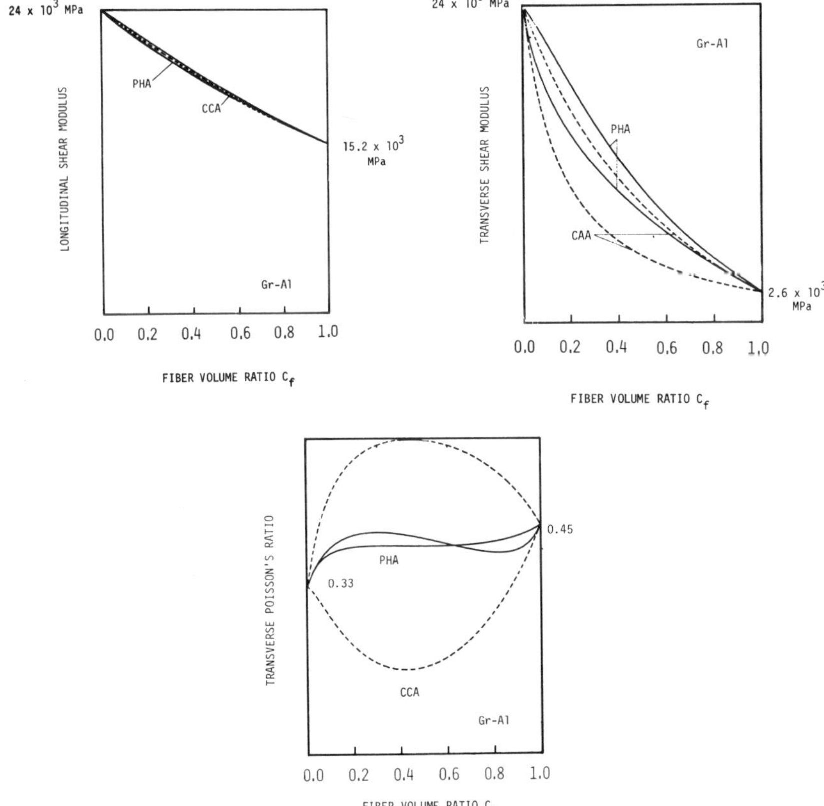

FIG. 11. Comparison of PHA and CCA bounds on elastic moduli for graphite–aluminum composites: graphite fibers and aluminum matrix.

Nagtegaal *et al.* [12], it is necessary to reduce the total number of constraints in the RVE domain. This has been achieved by choosing a refined mesh, and a reduced integration procedure. Reference 14 describes these and other aspects of the solution technique.

4. RESULTS

Space limitations prevent a more complete discussion of applications of the theory. Illustrations of results appear in Figs. 11 and 12 which show bounds on certain elastic moduli of fibrous graphite–aluminum and boron–aluminum systems. The present (PHA) results are plotted

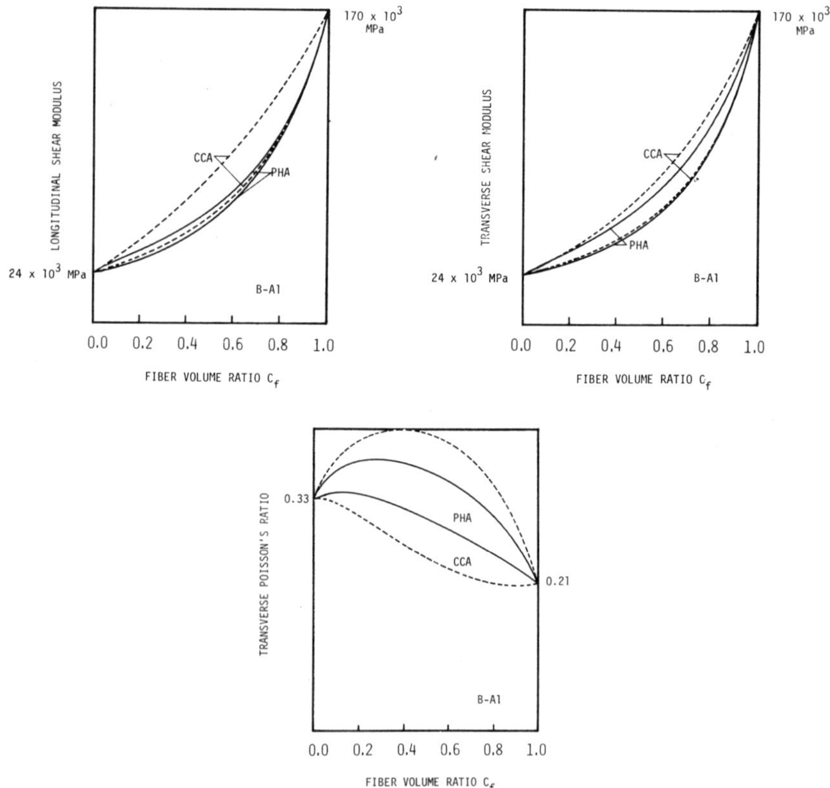

FIG. 12. Comparison of PHA and CCA bounds on elastic moduli for boron–aluminum composites: boron fibers and aluminum matrix.

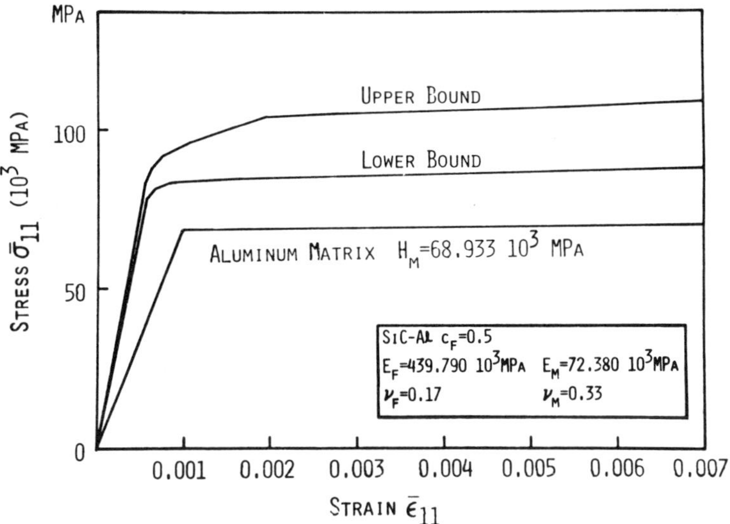

FIG. 13. Bounds on elastic–plastic response of a unidirectional SiC–Al composite for overall transverse tension in x_1-direction.

together with Hashin (CAA) bounds [5]. It is seen that the present approach provides generally closer bounds than the CAA model. Figure 12 shows upper and lower bounds. Figure 13 shows stress–strain curves of a fibrous medium under transverse normal stress. Results of both upper and lower bound calculations are presented; these can be readily utilized to determine bounds on the instantaneous transverse modulus.

ACKNOWLEDGEMENT

This work has been supported in part by the U.S. Army Materials and Mechanics Research Center, and by the U.S. Army Research Office.

REFERENCES

1. BUTLER, T. W. and E. J. SULLIVAN, JR. On the transverse strength of fiber-reinforced materials, *J. appl. Mech.*, **40** (1973), 523.

2. Drucker, D. C. Engineering and continuum aspects of high strength materials. In: *High Strength Materials* (Ed. V. F. Zackay), *Proc. 2nd Int. Sym. High Strength Materials*, June 1964, Wiley, New York (1965), 795–833.
3. Dvorak, G. J. and Y. A. Bahei-el-din. Elastic–plastic behavior of fibrous composites, *J. Mech. Phys. Solids*, **27** (1979), 51.
4. Dvorak, G. J. and Y. A. Bahei-el-din. Plasticity analysis of fibrous composites, *J. appl. Mech.*, **49** (1982), 327.
5. Hashin, Z. Analysis of properties of fiber composites with anisotropic constituents, *J. appl. Mech.*, **46** (1979), 543.
6. Hashin, Z. and B. W. Rosen. The elastic moduli of fiber reinforced materials, *J. appl. Mech.*, **31** (1964), 233.
7. Hashin, Z. and S. Shtrikman. A variational approach to the theory of the elastic behavior of multiphase materials, *J. Mech. Phys. Solids*, **11** (1963), 127.
8. Hill, R. Theory of mechanical properties of fiber-strengthened materials: I. Elastic behavior, *J. Mech. Phys. Solids*, **12** (1964), 199.
9. Hill, R. Theory of mechanical properties of fiber-strengthened materials: III. Self-consistent model, *J. Mech. Phys. Solids*, **13** (1965), 189.
10. Laws, N. On the thermostatics of composite materials, *J. Mech. Phys. Solids*, **21** (1973), 9.
11. Martin, J. B. *Plasticity*: *Fundamentals and General Results*, MIT Press, Cambridge, MA, 1975, 931.
12. Nagtegaal, J. C., D. M. Parks and J. R. Rice. On numerically accurate finite element solutions in the fully plastic range, *Comp. Methods appl. Mech. Engng*, **4** (1974), 153–177.
13. Strand, G. and G. J. Fix. *Analysis of the Finite Element Method*, Prentice-Hall, Englewood Cliffs, N.J., 1973.
14. Teply, J. L. Ph.D. Dissertation, University of Utah, 1983.
15. Veubeke, B. F. D. Displacement and equilibrium models in the finite element method. In: *Stress Analysis*, Chapter 9 (Eds O. C. Zienkiewicz and G. S. Holister), Wiley, New York, 1965.
16. Walpole, L. J. On the overall elastic moduli of composite materials, *J. Mech. Phys. Solids*, **17** (1969), 235.
17. O. C. Zienkiewicz. *The Finite Element Method*, 3rd edn, McGraw-Hill, New York, 1977.

36

Rock Plasticity

N. Cristescu

Department of Mechanics, University of Bucharest, Romania

ABSTRACT

The experimental results necessary to understand the kind of plasticity exhibited by rocks are presented. Generally for rocks the initial yield stress is practically zero, while various time effects are significant.

A viscoplastic constitutive equation is formulated. Mathematical definitions for dilatancy, compressibility and dilatancy threshold boundary are given. 'Elasticity' is to be found only along the hydrostatic line at high pressures and a 'volume elasticity' in the neighbourhood of the dilatancy threshold boundary. An example is given for granite. The model describes: creep, volumetric creep with compressibility and dilatancy, loading rate effect on stress–strain curves and on dilatancy and compressibility, etc. Failure prediction is envisaged. A time-independent elastic–plastic constitutive equation is also proposed for those rocks and conditions when time effects are negligible.

1. EXPERIMENTAL RESULTS

When trying to establish a constitutive equation for a new kind of material, generally we start with several diagnostic tests. The simplest test is certainly the uniaxial one, in which a cylindrical rock specimen is subjected to an axial stress, the lateral surface being stress-free. Both axial strain ε_1 and diameter strain ε_2 are measured. Thus we obtain two stress–strain curves σ_1–ε_1 and σ_1–ε_2. Typical curves of this sort are

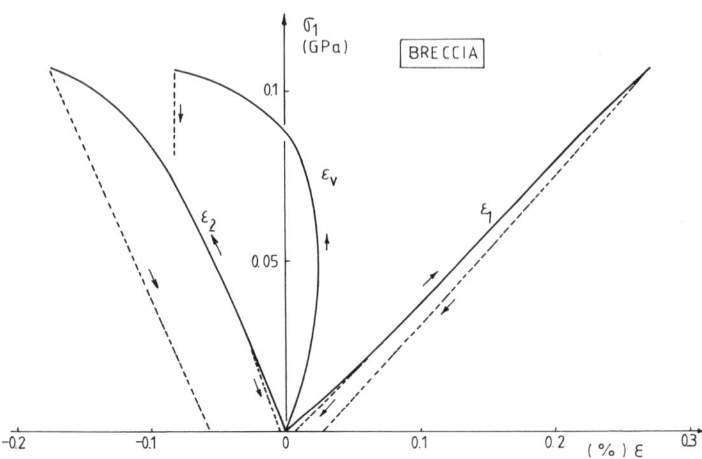

FIG. 1. Uniaxial stress–strain curves for breccia showing dilatancy.

shown on Fig. 1 for breccia. We can see that the σ_1–ε_1 curve is initially nonlinear, then more or less linear and finally at high stresses again nonlinear. Two unloading paths are also shown as dotted lines. An irreversible component of the strain ε_1 is thus revealed. Small loadings or unloadings are following approximately the same path only for stress states belonging to the middle part of this curve. The stress–strain curve σ_1–ε_2 has similar peculiarities, but the irreversible component of the strain is more significant. Since the volume deformation is defined by

$$\varepsilon_v = \varepsilon_1 + 2\varepsilon_2 \tag{1}$$

we can plot also the σ_1–ε_v curve. This last curve shows that at relatively small stresses, on the nonlinear portion of the σ_1–ε_v curve, the rock is compressible while at high stresses, on the other nonlinear portion of the curve, it is dilatant. At intermediate stress levels, on a very small portion, the curve σ_1–ε_v can possibly be approximated with the straight line $\sigma_1 = 3K\ \varepsilon_v$ with K standing for the bulk modulus. From this simple diagnostic test we come to the conclusion that for rocks 'elasticity' (in the sense of 'linear' and 'reversible' in a small cyclic loading–unloading test) is not to be found at small stresses, in the neighbourhood of the stress-free state, but somewhere at higher applied stresses.

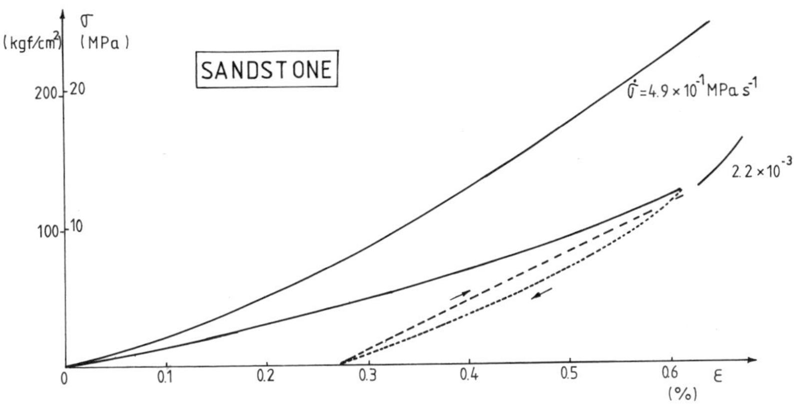

FIG. 2. Loading rate effect on uniaxial stress–strain curve for sandstone.

Another diagnostic test can be done by changing the loading rate in the uniaxial test. For instance in Fig. 2 are given the uniaxial stress–strain curves for sandstone, obtained at two loading rates. It follows that when the loading rate is increased the whole stress–strain curve is raised starting even from the origin. This result is typical for rocks: we do not have an initial 'linear elastic' portion of the stress–strain curve which would be rate-independent. On the other hand by unloading, a significant irreversible component of the strain is revealed. A typical loading–unloading hysteresis loop is also shown on the Fig. 2. Similar unloading paths are found starting from the smallest applied stresses.

The conclusion which results from the experimental data given above is that generally the yield stress of rock-like materials is essentially zero and that no elastic domain (in the sense used in metal plasticity) is surrounding the origin in the stress space.

The experimental data on Fig. 2 also show significant time effects. These can be checked also in creep tests. One typical creep curve for dolomite is given on Fig. 3. For any magnitude of the constant stress applied, after an initial sudden increase of the strain the rock deforms slowly with time. Even this initial strain increase is only partially reversible. Thus we do not have a 'yield stress' under which the irreversible deformation by creep would not be observed. This conclusion is true even if a high hydrostatic stress is superposed on a small deviatoric stress (cf. [22]). A major peculiarity of the creep curves shown on Fig. 3 is that for each constant applied stress the creep

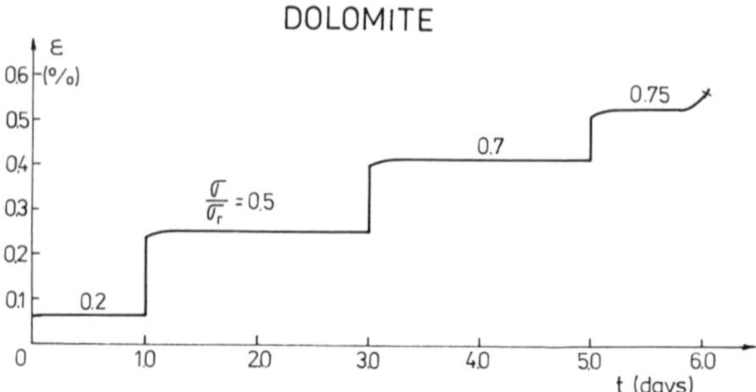

FIG. 3. Creep curves for dolomite.

deformation stops after a certain time interval; this is true however only for applied stresses which do not exceed about 0·6–0·7 from the stress producing failure (denoted here by σ_r). For higher stresses the deformation by creep continues with secondary and tertiary creep after which the specimen breaks. Another important peculiarity is that in the first minutes or hours after each reloading a significant 'fast' creep takes place, i.e. the deformation by creep in the first minutes or hours is a major portion of the total deformation obtained by creep.

In a similar way the creep curves for the diameter strain can be obtained and then using formula (1) the creep curves for the volumetric strain. Such curves are shown on Fig. 4 for a filler (an artificial rock used to fill up the underground opening) and two levels of applied stresses $\sigma_1 = 0{\cdot}79$ MPa and 3·15 MPa. It is interesting to observe that at the lower stress level shown first an elastic compressibility takes place followed by a fast creep producing dilatancy and finally by a slow creep producing compressibility. At higher stress level the creep of the volume produces dilatancy only. The oscillatory creep of the volumetric strain of various rocks was described by Cristescu [9]. Generally at lower stresses (as compared with the failure stress) the dominant phenomenon is the volume creep producing compressibility. At the intermediate stress level the volume creep is generally small. Finally at high stresses the volume dilatancy by creep is the dominant phenomenon. These are certainly the general features; some rocks may exhibit practically only dilatancy or only compressibility during uniaxial creep tests. If subjected also to a high hydrostatic stress generally all rocks are dilatant only.

The volume creep during tertiary creep of quartzite and granite was

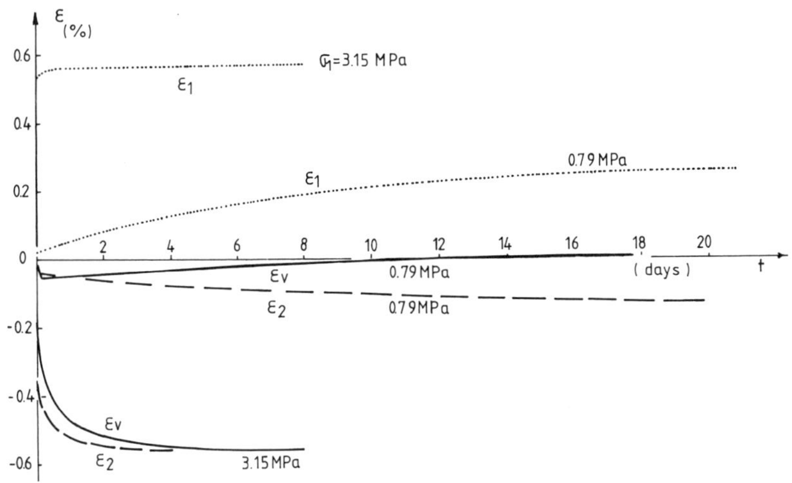

FIG. 4. Volumetric creep curves showing dilatancy followed by compressibility.

reported by Kranz and Scholz [17], the volumetric creep of sandstone was described by Cristescu [9] and that of cement concrete by Constantinescu and Cristescu [6].

In the previously described experiments the rock specimen was subjected to a certain deviatoric stress superposed on a hydrostatic stress. This information has to be supplemented by the reaction of the rocks when subjected to hydrostatic loadings alone. These are the confined tests in which the displacement of the lateral surface of the specimen is blocked and a very high pressure is transmitted axially. Various devices used for this purpose were described by Bridgman [4]. Volarovitch *et al.* [23] and others. We have done such experiments by increasing the axial stress in successive steps. At each quite sudden increase of the stress an instantaneous increase of the volumetric strain is observed; further during the period when the stress is held constant the volumetric strain continues to increase by creep. After a certain time interval the creep of the volume ceases. An unloading would reveal an irreversible portion of the volumetric strain though the slope of the unloading curve varies with the magnitude of the hydrostatic stress. As the hydrostatic applied stress is increased the property of the volume of the rock to deform by creep and generally to deform irreversibly is steadily diminishing. In other words above a certain quite high hydrostatic stress a small loading–unloading cycle would reveal reversible volumetric strains only. Generally the hydrostatic stresses involved in the mining applications are under this limiting hydrostatic stress. The creep of the volume strain in confined high pressure tests was described by the author for schist in ref. 8, for chalk in ref. 10 and for cement concrete in ref. 6.

2. TIME-DEPENDENT CONSTITUTIVE EQUATION

The time-dependent phenomena mentioned above can be described by a quasilinear constitutive equation (Cristescu [11]). Only compressive stresses (taken as positive) will further be considered. The rock will be assumed isotropic and homogeneous. Then assuming small displacements and material rotations we postulate the constitutive equation in the form (cf. refs. 7 and 13).

$$\dot{\varepsilon}_{ij} = \left[\frac{1}{3K(\sigma, \bar{\sigma}, \Delta)} - \frac{1}{2G(\sigma, \bar{\sigma}, \Delta)}\right]\dot{\sigma}\delta_{ij} + \frac{1}{2G(\sigma, \bar{\sigma}, \Delta)}\dot{\sigma}_{ij} + k(\sigma, \bar{\sigma}, \Delta)\left\langle 1 - \frac{W^{\mathrm{I}}(t)}{H(\sigma, \bar{\sigma}, \Delta)}\right\rangle \frac{\partial F(\sigma, \bar{\sigma}, \Delta)}{\partial \sigma_{ij}} \tag{2}$$

Here K and G are the elastic parameters, assumed to be variable and dependent on the mean stress, equivalent stress

$$\bar{\sigma} = \sqrt{\sigma_1^2 + \sigma_2^2 + \sigma_3^2 - \sigma_1\sigma_2 - \sigma_2\sigma_3 - \sigma_3\sigma_1} \tag{3}$$

and the third invariant Δ of the stress tensor. F is a viscoplastic potential dependent on the same arguments.

$$H(\sigma, \bar{\sigma}, \Delta) = W^{\mathrm{I}}(t) \tag{4}$$

is the equation of the stabilization boundary with

$$W^{\mathrm{I}}(T) = \int_0^T \sigma_{ij}(t)\dot{\varepsilon}_{ij}^{\mathrm{I}}(t)\,\mathrm{d}t \tag{5}$$

the stress work. Finally k is some kind of viscosity coefficient; more exactly E/k will be called the viscosity coefficient and will be given in

Poise, E being the Young's modulus. Superscript I stands for 'irreversible'.

The irreversible volumetric rate of deformation is therefore

$$\dot{\varepsilon}_{ij}^{I} = k\left\langle 1 - \frac{W^{I}(t)}{H(\sigma, \bar{\sigma}, \Delta)} \right\rangle \frac{\partial F(\sigma, \bar{\sigma}, \Delta)}{\partial \sigma_{ij}} \delta_{ij} \tag{6}$$

Thus the rock is by definition dilatant if

$$\frac{\partial F}{\partial \sigma_{ij}} \delta_{ij} < 0 \tag{7}$$

the opposite inequality defines the compressibility states. The states for which

$$\frac{\partial F}{\partial \sigma_{ij}} \delta_{ij} = 0 \tag{8}$$

are on the dilatancy threshold boundary.

The bracket $\langle\ \rangle$ from (2) or (6) has the meaning

$$\langle A \rangle = \begin{cases} A & \text{if} \quad A > 0 \\ 0 & \text{if} \quad A \leqslant 0 \end{cases} \tag{9}$$

Generally the dependence on the argument $1 - W^{I}/H$ in (2) may be nonlinear; the linear variant is a simplified version which seems quite reasonable.

All the constitutive coefficients involved in (2) can be determined from experimental data. Generally these experimental data are obtained in what is called 'triaxial' tests, but other types of experiments may also be necessary, as already mentioned in the previous section. Since in the triaxial tests only two distinct principal stress components can be measured, all the constitutive coefficients involved in (2) must be assumed to depend on two invariants only: the mean stress and the equivalent stress. The model can be improved if more complete experimental data become available.

The elastic moduli are determined from the slope of the unloading curves. A more difficult procedure is necessary to determine the stabilization boundary (4). An example is given in ref. 11. In principle various loading paths including the hydrostatic one are used to determine (5) and then (4). For instance using the experimental data for granite given by Brace [2], Volarovitch *et al.* [23] and Brace *et al.* [3]

we have determined the stabilization boundary in the form

$$H(\sigma,\bar{\sigma})=\left[\frac{a_o}{\sigma-\frac{\bar{\sigma}}{3}+a_1}+a_2\right]\left(\frac{\bar{\sigma}}{\sigma_*}\right)^4+\left[\frac{\left(\sigma-\frac{\bar{\sigma}}{3}-b_o\right)^2}{b_1}+b_2\right]\frac{\bar{\sigma}}{\sigma_*}$$

$$+\begin{cases} c_o\sin\left(\omega\dfrac{\sigma}{\sigma_*}+\varphi\right)+c_1 & \text{if } 0\leqslant\sigma\leqslant\sigma_o \\ c_o+c_1 & \text{if } \sigma_o\leqslant\sigma \end{cases} \tag{10}$$

with the following values of various constants:

$a_o=3{\cdot}833\times10^{-4}(\text{GPa})^2$, $a_1=9{\cdot}22\times10^{-3}$ GPa, $a_2=2{\cdot}62\times10^{-4}$ GPa
$b_o=6{\cdot}00\times10^{-3}$ GPa, $b_1=-969{\cdot}489$ GPa, $b_2=4{\cdot}51\times10^{-4}$ GPa
$\omega=493{\cdot}14748°$, $\varphi=-79{\cdot}8893°$, $c_o=1{\cdot}16454\times10^{-4}$ GPa,
$c_1=1{\cdot}14645\times10^{-4}$ GPa, $\sigma_o=0{\cdot}3445$ GPa, $\sigma_*=1$ GPa.

In the determination of $H(\sigma,\bar{\sigma})$ we have used the following loading paths: the hydrostatic one and several loading paths $\sigma_2=$ const. starting from a hydrostatic stress state. The curves $H(\sigma,\bar{\sigma})=$ const. are shown on Fig. 5 as dotted lines. From their shape we can conclude that the classical assumption $H(\sigma,\bar{\sigma})\equiv F(\sigma,\bar{\sigma})$ seems quite reasonable. With this assumption is plotted the dilatancy threshold boundary as a border line. D_c is the region where the rock is compressible while D_d is the region where the rock is dilatant. The behaviour of the volume is nearly linear elastic with practically no time effects in the neighbourhood of the dilatancy threshold boundary. Along the hydrostatic line for $\sigma\geqslant\sigma_o$ the behaviour of the volume is also linear elastic.

The last to be determined from the experimental data is the coefficient $k(\sigma,\bar{\sigma})$. It can be determined in uniaxial creep tests. From the author's experimental data for hard rocks the viscosity coefficient E/k is of the order 10^{14}–10^{15} Poise. For softer rocks such as rock salt or chalk it is smaller: 10^{12}–10^{13} Poise. The value of the viscosity coefficient at high hydrostatic stresses is not yet known.

Constitutive equation (2) can describe various time effects, some of them mentioned in the first section: creep, volumetric creep, the loading rate effect on dilatancy and on compressibility, the loading rate effect on the stress–strain curves, etc. For instance in Fig. 6 are given the curves of the volume creep in hydrostatic tests in which the mean stress is increased with increments $\Delta\sigma=0{\cdot}1$ GPa and afterwards is held

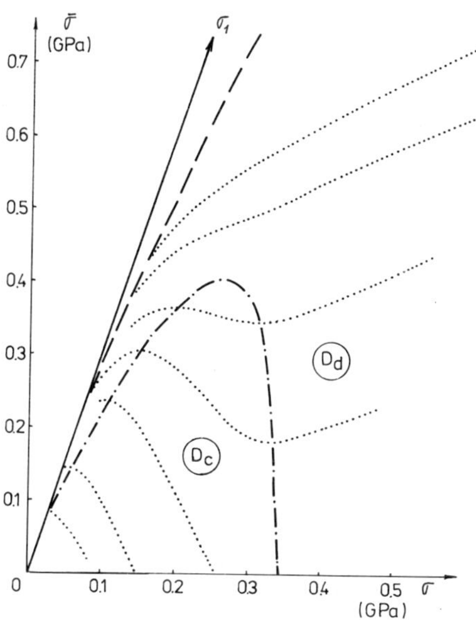

FIG. 5. Relaxation boundaries (dotted lines), dilatancy threshold boundary (border line), failure surface (interrupted line) for granite.

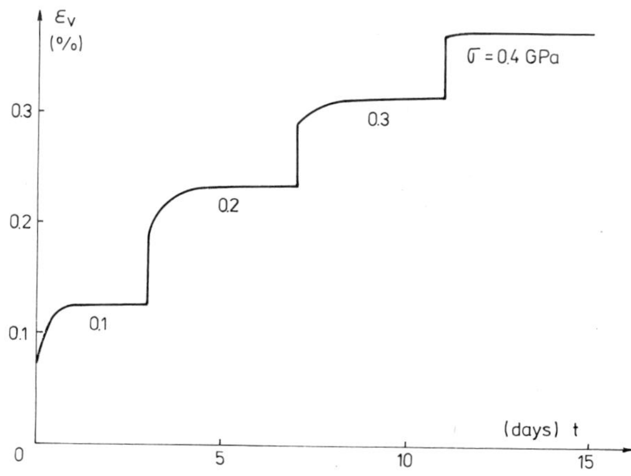

FIG. 6. Creep curves for volumetric strain in hydrostatic loadings.

constant for a long time until the volumetric strain is practically constant for 2 or 3 days. For stresses $\sigma > \sigma_o$ no more volume creep is possible, i.e. further loadings produce reversible volumetric strain only. The instantaneous responses in Fig. 6 are computed taking into account that the bulk modulus is variable with the mean stress as

$$K(\sigma, 0) = \begin{cases} K_o - K_1\left(1 - \dfrac{\sigma}{\sigma_o}\right)^4 & \text{if} \quad \sigma \leqslant \sigma_o \\ K_o & \text{if} \quad \sigma \geqslant \sigma_o \end{cases} \tag{11}$$

with $K_o = 59$ GPa and $K_1 = 48$ GPa.

As another example, in Fig. 7 are given the stress–strain curves σ_1–ε_1, σ_1–ε_2 and σ_1–ε_v obtained after a preliminary preloading with a hydrostatic mean stress of $\sigma = 0{\cdot}1$ GPa. Thus σ_1^R in the figure has the meaning of a relative stress $\sigma_1^R = \sigma_1 - 0{\cdot}1$. Four loading rates are considered corresponding to the following values of the mean loading rate: $\sigma_1 \simeq 0$ (border lines); $2{\cdot}8 \times 10^{-6}$ GPa s^{-1} (interrupted lines); $2{\cdot}8 \times 10^{-5}$ GPa s^{-1} (dotted lines) and $\dot{\sigma} \to \infty$ (full lines). The dilatancy threshold for this case is $\sigma_D = 0{\cdot}391$ GPa. The effect of the loading rate on compressibility and dilatancy is obvious from the figure. Only for very slow tests is there a significant effect of the loading rate on compressibility, while the effect on dilatancy is remarkable for both slow and fast tests.

Model (2) can be used to describe creep of rocks around underground structures; the creep of rocks around vertical shafts was studied by Massier and Cristescu [18] and around horizontal tunnels by

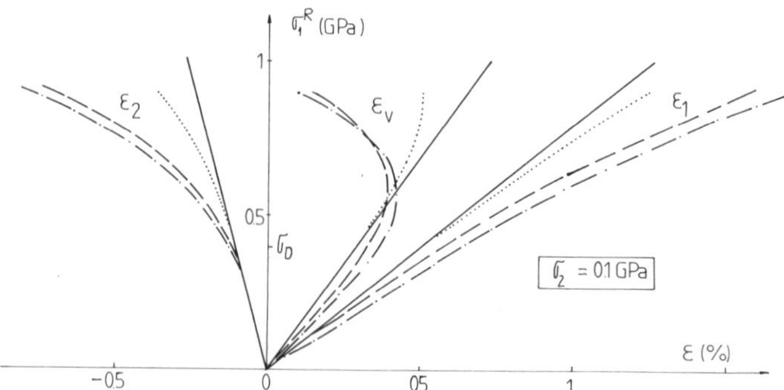

FIG. 7. Loading rate effect on stress–strain curves, on dilatancy and on compressibility.

Massier and Cristescu [19]. The model can be used also to predict the time of failure or of an earthquake if a failure condition is associated with the constitutive equation.

Another viscoplastic model aimed to describe the creep of rocks is due to Dragon and Mroz [15] and Gioda [16] who started from assumptions other than those given above.

3. TIME-INDEPENDENT CONSTITUTIVE EQUATION

In some cases, mainly for hard rocks, the time effects can be disregarded. Then a time independent constitutive equation can be used. Elastic–plastic constitutive equations of classical structures can be easily adapted for rocks. A number of variants have already been published. We mention here the approaches by Rice [20], Rudnicki and Rice [21], Bažant [1], Dorris and Nemat-Nasser [14] and others. In a distinct variant presented below we make the following assumptions [12]: the rock-like materials have a zero initial yield stress and the yield function has to describe both dilatancy and compressibility depending on the stress state to which the rock is subjected. A particular variant of this constitutive equation for cement concrete in uniaxial tests was given by Constantinescu and Cristescu [5].

Assuming again that the rock is isotropic and homogeneous and that displacements and material rotations are small the elastic–plastic constitutive equation can be written in the form

$$\dot{\varepsilon}_{ij} = \frac{\dot{\sigma}_{ij}}{2G} + \left(\frac{1}{3K} - \frac{1}{2G}\right)\dot{\sigma}\,\delta_{ij} + \frac{\left\langle \dfrac{\partial H}{\partial \sigma_{mn}}\,\dot{\sigma}_{mn} \right\rangle}{\dfrac{\partial F}{\partial \sigma_{kl}}\,\sigma_{kl}}\,\frac{\partial F}{\partial \sigma_{ij}} \quad \text{if} \quad H = W^{\mathrm{I}}$$

$$\dot{\varepsilon}_{ij} = \frac{\dot{\sigma}_{ij}}{2G} + \left(\frac{1}{3K} - \frac{1}{2G}\right)\dot{\sigma}\,\delta_{ij} \quad \text{if} \quad H < W^{\mathrm{I}} \tag{12}$$

Here the elastic parameters K and G as well as the loading function H and the plastic potential F are dependent on the mean stress, the equivalent stress and the third invariant of the stress tensor. W^{I} is the irreversible stress work.

If in order to give a numerical example function H is chosen in the form (10), then the yield surfaces are those represented by dotted lines in Fig. 5. From here it follows that $F = H$ is a very reasonable assumption, etc.

4. CONCLUSIONS

Dilatancy and compressibility, which are typical phenomena for rock-like materials, can be described by plastic, both time-dependent and time-independent, constitutive equations. The initial yield stress for rock-like materials is zero. 'Elasticity' is to be found along the hydrostatic line for $\sigma > \sigma_o$; this kind of elasticity can be called 'hydrostatic elasticity' for $\sigma > \sigma_o$. The behaviour of the volume is very close to the elastic one also in the neighbourhood of the dilatancy threshold surface where no significant time effects are observed. Inside this surface the rock is compressible, outside it is dilatant. Inside the irreversible dissipation is due mainly to pore collapse and microcrack closing, while outside the dominant irreversible dissipation mechanism is the friction along microcrack surfaces, and new microcrack formation. The relaxation boundary, or yield function respectively, can be determined by a standard procedure using a set of experimental data obtained in triaxial tests. All these concepts are certainly not path-independent. Thus it is hoped that these constitutive equations can be used for those loading paths of interest for mining applications. The model of unloading given above is to be improved in order to describe the nonlinear unloading, hysteresis loop, etc.

REFERENCES

1. Bažant, Z. P. Endochronic inelasticity and incremental plasticity, *Int. J. Solids Struct.*, **14** (1978), 691–714.
2. Brace, W. F. Some new measurements of linear compressibility of rocks, *J. Geophys. Res.*, **70** (1965), 391–398.
3. Brace, W. F., B. W. Paulding, Jr. and C. Scholz. Dilatancy in the fracture of crystalline rocks. *J. Geophys. Res.*, **71** (1966), 3939–3953.
4. Bridgman, P. W. *Physics of High Pressures*, Dover Publications, New York, 1949.
5. Constantinescu, M. and N. Cristescu, Anelastic properties of cement-concrete, *Rev. Roum. Sci. Techn. Méc. Appl.*, **27** (1982), 667–680.
6. Constantinescu, M. and N. Cristescu, Creep of rock-like materials, *Int. J. Engng Sci.*, **21** (1983), 45–49.
7. Cristescu, N. *Dynamic Plasticity*, North Holland, Amsterdam, 1967.
8. Cristescu, N. A Viscoplastic Constitutive Equation for Rocks. INCREST Preprints in Mathematics, No. 49/1979, Bucharest, 1979.
9. Cristescu, N. Rock dilatancy in uniaxial tests, *Rock Mechanics*, **15** (1982), 133–144.

10. CRISTESCU, N. *Mechanics of Rocks* (in Rumanian), Universitatea din Bucureşti, 1983.
11. CRISTESCU, N. Elastic–viscoplastic constitutive equations for rock-like materials, *Rock Mechanics* (1985) (submitted for publication).
12. CRISTESCU, N. Plasticity of compressible/dilatant rock-like materials, *Int. J. Engng Sci.* (1984) (in press).
13. CRISTESCU, N. and I. SULICIU. *Viscoplasticity*, Martinus Nijhoff, The Hague, 1982.
14. DORRIS, J. F. and S. NEMAT-NASSER. A plasticity model for flow of granular materials under triaxial stress states. *Int. J. Solids Struct.*, **18** (1982), 497–531.
15. DRAGON, A. and Z. MROZ. A model for plastic creep of rock-like materials accounting for the kinetics of fracture, *Int. J. Rock Mech. Min. Sci.*, **16** (1979), 253–259.
16. GIODA, G. A finite element solution of non-linear creep problems in rocks., *Int. J. Rock Mech. Min. Sci.*, **18** (1981), 35–46.
17. KRANZ, R. L. and C. H. SCHOLZ. Critical dilatant volume of rocks at the onset of tertiary creep, *J. Geophys. Res.*, **82** (1977), 4893–4898.
18. MASSIER, D. and N. CRISTESCU. In situ creep of rocks, *Rev. Roum. Sci. Techn. Méc. Appl.*, **26** (1981), 687–702.
19. MASSIER, D. and N. CRISTESCU. Creep of rocks around horizontal tunnels, (Submitted for publication), (1984).
20. RICE, J. R. On the stability of dilatant hardening for saturated rock masses. *J. Geophys. Res.*, **80** (1975), 1531–1536.
21. RUDNICKI, J. W. and J. R. RICE. Conditions for the localization of deformation in pressure-sensitive dilatant materials, *J. Mech. Phys. Solids*, **23** (1975), 371–394.
22. SASAJIMA, S. and H. ITO. Long-term creep experiment of rock with small deviator of stress under high confining pressure and temperature, *Tectonophysics*, **68** (1980), 183–198.
23. VOLAROVITCH, M. P., E. I. BAYUK, A. I. LEVYKIN and I. S. TOMASHEVSKAYA. *Physico-mechanical Properties of Rocks and Minerals at High Pressures and Temperatures* (in Russian), Nauka, Moscow, 1974.

37

On Constitutive Modelling of Dissipative Solids for Plastic Flow, Instability and Fracture

PIOTR PERZYNA

Institute of Fundamental Technological Research, Warsaw, Poland

ABSTRACT

The paper aims at the description of postcritical behaviour and fracture of dissipative solids during tensile tests. Ductile fracture is understood to be the final stage of the necking process and is preceded by nucleation of voids mostly due to plastic deformations and yield stress state as well as linkage of voids. An elastic–viscoplastic model of solids with internal imperfections is presented. An evolution equation for the internal imperfection parameter is postulated. The elastic–plastic response of the material is also discussed. The material functions and constants are determined from available experimental data. The synergetic nature of the fracture phenomenon is investigated. The theory of ductile fracture based on the evolution of internal imperfections is presented. The criterion of fracture is deduced. The initial-boundary-value problem is discussed.

1. INTRODUCTION

The main objective of the paper is to describe the postcritical behaviour and fracture of dissipative solids. A notion of postcritical behaviour means the response of a material of a body during the flow process when the onset of localization of plastic deformations has already taken place.

It has been proved that to describe the postcritical behaviour for dissipative solids we need very precise constitutive modelling and it is necessary to consider important interactive effects such as internal imperfections (nucleation and growth of voids), strain rate sensitivity, thermal effects, etc.

Ductile fracture may be treated as the final stage of the localization (necking) process and is preceded by nucleation of voids in the second phase growth of voids mostly due to plastic deformation and yield stress state and by coalescence of voids.

In this paper the constitutive modelling is developed within the framework of the rate type material structure with internal state variables.

We would like to introduce into the description some information about the final stage of the necking process. This information can be obtained by the modification of the assumed evolution equation for the imperfection parameter.

It is our conjecture that the final stage of the necking process depends on the evolution of all constitutive features of the material as well as on the boundary conditions and the shape of the body concerned.

In a previous paper [17] a simple model of an elastic–viscoplastic solid with internal imperfections (due to the nucleation, growth and diffusion of voids) was proposed. This model was justified by physical mechanisms of polycrystalline matter flow in some regions of temperature and strain rate changes. The identification procedure for all material functions and constants of the model was based on available mechanical test data and on metallurgical observations.

We modify the evolution equation of that model in such a way that we introduce the additional information about the final stage of the necking process. This modification is based on the precise metallurgical observations of nucleation and growth of voids during the necking process.

In Section 2 the discussion of experimental results concerning the postcritical behaviour of metals is presented. The main emphasis is on the nucleation and growth of voids during the necking process.

In Section 3 the description of an elastic–viscoplastic model with internal state variables is presented. An evolution equation for the internal imperfection parameter is postulated. Elastic–plastic response of the material is also discussed.

In Section 4 estimation of the material functions and constants is shown.

In Section 5 mechanisms of fracture are discussed. The influence of different effects on fracture phenomenon is investigated and the synergetic nature of fracture is shown. The theory of ductile fracture based on the evolution of internal imperfections is formulated.

In Section 6 the initial boundary-value problem is posed and a discussion of the solution obtained is presented.

Section 7 gives comments about a new criterion of ductile fracture.

2. POSTCRITICAL BEHAVIOUR OF SOLIDS

2.1. Localization of Plastic Deformation

In this paper we would like to confine our consideration to tensile tests. By postcritical behaviour of a specimen in the tensile test we understand the segment of inelastic flow process of duration $t_2 - t_1$, where t_1 is defined as the instant at which the onset of localization occurs and t_2 is the final instant at which the fracture phenomenon takes place. In Fig. 1 the trajectory of the inelastic flow process during the tensile test in the load–extension plane has been shown. The part of the trajectory between the supposed onset of necking and the

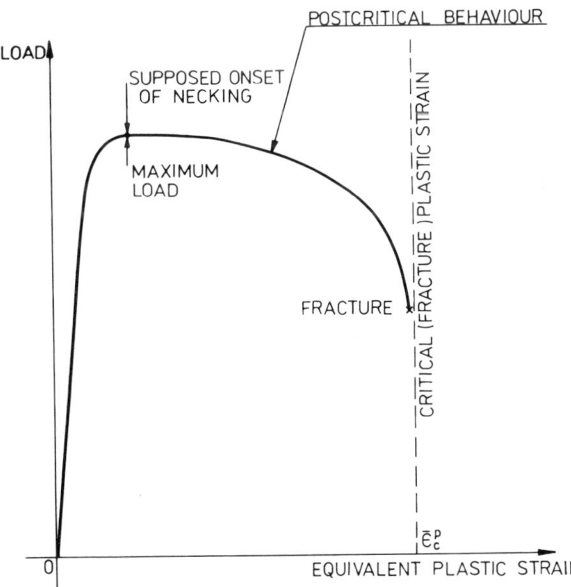

FIG. 1. Load–equivalent plastic strain curve for tensile test.

fracture point is related to the postcritical behaviour segment of the process.

As has been shown in a previous paper [17] the criteria of localization of plastic deformation during the flow process can be considered from two points of view. The first is purely of mathematical nature and can be achieved within the framework of investigation of discontinuous solution or bifurcation (branching) of the solution. The second is more an engineering approach and is based on the investigation of the load–extension curve during flow process.

We assume here the criterion of maximum load as a fundamental condition for the onset of instability by necking.†

It is a well known fact that this condition can be treated as an approximate criterion only. Because one can distinguish between incipient maximum load and incipient necking, these are not necessarily the same points on the load–extension curve.

It is noteworthy that experimental as well as theoretical studies of necking phenomena showed that some additional cooperative effects can influence the onset of instability by shifting the point of initiation of localization from the maximum load point on the load–extension curve.

2.2. Nucleation and Growth of Voids

Recent experimental investigations shedded new light on the real behaviour of the specimen during various postcritical stages of the flow process. First of all we should mention different kinds of experimental techniques used in those investigations. For instance, Bluhm and Morrissey [4] used, in addition to mechanically controlling the deformation of the specimen, the quartz crystal ultrasonic transducer mounted on the end of the specimen to continuously search for signs of the initial void appearance as well as metallographic observations to verify the ultrasonic techniques and to determine crack profile, the relationship to grain structure, and crack orientation. Fisher [6] performed investigations of various mechanical and material parameters on void formation at cementite particles in axisymmetric tensile specimens of spheroidized plain carbon steels by using optical and scanning

† A second basic mode of localization is the phenomenon of plastic instability in the direction of pure shear. A necessary condition for the localization of plastic deformations in the form of the plastic shear band is a maximum true flow stress criterion. For recent investigations in this subject see the review paper by Asaro [2].

electron microscopy. Similarly, Le Roy *et al.* [10] studied the nucleation and growth of voids during tensile straining in spheroidized carbon steel by using both an Instron machine and scanning electron microscopy.

From the investigations of Bluhm and Morrissey [4] we can follow the events (in the order in which things naturally happen) within the segment of the flow process corresponding to the postcritical behaviour of the specimen. After the incipient localization of plastic deformations is reached voids develop at inclusions in the central region of the neck. At this stage of the process voids nucleate and grow. With continued gross extension of the specimen the void volume expands leaving a matrix weakened and in the centre of the neck region coalescence of voids is observed. Compatibility of deformation then causes gross linkage of the central voids to form large macroscopically visible internal cracks which grow as the final shear lips are formed. At this stage of the tensile process in the neck region the solid is under a long range shear field and begins to shear in an overall manner. Finally the specimens either separate rapidly and approximately adiabatically forming the final shear lips, or the specimens may be controlled so that separation is a slow, essentially isothermal process. This latter overall shear separation is associated with the 'knee' of the load–deformation curve.

Le Roy *et al.* [10] have shown that voids nucleated both by the cracking of carbides and by their decohesion.† Once nucleated, the voids grow mainly in the tensile direction and only in the late stages of necking. Fracture occurs by the transverse linking of the elongated voids in a plane which is macroscopically normal to the tensile axis. Very few connected voids are observed below the fracture surface, which illustrates the highly localized and catastrophic character of the final coalescence process.

Le Roy *et al.* [10] have measured the void area fraction as a function of strain. Their results are shown in Fig. 2.

Fisher [6] has presented the quantitative void measurements during a tensile test. He measured the void volume fraction as a function of equivalent plastic strain for two kinds of steel (B and W types). His results are shown in Fig. 3.

Based on careful examination of the nucleation, growth and coalescence of voids and the accumulation of damage during tensile test and

† Compare here also suggestions presented by Gurland [7], and investigations conducted by Argon and Im [1] and Gurland and Fisher [8].

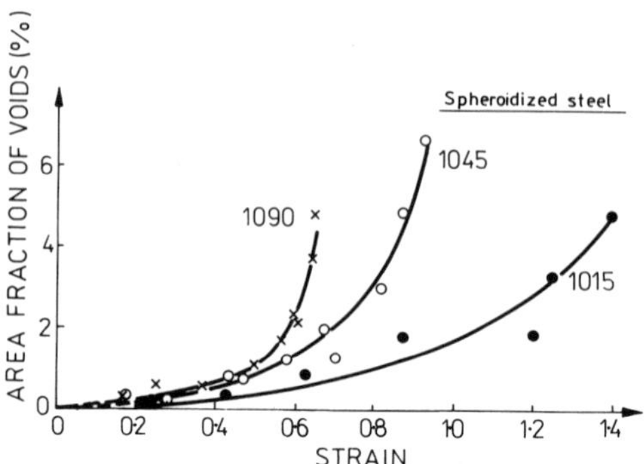

FIG. 2. Area fraction of voids as a function of strain for three kinds of steel (after Le Roy *et al.* [10]).

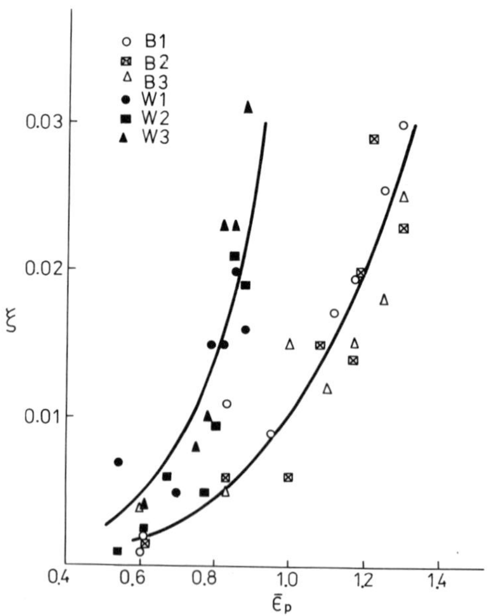

FIG. 3. Void volume fraction as function of equivalent plastic strain along specimen axis (after Fisher [6]).

the subsequent necking process as well as taking advantage of the physical suggestions we may assume a proper evolution equation for the scalar measure of the internal imperfection parameter [17].

3. ELASTIC–VISCOPLASTIC SOLIDS WITH INTERNAL IMPERFECTIONS

3.1. Constitutive Equations

In ref. 17 a simple model of an elastic–viscoplastic solid with internal imperfections was proposed. That model was developed within the internal state variable structure as well as within the rate type structure.

In the present paper we shall discuss the modified model of an elastic–viscoplastic solid with internal imperfections in the form of the rate type theory with internal state variables.†

Let us denote the symmetric rate of deformation tensor by **D** and postulate

$$
\begin{aligned}
&\mathbf{D}^{e} = \mathbf{D} - \mathbf{D}^{p} \\
&\mathbf{D}^{e} = \frac{1}{2G}\left[\overset{\nabla}{\boldsymbol{\sigma}} - \frac{\nu}{1+\nu}\,\mathrm{tr}\,\overset{\nabla}{\boldsymbol{\sigma}}\mathbf{I}\right] \\
&\mathbf{D}^{p} = \frac{\gamma_o}{\varphi}\left\langle \Phi\left[\frac{f(J_1, J_2', J_3', \xi)}{\kappa} - 1\right]\right\rangle \partial_{\boldsymbol{\sigma}} f
\end{aligned}
\tag{3.1}
$$

where $\overset{\nabla}{\boldsymbol{\sigma}}$ denotes the symmetric Zaremba–Jaumann rate of change of the Cauchy stress tensor $\boldsymbol{\sigma}$, G is the shear modulus, ν is the Poisson ratio, γ_o is the viscosity constant, φ is interpreted as the control function and is assumed to depend on $(I_2/I_2^s)-1$, $I_2=(II_{\mathbf{D}})^{1/2}$ is the second invariant of the rate of deformation tensor, $I_2^s=(II_{\mathbf{D}^s})^{1/2}$, $\mathbf{D}^s$ is the static rate of deformation tensor, Φ denotes the overstress viscoplastic function, f is the quasi-static yield function and is postulated in the form

$$f = f(J_2, J_2', J_3', \xi) \tag{3.2}$$

by J_1 we denote the first invariant of the Cauchy stress tensor $\boldsymbol{\sigma}$, J_2' and J_3' the second and third invariants of the stress deviator, respectively, κ plays a role of the material function and can be postulated in

† The advantages of the rate type formulation are sound when the constitutive structure is applied to the solution of the initial boundary-value problem.

the form

$$\kappa = \hat{\kappa}(\xi) \tag{3.3}$$

ξ denotes the imperfection internal state variable which is interpreted as the void volume fraction parameter, and the symbol $\langle [\ ] \rangle$ is understood according to the definition

$$\langle [\ ] \rangle = \begin{cases} 0 & \text{if} \quad f \leqslant \kappa \\ [\ ] & \text{if} \quad f > \kappa \end{cases} \tag{3.4}$$

We postulate the yield function as follows

$$f(\cdot) = J_2' \left[1 - (n_1 + \xi n_2) \frac{J_3'^2}{J_2'^3} + n_3 \xi \frac{J_1^2}{J_2'} \right] \tag{3.5}$$

and similarly

$$\kappa = \hat{\kappa}(\xi) = \kappa_o^2 (1 - n_4 \xi^{1/2})^2 \tag{3.6}$$

where n_1, n_2, n_3, n_4 and κ_o are the material constants.

The relations $(3.1)_3$ and (3.5) together with (3.6) give

$$\mathbf{D}^{\mathrm{p}} = \frac{\gamma_o}{\varphi} \left\langle \Phi \left[\frac{J_2' - (n_1 + \xi n_2) \dfrac{J_3'^2}{J_2'^2} + n_3 \xi J_1^2}{\kappa_o^2 (1 - n_4 \xi^{1/2})^2} - 1 \right] \right\rangle \times (z_1 \mathbf{I} + z_2 \mathbf{S} + z_3 \mathbf{S}^2) \tag{3.7}$$

where

$$\begin{aligned} z_1 &= \frac{1}{\kappa_o} \left[2 n_3 \xi J_1 + \tfrac{4}{3}(n_1 + \xi n_2) \frac{J_3'}{J_2'} \right] \\ z_2 &= \frac{1}{\kappa_o} \left[1 + 2(n_1 + \xi n_2) \frac{J_3'^2}{J_2'^3} \right] \\ z_3 &= \frac{1}{\kappa_o} \left[-2(n_1 + \xi n_2) \frac{J_3'}{J_2'^2} \right] \end{aligned} \tag{3.8}$$

It is noteworthy that eqns. (3.1) lead to the evolution equation for the stress tensor $\boldsymbol{\sigma}$ in the following form

$$\frac{1}{2G} \left[\overset{\triangledown}{\boldsymbol{\sigma}} - \frac{\nu}{1+\nu} \operatorname{tr} \overset{\triangledown}{\boldsymbol{\sigma}} \mathbf{I} \right] = \mathbf{D} - \frac{\gamma_o}{\varphi} \left\langle \Phi \left[\frac{f(J_1, J_2', J_3', \xi)}{\kappa} - 1 \right] \right\rangle \partial_{\boldsymbol{\sigma}} f \tag{3.9}$$

3.2. Evolution Equation for the Volume Fraction Parameter

It has been postulated that internal imperfections in solids are generated by the nucleation and growth of voids and by the accommodated transport phenomena during the inelastic flow process. So, the evolution equation for the imperfection parameter ξ can be assumed to be as follows

$$\dot{\xi} = (\dot{\xi})_{\text{transport}} + (\dot{\xi})_{\text{nucleation}} + (\dot{\xi})_{\text{growth}} \tag{3.10}$$

The transport phenomena in solids are important only at elevated temperatures and then can be approximated by a diffusional term†

$$(\dot{\xi})_{\text{transport}} = (\dot{\xi})_{\text{diffusion}} = D_0 \nabla^2 \xi(X, t) \tag{3.11}$$

where D_o is the diffusion constant (at constant temperature) and ∇^2 denotes the Laplacian operator.

Physical considerations suggest that the nucleation of voids is connected with the inelastic power and with the rate of the first invariant of the stress tensor. Based on this suggestion we postulate (cf. refs. 13, 17, 22)

$$(\dot{\xi})_{\text{nucleation}} = \frac{h}{1-\xi} \operatorname{tr}(\boldsymbol{\sigma}\mathbf{D}^{\text{p}}) + l\dot{J}_1 \tag{3.12}$$

where h and l denote the nucleation material functions.

Similarly the growth of voids during the inelastic flow process is mainly implied by the inelastic strain rate, so we can assume as in ref. 17

$$(\dot{\xi})_{\text{growth}} = (1-\xi) \operatorname{tr}(\boldsymbol{\Xi}\mathbf{D}^{\text{p}}) \tag{3.13}$$

where $\boldsymbol{\Xi}$ is the matrix of the material functions.

As a result of these assumptions we have the evolution equation for the void fraction parameter ξ in the form

$$\dot{\xi} = D_o \nabla^2 \xi + \frac{h}{1-\xi} \operatorname{tr}(\boldsymbol{\sigma}\mathbf{D}^{\text{p}}) + l\dot{J}_1 + (1-\xi) \operatorname{tr}(\boldsymbol{\Xi}\mathbf{D}^{\text{p}}) \tag{3.14}$$

In the following considerations which are assumed for room temperature we shall postulate a particular form of the evolution equation (3.14), namely

$$\dot{\xi} = \frac{h}{1-\xi} \operatorname{tr}(\boldsymbol{\sigma}\mathbf{D}^{\text{p}}) + l\dot{J}_1 + (1-\xi)\Xi \operatorname{tr} \mathbf{D}^{\text{p}} \tag{3.15}$$

In this particular form the diffusional term is neglected and instead of the matrix $\boldsymbol{\Xi}$ there is postulated a scalar material growth function Ξ.

† For a thorough discussion of the diffusion accommodated process in dissipative solids see Ashby and Verrall [3], also refs. 15–17.

3.3. Elastic–Plastic Response

By introducing the control function φ the model proposed can describe the properties of a material in a range of strain rates near the static value $I_2 = I_2^s$.

Let us assume that the control function φ has properties as follows

$$\lim_{I_2 = I_2^s} \varphi = 0 \quad \text{and} \quad \varphi(\cdot) = 0 \quad \text{for} \quad I_2 < I_2^s \tag{3.16}$$

then the model proposed satisfies also the requirement that during the deformation process in which the effective strain rate is equal to the static value the response of a material becomes elastic–plastic.

For the limiting case, when $I_2 \leq I_2^s$, we have the evolution equations in the form

$$\begin{aligned} \mathbf{D}^{\mathrm{p}} &= \Lambda\, \partial_{\boldsymbol{\sigma}} f = \Lambda(z_1 \mathbf{I} + z_2 \mathbf{S} + z_3 \mathbf{S}^2) \\ \dot{\xi} &= \frac{h}{1-\xi} \operatorname{tr}(\boldsymbol{\sigma} \mathbf{D}^{\mathrm{p}}) + l \dot{J}_1 + (1-\xi) \operatorname{tr}(\boldsymbol{\Xi} \mathbf{D}^{\mathrm{p}}) \end{aligned} \tag{3.17}$$

for

$$f(\cdot) = \kappa \quad \text{and} \quad \operatorname{tr}(\partial_{\boldsymbol{\sigma}} f \dot{\boldsymbol{\sigma}}) > 0 \tag{3.18}$$

The parameter Λ can be determined from the condition

$$\dot{f} = \dot{\kappa} \tag{3.19}$$

The evolution equation for the plastic deformation in index notation can be written as in ref. 17.

$$\mathbf{D}_{ij}^{\mathrm{p}} = \frac{\lambda^2}{H} P_{ij}^* Q_{kl}^* \overset{\nabla}{\sigma}_{kl} \tag{3.20}$$

where

$$\begin{aligned} P_{ij}^* &= \frac{1}{2\tau_e}(s_{ij} + \beta_1 s_{ik} s_{kj}) + \frac{\beta_2}{3} \delta_{ij} \\ Q_{kl}^* &= \frac{1}{2\tau_e}(s_{kl} + \mu_1 s_{km} s_{ml}) + \frac{\mu_2}{3} \delta_{kl} \\ \tau_e &= \sqrt{J_2'} \\ \beta_1 &= -2(n_1 + \xi n_2) J_2' J_3' [J_2'^3 + 2(n_1 + \xi n_2) J_3'^2]^{-1} \\ \beta_2 &= 3 n_3 \xi J_1 J_2' + 2(n_1 + \xi n_2) J_3' \left\{ \tau_e \left[1 + 2(n_1 + \xi n_2) \frac{J_3'^2}{J_2'^3} \right] \right\}^{-1} \end{aligned}$$

$$\mu_1 = \beta_1$$

$$\mu_2 = \beta_2 + \frac{\frac{1}{2}\left[n_3 J_1^2 - n_2 \frac{J_3'^2}{J_2'^2} - (n_4 - \xi^{-1/2}) n_4 \kappa_0^2\right] l}{\tau_e\left[1 + 2(n_1 + \xi n_2)\frac{J_3'^2}{J_2'^3}\right]}$$

$$\lambda = 2\tau_e\left[1 + 2(n_1 + \xi n_2)\frac{J_3'^2}{J_2'^3}\right]$$

$$H = \left[n_4 \kappa_0^2 (n_4 - \xi^{-1/2}) + n_2 \frac{J_3'^2}{J_2'} - n_3 J_1^2\right]$$

$$\times\left[(1-\xi)\Xi_{ij} + \frac{h}{1-\xi}\sigma_{ij}\right]\frac{\partial f}{\partial \sigma_{ij}} \tag{3.21}$$

4. DETERMINATION OF THE MATERIAL FUNCTIONS AND CONSTANTS

4.1. Mechanical Test Data

To establish the material functions and constants we shall consider a particular case of the constitutive relations proposed. Let us assume in eqns. (3.5) and (3.6)

$$n_1 = n_2 = 0, \qquad n_3 = n, \qquad n_4 \neq 0 \tag{4.1}$$

then

$$f(\cdot) = J_2'\left(1 + n\xi\frac{J_1^2}{J_2'}\right)$$
$$\kappa = \kappa_0^2(1 - n_4\xi^{1/2})^2 \tag{4.2}$$

and

$$\mathbf{D}^{\mathrm{p}} = \frac{\gamma_o}{\varphi}\left\langle \Phi\left[\frac{J_2'\left(1 + n\xi\frac{J_1^2}{J_2'}\right)}{\kappa_o^2(1 - n_4\xi^{1/2})^2} - 1\right]\right\rangle \frac{1}{\kappa_o}(2n\xi J_1 \mathbf{I} + \mathbf{S}) \tag{4.3}$$

The material constant n can be determined based on the comparison of the Gurson [9] solution with the present proposition. To do this we used the yield condition for a voided solid in the form

$$\frac{J_2'}{\kappa_o^2} + n\xi\frac{J_1^2}{\kappa_o^2} = (1 - n_4\xi^{1/2})^2 \tag{4.4}$$

If we assume that for $J_2' = 0$ the Gurson solution coincides with our prediction then we have the condition

$$n\xi\left(\frac{J_1}{\kappa_o}\right)^2 = (1 - n_4\xi^{1/2})^2 \tag{4.5}$$

This condition gives the result

$$n = \frac{(1 - n_4\xi^{1/2})^2}{\xi(J_1/\kappa_o)^2} \tag{4.6}$$

in which J_1/κ_o is given a value from the Gurson solution (e.g. for $\xi = 0{\cdot}02$ we have $(J_1/\kappa_o) = 4{\cdot}52$) and n_4 can be determined from the fracture condition (this gives $n_4 = 2{\cdot}00$). So, we have $n = 1{\cdot}2587$.

The constants γ_o and κ_o, the value I_2^s and the material functions φ and Φ may be determined by comparison of the results obtained from the evolution equation (4.3) for $\xi = 0$ with the experimental data obtained for the dynamical test under combined loading. In previous papers [14, 17] this procedure of specifying the material functions and constants was performed based on experimental data obtained by Lindholm [11] for mild steel. The results obtained were as follows†

$$\kappa_o = 21{\cdot}50 \text{ ksi}, \ \gamma_o = 25{\cdot}82374 \text{ s}^{-1}, \ I_2^s = 10^{-6}\text{ s}^{-1},$$

$$\Phi = \left[\frac{(J_2')^{1/2}}{\kappa_o} - 1\right]^5, \ \varphi = \left(\frac{I_2}{I_2^s} - 1\right)^5. \tag{4.7}$$

4.2. Metallurgical Observation Results

Let us assume that in the evolution equation (3.15) for the void volume fraction the material function l vanishes, then we have

$$\dot{\xi} = \frac{h}{1-\xi}\operatorname{tr}(\boldsymbol{\sigma}\mathbf{D}^{\mathrm{p}}) + (1-\xi)\Xi \operatorname{tr}\mathbf{D}^{\mathrm{p}} \tag{4.8}$$

in which there are two material functions, namely the nucleation function h and the growth function Ξ.

To determine these material functions we shall use the metallurgical observation data obtained by Fisher [6] for two kinds of steel as have been shown in Fig. 3.

The procedure for the determination of the material functions h and Ξ is based on the assumption that the distribution of the stress in the

† This procedure of establishing the material functions and constants has been generalized to the elastic-work hardening viscoplastic material in ref. 21.

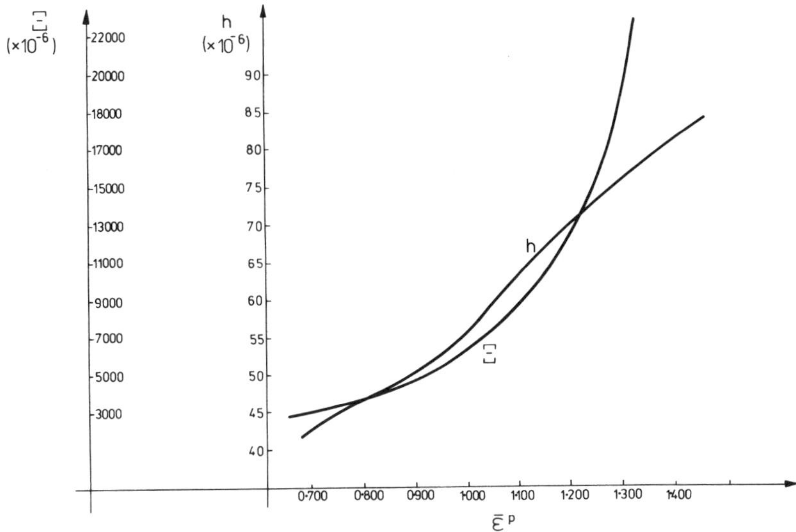

FIG. 4. The material functions h and Ξ determined on the basis of Bridgman's distribution of stress.

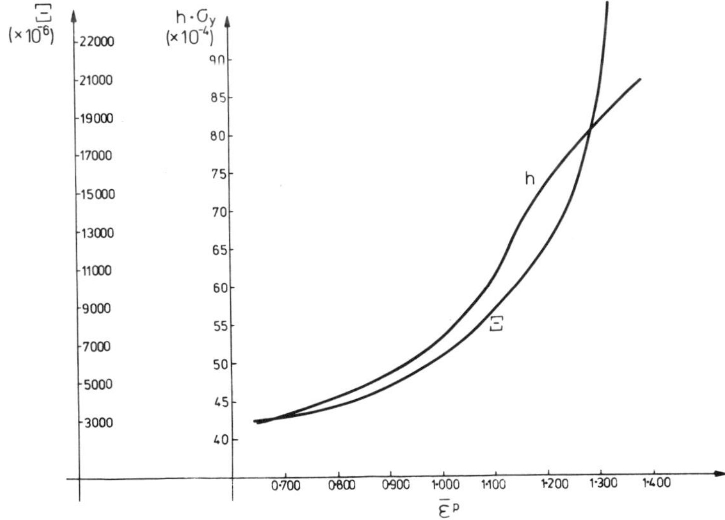

FIG. 5. The material functions h and Ξ determined on the basis of Needleman's distribution of stress.

neck is given. We shall take advantage here of the analytical solution given by Bridgman [5] as well as of the numerical solution presented by Needleman [12]. The results obtained are shown in Fig. 4 for Bridgman's solution and in Fig. 5 for Needleman's solution.†

5. THEORY OF FRACTURE PHENOMENON

5.1. Synergetic Nature of Fracture

From previous considerations it is clear that to describe fracture phenomenon we need very precise constitutive modelling and must take into account important interactive effects such as plastic deformation, internal imperfections induced by nucleation, growth and linkage of voids, strain rate sensitivity, thermal effects, etc.

The main idea is to treat the fracture phenomenon as a final stage of the inelastic flow process and to describe the dependence of fracture upon the evolution of the constitutive structure of solids.

On the other hand it would be unrealistic to include in the description all the effects observed experimentally. Constitutive modelling to some extent is understood as a reasonable choice of effects which are most important for explanation of the phenomenon described.

During the postcritical behaviour of dissipative solids it can be observed (cf. the experimental results obtained in tensile tests by Bluhm and Morrissey [4] for steel and copper) that two interactive effects, namely the plastic deformation (together with the localization phenomenon) and internal imperfections (generated by nucleation, growth and linkage of voids) play the main role in the true explanation of the final stage of the flow process.

We can also observe synergetic effects which give a definite increase in the results induced by these two interactive phenomena. This synergetic effect is especially pronounced at a final stage of the flow process, that is when the fracture mechanism operates.

To describe this synergetic nature of interactive phenomena we have to introduce some important information about the final stage of the flow process directly into the evolution equations and into the constitutive relations. That is the reason why the constitutive modelling for fracture is of great importance.

† The procedure of the determination of the material functions h and Ξ has been discussed in detail in ref. 19.

5.2. Constitutive Modelling for Fracture

From the analysis of experimental observations presented by Fisher [6] (Fig. 6) we can deduce for each kind of two steels investigated the critical plastic deformation $\bar{\varepsilon}^{\mathrm{p}}_{\mathrm{c}}$ which is understood in such a way that during the inelastic flow process when $\bar{\varepsilon}^{\mathrm{p}}$ tends to $\bar{\varepsilon}^{\mathrm{p}}_{\mathrm{c}}$ then the volume fraction parameter ξ tends to its upper bound, i.e. $\xi \rightarrow 1$.

This suggestion is also confirmed by the experimental observations for the area fraction parameter presented by Le Roy *et al.* [10] (Fig. 7).

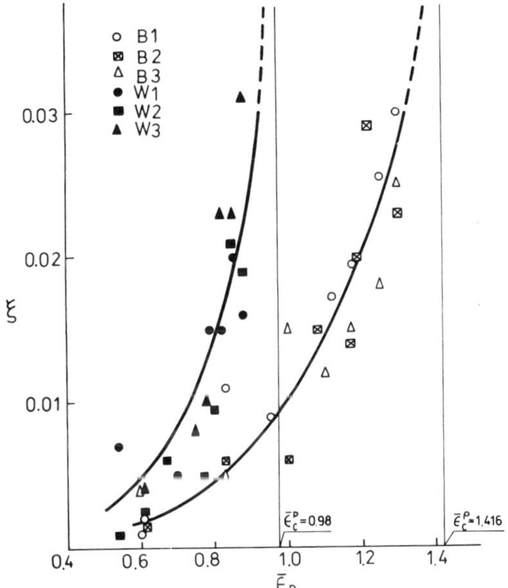

FIG. 6. The critical plastic deformation based on Fisher's results.

It would be reasonable to introduce this important feature of the flow process directly to the evolution equation for the volume fraction ξ. To do this we need to modify the evolution equation proposed (eqn. (3.15)).

There is no unique way to achieve this modification.† We shall

† Recently Tvergaard [23, 24] presented a conception of modification of the evolution equation for the volume fraction parameter. It seems however that Tvergaard's method of modification has no clear physical foundations. Very recently Tvergaard and Needleman [25] introduced a new method of describing the fracture phenomenon based only on the modification of the yield condition.

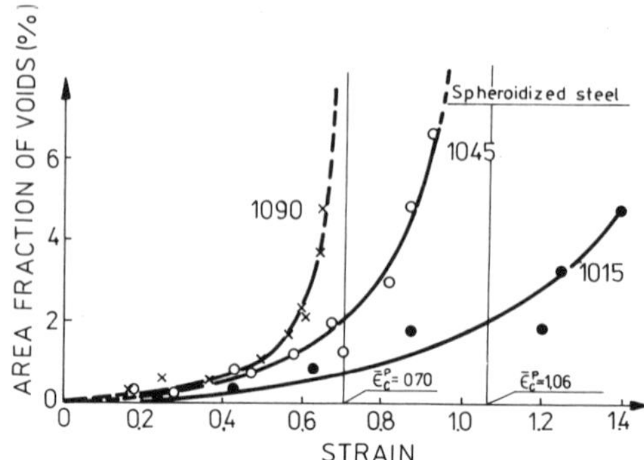

FIG. 7. The critical plastic deformation based on the Le Roy *et al.* results.

mention possible different methods of modification of the evolution equation for the volume fraction parameter ξ:

(i) By introducing the control function

$$\psi = \psi\left(1 - \frac{\bar{\varepsilon}^{\mathrm{p}}}{\bar{\varepsilon}^{\mathrm{p}}_{\mathrm{c}}}\right) \tag{5.1}$$

where $\bar{\varepsilon}^{\mathrm{p}} = (II_{e^{\mathrm{p}}})^{\frac{1}{2}}$ and the function ψ has the property: $\psi(0) = 0$. The value $\bar{\varepsilon}^{\mathrm{p}}_{\mathrm{c}}$ is called the critical equivalent plastic deformation.

(ii) By assuming that the material functions h, l and Ξ are such that

$$\bar{\varepsilon}^{\mathrm{p}} \rightarrow \bar{\varepsilon}^{\mathrm{p}}_{\mathrm{c}} \Rightarrow \xi \rightarrow 1 \tag{5.2}$$

(iii) By specifying the small parameter from the material functions h, l and Ξ and by application of the perturbation procedure with the assumption that $\xi \rightarrow 1$ when $\bar{\varepsilon}^{\mathrm{p}} \rightarrow \bar{\varepsilon}^{\mathrm{p}}_{\mathrm{c}}$.

Let us focus here on the first method of modification and write the evolution equation (3.15) in the new form

$$\dot{\xi} = \frac{h}{1-\xi}\,\mathrm{tr}\,(\boldsymbol{\sigma}\mathbf{D}^{\mathrm{p}}) + \frac{l}{\psi\left(1 - \dfrac{\bar{\varepsilon}^{\mathrm{p}}}{\bar{\varepsilon}^{\mathrm{p}}_{\mathrm{c}}}\right)}\,\dot{J}_1 + \frac{\Xi}{\psi\left(1 - \dfrac{\bar{\varepsilon}^{\mathrm{p}}}{\bar{\varepsilon}^{\mathrm{p}}_{\mathrm{c}}}\right)}\,(1-\xi)\,\mathrm{tr}\,\mathbf{D}^{\mathrm{p}} \tag{5.3}$$

with the property for the function ψ that $\psi(0) = 0$.

We know also from experimental observations of evolution of voids that for mild steel the coalescence of voids takes place at a value of ξ around 0·15 and fracture (by shearing) occurs at value $\xi = 0{\cdot}25$.

We shall introduce this information directly by the yield condition, namely by proper choice of the constant n_4.

5.3. Fracture Criterion Based on Evolution of Internal Imperfections

Let us consider the inelastic flow process (it can be understood as a tensile test) for which

$$\bar{\varepsilon}^{\mathrm{p}} \rightarrow \bar{\varepsilon}^{\mathrm{p}}_{\mathrm{c}} \Rightarrow \xi \rightarrow 1 \tag{5.4}$$

where $\bar{\varepsilon}^{\mathrm{p}}_{\mathrm{c}}$ denotes the critical equivalent plastic deformation. The interval $[0, 1]$ is the domain of ξ parameter, i.e. $\xi \in [0, 1]$.

We postulate that there exists the interval $[\xi^{\mathrm{L}}, \xi^{\mathrm{F}}]$ with the property

$$[\xi^{\mathrm{L}}, \xi^{\mathrm{F}}] \subset [0, 1] \tag{5.5}$$

where ξ^{L} is the value of the volume fraction parameter which corresponds to the incipient linkage of voids and ξ^{F} is the value at which the incipient fracture occurs (Fig. 8).

During the segment of the flow process which corresponds to the interval $[\xi^{\mathrm{L}}, \xi^{\mathrm{F}}]$ or $[\bar{\varepsilon}^{\mathrm{p}}_{\mathrm{L}}, \bar{\varepsilon}^{\mathrm{p}}_{\mathrm{F}}]$ (Fig. 8) the coalescence of voids plays the most important role and at the end of this segment when $\bar{\varepsilon}^{\mathrm{p}} = \bar{\varepsilon}^{\mathrm{p}}_{\mathrm{F}}$ and $\xi = \xi^{\mathrm{F}}$ the shear phenomenon takes place. This instantaneous shearing is understood to be a fracture mechanism.

It is noteworthy that the interval $[\xi^{\mathrm{L}}, \xi^{\mathrm{F}}]$ corresponds to very small changes of plastic deformation. In other words the difference between the plastic deformation $\bar{\varepsilon}^{\mathrm{p}}_{\mathrm{L}}$ corresponding to the incipient linkage of voids and the plastic deformation $\bar{\varepsilon}^{\mathrm{p}}_{\mathrm{F}}$ corresponding to the incipient fracture is very small (Fig. 8).

For $\xi = \xi^{\mathrm{F}}$ the catastrophe takes place, that is

$$\hat{\kappa}(\xi)\big|_{\xi = \xi^{\mathrm{F}}} = 0 \tag{5.6}$$

then the material loses its stress-carrying capacity (Fig. 8).

The condition (5.6) describes the main feature observed experimentally that the load tends to zero at the fracture point. It also leads directly to the particular relation

$$\lim_{\xi \rightarrow \xi^{\mathrm{F}}} \hat{\kappa}(\xi) = \lim_{\xi \rightarrow \xi^{\mathrm{F}}} \kappa_{\mathrm{o}}^2 (1 - n_4 \xi^{1/2})^2 = 0 \tag{5.7}$$

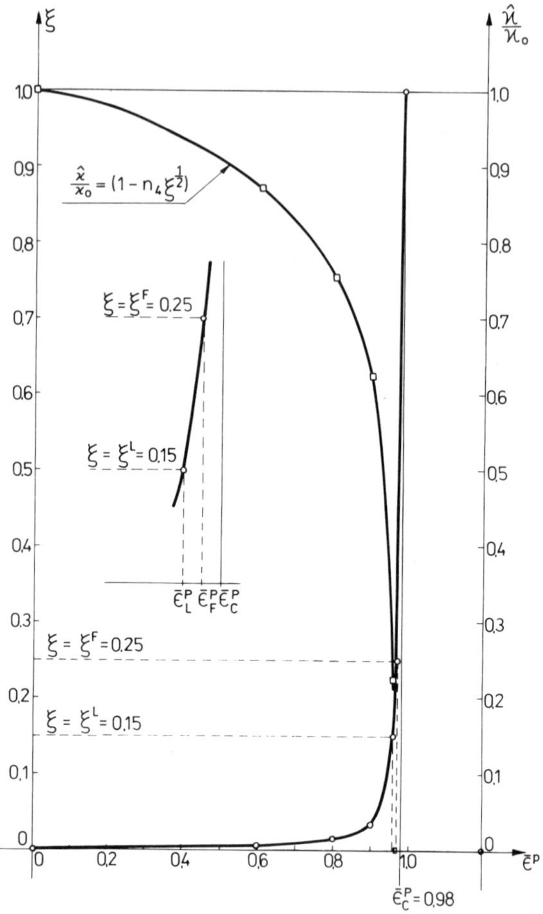

FIG. 8. Explanation of the fracture criterion based on the evolution of internal imperfections for W type steel.

hence we can obtain the material constant

$$n_4 = (\xi^{\mathrm{F}})^{-1/2} \tag{5.8}$$

which is crucial for the description of the fracture phenomenon.

For instance, for W type steel we have from metallurgical observations

$$[\xi^{\mathrm{L}}, \xi^{\mathrm{F}}] = [0{\cdot}15, 0{\cdot}25] \tag{5.9}$$

and then $n_4 = 2{\cdot}00$.

So, we are now in a position to formulate the criterion of fracture. During the tensile flow process the fracture occurs when

$$\bar{\varepsilon}^{\mathrm{p}} = \bar{\varepsilon}^{\mathrm{p}}_{\mathrm{F}} \Rightarrow \xi = \xi^{\mathrm{F}} \tag{5.10}$$

which leads to the condition $\hat{\kappa}(\xi)\,|_{\xi=\xi^{\mathrm{F}}} = 0$,

It is noteworthy that the criterion of fracture in the form

$$\bar{\varepsilon}^{\mathrm{p}} = \bar{\varepsilon}^{\mathrm{p}}_{\mathrm{F}} = \mathrm{const} \tag{5.11}$$

is also acceptable for an elastic–viscoplastic model of solids provided the process considered has not got large strain rates.

However the experimental investigations on iron presented by Wray [26] have shown that even for very small strain rates but at elevated temperature the fracture equivalent plastic deformation depends on the rate of deformation as well as on temperature, i.e.

$$\bar{\varepsilon}^{\mathrm{p}} = \bar{\varepsilon}^{\mathrm{p}}_{\mathrm{F}}(\vartheta, \mathbf{D}^{\mathrm{p}}) \tag{5.12}$$

On the other hand the theoretical analysis on ductile fracture of plain carbon steel at room temperature presented in ref. 20 proved that the assumption (5.11) is valid up to a strain rate of the order of 10^3. If the strain rate is larger than 10^3 then for constant temperature the criterion of fracture (5.11) has to be replaced by

$$\bar{\varepsilon}^{\mathrm{p}} = \bar{\varepsilon}^{\mathrm{p}}_{\mathrm{F}}(\mathbf{D}^{\mathrm{p}}) \tag{5.13}$$

6. INITIAL BOUNDARY-VALUE PROBLEM

6.1. Formulation of the Problem

Let us study the tensile deformation of a circular bar of initial length $2L_{\mathrm{o}}$ and initial radius R_{o}. The problem is described in the cylindrical co-ordinates r, θ, z. It has been assumed that the problem is axisymmetric and additionally that the deformations are symmetric about the mid-plane $z = 0$. To the ends of the specimen is applied the constant velocity $\dot{U}$ and additionally the ends are assumed to remain shear free.

The material of the specimen is assumed to be elastic–plastic with internal imperfections, the yield condition has the form (4.4) and the rate of plastic deformation is given by eqn. (3.20) with $n_1 = n_2 = 0$.

The evolution equation for the volume fraction parameter ξ is assumed in the modified form (5.3) with $l = 0$.

The material functions h and Ξ and the material constants n, n_4 and κ_{o} are assumed to have been determined in Section 4.

The problem is treated as quasi-static and the initial values are assumed as follows

$$
\begin{gathered}
t=0,\ r\in[0, R_o],\ z\in[0, L_0] \\
u(r, z, 0)=0 \\
\boldsymbol{\sigma}(r, z, 0)=0 \\
\xi(r, z, 0)=\xi^{o}(r, z)
\end{gathered} \tag{6.1}
$$

where u denotes the displacement vector.

6.2. Discussion of the Results

The solution of the problem for two different descriptions of the volume fraction parameter ξ for W type steel is shown in Fig. 9. In both cases considered the load has the property that it tends to zero when the equivalent plastic deformation tends to the critical value which for W type steel is assumed to be $\bar{\varepsilon}^{p}_{c}=0{\cdot}98$.

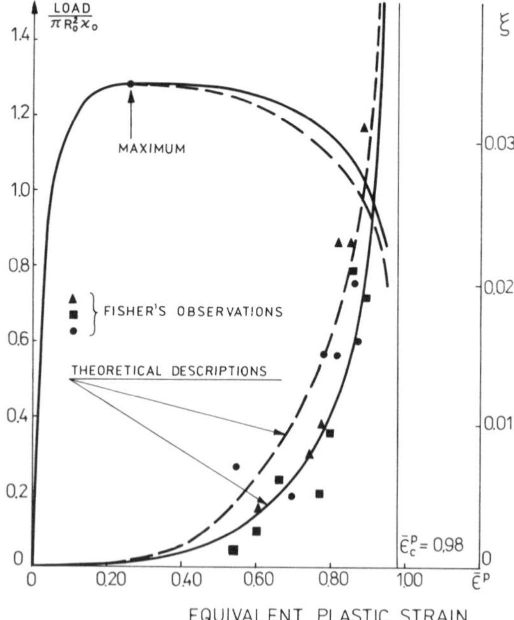

FIG. 9. Solution of the initial boundary-value problem based on the modified evolution equation for the volume fraction parameter.

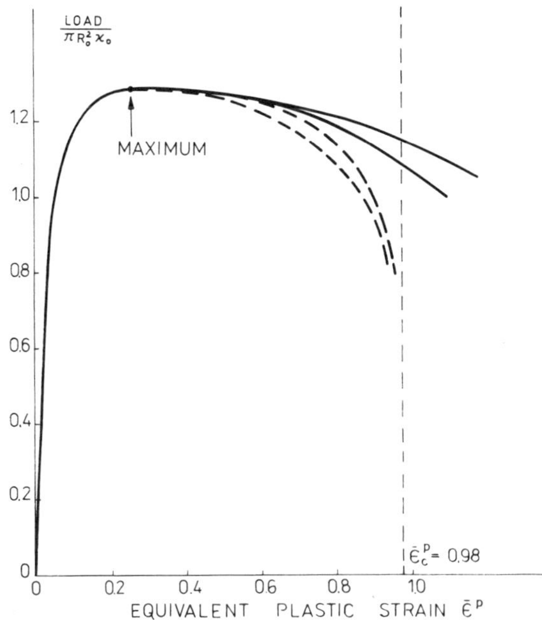

FIG. 10. Comparison of the results: —— results for unmodified evolution equation; – – – – results for modified evolution equation.

If the evolution equation for the volume fraction ξ is assumed to be in the unmodified form (4.8) and n_4 is supposed to equal unity [17] then the solution of the problem for two similar descriptions of the volume fraction parameter ξ does not have the property that the load tends to zero when the equivalent plastic deformation $\bar{\varepsilon}^p$ tends to the critical value $\bar{\varepsilon}_c^p = 0{\cdot}98$ (cf. Fig. 10).

7. CONCLUSIONS AND COMMENTS

First of all it is noteworthy that the theory of fracture presented has been inspired by the brilliant experimental investigations carried out by Bluhm and Morrissey [4], Le Roy *et al.* [10] and Fisher [6]. All these experimental works have given a deep understanding of real features of the tensile deformation process and have given many important measurements needed for development of the theoretical description.

We hope that the new theory of fracture proposed is sufficiently simple in its nature that it can be applicable to the solution of initial boundary-value problems.

The crucial idea in this theory is the very efficient interpretation of the internal state variable ξ. The assumption that ξ stands for the volume fraction allowed us to base all considerations on good physical foundations and to use all available experimental observations.

The criterion of fracture proposed for tensile deformation problems does depend on the entire evolution of the constitutive structure of solids. This has been achieved mainly due to careful analysis of the evolution of the volume fraction parameter ξ during the tensile inelastic flow process.†

REFERENCES

1. ARGON, A. S. and J. IM. Separation of second phase particles in spheroidized 1045 steel, CU-0.6 pct Cr alloy, and maraging steel in plastic straining, *Metall. Trans.*, **6A** (1975) 839–851.
2. ASARO, R. J. Micromechanics of crystals and polycrystals, Brown University Report, April 1982.
3. ASHBY, M. F. and R. A. VERRALL. Diffusion-accommodated flow and superplasticity, *Acta Met.*, **21** (1973), 149–163.
4. BLUHM, J. I. and R. J. MORRISSEY. Fracture in a tensile specimen, *Proc. First Int. Conf. on Fracture*, Sendai, Japan, September 1965 (Eds. T. Yokobori, T. Kawasaki and J. L. Swedlow), Vol. 3, 1739–1780.
5. BRIDGMAN, P. W. The stress distribution at the neck of a tension specimen, *Trans. A.S.M.*, **32** (1943), 553–574.
6. FISHER, J. R. Void nucleation in spheroidized steels during tensile deformation, Ph.D. Thesis, Brown University, 1980.
7. GURLAND, J. Observations on the fracture of cementite particles in a spheroidized 1·05% C steel deformed at room temperature, *Acta Met.*, **20** (1972) 735–741.
8. GURLAND, J. and J. R. FISHER. Void nucleation in spheroidized carbon steels, *Mater. Sci.* (May 1981), 185–202.
9. GURSON, A. L. Continuum theory of ductile rupture by void nucleation and growth, *J. Engng Mater. Technol.*, **99** (1977) 2–15.
10. LE ROY, G., J. D. EMBURY, G. EDWARD and M. F. ASHBY. A model of ductile fracture based on the nucleation and growth of voids, *Acta Met.*, **29** (1981) 1509–1522.
11. LINDHOLM, U. S. Dynamic deformation of metals. In: *Behaviour of Materials under Dynamic Loads*, ASME, New York, 1965, 42–61.

† A thorough analysis of the evolution equation for the volume fraction parameter in the entire domain is given in a forthcoming paper [18].

12. NEEDLEMAN, A. A numerical study of necking in circular cylindrical bars, *J. Mech. Phys. Solids*, **20** (1972), 111–127.
13. NEEDLEMAN, A. and J. R. RICE. Limits to ductility set by plastic flow localization. In: *Mechanics of Sheet Metal Forming* (Eds. D. P. Koistinen and N. M. Wang), Plenum Publ. Co., New York, 1978, 237–265.
14. PERZYNA, P. Modified theory of viscoplasticity. Application to advanced flow and instability phenomena, *Arch. Mech.*, **32** (1980), 403–420.
15. PERZYNA, P. Stability phenomena of dissipative solids with internal defects and imperfections, *XVth IUTAM Congress*, Toronto, August 1980; in *Theoretical and Applied Mechanics* (Proc. Eds. F. P. J. Rimrott and B. Tabarrok), North-Holland, Amsterdam, 1980, 369–374.
16. PERZYNA, P. Stability problems for inelastic solids with defects and imperfections, *Arch. Mech.*, **33** (1981), 587–602.
17. PERZYNA, P. Stability of flow processes for dissipative solids with internal imperfections, submitted to *ZAMP*, 1983.
18. PERZYNA, P. and A. DRABIK. Analysis of the evolution equation describing the postcritical behaviour of dissipative solids, *Arch. Mech.* (1983), in press.
19. PERZYNA, P. and Z. NOWAK. Evolution equation for void fraction parameter in necking region, *Arch. Mech.* (1983), in press.
20. PERZYNA, P. and R. B. PECHERSKI. Analysis of strain rate effects on ductile fracture of metals, *Arch. Mech.*, **35** (1983), 287–301.
21. PERZYNA, P. and R. B. PECHERSKI. Modified theory of viscoplasticity. Physical foundations and identification of material functions for advanced strains, *Arch. Mech.*, **35**, (1983), in press.
22. SAJE, M., J. PAN and A. NEEDLEMAN. Void nucleation effects on shear localization in porous plastic solids, *Int. J. Fract.* **19** (1982), 163–182.
23. TVERGAARD, V. Influence of void nucleation on ductile shear fracture at a free surface, *J. Mech. Phys. Solids*, **30** (1982), 399–425.
24. TVERGAARD, V. Material failure by void coalescence in localized shear bands, *Int. J. Solids Struct.*, **18** (1982), 659–672.
25. TVERGAARD, V. and A. NEEDLEMAN. Analysis of the cup-cone fracture in a round tensile bar, The Technical University of Denmark Report No. 264, June 1983.
26. WRAY, P. J. Tensile failure of austenitic iron at intermediate strain rates, *Metall. Trans.* **6A** (1975), 1379–1391.

Section 7
APPLICATIONS OF PLASTICITY

38

On Problems of Plastic Flow of Metals

J. M. ALEXANDER*

Department of Mechanical Engineering, University College of Swansea, Swansea, UK

ABSTRACT

The main applications of the classical theory of plasticity are to be found in situations involving large deformation, which inevitably leads to consideration of metal manufacturing processes. Problems of fracture, creep and instability often occur in certain metal forming processes such as cropping, creep forming and high speed deformation but are neglected in the classical theory. Although there is a general belief that plasticity theory can give good estimates of the behaviour of a hypothetical material in which the effects of temperature and strain rate are neglected, this is not necessarily true and examples are given. It is pointed out that, since even the hypothetical material of plasticity theory generally requires numerical solution procedures, it is sensible to include elasticity, temperature and strain rate effects so as to give results which will be really useful in industrial practice. Attention is given to modern 'precision' and 'near-to-net shape' metal forming processes in an attempt to classify the ranges of homologous temperature and strain rate characterizing such processes, which are now of most importance to industry. This leads to a discussion of constitutive laws, typical examples being given and their limitations pointed out. Problems of instability and fracture which occur during metal forming are discussed, drawing attention to the lack of consideration which has been given to such problems by theoreticians. Finally a general discussion is given of the need for some new directions to be taken in theoretical plasticity, particularly in giving more realistic properties to the hypothetical materials used.

* Present address: University of Surrey, Guildford, UK.

1. INTRODUCTION

During the past forty years or so, the mathematical theory of plasticity has been developed and applied to many technological problems, particularly those concerned with the deformation of metals during processing. Since metals are inherently strong and abundant, they are inevitably chosen as the basic material from which engineering components are manufactured, although other materials such as polymers, ceramics and composites are increasingly being developed to replace metals in certain applications, where they offer advantages. During the manufacture of such materials plasticity also plays a part, and even in the actual use of the manufactured artefact, whatever the material of which it is made, plastic flow may occur which requires study in order to understand and be able to predict its effects. A particular example of this is in the creep of materials under stress, which must be allowed for in the design of components. In a similar way the processes leading to fracture, particularly under the repeated or reversed stressing conditions of fatigue, must be studied and understood so that engineering components can be designed to be safe.

In all these problems, in fact, the effects of temperature and rate of deformation are of paramount importance but are often neglected. The purpose of the present article is to concentrate mainly on metal forming processes and to consider to what extent such phenomena can or should be included in any analyses of plastic flow which are developed. Apart from temperature and strain rate effects, there are also such considerations as: inertia effects, particularly in the shocks or strain waves generated under impact conditions; fracture; anisotropy; the need to include elastic recovery and compressibility in large elastic–plastic deformations; and thermomechanical interactions. In addition to these material effects there is also the extremely difficult problem of predicting and describing the surface friction and consequent interfacial stresses which must exist at every interface between tools and workpiece. This depends on the developing shape of the workpiece and on the microstructure of surfaces of tools and continually changing workpiece.

2. PLASTICITY THEORY

Although there is a general belief that the classical mathematical theory of plasticity, in which many of the phenomena discussed in the introduction are neglected, can give good estimates of the behaviour to be expected during metal flow, this is not necessarily true. Two of the most important effects which should not be neglected are elasticity and work-hardening. Thus the concept of the elastic–plastic work-hardening material is embraced within the classical theory, as developed and described by Hill [1].

However, the number of problems which can be solved analytically for even the simple elastic–plastic work-hardening material are very few and numerical methods generally have to be used, as for example by Alexander and Ford [2] on the expansion of a hole in a plate and nowadays based on the modern computer. In the past, before modern computers were available, even elasticity and work-hardening were neglected, leading to the concepts of plastic–rigid material and limit analysis. The further assumption of plane strain conditions allowed the development of slip line field theory which has been a powerful method for predicting the flow of this hypothetical rigid–plastic material under plane strain conditions. Metal-forming processes such as cogging (flat forging), rolling, shearing, blanking and punching, ironing and machining *do* occur substantially under plane strain conditions but an even greater number occur under axisymmetrical or complex three-dimensional conditions for which the slip line field theory really does not apply at all. From the point of view of predicting loads, limit analysis methods which lead to upper and lower bound loads has been a powerful technique. However, because it is easier to visualize kinematically admissible velocity fields (upper bounds) than statically admissible stress fields (lower bounds), researchers have almost invariably concentrated on deriving upper-bound solutions. In my view, this has been a serious deficiency in the proper application of limit analysis and I am going to illustrate the point with a typical example, involving a novel process for the upsetting of a rod against a flat surface under an external pressure (injection upsetting) which my colleague Dr B. Lengyel and I devised many years ago and which is illustrated in Fig. 1. In 1965 Alexander and Lengyel [3] produced the upper and lower bound solutions illustrated in Fig. 2, the upper bound solution being represented by the equation:

$$p_{\mathrm{u}} = \left[1 + a + \frac{4}{3a} + \ln R\right]k + q \tag{1}$$

where $a = R_{\mathrm{o}}/h$, $R = (R_1/R_{\mathrm{o}})^2$, k = yield shear stress.

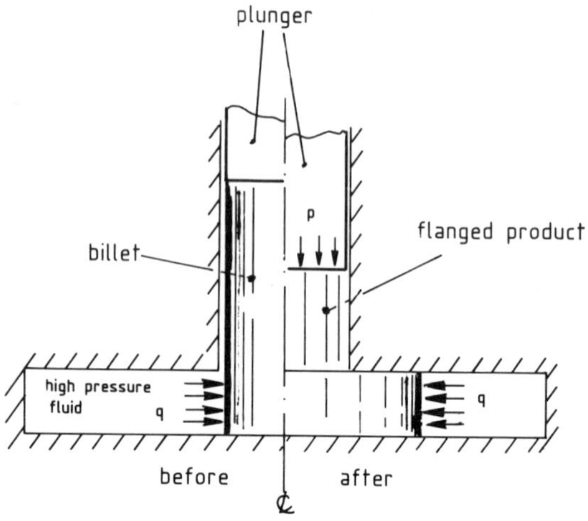

FIG. 1. Injection upsetting.

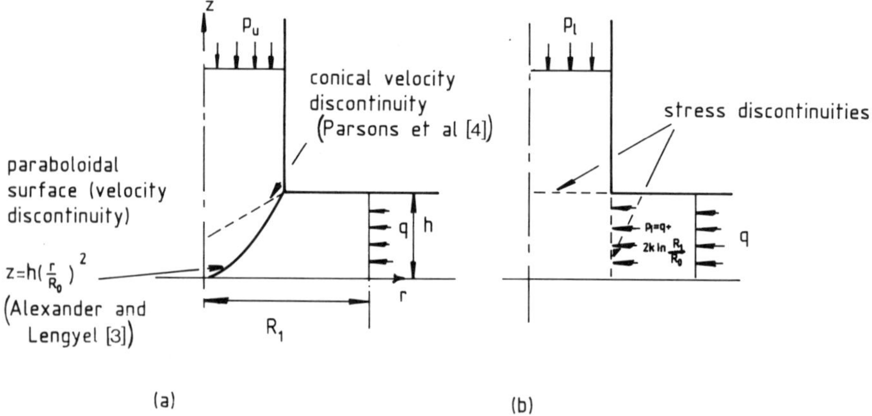

FIG. 2. Upper and lower bound pictures for injection upsetting. (a) Upper bounds; (b) lower bound (Alexandèr and Lengyel [3]).

The lower bound solution is given by the equation:

$$p_l = (\sqrt{3} + \ln R)k + q \tag{2}$$

We also carried out extensive experimental work on a number of materials which indicated that a solution based on the mean of our upper and lower bounds, together with the yield criterion of Tresca,

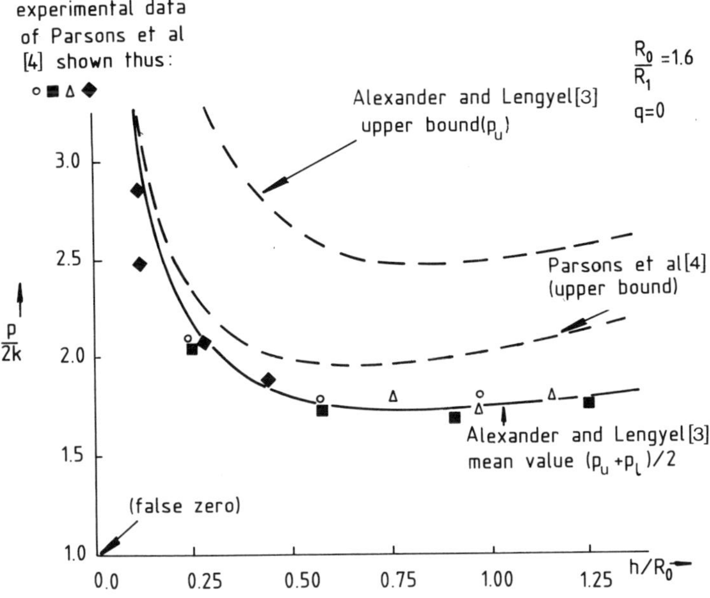

FIG. 3. Comparison between limit analysis and experimental results.

gave very accurate predictions of our experimental results, i.e.

$$p = \frac{\sqrt{3}}{2} \frac{(p_u + p_l)}{2} = \frac{\sqrt{3}}{2} \left[\frac{1+\sqrt{3}}{2} + \frac{a}{2} + \frac{2}{3a} + \ln R \right] k + q \qquad (3)$$

In 1973, Parsons *et al.* [4] used a different upper bound picture, as shown in Fig. 2(a), and produced a better (lower) *upper* bound solution, as indicated in Fig. 3 (for $q = 0$). They also obtained extensive experimental results as shown, from which it can be seen that a much better prediction of their own experimental results is given by the much simpler 'mean' solution of Alexander and Lengyel. This illustrates that even the existing simple mathematical techniques of limit analysis in plasticity theory are often not being used to their best advantage, because lower bound solutions are neglected.

From this brief discussion it is apparent that the classical mathematical theory of plasticity is unfortunately rather inadequate for describing many of the features characterizing actual metal forming processes. The same observation is also true for materials other than metals.

Since even the classical theory requires sophisticated numerical computational techniques such as the finite element method for the solution of all but the simplest of geometrical configurations it seems worthwhile examining how it can be modified to include the more important effects discussed in the introduction. To this end, it is proposed now to categorize the more important metal forming processes, particularly those involved in either the modern 'precision forming' or 'near-to-net-shape forming' techniques which are being currently developed to eliminate or reduce the waste of material and energy involved in machining away massive amounts of metal. In this discussion, some of which derives from a keynote lecture to be given to the Forming Group of C.I.R.P. at their General Assembly in Great Britain in August 1983 [5] an attempt will be made to set limits on the temperatures and strain rates found in the particular process involved.

3. TEMPERATURE AND STRAIN RATE EFFECTS

3.1. Classification

In any discussion of metal-forming processes aimed at examining the effects of temperature and the rate of straining, it is advisable to recall the too often neglected terms 'hot working' and 'cold working' (Professor W. Johnson F.R.S. (private communication)) and recognize the importance of some of the characteristics of the metallic structure implied by these simple phrases. Brief definitions of them are as follows:

Hot working: plastic deformation of metal at high homologous temperature and at a rate of straining low enough to avoid significant strain-hardening and to allow recovery and/or recrystallization to occur without excessive grain growth.

Cold working: plastic deformation of metal at low homologous temperature and at a rate of straining high enough to avoid significant recovery and/or recrystallization and to allow moderate strain hardening to occur, without fracture.

The interplay and delicate balance between homologous temperature T_H (= absolute actual temperature/absolute melting temperature) and effective strain rate $\dot{\bar{\varepsilon}}$ is evident from these definitions. The importance of metallurgical phenomena such as phase changes and the movement on the atomic scale of vacancies and dislocations is also obvious. If all the heat of deformation is allowed to escape the deformation will be 'isothermal'; if none of it is allowed to escape deformation is 'adiabatic'.

A recent survey commissioned by the S.E.R.C. (Science and Engineering Research Council) and the D.O.I. (Department of Industry) in the U.K. to look into the possible benefits of precision or near-to-net shape forming has been made by Waterman and Neale [6] in which such processes have been categorized under the headings of (1) casting, (2) forging and/or extrusion, (3) sheet metal forming, (4) powder forming and (5) deposition forming. These categories of process will now be considered in some detail, in view of their importance in the present context. Further categories which could be added are (6) machining and shearing, (7) explosive forming and finally one of the main conventional processes for producing semi-finished stock, namely (8) rolling, and these will also be discussed. Several other modern precision forming processes exist, e.g. rotary metal working processes such as spinning and roll-forming but the aforementioned categories cover the main ranges of temperature and strain rate found.

3.1.1. *Casting*

In general, plastic flow (as normally understood for solids) does not occur in casting processes during which hot liquid metal is poured into moulds. However, recent techniques for the plastic deformation of metals in the 'mashy state' are being developed in Japan, for example by Kiuchi *et al.* [7]. At such high temperatures high rates of straining may be expected to induce sufficient adiabatic heating to liquefy the metal completely, so there is a certain limitation on the rate of working for that reason. From their results these researchers observed that T_H ranged from 0·9 to 1·0 for the non-ferrous metals they investigated, as between the fully solid and fully liquid states. Testing their metals at a constant strain rate of $0{\cdot}2\,s^{-1}$ they found that there was a continual reduction in flow stress for the completely solid metal as T_H was raised to 0·9. Further increase of T_H above 0·9 resulted in dramatic decreases of flow stress, the metal being partly molten.

3.1.2. *Forging and/or Extrusion*

Hot forging is traditionally the oldest of the metal-working crafts, as practised on a small scale by the village blacksmith, carried out at homologous temperatures typically of about 0·7. The process takes place at such high temperatures that the rate of straining can be quite high with the process still being able to be classified as one of hot working. Hot forging with very large drop hammers or hydraulic presses is standard conventional practice and strain rates from 100 to 200 s^{-1} are fairly typical, corresponding to tup-anvil approach speeds of around 10 m/s.

Perhaps the most interesting of the modern plastic deformation processes are those of cold forging and cold extrusion, usually carried out at room temperature ($T_H \simeq 0{\cdot}16$–$0{\cdot}32$). These processes are very beneficial from the point of view of material and energy conservation and therefore of cost-saving, particularly when compared with machining from the solid and they also generally confer better mechanical properties on the final product because of work-hardening. For metals like steel the forces involved are very high and development of the processes has depended on the parallel development of very hard tool steels and 'hard metals' (cemented carbides such as tungsten carbide). To avoid brittle fracture of the tools and dies hammers are not used and deformation is usually achieved by mechanical or hydraulic presses having tool approach speeds of about 0·1 m/s, giving strain rates of around 1·0 s^{-1}. Strain rates in cold extrusion depend very significantly on the extrusion ratio R. The mean strain rate is approximately given by Feltham's [44] expression:

$$\dot{\bar{\varepsilon}} = (6U/D) \ln R \tag{4}$$

where U is the approach velocity and D is the billet diameter. Thus, if $U = 0{\cdot}1$ m/s, $D = 0{\cdot}1$ m and $R = 10$ (all typical values), a typical strain rate is approximately 14 s^{-1}. In general, strain rates in this process vary from about 10 to 100 s^{-1}.

These processes have been developed extensively in Germany, the U.S.A. and Japan (roughly in that order) and it is only recently that industry in the U.K. has begun to appreciate their importance (with one or two notable exceptions). Because of the high forces required for the cold forging or extrusion of steel and its alloys, much research has been and is currently being applied to developing the process of 'warm forging'. For steels this takes place at a homologous temperature of approximately 0·55 and at about the same strain rates as for cold forging. Temperatures are around 700°C, which immediately introduces problems of adequate die lubrication in view of the difficulty of finding lubricants which will not break down above temperatures of about 400°C. Also, the process tends towards hot working involving recovery and recrystallization, which takes away the benefit of strain hardening characteristic of cold forging.

A recent development in forging has been the technique of 'isothermal forging' or 'creep forming'. As the names imply, this involves heating both the workpiece and the tools up to the same high temperature and then applying the tools to the workpiece under a relatively low load. Deformation will then take place at an extremely low rate, at strain rates of approximately $10^{-4}\,s^{-1}$ and homologous temperatures of about 0·63. This process is particularly suited to metals such as titanium and its alloys which exhibit 'superplasticity' at high T_H, tending to flow with Newtonian viscosity. A constitutive equation often used for metals is:

$$\bar{\sigma} = C\dot{\bar{\varepsilon}}^n \tag{5}$$

where $\bar{\sigma}$ = effective stress, $\dot{\bar{\varepsilon}}$ = effective strain rate and C and n are assumed to be constants. Actually they must obviously be functions of $\dot{\bar{\varepsilon}}$ and T_H, and typical values of n for metals lie in the range 0·1–0·3. As n approaches unity, Newtonian viscosity is approached and the metal flows like a liquid. This occurs for titanium as it is heated up to about 900°C.

The conventional hot extrusion of large billets at homologous temperatures of about 0·7 and mean effective strain rates of about 10–30 s^{-1} to form long prismatic sections of semi-finished product in the form of rod, bar or sections is a well-known process.

Friction between the tools and the workpiece has always been a problem and has encouraged the use of indirect extrusion, in which the extrusion die is pushed into the billet, thus eliminating movement and hence friction between the billet and the inner container wall. Friction can also be substantially eliminated by using 'hydrostatic extrusion', most easily carried out cold because of the difficulty of finding suitable fluids for high temperature. Recently, however, Japanese researchers and industrialists have developed and applied the hot hydrostatic extrusion of steel at homologous temperatures of about 0·7 and mean effective strain rates of about 10, with some success (notably at Kobe Steel).

3.1.3. *Sheet Metal Forming*

Most sheet metal forming in industry takes place at room temperature in large hydraulic or mechanical presses, with relatively low approach speeds of the tools (≃0·1 m/s). It is essentially a precision forming process since the final shape is produced without any machining, apart from the drilling or blanking out of holes. Straining of the metal is confined to plastic bending and stretching which takes place over the surface of the sheet. Maximum strain rates occur at the well-known zones of sudden thinning where the sheet is drawn over a die radius under tension and can range between 10 and 100 s^{-1}, typically. The process is usually operated at room temperature ($T_H \simeq 0{\cdot}16$ for steels) but recently better performance has been obtained on steels at higher temperatures ($T_H \simeq 0{\cdot}44$) by Wilson [8]. Conversely, better performance has been obtained at low (cryogenic) temperatures ($T_H \simeq 0{\cdot}075$) for face centred cubic metals (aluminium and copper) by Kobayashi *et al.* [9].

3.1.4. *Powder Forming*

Powder metallurgy has existed for many years and covers a great range of powder processing methods. One of the main features of the process is the plastic deformation of the particles by either cold or hot compaction. Cold compaction produces a 'green' compact of little strength which must be sintered to give the necessary strength to the final component. Hot compaction can provide both plastic deformation and some degree of sintering. Compaction can be achieved either by mechanical pressing or by isostatic pressure, the main objective being to approach the theoretical 100% density of the homogeneous metal. The metal particles can be either elemental or alloys. Estimation of the strain rate during particle deformation is extremely difficult and the temperature of deformation can range from room temperature to hot working temperatures.

3.1.5. *Deposition Forming*

Some aerospace type components are of such awkward shape that it is impossible to forge them between closed dies because of the difficulty of removing them from the dies. Because such components must be made from forged material to withstand the high stresses imposed upon them in service, they are machined from solid bars or slabs, often with the loss of up to 98% of the stock material, as swarf (e.g. 'Nimonic' gas turbine rotors).

The idea of building up complex shapes by simultaneous deposition and forming has been developed by Alexander and Dhillon [10] and Singer [11]. In the former method, metal inert gas (MIG) welding followed immediately by rolling or forging is used. In the latter, deposition is by spraying of liquid metal under controlled atmosphere, with simultaneous peening by a recirculated stream of small balls.

Some combination of both methods is now being investigated, with considerable success.

Estimation of temperatures and strain rates in these processes is again extremely difficult. Homologous temperatures around 0·9 obviously apply. Strain rates in rolling or forging by small pneumatic hammers will be in the region of 200–600 s^{-1}. Strain rates in peening do not seem to have been measured, although impact velocities of about 100 m/s are mentioned by Al-Obaid *et al.* [12], so that surface strain rates of more than 1000 s^{-1} could be occurring.

3.1.6. *Machining and Shearing*

In both of these processes, generally carried out at room temperature, deformation is concentrated into a narrow band of intense shearing within the material. The deformation patterns for both processes are illustrated in Fig. 4. From Fig. 4(b) it can be seen that the shear strain suffered by the material within the shear zone of average breadth b is:

$$\gamma = \frac{\delta}{b} \tag{6}$$

and the shear strain rate is:

$$\dot{\gamma} = \frac{d\gamma}{dt} = \frac{1}{b}\frac{d\delta}{dt} = \frac{v}{b} \tag{7}$$

where v is the velocity of the shear.

Evidently, as b tends to zero (which often occurs as the clearance between punch and die is decreased and also for brittle metals), the

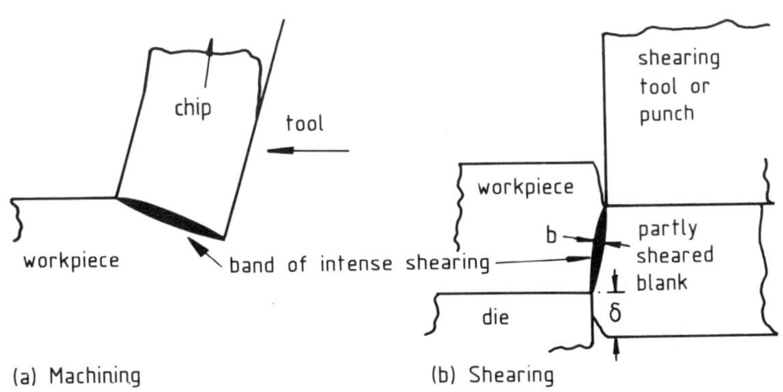

FIG. 4. Machining and shearing.

shear strain rate tends to approach infinity. The same occurs in machining where the width of the shear zone depends upon a number of factors but is generally very small.

Although both of these processes take place at room temperature, the concentration of strain within an intense band of shear, where all the deformation work occurs, results in some increase of temperature due to adiabatic heating. This is rapidly dissipated, however, either by being carried away with the chip in machining or the blank in shearing.

Instantaneously, therefore, the machining and blanking process takes place at moderately high temperature and very high strain rates, the magnitudes of which depend on many factors such as tool velocity, depth of cut (or clearance) and width of shear zone, all of which are difficult to quantify. Nevertheless, machining has been used by several workers as a convenient way of testing metals at very high rates of strain (of the order of $1000\,s^{-1}$).

3.1.7. *Explosive Forming*

Explosion is now being used to form metals in a number of different ways. Quite dissimilar metals can be joined together by explosive welding, where impact velocities ranging from 200 to 700 m/s have been mentioned. The resulting stress waves which occur lead to strain rates which are not known but are so high that the metals apparently melt at their interface and behave as fluids under the high pressures generated.

In the explosive forming of sheet into dies the explosive is usually detonated in water which transmits a shock wave to the sheet so that it impacts the dies at velocities from about 20–200 m/s. Deformation is by bending and stretching under tension in the plane of the sheet so that strains are similar to those already discussed under 'Sheet metal forming'. Again, maximum strain rates will occur at zones of sudden thinning but will be considerably higher than in conventional sheet forming due to the higher impact speeds of the sheet on the die, ranging from 200 to 2000 m/s. The temperature is generally ambient except at the zones of thinning where some small increase may occur. Ezra [13] has mentioned strain rates ranging from 100 to $1000\,s^{-1}$ in this process.

3.1.8. *Rolling*

The rolling process for flat sheets or slabs is relatively easy to describe with simple and realistic mathematical models. A simple equation for the mean strain rate was developed by Ford and Alexander [14] which gives reasonable estimates, namely:

$$\dot{\bar{\lambda}} = \frac{v_n}{\sqrt{(Rh_1)}} \sqrt{r} \left[1 + \frac{r}{4}\right] \tag{8}$$

where $\dot{\bar{\lambda}}$ is the mean strain rate in plane strain ($=2/\sqrt{3}\dot{\bar{\varepsilon}}$, where $\dot{\bar{\varepsilon}}$ is the mean *effective* strain rate); v_n = peripheral speed of the roll; R = undeformed roll radius; h_1 = initial strip thickness; r = pass reduction = $(h_1 - h_2)/h_1$.

For a typical 50% reduction in hot rolling, $\dot{\bar{\lambda}} \simeq 0{\cdot}8 v_n/\sqrt{(Rh_1)}$. For a roll of 0·5 m diameter rotating at 300 rev/min and rolling a slab of 0·1 m thickness (typical values in hot rolling), $\dot{\bar{\lambda}} \simeq 40\ \mathrm{s}^{-1}$. In the high speed rolling of thin strip, say 1 mm thick, the strip speed can be as high as 30 m/s. If the reduction is 10% and the roll 0·2 m diameter rotating at 3000 rev/min, $\dot{\bar{\lambda}} \simeq 1000\ \mathrm{s}^{-1}$. Thus strain rate can vary from about 40 up to $1000\ \mathrm{s}^{-1}$ in rolling.

A convenient summary of the ranges of T_H and strain rate in processes (1)–(8) is given in Table I.

TABLE I
Typical ranges of T_H and strain rate

Process	*Range of T_H*	*Approximate range of strain rate* (s^{-1})
Casting	0·9–1·0	0·0–0·4
Hot forging	0·7	100–200
Hot extrusion	0·7	10–20
Cold forging and extrusion	0·16–0·32	10–100
Warm forging	0·55	10–100
Creep forming	0·63+	10^{-4}
Sheet metal forming	0·16–0·32	10–100
Powder forming	?	?
Deposition forming	0·9	200 to 1 000
Machining, shearing	0·16–0·32	⩾1000
Explosive forming	0·16–0·32	100–1 000
Hot rolling	0·7	40–100
Cold rolling	0·16–0·32	100–1 000

4. CONSTITUTIVE EQUATIONS

4.1. Tests

In recent years considerable advances have been made in the mathematical modelling of many of the modern metal working processes, including factors such as equilibrium conditions, strain compatibility, elastic deformation of the tools, friction at tool–work interfaces and material behaviour (in increasing order of complexity). Undoubtedly the most intractable of these factors is material behaviour and yet it is likely to have the most effect in any attempt to set up a realistic model of a given process, as can be seen by examining any set of tensile tests obtained over a wide range of temperature and/or strain rate.

Having just considered the ranges of temperature and strain rate which can exist in some of the more important modern metal working processes it is therefore relevant to examine how such effects can be included in constitutive relations set up to describe material behaviour. A considerable amount of both experimental and theoretical work has been carried out on these effects in the past and a brief review of some of that work will now be given. The most important testing methods used have been tension, compression and torsion and much of the work has been on steels because of their strength and importance in the manufacture of engineering components.

4.1.1. *Tension Tests*

Although the traditional tensile test is easy to visualize and simple to perform, it suffers a number of disadvantages which limit its usefulness. The main problem is tensile plastic instability, which occurs at overall strains of between about 5 and 30%, depending on the metal. Further straining leads to non-uniform deformation, all strain being concentrated in the necked portion of the test piece and resulting in a complex triaxial system of tensile stresses which renders interpretation of the test results very difficult. Thomsen *et al.* [15] developed a modified tension test aimed at overcoming this difficulty, by interrupting the test at intervals and re-machining the specimen to remove the necked-down section before continuing with the test. Marion [16] has recently developed a high temperature tensile testing facility and he describes in some detail the methods adopted for the measurement and control of temperature, temperature gradient and strain but makes no mention of the problem of 'necking'. Nicholas [17] has used a split Hopkinson bar for the high strain rate tensile testing of aluminium, achieving strain rates up to $1000\,s^{-1}$ (at room temperature only).

4.1.2. *Compression Tests*

By its nature, compression allows significantly higher strains to be attained before any form of instability intervenes, although friction at the platen–specimen interface inevitably leads to non-uniform deformation albeit stable. The first really definitive tests were carried out by Orowan [18] using a 'cam plastometer', which ensured that the strain rate ($\dot{\varepsilon} = v/h$, where v = downward velocity of platen, h = height of specimen) remained sensibly constant during the test itself. Recently, Luton *et al.* [19] have described the modifications required to produce constant strain rate on an Instron testing machine, by using an analogue function generator. Uniaxial compression testing by using a drop hammer was used by both Suzuki *et al.* [20] and Douglas and Altan [21]. The report by Suzuki *et al.* is very comprehensive, comprising over 100 pages of data for both ferrous and non-ferrous metals. They used a cam plastometer and also a drop hammer, measuring the flow stress of 21 non-ferrous metals and alloys and 49 steels (carbon, low alloy and stainless) in the hot-working range of temperature for each material and ranging over strain rates from $0{\cdot}1\ s^{-1}$ up to $650\ s^{-1}$. The ranges of strain rate and homologous temperature they actually used are shown in Table II.

Previous research workers had measured properties over more restricted ranges, for example Alder and Phillips [22] used a cam plastometer in the strain rate range $1–40\ s^{-1}$ and up to a strain of 0·5 in uniaxial compression to measure the flow stresses of aluminium ($T_H = 0{\cdot}09–0{\cdot}88$), copper ($T_H = 0{\cdot}21–0{\cdot}87$), and 0·17% carbon steel ($T_H = 0{\cdot}7–0{\cdot}85$). Inoue [23] measured the flow stress of 15 steel alloys

TABLE II
Compression test conditions (Suzuki *et al.* [20])

Metal	*Melting temp., °C*	*Actual temp. range, °C*	*Range of T_H*	$\dot{\varepsilon}\ s^{-1}$ *Ranges shown*
Aluminium	660	−75 to 650	0·21 to 0·99	
'Duralumin'	640	200 to 500	0·52 to 0·85	
Zinc	420	−75 to 300	0·29 to 0·83	
Magnesium	650	18 to 500	0·32 to 0·84	0·1–650
Titanium	1 670	18 to 900	0·15 to 0·60	
Copper	1 083	18 to 900	0·21 to 0·87	
Copper alloys	950	18 to 900	0·24 to 0·96	
Steel alloys	≃1 450	0 to 1 200	0.16 to 0.85	0·2–650

in tension up to a maximum strain of 0·25 over a range of strain 1–100 s^{-1} and in a T_H range of 0·62–0·85. Cook [24] carried out compression tests on 12 steels up to a strain of 0·5, strain rate = 12–100 s^{-1}, T_H = 0·68–0·85. Arnold and Parker [25] also used a cam plastometer to determine the flow stress of aluminium alloys at strain rates from 1 to 30 s^{-1} and T_H range 0·63–0·9. Bailey and Singer [26] carried out plane strain compression tests to determine flow stress curves for several non-ferrous metals at strain rates ranging from 0·4 to 311 s^{-1}, e.g. aluminium T_H = 0·32–0·94, Al–5·7% Zn alloy T_H = 0·74–0·9, Al–4·2% Cu alloy T_H = 0·63–0·85, lead T_H = 0·49–0·96.

The main difficulties encountered in all compression tests are (1) friction at the platen–specimen interfaces, (2) 'barrelling', or other instabilities such as buckling, of the specimen. Cook and Larke [27] compressed solid cylinders of varying aspect ratios (h/d ratios) in order to extrapolate their results to an infinite height/diameter (h/d) ratio to estimate the zero-friction flow stress. This has been one of the most successful ways of estimating flow stresses, although h/d ratios are limited by buckling of the specimen. For processes such as rolling, the plane strain compression test developed by Watts and Ford [28] is unsurpassed, giving directly the flow stress in plane strain ($2k = 2Y/\sqrt{3} \simeq 1{\cdot}155Y$). Slip line field solutions for this test are well known and, if the ratio platen width/strip thickness is kept within the limits of 2 and 4, frictional effects are negligible and instabilities such as buckling do not occur.

4.1.3. *Torsion Tests*

Many investigators have used torsion testing as a convenient method for obtaining known stresses in the test specimen. Variation of the strain and strain-rate with radius is obviously a problem but can be minimized by using tubular specimens and McQueen and Jonas [29] have achieved strains of order 20. Another problem is the possible buckling of thin-walled tubes twisted to such large strains and also the plastic deformation of tubes tends to be concentrated at one location along the length of the specimen (a phenomenon akin to 'necking' in tension) so that exact determination of strain, strain-rate and stress becomes difficult.

4.2. Modelling

Tests of the various types just outlined have enabled determination of a wealth of data, usually presented as families of curves of flow stress versus strain at various constant strain rates and/or constant temperatures. Constant strain rate during compression testing can really only be achieved easily by using a 'cam plastometer' but the variation of strain rate during the stroke of a mechanical or hydraulic press is not great, since the parameter $\dot{\bar{\varepsilon}} = v/h$ does not vary greatly. (For most deformation processes of this type v and h have their maximum values at the start of the test and drop off roughly in proportion to one another.)

It is much more difficult to keep the temperature constant during testing, particularly at high strain rates. Most investigators simply rely upon measuring temperature accurately throughout any test and then interpolating between the various tests to try to determine flow stress curves at both constant strain rate and constant temperature. The stress required to produce flow in a deforming solid is a complex function of the strain rate, temperature, total strain and past deformation history so that it is not to be expected that any constitutive relation based on an equation of state could provide a full description. For most metals the well-known Prandtl–Reuss equations provide an adequate relationship between stress and incremental strain but determination of material parameters is difficult. Certain approximations are useful provided they are not applied to situations where there are large changes in stress state or temperature during flow.

Different metals exhibit stress–strain curves whose shapes are strongly influenced by temperature. This is illustrated in Fig. 5. Curve A is usually found at low temperatures and shows continuous work-hardening to fracture. Curve B often occurs in aluminium and structurally similar alloys and exhibits work-hardening followed by a period of

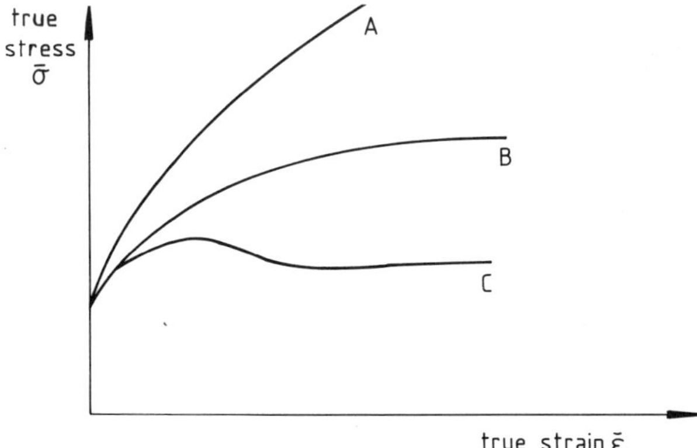

FIG. 5. Typical stress–strain curves.

constant stress deformation (the 'steady state') where work-hardening is balanced by structural recovery. Curves of type C are found in many steels at high temperatures in the austenitic range. Work-hardening is followed by a period of decreasing stress with strain before an eventual steady state is reached. The stress decrease has been associated with dynamic recrystallization.

Figure 5 represents the results of hypothetical tests carried out supposedly under conditions of constant strain rate. An alternative test procedure for the characterization of stress/strain-rate relationships at various temperatures is to measure strain as a function of time, at constant stress. At low temperatures, such tests result in the well-known 'creep curves' where the strain rate shows a continuous decline with time t as shown by curve (a) in Fig. 6 so that the total creep strain (ε_c) can be approximately represented by the equation:

$$\varepsilon_c = \varepsilon_0 + \alpha \ln(1 + \gamma t) \tag{9}$$

where ε_o, α and γ are appropriate constants. At higher temperatures, the shape of the creep curve changes to that shown in Fig. 6, curve (b), which possesses an initially declining creep rate (the primary stage) followed by a period of approximately constant strain rate (the secondary stage) and, eventually, an increasing creep rate (the tertiary stage) leading to fracture. The creep rate in the secondary stage at any temperature and stress corresponds closely with the strain rate in steady state deformation at the same temperature measured by the constant strain rate tests sketched in Fig. 5.

In many cases, the constitutive relationships which have been used to find values for the constants in the Prandtl–Reuss (or Levy–Mises

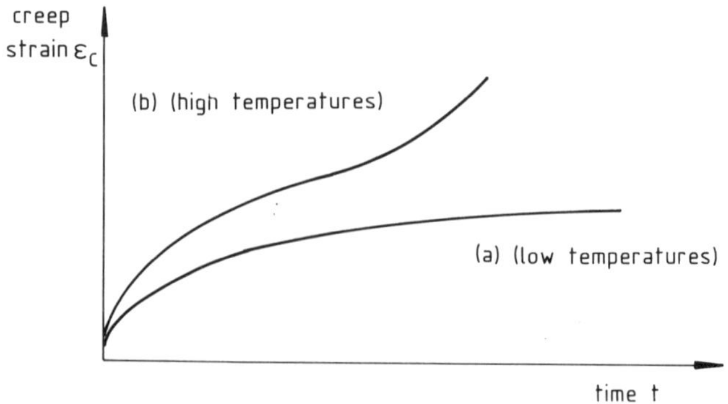

FIG. 6. Typical creep curves.

for large strains) equations have referred exclusively to the steady-state conditions. At low stresses, the effective stress/effective strain rate relationship for steady-state deformation can be adequately described by the inverse of eqn. (5), viz:

$$\dot{\bar{\varepsilon}} = B\bar{\sigma}^m \tag{10}$$

where m, the creep rate exponent, usually lies between 4 and 12 (i.e. $n = 1/m$ lies between 0·0833 and 0·25). This is usually referred to as 'power law' creep.

By plotting the natural logarithm of $\dot{\bar{\varepsilon}}$ against the reciprocal of the absolute temperature ($1/T$) it is found to be proportional to $-Q/RT$ where Q is an Activation Energy for Creep (measured in units J mol^{-1}) and R is the Universal Gas Constant (8·31 J mol^{-1} K^{-1}). The creep rate thus increases exponentially with temperature; typically, an increase of 20°C in temperature can double creep rate. Combining these two dependencies gives a constitutive relationship of the form:

$$\dot{\bar{\varepsilon}} = A\bar{\sigma}^m \exp(-Q/RT) \tag{11}$$

and this is a possible constitutive relationship which might be useful for representing the flow stress dependence upon strain rate and temperature, viz:

$$\bar{\sigma} = \left[\frac{\dot{\bar{\varepsilon}}}{A \exp(-Q/RT)}\right]^{1/m} = C\dot{\bar{\varepsilon}}^n \exp(Q/RT) \tag{12}$$

For pure metals n is usually between 0·25 and 0·0167 but at very low stress, n tends to 1 and deformation exhibits almost Newtonian viscosity, as mentioned previously. The activation energy Q is usually attributed to self-diffusion but at low homologous temperatures, lower values are observed. For complex alloys very small values of n (0·1–0·0833) and large values of Q (several times the value for self-diffusion) are found. These various values of n and Q have been discussed in terms of the different micromechanisms of deformation operating in different stress regimes. To provide a visual description of these different types of behaviour, the concept of *deformation mechanism maps* has gained increasing acceptance. These maps represent the deformation behaviour in stress–temperature space with stress normalized by the appropriate elastic modulus and absolute temperature normalized by the absolute melting temperature, as illustrated in Fig. 7.

One interesting modern approach is that of Ashby and Verrall [30] described by Ashby and Frost [31]. The strain rate of a plastically deforming crystalline solid is dependent upon a number of basic atomic

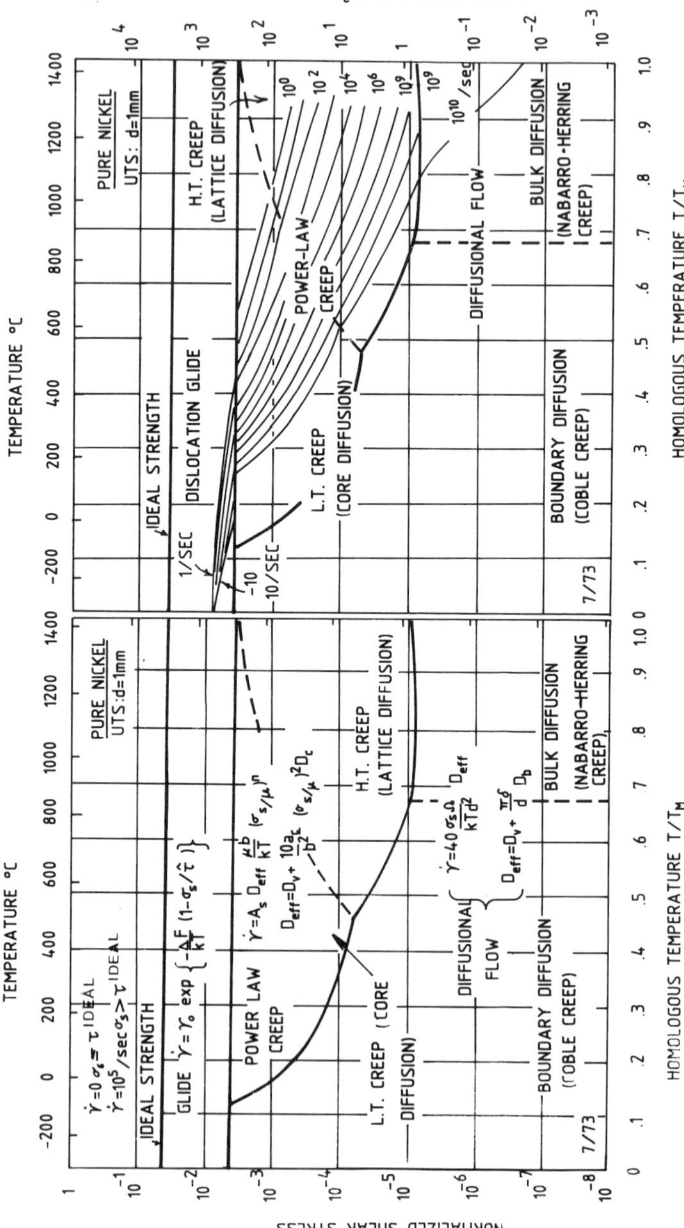

FIG. 7. Deformation mechanism maps (Figs. 4.1 and 4.2 of Ashby *et al.* [31]).

processes such as: dislocation motion, diffusion, grain boundary sliding, twinning or a phase transformation. Such processes can combine to give at least 12 distinctive deformation mechanisms, e.g. yielding (or slip), power law creep, Nabarro–Herring creep, Coble creep, superplastic flow, etc.

Deformation mechanism maps for any polycrystalline material can be constructed in stress–temperature space, showing the area of dominance of each flow mechanism for each of which a rate equation exists, linking shear strain rate $\dot{\gamma}$ to shear stress τ, temperature T and to 'structure' (including all parameters describing its atomic structure, e.g. bonding, crystal class, defect structure, grain size, dislocation density and arrangement, solute or precipitate concentration, etc.). A typical example of such a map is shown in Fig. 7 {Figs. 4.1 and 4.2 of Ashby *et al.* [31]} on which typical equations have been shown, relating the normalized shear stress τ/μ, where μ = shear modulus, to homologous temperature. Kelly [32] has shown that τ cannot exceed τ_{ideal} ($\simeq\mu/20$), shown by the line marked 'ideal strength' in the figure, which is for pure nickel, but polycrystalline metals deform at stresses much less than this by dislocation glide and/or movement of vacancies. At stresses near to the yield stress ($\simeq\mu/1000$) deformation can occur simply by dislocation glide. At lower stresses, deformation is still possible by special dislocation movement processes and, at very low stresses where n approaches 1, deformation can occur simply by stress directed vacancy flow. In these cases, preferential diffusion paths such as grain boundaries and dislocation lines become very important and it is such short-circuiting diffusion paths which lead to the observed low activation energies. The net strain rate at temperature T and stress τ is some superposition of the several mechanisms described. Dislocation creep superimposes roughly additively with diffusional flow but not with glide, so it is customary to choose always the faster rate. Equations are of the form:

$$\dot{\gamma}_1 = \dot{\gamma}_o \exp\left[-\frac{\Delta F}{kT}(1-\tau/\tau_0)\right]$$

$$\dot{\gamma}_2 = \dot{\gamma}_p \exp\left[-\frac{\Delta F}{kT}\right]$$

$$\dot{\gamma}_3 = \frac{A' D_{eff} \mu b}{kT}\left[\frac{\tau}{\mu}\right]^m$$

$$\dot{\gamma}_4 = 42\frac{\tau\Omega}{kTd^2}\left[D_v + \frac{\pi\delta}{d} D_b\right]$$

where $\dot{\gamma}_o$ and $\dot{\gamma}_p$ are a number of mobile dislocation segments, ΔF is total free energy required to overcome a dislocation obstacle, τ_o is the flow stress at $0\ K = \mu b/l$ where μ = shear modulus, b = Burgers vector, l = obstacle spacing; $\dot{\gamma}_o$ and $\dot{\gamma}_p$ are typically limiting strain rates of the order 10^6–$10^8\ s^{-1}$. A' and m are material constants, D_{eff} is a diffusion coefficient, as are D_v and D_b and $D_{eff} = D_v f_v + D_c f_c$ where D_c is a diffusion coefficient, D_v is the lattice diffusion coefficient and f_v and f_c are the fractions of atomic sites associated with each diffusion type. The atomic volume is Ω, d the grain size and δ the effective thickness of a boundary for diffusional transport.

To summarize, the net strain rate of a polycrystal subjected to a shear stress τ at a temperature T is approximately:

$$\dot{\gamma}_{net} = \dot{\gamma}_4\,(\textit{diffusional flow}) + \text{Greatest of} \begin{bmatrix} \gamma_3\,(\textit{dislocation creep}) \\ \text{Least of } \dot{\gamma}_1 \text{ or } \dot{\gamma}_2\,(\textit{glide}) \end{bmatrix}$$

and this information constitutes the deformation mechanism map. Examples of other maps and the methods of using them are described by Ashby *et al.* [31].

In some forming processes, where very large strain rate ranges may be encountered, the description of steady state behaviour given in eqn. (10) is inadequate. At very high stress, an exponential stress dependence on strain rate is more appropriate and the power law and exponential variation can be combined as:

$$\dot{\bar{\varepsilon}} = A\{\sinh(\alpha\bar{\sigma})\}^m \exp(-Q/RT) \tag{13}$$

This relationship has been shown to give a good fit with experimental data for aluminium alloys over many orders of magnitude change in strain rate. However, the description is still of ‘steady state’ behaviour and there appears to be no good evidence that it can be used as an equation of state relationship when stress or temperature change sharply. This problem has been approached in two ways.

One procedure is to define an equation of state which contains an internal variable describing the structure of the material. If the variation of this internal variable with change in stress, temperature and time can be established, an adequate constitutive equation may result. Numerous attempts have been made to provide such a description but they all suffer from the disadvantage that large numbers of sophisticated tests have to be conducted to establish the internal variables. Thus not only constant stress and constant strain rate tests are required but also complex relaxation and other stress and temperature cycling experiments.

An alternative approach has recently been proposed for creep analysis by Evans *et al.* [33]. This suggests that true steady state creep may not exist but that it is merely a point of inflexion on a curve which is the sum of decelerating and accelerating components. The creep curve is described by the equation:

$$\varepsilon = \theta_1\{1-\exp(-\theta_2 t)\}+\theta_3\{\exp(\theta_4 t)-1\} \tag{14}$$

and the various θ-values can be shown to vary systematically with stress and temperature. In this theory, the m values of eqn. (10), which form the basis of deformation maps, are simply a coincidental outcome of the variation of the θ's with stress. The method has already been applied usefully to the long term prediction of creep life and the stress dependence of the θ's seems to be such as to predict an exponential dependence of stress on the deformation rate at high stresses. Many of these developments are very recent but the lack of good constitutive relationships is proving to be such a major obstacle to the application of modern numerical procedures to forming processes that considerable effort must continue to be maintained, to provide adequate models.

5. INSTABILITY AND FRACTURE

The failure of materials during the actual process of forming is not uncommon. For example, sheet can fail in biaxial tension by tensile plastic instability (necking). This was a problem studied extensively over the years by the late Professor H. W. Swift [34] and later by Professor P. B. Mellor [35]. There is an excellent review of the subject in the book by Johnson and Mellor [36] from which it can be seen that this mode of failure is reasonably well understood, although not for anisotropic materials. Keeler [37] developed his idea of the Forming Limit Diagram (FLD) as a convenient way of studying the limits to the formability of sheet metal under biaxial stress, as illustrated in Fig. 8.

This diagram is now widely used and is based on plotting the maximum principal surface strain as ordinate against the minimum principal surface strain as abscissa. In the left hand upper quadrant of the diagram these strains are of opposite sign and failure is by necking. In the upper right hand quadrant both strains are of the same sign and failure is usually by tearing and rupture. Characteristic curves of the

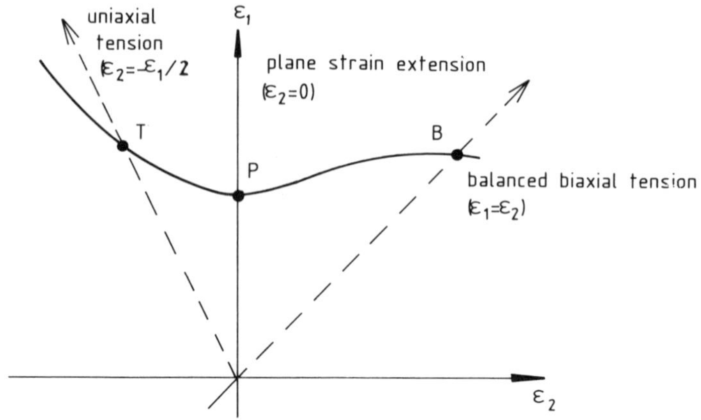

FIG. 8. A typical forming limit diagram (FLD).

type illustrated by TPB can be determined for any particular material, combinations of strain lying above it giving visible localized strain whilst those below do not. Point T corresponds with uniaxial tension ($\varepsilon_2=-\varepsilon_1/2$), point P with plane strain ($\varepsilon_2=0$) and point B with equal biaxial tension ($\varepsilon_2=\varepsilon_1$). Experimental work has revealed the value of FLD's for sheet forming and similar diagrams can be constructed for the surface strains occurring in the bulk compression of billets in production processes, again producing dangerous areas in the lower quadrants of the diagram, in which compressive instability or fracture may occur.

Undoubtedly the major problem in plasticity today is that of predicting when fracture will occur. Fracture mechanics, Griffith's theory and extensions of it into the small-strain elastic–plastic range have been very successful in enabling better understanding of elastic fracture. Similar approaches for large strain deformation processes have been less successful and this is an area requiring much more research and study. Two of the most active researchers in Britain on this topic are Professors J. G. Williams [43] on polymers and A. G. Atkins and T. W. Mai [38] on metals, both of whom are shortly publishing books on the subject, with Ellis Horwood Ltd. of Chichester.

Historically, materials (particularly metals) have broadly been classified as either 'brittle' or 'ductile' but this is a great over-simplification of the situation in respect of their resistance to fracture. Bridgman showed many years ago that ostensibly brittle materials like rock, marble or concrete could be made to behave in a quite ductile fashion, for example exhibiting the 'necking' phenomenon, under large enough hydrostatic pressure, so that the 'brittleness' of a material is certainly not an inherent property. It is well known that the success of most metal forming processes depends upon their being mainly compressive; thus, for example, it is possible to draw a work-hardened copper wire through a die to give it 50% extension whereas it might only extend by about 10% before necking and fracturing, in simple tension. The small taper angle of the die introduces very high compressive stresses in the plastically deforming wire within the die, thus preventing fracture. Conversely, the presence of a high stress concentration such as a sharp notch, crack or neck in a tensile specimen will embrittle the specimen and cause it to fracture at a much lower strain due to the hydrostatic tension resulting from the notch effect.

All this is well known qualitatively but little or no quantitative theory exists, certainly not of the sophistication of Griffith's linear elastic fracture mechanics and this problem needs the attention of both theoretical plasticians and materials scientists. As already mentioned, Ashby has pioneered the use of deformation mechanism maps, with the concept of admissible areas, depending upon the various micromechanisms involved in the deformation. Similarly he has studied cleavage and fracture of polycrystalline metals, which are either intergranular or transgranular and brittle or ductile depending upon the interaction of dislocations, inclusions, voids and other defects, giving characteristic appearances to fracture surfaces of the type described by Rao and Alexander [39] (for deformation in a shearing or 'cropping' process).

Ashby has recognized that a solid may fail by any one of a number of competing micromechanisms and has tried to summarize the fracture behaviour of a given material subjected to a given deformation pattern by a fracture mechanism map. These maps are similar to those illustrated in Fig. 7 and are constructed by assembling fracture observations and data, usually from tests in uniaxial tension. The map has zones within it relating to intergranular fracture (e.g. at high temperature and low stress (for tungsten)), transgranular fracture (e.g. at low T_H high σ), dynamic fracture (high stress, all temperatures), ductile fracture (at intermediate stress and temperature), and rupture (at high temperature).

In his Keynote Lecture at the most recent conference on high strain rates, Johnson [40] referred to Ashby's work and drew attention to the

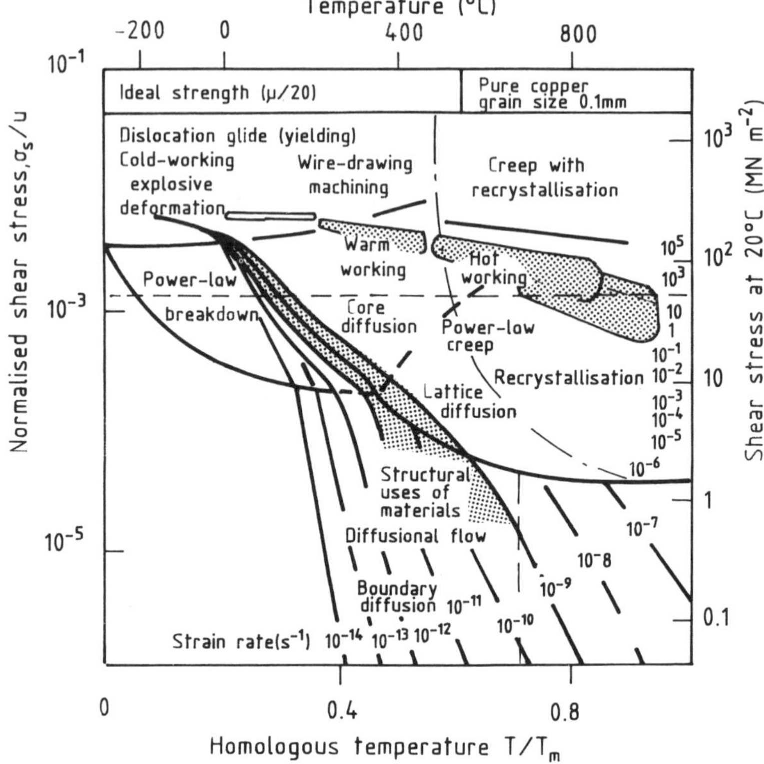

FIG. 9. Forming operations on a deformation map for pure copper.

importance of these maps, for describing and illustrating both deformation and fracture mechanisms. In Figs. 9 and 10 are reproduced Figs. 9 and 10 of Johnson's paper, which show the way in which various zones of the maps of Fig. 7 can be designated for different deformation and fracture processes.

The deformation mechanism map of Fig. 9 relates to pure copper of grain size 0·1 mm and is complicated by the presence of a regime of recrystallization and a limit of flow stress above which simple power-law constitutive relations break down. The fracture mechanism map of Fig. 10(a) is for tungsten of 1 mm grain size and the types of fracture referred to in the various zones of the map are illustrated in Fig. 10(b). It should be noted that the ordinates of Ashby's deformation maps

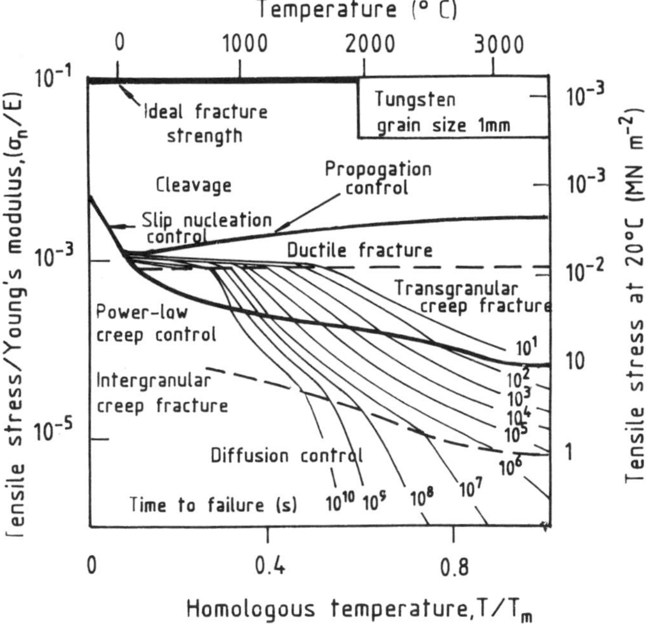

FIG. 10a. A 'mathematical-model'-based fracture map for tungsten.

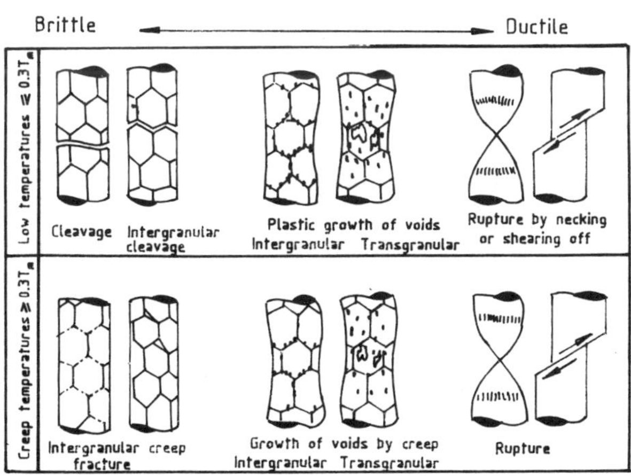

FIG. 10b. Simple classifications and illustrations of fracture mechanisms.

have normalized tensile stresses (σ/E) as ordinates, which seems very sensible in view of the role played by tensile hydrostatic stress in fracture. However, Johnson quotes Ashby's remark that fracture maps are as yet new and somewhat over-simplified.

It should perhaps also be mentioned here that there have been other conferences recently at which problems of constitutive relationships for both flow and fracture have been discussed. One of these was held in 1981 at Stanford University [41], devoted to large deformation plasticity and split into four divisions: macroscopic measurements, micromechanisms, constitutive relations and computing. This conference was under the general chairmanship of Professor E. H. Lee and the proceedings contain many useful papers and also valuable accounts of the detailed discussions at the conference meetings.

The most recent conference on these topics seems to be that held in 1983 at the University of Arizona in Tucson [42], comprising about 100 papers covering seven main topics: General Theory; Metals and Composites; Geological Materials; Discontinuous Media; Interfaces; Joints; Concrete; Granular Materials and Aggregates; Implementation and Evaluation. The aim of that conference was to bridge the well-known gap between theoretical developments and engineering applications. There were only nine papers devoted to metals (apart from one or two of the general papers) and they included no mention of either deformation or fracture maps. The papers appear mostly to be devoted to detailed research findings and seem to the writer to fall far short of achieving the 'gap-bridging' objective mentioned.

6. DISCUSSION AND CONCLUSIONS

In attempting to highlight current problems of plasticity as applied to the forming of metals it has become apparent to me that many researchers have devoted their attention to very restricted academic studies which contribute very little to our overall understanding of the particular class of problem involved. Thus, in the field of limit analysis, researchers have concentrated on upper bound solutions, presumably because they are easy to visualize and to calculate and give a safe (upper) estimate of the loads or power required from the machine needed to effect the deformation. This is really not very satisfactory—more and more realistic estimates of the various parameters are necessary to analyse modern manufacturing processes and it would be better if the correct mathematical procedures laid down for limit analysis and load-bounding techniques were adopted, as shown by my first example.

The same comment is true, to some extent, in relation to the problem of predicting the behaviour in hot working where temperature and strain rate effects predominate. To be able to include the complex variation of flow stress throughout the material volume, due to the similarly complex and interdependent variation of strain rate and temperature in a hot working process, it is now desirable and usually necessary to use the iterative techniques of the finite element method made possible by the ability of the modern computer to manipulate large matrices. Whether or not expensive FEM programs can compete with carrying out actual experiments is an important subject for debate. Without doubt, the development and possession of a good finite element program enables many solutions to be obtained at a fraction of the cost of carrying out the necessary factorial experiment required to obtain the data experimentally. In order to apply any finite element program in a hot-working situation extensive flow stress data are required, however, and the situation here is very bleak. Again, researchers have seemingly concentrated on formulae of the type characterized by my equations (10) and (11) without taking any account of the history of the deformation and its effects on the metallurgical structure. Personally, I have great sympathy for Ashby's approach and his attempts to characterize the flow stress in terms of the micromechanisms of interacting defects, because this is essentially a mechanistic approach: but it is extremely complicated and very difficult to apply in any real engineering situation.

The same is apparently true of the problems of fracture in large deformation processes. We are well aware of the effect of the hydrostatic stress (compressive promoting 'ductility', tensile promoting 'brittleness' and hence fracture) but there seems to be no method as yet of introducing it quantitatively into any theory of fracture. We have 'fracture mechanics' but this is essentially restricted to small strains and needs adapting to the large deformation problem. Again, for fracture as for flow stress, Ashby's techniques help to indicate and explain the role of micromechanisms but not quantitatively and with no account taken of the importance of the hydrostatic component of any complex stress state.

To summarize, I conclude that there is still a great deal of research work to be done but on a more general basis than has been the case up to now. We need a few distinguished researchers to take an overall view of these problems, thereby endeavouring to produce more realistic mathematical models for use by the people who have to design and carry out large deformation metal forming processes in practice.

ACKNOWLEDGEMENTS

I am grateful to Professor J. G. Lenard of the Department of Mechanical Engineering, University of Brunswick, Canada, for contributing to the review on 'constitutive laws'. This forms part of the text of a successful application which we made to N.A.T.O. for funds to support our investigation of steels in collaboration with Professor K. Lange, University of Stuttgart, West Germany and Professor B. Kaftanoglu, Department of Mechanical Engineering, Middle East Technical University, Turkey.

I should also like to thank my colleagues Professor B. Wilshire and Dr R. W. Evans of the Department of Metallurgy and Materials Technology at University College of Swansea for their advice on the same topic.

Finally, I thank Professors W. Johnson and E. H. Lee for their helpful comments and for drawing my attention to the conferences I have mentioned (refs. 40 and 41).

REFERENCES

1. Hill, R. *The Mathematical Theory of Plasticity*, Clarendon Press, Oxford, 1950.
2. Alexander, J. M. and H. Ford. On expanding a hole from zero radius in a thin infinite plate, *Proc. R. Soc. A*, **226** (1955) 543.
3. Alexander, J. M. and B. Lengyel. On the cold extrusion of flanges against high hydrostatic pressure, *J. Inst. Metals*, **93** (1965), 137.
4. Parsons, B., P. R. Milner, and B. N. Cole. Study of the injection upsetting of metals, *J. Mech. Engng Sci.*, **15** (1973), 410.
5. Alexander, J. M. On the economics of alternative methods of metal processing, *Annals C.I.R.P.*, **32**(2) (1982).
6. Waterman, N. A. and M. J. Neale. A survey of applications in industry for the precision forming of metal components and the related needs for research, Michael Neale and Associates Ltd., with S.E.R.C., 1982, Report TRS 284.
7. Kiuchi, M., S. Sugiyama and K. Arai. Study of metal forming in the mashy state, *Proc. 20th Int. M.T.D.R. Conf.* 1979, 71.

8. WILSON, D. V. Shaping things to come—a metallurgist's view, *Sheet Metal Industries*, Part 1, **57**(12) (1980), 1054, Part 2, **58**(3) (1981), 115, Part 3, **58**(4) (1981), 313.
9. KOBAYASHI, M., A. KAMADA, T. TERABAYASHI, and H. ASAO. Press formability of face-centred cubic metals at cryogenic temperatures, *Proc. 20th Int. M.T.D.R. Conf.*, 1979, 239.
10. ALEXANDER, J. M. and M. S. DHILLON. On the deposition forming of steels, *Annals C.I.R.P.*, **28**(1) (1979) 177.
11. SINGER, A. R. E. Simultaneous Spray Deposition and Peening of Metal, U.K. Patent No. 1605035.
12. AL-OBAID, Y. F., S. T. S. AL-HASSANI and T. ALP. A simplified approach to shot-peening mechanics, *Proc. 2nd Machine Design Conf.* (*Cairo*), 1982.
13. EZRA, A. A. The explosive forming of metals, *Proc. C.I.R.P. Int. Conf. on Manufacturing Technology*, 1967, A.S.T.M.E., 775.
14. FORD, H. and J. M. ALEXANDER. Simplified hot rolling calculations, *J. Inst. Metals*, **92** (1963), 397.
15. THOMSEN, E. G., A. H. SHABAIK, and S. SOHRABPOUR. *Engng. Mater. Technol.*, **99** (1977), 252.
16. MARION, R. H. A short-time, high temperature mechanical testing facility, *J. Testing and Evaluation*, **6** (1978), 3.
17. NICHOLAS, T. Tensile testing of materials at high ratio of strain, *Exp. Mech.*, **21** (1981), 177.
18. OROWAN, E. The calculation of roll pressure in hot and cold flat rolling, *Proc. I. Mech. E.*, **150** (1943), 140.
19. LUTON, M. J., J-P.A. IMMARIGEON and J. J. JONAS Constant true strain rate apparatus for use with Instron testing machines, *J. Physics*, **7** (1974), 862.
20. SUZUKI, H., S. HASHIZUME, Y. YABUKI, Y. ICHIHARA, S. NAKAJIMA and K. KENMOCHI. Studies on the flow stress of metals and alloys, *Inst. Ind. Science* (*Tokyo*), **18**(3) (1968), Serial No. 117, 139.
21. DOUGLAS, J. R. and T. ALTAN. Flow stress determination for metals at forging rates and temperatures, *J. Engn. Ind.* (*A.S.M.E.*), **97**(B) (1975), 66.
22. ALDER, J. F. and V. A. PHILLIPS. The effect of strain rate and temperature on the resistance of aluminium, copper and steel to compression, *J. Inst. Metals*, **83** (1954), 80.
23. INOUE, K. Studies on the hot-working strength of steels, *Tetu-to-Hagane*, **41** (1955), 506–515, 593–601.
24. COOK, P. M. True stress–strain curves for steel in compression at high temperatures and high strain rates, for application to the calculation of load and torque in hot rolling, *Proc. Conf. on Properties of Materials at High Strain Rates*, I. Mech. E., 1957, 86.
25. ARNOLD, R. R. and R. J. PARKER. Resistance to deformation of aluminium and some aluminium alloys: Its dependence on temperature and rate of deformation, *J. Inst. Metals*, **88** (1959) 255.
26. BAILEY, A. J. and A. R. E. SINGER. A plane-strain cam plastometer for use in metal-working studies, *J. Inst. Metals*, **92** (1963), 288.

27. Cook, M. and E. C. Larke. Resistance of copper and copper alloys to homogeneous deformation in compression. *J. Inst. Metals*, **71** (1945), 371.
28. Watts, A. B. and H. Ford. An experimental investigation of the yielding of strip between smooth dies, *Proc. I. Mech. E.*, **1B** (1952), 448.
29. McQueen, H. J. and J. Jonas. *Proc. Symposium on the Relation between Theory and Practice in Metal Forming*, Plenum Press, New York, 1970, 393.
30. Ashby, M. F. and R. A. Verrall. *Acta. Met.*, **21** (1973), 149.
31. Ashby, M. F. and H. J. Frost. Constitutive equations in plasticity (Ed. A. S. Argon), M.I.T. Press, 1975, 117.
32. Kelly, A. *Strong Solids*, Clarendon Press, Oxford, 1966.
33. Evans, R. W., J. D. Parker, and B. Wilshire. *Recent Advances in the Creep and Fracture of Engineering Materials and Structures*, Pineridge Press, Swansea, 1982, 135.
34. Swift. H. W. Plastic instability under plane stress, *J. Mech. Phys. Solids* **1** (1952), 1.
35. Mellor, P. B. Plastic instability in tension, *The Engineer*, **209** (1960), 517.
36. Johnson, W. and P. B. Mellor. *Engineering Plasticity*, Van Nostrand Reinhold, London, 1973, 243 et seq. (first published by D. Van Nostrand, 1962, 170 et seq.).
37. Keeler, S. P. Determination of forming limits, *Sheet Metal Industries*, **42** (1965), 683.
38. Atkins, A. G. and T. Mai. *Elastic and Plastic Fracture*, to be published by Ellis Horwood Ltd., Chichester, 1985.
39. Rao, R. S. and J. M. Alexander. Fractographic study of mechanisms of fracture in cropping thin-walled steel tubes, *Metals Technology*, **10** (1983), 14.
40. Johnson, W. Applications: processes involving high strain rates, Keynote Paper, Session 4, *High Strain Rate Conference*, Oxford, Easter, 1979. Inst. Phys. Conf. Ser, 1979, No. 47, Chapter 4, 337–359.
41. Lee, E. H. and R. L. Mallett (Eds). Plasticity of metals at finite strain: theory, computation and experiment, *Proc. Research Workshop held at Stanford University, July* 1981. published in 1982 by the Division of Applied Mechanics, Stanford University and the Department of Mechanical Engineering, Rensselaer Polytechnic Institute, New York 12181, U.S.A.
42. Desai, C. S. and R. H. Gallagher (Eds). Constitutive laws for engineering materials: theory and application, *Proc. International Conference held in University of Arizona, Tucson, January* 1983, published in 1983 and sponsored by the Department of Civil Engineering and Engineering Mechanics, University of Arizona, Tucson, Arizona, U.S.A.
43. Williams. J. G. *Fracture Mechanics of Polymers*, Ellis Horwood Ltd., Chichester, 1983.
44. Feltham, P. Extrusion of metals, *Metal Treatment*, **23** (1956), 440.

39

An Aperçu of Superplastic Forming

J. ARGYRIS and J. ST. DOLTSINIS

Institut für Statik und Dynamik der Luft- und Raumfahrtkonstruktionen, University of Stuttgart, Federal Republic of Germany

ABSTRACT

Manufacturing processes rely increasingly on superplastic properties of certain structural materials, such as titanium alloys, and gain significant importance, in particular, in the aircraft and spacecraft industry.

This contribution presents a computer oriented review and analysis of superplastic forming processes. The modelling of the material response subject to superplastic deformations is tackled first, and relies on recent experimental findings with respect to rate of deformation and grain growth effects. Multiaxial constitutive relations are based on the homogeneous natural definition of stress and rate of deformation [1, 2].

In order to develop computational techniques within the framework of finite element discretisations it is necessary to consider critically the observance of the incompressibility constraint. Nonlinear solution methods for the velocity field are examined as well as the temporal approximation of the unsteady deformation process. The numerical characteristics of the proposed techniques are demonstrated.

A number of examples illustrating superplastic forming processes are investigated and compared, where possible, with experimental data. In this context unsteady contact with the die, frictional sliding, and the forming of an actual structural component are also examined.

1. NATURAL DESCRIPTION OF SUPERPLASTIC MATERIAL BEHAVIOUR

1.1. Uniaxial Considerations

Superplasticity is, in accordance with ref. 13, a deformation process producing high, essentially neck-free, inelastic elongations in metallic materials under tension. In particular, the manifestation known as structural superplasticity can be induced in materials possessing a stable, ultrafine grain size at a temperature of deformation $T \geqslant 0{\cdot}4\ T_M$, T_M denoting the absolute melting temperature of the material.

Fundamental to superplastic deformation is the dependence of the so-called flow stress on the rate of deformation. Under uniaxial conditions, the variation of the flow stress σ (true or Cauchy stress) as a function of the rate of deformation δ, is commonly given in the form of a logarithmic stress rate of deformation relation or a diagram, as illustrated in Fig. 1 for a Ti-6Al-4V alloy. The grain size d of the material represents a parameter determining the relevant log σ, log δ curve. As a result, if the grain size varies during the process of deformation, it is necessary to account for the functional dependence

$$\sigma = f(\delta, d) \tag{1}$$

Relation (1) has to be supplemented by an evolution law for the grain size. To this purpose the experimental observation of ref. 8 may be

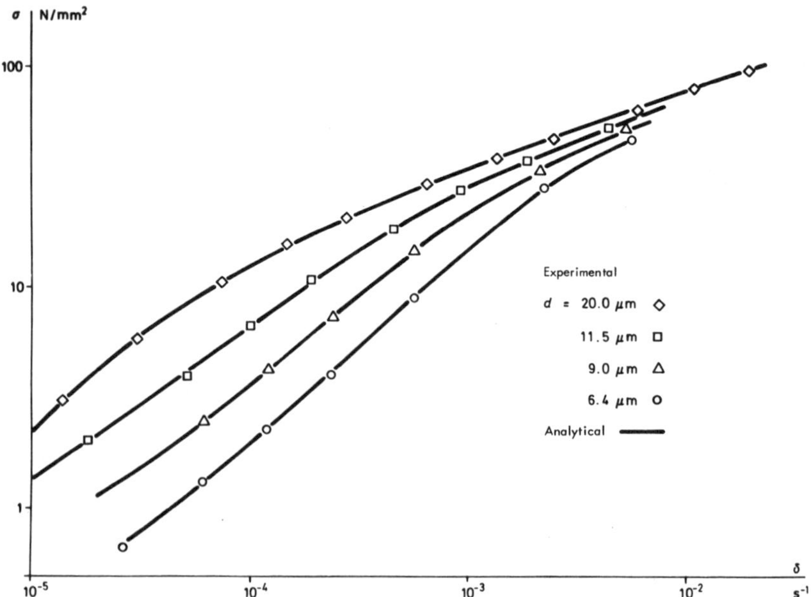

FIG. 1. Dependence of flow stress σ on rate of deformation δ (Ti-6Al-4V, $T = 1200$ K).

interpreted to yield a rate of grain growth in the form

$$\dot{d} = g(t, \delta) \tag{2}$$

The grain size may then be obtained by integration of (2) with respect to time for a given history of the rate of deformation.

In the literature a relation of the form

$$\sigma = k\delta^m d^n \tag{3}$$

is usually proposed in order to establish the dependence of the flow stress σ on the rate of deformation δ and the grain size d. Strictly, expression (3) may be considered as a local approximation of (1). In this case the numerical factor k may be put as

$$k = \sigma_o \, \delta_o^{-m} d_o^{-n} \tag{4}$$

where σ_o, δ_o, d_o appertain to the location to which the approximation is referred. The exponents m and n of (3) are also based on this location and are given by

$$m = \left(\frac{\partial \log \sigma}{\partial \log \delta}\right)_o \tag{5}$$

and

$$n = \left(\frac{\partial \log \sigma}{\partial \log d}\right)_o \tag{6}$$

The local approximation (3) is mainly relevant for considerations on the stability of flow. On the other hand, an analysis of superplastic deformation processes comprising a wide range of variation in the rate of deformation depends on the global description of (1). For the specific case of a Ti-6Al-4V superplastic alloy, ref. 9 proposes for (1) a power series expression up to the fourth order in the form

$$\log \sigma = a_o + a_1 \log \delta + a_2(\log \delta)^2 + \cdots + a_p(\log \delta)^p \tag{7}$$

This expression is intended for the analytical description of data obtained from tests at constant grain size. Alternatively, relation (7) may be interpreted as

$$\sigma = K\,\delta^M \tag{8}$$

with

$$K = 10^{a_o} \tag{9}$$

and

$$M = a_1 + a_2 \log \delta + \cdots + a_p(\log \delta)^{p-1} \tag{10}$$

where the coefficients $a_o \ldots a_p$ vary with the grain size d.

An analytical description of the grain growth kinetics of the Ti-6Al-4V alloy derived from experimental data is presented in ref. 8 and reads

$$\frac{d}{{}^{\circ}d} = \left(\frac{t}{{}^{\circ}t}\right)^{N} \tag{11}$$

where the exponent

$$N = N(\delta) \tag{12}$$

and ${}^{\circ}d$ denotes the initial grain size at time ${}^{\circ}t$. A comparison of the proposed approximation with test data is given in Fig. 2. Relation (11) is strictly applicable to processes taking place at a constant rate of deformation δ. Consequently, the associated grain growth law in the form of (2) is obtained as the time rate of (11) for $\delta = \text{const}$ as

$$\frac{\dot{d}}{{}^{\circ}d} = \frac{N}{{}^{\circ}t}\left(\frac{t}{{}^{\circ}t}\right)^{N-1} \tag{13}$$

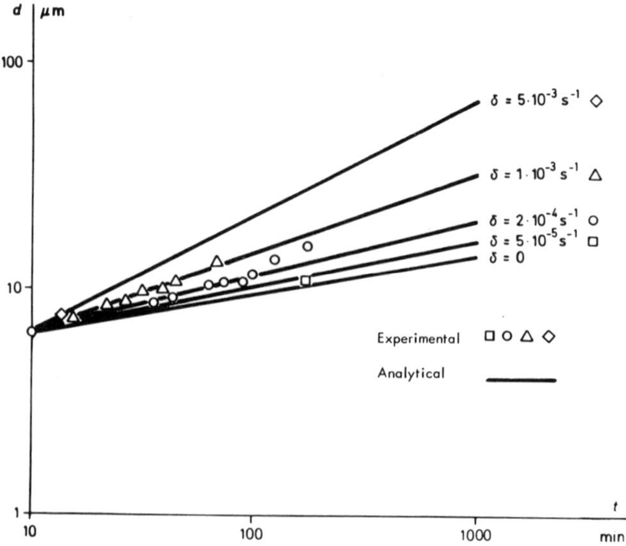

FIG. 2. Grain growth kinetics (Ti-6Al-4V, ${}^{\circ}d = 6{\cdot}4\ \mu\text{m}$, $T = 1200$ K).

Integration of (13) based on the variation of N with the history of the rate of deformation furnishes the current grain size d as the actual process develops.

1.2. Multiaxial Conditions

Proceeding next to the formulation of the constitutive relations appertaining to the general three-dimensional deformation of the material, we follow the natural approach of refs 1 and 2. The six natural directions α to ζ are defined along the edges of an arbitrary reference tetrahedron as shown in Fig. 3. The associated natural rate of deformation (cf. Fig. 4) may either be of the total type [1, 2]

$$\boldsymbol{\delta}_t = \{\delta_t^\alpha \ \delta_t^\beta \ \delta_t^\gamma \ \delta_t^\delta \ \delta_t^\varepsilon \ \delta_t^\zeta\} \tag{14}$$

or of the component type

$$\boldsymbol{\delta}_c = \{\delta_c^\alpha \ \delta_c^\beta \ \delta_c^\gamma \ \delta_c^\delta \ \delta_c^\varepsilon \ \delta_c^\zeta\} \tag{15}$$

Also

$$\boldsymbol{\delta}_t = \mathscr{A}\boldsymbol{\delta}_c \tag{16}$$

where the matrix

$$\mathscr{A} = [\cos^2(\theta, \varphi)], \qquad \theta = \alpha \ldots \zeta, \ \varphi = \alpha \ldots \zeta \tag{17}$$

defines the relative position of the natural directions.

Analogously the natural stresses

$$\boldsymbol{\sigma}_c = \{\sigma_c^\alpha \ \sigma_c^\beta \ \sigma_c^\gamma \ \sigma_c^\delta \ \sigma_c^\varepsilon \ \sigma_c^\zeta\} \tag{18}$$

and

$$\boldsymbol{\sigma}_t = \{\sigma_t^\alpha \ \sigma_t^\beta \ \sigma_t^\gamma \ \sigma_t^\delta \ \sigma_t^\varepsilon \ \sigma_t^\zeta\} \tag{19}$$

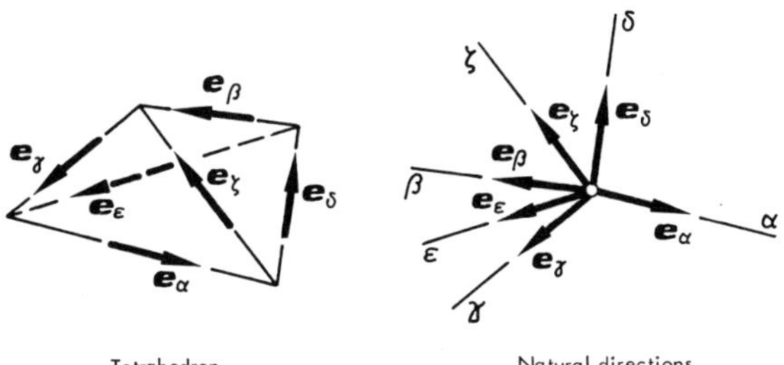

FIG. 3. Tetrahedron element and natural directions.

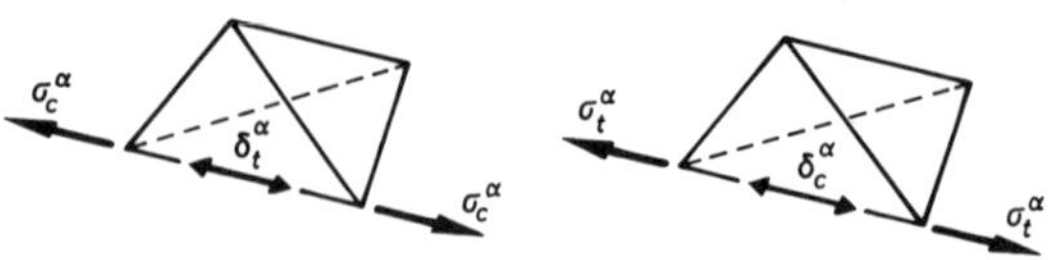

FIG. 4. Corresponding natural stresses and rates of deformation.

are interrelated by

$$\boldsymbol{\sigma}_c = \mathscr{A}^{-1}\boldsymbol{\sigma}_t \tag{20}$$

Natural stresses correspond to the natural rate of deformation satisfying the invariance of the virtual rate of work

$$\boldsymbol{\delta}_t^t\boldsymbol{\sigma}_c = \boldsymbol{\delta}_c^t\boldsymbol{\sigma}_t \tag{21}$$

and are in this case of the Cauchy type.

It is convenient to separate the stress state into hydrostatic and deviatoric components. Considering first total stresses, we write

$$\boldsymbol{\sigma}_t = \boldsymbol{\sigma}_{tH} + \boldsymbol{\sigma}_{tD} \tag{22}$$

with the hydrostatic part

$$\boldsymbol{\sigma}_{tH} = \tfrac{1}{3}\boldsymbol{E}_6\boldsymbol{\sigma}_c \tag{23}$$

where

$$\boldsymbol{E}_6 = \boldsymbol{e}_6\boldsymbol{e}_6^t, \ \boldsymbol{e}_6 = \{1 \ \ 1 \ \ 1 \ \ 1 \ \ 1 \ \ 1\} \tag{24}$$

and the deviatoric part

$$\boldsymbol{\sigma}_{tD} = [\mathscr{A} - \tfrac{1}{3}\boldsymbol{E}_6]\boldsymbol{\sigma}_c \tag{25}$$

The two contributions to the component stress

$$\boldsymbol{\sigma}_c = \boldsymbol{\sigma}_{cH} + \boldsymbol{\sigma}_{cD} \tag{26}$$

follow then by application of (20) from the corresponding terms of the total stress in (22).

The decomposition of the rate of deformation $\boldsymbol{\delta}$ into volumetric and deviatoric parts proceeds along the same lines. For an incompressible or isochoric material, the volumetric rate of deformation must vanish. This yields for the incompressibility condition the expression

$$\boldsymbol{e}_6^t\boldsymbol{\delta}_c = \boldsymbol{e}_6^t\mathscr{A}^{-1}\boldsymbol{\delta}_t = 0 \tag{27}$$

Assuming the material possesses viscous properties, one obtains the total natural stress

$$\boldsymbol{\sigma}_t = \boldsymbol{\sigma}_{tD} + \boldsymbol{\sigma}_{tH} = 2\mu\boldsymbol{\delta}_t - p\boldsymbol{e}_6 \tag{28}$$

Here the deviatoric stress arises from the viscosity coefficient μ of the material whilst the hydrostatic stress derives from the static pressure p. Ultimately, (28) is expressed as a relation between the component natural stress and the corresponding total natural rate of deformation. Thus,

$$\boldsymbol{\sigma}_c = \mathscr{A}^{-1}[2\mu\boldsymbol{\delta}_t - p\boldsymbol{e}_6] \tag{29}$$

A connection with the aforementioned uniaxial case is achieved by introducing the von Mises equivalent stress $\bar{\sigma}$, defined in natural terms as

$$\bar{\sigma}^2 = \tfrac{3}{2}\boldsymbol{\sigma}^{\mathrm{t}}_{cD}\boldsymbol{\sigma}_{tD} \tag{30}$$

and the equivalent rate of deformation $\bar{\delta}$, defined by

$$\bar{\delta}^2 = \tfrac{2}{3}\boldsymbol{\delta}^{\mathrm{t}}_t\boldsymbol{\delta}_c \tag{31}$$

where $\boldsymbol{\delta} \equiv \boldsymbol{\delta}_D$, in the present incompressible case. Using the deviatoric part of the constitutive relation (28) or (29), one obtains for the equivalent quantities

$$\bar{\sigma} = 3\mu\bar{\delta} \tag{32}$$

On the other hand, uniaxial constitutive equations may be interpreted as a relation between the equivalent stress and the equivalent rate of deformation,

$$\bar{\sigma} = f(\bar{\delta}) \tag{33}$$

Comparison between (32) and (33) yields the viscosity coefficient

$$\mu = \frac{f(\bar{\delta})}{3\bar{\delta}} = \mu(\bar{\delta}) \tag{34}$$

In order to apply standard computational procedures to the treatment of incompressible processes, the so-called penalty approach is used. Thereby the isochoric condition (27) is relaxed and the pressure p is related to the volumetric rate of deformation as follows

$$p = -\bar{\kappa}\boldsymbol{e}^{\mathrm{t}}_6\boldsymbol{\delta}_c = -\bar{\kappa}\boldsymbol{e}^{\mathrm{t}}_6\mathscr{A}^{-1}\boldsymbol{\delta}_t \tag{35}$$

where

$$\bar{\kappa} = \frac{2\mu}{3}\frac{1+\bar{\nu}}{1-2\bar{\nu}} \rightarrow \infty \tag{36}$$

represents the penalty parameter in the form of a modulus of viscous compressibility as a function of a pseudo Poisson's ratio $\bar{\nu}$. It is put here in a form modelled on the well-known elastic bulk modulus. Then, the strictly incompressible constitutive relation (29) assumes, in the penalty approach, the form

$$\boldsymbol{\sigma}_c = 2\mu\left[\mathcal{A}^{-1} + \frac{\bar{\nu}}{1-2\bar{\nu}}\mathcal{A}^{-1}\mathbf{E}_6\mathcal{A}^{-1}\right]\boldsymbol{\delta}_t \tag{37}$$

and

$$\bar{\nu} \rightarrow \tfrac{1}{2} \tag{37a}$$

may be used as an alternative penalty parameter.

2. NUMERICAL SOLUTION METHODS FOR SUPERPLASTIC DEFORMATION PROCESSES

2.1. Governing Equations

The following considerations concern the motion of viscous incompressible media as arising in superplastic materials. In particular, we deal with slow viscous processes where inertia effects may be disregarded and the motion is quasi-static but not necessarily steady. In this case the motion satisfies the equilibrium condition

$$\boldsymbol{S} = \boldsymbol{R} \tag{38}$$

between the resultants of the stresses $\boldsymbol{S}$ in the medium as discretised by finite elements, and the applied loads $\boldsymbol{R}$.

In an incompressible medium, the static pressure p appearing in the constitutive relations (28) and (29) as a part of the stress cannot be derived from the rate of deformation $\boldsymbol{\delta}$. The latter being purely deviatoric must satisfy the isochoric condition (27). As a consequence, both the velocity field and the pressure field must appear as the unknown variables in the solution of the problem. Therefore, they must be considered as independent variables within a finite element discretisation and furnish the nodal stress resultants

$$\boldsymbol{S} = \boldsymbol{C}\boldsymbol{V} - \boldsymbol{H}\boldsymbol{p} \tag{39}$$

in dependence of the nodal velocities $\boldsymbol{V}$ and the pressures $\boldsymbol{p}$ of the discretised medium. In particular, $\boldsymbol{C}$ is termed the viscosity matrix and $\boldsymbol{H}$ the hydrostatic matrix. Equation (39) represents the discretised form of the constitutive relations (28), (29) and must be supplemented by the discretised form of the incompressibility condition (27), which reads

$$\boldsymbol{H}^{t}\mathbf{V} = \boldsymbol{o} \tag{40}$$

The set (38), (39) and (40) presents a complete system of equations governing slow incompressible viscous flow of the medium within the finite element methodology [3]. In this context, we observe that the pressure is a variable akin to the stress. As a result, a consistent approximation of the pressure distribution within a finite element must be of one order lower than the approximation of the velocity field [4]. However, convergence and stability requirements with respect to the discretisation impose certain additional limitations on the representation of the pressure field as against the associated velocity field. For these aspects the reader may consult [12] where additional references are quoted.

Considering now the penalty approach to the solution of the incompressible problem, we observe that (39) becomes

$$\boldsymbol{S} = \bar{\boldsymbol{C}}\mathbf{V} \tag{41}$$

Here the stress resultants $\boldsymbol{S}$ depend solely on the nodal velocities $\mathbf{V}$. Equation (41) is a consequence of the relaxed incompressible constitutive relation (37) and is seen to require the matrix $\bar{\boldsymbol{C}}$ which includes the influence of the slightly compressible behaviour as assumed for the viscous medium. Despite the interdependency of the pressure and velocity fields as postulated by the constitutive relations it is feasible to approximate the two variables independently and select to this purpose a mixed finite element formulation (for example, ref. 14). Alternatively, the finite element approach may be based exclusively on an approximation of the velocity field. In this case, the requirement of a lower order approximation of the pressure distribution is replaced by reduced integration techniques applied to the volumetric response of the finite element model [10].

In conclusion, we observe that whatever method is chosen among the three aforementioned techniques—summarily described as the incompressible formulation, the penalty method with mixed approximation and the penalty method with reduced integration—we ultimately find that the results differ mainly with respect to the pressure field.

2.2. Determination of the Velocity Field

In this section we deal exclusively with the penalty approach, in which case (41) supplies the stress resultants as functions of the nodal velocities. The equilibrium condition (38) may then be written in the form

$$\boldsymbol{F}(\mathbf{V}) = \boldsymbol{S}(\mathbf{V}) - \boldsymbol{R} = \boldsymbol{o} \tag{42}$$

which defines the velocities $\mathbf{V}$.

In the case of a non-Newtonian viscous material, (42) is nonlinear in the velocities. It may be solved by application of the Newton method. To this purpose, one starts the ith iteration cycle by injecting an estimate $\mathbf{V}_i$ in (42) and deducing a search direction

$$\boldsymbol{d}_i = -\left[\frac{\mathrm{d}\boldsymbol{F}}{\mathrm{d}\mathbf{V}}\right]_i^{-1} \boldsymbol{F}_i = -\boldsymbol{G}_i^{-1}\boldsymbol{F}_i \tag{43}$$

where

$$\boldsymbol{G} = \frac{\mathrm{d}\boldsymbol{F}}{\mathrm{d}\mathbf{V}} = \frac{\mathrm{d}\boldsymbol{S}}{\mathrm{d}\mathbf{V}} \tag{44}$$

is the gradient matrix of the function $\boldsymbol{F}(\mathbf{V})$ and is to be detailed below. From the first estimate $\mathbf{V}_i$ a new estimate $\mathbf{V}_{i+1}$ for the velocities is obtained via

$$\mathbf{V}_{i+1} = \mathbf{V}_i + s_i\boldsymbol{d}_i \tag{45}$$

where the scalar s is usually set to $s = 1$. For a closer approximation we may determine s_i by a line search in accordance with the condition

$$\boldsymbol{d}_i^t\boldsymbol{F}(\mathbf{V}_i + s_i\boldsymbol{d}_i) = 0 \tag{46}$$

which involves considerable computational effort.

Following this brief recapitulation of the Newton technique we derive the gradient matrix of the system as required in the computational process. Substituting expression (41) for the stress resultants in (44)

$$\boldsymbol{G} = \frac{\mathrm{d}\boldsymbol{S}}{\mathrm{d}\mathbf{V}} = \bar{\boldsymbol{C}} + \tilde{\boldsymbol{C}} \tag{47}$$

where the incremental matrix $\tilde{\boldsymbol{C}}$ reflects the dependence of the viscosity matrix $\bar{\boldsymbol{C}}$ on the velocities $\boldsymbol{V}$ as detailed in ref. 2.

A valid criticism of the Newton technique arises from the uneconomical necessity of setting up a new inverse gradient matrix in each iteration cycle. A possible evasion of this difficulty may be achieved by application of an approximate gradient matrix, whereby

$$\boldsymbol{G} \Leftarrow \bar{\boldsymbol{C}} \tag{48}$$

appears an obvious choice. This initial approximation may then be updated during the course of the iteration process in accordance with the so-called Broyden–Fletcher–Goldfarb–Shanno (BFGS) method described in ref. 11. In fact, this proposed update may be implemented directly on the inverse of the gradient matrix in (43) within the scheme

$$\boldsymbol{G}_i^{-1} = [\boldsymbol{I} + \boldsymbol{w}_i \boldsymbol{u}_i^{\mathrm{t}}]\boldsymbol{G}_{i-1}^{-1}[\boldsymbol{I} + \boldsymbol{u}_i \boldsymbol{w}_i^{\mathrm{t}}] \tag{49}$$

where

$$\boldsymbol{u}_i = \left[1 - s_{i-1}\left(\frac{\boldsymbol{d}_{i-1}^{\mathrm{t}} \boldsymbol{F}_{\Delta i}}{\boldsymbol{F}_{i-1}^{\mathrm{t}} \mathbf{V}_{\Delta i}}\right)^{1/2}\right]\boldsymbol{F}_{i-1} - \boldsymbol{F}_i \tag{50}$$

$$\boldsymbol{w}_i = \frac{1}{\mathbf{V}_{\Delta i}^{\mathrm{t}} \boldsymbol{F}_{\Delta i}} \mathbf{V}_{\Delta i} \tag{51}$$

and

$$\mathbf{V}_{\Delta i} = \Delta_i \mathbf{V} = \mathbf{V}_i - \mathbf{V}_{i-1} \tag{52}$$

$$\boldsymbol{F}_{\Delta i} = \Delta_i \boldsymbol{F} = \boldsymbol{F}_i - \boldsymbol{F}_{i-1} = \boldsymbol{F}(\mathbf{V}_i) - \boldsymbol{F}(\mathbf{V}_{i-1}) \tag{53}$$

Alternatively, the condition of equilibrium (38) and expression (41) for the stress resultants may be rearranged in the form

$$\boldsymbol{S} = \bar{\boldsymbol{C}}(\mathbf{V})\mathbf{V} = \boldsymbol{R} \tag{54}$$

A standard method for the solution of the nonlinear equation (54) applies the iteration procedure

$$\mathbf{V}_{i+1} = \bar{\boldsymbol{C}}_i^{-1} \boldsymbol{R} \tag{55}$$

with

$$\bar{\boldsymbol{C}}_i = \bar{\boldsymbol{C}}(\mathbf{V}_i) \tag{56}$$

This supplies an improved estimate $\mathbf{V}_{i+1}$ following the initial estimate $\mathbf{V}_i$ when entering the ith iteration cycle.

Despite its simple structure, the iterative technique proposed in (55) is known to be subject to convergence limitations as discussed in ref. 2. It turns out that the constituent matrices of the gradient matrix $\boldsymbol{G}$ in (47) enter into a convergence criterion of the form

$$\|\bar{\boldsymbol{C}}^{-1}\tilde{\boldsymbol{C}}\| < 1 \quad \text{for convergence} \tag{57}$$

In this context it is interesting to consider a uniaxial viscous specimen subject to tension. Let us assume a constitutive relation of the power form,

$$\sigma = k\delta^m = 3\mu\delta \tag{58}$$

where the second expression reproduces (32) with a viscosity coefficient

$$\mu = \frac{k}{3}\,\delta^{m-1} \tag{59}$$

If (58) is solved by application of the iterative scheme (55), convergence is shown in ref. 2 to hold within the range

$$0 < m < 2 \quad \text{for convergence} \tag{60}$$

of the exponent m. Furthermore, the lower limit $m = 0$ is found to coincide with the limiting case of an inviscid, perfectly plastic material response.

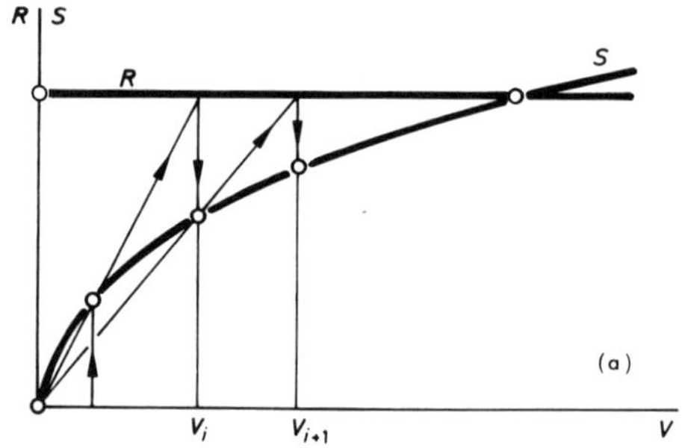

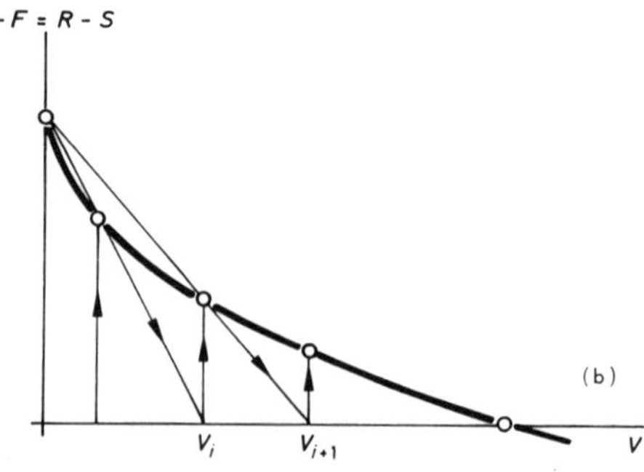

FIG. 5. Successive approximation. (a) Standard scheme; (b) quasi-Newton scheme.

The same uniaxial problem may serve as a common criterion of the numerical properties of the proposed three different solution methods. In particular, the standard iteration method of (55) may be interpreted for a single unknown as the secant method of Fig. 5(a). Figure 5(b) demonstrates this technique in the guise of a quasi-Newton procedure applied to (42). Finally, the BFGS technique relies also on a secant method but makes use of the results of two successive iterations, as shown in Fig. 6(a). The last demonstration should be set against the original Newton procedure given in Fig. 6(b).

The numerical solution of (58) has been carried out by the three techniques for various values of the exponent m but within the range $0<m<2$ of the standard iterative method. Throughout this nonlinear investigation, the solution of the linear case, $m=1$, yields in each case an initial estimate for δ. Figure 7 indicates for all three techniques the number of iterations NIT required for convergence in dependence of m. The iteration procedure was terminated at a relative error of 10^{-3}.

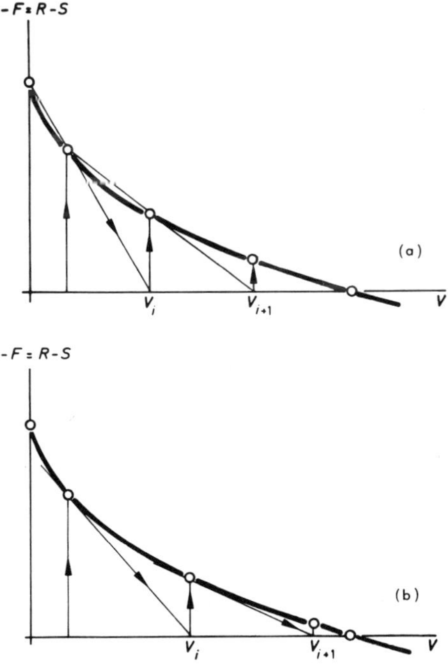

FIG. 6. (a) Quasi-Newton (BFGS) and (b) Newton iteration.

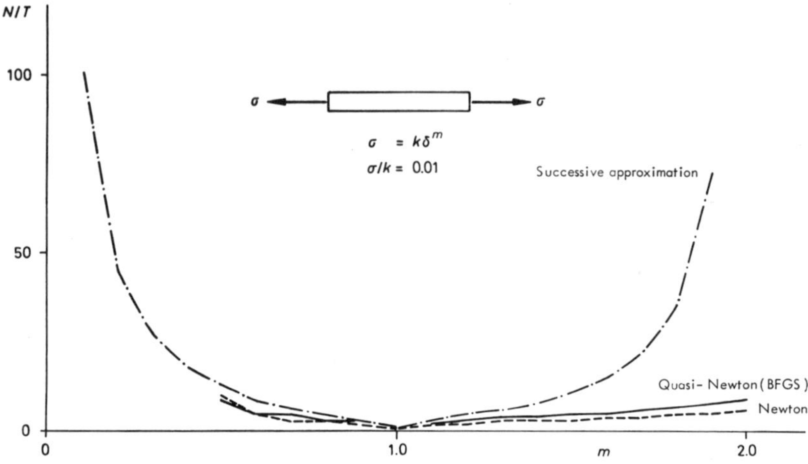

FIG. 7. Convergence behaviour in a uniaxial case.

We observe that the standard iterative scheme (55) converges within the limits set by (60). However, the quasi-Newton procedure which relies on the initial approximation (48) to the gradient is seen to reduce considerably with increasing nonlinearity the number of iterations and to approach the behaviour of the original Newton method. It is noteworthy that the two latter procedures cease to converge for $m < 0{\cdot}5$ as a result of the initial estimate used for δ.

2.3. Motion During the Forming Process

Once the velocity field is known as a function of the time t, the motion of the material, i.e. the positions occupied successively by the particles, is obtained by integration of the velocity field with respect to time. Inevitably, the time integration is performed in an incremental manner. Let the finite element mesh occupied by the material at time ${}^{a}t$, the beginning of a time increment, be specified by the co-ordinates ${}^{a}\mathbf{X}$ of the mesh nodal points with respect to a cartesian frame fixed in space. The subsequent location of the nodal points at the end of the current interval

$$ {}^{b}t = {}^{a}t + \tau \tag{61} $$

is given by

$$ {}^{b}\mathbf{X} = {}^{a}\mathbf{X} + \int_{{}^{a}t}^{{}^{b}t} \mathbf{V}\, dt = {}^{a}\mathbf{X} + \tau^{\zeta}\mathbf{V} \tag{62} $$

where the parameter

$$\zeta = \frac{t - {}^a t}{{}^b t - {}^a t} = \frac{1}{\tau}(t - {}^a t),\ 0 \leqslant \zeta \leqslant 1 \tag{63}$$

specifies a certain instant within the time interval (61).

If the velocity $\mathbf{V}$ of the change of geometry is assumed to vary linearly within the time interval, then

$$^{\zeta}\mathbf{V} = (1-\zeta)^a\mathbf{V} + \zeta^b\mathbf{V} \tag{64}$$

Substitution of (64) in (62) provides in this simple case the geometry of the discretised body at the end of the time increment,

$$^b\mathbf{X} = {}^a\mathbf{X} + (1-\zeta)\tau^a\mathbf{V} + \zeta\tau^b\mathbf{V} \tag{65}$$

The velocities $\mathbf{V}$ are governed by the equations of Section 2.1. Application of these relations requires an explicit reference to the current geometry. As a result (42) has to be expressed as

$$\boldsymbol{F}(\mathbf{X}, \mathbf{V}) = \mathbf{S}(\mathbf{X}, \mathbf{V}) - \boldsymbol{R}(t, \mathbf{X}) = \boldsymbol{o} \tag{66}$$

This relation governs, in conjunction with the integration scheme (65), the motion of the deforming body on the basis of the penalty approach to the incompressibility condition.

An appraisal [2] of the numerical stability of the integration scheme (65) as against the time interval τ indicates that the matrix

$$\mathbf{N} = \left[\frac{\partial \boldsymbol{F}}{\partial \mathbf{V}}\right]^{-1} \frac{\partial \boldsymbol{F}}{\partial \mathbf{X}} = \boldsymbol{F}_{,\mathrm{V}}^{-1}\boldsymbol{F}_{,\mathrm{X}} \tag{67}$$

plays a fundamental role. In particular, its spectral norm

$$\rho = \rho(\mathbf{N}) \tag{68}$$

limits the time step for conditional stability to

$$\tau < \frac{2}{{}^a\rho - ({}^a\rho + {}^b\rho)\zeta} \quad \text{and} \quad 0 \leqslant \zeta \leqslant \frac{{}^a\rho}{{}^a\rho + {}^b\rho} \tag{69}$$

Unconditional stability is achieved when

$$\tau \rightarrow \infty \quad \text{and} \quad \frac{{}^a\rho}{{}^a\rho + {}^b\rho} \leqslant \zeta \leqslant 1 \tag{70}$$

Figure 8 demonstrates the lower limit of ζ with respect to the unconditional stability of (70), as a function of an increasing quotient ${}^b\rho/{}^a\rho$. For

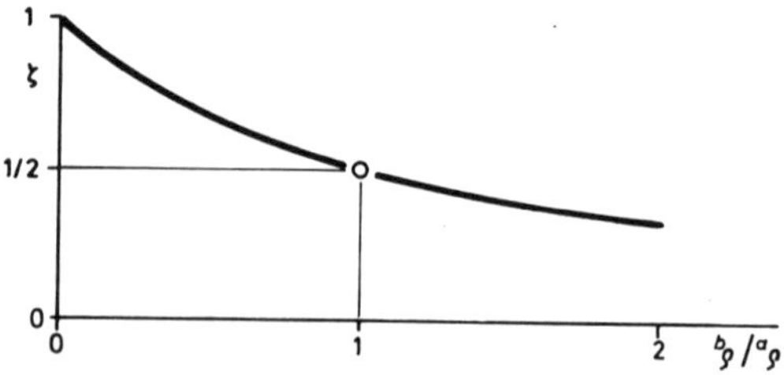

FIG. 8. Unconditional stability limit.

${}^b\rho/{}^a\rho = 1$ one recovers the unconditional stability limit $\zeta = 1/2$ appertaining to a prescribed linear variation of the geometry $\boldsymbol{X}$ instead of the presently adopted scheme. Figure 9 illustrates the stability condition (69) for the case of a constant spectral norm ρ within the interval τ and also shows the oscillation frontier beyond which solutions, or rather their perturbations, start to change their sign in the course of the incremental computation.

Considering next techniques for solving (65), we begin with the method of successive solutions whereby the result of iteration i is used to predict the solution at iteration $i+1$ in accordance with

$$ {}^b\mathbf{X}_{i+1} = {}^a\mathbf{X} + (1-\zeta)\tau\, {}^a\mathbf{V} + \zeta\tau\, {}^b\mathbf{V}_i \tag{71} $$

for ${}^b\mathbf{X}_{i+1}$, and

$$ \boldsymbol{F}({}^b\mathbf{X}_i, {}^b\mathbf{V}_i) = \boldsymbol{S}({}^b\mathbf{X}_i, {}^b\mathbf{V}_i) - \boldsymbol{R}({}^bt, {}^b\mathbf{X}_i) = \boldsymbol{o} \tag{72} $$

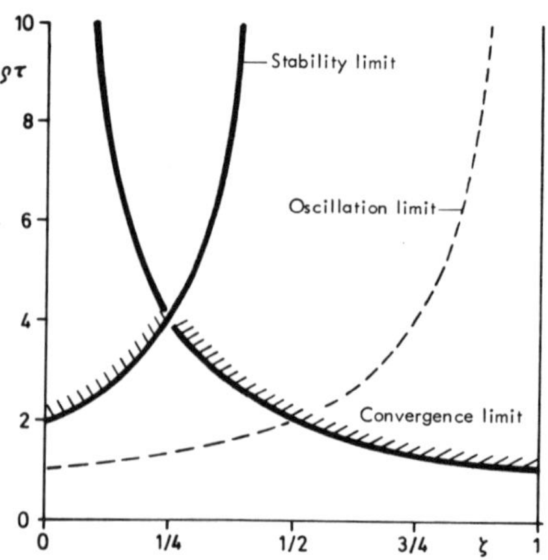

FIG. 9. Limitation of time increment.

for ${}^{b}\boldsymbol{V}_i$. Convergence considerations developed in ref. 2 rely on the matrix $\boldsymbol{N}$ of (67). The spectral norm ρ of the matrix $\boldsymbol{N}$, (68), limits in this case the time interval in accordance with the rule

$$\tau < \frac{1}{\zeta \rho^{b}} \qquad \text{for convergence} \tag{73}$$

We observe in Fig. 9 the growing importance of the convergence limit with increasing ζ, as against the stability limit.

In order to overcome the aforementioned limitation imposed on the time step when solving (71) and (72) successively, a simultaneous method of solution may be applied. To this end, we recall that the variable geometry $\boldsymbol{X} = {}^{b}\boldsymbol{X}$ of (66) may be expressed via (65) in terms of the unknown velocity $\boldsymbol{V} = {}^{b}\boldsymbol{V}$. In this way, only ${}^{b}\boldsymbol{V}$ appears in (66) and may be determined via the Newton procedure described in Section 2. In the present case, $\boldsymbol{F}$ in (43) is that of (66) and accordingly the gradient matrix of the system is expressed as

$$\boldsymbol{G} = \frac{\mathrm{d}\boldsymbol{F}}{\mathrm{d}\boldsymbol{V}} = \frac{\partial \boldsymbol{F}}{\partial \boldsymbol{X}} \frac{\mathrm{d}\boldsymbol{X}}{\mathrm{d}\boldsymbol{V}} + \frac{\partial \boldsymbol{F}}{\partial \boldsymbol{V}} = \zeta \tau \boldsymbol{F}_{,X} + \boldsymbol{F}_{,V} \tag{74}$$

in which use is made of (65) when relating the variable geometry to the unknown velocities. The velocity at the end of the time interval is computed in accordance with (45), and the associated geometry via (65), before the subsequent iteration cycle can be started. The outlined procedure may alternatively be stated in terms of the unknown geometry. Furthermore, a convenient estimate of the gradient matrix may be obtained in conjunction with the quasi-Newton or BFGS technique described in Section 2.2. For this reason, the constituents of the gradient matrix $\boldsymbol{G}$ in (74) are identified below and yield also the constituent matrices of $\boldsymbol{N}$ in (67) which are required for an appraisal of stability and convergence. To this purpose, we obtain by a differentiation of (66),

$$\boldsymbol{F}_{,V} = \frac{\partial \boldsymbol{F}}{\partial \boldsymbol{V}} = \frac{\partial \boldsymbol{S}}{\partial \boldsymbol{V}} = \bar{\boldsymbol{C}} + \tilde{\boldsymbol{C}} \tag{75}$$

in which we use (47), and

$$\boldsymbol{F}_{,X} = \frac{\partial \boldsymbol{F}}{\partial \boldsymbol{X}} = \frac{\partial \boldsymbol{S}}{\partial \boldsymbol{X}} - \frac{\partial \boldsymbol{R}}{\partial \boldsymbol{X}} = \boldsymbol{K}_G - \boldsymbol{K}_L \tag{76}$$

In (76) the matrix

$$\boldsymbol{K}_G = \frac{\partial \boldsymbol{S}}{\partial \boldsymbol{X}} \tag{77}$$

reflects changes of the stress resultants with varying geometry at constant velocity. The matrix $\boldsymbol{K}_G$ which corresponds to constant stress is known as the geometric stiffness of the discretised body. Its definition is independent of the particular material, be it viscous as in the present case or elastic as in ref. 5. Also the matrix

$$\boldsymbol{K}_L = \frac{\partial \boldsymbol{R}}{\partial \boldsymbol{X}} \tag{78}$$

accounts for changes of nonconservative loads acting on the body, as a result of the varying geometry [6].

Substitution of (76) and (75) in (74) yields the gradient matrix of the Newton method in the form

$$\boldsymbol{G} = \bar{\boldsymbol{C}} + \tilde{\boldsymbol{C}} + \zeta\tau[\boldsymbol{K}_G - \boldsymbol{K}_L] \tag{79}$$

which may conveniently be approximated for application within a quasi-Newton technique. Furthermore, substitution of (76) and (75) in (67) leads to

$$\boldsymbol{N} = [\bar{\boldsymbol{C}} + \tilde{\boldsymbol{C}}]^{-1}[\boldsymbol{K}_G - \boldsymbol{K}_L] \tag{80}$$

Expression (80) determines the numerical properties of the approximate integration in time.

A typical uniaxial specimen under constant compressive stress is shown in Fig. 10 and appears well suited for a verification of the proposed algorithmic schemes. By virtue of (32) one obtains first an expression governing the rate of change of length $\dot{l}/{}^{\circ}l$. Its subsequent numerical integration relies on (65) and yields $l/{}^{\circ}l$. The illustrated computations presume a constant μ and demonstrate the influence of the parameter ζ on the accuracy of the numerical results. Stability and convergence are examined in conjunction with the scheme of Fig. 9 assuming $\rho = \sigma/3\mu = \text{const}$. The predicted limits for the onset of oscillatory and unstable behaviour of the solution may easily be verified. Furthermore, the convergence limit of the method of successive solutions as presented in (71), (72) may also be investigated. In the current linear case the simultaneous treatment via the Newton procedure furnishes of course the solution immediately.

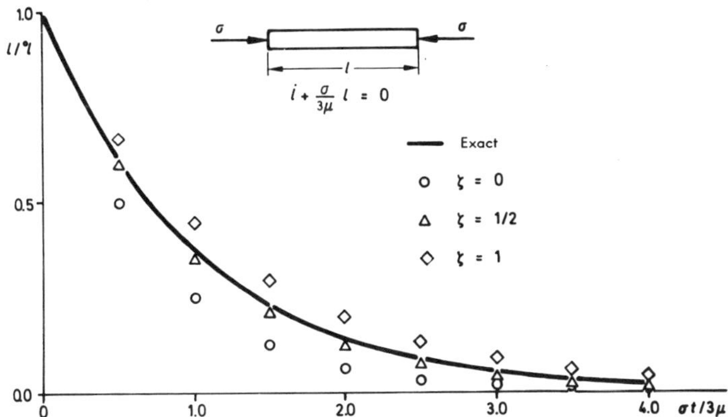

FIG. 10. Accuracy of numerical integration in a uniaxial case.

3. APPLICATIONS

3.1. Superplastic Bulging of Circular Membranes

We first discuss the free forming of circular membranes using the developed algorithms and comparing with the experimental data of ref. 7.

Following Fig. 11 two materials are examined, a titanium alloy and a stainless steel. The lower part of the figure shows the initial flat geometry of the membranes, their discretisation by four-node axisymmetric elements based on a reduced integration of their volumetric response and the boundary conditions. Radial motion is suppressed on the periphery, also motion against the die is not allowed. The relevant constitutive relations are quoted in the figure and correspond to non-Newtonian viscous behaviour. The upper part of Fig. 11 demonstrates the configurations achieved after 30 min. The corresponding forming pressures increase as indicated in Fig. 12. The use of auxiliary surface elements facilitates the application of the pressure loading on the lower face of the membrane.

In Fig. 13, the distribution of the logarithmic strain across the thickness of the membrane is shown for the two materials, and compared with the experimental data of ref. 7. It is seen that numerical results for the titanium alloy yield better agreement with experimental data when not only the periphery but the entire edge contacted by the die is assumed not to move radially (rigid edge). We also observe that

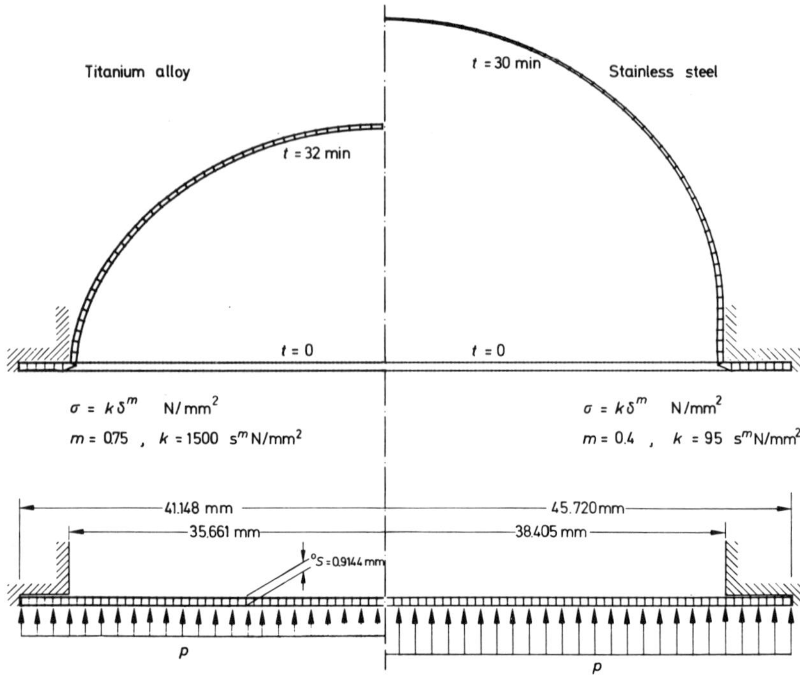

FIG. 11. Superplastic bulging of circular membranes.

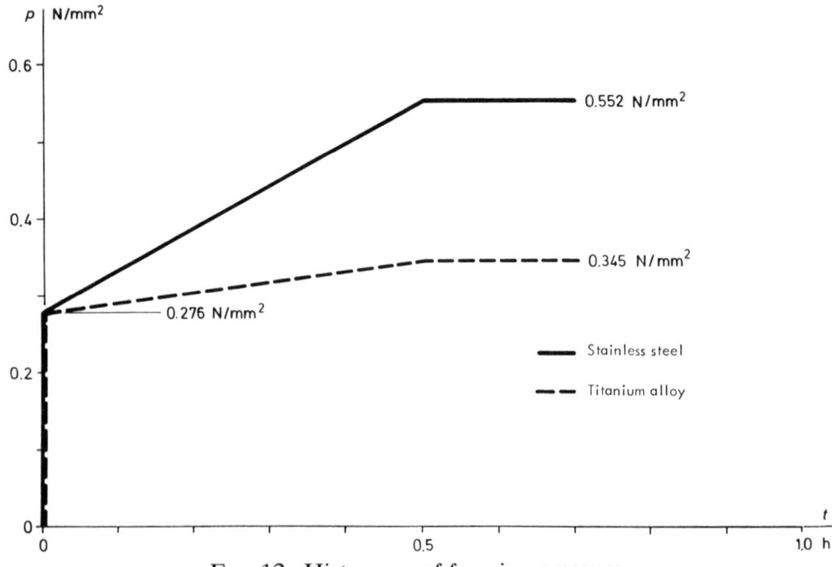

FIG. 12. Histogram of forming pressure.

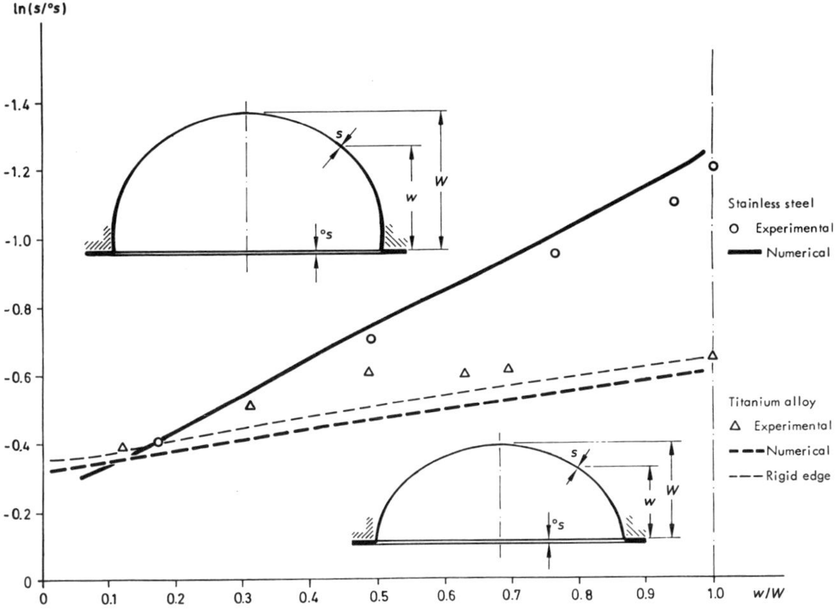

FIG. 13. Variation of thickness strain.

the single element positioned around the axis of revolution requires, in contrast to the other elements, application of the complete integration rule. This prevents extensive thinning in the cases investigated here.

3.2. Superplastic Sheet Forming onto Different Dies

We now consider variations of the first example. They concern the forming of a sheet onto dies of alternative shapes and involve frictionless, non-steady contact between the deforming sheet and the die. All forming processes are investigated for a titanium alloy, the constitutive assumptions being identical with those given in Fig. 11 of the preceding example.

The die considered first has a spherical form. Description of this problem and results of the computations are presented in Fig. 14. On the left-hand side of the figure, the initial geometry of the flat sheet and the geometry of the spherical die are illustrated. The discretisation of the sheet by finite elements corresponds to that in the first example. Radial motion is suppressed on both the periphery of the circular sheet

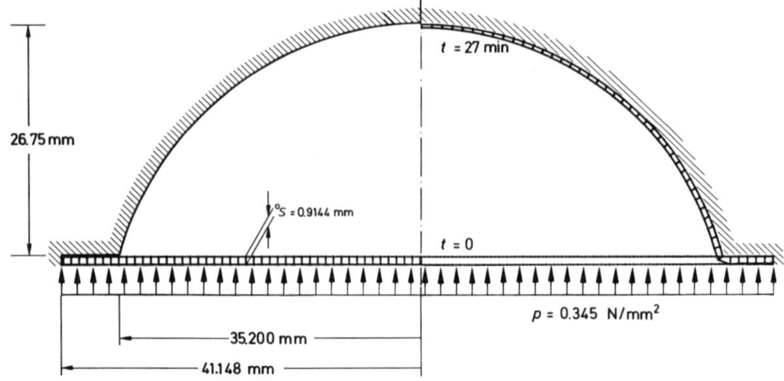

FIG. 14. Superplastic forming of a sheet onto a spherical die (titanium alloy).

and the part initially in contact with the die. Maximum pressure is applied in the present case right from the start of the computations. In the course of the forming process the sheet contacts the inner surface of the die and is subsequently allowed to slide along it without friction. Full contact with the die is achieved after 27 min of forming, as indicated on the right-hand side of Fig. 14 which illustrates also the

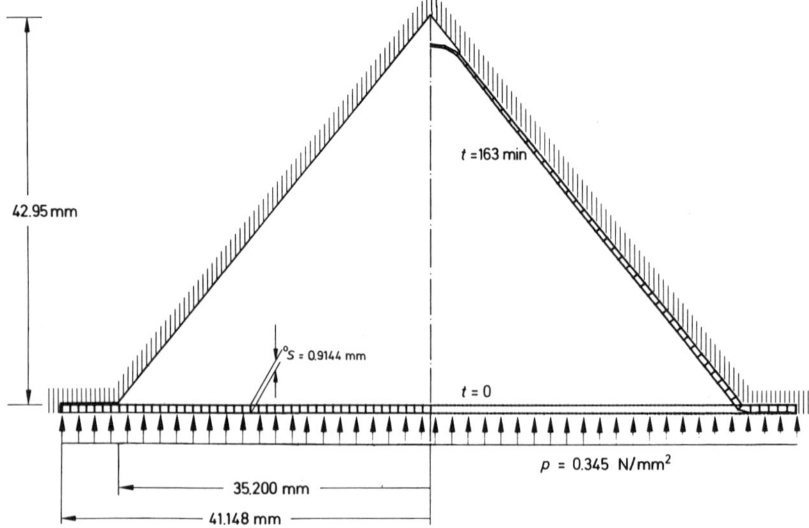

FIG. 15. Superplastic forming of a sheet onto a conical die (titanium alloy).

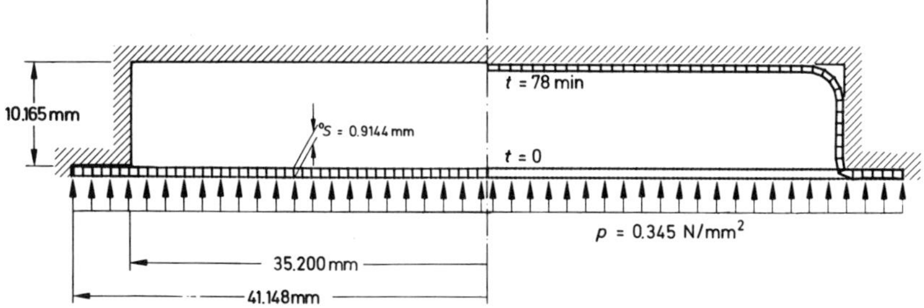

FIG. 16. Superplastic forming of a sheet onto a cylindrical die (titanium alloy).

variation of sheet thickness at the ultimate state as against the initial uniform thickness of the material.

Figure 15 demonstrates an alternative forming process onto a conical die. The left-hand side of the figure presents the data of the problem and details the die geometry. The final configuration following 163 min of forming time is depicted on the right-hand side of Fig. 15. The wide variation of the sheet thickness in the deformed state is clearly indicated.

As a final variant we consider in Fig. 16 the forming of the sheet onto a cylindrical die. The shape of the die is defined on the left-hand side of the figure whilst the right-hand side shows the deformed sheet after 78 min of process time. It is remarkable that the thickness of the sheet is almost constant on the top of the ultimate configuration.

3.3. Superplastic Forming of a Structural Component

The last example considered here was first investigated in ref. 1 on a different constitutive foundation and is taken from industrial practice in Aerospace Engineering. The complex forming process is illustrated in Fig. 17 and demonstrates the manufacture of a conical structural component starting with the virgin material in the form of a circular plate. The superplastic forming process requires a period of several hours and the application of a low pressure on the preheated Ti-6Al-4V material which deforms slowly towards the die. While the inner surface of the component is subjected to the forming pressure, its outer surface progressively comes into contact with the die and slides along it.

The geometry of the undeformed material is specified in Fig. 18 together with the geometry of the die. This figure demonstrates also

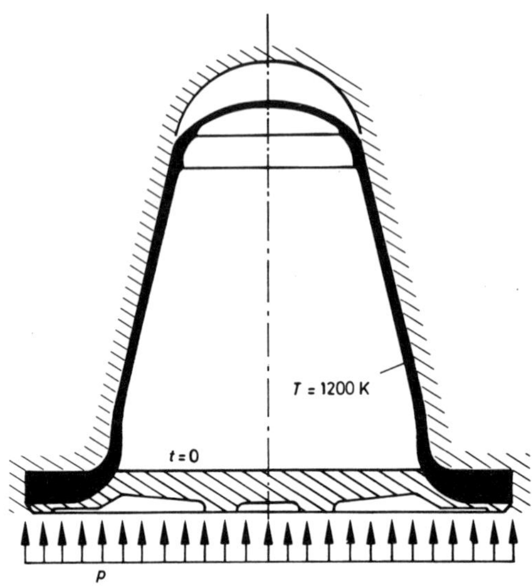

FIG. 17. Superplastic forming of a structural component in Ti-6Al-4V alloy.

the discretisation with four-node axisymmetric finite elements based on a reduced integration of the volumetric response. However, in order to prevent extensive local thinning (cf. ref. 1), a complete integration rule is applied to the single element positioned on the axis of revolution. Auxiliary surface elements are used for the application of the pressure on the lower face of the plate, which becomes the inner surface of the ultimate component. The time history of the applied pressure is depicted in Fig. 19. The pressure attains its final intensity of $1{\cdot}4\ \text{N/mm}^2$ within 4 h.

The uniaxial properties of the Ti-6Al-4V superplastic alloy at the temperature of operation of $T = 1200$ K are described in refs. 8 and 9. Figure 1 indicates the variation of the flow stress as against the rate of deformation obtained from tests at constant grain size. The data are approximated by the power series expression (7) taken to the fourth order with coefficients varying with the grain size. The required viscosity coefficient is obtained via (34) for given values of the equivalent rate of deformation $\bar{\delta}$ and of the grain size d. The experimental grain growth kinetics of the material are reproduced in Fig. 2 for an initial grain size of ${}^{\circ}d = 6{\cdot}4\ \mu\text{m}$. In the evolution of the deformation the grain size is determined by integration of (13) starting with the grain size ${}^{\circ}d$

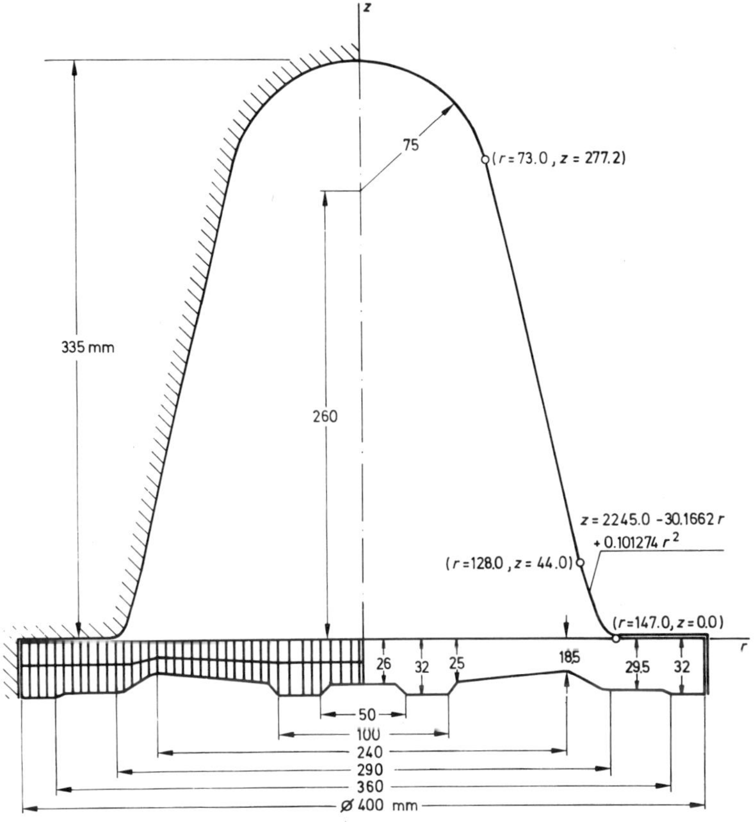

FIG. 18. Geometry and finite element discretisation.

and an initial time $^{\circ}t = 10$ min. Following ref. 8 the exponent N is taken to vary in (13) as

$$N = 1{\cdot}8(\bar{\delta} + 0{\cdot}00005)^{0{\cdot}237}$$

where the dimension of the equivalent rate of deformation $\bar{\delta}$ is s^{-1}.

The numerical analysis of the superplastic forming process is performed using time increments of $\tau = 600$ s in conjunction with an explicit integration scheme, $\zeta = 0$. For an initial grain size of $^{\circ}d = 9\ \mu$m results are first obtained assuming that the material slides along the inner surface of the die without friction. Figure 20 shows the shape of the component at different stages of the forming process. Full contact

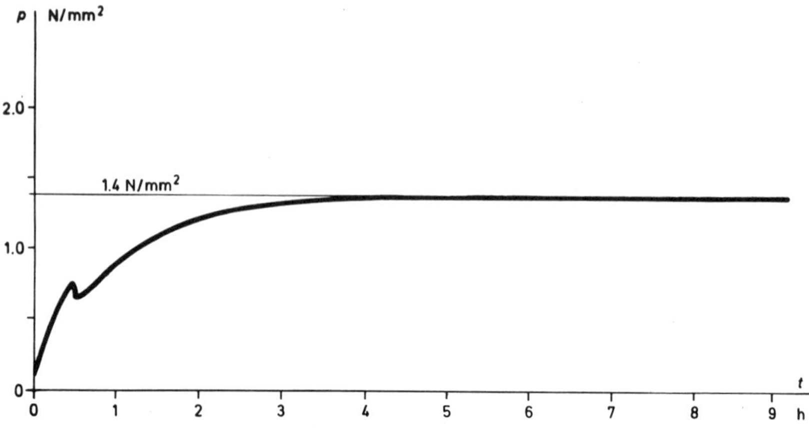

FIG. 19. Time history of forming pressure.

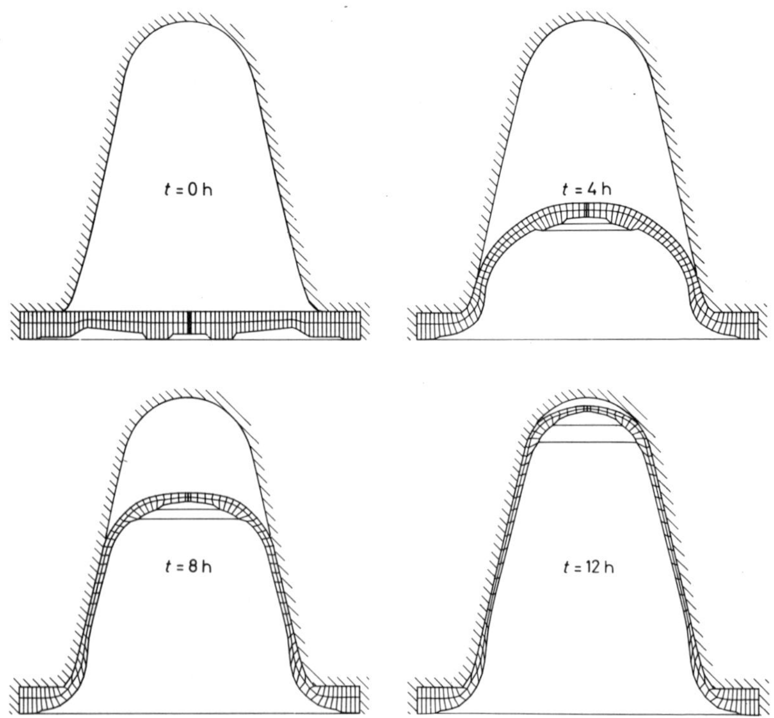

FIG. 20. Different stages of forming process (no friction).

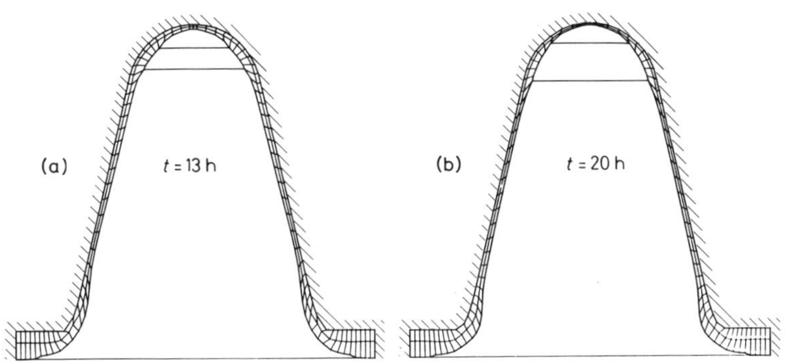

FIG. 21. Comparison of final shapes. (a) Without friction; (b) with friction.

with the given contour is attained after a process time of $t = 13{\cdot}0$ h which is in full agreement with the experimental evidence. In an additional computation, Coulomb friction between the deforming material and the die is accounted for, assuming a frictional coefficient $c = 0{\cdot}3$. Friction has a considerable influence on the forming process, full contact with the contour now requiring a time of $t = 20$ h. Also the shape, or rather the thickness distribution of the component, differs

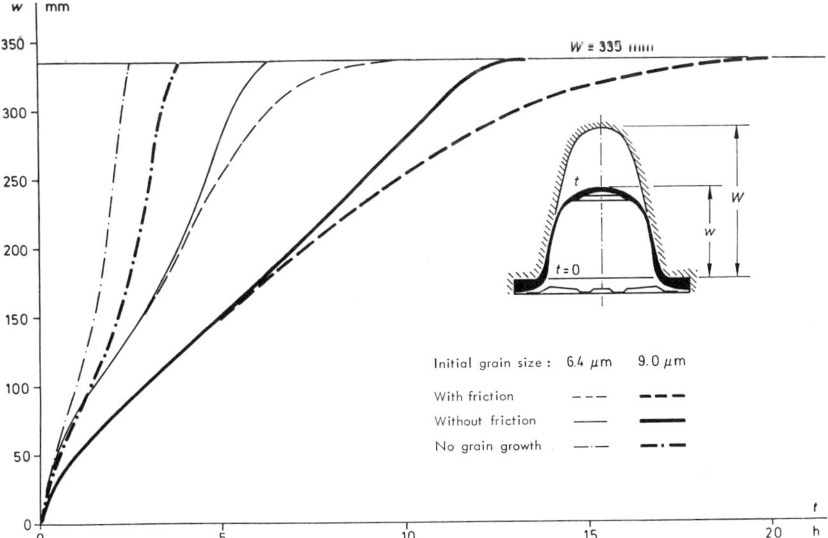

FIG. 22. Elevation of the apex of the specimen during the process of forming.

now from that of the component formed without friction. This is clearly demonstrated in Fig. 21 by comparison of the respective final shapes. In Fig. 22, the elevation of the apex of the component is depicted as a function of the process time. Besides the indicated influence of friction, the effect of an imposed constant grain size (taken to be $d = {}^{\circ}d = 9\ \mu m$) is also demonstrated in the figure for the frictionless case. The figure summarises also the results of an additional investigation for an initial grain size of ${}^{\circ}d = 6{\cdot}4\ \mu m$.

REFERENCES

1. ARGYRIS, J., J. ST. DOLTSINIS and H. WÜSTENBERG. Analysis of thermoplastic forming processes—Natural approach, *Computers and Structures*, **19** (1984).
2. ARGYRIS, J. and J. ST. DOLTSINIS. A primer on superplasticity in natural formulation, *Computer Meth. Appl. Mech. Engng*, **46** (1984), 83–131.
3. ARGYRIS, J. and G. MARECZEK. Finite element analysis of slow incompressible viscous fluid motion, *Ing. Archiv*, **43** (1974), 92–109.
4. ARGYRIS, J., J. ST. DOLTSINIS, W. C. KNUDSON, J. SZIMMAT, K. J. WILLAM and H. WÜSTENBERG. Eulerian and Lagrangian techniques for elastic and inelastic deformation processes, TICOM 2nd Int. Conf., Austin, Texas, 1979. In: *Computational Methods in Nonlinear Mechanics* (Ed. J. T. Oden), North-Holland, Amsterdam, 1980, 13–66.
5. ARGYRIS, J., P. C. DUNNE, TH. ANGELOPOULOS and B. BICHAT. Large natural strains and some special difficulties due to nonlinearity and incompressibility in finite elements, *Computer Meth. Appl. Mech. Engng* **4** (1974), 219–278.
6. ARGYRIS, J., K. STRAUB and SP. SYMEONIDIS. Static and dynamic stability of nonlinear elastic systems under nonconservative forces—Natural approach, Fenomech '81, *Computer Meth. Appl. Mech. Engng*, **32–34** (1982), 59–83.
7. CORNFIELD, G. C. and R. H. JOHNSON. The forming of superplastic sheet metal, *Int. J. mech. Sci.*, **12** (1970), 479–490.
8. GHOSH, A. K. and C. H. HAMILTON. Mechanical behaviour and hardening characteristics of a superplastic Ti-6Al-4V alloy, *Metall. Trans.*, A, **10A** (1979), 699–706.
9. GHOSH, A. K. and C. H. HAMILTON. Influences of material parameters and microstructure on superplastic forming, *Metall. Trans.* A, **13A** (1982), 733–743.
10. MALKUS, D. S. and T. J. R. HUGHES. Mixed finite element methods—reduced and selective integration technique: a unification of concepts, *Computer Meth. Appl. Mech. Engng.*, **15** (1978), 63–81.
11. MATTHIES, H. and G. STRANG. The solution of nonlinear finite element equations, *Int. J. Num. Meth. Engng*, **14** (1979), 1613–1626.

12. ODEN, J. T. RIP-Methods for Stokesian flows. In: *Finite Elements in Fluids*, Vol. 4 (Ed. R. H. Gallagher *et al.*), John Wiley, New York, 1982.
13. PADMANABHAN, K. A. and G. J. DAVIES. *Superplasticity*, Springer, Berlin, 1980.
14. TAYLOR, R. L. and O. C. ZIENKIEWICZ. Mixed finite element solution of fluid flow problems. In: *Finite Elements in Fluids*, Vol. 4 (Ed. R. H. Gallagher *et al.*), John Wiley, New York, 1982.

40

Common Defects in the Processing of Metals and Composite Materials

W. JOHNSON

Formerly Department of Engineering, University of Cambridge, Cambridge, UK

and

A. G. MAMALIS

Department of Mechanical Engineering, National Technical University of Athens, Athens, Greece

ABSTRACT

A wide range of physical defects which occur in various metal working processes is surveyed. These defects are treated descriptively but theoretical plasticity, e.g. slip-line field theory and upper bound techniques, are employed to explain some of them. The major defects arising during fabrication and further processing of composite materials are also surveyed. The kinds of composites reviewed are fibre-reinforced materials, metal–matrix composites, ceramic–matrix composites and bonded (clad, rolled) materials. The list of defects described is not exhaustive but the phenomena treated exemplify the practical value of knowledge of the principles of metal plasticity theory and the difficulties encountered in composite material processing.

1. INTRODUCTION

Researchers into the mechanics of metal processing are mostly devoted to finding the load to perform an operation. This is indeed important for work-scheduling, i.e. deciding which presses or machines are capable of applying any requisite force or power. However, there are occasions when other features may be even more important and deserving of attention. Such is the case concerning the defects and the limitations to processing which can and do arise in products; certainly, until very recently, they have tended to draw little attention from academic workers even though their economic consequences can be great.

The range of material working and fabrication defectiveness embraces [1]:

(i) The occurrence of defects due to interaction between the workpiece material, the tooling, the friction between the latter and the process-geometry:
(ii) some forms of metallurgical structure which result from purely mechanical action;
(iii) the limits of performance imposed by the material properties themselves with a given tooling and stressing system;
(iv) elastic springback and generated residual stresses.

Interaction of the above-mentioned features during material processing makes it difficult to account precisely for the defects met in terms of the mechanics: certain defects are associated with particular processes whilst some defects are peculiar to some materials.

In the following sections a wide range of defects which occur during various metal working processes and defects arising during the processing of composite materials is documented. Table I, most of which is reproduced from ref. 2, lists the major defects that arise in both bulk and sheet metal forming operations, whilst Table II, reproduced from ref. 3, presents a separate list of some surface defects that are encountered in either one particular process or a number of them. In Table III the principal defects associated with composite material fabrication and processing, as reviewed in ref. 4, are summarised.

Readers are referred to the work of Johnson and Mamalis [2] and Johnson and Ghosh [4] for an extensive list of references to the above mentioned defects. Further treatments of defects in metal processing can be found in various metal working books (see those listed in ref. 2 and also refs. 5 and 6) and in the papers of conferences considering defects due to metal working (see for example refs. 7 and 8).

Metal working defects of metallurgical origin, i.e. pipes, seams and segregation, etc., in ingots, are not reviewed in this paper but an extensive treatment of such defects can be found in the book by Engel and Klingele [9].

TABLE I
Common physical metal processing defects [2]

ROLLING
Flat- and Section-Rolling
Edge cracking
Transverse – fire cracking
Alligatoring (crocodiling)
Fish-tail
Folds, laps
Flash, fins
Laminations
Ridges – spouty material
Ribbing
Sinusoidal fracture
Zippering
Cross-, Transverse- and Helical Rolling
Central cavity (axial or annular fissure)
Overheated ball bearing
Roll mark
Folding (or seaming), laps, fringes
Squaring
Necking
Triangulation and triangular fish-tail
Ring-rolling
Cavities
Fish-tail
Edge cracking
Straight-sided forms

FORGING
Open- and Closed-Die Forging, Upsetting, Indentation
Longitudinal cracking
Hot tears and tears
Edge cracking
Central cavity
Centre bursts
Cracks due to t.v.ds and thermal cracks
Folds, laps
Flash, fins
Laminations
Orange peel
Shearing fracture
Piping
Rotary Forging
Mushrooming
Central fracture
Flaking
High Energy Rate Forging
Piping
Dead metal region
Laps
Turbulent metal flow

EXTRUSION-PIERCING
Christmas tree (fir tree)
Hot-, cold-shortness
Radial and circumferential cracking
Internal cracking
Central burst (chevrons)
Piping (cavity formation)
Sucking-in
Corner lifting
Skin inclusions (side and bottom of the billet)
Longitudinal streaks
Laps
Laminated fractures
Mottled appearance
Extrusion defect
Impact Extrusion
Multiple tensile 'necks'
Thermal break-off

DRAWING OF ROD, SHEET, WIRE AND TUBE
Internal bursts (cup and cone chevron)
Transverse surface cracking
Chips of metal
Poor surface finish
Folding and buckling
Fins, laps
Chatter (vibration) marks
Season cracking
Island-like welding

TABLE I (*contd.*)

DEEP-DRAWING Wrinkling Puckering Tearing (necking) Edge cracking Orange peel Stretcher-strains (Lüders lines) Earing	Eyes, ears, warts, beards, tongues
BENDING AND CONTOUR FORMING Cracking Wrinkling Springback	SPINNING, FLOW TURNING, SHEAR FORMING Springback Wall-fracture (shear splitting and circumferential splitting) Wrinkling Buckling Back-extrusion (over-reduction) Under-reduction
HOLE FLANGING Lip formation Petal formation Plug formation	PEEN FORMING, BALL FORMING Overlapping dimples Orange peel Surface tearing Break-up of surface grains Intergranular cracking Wrinkling Microfissures Folds
BLANKING AND CROPPING Distortion of the part (doming and dishing) Cracking Martensitic lines	

TABLE II
Surface defects in massive and sheet metal forming [3]

Alligator skin	Overlapping (laps)
Blisters	Peeling (orange peel, pebbles)
Burnt surface	Pickle pitting
Checks marks	Pinchers
Cold laps	Pinheads (blisters)
Cold shut	Pitting
Crowfeet (check marks)	Ragging marks
Dark areas	Roakes
Die lines	Rolling mill laps
Edge checks	Rough surface
Fishhooks (check marks)	Scratches
Folds	Seams
Galls	Shearing defects
Grease spots	Slivers
Greasy surface	Snakes
Hair lines (cracks)	Spangled surface (mottled surface)
Hair seams (seams)	Stretcher strains
Mottled surface	White spots on steel ingots

TABLE III
Physical defects in composite material fabrication [4]

Defect	Stage
FIBRE-REINFORCED PLASTICS	
Incomplete impregnation of fibre Incomplete cure of resin Poor wetting and subsequent poor adhesion of fibre to matrix Bubbles Voids Delaminations Broken strands Loose ends of fibres Knotted strands Wrinkled strands and crevices Crazing cracks Local resin-rich areas	Manufacture
Severe delamination Concealed cuts Rupture of resin starved layers	Forming
Fibre pull-out Fibre–matrix debonding	Tensile loading
Splitting Buckling	Compressive loading
Transverse cracking of fibres Parallel splitting of laminates	Static and dynamic piercing
Delamination Translaminal cracking perpendicular to fibres Spalling	Impact
Surface flaws (step, hole, ripple, branch, fissure, crack) of whiskers	
METAL–MATRIX COMPOSITES	
Incompatibility of fibre and matrix Poor wettability Reaction between fibre and matrix Inadequate percolation of the matrix material to properly surround the fibres Voids Porosity Matrix–filament debonding	Manufacture
Axial densification Density gradients Delaminations Filaments to break Filament buckles Formation and flattening of ribbons	Powder compaction
Flaky area of carbon fibre	Fibre-coating

TABLE III (*contd.*)

METAL–MATRIX COMPOSITES (*contd.*)	
Void formation at the poles of fibres Transmatrix cracks Decohesion of a matrix–fibre interface Brooming or crushing of component ends	Forging
Transverse cracking Rough 'tree-bark' surface finish Delaminations Fibre rupture as multiple necking	Rolling
Transverse cracking Rough 'tree-bark' surface finish Delaminations Burst Split into bundles of metal-coated fibres Microcracks	Extrusion
Bamboo shape defect Breakage of continuous fibres or filament Voids Debonding	Drawing of bar, rod and wire
Filament breakage Splitting	Bending
Brittle fracture	
SANDWICH, CLAD/BONDED MATERIALS	
Poor bonding Bond breakage Warping at high temperatures Edge delamination Destruction of the flyer plate	Manufacture
Wrinkling Earing Wall fracture Delamination or bowing	Deep-drawing and bending
Surface defects	Abrasive wear
COATED MATERIALS	
Poor coating Thinning and chipping at the corners and edges Bubbles and specks due to excessive pickling Fish scale, ruptures, flakes, peeling off due to absorbed hydrogen between coating and base material Severe oxidation Substrate Voids Porosity Overspraying	Manufacture

TABLE III (*contd.*)

COATED MATERIALS (*contd.*)	
Surface marks	Sheet forming
Wrinkling	Sheet forming
Earing	Sheet forming
Springback	Sheet forming

2. MASSIVE FORMING

2.1. Rolling Processes

Successful rolling practice requires a careful balance of such factors as the control of temperature, intermediate annealing, soundness of the ingot, control of scaling, lubrication, the condition of roll surfaces and angular speed and mill stiffness [2].

The common principal defects arising in rolling processes are summarised in Table I and an extensive study of those defects is given in ref. 2. Some of these defects occurring in conventional and non-conventional rolling processes are outlined below.

Failures occurring during such compression-like working processes are predominantly caused by so-called 'secondary' tensile stresses. Fracturing is observed in the rolling of slabs at the edges (edge-cracking) where longitudinal tensions (or strains) may develop under certain rolling conditions or along the centre-plane leading to alligatoring.† As pointed out [2], the formation of internal or surface cracks during the rolling of heavy ingots in a blooming mill, when the reduction per pass is small, is basically encouraged by inhomogeneous deformation.

Surface defects induced during hot and cold flat and section rolling have been reported; see refs. 10–13. In ref. 13 the types of defects encountered in the cold rolling of strip and sheet are defined and usefully classified.

Edge cracking was examined by Schey [14] and three causes adduced for its occurrence, viz. limited ductility, variation of the stresses across the width of the rolled material and uneven deformation at the edges. In subsequent passes, 'overhanging' material is not directly compressed but forced to elongate and therefore subjected to longitudinal tensile stresses leading to cracking. The initiation and propagation

† This is fracture at about mid-thickness level. Reference 35 is much concerned with centre-plane through thickness fracture, called in-zippering.

of edge cracks in cold rolling have been examined by Dodd and Boddington [15], but see also the work on edge cracking in sandwich rolling reported in ref. 16.

Unwanted end shapes can be developed particularly in the rolling of slabs and blooms and can have important consequences for production costs. The form encountered at the front end of an initially square-ended ingot or slab after rolling is termed overhang and that at the rear when well developed combines overlap and fish-tail. (The former is due to folding over of the ingot head and the latter to tail end-folding in the direction of its thickness for overlap whilst the width effect produces the fish-tail) (see Figs 1a and 1b). These ends are cut off, constituting a crop-loss and may account for about 5% of a total throughput of bloom or slab weight. Even a small average percentage reduction in the crop loss of the annual volume of production of a mill leads to substantial material and monetary savings. Some experimental results and useful conclusions about these end defects as a result of experiments with Plasticine are given in refs. 17 and 18. Clearly, losses are capable of reduction by changing ingot bottom shape and by starting with a convex or cone-bottom ingot [19]. N.K.K., Japan are reported to have reduced their crop losses substantially after modelling the phenomenon with Plasticine. Slip-line fields and upper bound solutions for overlap developments have been suggested [1], [18].

In rolling a round or a section a type of fold develops if a flash or fin is formed in one pass and pressed into the metal during the subsequent pass [2]. The promotion of these defects is aided by overfilling or

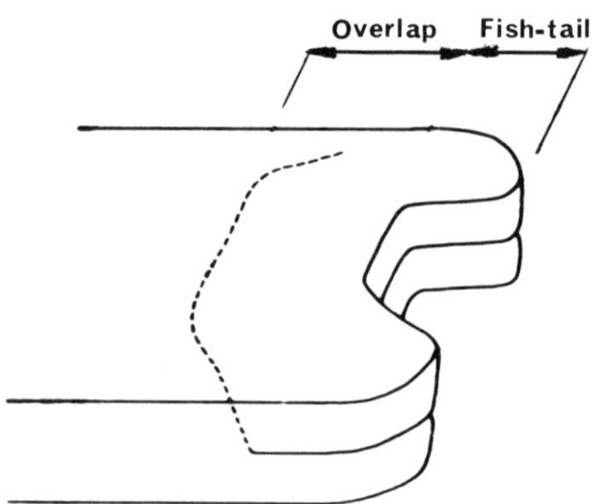

FIG. 1a. Definitions of overlap and fish-tail in slabbing.

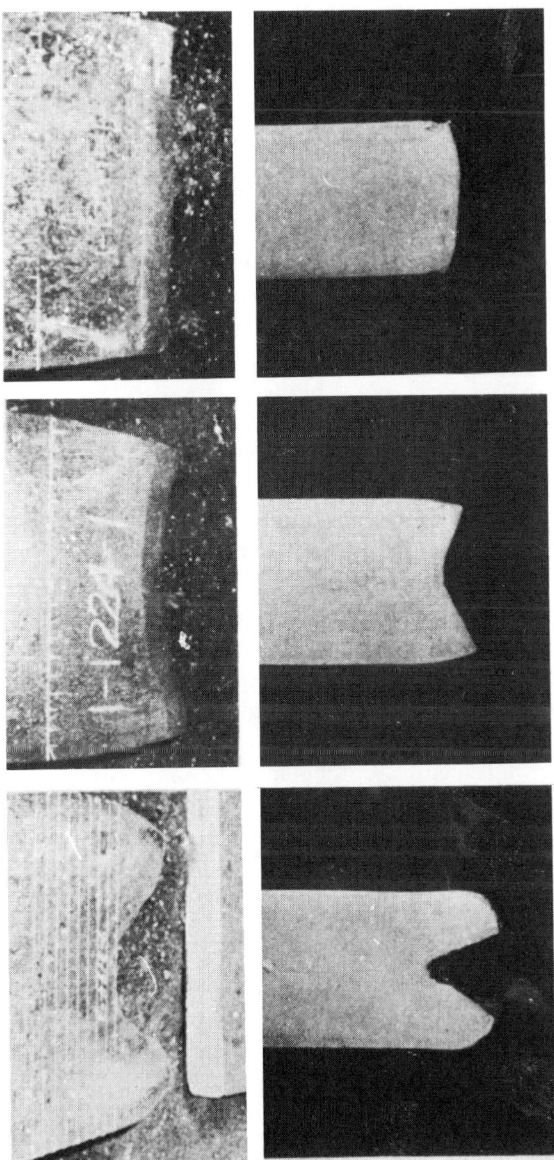

FIG. 1b. Overlap development with reduction in actual slabs (left) and plasticine models (right) [17].

underfilling passes. In overfilling—'too much metal enters the pass'—there is a tendency to cause finning into the roll gap and to result in rolled-in laps when the bar is turned through 90° on the next pass, whilst underfilling the pass may lead to later passes turning sideways ('drunken' passes). Perhaps most theoretical attention to profile rolling is that given by Tarnovski *et al.* [20], but see also Wusatowski [21].

Central cracking is a common defect in light pass rolling, wedge-rolling, helical or transverse-rolling and in rotary forming operations; an axial fissure is observed in two roll transverse rolling or an annular one in the three-roll process. Large inclusions in the central region of the billet strongly predispose the material to cracking. Another cause of cracking is believed to be excessive 'kneading' of a section after forming, due to excess metal being trapped in this section [2].

The reasons for axial cavity formation or annular fissure development may, to some extent, be understood with the help of plane strain slip-line field solutions for forging. The approach of opposed (two) rigid flat-ended indenters will tend to open (due to heavy shearing) any existing or incipient crack and to cause central voids to be formed where the cohesive strength of the structure is exceeded in the presence of plastic straining, especially if the ratio of the die width to distance apart is sufficiently small. (This geometry will cause tensile stresses towards the middle of the product.) Nasmyth introduced his vee-anvil to avoid this defect (though in fact the defect location may simply be shifted) in effect by using three indenters or dies. Many investigations into central defects have been reported in the USSR and principal contributions to the literature are briefly listed in ref. 2.

When rolling profiled T-shaped rings at large total reductions of ring

Fig. 2. Cavity formation in rolling profiled rings [22].

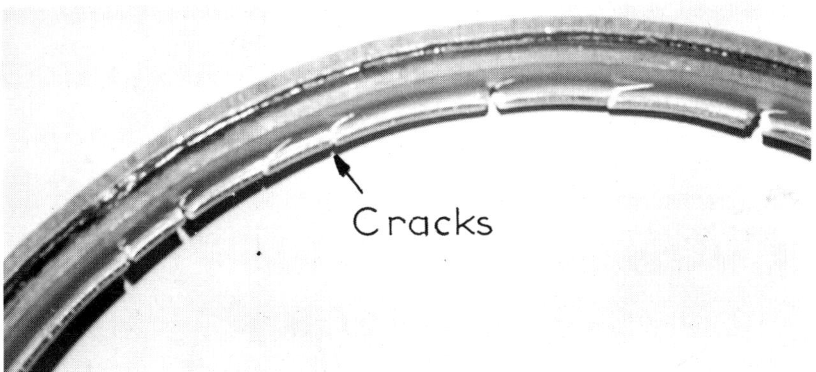

FIG. 3. Edge cracking in L-shaped ring-rolling (after Dong Yol Yang; unpublished work).

wall thickness, it was found that for certain rolling conditions a specific form of cavity formation arises; separation of the material from the main roll occurs where initially it was in contact (see Fig. 2). Accounting for this defective ring-rolling phenomenon is attempted together with some quantitative assessments by analogy with the piping defect well known in extrusion [22].

Fish-tail formation, i.e. an irregular and non-rectangular spread profile with concave edges (see Fig. 2), and edge cracking (see Fig. 3) are defects also developed in ring-rolling.

2.2. Forging

Cracking, folds, laps and improper sections are the three main groups of defects observed in forged products (see Table I and ref. 2). Some of these defects are outlined below.

Among other factors governing surface and internal cracking are the ductility of the material, temperature and stress field (and the magnitude of the local hydrostatic pressure) around the position of the crack.

Longitudinal cracking on the curved surface of a billet when upsetting is due to the presence of secondary tensions. Thomason's work [2] studying longitudinal surface defect influence on ductility in cold heading and upsetting is noteworthy; very usefully, an index of surface quality for the cold heading of wire was proposed; see also ref. 23.

A centre-cavity may be formed in a forged round billet similar to that already mentioned concerning cross-rolling, because of the development of tensile stresses. The introduction of reliable criteria for the onset of void formation or crack initiation in ductile metals, in regions of complex stress, should soon make it possible to calculate or anticipate the site of creation of these costly internal fractures in metal forming processes; see the work by Kuhn [24].

To combat centre-bursting in open-die forging, flat dies may be replaced by curved or so-called swaging dies which introduce favourable lateral compression; refer to the remarks on Nasmyth above. The extrusion-forging of materials possessing a limited ductility is, therefore, more likely to be successful than closed-die forging, which in turn is superior to open-die forging [2].

Further, when using dies of appropriate shape to develop a system of

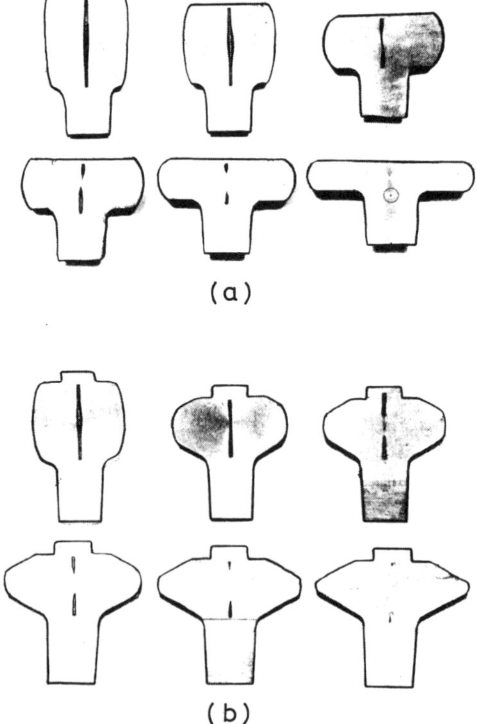

FIG. 4. Closing of cavities in forging between (a) flat tools and (b) dished tools [2].

compressive stresses, it is possible to close existing internal cavities (e.g. shrinkage cavities in ingots) during upsetting operations and to produce good material properties at the healed position; this has been investigated by Tomlison and Stringer ([2] and [36]) (see Fig. 4). Dies with dished working faces have been suggested.

Folding due to buckling is a common mode of failure in upsetting and heading operations, e.g. in upsetting the head of a cartridge case. However, buckling is not the only reason for folds and laminations.

Folds at the edges of hammer-forged products originate frequently from too small reductions per pass when local deformation occurs near the surface layers. (Compare this with the rolling of ingots mentioned earlier.) In a subsequent heavy reduction, the metal may spread from both surfaces over the centre part and form folds or laminations. Folds (or laps) are also formed whenever metal folds over itself during die forging, i.e. a flash or fin is formed in one stroke and pressed into the metal during a subsequent one [2].

2.3. Extrusion, Piercing and Wire Drawing

A comprehensive survey of defects occurring during extrusion and piercing is given in the book by Johnson and Kudo [25]; slip-line field and upper bound techniques are extensively used to throw light on the phenomena occurring. Cracking (surface and internal), sinking-in or contracting and skin-inclusion defects are the three main groups into which the defects may be classified. A detailed account of extrusion and wire drawing defects is given in ref. 2 (see also Table I). Some of the reported defects are outlined below.

Central-bursts (or chevrons), i.e. internal arrow-shaped defects, are occasionally encountered in the cold extrusion and wire drawing of round bars of steel. An upper bound approach to formulate a central-burst fracture criterion has been made by Avitzur [26]; see also the work by Lee and McMeeking [27].

A well-known defect occurring at the rear end of extruded products is that of a sinking-in of the material, i.e. the beginning of piping or cavity formation, on the bottom of a slug; it occurs during the final unsteady state phase of extrusion when the slug thickness has become greatly reduced (see Fig. 5). Upper bounds for the prediction of this defect have been suggested [25].

It was shown by Hill [28] that a defect can be created by indenting strip resting on a rigid foundation with a wedge-shaped or a flat-ended die. Adapting the slip-line field used by Hill, Dodd and Kudo [1] have

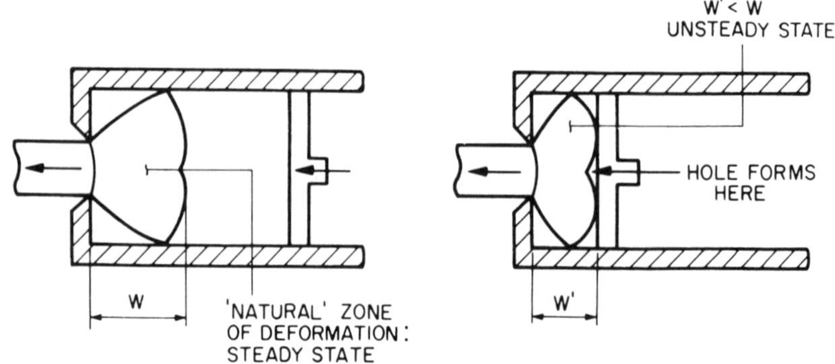

FIG. 5. Extrusion through square dies showing how the steady state and the unsteady state of deformation depend on billet length.

recently demonstrated how a sizing problem may arise in a tube container extrusion situation. (Compare also with similar cavity formation occurring in ring-rolling mentioned above.)

Defects characterised by the inclusion or surface skin or layers in a billet body (see Fig. 6) are examined in detail in ref. 25, where associated explanatory slip-line field solutions are given.

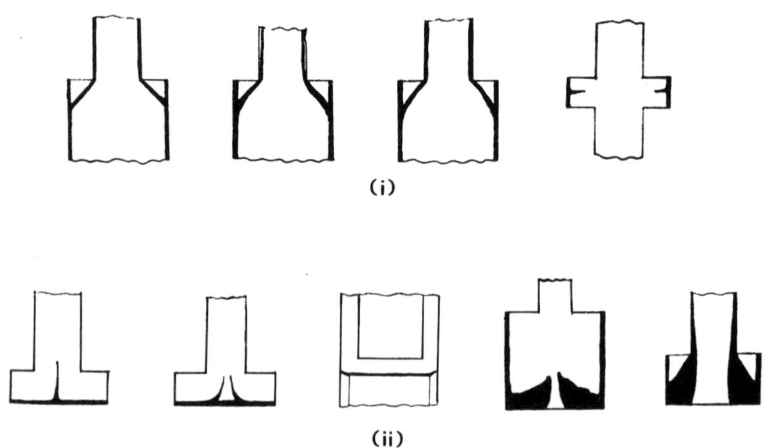

FIG. 6. Surface skin of layer inclusions from (i) billet side and (ii) billet bottom [25].

3. SHEET METAL FORMING

There are three types of end point or limiting condition for sheet metal transversely loaded over some large fraction of its area and transmitting biaxial tensile stress in its own plane in some regions, in deep-drawing and pressing. These are [2]:

(1) wrinkling or compressive instability in regions of high compressive stress;
(2) local tensile instability in regions where a line of zero extension can exist;
(3) failure in regions of biaxial tensile strain.

Because sheet forming is largely a plane stress operation, the above three conditions may be associated with a portion of a yield surface as in Fig. 15 of ref. 2. Forming limit diagrams based on the dimensions of the initial blank and the tool have been suggested by various investigators.

The mechanical defects which arise during various sheet forming processes range from such common principal defects as springback, wrinkling, buckling, necking and fracture and surface marks to ones of specialised nature such as intergranular cavitation and void formation, shape inaccuracy defects in large automotive panels, decohesion and localised shear-band formation (see refs. 2 and 3, and Tables I and II). A selection of common defects in typical deep-drawn products is shown in Fig. 7, taken from the book of Eary and Reed [29]. These comprise, among other things, buckling due to large compressive strains, tooling marks, fractures due to excessive tensile stress and earing due to planar anisotropy.

Concerning sheet metal forming defects, the interested reader may consult the proceedings of the Biennial Congresses of the International Deep Drawing Research Group, where much work is devoted to the understanding of the mechanics of sheet metal forming; see for example ref. 3. (The book, *Metallurgy of Deep Drawing and Pressing*, by J. D. Jevons, Chapman and Hall Ltd, 1940, though now rather old is recommended for its three chapters on Defects and Difficulties.)

Springback when press-forming shallow automotive panels of high strength steel sheets is of the utmost importance. Restriking (a method of correcting or compensating for springback carried out by 'bottoming' the punch in the die so as to produce a coining action) components

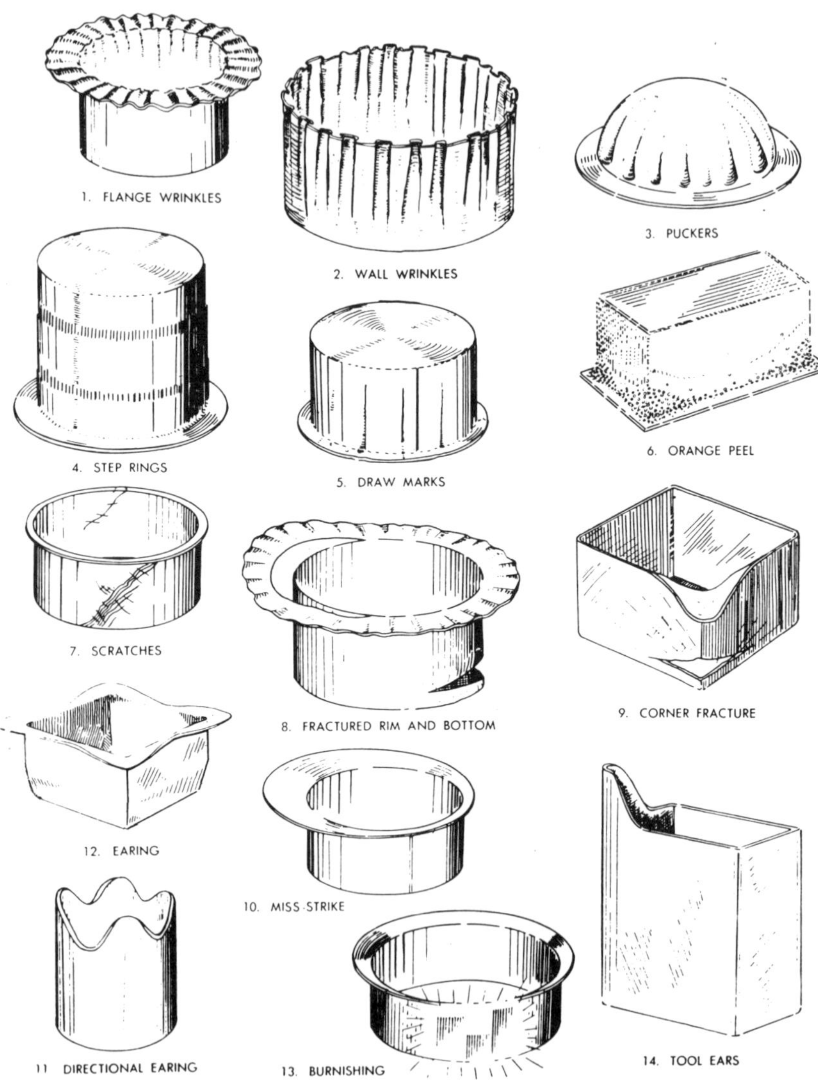

FIG. 7. Some deep-drawing defects [29].

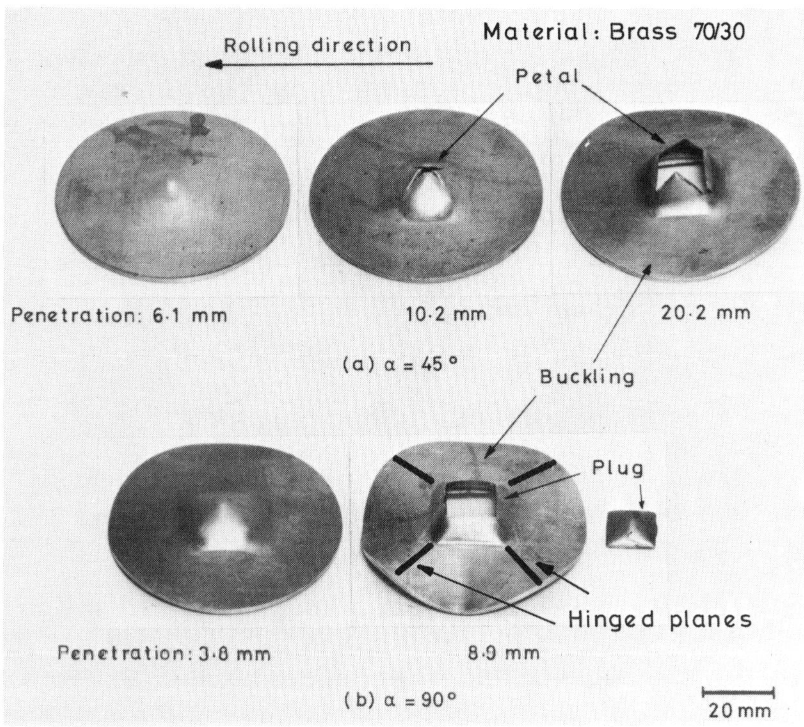

FIG. 8. Showing petal formation, plug formation and buckling of plate in deformed brass 70/30 specimens for a punch semi-angle (a) 45° and (b) 90° [32].

is shown to lower substantially the magnitude of springback.† As shown in ref. 3, springback is reduced with increasing blank-holder force in stretch-bending tests; see also ref. 30.

Relating to the shape inaccuracy of large panels possessing complex contours, such geometrical surface defects as surface warp, wrinkling and surface deflection have been reported [3] (see Table II). These defects are attributed to non-uniform stress distribution in the plane of the sheet; in addition, the inherent difficulties encountered in bending

† Heavy bottoming also confers closer adherence to die shape; see Johnson, W. and T. X. Yu, Experiments in the Cylindrical Bending of Metal Strips, *Metals Technology*, 1983 (in press).

a sheet of metal under tension were pointed to as major factors giving rise to shape fixing defects.

References 31 and 32 report on the failure modes obtained in the piercing and hole-flanging of sheet metals. These occur either by petalling, plugging or lip-fracture (see Fig. 8). In the context of the hole-expanding limit test, it has been pointed out that the above limit is generally lowered by the presence of microcracks at the hole-edge produced during the punching of a hole; there a strain concentration exists which causes early failure [3].

The common defects observed in the peen-/ ball-forming and shear forming processes are listed in Table I.

4. COMPOSITE MATERIAL FABRICATION

A wide range of common principal defects that arise in composite material production and subsequent forming is presented in ref. 4. The kinds of composites treated are:

(1) fibre-reinforced plastics,
(2) metal–matrix composites,
(3) clad/bonded materials and
(4) coated composites.

An extensive list of references is provided in ref. 4 that will furnish interested readers with a substantial background.

In Table III we have attempted to summarise the physical defects associated with composite material fabrication and forming as reported in Ref. 4.

Figure 9, reproduced from ref. 33, shows damage regions occurring in bands parallel to the direction of maximum shear stress during the forging of thorium-coated tungsten wires embedded in a fine-grain nickel-base superalloy composite.

The principal defect encountered in the deep-drawing and stretch-forming of coated sheets has been recognised to be the flaking or peeling off of the coat material. Flaking limit strain diagrams have been suggested [3]. Additionally, in forming coated sheets, especially into box-shaped panels, there is the possibility of a loss of coat material at the corners. Testing procedures to evaluate the adherence properties of coatings have been also suggested [3]; see also ref. 34.

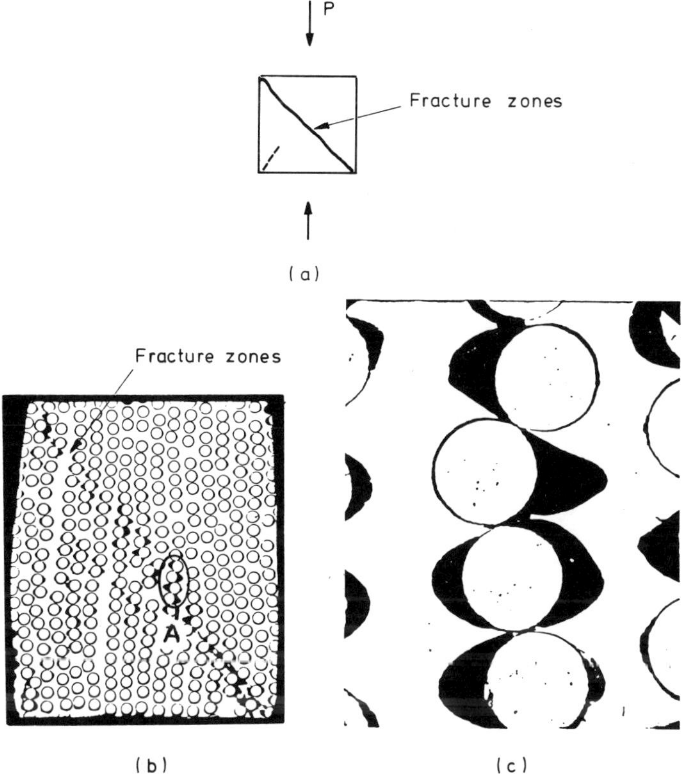

FIG. 9. Zones of severe damage in a section normal to the direction of fibre alignment for a forged nickel-base superalloy fibre-reinforced composite. (a) A schematic diagram of the fracture zones. (b) A micrograph of a polished specimen. (c) Magnified view of damaged region A [33].

5. CONCLUSION

This paper is written to give an interpretive and (tentatively) consolidated view of material processing defects, for the general interest of the metal working engineer, and for improving the academic teaching of the subject of material processing mechanics at a practical level. It is hoped that directing attention to a wide spectrum of defects will suggest to those concerned with developing new processes, principles for avoiding them. It is also our hope that the catalogue of defects described will quicken interest in creative and successful material processing practice and promote better interpretation through theoretical analyses.

6. ADDENDUM

Reference 35 became available after completing this paper.

REFERENCES

1. Johnson, W. A short review of metal forming mechanics and some processes of current research interest, *Proc. 3rd Int. Conf. on Mechanical Behaviour of Materials*, Cambridge 1979, Pergamon Press, Oxford, 1980, 167–225.
2. Johnson, W. and A. G. Mamalis. A survey of some physical defects arising in metal working processes, *Proc. 17th Int. Machine Tool Design and Research Conference*, Birmingham 1976, Macmillan, London, 1977, 607–621.
3. Ghosh, S. K. Principally on sheet metal forming defects as described in the eleventh biennial congress of the international deep-drawing research group, *Int. J. Mech. Sci.*, **22** (1981), 195–211.
4. Johnson, W. and S. K. Ghosh. Some physical defects arising in composite material fabrication, *J. Mater. Sci.*, **16** (1981), 285–301.
5. ASM., *Metals Handbook*, 8th edn, Vol. 4 Forming, Metals Park, Ohio, 1973.
6. U.S. Steel, *The Making, Shaping and Treating of Steel*, 9th edn, Pittsburgh, 1971.
7. *54th Symposium on Metals Plasticity* (Defects in Metal Working Processes), Papers 1–6, Osaka, Japan, 1976.
8. *4th International Conference on Fracture*, Fracture 1977 (Ed. D. M. R. Taplin), Waterloo, Canada, 1977.
9. Engel, L. and H. Klingele, *Rasterelektronmikroskopische Untersuchungen von Metallschäden*, Carl Hanser Verlag, Müchen, 1974. (Available in English as, *An Atlas of Metal Damage*, Wolfe Science Books. London 1981, pp. 271.)
10. Bading, W., P. Funke and T. Kootz, Einfluss der Stichplangestaltung und der Rohbrammenabmessungen auf das Ausbringen beim Brammenwalzen, *Stahl und Eisen*, **97** (1977), 1307–1314.
11. Malygin, R. Z. *et al.* Einfluss der Erwärmung und Verformung auf die Tiefe von Oberflächenfehlern beim Walzen von legiertem Stahl, *Stal*, **38** (1978), 156–157.
12. Maddison, R. Surface Defects Induced during Hot Rolling of Steel Rod, Ph.D. Thesis, University of Bradford, England, March, 1978.

13. GRÜNHOFER, H. G. *et al. Oberflächenfehler an kaltgewalztem Band und Blech*, Verlag Stahleisen, Düsseldorf, 1967.
14. SCHEY, J. A. Prevention of edge cracking in rolling by means of edge restraint, *J. Inst. Metals*, **94** (1966), 193–200.
15. DODD, B. and P. BODDINGTON. The causes of edge cracking in cold rolling, *J. Mech. Working Technology*, **3** (1980), 239–252.
16. AFONJA, A. A. and D. H. SANSOME. Edge cracking in sandwich rolling, *J. Mech. Working Technology*, **3** (1979), 77–83.
17. IKUSHIMA, H. *et al.* Plasticine Model Test on Slabbing-Shape of Slab Ends, Nippon Kokan Technical Report-Overseas, July 1975, 21–29.
18. (a) MIHARA, Y. and W. JOHNSON. Crop loss: front and back end deformation during slab and bloom rolling, *Metallurgia and Metal Forming*, **44** (1977), 332–339.
(b) STAHLBERG, U., J. O. SODERBERG and A. WALLERO. Overlap at the back and front end in slab ingot rolling, *Int. J. Mech. Sci.*, **23** (1982), 243–252.
19. PARKINS, R. N. *Mechanical Treatment of Metals*, Allen and Unwin Ltd, London, 1968.
20. TARNOVSKI, I. J., A. N. SKOROHODOV and B. M. ILYKOVICH. *The Elements of Theory of Complex Profile Rolling*, Metallurgy, Moscow, 1972.
21. WUSATOWSKI, Z. *Fundamentals of Rolling*, Pergamon Press, Oxford, 1969.
22. MAMALIS, A. G., J. B. HAWKYARD and W. JOHNSON. Cavity formation in rolling profiled rings, *Int. J. Mech. Sci.*, **17** (1975), 669–672.
23. JENNER, A. and B. DODD. Cold upsetting and free surface ductility, *J. Mech. Working Technology*, **5** (1981), 31–43.
24. KUHN, H. A. Forming limit criteria-bulk deformation processes. In: *Advances in Deformation Processing* (Eds. J. J. Burke and V. Weiss), Plenum Press, New York, 1978, 159–186.
25. JOHNSON, W. and H. KUDO. *The Mechanics of Metal Extrusion*, Manchester University Press, Manchester, 1962.
26. AVITZUR, B. Analysis of central bursting defects in extrusion and wire drawing, *Trans. ASME, J. Engng Industry*, **90** (1968), 79.
27. LEE, E. H. and R. M. MCMEEKING. Concerning analysis of central burst in metal forming, *Trans. ASME, J. Engng Industry*, **100** (1978), 386–387.
28. HILL, R. On the mechanics of cutting metal strips with knife-edged tools, *J. Mech. Phys. Solids*, **1** (1953), 265–270.
29. EARY, D. F. and F. A. REED. *Techniques of Press-working Sheet Metal*, Prentice-Hall, Englewood Cliffs, N.J., 1974.
30. JOHNSON, W. and A. N. SINGH. Elastic springback in circular blanks formed by bending with a hemi-spherical punch and die, *Metallurgia and Metal Forming*, **47** (1980), 275–280.
31. JOHNSON, W., N. R. CHITKARA, A. H. IBRAHIM and A. K. DASGUPTA. Hole-flanging and punching of circular plates with conically headed cylindrical punches, *J. Strain Analysis*, **8** (1973), 228–241.
32. JOHNSON, W. and A. G. MAMALIS. The perforation of circular plates with four-sided pyramidally-headed square-section punches, *Int. J. Mech. Sci.*, **20** (1978), 849–866.

33. MAMALIS, A. G., W. WALLACE, A. KANDEIL, M. C. DE MALHERBE and J.-P. A. IMMARIGEON. Spread and fracture patterns in forging fibre-reinforced composites, *J. Mech. Working Technology*, **5** (1981), 15–30.
34. LAWRENZ, K. J. Untersuchungen über das Tiefziehen kunststoffbeschichteter Stahlbleche, *Industrie Anzeiger*, **99** (29), (1977), 515–516.
35. SHERBY, O. D., S. DE JESUS, T. OYAMA, and E. MILLER. Metal forming defects at large strain deformation at intermediate temperatures. In: *Plasticity of Metals At Finite Strain*, (Ed. E. H. Lee and R. L. Mallett), Division of Applied Mechanics, Stanford University, 3–15.
36. CHAABAN. A. A. and J. M. ALEXANDER. A study of the closure of cavities in swing forging, 17*th M.T.D.R. Conf.*, Macmillan, 1977, 633–646 (see also 623–632).

41

Limit Analysis and Design: An Up-to-date Subject of Engineering Plasticity

M. A. SAVE

Faculté Polytechnique, Mons, Belgium

ABSTRACT

Though the theories and methods of plastic limit analysis and optimal design are now well developed and widely known, much remains to be done to arrive at the completion of practical tools of engineering plasticity. In most cases, theoretical progress is still to be complemented in order to achieve practical applicability and evaluate the influence of assumptions or neglected facts. To illustrate this point of view, we first outline some examples of such progress in optimal design of grids, and in limit analysis and optimal design of circular plates and cylindrical shells with stepwise varying thickness. We then point out various aspects of the solutions that necessitate complementary studies, most often with some experimental part. This situation should be faced, maybe through some kind of international co-ordination of applied research in the field, if plasticity today is to become engineering plasticity tomorrow.

1. INTRODUCTION

Though it can be agreed that 'the plasticity theory is well developed regarding:

(a) the mathematical description of the rate independent behavior;
(b) methods of solutions concerning perfect plasticity; and

(c) applications of perfect plasticity in metal forming and structural engineering'.†

and that progress in the theory of plasticity today necessitates consideration of various couplings, cyclic and dynamic effects, anisotropy etc., it is believed that the situation in engineering plasticity is quite different, and that much remains to be done by 'plasticians' if they want to see the results of their work become useful practical engineering tools. For this reason, the present paper aims at reviewing some important progress made during the last decade in classical plastic limit analysis and optimal plastic design, with some emphasis on more recent (unpublished) works of the author and his co-workers, and discussing the conditions of practical applicability of the considered results.

Section 2 is devoted to optimal plastic design. It describes generalizations of optimality criteria, including layout optimization with particular reference to the work of Prager and Rozvany. Typical applications of beams, grids, arch-grids and reinforced concrete plates are briefly quoted.

Section 3 is devoted to circular metal plates, metal cylindrical shells and reinforced concrete cylindrical shells. It first considers limit analysis of Tresca plates with piecewise constant thickness, for various loadings and boundary conditions. The plate with reinforcing rings is treated as a particular case, and inadequacy of the 'smearing-out' technique is pointed out. Minimum-volume design is then considered, for both sandwich and solid plates, with given or with optimized division radii.

Limit analysis of axisymmetrical loaded cylindrical Tresca sandwich shells is presented, where reinforcing rings with rectangular or T cross-section are treated with due account for their large curvature.

Vertical cylindrical tanks filled with liquid and made of rings of piecewise constant thickness are studied, and a practical design procedure is suggested. Finally, cylindrical reinforced concrete shells designed for minimum-volume of reinforcing steel are discussed.

In the fourth section, we reconsider some of the examples treated in the two first parts, and discuss the practical applicability of the solutions obtained. It is shown that a variety of supplementary information is needed, before there is enough confidence for these results to be used in practice.

† Plasticity today: First announcement.

Subjects to be clarified, experimentally in many cases, are: effects of big increase of the thickness, of concentration of beams in continuum-like optimal grids and in optimal solid plates, service behavior and post-limit behavior of optimal plastic designs, especially of reinforced concrete plates and shells, instability of compressed ribs, etc. We conclude that, in parallel with newer and more sophisticated problems of theoretical plasticity, classical plastic limit analysis and optimal design and their connected problems, remain important aspects of today's engineering plasticity, and we advocate international co-ordination of structural engineering research in this field.

2. OPTIMAL DESIGN FOR ASSIGNED LOAD FACTOR AT PLASTIC COLLAPSE

Initiated by the pioneering paper of Drucker and Shield in 1956 [1], optimal limit design has since made enormous progress: optimality criteria and examples of applications successively included: multiple loading [2] and movable loads [3–7], generalized convex [8, 9] and non-convex [10] cost functions (instead of simply volume or weight), piecewise varying (or partially pre-assigned) design variables and upper and lower bounds on the design variables [11] (so-called technological constraints) and optimal location of discontinuities or supports [12]. Simultaneously, layout optimization of concrete plate reinforcement, grids and arch-grids developed very rapidly, mostly due to Prager and Rozvany [12–15]. Most applications based on the optimality criteria so derived are analytical, even if computer-aided. Some plate problems have been tackled by a basically numerical approach (by finite elements) using the optimality criterion as the test for design improvement. Most numerical optimal plastic designs however have been treated by mathematical programming methods, resulting in a large number of scientific and technical papers [16–22]. Figures 1 and 2 show typical examples of minimum-volume designs.

Hence, we see that we can have presently at our disposal a large number of solutions or solution methods for optimal plastic design. It is thus worth trying to find why they are used so rarely in practice. Not only is it so for plastic optimal design, but also, very often, for simple plastic analysis or for one-variable design. This question will be reconsidered in Section 4, after some practically oriented plate and shell examples will have been treated in Section 3.

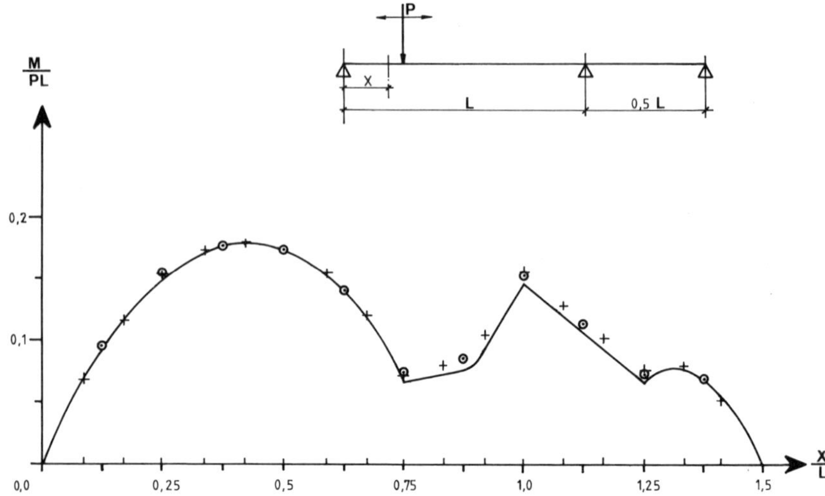

FIG. 1. Minimum-volume design for one movable load [7]. Distribution of the yield moments: ——, obtained analytically; ⊙, calculated by L.P. (11 critical sections); +, calculated by L.P. (17 critical sections).

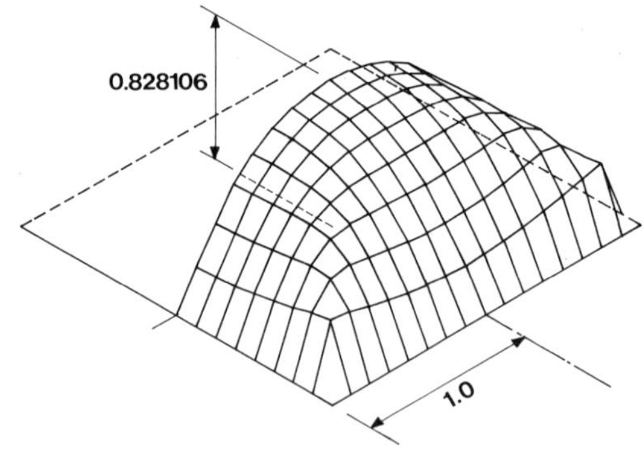

FIG. 2. Minimum-volume design of an arch-grid in which arches are subjected solely to axial forces under vertical loading [15].

3. CIRCULAR PLATES WITH PIECEWISE-CONSTANT THICKNESS

Axisymmetrically loaded and supported solid circular and annular plates with piecewise constant thickness have been treated in detail by Lamblin in his thesis [23]. The yield condition of Tresca is used, and computer-aided analytical complete solutions are obtained for any number of constituting rings, simply supported with built-in edges, under uniformly distributed annular loads and concentrated ring loads. Results are summarized in Fig. 3 in the case of the uniformly loaded built-in plate made of n equally wide rings. In the figure, M_0 stand for the yield moment, and $n = \infty$ corresponds to the physically orthotropic plate, as considered by Olszak and Sawczuk [24]. Figure 4 corresponds to the same example but with conditions of simple support. The theoretical predictions have been compared to earlier experiments by Save [25] on plates with 4, 8, 10 and 16 rings and new tests by Lamblin [23] on plates with only 2 or 4 rings. The general conclusions are as follows: if we denote by μ the 'slenderness ratio' of the plate obtained by dividing the diameter by the average thickness, experiments support the theory within a few per cent on the safe side for $20 \leqslant \mu \leqslant 30$. For larger values of μ, the 'knee' in the pressure-vs-deflection diagram tends to flatten, and cannot be used anymore for experimental definition of a limit pressure beyond $\mu = 40$. Nevertheless, the theoretical limit pressure is always smaller than the experimental pressure corresponding to a ratio of the central deflection to the diameter of 2%, up to $\mu = 45$. If we accept this deflection criterion as a substitute for the lacking experimental limit load as done in some codes [26], the theory is physically significant and acceptably accurate in the range $20 \leqslant \mu \leqslant 45$. In particular it shows that the 'smearing-out' technique that uses an equivalent orthotropic plate can be very much in error (of the order of 50% or more) either on the safe or the unsafe side (see Figs. 3 and 4), for small numbers of rings (4 and less). The theory can be used to determine how to rationally strengthen a circular plate, that is to increase its limit load as much as possible with the smallest volume of supplementary rings. This problem leads us naturally to that of minimum-volume design.

Starting with a sandwich plate with given division radii, Lamblin [23] has shown that collapse mechanisms can be found that satisfy the optimality criterion and correspond by the normality law to the stress regimes used in his limit analysis procedure. Hence, computer-aided analytical solutions could be obtained for the absolute minimum-volume designs of circular sandwich plates with various loadings and boundary conditions, the number of component rings ranging from 2 to 6. When the division radii were later optimized, or when increasing the number of rings to infinity in order to recover the design with continuously varying sheet thickness, it becomes necessary to set an upper bound on the yield moment in order to avoid creating very large

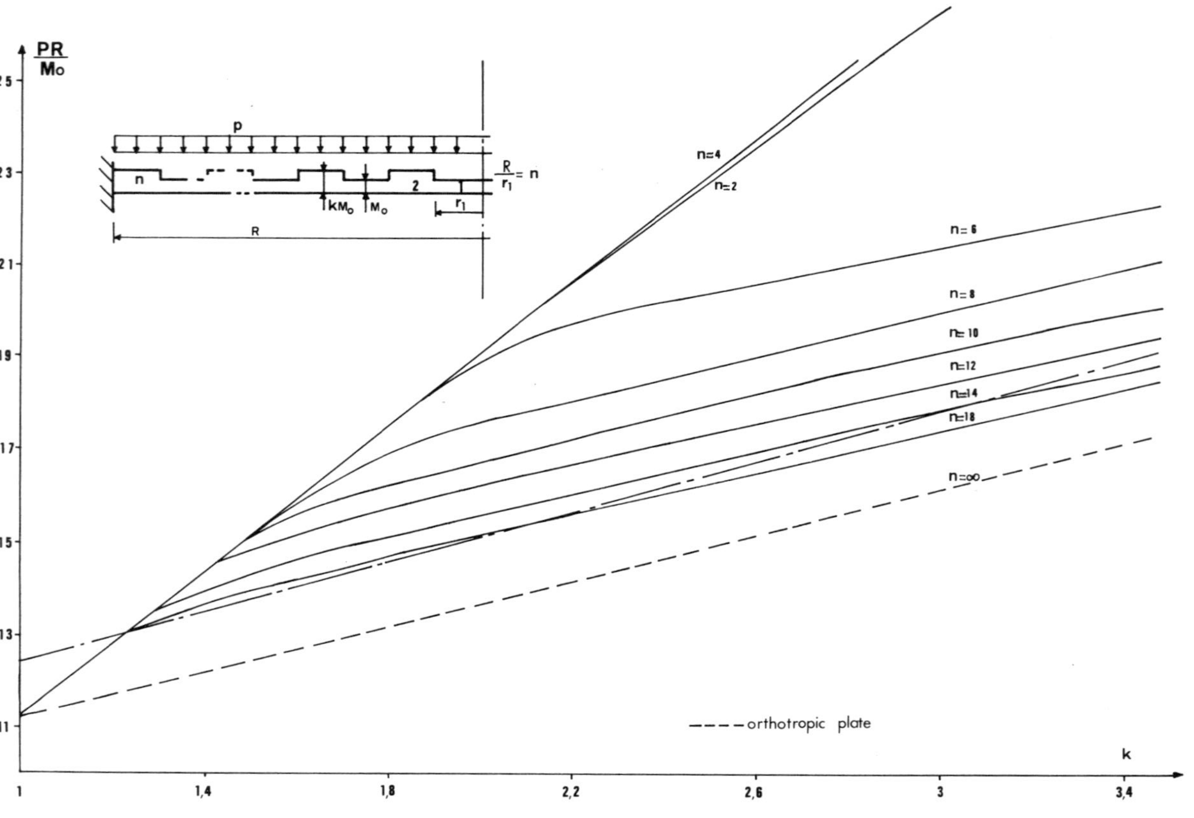

FIG. 3. Limit pressure of built-in circular plate [23].

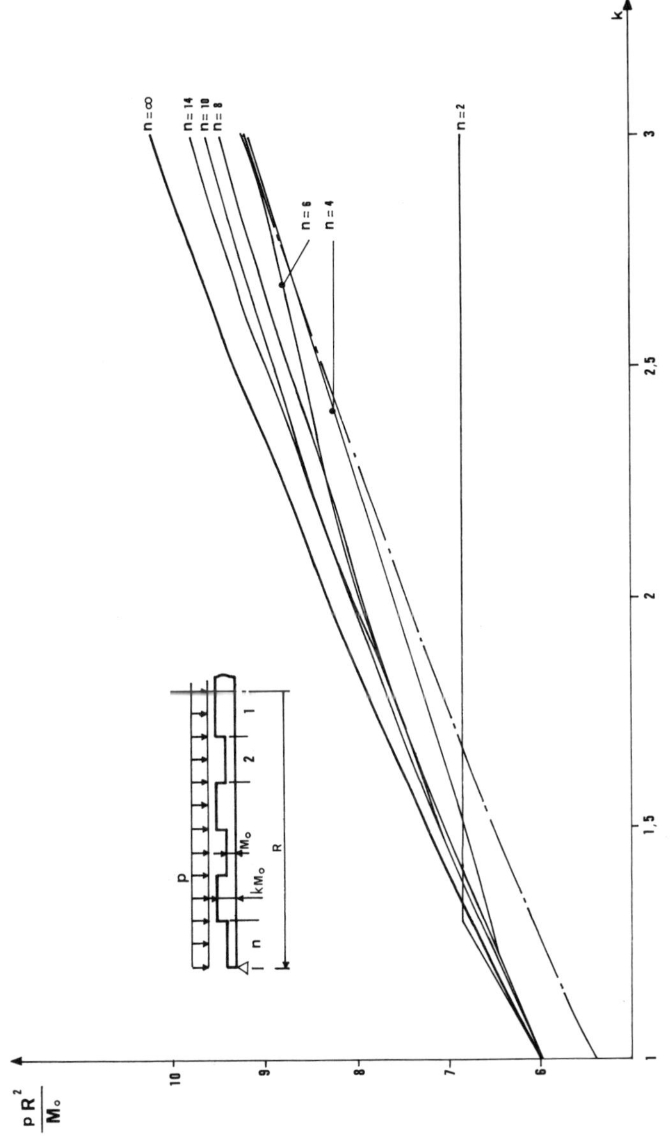

FIG. 4. Limit pressure of simply supported circular plate [23].

thickness in rings adjacent to central holes. Real metal plates however are not sandwich but solid plates, for which it is known that the minimum-volume criterion is but a necessary condition of local optimum. Let us nevertheless apply it to simple supported plates, and discuss later the usefulness of the results. Taking the yield moments M_{oi} in the various rings as the design variables, it can be shown that the minimum-volume distributions of M_{oi} of the sandwich plates with the same loading satisfy the optimality condition for the solid plate, provided the division radii are given. Obviously, the identical designs M_{oi} give different volumes for the solid and sandwich plates and, hence, different sets of optimal division radii. For example, in the case of the uniformly loaded circular plate made of two equal rings, we have $M_{o1}=0{\cdot}1875pR^2$ and $M_{o2}=0{\cdot}1458pR^2$. But if we optimize the division radius r_1, we find $r_1=0{\cdot}75R$, yielding $M_{o1}=0{\cdot}1901pR^2$ and $M_{o2}=0{\cdot}0963pR^2$ for the sandwich plate, whereas minimization of the volume of a solid plate with respect to r_1 gives $r_1=0{\cdot}806R$ and corresponding yield moments $M_{o1}=0{\cdot}1877pR^2$ and $M_{o2}=0{\cdot}0794pR^2$ (Hopkins and Prager [27]).

If the number of rings increases without limit, we recover the corresponding functions $M_o(r)$ obtained for sandwich plates, but the savings with respect to the constant-thickness plate are different; for example, it is 18% for the uniformly loaded solid plate, compared to 25% for the corresponding sandwich plate. Examples of circular plates with 2–6 rings, subjected to circular or annular loadings, can be found in Lamblin *et al.* [28]. As emphasized at the beginning of this section, the designs considered here furnish relative optima. Hence it may be expected that other designs exist that exhibit smaller costs. For example, a two-ring plate uniformly loaded and simply supported can be made of a very narrow and strong external ring in which the internal ring is 'built-in'. As shown by Kozlowski and Mroz [29], if one is prepared to accept $r_1>0{\cdot}97R$ and $M_{o2}>39{\cdot}5\ M_{o1}$ (!), a volume smaller than that of the classical design of Hopkins and Prager [27] can be achieved. If one recalls that this latter design saves only 6·87% of volume with respect to the constant thickness plate, whereas transforming the simple support of this plate into a built-in edge multiplies its limit load by 1·88 (see Save and Massonnet [30]), the result of Kozlowski and Mroz was to be expected, even with some cost assigned to the fixed support so obtained.

Another way of decreasing the volume is the use of stiffeners, as shown by these authors in the same paper. The volume can even be made theoretically to vanish with an infinite number of infinitely thin and infinitely high stiffeners, as first pointed out by Brotchie [31]. In order to take advantage of this possibility, we must on the one hand admit the use of functions with an unlimited number of discontinuities and on the other hand set an upper bound on the height of the stiffeners. Such a procedure is discussed by Rozvany *et al.* [32] and applied to the simply supported circular Tresca plate uniformly loaded. For example, setting the upper bound $M_o^+ = M_{oc}$, where M_{oc} is the yield moment of the solid plate with constant thickness and same limit load, it is found that the design with the smallest volume is formed of a solid central region ($0 \leqslant r \leqslant 0{\cdot}5014R$) with a variable thickness such that $M_o(r) < M_o^+$, surrounded by an infinity of radial beams (stiffeners without plate) of maximum height $t = (4M_o^+/\sigma_o)^{1/2}$, and of variable breadth such that $M_o(r) = M_r(r)$. The saving obtained is 35%, compared to 18% given by the design with a smooth function $M_o(r)$. However, without mentioning the machining cost, this material saving of 35% is likely to be reduced to some extent by the necessity of approximating the considered design with the use of a finite number of radial beams and, for application of the load, of a non-vanishing plate even in the outer region.

Consider now the uniformly loaded built-in plate. The same plastic regimes will be used as for the corresponding sandwich plate, but the radius r_o separating the regions of application of these two regimes is different because it is now dependent on the type of loading. Indeed, it is determined from kinematic continuity conditions applied to a collapse mechanism satisfying the optimality criterion. But the latter contains $M_o^{-1/2}(r)$, which in turn depends on the type of loading. Hence, the optimal designs of corresponding sandwich and solid plates are not identical any more. A successive approximation procedure similar to that used for the sandwich plate could be applied. In order to avoid these supplementary lengthy and tedious calculations, we shall satisfy ourselves with the approximations given by the optimal designs M_{oi} ($i = 1, \ldots, n$) of the sandwich plates. They are able to support the load but, as a rule, do not satisfy the optimality criterion of the solid plate and, hence, have a volume in excess of the (relative) minimum.

The benefit of these approximations to the optimal designs will be evaluated by reference to the exact minimum-volume design obtained in the case of continuous smooth variation of M_o with r by Onat *et al.* [33].

Uniform loading corresponds to $r_o = 0{\cdot}664R$. The volume is

TABLE I

Built-in solid plate uniformly loaded. $r_1, r_2, \ldots, r_{n-1}$ are optimal division radii for sandwich plate.

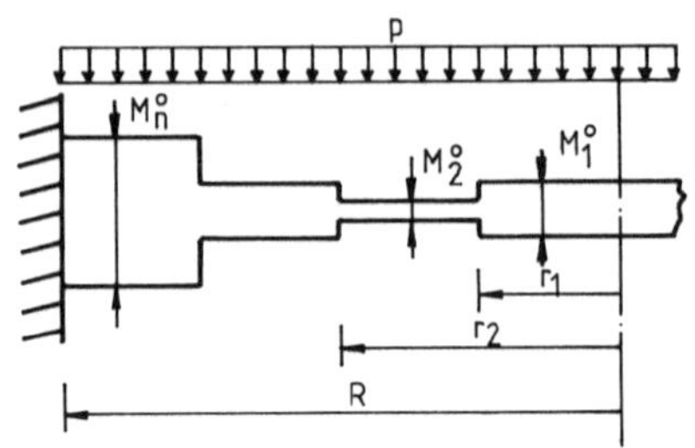

	Number of rings					
	1	2	3	4	5	∞
r_1/R	—	0·86	0·816	0·49	0·498	—
r_2/R	—	—	0·915	0·801	0·774	—
r_3/R	—	—	—	0·91	0·86	—
r_4/R	—	—	—	—	0·934	—
r_5/R	—	—	—	—	—	—
G %	0	6·95	9·26	11·9	13·1	23

evaluated by integration, with $t = 2(M_o/\sigma_o)^{1/2}$, and the saving is found to be 23%.

Tables I and II show that the design of approximate optimality (yield moments of the sandwich plate) gives very rapidly a large amount of maximum savings, either with the optimal division radii of the sandwich plate, or upon separate optimization of these radii by the grid method for the solid plate, a computation that proves thus not useful for practical application.

4. METAL CYLINDRICAL SHELLS

Though several books dealing with limit analysis of shells appeared after Hodge's *Limit Analysis of Structures* of 1959, the simply supported shell under uniform pressure and no end load was, surprisingly enough, not treated until 1975 [34]. Such 'forgotten' problems certainly slow down application of engineering plasticity. Also the treat-

TABLE II
Built-in solid plate uniformly loaded. Optimal division radii obtained by the grid method.

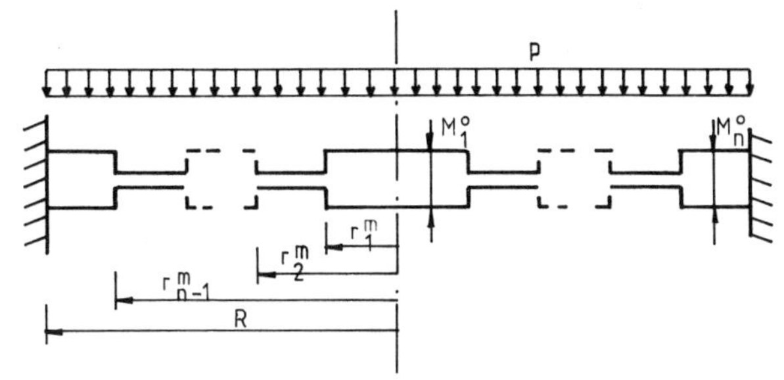

		Number of rings					
		1	2	3	4	5	∞
$x_i = \frac{r_i^m}{R}$	x_1	—	0·831	0·507	0·517	0·448	
	x_2	—	—	0·826	0·763	0·573	
	x_3	—	—	—	0·886	0·763	
	x_4	—	—	—	—	0·885	
	x_5	—	—	—	—	—	
$G = \frac{v_c - v}{v}$	%	0	7·1	9·33	12·15	13·19	23

ment of ring stiffeners was as a rule done by the smearing-out technique [35–39], which, as stiffened plates, will give bad results in the case of a small number of strong stiffeners. In ref. 34 and in subsequent papers with co-workers [40–42], Guerlement has given computer-aided analytical solutions of the limit behavior of (a) simply supported cylindrical shells with one central stiffening ring, with and without axial load; (b) infinitely long cylindrical shell with regularly spaced stiffening rings, with and without end-loads. The linearized Tresca yield condition is used, and due account is taken of the curvature of the stiffeners that are treated in a biaxial state of stress. Stiffeners of rectangular and T cross-section are studied, the latter giving rise, by extension of the flanges, to the case of the double-layer shell. In case (b) effects of end supports are also treated in an approximate manner. Some typical results are depicted in Figs. 5 and 6.

Another example of important practical interest is the cylindrical

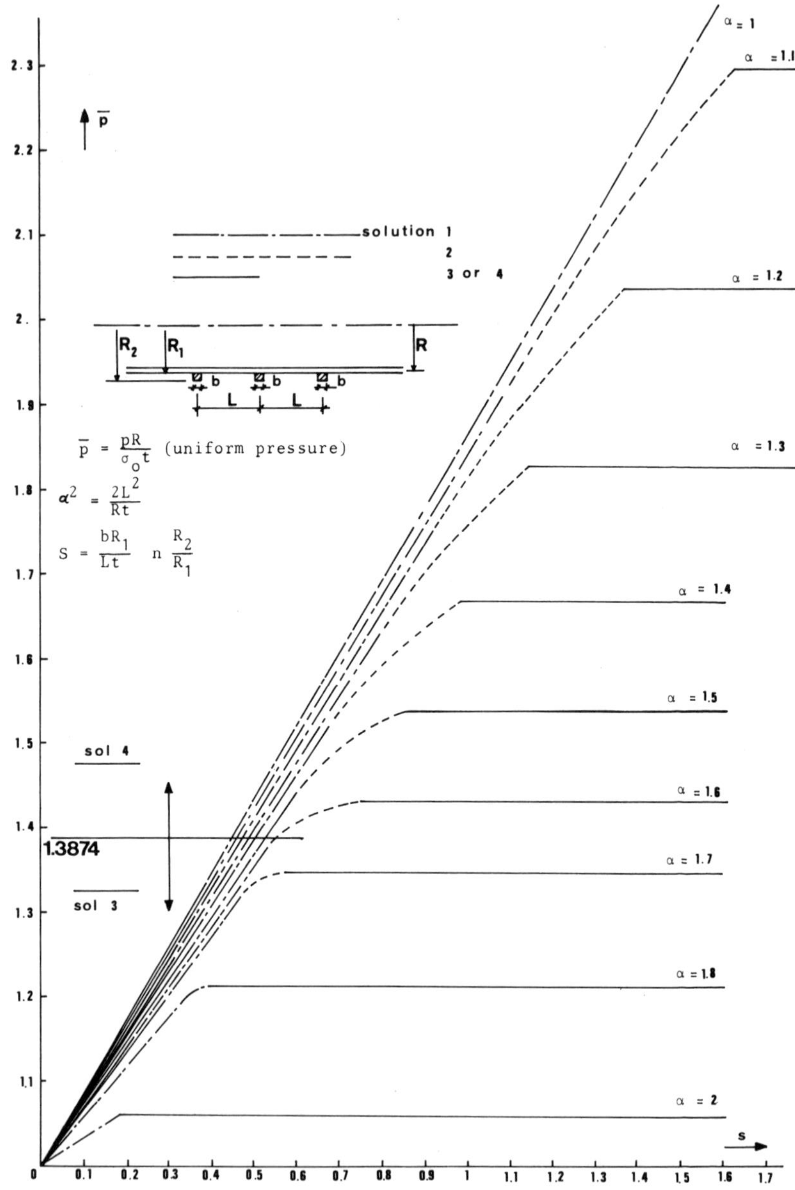

FIG. 5. Limit pressure of infinite cylindrical shell with equally spaced stiffeners [34].

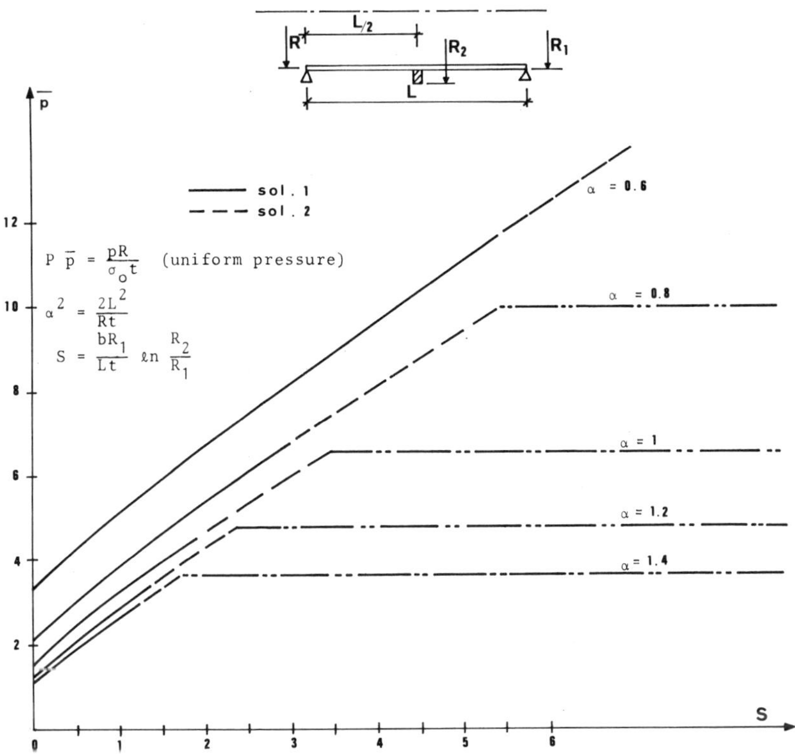

FIG. 6. Limit pressure of simply supported shell with one central stiffener [34].

tank with vertical axis subjected to hydrostatic pressure and made of several rings of different thickness. The API Standard 650 prescribes the design of such a tank on the basis of a limited circumferential membrane stress at an empirically given level in each ring (one foot above the lower end). This is likely to be a non-economic design procedure, without speaking of its crude aspect. Hence, a 'true lower bound' solution as accepted by the A.S.M.E Code should save some material and give better knowledge of the real safety with respect to collapse. As a first step toward this goal, analytical limit analysis of a cylinder of uniform thickness subjected to hydrostatic loading was achieved for all possible boundary conditions and shell geometries by Cinquini *et al.* [43]. For the case of several rings of different thicknesses, it is then suggested we adopt the following procedure:

> each intermediate ring is treated as being free at both ends (an assumption on the safe side) and subjected to a linearly varying radial pressure. Its thickness is determined from the graph of Fig. 7, L, R and T being the length, radius and thickness of the ring, respectively. Obviously, a safety factor s with respect to plastic collapse must be applied to the service pressure.

The analytical results of Fig. 7 are due to Cinquini [44]. The thickness of the upper ring is determined from Fig. 8 [43]. The analytical solution for the lower ring is likely to be lengthy and tedious because the support conditions at the end necessitate, when varying the shell parameter α, consideration of a variety of stress profiles. Hence, graphs similar to that of Figs. 7 and 8 are presently being constructed by a finite element approach using the SAP 7 program. With such a practical procedure, an interactive computer program could be conceived that allows variation of the lengths of the rings to achieve some kind of weight minimization. Of course, the ideal goal is to make available a program of minimum-weight design, including optimization

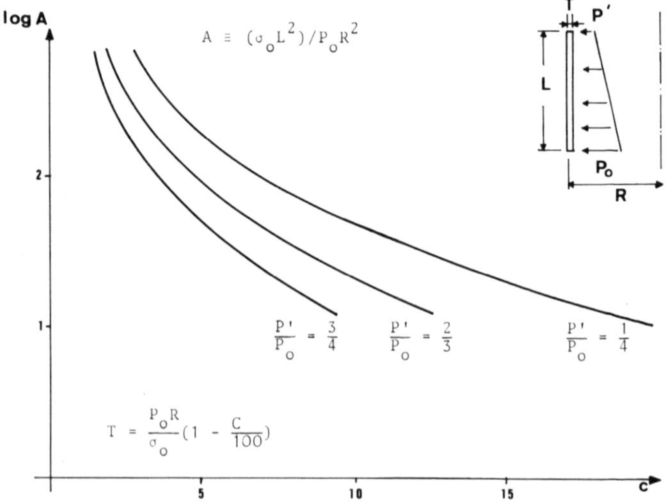

FIG. 7. Cylindrical shell with free ends, subjected to linearly varying pressure [44].

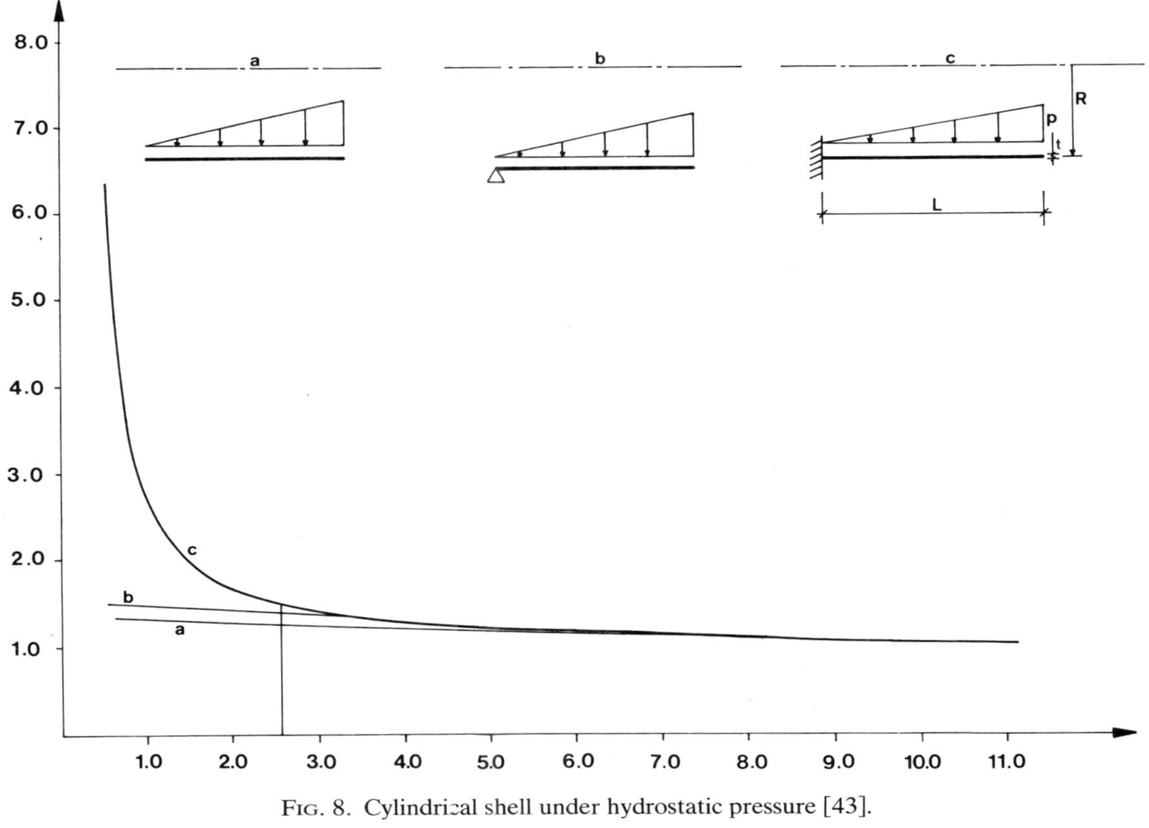

FIG. 8. Cylindrical shell under hydrostatic pressure [43].

of the ring length. Recent progress in plastic optimal design of cylinders [42, 45, 46], either by optimality criterion or mathematical programming, gives reasons for being optimistic if efforts are pursued.

Though promising results in reinforcement minimization of cylindrical concrete tanks have been obtained [45], their practical use necessitates obviously much complementary theoretical and experimental work.

5. NECESSARY COMPLEMENTARY RESEARCHES

Multipurpose optimal structural design is now reconciling elastic and plastic design philosophies: typical is the beam used in bending that is designed optimally for assigned elastic deflection under service load and assigned load factor at plastic collapse [47], the design being in general influenced by both behavioral constraints. Though the numerical approach will be most often the only tractable one, it may prove useful and efficient to solve analytically the more simple but frequently encountered problems. Very little has been done to date in this direction. In the case of grids in bending, optimal design for practically all types of loadings and boundary conditions has been achieved by Rozvany [12]. Applicability to reinforced concrete plate is immediate. The solutions are however continuum-like grids, made of infinite sets of infinitely narrow beams. Discretization is compulsory for practical use, together with some simplification of the theoretical distributions of bars (or beams) or some adaptations to comply with the requirements of Codes. How much these modifications will alter the optimum is important to know. More important however is to be certain of the acceptable behavior of an optimum plastic design with respect to the other relevant behavioral constraints (deflection, cracking of concrete, instabilities) when, for some reasons, a multipurpose optimal design has not been done. In optimality designed circular plates with piecewise varying thickness, variations in yield moment are assumed to follow strictly that of the thickness, in contradiction with the real physical situation. How much this assumption influences the limit load of plates with a small number of rings of very different thickness should be studied by three-dimensional (finite element) analysis.

The post-yield behavior of optimal plastic design is also of importance: the predicted instabilities (see for example ref. 48) might be too pessimistic because they neglect material work-hardening.

In all the problems quoted above, and in many others, experimental results on practically adapted optimum plastic designs are urgently needed.

The situation is similar in limit analysis of plates and shells, despite the existence of a certain amount of experimental work.

The conclusion is obvious: despite the well established theories of plastic limit analysis and optimal design, much theoretical and experimental work in plasticity today remains to be done to render available the tools needed for practical engineering plasticity. Organized international co-operation in this field would certainly be most valuable.

REFERENCES

1. Drucker, D. C. and R. T. Shield. Design for minimum weight, *Proc. 9th Int. Congr. Appl. Mech.*, Brussels, **5** (1956), 212–222.
2. Shield, R. T. Optimum design method for multiple loading, *ZAMP*, **14** (1963), 38–45.
3. Save, M. A. and W. Prager. Minimum weight design of beams subjected to fixed and moving loads, *J. Mech. Phys. Solids*, **11** (1963), 38–45.
4. Save, M. A. and R. T. Shield. Minimum weight design of sandwich shells subjected to fixed and moving loads, *Proc. 11th Int. Congr. Appl. Mech.* (Munich 1964), Springer, Berlin, 1966, 341–349.
5. Lamblin, D. O. and M. A. Save. Minimum volume plastic design of beams for movable loads, *Meccanica*, **6** (1971), 157–163.
6. Lamblin, D. O. Minimum weight plastic design of continuous beams subjected to one single movable load, *J. Struct. Mech.*, **1** (1972), 133–157.
7. Sacchi, G., G. Maier and M. A. Save. Limit design of frames for movable loads by linear programming. In: *Optimization in Structural Design.* (Ed. A. Sawczuk and Z. Mroz), IUTAM Symp., Warsaw, 1973, Springer, Berlin, 1975.
8. Prager, W. and R. T. Shield. The general theory of optimal plastic design, *J. appl. Mech.*, **34** (1967), 184–186.
9. Save, M. A. A unified formulation of the theory of optimal plastic design with convex cost functions, *J. Struct. Mech.*, **1** (1972) 267–276.
10. Rozvany, G. I. N. Non-convex structural optimization problems, *Proc. ASCE*, **99**, EM1 (1973), 243–248.
11. Prager, W. Introduction to structural optimization. In: *CISM Lecture Notes*, Springer, Vienna, 1974.
12. Rozvany, G. I. N. *Optimal Design of Flexural Systems*, Pergamon Press, Oxford, 1976.
13. Prager, W. and G. I. N. Rozvany. Optimization of structural geometry. In: *Dynamical systems. Proc. Int. Symp., Florida* (Ed. Bednarek), 1976, Academic Press, New York, 1977, 265–294.
14. Rozvany, G. I. N. and R. D. Hill. The theory of optimal load transmission by flexure, *Adv. appl. Mech.*, **16** (1976), 183–306.

15. ROZVANY, G. I. N. and W. PRAGER. A new class of structural optimization problems: optimal archgrids, *Comp. Meth. appl. Mech. Engng*, **19** (1970), 127–150.
16. MUNRO, J. Optimal plastic design of frames. In: *Engineering Plasticity by Mathematical Programming*, Chapter 7, Proc. of the NATO Advanced Study Institute, University of Waterloo (Ed. M. Z. Cohn and G. Maier), Waterloo, Ontario, 1977, 135–171.
17. MAIER, G., R. SRINIVISAN and M. A. SAVE. On limit design of frames using linear programming, *Proc. Int. Symp. Computer-Aided Struct. Design*, University of Warwick, 1972.
18. SMITH, D. L. Plastic limit analysis and synthesis of structures by linear programming, Ph.D. Thesis, University of London, 1974.
19. ZAVELANI-ROSSI, A. Optimal design of discretized continua. In: *Engineering Plasticity by Mathematical Programming*, Chapter 10, Proc. of the NATO Advanced Study Institute, University of Waterloo (Ed. M. Z. Cohn and G. Maier), Waterloo, Ontario, 1977, 223–238.
20. ZAVELANI, A., G. MAIER and L. BINDA. Shape optimization of plastic structures by zero-one programming. In: *Optimisation in Structural Design*, (Ed. A. Sawczuk and Z. Mroz), IUTAM Symposium, Warsaw, 1973. Springer, Berlin, 1975.
21. ZAVELANI, A. A compact linear programming formulation for optimal design in plane stresses, *J. Struct. Mech.*, **2** (1973), 301–324.
22. MAIER, G., A. ZAVELANI and D. BENEDETTI. A finite element approach to optimal design of plastic structures in plane stress, *Int. J. Num. Methods Engng*, **4** (1972), 455–473.
23. LAMBLIN, D. O. Analyse et dimensionnement plastique de 'coût' minimum de plaques circulaires, Doctoral Thesis, Faculté Polytechnique de Mons, 1975.
24. OLSZAK, W. and A. SAWCZUK. Théorie de la capacité portante des constructions non-homogènes et orthotropes, Suppl. to Annales de l'Inst. Tech. Bat. et Trav. Pub., Paris, 149, May, 1960.
25. SAVE, M. A. Vérification expérimentale de l'analyse plastique des plaques et des coques en acier doux. (Experimental verification of plastic limit analysis of mild steel plates and shells), C.R.I.F. Report, M.T. 21, February, 1966.
26. Belgian Code N.B.N-B-51-001.
27. HOPKINS, H. and W. PRAGER. Limits of economy of material in plates, *J. appl. Mech.*, **22** (1955), 372–374.
28. LAMBLIN, D. O., M. A. SAVE and G. GUERLEMENT. Solutions de dimensionnement plastique de volume minimal de plaques circulaires pleines et sandwich en présence de contraintes technologiques, *J. Mec. Théor. Appl.* (to be published).
29. KOZLOWSKI, W. and Z. MROZ. Optimal design of solid plates, *Int. J. Solids Struct.*, **5** (1969), 781–794.
30. SAVE, M. A. and C. E. MASSONNET. *Plastic Analysis and Design of Plates, Shells and Disks*, North Holland, Amsterdam, 1972.
31. BROTCHIE, J. F. Discussion to the paper by G. J. Megarefs. *J. Engng Mech. Div. Am. Soc. Civ. Engrs*, **93** (1967), 173–175.

32. Rozvany, G. I. N., N. Olhoff, K. T. Cheng and J. Taylor. On the solid plate paradox in structural optimization, *J. Struct. Mech.*, **10** (1982), 1–32.
33. Onat, E. T., R. T. Shield and W. Schumann. Design of circular plates for minimum weight, *ZAMP*, **8** (1957), 485–499.
34. Guerlement, G. Contribution à l'analyse limite des coques cylindriques, Doctoral Thesis, Faculté Polytechnique de Mons, 1975.
35. Paul, B. Limit loads of clamped shells with a reinforcing ring (PIBAL report, no. 424, December 1958).
36. Nemirovsky, Y. V. and Y. N. Rabotnov. Limit analysis of ribbed plates and ribbed shells. In: *Non Classical Shell Problems* (Eds W. Olszak and A. Sawczuk) North Holland, Amsterdam, 1964, 786–807.
37. Capurso, M. and A. Gandolfi. Sul collasso rigido plastico dei gusci nervati di rivoluzione. Internal Report. Istituto di Tecnica delle Costruzione, Napoli, 1967.
38. Capurso, M. and A. Gandolfi. Sul collasso plastico dei tubi circolari nervati soggetti a pressione es forzo assiale. Internal Report. Istituto di Tecnica delle Costruzioni, Napoli, 1967.
39. Biron, A. and A. Sawczuk. Plastic analysis of rib reinforced cylindrical shells, *J. appl. Mech.*, **34** (1967), 37–42.
40. Guerlement, G., D. O. Lamblin and A. Save. Limit analysis of a cylindrical shell with reinforcing rings (to be published).
41. Cinquini, C., G. Guerlement and D. O. Lamblin. Shell-stiffener interaction. Application to simply supported cylindrical shell under uniform pressure (to be published).
42. Cinquini, C. and M. Kouam. Optimal plastic design of stiffened shells. *Int. J. Solids Struct.*, **19** (1983), 773–783.
43. Cinquini, C., D. O. Lamblin and G. Guerlement. Limit analysis of circular cylindrical shells under hydrostatic pressure (to be published in *J. Struct. Mech.*).
44. Cinquini, C. Limit analysis and optimal plastic design of circular cylindrical shells. Euromech Coll., Siegen, October 1982.
45. Igic, T. Contribution au dimensionnement optimal des structures, Doctoral Thesis, Université de Nis, 1981.
46. Kouam, M. Contribution à l'analyse limite et au dimensionnement optimal des coques cylindriques, Doctoral Thesis, Faculté Polytechnique de Mons, 1983.
47. Save, M. A. and T. Igic. Exemples de poutres optimales à deux fonctions, *J. Mec. Théor. Appl.* **1** (1982), 311–321.
48. Mroz, Z. and A. Gawecki. Post-yield behaviour of optimal plastic structures. In: *Optimization in Structural Design*, IUTAM Symp. Warsaw 1973 (Ed. A. Sawczuk and Z. Mroz), Springer, Berlin, 1975, 518–540.

42

Dynamic Plastic Response of Structures

S. KALISZKY

Department of Mechanics, Technical University, Budapest, Hungary

ABSTRACT

The purpose of this paper is to present a review of the basic principles and methods of the overall inelastic response of structures subjected to short-time, high intensity dynamic pressure. The review concentrates on the approximate methods based on the simple rigid–plastic idealization; in the second part, however, problems including elastic and viscous effects and large deformations are also discussed. In the solutions pulse and impulsive loadings are taken into consideration.

In course of the theoretical survey the numerical analysis of a two-mass discrete beam model is discussed. This simple model provides a clear insight into the physical background of the theorems and approximations.

The approximate solutions presented can form the basis of rapid calculations and provide reliable information about the dynamic plastic response of structures.

1. INTRODUCTION

Short time, high intensity dynamic pressure can cause stress waves as well as an overall structural response. It is customary practice to separate these two phenomena, because their time durations usually differ by a few orders of magnitude.

In the overall inelastic dynamic response the following phenomena can play an important role: plastic deformations, elastic deformations, strain rate sensitivity, strain hardening, finite deflections. In a great number of studies rigid-perfectly plastic material is assumed and the effect of finite deflections is neglected. This idealization leads to significant simplifications and yet provides reliable information. The assumptions of this simple theory are, however, to some extent contradictory, since the elastic deformations can only be neglected when the external energy input is several times greater than the total elastic strain energy capacity. On the other hand, when the transverse deflections are larger than the corresponding thickness of the structure, the geometry changes can introduce in-plane or membrane forces which might exercise a significant effect. Similarly, the dynamic buckling or unstable behaviour of structures can also have an important influence.

The strain rate sensitive behaviour exerts a strengthening effect, which at higher strain rates can be significant. Therefore the rigid–plastic approximation might require further refinements. The strain hardening is relatively unimportant compared to other effects; thus, it can usually be neglected.

The purpose of this article is to present a review of the basic principles and methods of the overall inelastic response of structures subjected to short-time, high intensity dynamic pressure. Because of the limited extent no comprehensive survey of methods and literature is attempted. The first part of the review concentrates on the approximate methods based on the simple rigid–plastic idealization; in the second part, however, the problems including elastic and viscous effects and large deformations will also be discussed. The problems of repeated and periodical loadings, the description of proofs of the theorems and the discussion of computer solutions are beyond the scope of this article.

In the course of theoretical survey the analysis of a two-mass discrete beam model will be presented. This simple model provides a clear insight into the physical background of the theorems and makes possible the comparison of the applications and results of the different methods.

2. FUNDAMENTAL RELATIONSHIPS

In the following, one- and two-dimensional structures will be considered and the following notation will be used (Fig. 1).

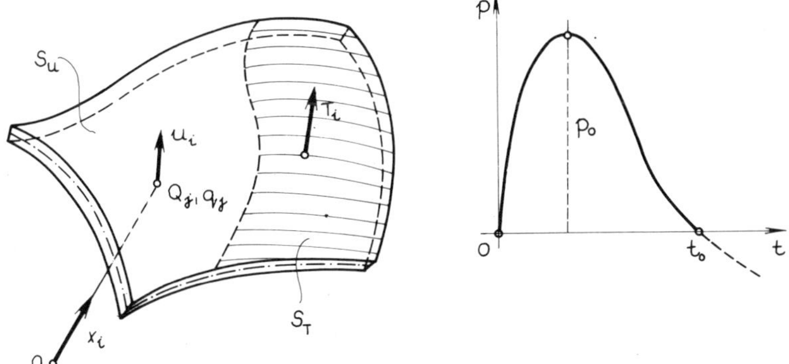

FIG. 1. The dynamically loaded structure.

x_i, t: space and time variables $(i = 1, 2, 3)$;
$m(x_i)$: mass density per unit length or surface;
$u_i(x_i, t)$: displacement field;
$q_j(x_i, t)$; $Q_j(x_i, t)$: generalized strain and stress fields

$$(j = 1, 2, \ldots, n);$$

$T_i(x_j, t)$: dynamic surface traction;
$p_o(t)$; $T_i^o(x_j)$: load parameter and load distribution function;
$v_i^o(x_j)$: initial velocity field;
$f(Q_j, x_i)$: yield function;
p_k; p_c: kinematically admissible and collapse load multipliers;
p_o; t_o: peak value and duration of the pulse load;
t_f: response time.

The dot denotes derivation with respect to the time. On a part of the total domain S of the structure denoted by S_T, dynamic surface tractions $T_i(x_j, t)$ are given and initial velocities $v_i^o(x_j)$ can also be prescribed. On the remainder of S, denoted by S_u, the velocities $\dot{u}_i(x_j, t)$ are set equal to zero and the initial displacements $u_i(x_j, 0)$ are taken to be zero everywhere. The weight of the structure is disregarded.

The primary aim of the investigation is to calculate the response time t_f at which the structure ceases to move and to determine the final (permanent) displacements: $u_i^f(x_j) = u_i(x_j, t_f)$. Besides, in some cases the complete determination of the state variables u_i, $\dot{u}_i$, q_j, $\dot{q}_j$ and Q_j is also required. Usually two loading cases are distinguished.

In case of pulse loading the dynamic surface tractions are specified in a separated-variable form as

$$T_i(x_j, t) = p(t)T_i^{\circ}(x_j) \tag{1}$$

The most widely used pulse loading is the 'square' pressure defined as

$$\begin{aligned} p &= p_{\circ} \quad \text{if} \quad 0 \leqslant t \leqslant t_{\circ} \\ p &\equiv 0 \quad \text{if} \quad t > t_{\circ} \end{aligned} \tag{2}$$

Impulsive loading can be applied when $p(t)$ is non-zero for a time period which is very short compared to the response time of the structure. Then, introducing the impulse

$$I_{\circ} = \int_{\circ}^{t_{\circ}} p(t)\,\mathrm{d}t \tag{3}$$

represented by the pressure, eqn. (1) can be replaced by a delta function:

$$T_i(x_j, t) = T_i^{\circ}(x_j) I_{\circ} \delta(t) \tag{4}$$

Here $\delta(t)$ is the unit Dirac delta function. Then the solution of the problem can be reduced to the dynamic analysis of an unloaded structure with prescribed initial velocity field:

$$\begin{aligned} &T_i(x_j, t) = 0 && \text{on} \quad S_T \\ &u_i(x_j, 0) = 0 && \text{on} \quad S \\ &\dot{u}_i(x_j, 0) = v_i^{\circ}(x_j) = \frac{T_i^{\circ}(x_j)}{m(x_j)} I_{\circ} && \text{on} \quad S_T \end{aligned} \tag{5}$$

For any kinematically admissible strain rate and acceleration field $\dot{u}_i^{\mathrm{k}}$, $\dot{q}_j^{\mathrm{k}}$, $\ddot{u}_i^{\mathrm{k}}$, $\ddot{q}_j^{\mathrm{k}}$ and for any dynamically admissible stress and acceleration field Q_j^{d} and $\ddot{u}_i^{\mathrm{d}}$ the principle of virtual velocities is formulated as

$$\int_{S_T} T_i \dot{u}_i^{\mathrm{k}}\,\mathrm{d}S - \int_S m \ddot{u}_i^{\mathrm{d}} \dot{u}_i^{\mathrm{k}}\,\mathrm{d}S = \int_S Q_j^{\mathrm{d}} \dot{q}_j^{\mathrm{k}}\,\mathrm{d}S \tag{6}$$

In the simple rigid–plastic theory of structures it is assumed that the structure is composed of a strain, strain-rate and temperature independent rigid-perfectly plastic material and the effect of changes in geometry is neglected. Introducing a convex yield function $f(Q_j, x_i)$ the constitutive equation has the form

$$\dot{q}_j = \dot{\lambda}\frac{\partial f}{\partial Q_j} \tag{7}$$

where $\dot{\lambda}(x_i, t) \geqslant 0$ is the scalar multiplier of plastic strain rates.

Because of the convexity of the yield surface $f(Q_j) = 0$ the specific rate of dissipation energy function D is uniquely determined if $\dot{q}_j$ is given:

$$D = Q_j\dot{q}_j = D(\dot{q}_j) \tag{8}$$

From that

$$Q_j = \frac{\partial D}{\partial \dot{q}_j} \tag{9}$$

Considering a quasi-static one parameter loading defined by eqn. (1) the corresponding kinematically admissible load multiplier is defined by

$$p_k = \frac{\int_S Q_j^k \dot{q}_j^k \, dS}{\int_S T_i^o \dot{u}_i^k \, dS} \tag{10}$$

3. TIME HISTORY ANALYSIS

3.1. General Notes

The basic problem of time history analysis of dynamically loaded structures is that of determining the acceleration field $\ddot{u}_i(x_j, t)$ at an instant t, when the velocity field $\dot{u}_i(x_j, t)$ is known. Solving this problem the entire dynamic solution can be obtained by considering successive instants in which the velocity fields are determined from the velocities and accelerations at preceding instants.

In the case of rigid–plastic materials difficulties arise from the possibility of time discontinuities in the acceleration field $\ddot{u}_i(x_j, t)$. Besides, in the rigid regions, where $q_j = 0$, but in time t under consideration plastic flow starts to occur, the stresses cannot be determined from eqn. (9). In the time history analysis these phenomena should be taken into consideration. For that purpose, among others two extremum principles will be described next.

3.2. Kinematic Principle

Suppose that at an instant t the state variables u_i, $\dot{u}_i$ and $\dot{q}_j$ of a rigid-perfectly plastic structure are known and consider kinematically admissible velocity fields $\dot{u}_i^k(x_j, t)$ defined for $t_1 \geqslant t$, which fulfil the condition $\dot{u}_i^k = \dot{u}_i$ at time t. From the function $\dot{u}_i^k$ the velocity and acceleration fields $\dot{q}_j^k$, $\ddot{q}_j^k$ and $\ddot{u}_i^k$ can be derived. Then Tamuzh's principle, which is the generalization of Gauss's principle of least constraint to rigid–plastic problems, states that the functional

$$J(\ddot{u}_i^k) = \frac{1}{2}\int_S m\ddot{u}_i^k\ddot{u}_i^k \, dS - \int_{S_T} T_i\ddot{u}_i^k \, dS + \int_{S_1} Q_j^k\ddot{q}_j^k \, dS + \int_{S_2} Q_j^*\ddot{q}_j^k \, dS \tag{11}$$

takes on its least value when $\ddot{u}_i^k = \ddot{u}_i$ [24]. In the functional (11) S_1 denotes the region where at the moment t when $\dot{q}_j = \dot{q}_j^k \neq 0$ and $D(\dot{q}_j^k)$ has continuous derivatives, while S_2 is the region where $\dot{q}_j = \dot{q}_j^k = 0$ or $\dot{q}_j = \dot{q}_j^k \neq 0$, but $D(\dot{q}_j^k)$ has discontinuous derivatives. In S_1 the Q_j^k is defined by eqn. (9) and in S_2 Q_j^* is associated with $\ddot{q}_j^k$ through eqn. (7) if q_j^k were a strain rate.

Using piecewise linear yield functions Capurso extended Tamuzh's principle and then the determination of the acceleration field becomes a quadratic programming problem [1].

3.3. Dynamic Principle

Consider again the same problem, and let $\ddot{u}_i^d(x_j, t)$ and $Q_j^d(x_i, t)$ denote dynamically admissible, acceleration and stress fields for $t_1 > t$ which satisfy the yield conditions in each point of the structure. Then, the principle proved by Martin states that the functional

$$J(\ddot{u}_i^d) = \int_S m\ddot{u}_i^d\ddot{u}_i \, dS \tag{12}$$

takes on its least value when $\ddot{u}_i^d = \ddot{u}_i$ [12].

3.4. Simple Dynamic Model

The fundamental ideas of structural dynamics will be illustrated by means of a two-degree-of-freedom model used already by other authors. The model is a simply supported, rigid-perfectly plastic, massless beam with a plastic moment M_o (Fig. 2). Two equal concentrated masses M are attached to the beam at the points $\alpha = 1, 2$. The dynamic

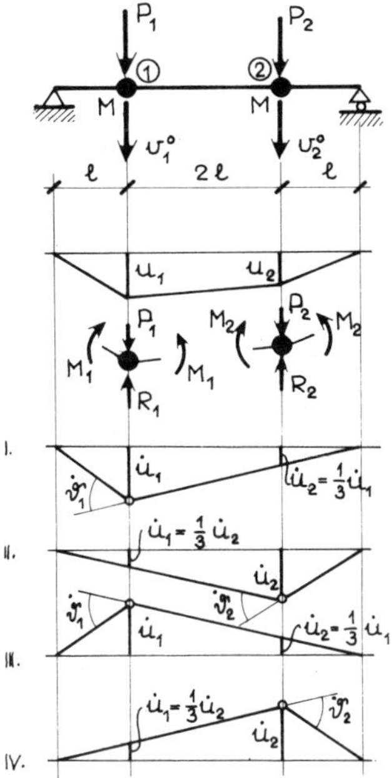

FIG. 2. Two-degree-of-freedom model.

equations of the masses and the equilibrium and compatibility equations of the basic yield mechanisms I–IV shown in Fig. 2 have the form:

$$P_\alpha - R_\alpha = M\ddot{u}_\alpha, \qquad (\alpha = 1, 2) \tag{13}$$

$$\boldsymbol{\phi} = \mathbf{N}R_\alpha - \mathbf{k} = \mathbf{0} \tag{14}$$

$$\dot{u}_\alpha = \mathbf{G}\dot{\theta}_\alpha \tag{15}$$

where

$$\mathbf{N} = \begin{bmatrix} 1 & \frac{1}{3} \\ \frac{1}{3} & 1 \\ -1 & -\frac{1}{3} \\ -\frac{1}{3} & -1 \end{bmatrix}, \qquad \mathbf{k} = \frac{4M_o}{3l}\begin{bmatrix} 1 \\ 1 \\ 1 \\ 1 \end{bmatrix}, \qquad \boldsymbol{G} = \frac{l}{4}\begin{bmatrix} 3 & 1 \\ 1 & 3 \end{bmatrix}$$

The safe feasible region of the internal forces R_α corresponding to eqn. (14) and to the yield mechanisms of Fig. 2 are bounded by four straight

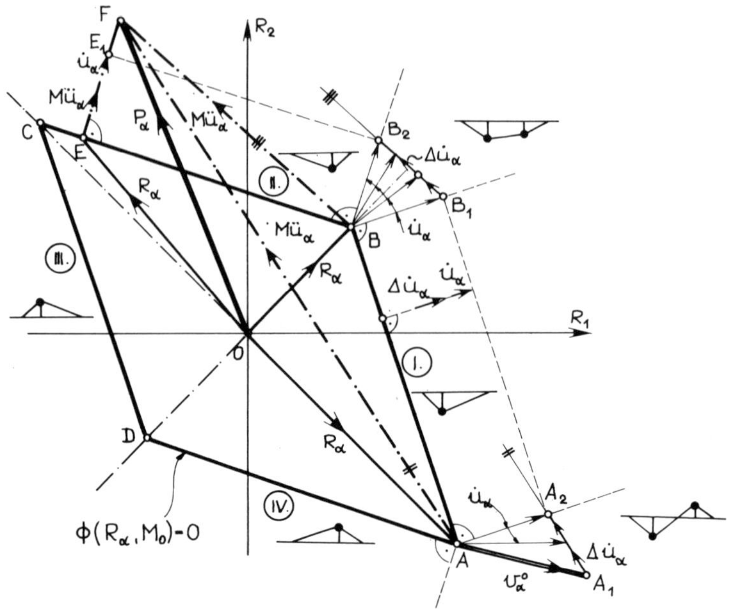

FIG. 3. Feasible domain of the two-degree-of-freedom model.

lines (Fig. 3). For this limit curve the normality rule holds, thus

$$\dot{u}_\alpha = \dot{\Lambda}\frac{\partial \mathbf{\Phi}}{\partial R_\alpha}\ \dot{\Lambda}\mathbf{N} \tag{16}$$

where $\dot{\Lambda} \geqslant 0$ is a scalar multiplier.

The energy dissipation function of the beam is

$$D = \frac{M_o}{2l}[|3\dot{u}_1 - \dot{u}_2| + |3\dot{u}_2 - \dot{u}_1|] \tag{17}$$

According to eqns. (9) and (13)

$$R_\alpha = \frac{\partial D}{\partial \dot{u}_\alpha}, \qquad M\ddot{u}_\alpha = -\frac{\partial}{\partial \dot{u}_\alpha}[D - P_\alpha \dot{u}_\alpha] \tag{18}$$

Thus, the acceleration vector can be shown as the gradient of the function in the square bracket. The level curve of this function can be drawn in the velocity space. Figure 4(c) shows the special case when $P_\alpha = 0$.

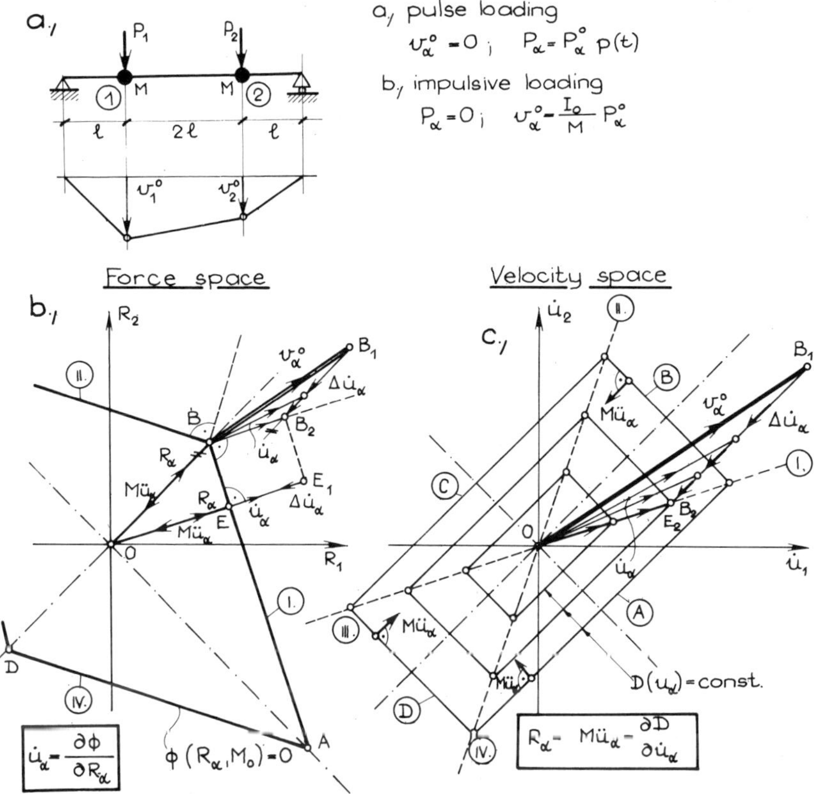

FIG. 4. Force and velocity spaces representing the response of the two-degree-of-freedom model under impulsive loading.

Consider the problem when the beam is loaded by the forces $P_\alpha = \text{const.}$ and has an initial velocity v_α°. At the instant $t = 0$ the internal forces and accelerations are determined by the triangle OAE. Thereafter, when the velocity $\dot{u}_\alpha$ reaches the direction AA_2 and BB_2 respectively a jump takes place since states that correspond to the points of the lines between the corners of the limit curve are not feasible. At the point E $\ddot{u}_\alpha$ will remain constant and $\dot{u}_\alpha$ will change only in magnitude but not direction. Thus, during the further response the shape of the velocity field remains constant. From the example one can state that the acceleration field may be discontinuous in time and at constant external load the response tends toward a stationary motion, which corresponds to a mode shape of the structure.

In case of impulsive loading, i.e. when the external load $P_\alpha = P^o_\alpha p(t)$ is replaced by an initial velocity field: $v^o_\alpha = (I_o/M)P^o_\alpha$, similar statements can be made (Fig. 4(b)). In the velocity space the energy level lines are rectangles and the velocity increments $\Delta\dot{u}_\alpha$ are normal to these surfaces (Fig. 4(c)). Reaching the corner B_2 the velocity vector $\dot{u}_\alpha$ will change only in magnitude and a mode form motion will take place until $\dot{u}_\alpha$ diminishes to zero.

Assuming the numerical values $P^o_\alpha(1, 2/3)$ and following the different stages of the response for the response time and for the final permanent displacements of the beam the following results can be obtained:

$$\left.\begin{aligned} t_f &= 0{\cdot}91667\frac{I_o l}{M_o} \\ u_{1f} &= 0{\cdot}47917\frac{I_o^2 l}{MM_o}, \quad u_{2f} = 0{\cdot}2430\frac{I_o^2 l}{MM_o} \end{aligned}\right\} \tag{19}$$

4. MODE SOLUTION

4.1. Basic Concepts

It was shown that the dynamic response of rigid-perfectly plastic structures subjected to a constant external pressure or initial velocity tends toward a stationary mode form motion. This mode form motion is defined as

$$\dot{u}_i^*(x_j, t) = \psi(t)\ddot{u}_i^*(x_j) = \dot{W}(t)u_i^m(x_j) \tag{20}$$

where $\psi(t)$ and $\dot{W}(t)$ are linear functions of time and $u_i^m(x_j)$ is the mode shape of the structure.

The concept of the mode solution proposed by Martin and Symonds [13] is based on two approximations:

(1) The pulse loading is replaced by impulsive loading with the initial velocity field $v_i^o(x_j)$ defined by eqn. (4);

(2) the initial velocity field $v_i^o(x_j)$ is replaced by another equivalent initial velocity field, which corresponds to a mode shape. This is expressed in the form:

$$\dot{u}_i^*(x_j, 0) = \dot{W}(0)u_i^m(x_j) = V_o u_i^m(x_j) \tag{21}$$

It means that the mode approximation solves a different initial value problem and therefore violates the original initial conditions. The advantage of this approximation is that introducing the modified initial conditions the structure undergoes solely mode form motion which can be solved in a simple closed form. The only problem is to select the mode shape $u_i^m(x_j)$ and the parameter V_o.

4.2. Solution Technique

Supposing that the mode shape $u_i^m(x_j)$ used for the mode solution has already been selected the next problem is to find the initial velocity. V_o can be chosen so as to provide a 'best fit' to the given initial velocity field in a mean square sense, i.e. to minimize

$$\Delta_o(V_o) = \frac{1}{2}\int_S m(v_i^o - V_o u_i^m)(v_i^o - V_o u_i^m)\,dS \tag{22}$$

This gives

$$V_o = \frac{\int_S m v_i^o u_i^m\,dS}{\int_S m u_i^m u_i^m\,dS} \tag{23}$$

Using this value the response time t_f and the parameter $W_f = W(t_f)$ can be easily obtained. Omitting the details we obtain:

$$t_f = \frac{\int_S m v_i^o u_i^m\,dS}{\int_S Q_j^m q_j^m\,dS} = \frac{I_o}{p_k} \tag{24}$$

$$W_f = W(t_f) = \frac{\left[\int_S m v_i^o u_i^m\,dS\right]^2}{2\int_S m u_i^m u_i^m\,dS \int_S Q_j^m q_j^m\,dS} = \frac{1}{2}K\frac{I_o^2}{p_k} \tag{25}$$

where p_k is the kinematically admissible load multiplier associated with the displacement field $u_i^m(x_j)$ and defined by eqn. (10) and

$$K = \frac{\int_S T_i^o u_i^m\,dS}{\int_S m u_i^m u_i^m\,dS} \tag{26}$$

The mode shapes comprise a class of the kinematically admissible displacement fields at which at constant external loads the structure will reach one or the other of these modes. For the selection of the mode shape which leads to the best approximation different rules and methods have been suggested [23].

4.3. Example

Let us consider again the impulsively loaded simple model shown in Fig. 4. The actual response of the structure corresponding to the initial velocity v°_{α} is characterized by the line $B_1 - B_2 - O$ in Fig. 5. Applying the mode approximation and assuming that the mode shape is given by the line O–G the 'best choice' of V_{o} defined by eqn. (23) is shown by the distance $\overline{OG}$.

Let us solve the numerical example shown in Fig. 4. Then $v^{\circ}_{\alpha} = I_{o}/M$ (1, 2/3) and $u^{m}_{\alpha}(1, 1/3)$ and from eqn. (3)

$$V_{o} = \frac{Mv^{\circ}_{\alpha}u^{m}_{\alpha}}{Mu^{m}_{\alpha}u^{m}_{\alpha}} = 0{\cdot}91667\,\frac{I_{o}}{M} \tag{27}$$

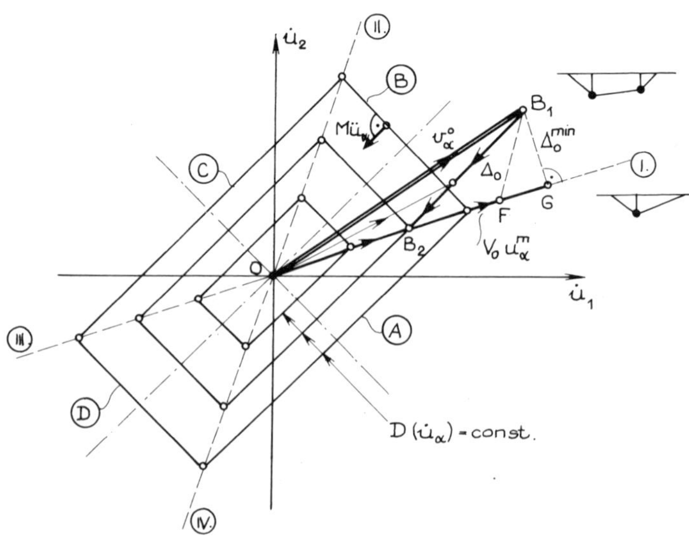

FIG. 5. Mode solution of the two-degree-of-freedom model under impulsive loading.

Further, using eqns. (24) and (25) the following results can be obtained

$$t_f = \frac{Mv_\alpha^o u_\alpha^m}{Q_\alpha^m q_\alpha^m} = 0{\cdot}91667 \frac{I_o l}{M_o} \tag{28}$$

$$u_{1f} = W_f = \frac{[Mv_\alpha^o u_\alpha^m]^2}{2[Mu_\alpha^m u_\alpha^m][Q_\alpha^m q_\alpha^m]} = 0{\cdot}50417 \frac{I_o^2 l}{M_o M} \tag{29}$$

$$u_{2f} = \tfrac{1}{3} W_f = 0{\cdot}1680 \frac{I_o^2 l}{M_o M} \tag{30}$$

5. KINEMATICAL SOLUTION

5.1. Basic Concepts

The response of rigid–plastic structures subjected to pulse or impulsive loading generally is in two stages: first a nonstationary motion takes place but later the response tends toward a stationary motion. The aim of the kinetical solution is to replace these two stages by a single stationary motion [5]. To achieve this motion one has to impose on the structure a kinematically admissible displacement field expressed in a separated-variable form as

$$u_i^o(x_j, t) = W(t) u_i^k(x_j) \tag{31}$$

Here $u_i^k(x_j)$ denotes *any* postulated kinematically admissible displacement field and $W(t)$ is an unknown displacement parameter function.

Considering a rigid-perfectly plastic structure subjected to a pulse loading and using eqn. (31) the determination of $W(t)$ can be reduced to the solution of the differential equation

$$\ddot{W}(t) = K[p(t) - p_k] \tag{32}$$

Here

$$K = \frac{\int_S T_i^o u_i^k \, dS}{\int_S m u_i^k u_i^k \, dS} \tag{33}$$

and p_k denotes the kinematically admissible load multiplier associated with the displacement field $u_i^k(x_j)$ and the quasi-static loading $T_i^o(x_j)$

(see eqn. (10)). Equation (32) might be considered as the dynamic equation of a single mass point. Thus, the kinematical approximation reduces the solution of the problem to the solution of the equivalent quasi-static problem and to the analysis of a one-degree-of-freedom system.

5.2. Response Time and Final Displacements

Assuming the initial conditions $u_i(x_j, 0) = \dot{u}_i(x_j, 0) = 0$ the solution of eqn. (32) can be expressed in the following form [5, 8]:

$$t_f = \frac{I(t_f)}{p_k} \tag{34}$$

$$W_f = W(t_f) = KS_o\left[\frac{p_e}{p_k} - 1\right] \tag{35}$$

Here

$$I(t_f) = \int_o^{t_f} p(t)\,dt, \qquad I_o = \int_o^{t_o} p(t)\,dt$$

$$S_o = \int_o^{t_o} tp(t)\,dt, \qquad p_e = \frac{I_o^2}{2S_o} \tag{36}$$

One has to note that the validity of eqn. (35) is restricted only to the loading cases when the following conditions are simultaneously fulfilled [8]:

$$\begin{aligned} p(0) \geqslant p_k, \qquad I_o \geqslant p_k t_o \\ p(t) \equiv 0 \quad \text{if} \quad t > t_o \end{aligned} \tag{37}$$

The accuracy of the kinematical approximation depends on the proper choice of the kinematically admissible displacement field. It has been proved that eqn. (34) gives a lower bound for the response time therefore the best lower bound can be obtained if p_k equals the collapse load multiplier p_c. For the final displacements similar statements cannot be proved. Still, even in this case the use of the collapse mechanism obtained under quasi-static conditions is suggested to obtain the best approximation.

It can be proved that when in the mode solution and in the kinematical solution the same kinematically admissible displacement fields are used, then both methods yield the same results. It means that the kinematical solution is a more general method which is equally suitable for the analysis of pulse and impulsive loading problems and in case of impulsive loading it can provide as a special case the mode solution.

5.3. Example

Consider again the model shown in Fig. 4. Assuming a pulse loading the exact solution of the response is illustrated in Fig. 6(a). Here while the load is decreasing the velocity vector $\dot{u}_\alpha$ rotates from the direction AA_1 to AA_3 and the model performs a nonstationary motion. After this stage, however, the model undertakes stationary motion which corresponds to the yield mechanism I of Fig. 2.

Applying the kinematical approximation first the quasi-static solution of the problem has to be determined (Fig. 6(b)). This is defined by the intersection of the vector P_α with the limit surface $\Phi(R_\alpha, M_o) = 0$. The length OC gives the magnitude of the collapse load multiplier p_c and the vector normal to the line $\overline{AB}$ corresponds to the collapse mechanism $u_i^\circ(x_j)$ which forms the basis of the solution. Since $\ddot{u}_i^\circ$ has to be also normal to the line $\overline{AB}$ during the entire response therefore, at

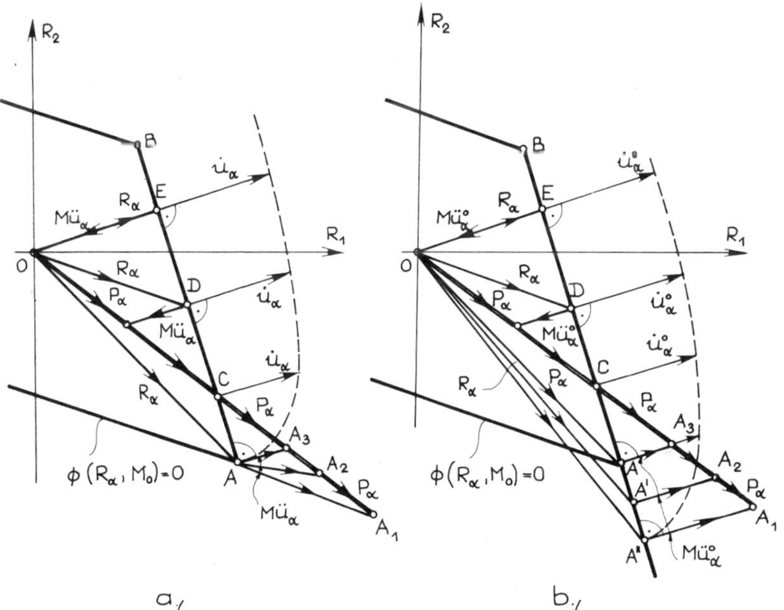

FIG. 6. (a) The exact and (b) the kinematical solutions of the two-degree-of-freedom model under pulse loading.

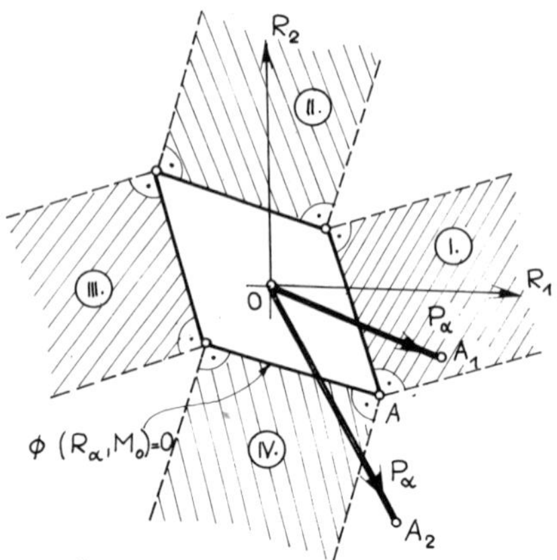

FIG. 7. The domains when the kinematical solution gives exact (shaded regions) or approximate (unshaded regions) results.

the instant $t=0$ the vector of R_α intersects the limit curve. This stage is valid until the load intensity decreases to the value OA_3. Thereafter R_α remains inside the limit curve and the motion of the model will be identical to that described in the exact solution.

We can conclude that at the kinematical solution in the first stage of the motion the yield condition might be violated. This occurs in those cases when the peak value of the pulse load is large and the end point of P_α does not remain inside the shaded regions I, II, III, IV shown in Fig. 7. Otherwise by the use of the kinematical approach an exact solution can be obtained.

Finally, let us consider the numerical values $P_\alpha^o(1, 2/3)$ and the collapse mechanism shown by I in Fig. 2. Then the corresponding kinematically admissible multiplier $p_k = 1{\cdot}0909\ (M_o/l)$.

Using the 'square' load parameter function defined by formula (2) eqns. (35) and (36) yield the following results

$$t_f = 0{\cdot}91667 \frac{I_o l}{M_o} \tag{38}$$

$$W_f = 0{\cdot}50417 \frac{I^2 l}{M_o M} - 0{\cdot}550 I_o t_o \tag{39}$$

In the case of impulsive loading, i.e. when $t_o \to 0$, these results are identical with those obtained by the mode solution.

6. BOUNDING THEOREMS

A number of theorems have been published in the literature to obtain bounds on the response time and on the final displacements of impulsively loaded structures (e.g. refs. 20, 11). Here only the theorem based on the simple rigid–plastic theory will be briefly described [14].

The expression (24) obtained by the mode solution and given in the form

$$t_f \geqslant \frac{I_o}{p_k} \tag{40}$$

is a lower bound on the response time of impulsively loaded structures. Using the kinematical solution this theorem can be generalized for pulse loading [5]:

$$t_f \geqslant \frac{I(t_f)}{p_k} \tag{41}$$

On a single final displacement at a point A in the direction n Martin derived the following upper bound [10]:

$$u^f_{An} \leqslant \frac{1}{2P_s} \int_S m v_i^o v_i^o \, dS \tag{42}$$

Here P_s is a statically admissible multiplier of a concentrated virtual force acting at the point A in the direction n on the structure. The best upper bound can be obtained when P_s assumes its maximum value which is the collapse load multiplier P_c.

Considering the impulsively loaded model and assuming the numerical values $v^o_\alpha = I_o/M$ (1, 2/3) let us apply a virtual vertical force at the point 1. The collapse load intensity of this force is $p_c = (4/3)(M_o/l)$, such that using expression (42) the following upper bound can be calculated on the final vertical displacement at the point 1:

$$u_{1f} \leqslant \frac{1}{p_c}\left[\frac{1}{2} M v^o_\alpha v^o_\alpha\right] = 0{\cdot}54167 \frac{I_o^2 l}{M M_o} \tag{43}$$

7. OPTIMAL DESIGN

7.1. Basic Concepts

Only a few papers have so far been published in the literature which deal with the optimal design of plastic structures exposed to dynamic pressure [19, 9]. In the following an approximate method will be presented for the optimal design of perfectly plastic inhomogeneous structures subjected to pulse or impulsive loading. The solution is based on the kinematical approach [2].

It is assumed that the surface S of the rigid-perfectly plastic inhomogeneous structure under consideration can be subdivided into regions S_k $(k = 1, 2, \ldots, n)$ in which the density (ρ_k), the specific cost (c_k) and the yield stress (σ_y^k) are given constants. The size of S_k and the thickness of the structure (t_k), which is also constant in each region, are design variables restricted by the design constraints

$$S'_k \subset S_k \subset S''_k, \qquad t'_k \leqslant t_k \leqslant t''_k \tag{44}$$

Here S'_k, S''_k, t'_k, t''_k are prescribed bounds for S_k and t_k. Assuming different kinematically admissible displacement fields

$$u^{\circ}_{ih}(x_j, t) = W_h(t) u^{k}_{ih}(x_j), (h = 1, 2, \ldots, m) \tag{45}$$

which are extended to the total domain S and specifying for each displacement mode an allowable parameter W_{oh}, the design condition is expressed in the form

$$W_{fh} \leqslant W_{oh}, \qquad (h = 1, 2, \ldots, m) \tag{46}$$

where W_{fh} is defined by eqn. (35). Finally, for the objective function we shall choose the cost of the structure, which by making use of the summation convention has the form

$$C = c_k t_k S_k, \qquad (k = 1, 2, \ldots, n) \tag{47}$$

Then, in the case of optimal design the sizes of S_k and t_k are to be determined such that S_k, t_k and W_{fh} do not exceed the constraints (44) and (45) and the cost of the structure be a minimum. In brief,

minimize

$$C = c_k t_k S_k, \qquad (k = 1, 2, \ldots, n) \tag{48}$$

subject to

$$S'_k \subset S_k \subset S''_k, \qquad t'_k \leqslant t_k \leqslant t''_k \tag{49}$$

$$W_{fh} = K_h S_o \left[\frac{p_e}{p_{kh}} - 1\right] \leqslant W_{oh}, (h = 1, 2, \ldots, m) \tag{50}$$

In eqns. (48)–(50) p_{kh}, K_h, p_e and S_o are defined by eqns. (10), (33) and (36). It can be seen that we have a mathematical programming problem.

7.2. Example

Consider again the simple model shown in Fig. 4(a). Now we assume that the massless beam has built-in ends and is subjected to square-type pulse loading defined by eqn. (2) and by the force distribution $P^{\circ}_{\alpha}(1, 2/3)$. The allowable final displacements for both mass points are identical and denoted by W_o.

The beam is subdivided by the masses into three regions S_1, S_2 and S_3 (Fig. 8). Now these regions are fixed and we assume that in each region the plastic moments M_{01}, M_{02} and M_{03} are constants and that the cost per unit length of the beam is proportional to these plastic moments. Then, the cost of the entire beam can be expressed in the form

$$C = M_{01} + 2M_{02} + M_{03} \tag{51}$$

For the plastic moments the following restrictions are prescribed:

$$M_{01} \geqslant 0, \qquad M_{02} \geqslant 0, \qquad M_{03} \geqslant 0 \tag{52}$$

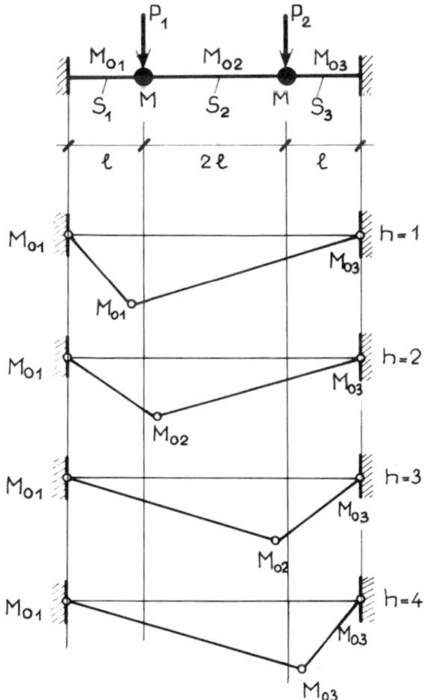

FIG. 8. Optimal design of the two-degree-of-freedom model under pulse loading.

In order to express the final displacements four kinematically admissible displacement fields will be considered (see Fig. 8). The corresponding kinematically admissible load multipliers are

$$p_{k1} = \frac{3}{11l}(7M_{01} + M_{03})$$

$$p_{k2} = \frac{3}{11l}(3M_{01} + 4M_{02} + M_{03})$$

$$p_{k3} = \frac{3}{11}(M_{01} + 4M_{02} + 3M_{03})$$

$$p_{k4} = \frac{3}{11}(M_{01} + 7M_{03})$$

From eqn. (33) $K_1 = K_2 = 11/10M$ and $K_3 = K_4 = 9/10M$ and applying formula (50) for each yield mechanism the following inequalities

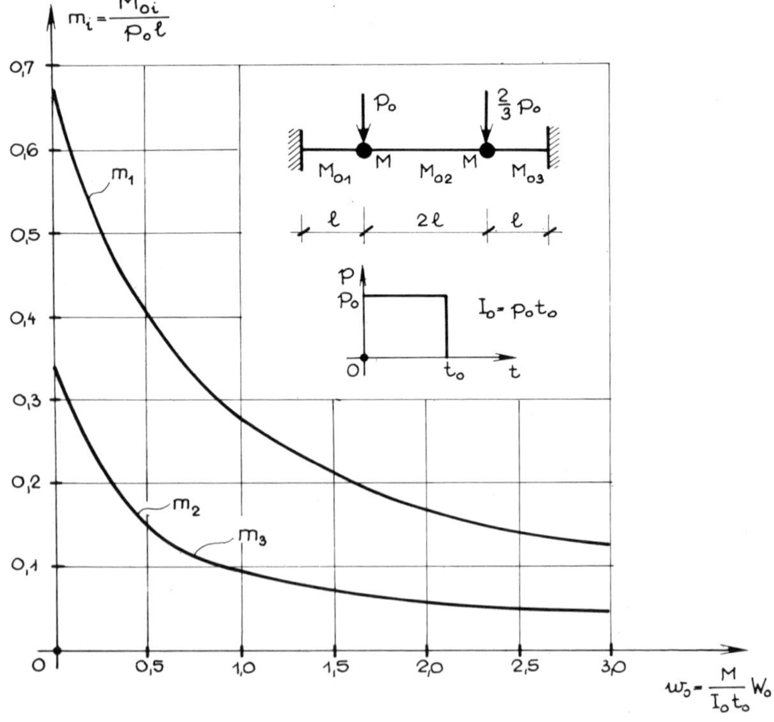

FIG. 9. Optimal solution of the two-degree-of-freedom model under pulse loading.

can be obtained:

$$\left.\begin{aligned}7m_{01}+\qquad\qquad m_{03}&\geq\frac{2\cdot016}{w_o+0\cdot550}\\3m_{01}+4m_{02}+m_{03}&\geq\frac{2\cdot016}{w_o+0\cdot550}\\m_{01}+4m_{02}+3m_{03}&\geq\frac{1\cdot35}{w_o+0\cdot450}\\m_{01}+\qquad\qquad 7m_{03}&\geq\frac{1\cdot35}{w_o+0\cdot450}\end{aligned}\right\}\qquad(53)$$

Here $m_{0k}=M_{0k}/(p_o l)$ and $w_o=M/(I_o t_o)W_o$ are dimensionless variables. To find the optimal solution for the plastic moments M_{0k} one has to minimize the object function (51) subject to the constraints (52) and (53). The results obtained for different values of the allowable final displacements W_o and impulses I_o are plotted in Fig. 9.

8. EFFECT OF STRAIN RATE SENSITIVITY

8.1. Constitutive Law

In most dynamic problems the rigid–viscoplastic behaviour of structures is represented by the constitutive law proposed by Perzyna [17]. This can be expressed in the following special form:

$$\left.\begin{aligned}\dot{q}_j&=\dot{q}_o\left[\frac{\psi(Q_j)}{Q_o}-1\right]^n\frac{\partial\psi}{\partial Q_j}\quad &&\text{for}\quad \psi(Q_j)\geq Q_o\\\dot{q}_j&=0 &&\text{for}\quad \psi(Q_j)<0\end{aligned}\right\}\qquad(54)$$

In case of simple bending

$$Q_j=M,\ Q_o=M_o,\ \dot{q}_j=\dot{\kappa},\ \dot{q}_o=\dot{\kappa}_o=\frac{2\dot{\varepsilon}_o}{H}\left(1+\frac{1}{2n}\right)^n$$

and hence

$$\left.\begin{aligned}\dot{\kappa}&=(\text{sign }M)\dot{\kappa}_o\left[\frac{|M|}{M_o}-1\right]^n\\ &\text{or}\\ M&=(\text{sign }\dot{\kappa})M_o\left[1+\left(\frac{|\dot{\kappa}|}{\dot{\kappa}_o}\right)^{1/n}\right]\end{aligned}\right\}\quad\text{for}\quad |M|\geq M_o\qquad(55)$$

$$\dot{\kappa}=0\qquad\qquad\text{for}\quad |M|<M_o$$

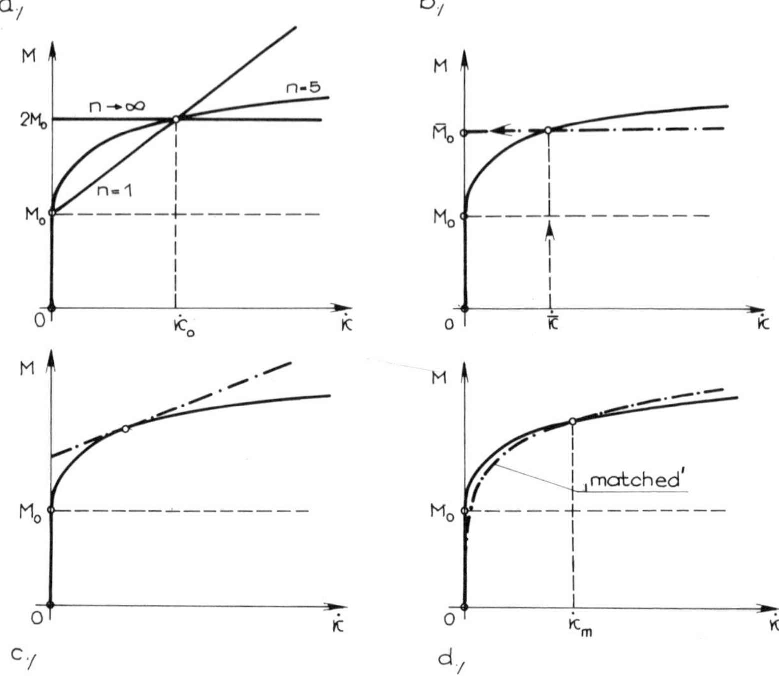

FIG. 10. Approximation of the constitutive law of strain rate sensitive structures.

Here $\dot{\kappa}$ is the curvature rate, H and M_o are the thickness and the plastic moment of the beam and n and $\dot{\varepsilon}_o$ denote viscous constants (e.g. for mild steel $\dot{\varepsilon}_o = 40\ \mathrm{s}^{-1}$ and $n = 5$ [20]). When $n = 1$ we obtain the case of linear viscosity, while when $n \to \infty$ the behaviour approaches that of rigid–plastic material with doubled yield stress (Fig. 10(a)).

In the following a few methods will be outlined for the approximate analysis of dynamically loaded rigid–viscoplastic structures.

8.2. Linearization of the Constitutive Law

The simplest idea of approximation is that the exponential function of the one-dimensional constitutive law is replaced by a straight line with zero slope, which provides a reasonable approximation for higher strain rates (Fig. 10(b)). Based on this idea Perrone elaborated a quasi-iteration method in which first the rigid-perfectly plastic dynamic problem has to be solved [16]. Then, at the strain rate at which the correction is computed a maximum or a time average magnitude of strain rates has to be determined (see $\bar{\dot{\kappa}}$ in Fig. 10(b)). Finally, using the modified yield stress (see $\bar{M}_o$ in Fig. 10(b)) the rigid-perfectly plastic solution can be corrected.

The nonlinear function of the one-dimensional constitutive law can be replaced by a straight line which is either tangent to the curve representing the function or intersects it at a chosen point (Fig. 10(c)). The general constitutive law, however, can be only partially linearized in this manner, since it involves $\psi(Q_j)$ described by non-linear functions of the generalized stresses. A full linearization is obtained if a suitably modified statically admissible system is introduced. Then the governing equations become similar to those of vibrational problems of elastic structures.

8.3. Mode Solution

Inhomogeneous stress–strain rate relations do not allow mode form solutions in the sense of separated variable forms that hold for the entire motion. Physically this is due to the fact that the interfaces between the rigid and deformable regions of the structure are not fixed during the response. This difficulty may be overcome by the use of a homogeneous viscous rather than a rigid–viscoplastic constitutive law, replacing (54) and (55) by a 'matched' viscous representation as below [22]:

$$\dot{\kappa} = (\text{sign } M)\dot{\kappa}'_o\left[\frac{|M|}{M_o}\right]^{n'} \quad \text{or} \quad M = (\text{sign } \dot{\kappa})\left[\frac{|\dot{\kappa}|}{\dot{\kappa}'_o}\right]^{1/n'} \tag{56}$$

where $M'_o = \mu M_o$, $\dot{\kappa}'_o = \dfrac{4\dot{\varepsilon}_o}{H}$ and $n' = \nu n$. Choosing $\dot{\kappa}_m = \beta\dot{\kappa}_o$ as 'matching' point (Fig. 10(d)) μ and ν are determined by the relationships

$$\mu = \frac{1+\beta^{1/n}}{\beta^{1/n'}}, \qquad \nu = \frac{1+\beta^{1/n}}{\beta^{1/n}} \tag{57}$$

Introducing the homogeneous 'matched' viscous constitutive law (56) the impulsively loaded structures have no rigid regions and can undergo permanent mode form motions. Therefore the mode solution described in Section 4 can be generalized for these problems.

8.4. Kinematical Solution

The kinematical solution of rigid-perfectly plastic structures presented in Section 5 has been generalized for rigid–viscoplastic behaviour [6]. The idea of this further approximation is to replace in eqn. (32) the kinematically admissible multiplier p_k obtained by the simple rigid–plastic theory by p_k^v which expresses the resistance displayed by the rigid–viscoplastic structure. While p_k is a constant, p_k^v can be expressed in terms of the velocity of the displacement parameter function W. Using the constitutive law (54) p_k^v can be written in the following general form:

$$p_k^v(\dot{W}) = p_k(1 + a\dot{W}^{\alpha}) \tag{58}$$

where the constants a and α are obtained from the quasi-static solution of the rigid–viscoplastic problem. Substituting (58) in eqn. (32) the differential equation of motion has a modified form

$$\ddot{W}(t) = K[p(t) - p_k(1 + a\dot{W}^{\alpha})] \tag{59}$$

Let us apply the kinematical approximation to the simple rigid–viscoplastic model subjected to impulsive loading defined by the initial velocity $v_\alpha^o = \dfrac{I_o}{M} P_\alpha^o$, where $P_\alpha^o(1, 2/3)$. Select the kinematically admissible displacement field to the yield mechanism I of Fig. 2. Then, the kinematically admissible load multiplier of the rigid–viscoplastic model associated with the assumed yield mechanism can be calculated by making use of eqn. (55):

$$p_k^v = \frac{12}{11}\frac{M_o}{l}\left[1 + \left(\frac{4}{3}\frac{H}{l}\right)^{1/n}\left(\frac{\dot{w}}{\dot{\kappa}_o}\right)^{1/n}\right] \tag{60}$$

Here H is the thickness of the beam, $\dot{w} = \dot{W}/H$ and

$$\dot{\kappa}_o = \frac{2\dot{\varepsilon}_o}{H}\left(1 + \frac{1}{2n}\right)^n$$

Let us assume the numerical values $n = 5$, $\dot{\varepsilon}_o = 40\ \mathrm{s}^{-1}$ and $H/l = 0{\cdot}25$ and modify the differential equation (59) such that $p(t) = 0$ and an equivalent initial velocity $V_o = KI_o$ is taken into account. Then, for the problem under consideration the following equation is derived:

$$\frac{\mathrm{d}^2 w}{\mathrm{d}t^2} + 0{\cdot}9917\frac{v_o^2}{\lambda}\left[1 + 0{\cdot}3036\left(\frac{\mathrm{d}w}{\mathrm{d}t}\right)^{1/5}\right] = 0 \tag{61}$$

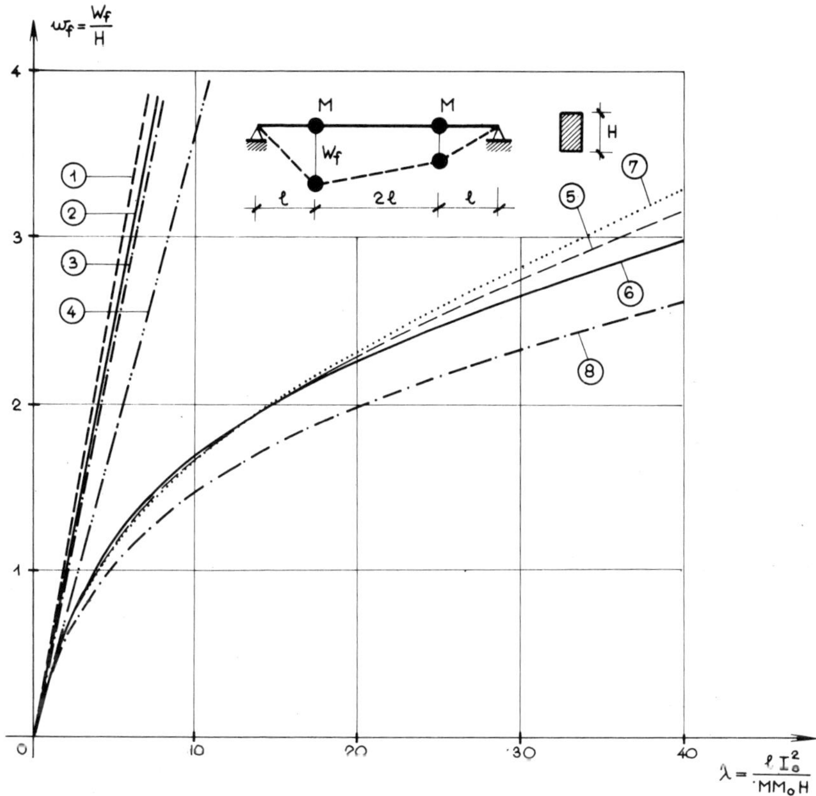

FIG. 11. Solutions based on different methods of the two-degree-of-freedom model under impulsive loading.

where

$$w = \frac{W}{H}, v_o = \frac{V_o}{H} \quad \text{and} \quad \lambda = \frac{I_o^2 l}{MM_oH} \tag{62}$$

This nonlinear differential equation has been solved numerically. The final displacements at the point 1 of the model are presented in Fig. 11 by curve 4. The curves 1, 2 and 3 show the results given by eqns. (43) (bounding theorem), (31) and (39) (mode solution and kinematical solution) and (19) (exact solution), respectively.

9. EFFECT OF LARGE DISPLACEMENTS

9.1. Mode Form Solution

It can readily be seen that when nonlinear terms for finite deflections appear in the field equations it becomes impossible to obtain full solutions in mode form. Hence, the mode form solution cannot be directly applied for the analysis of structures undergoing large deformations.

It is possible to determine such quasi-mode form solutions that start out as simple mode form solutions in the small deflection range and take account of the effects of large deformations as the response continues. In these solutions the minimum Δ_o choice can be applied.

The aim of the 'instantaneous mode form' solution suggested by Symonds and Chon [23] is to apply a sequence of instantaneous mode solutions in separated-variable form for subsequent time intervals. These satisfy the field equations at each stage but they do not provide a smoothly connected system of solutions since continuity is imposed only at certain points of the structure. This method should provide a satisfactory simple approximation at least for a structure whose configuration does not change significantly during the response.

9.2. Kinematical Solution

The kinematical solution can be generalized for structures undertaking moderately large displacements [6]. The idea of this further approximation is to replace in eqn. (32) the kinematically admissible multiplier p_k obtained by the simple rigid–plastic theory by another one denoted by p_k^1, which expresses the resistance displayed by the structure undertaking large deflections. While p_k is a constant, p_k^1 can be expressed in terms of the displacement parameter function: $p_k^1 = p_k^1(W)$. In most cases this function can be written in the following general form:

$$p_k^1(W) = p_k(1 + bW^\beta) \tag{63}$$

Here the constants b and β are obtained from the quasi-static solution of the problem taking into consideration the effects of deflections. Substituting (63) into eqn. (32) the differential equation of motion has a modified form

$$\ddot{W}(t) = K[p(t) - p_k(1 + bW^\beta)] \tag{64}$$

By the combination of eqns. (59) and (64) the kinematical approach can be extended to the dynamic analysis of problems at which both the effect of strain rate sensitivity and the influence of large displacements are taken into consideration. Then the differential equation of motion has the following general form:

$$\ddot{W}(t) = K[p(t) - p_k(1 + a\dot{W}^{\alpha})(1 + bW^{\beta})] \tag{65}$$

9.3. Bounding Theorem

The bounding theorem described below was proposed by Ploch and Wierzbicki for impulsively loaded rigid-perfectly plastic structures undergoing moderately large deformations [18].

Let us apply a quasi-static concentrated virtual force P_s on the structure at the point A in the direction n and denote the corresponding statically admissible stress and displacement fields by σ_{ij}^s and u_i^s. Now the system P_s, σ_{ij}^s, u_i^s has to satisfy the equilibrium equations and yield conditions taking into consideration the second order effects caused by the assumed displacements u_i^s. Then, an upper bound on a single final displacement at the point A in the direction n can be obtained as

$$u_{An}^f \leqslant \frac{1}{P_s}\left[\frac{1}{2}\int_S mv_i^o v_i^o \, dS + \frac{1}{2}\int_V \sigma_{ij}^s u_{i,j}^s u_{j,i}^s \, dV\right] \tag{66}$$

where v_i^o is the initial velocity field corresponding to the impulsive loading. When the effect of large deformations is neglected the second term of the right-hand side equals zero and formula (66) gives the bound (42) derived by Martin for structures with small deformations.

9.4. Example

Consider the two-mass model of Fig. 4 subjected to impulsive loading defined by the initial velocity $v_\alpha^o = \dfrac{I_o}{M} P_\alpha^o$, where $P_\alpha^o(1, 2/3)$. Suppose that the beam is fully restrained from axial motion and for its rectangular cross-section the yield condition is

$$f = \frac{|M|}{M_o} + \left(\frac{N}{N_o}\right)^2 - 1 = 0$$

Here $M_o = H^2/4\ \sigma_o$ and $N_o = H\sigma_o$ denote the fully plastic moment and axial force, respectively, and H is the height of the beam.

(a) First let us apply the kinematical solution and assume the kinematically admissible velocity field shown in Fig. 12(a). Then, from Fig. 12(b) for the kinematically admissible load multiplier the following

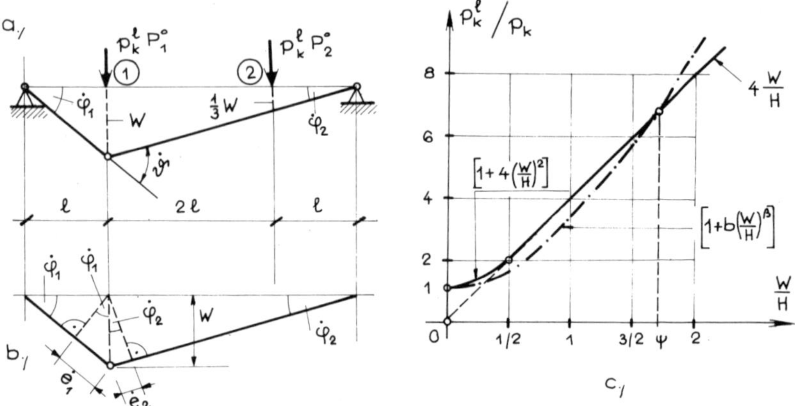

FIG. 12. Large deflections of the two-degree-of-freedom model.

results can be obtained (see Fig. 12(c))

$$p_k^1 = p_k\left[1+4\left(\frac{W}{H}\right)^2\right], \quad \text{if} \quad \frac{W}{H} \leqslant \frac{1}{2} \tag{67}$$

$$p_k^1 = 4p_k\frac{W}{H}, \quad \text{if} \quad \frac{W}{H} \geqslant \frac{1}{2} \tag{68}$$

Here $p_k = \dfrac{12}{11}\dfrac{M_o}{l}$.

In the range of larger deflections a good approximation can be expected by the use of eqn. (68). Then the differential equation (64) becomes

$$\frac{d^2w}{dt^2} + 3{\cdot}967\frac{v_o^2}{\lambda}w = 0 \tag{69}$$

and after its solution the final value of the displacement parameter function is

$$w_f = \frac{W_f}{H} = 0{\cdot}5021\sqrt{\lambda}\,; \; \lambda = \frac{I_o^2 l}{MM_oH} \tag{70}$$

This result is illustrated in Fig. 11 by curve 5.

(b) Equations (67) and (68) can be approximated by the single formula

$$p_k^1 = p_k\left[1+2\left(\frac{W}{H}\right)^2\right] \tag{71}$$

Then the differential equation (64) has the form

$$\frac{d^2w}{dt^2}+0{\cdot}9917\frac{v_o^2}{\lambda}[1+2w^2]=0 \tag{72}$$

The results of the numerical solution of this nonlinear differential equation are given in Fig. 11 by curve 6.

(c) Solving the same problem by the use of the bounding theorem described in Section 9.3 the following upper bound is obtained:

$$w_f=\frac{W_f}{H}=0{\cdot}5204\sqrt{\lambda} \tag{73}$$

This upper bound is illustrated in Fig. 11 by curve 7.

(d) Finally consider the problem when both the effect of strain rate sensitivity and the influence of large deformations are taken into consideration. Using the kinematical approach and formula (65) by the combination of eqns. (61) and (72) the following differential equation is obtained

$$\frac{d^2w}{dt^2}+0{\cdot}9917\frac{v_o^2}{\lambda}[1+2w^2]\left[1+0{\cdot}3036\left(\frac{dw}{dt}\right)^{1/5}\right]=0 \tag{74}$$

The numerical solution of this nonlinear differential equation leads to the final displacements presented in Fig. 11 by curve 8.

10. EFFECT OF ELASTIC DEFORMATIONS

10.1. General Remarks

The exact dynamic analysis involving elastic and plastic deformations is far more difficult than the elastic, the rigid–plastic or the rigid–viscoplastic solutions. In the elasto–plastic structure several plastic regions might develop which increase or decrease in size during the response. This complicated behaviour can be followed only by numerical methods. Now two simple approximate methods will be briefly described.

10.2. Mode Form Solution

The concept of this approach, proposed by Symonds [21], is to divide the response of the structure into two sharply separated phases; in the first phase the structure performs solely linear elastic deformations while in the second phase a rigid–plastic behaviour is assumed.

The linear elastic phase is treated by the elastic dynamic equations. This phase is terminated at time t_1 when a global criterion of plastic yielding is satisfied. For this purpose a 'global average' stress has to be calculated from the elastic stresses in the regions where plastic flow occurs.

In the rigid–plastic phase an appropriate mode form has to be assumed expressed in a separated-variable form $W(t)u_i^{\mathrm{m}}(x_j)$. The initial value of W is taken so that the displacement at a chosen point of main interest in the structure equals the displacement at this point given by the elastic solution at time t_1. The initial velocity amplitude V_{o} is determined from the elastic velocity field $\dot{u}_i^{(1)}(x_j, t_1)$ by minimizing the 'difference' function Δ_{o} as it was described in Section 4.2. Finally, the response time and the maximum displacements of the elasto–plastic structure are equal to the sum of the corresponding values of the elastic and plastic phases.

In the described method the effect of strain rate sensitivity can also be taken into consideration and the solution has been extended to the analysis of the phase of large plastic deformations as well.

10.3. Kinematical Solution

The idea of the solution is again to impose a kinematically admissible displacement field $u_i^{\mathrm{k}}(x_j)$ on the structure but now the regions of the structure which perform deformations are elasto–plastic rather than perfectly plastic [7]. Then, during the entire response the structure undertakes stationary displacements expressed in the form $u_i^{\mathrm{o}}(x_j, t) = W(t)u_i^k(x_j)$. For example in case of frames the kinematically admissible displacement field is obtained by introducing elasto–plastic hinges while the other parts of the frame are assumed to be perfectly rigid. The elastic properties of the hinges can be approximately calculated from the elastic stiffnesses of the neighbouring parts of the frame while the plastic characteristics are determined by the plastic moments of the cross-sections. This frame undertakes only stationary displacements during the entire response and forms a one-degree-of-freedom system. Still, the states of the hinges are changing; some of them are in an elastic and others in a plastic stage during the response, consequently the resistance displayed by the system is not constant. Therefore a step-by-step time history analysis is needed for the solution of the problem.

The computational work can be further reduced by introducing the concept of 'average global' elastic stiffness of the structure. Let us

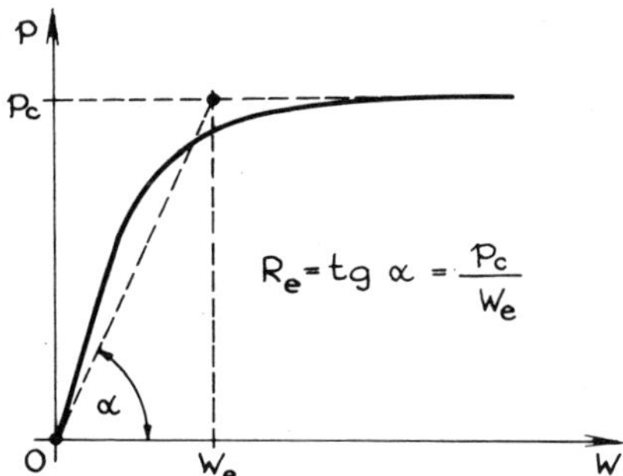

FIG. 13. Linearization of the quasi-static response of an elasto–plastic structure.

suppose that the elasto–plastic solution of the structure subjected to the quasi-static loading $pT_i^o(x_j)$ is known and the relationship between the load parameter p and the displacement W of a chosen point of the structure is also given (Fig. 13). Then the 'global average' stiffness of the elasto–plastic structure can be chosen $R_e = p_c/W_e$, where p_c is the collapse load parameter and W_e is an appropriately chosen value. Now superimposing the collapse mechanism on the structure two phases of motion can be distinguished which are described by the following two differential equations:

if $W < W_e$, then

$$\ddot{W}(t) = K\left[p(t) - \frac{p_c}{W_e}\,W(t)\right] \tag{75}$$

if $W \geqslant W_e$, then

$$\ddot{W}(t) = K[p(t) - p_c] \tag{76}$$

The solution of these differential equations provides the maximum displacements of the elasto–plastic structure under consideration.

10.4. Example

To illustrate the application of the kinematical method consider the simple model of Fig. 4. Now the ends of the beam are built-in and the constant flexural stiffness is denoted by EJ (Fig. 14(a)). The pulse load $pP_\alpha^o(1, 2/3)$ is replaced by impulsive loading characterized by the impulse I_o. The relationship between the load multiplier p and the displacement W of the point 1 determined under quasi-static conditions is

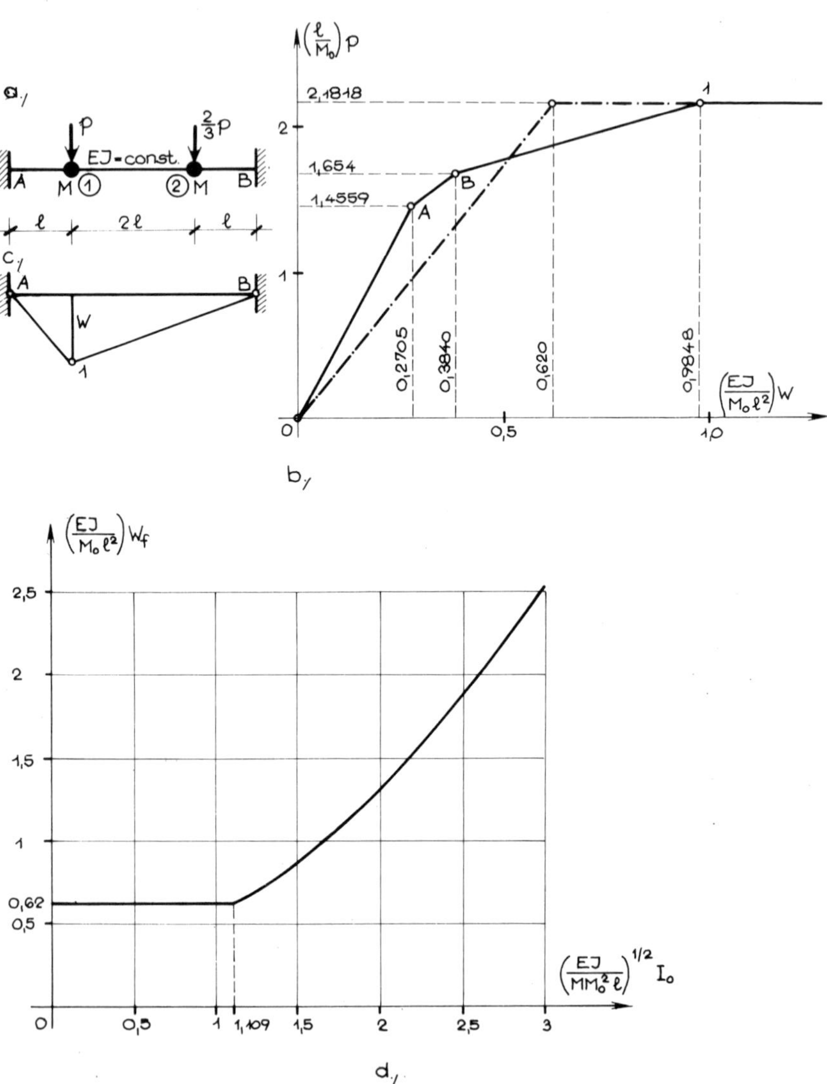

FIG. 14. Kinematical solution of the elasto–plastic two-degree-of-freedom model under pulse loading.

shown in Fig. 14(b). To obtain the 'average global' elastic stiffness of the model we chose $W_e = 0{\cdot}620\ (M_o l^2/EJ)$ and then $R_e = (p_c/W_e) = 3{\cdot}519(EJ/l^3)$. Using the collapse mechanism of Fig. 14(c) and eqns. (75) and (76) we obtain the differential equations of motion which can be solved analytically. The results are plotted in Fig. 14(d).

11. CONCLUDING REMARKS

This paper has attempted to present a review of the most important effects and methods of the dynamic plastic response of structures subjected to high intensity short-time pressure. Besides the plastic deformation the effects of strain rate sensitivity and elastic deformations and the influence of large deflections have been discussed and the principles and the simple approximate solutions due to these problems have also been described.

The material strain hardening, the large transverse shear forces and the rotatory inertia might also influence the dynamic plastic behaviour of structures and in certain cases the plastic buckling or unstable behaviour of the structure might play a dominant role in the dynamic response [15, 14, 3]. The description of these phenomena was beyond the scope of this article.

Finally, it is to be noted that a large number of experimental results and computer methods and solutions are available in the literature [4]. This text has great theoretical and practical importance; its survey should form, however, the subject of a separate article.

REFERENCES

1. Capurso, M. Extended displacement bound theorems for continua subjected to dynamic loading, *J. Mech. Phys. Solids*, **23** (1975), 113–122.
2. Heinloo, M. and S. Kaliszky. Optimal design of dynamically loaded rigid-plastic structures. Application: thick-walled concrete tube, *J. Struct. Mech.*, **9** (1981), 235–251.
3. Jones, N. Response of Structures to Dynamic Loading, Inst. Phys. Conf. Ser. No. 47. Chapter 3.
4. Jones, N. A literature review of the dynamic plastic response of structures, *Shock and Vibration Digest*, **7** (1975), **10** (1978), **13** (1981).
5. Kaliszky, S. Approximate solutions for impulsively loaded inelastic structures and continua, *Int. J. Non-Linear Mech*, **5** (1970), 143–158.
6. Kaliszky, S. Large deformations of rigid-viscoplastic structures under impulsive and pressure loading, *J. Struct. Mech.*, **1** (1973), 295–317.
7. Kaliszky, S. and V. T. Duong. Approximate analysis of elasto–plastic

structures under dynamic pressure (in Hungarian), *Építés-Építészettudomány*, **XVI**(1–2) (1984), 25–38.
8. KRAJCINOVIC, D. On approximate solutions for rigid-plastic structures subjected to dynamic loading, *Int. J. Non-Linear Mech.*, **7** (1972), 571–575.
9. LEPIK, U. and Z. MRÓZ. Optimal design of plastic structures under impulsive and dynamic pressure loading, *Int. J. Solids Struct.*, **13** (1977), 657–674.
10. MARTIN, J. B. A displacement bound technique for elastic continua subjected to a certain class of dynamic loading, *J. Mech. Phys. Solids*, **12** (1964), 165–175.
11. MARTIN, J. B. Displacement bounds for dynamically loaded elastic structures, *J. Mech. Engng Sci.*, **10** (1968), 213–218.
12. MARTIN, J. B. Extremum principles for a class of dynamic rigid–plastic problems, *Int. J. Solids Struct.*, **8** (1972), 1185–1204.
13. MARTIN, J. B. and P. S. SYMONDS. Mode approximations for impulsively loaded rigid–plastic structures, *J. Engng Mech. Div. Proc. ASCE*, **92**, No. EM5 (1966), 43–66.
14. NONAKA, T. Shear and bending response of a rigid–plastic beam to blast-type loading, *Ing. Arch.*, **46** (1977), 35–52.
15. OLIVEIRA, J. G. and N. JONES. A numerical procedure for the dynamic plastic response of beams with rotatory inertia and transverse shear effects, *J. Struct. Mech.*, **7** (1979), 193–230.
16. PERRONE, N. On a simplified method for solving impulsively loaded structures of rate-sensitive materials, *J. appl. Mech.*, **32** (1965), 489–492.
17. PERZYNA, P. The constitutive equations for rate sensitive plastic materials, *J. appl. Math.*, **20** (1963), 321–331.
18. PLOCH, J. and T. WIERZBICKI. Bounds in large inelastic deformations of dynamically loaded structures (in Polish), *Proc. 3rd Symposium Dynamic Stability of Structures.*
19. REITMAN, M. I. and G. S. SHAPIRO. Optimization of structures under dynamic loading (in Russian), *Conference on Problems of Optimization in the Mechanics of Deformable Bodies*, Vilnius, 1974.
20. SYMONDS, P. S. and T. WIERZBICKI. On an extremum principle for mode form solutions in plastic structural dynamics, *J. appl. Mech.*, **42** (1975), 630–640.
21. SYMONDS, P. S. Elastic, finite deflection and strain rate effects in a mode approximation technique for plastic deformations of pulse loaded structures, *J. Mech. Engng Sci.*, **22** (1980), 189–197.
22. SYMONDS, P. S. and C. T. CHON. Approximation techniques for impulsive loading of structures of time-dependent plastic behavior with finite deflections, *Proc. Oxford Conf. Mechanical Properties of Materials at High Strain Rates*, Institute of Physics Conference Series No. 21, Dec. 1974, 299–315.
23. SYMONDS, P. S. and C. T. CHON. On dynamic plastic mode form solutions, *J. Mech. Phys. Solids*, **26** (1978), 21–35.
24. TAMUZS, B. P. On a minimum principle in the dynamics of rigid–plastic solids (in Russian), *Prikl. Mat. Mekh.*, **26** (1962), 715–722.

43

An Approach to the Limit Design of Masonry Arches under Seismic Loads

VINCENZO FRANCIOSI

Istituto di Scienza delle Costruzioni-Ingegneria, University of Naples, Italy

ABSTRACT

Attention is called to the fact that Heyman's suggestion (a masonry arch collapses owing to a mere impossibility of equilibrium) is particularly near the truth when the thrust line of living loads, e.g. seismic loads, is very different from the axis of the arch. It is advisable to consider the possibility of using slides; but this is especially true when the loads are alternating, as is the case during an earthquake. The primal-dual method can be used, but a program based on the contiguous classes procedure requires a rather small memory. Standard and non-standard materials have been considered. Some examples have been given; from these it transpires that by using slides the collapse factor can be substantially reduced.

1. THE 'BOUNDLESS YIELD STRESS' PROCEDURE: ITS VALIDITY AND ITS LIMITATIONS

In 1966, while studying the statics of Gothic cathedrals, J. Heyman [7] proposed a simplification of the plastic design of masonry arches by substituting the two tangents to the curve in the single point O (Fig. 1) for the yield curve Λ in the NM plane. This is the same as supposing there is no limit to the increase of yield stress of the masonry. Thus the problem is once more one of mere mechanics and therefore we can call the consequent design procedure 'boundless yield stress'.

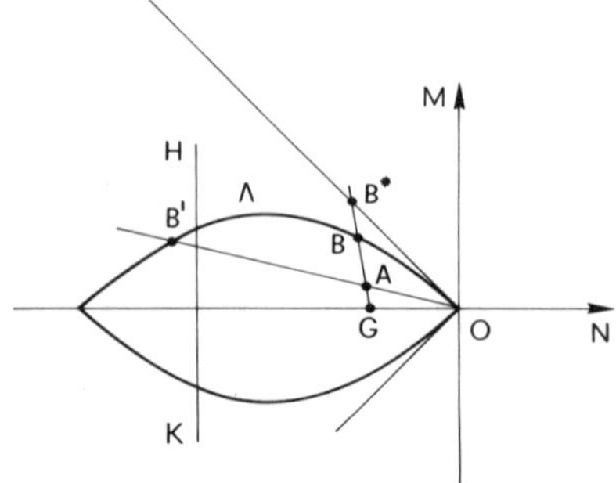

FIG. 1. True and Heyman's yield curve.

The collapse factor k_r is insignificant if the dead and living loads increase together. Thus occurs when we study a retaining wall, or a slope, employing Coulomb's criterion without cohesion: the only information we can obtain is whether or not the set of loads will cause the collapse of the arch. If (and only if) there is at least one thrust line lying entirely within the arch (statically admissible line of thrust) collapse cannot occur, and the set of forces can increase beyond any limit.

On the other hand, the collapse factor regains its full interest when the dead loads do not vary, i.e. when only the living loads increase by multiplication by only one factor. In this case, the importance of this factor is much greater. In fact, living loads are aleatory, while the nature of dead loads is deterministic. This is very evident, for instance, in the case of seismic equivalent loads.

Heyman's proposal appears unacceptable in the first case, but fortunately appears acceptable in the second. In fact, for what concerns the four plastic hinges of the mechanism, in the orthodox limit design in the first case the yield point B lies in the region of Λ farthest from the M-axis, while, in the second case, the yield point lies in the nearest region, and it is of no great consequence if we assume $B = B^*$.

Besides, in classic limit design practice, the elastic domain is limited on the left side by a vertical straight line HK, to bound the strain compression. Thus it is evident that Heyman's proposal results in a yield curve not much different from the classic one.

We remark that the error is smaller when the vector GA is near the vertical, i.e. when the internal forces caused by the living loads show a considerable degree of eccentricity. We can say that results become more and more acceptable as the 'shape' of living loads differs more and more from that of dead loads. Therefore, results are not reliable when, for example, the living forces consist of a uniform distributed load that engages the arch almost from one end to the other. Vice versa, results correspond very closely to what actually occurs in reality under antisymmetric loads (either horizontal or vertical), as for instance in an earthquake, or when a strip of uniformly distributed load lies on only half a span.

The above considerations are valid also for reinforced concrete arches, if the sections are ideally reinforced, or, rather, underreinforced. Comparative calculations carried out on a 100-m span bridge arch with thin walled symmetrically reinforced cross-section, and with a steel area (superior or inferior) corresponding to 0·5% of the section's total area, under a strip of uniform load distributed on one third of the span, gave an error which in practice is negligible [5].

2. EXCEEDED FRICTION COLLAPSE IN PRESENCE OF STANDARD MATERIALS

The kinematic procedure implies the indestructibility of shallow arches since there is no possibility of a 'consistent' mechanism [7, 8, 9]. This absurdo goes hand in hand with another according to which no variation of temperature can be tolerated by the structure [6]. Both these absurdities prove to be mere paradoxes if we consider that a slide between two voussoirs is possible once the slope of the thrust line to the tangent to the axis of the arch is greater than the angle of friction [4]. We can observe that this is the same as considering or disregarding the influence of the shear stresses in the strength control of the cross-section in elastic range. In both cases two ratios f/l and h/f play the leading role.

The general normality condition of plastic theory is trivially satisfied in the orthodox limit design when we assimilate the elementary trunk to a set of bars not capable of bearing tensile stress. The normality condition is likewise trivially satisfied in the kinematic procedure. Taking into account shearing forces, the normality condition must be postulated in both cases. In reality this condition is not always respected. If this condition is valid, the static and kinematic theorems of limit design also are valid. The slide plastic joint (sliding block, or

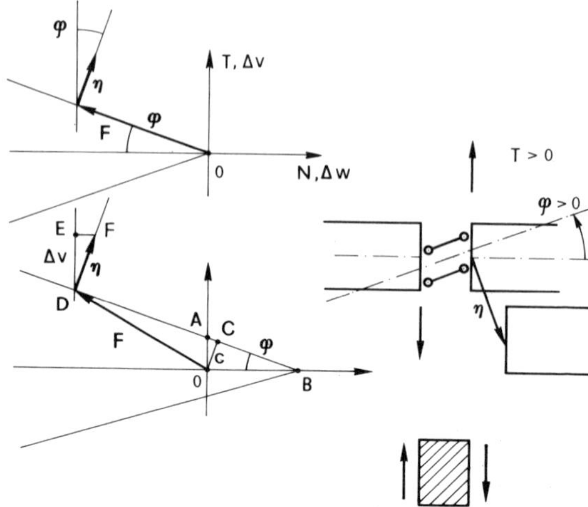

FIG. 2. Slide plastic joint ('pattino').

'pattino') is shown in Fig. 2. We must point out that, in the absence of concentrated loads, the mechanism can only present 'pattini', or hinges formed in the intrados or the extrados. In fact, if the thrust line has no single points, it cuts the cross-section at an internal point and thus only a 'pattino' can be formed. Otherwise it meets the section in the intrados or extrados, necessarily as tangent to the edge. In this case only a hinge can result.

The 'pattino', like the hinge, is a unilateral joint and can be one of two kinds. If $T>0$, as in Fig. 2, the right side will drop down with respect to the left side ($\Delta v>0$). With dilatancy this slide is compatible with two bars whose axes 'a' are parallel and form an angle φ with the tangent to the axis of the arch (Fig. 2). Hence, the 'pattino' must have a slope $\pm\varphi$, and must allow a slide $\Delta v \gtrless 0$.

In absence of cohesion, the work done by the internal forces acting on the two faces assembled by the 'pattino' is zero; in presence of cohesion the work is

$$L^* = -F \, . \, \eta = -C\eta$$

for any and every value of F.

In absence of cohesion—and, for safety, this hypothesis must be considered—the collapse factor that multiplies all the loads is still insignificant. The analogy with retaining walls is more than evident.

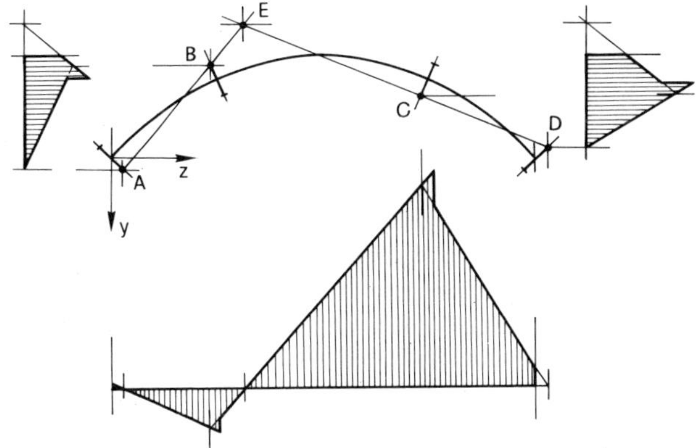

FIG. 3. Four hinges mechanism.

The mechanism can have four hinges (Fig. 3). If every possibility of slides is excluded this is the only mechanism possible. If the arch is slender and not too shallow, this is the true mechanism, even though the angle of friction is rather small. Hence, it is useful to start from this mechanism, also because the safety factor can only be one and an interval thus obtained (taking into account friction) is valid and often acceptable.

Other kinds of mechanisms have one 'pattino' and three hinges (Fig. 4) or two 'pattini' and two hinges (Fig. 5). In all there are nine of

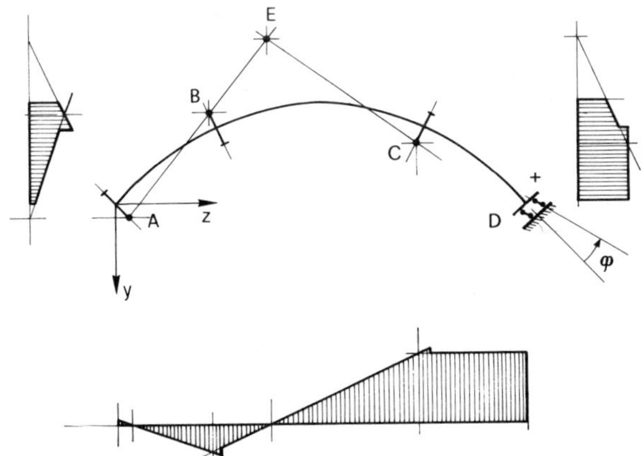

FIG. 4. Three hinges and one slide joint mechanism.

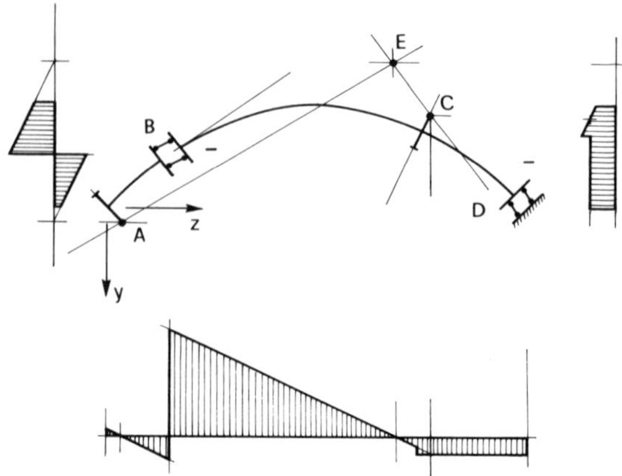

FIG. 5. Two hinges and two slide joints mechanism.

these mechanisms. Finally, there is the mechanism with three 'pattini' (Fig. 6), that has no hinges since three 'pattini' are sufficient to change the structure into a mechanism.

The classic procedure concerning possible classes of mechanisms can be used. We begin by choosing a mechanism, preferably the first one with four hinges, and obtain the kinematically admissible factor when imposing the equilibrium conditions. Then we calculate the values of the three redundant quantities X_1 X_2 X_3 referred to the centroid of the

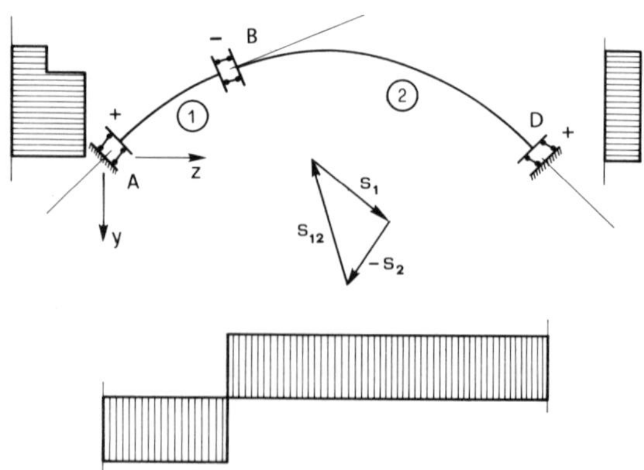

FIG. 6. Three slide joints mechanism.

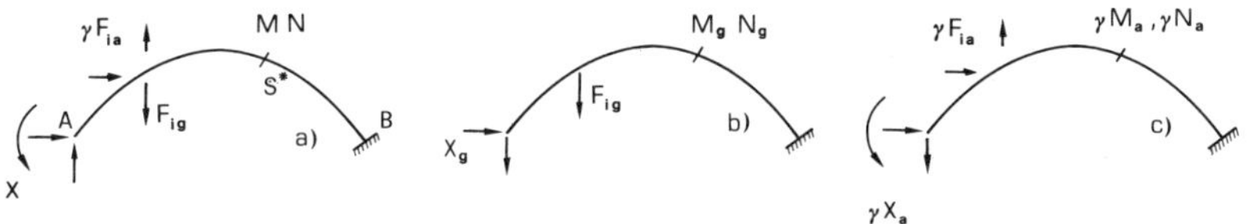

FIG. 7. The equivalent console.

left support and those of the internal forces N, M, T in all the cross-sections (Fig. 7(a)).

Let N_g, M_g, T_g be the set of internal forces (Fig. 7(b)) statically admissible and in equilibrium with the dead forces F_{ig}. Under the assumption that dead loads cannot cause collapse, at least one such set must exist. The set looked for can be obtained by drawing the thrust line of the dead forces that touches three points lying on three cross-sections, e.g. the centroids of the end sections and of the keystone.

The internal forces $C-C_g$ are in equilibrium with the living forces γF_{ia} (Fig. 7(c)). The internal forces $C_a=(C-C_g)/\gamma$ are in equilibrium with the living forces F_{ia}, and the internal forces $C_g+(\gamma/r)C_a$ are in equilibrium with the external forces $F_{ig}+(\gamma/r)F_{ia}$.

In every cross-section (Fig. 8), where the point P_m(NM) (or the point P_t(NT)) lies outside the yield bilatera NM (or NT), we calculate the ratio $r_m=P_gP_m/P_gP'_m$ (or $r_t=P_gP_t/P_gP'_t$). Let $\bar{r}$ be the highest value of 'r'; the set of internal forces $C_g+(\gamma/\bar{r})C_a$ is in equilibrium with the external forces $F_{ig}+(\gamma/\bar{r})F_{ia}$, and is statically admissible. The number $\gamma/\bar{r}$ is a statically admissible factor and we can write

$$\frac{\gamma}{\bar{r}}\leqslant\kappa_r\leqslant\gamma \tag{1}$$

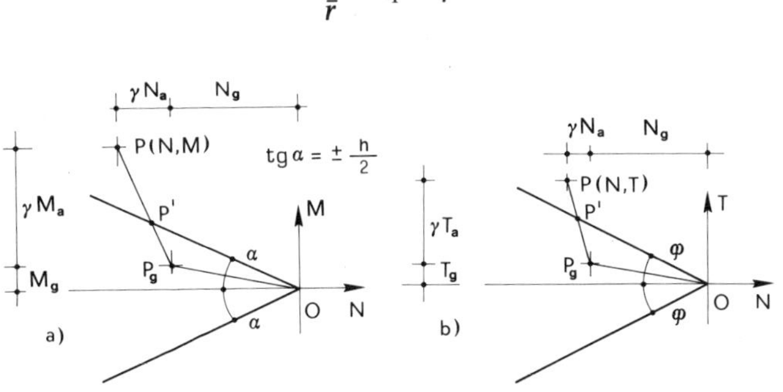

FIG. 8. Study of the safety interval.

We must bear in mind the two upper values of r_m under $M \gtrless 0$, and the two upper values of r_t under $T \gtrless 0$, within the abscissae of the relative sections. In fact, in the subsequent trial, intended to reduce the interval (1), it will be convenient to consider a mechanism that presents a hinge formed in the extrados, specifically in the section where the highest r_m under $M>0$ is found, either a hinge formed in the intrados, in the section where the highest r_m under $M<0$ is found, or a 'pattino' with $\varphi=(ta)>0$ where the highest r_t under $T>0$ is found; the fourth possibility is with $\varphi=(ta)<0$ where the highest r_t under $T<0$ is found. Of course, it is not necessary that all the four situations are attained.

3. LINEAR PROGRAMMING IN PRESENCE OF STANDARD MATERIALS

The collapse factor k_r can also be obtained by linear programming. If n is the number of the elementary trunks, in each of the $n+1$ assembly sections the following inequalities hold:

$$\frac{M_x+M_g+kM_a}{N_x+N_g+kN_a} \begin{array}{l} \leqslant h/2 \\ \geqslant -h/2 \end{array} \qquad \frac{T_x+T_g+kT_a}{N_x+N_g+kN_a} \begin{array}{l} \leqslant \operatorname{tg}\varphi \\ \geqslant -\operatorname{tg}\varphi\,; \end{array} \tag{2}$$

the cantilever built in B is an isostatic structure and C_g, C_a and C_x are the internal forces caused by the dead loads, the living loads and the redundant quantities $X_1X_2X_3$ acting on the end section A. The inequalities (2) constitute 4 $(n+1)$ linear inequalities involving the four unknown quantities $X_1X_2X_3k$, and k is the target. Therefore, we must operate on a 5×4 $(n+1)$ matrix and this can be very onerous if n exceeds a certain limit. Using the same engagement of memory space, the admissible classes procedure permits a much larger subdivision into elementary trunks and hence a greater precision. Moreover (see Section 5), if the material cannot be considered standard, the factor k_r must be calculated as the minimum γ value.

For this kind of calculus, we can use the program elaborated for admissible classes procedure with only slight corrections, while linear programming, as set out in this paragraph, gives the maximum ψ, which does not a fortiori correspond to the collapse factor k_r, as the static theorem is no longer valid for non-standard materials.

4. SOME EXAMPLES

(a) The object of this examination is an ancient bridge (Fig. 9) of the Aosta valley. The bridge in question has a span of 31·4 m, and a rise of 11·4 m. It was built by the Romans along one of the consular roads leading to Gaul, near the present-day town of Pont Saint Martin.

In our study the structure has been subjected to equivalent seismic forces laid down by the Italian seismic standards established for masonry structures after the earthquake in Southern Italy (1980). The horizontal loads are kF_i, while the vertical loads are $\frac{k}{2}F_i$, where F_i are the weights of the trunks and

$$k = 4\frac{S-2}{100} \tag{3}$$

S being the degree of seismicity ($2 \leqslant S \leqslant 12$).

The design in elastic range must consider that the structure cannot tolerate any tensile stress and so the cross-sections are reduced progressively as the value of k increases. From this research [1], it results that the compressive stress of 300 kg cm^{-2} is reached when $k = 0{\cdot}118$ (Fig. 10).

The $k \rightarrow \sigma$ curve shows a vertical asymptote in $k = 0{\cdot}140$.

When working on a mechanism with four hinges, limit design in the kinematic form gives the results shown in Fig. 9. We notice that

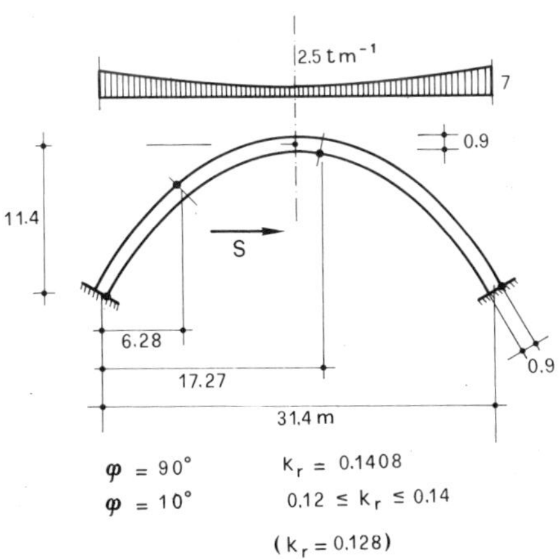

FIG. 9. Masonry arch under seismic loads.

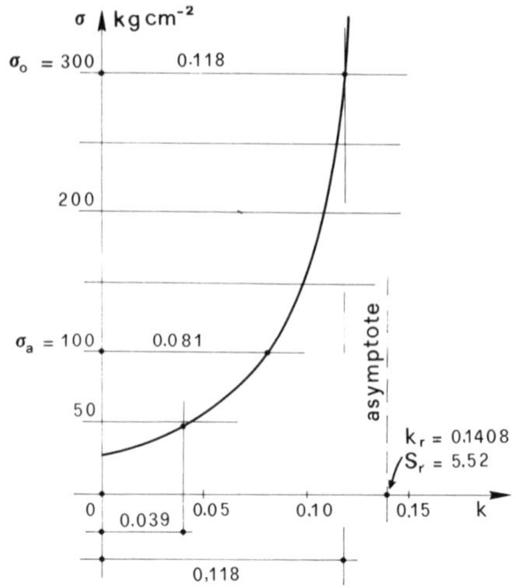

FIG. 10. Study of collapse factor via elastic procedure.

$k = 0{\cdot}1408$ and the thrust line lies entirely within the arch. If we do not consider the possibility of sliding, the results are surprisingly close to those obtained in the elastic range, either for the value of k_r that coincides with the asymptote's abscissa in the $k \rightarrow \sigma$ diagram (Fig. 10), or for the position of the hinges. But the value $\varphi = 10°$ fixed for the angle of friction reduces the value of k_r. This characteristic is contained in the interval

$$0{\cdot}1212 \leqslant k_r \leqslant 0{\cdot}1408$$

In fact, $r_{tmax} = 1{\cdot}161$ (under $T > 0$) at the abscissa 0. If we introduced at $z = 0$ a 'pattino' inclined at $+\varphi$, and if the three residual hinges remain, we have

$$0{\cdot}1292 \leqslant k_r \leqslant 0{\cdot}1278$$

and thus collapse will occur according to such a mechanism for $k_r = 0{\cdot}128$.

(b) The squat and shallow arch of Fig. 11, 40 m span and 4 m rise, does not collapse if the possibility of slides is disregarded. The angle of friction is $\varphi = 5°$. The living loads are the same as those of the preceding example.

A first trial has been carried out by introducing three hinges at the

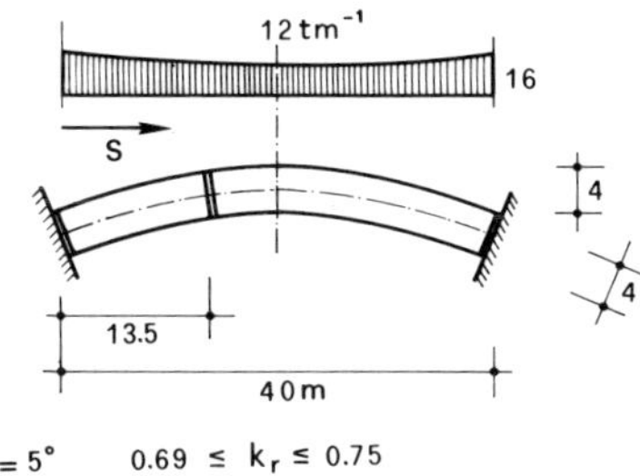

FIG. 11. Shallow masonry arch.

abscissae = 0, 4·5 m, 24 m, and 'pattino' at the right end section, obtaining

$$0{\cdot}4937 \leqslant k_r \leqslant 1{\cdot}1299$$

We also have $r_{tmax} = 1{\cdot}684$ under $T < 0$ at the abscissa $z = 16{\cdot}5$ m, and $r_{tmax} = 2{\cdot}288$ under $T > 0$ at the abscissa $z = 0$.

A second trial is carried out by introducing three 'pattini' ($\varphi + - +$) at the extremities and at the abscissa $z = 16{\cdot}5$ m, obtaining

$$0{\cdot}6861 \leqslant k_r \leqslant 0{\cdot}7609$$

We also have $r_{tmax} = 1{\cdot}109$ under $M < 0$ at the abscissa $z = 24{\cdot}5$ m.

The third trial is carried out by displacing the central 'pattino' at the abscissa $z = 13{\cdot}5$ m and we obtain

$$0{\cdot}6868 \leqslant k_r \leqslant 0{\cdot}7493 \tag{4}$$

whilst $r_{mmax} = 1{\cdot}09$ at the abscissa $z = 24{\cdot}5$ m.

All the mechanisms with at least one hinge give intervals larger than (4). This suggests a collapse mechanism where the first and third 'pattino' are not at the ends but very near to them. However, the interval (4) being very narrow is quite acceptable.

(c) The flying buttress of Fig. 12 is subjected to seismic loads from the left side. Using the mechanism with three 'pattini', the central one of which is at the abscissa $z = 6$ m, we have

$$0{\cdot}6756 \leqslant k_r \leqslant 0{\cdot}6907 \tag{5}$$

with $r_{mmax} = 1{\cdot}022$, under $M < 0$, at the abscissa $z = 18{\cdot}5$ m.

If the seismic forces act from the right side, we consider the arch of Fig. 13. The mechanism with three 'pattini', the central one of which is at

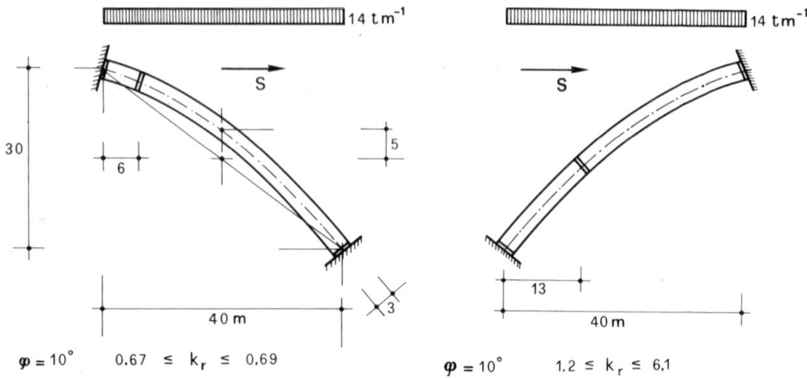

FIG. 12. Flying buttress.

FIG. 13. Flying buttress of Fig. 12 under reversed seismic loads.

the abscissa $z = 13$ m, gives

$$1{\cdot}2037 \leqslant k_r \leqslant 6{\cdot}1475$$

and there is no need for a second test because (5) establishes that $k_r \leqslant 0{\cdot}6907$ and therefore collapse is determined by the earthquake from the left.

5. THE CASE OF NONSTANDARD MATERIALS

Let us consider nonstandard materials, particularly materials of zero dilatancy.

The axis of the 'pattini' (Fig. 14) and the tangent to the axis of the arch are parallel. The slide $\Delta\eta_i$ implies two shear forces $\pm T_i$ acting on the faces jointed by the 'pattino'. The sign of these forces is such that

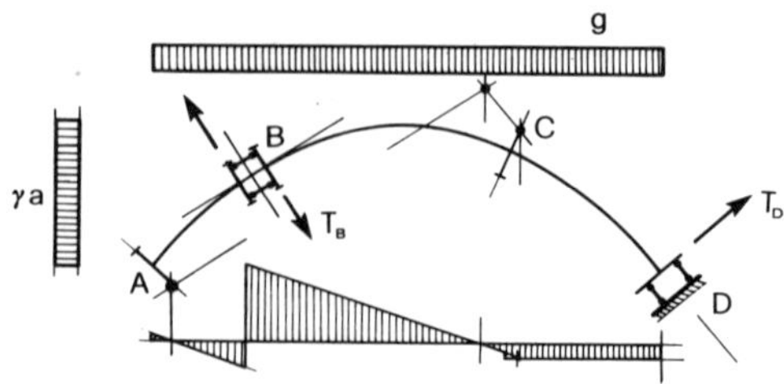

FIG. 14. The case of nonstandard material.

$T_i\ \Delta\eta_i < 0$ and the moduli are

$$T_i = |N_i|\ \mathrm{tg}\ \varphi_i \tag{6}$$

The equilibrium condition is

$$L_g + \gamma L_a + \sum T_i \Delta\eta_i = 0 \tag{7}$$

The slides $\Delta\eta_i$, as well as L_g and L_a, are determined if the value of the Lagrangian parameter is fixed. But (7) alone cannot give the value of γ.

However, N_i is a linear function of γ that we can establish. For example, N_B can be obtained (Fig. 15) by changing the 'pattino' in B with a joint that permits only the forces N_B to work, and by modifying the other 'pattini' so that the work of the forces N_i and T_i is zero in every joint. In order to do so, it is sufficient to turn the bars of the 'pattino' until their axes become orthogonal to the resultant of N and T. Hence we have:

$$L'_g + \gamma L'_a + N_B \Delta' \xi_B = 0$$

from which

$$N_B = -\frac{L'_g}{\Delta'\xi_B} - \gamma \frac{L'_a}{\Delta'\xi_B} = N_{Bg} + \gamma N_{Ba}. \tag{8}$$

From (6) and (8) the following results:

$$T_i = -\frac{\Delta v_i}{|\Delta v_i|} |N_{ig} + \gamma N_{ia}|\ \mathrm{tg}\ \varphi_i = \frac{\Delta v_i}{|\Delta v_i|} (N_{ig} + \gamma N_{ia})\ \mathrm{tg}\ \varphi_i$$

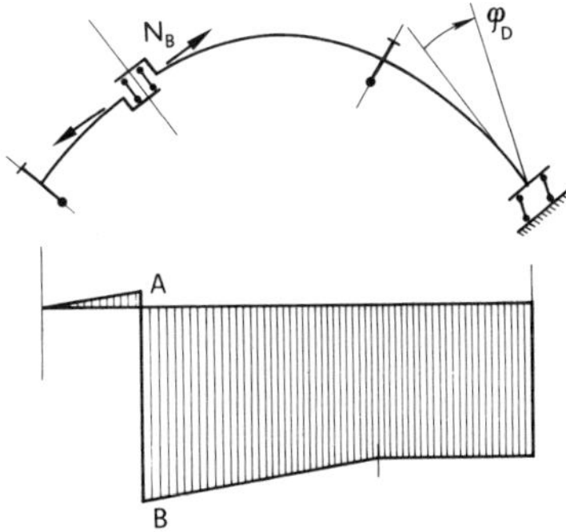

FIG. 15. Search for the N_B in the previous example.

therefore (7) can be written as follows:

$$L_g + \gamma L_a + \sum \frac{\Delta v_i}{|\Delta v_i|} N_i \operatorname{tg} \varphi_i + \gamma \sum \frac{\Delta v_i}{|\Delta v_i|} N_{ia} \operatorname{tg} \varphi_i = 0$$

from which

$$\gamma = -\frac{L_g + \sum \frac{\Delta v_i}{|\Delta v_i|} N_{ig} \operatorname{tg} \varphi_i}{L_a + \sum \frac{\Delta v_i}{|\Delta v_i|} N_{ia} \operatorname{tg} \varphi_i} \tag{9}$$

REFERENCES

1. AUCIELLO, N., A. DE ROSA and V. FRANCIOSI. Il procedimento delle tensioni ammissibili nella verifica degli archi e delle volte in muratura in zona sismica, *Autostrade* **9** (1982), 35–47.
2. DELBECQ, J. M. Analyse de la stabilité des vôutes en maçonnerie par le calcul à la rupture, *J. Méc. Théor. Appl.*, **1** (1982), 91–121.
3. DRUCKER, B. C. Coulomb friction, plasticity and limit loads, *J. appl. Mech.*, **3** (1954), 71–74.
4. FRANCIOSI, V. L'attrito nel calcolo a rottura delle murature, *Giorn. Genio Civile*, **3** (1980), 215–234.
5. FRANCIOSI, V. La verifica a rottura degli archi in conglomerato debolmente armati. Corso di perfezionamento per le costruzioni in c.a., Milano, 1983.
6. FRANCIOSI, V. and L. NUNZIANTE. The masonry arch under some types of variable actions, Atti dell'Istituto di Scienza delle Costruzioni dell'Università di Catania, no. 13/83.
7. HEYMAN, J. The stone skeleton, *Int. J. Solids Struct.*, **2** (1966), 249–279.
8. HEYMAN, J. The safety of masonry arches, *Int. J. Mech. Sci.*, **2** (1969), 363–385.
9. HEYMAN, J. *Plastic Design of Frames*, Vol. 2, Chap. IV, University Press, Cambridge, 1971.
10. MASSONNET, CH. and M. SAVE. *Calcul Plastique des Constructions*, 3rd edn, Nelissen, Liege, 1976, 599.
11. SACCHI, G. *et al. Comportamento statico e sismico delle strutture murarie*, CLUP, Milano, 1982.

Index